Handbook of Experimental Pharmacology

Volume 95/II

Peptide Growth Factors and Their Receptors II

Contributors

J.F. Battey, B. Beutler, L. Bonewald, R.L. Cate, M.V. Chao, P.K. Donahoe, K. Elgjo, J. Folkman, T. Graf, R. Grosse, M.E. Gurney, V.K.M. Han, C.-H. Heldin, A. Hsueh, M. Klagsbrun, O.D. Laerum, P. Langen, A.-M. Lebacq-Verheyden, D.C. Lee, A. Leutz, L.A. Liotta, D.T. MacLaughlin, G.R. Martin, K. Miyazono, D. Monard, M.A.S. Moore, D.E. Mullins, G.R. Mundy, C.F. Nathan, W.R. Paukovits, W.E. Paul, J. Pfeilschifter, D.B. Rifkin, C. Rivier, A.C. Sank, E.A. Sausville, E. Schiffmann, M.L. Stracke, J. Trepel, W. Vale, J. Vilček, E.S. Vitetta, S.M. Wahl, H.L. Wong, J. Yu

Editors

Michael B. Sporn and Anita B. Roberts

Springer-Verlag
Berlin Heidelberg New York
London Paris Tokyo
Hong Kong

MICHAEL B. SPORN, M. D.

ANITA B. ROBERTS, Ph. D.

Laboratory of Chemoprevention
National Cancer Institute
Bethesda, MD 20892
USA

With 75 Figures

ISBN-13:978-3-642-74783-0 e-ISBN-13:978-3-642-74781-6
DOI: 10.1007/978-3-642-74781-6

Library of Congress Cataloging-in-Publication Data. Peptide growth factors and their receptors/contributors, K.-I. Arai ... [et al.]; editors, Michael B. Sporn and Anita B. Roberts. p. cm. – (Handbook of experimental pharmacology; v. 95) Contributors for v. 2, J.F. Battey and others. Includes bibliographical references. ISBN-13:978-3-642-74783-0
1. Growth factors. 2. Growth factors — Receptors. I. Arai, Ken-Ichi. II. Sporn, Michael B. III. Roberts, Anita B. IV. Battey, James F. V. Series. [DNLM: 1. Cell Communication. 2. Growth Substances. 3. Peptides. 4. Receptors, Endogenous Substances. W1 HA51L v. 95/QU 68 P42404] QP905.H3 vol. 95 [QP552.G76] 615'.1 s — dc20 [574.87'6] DNLM/DLC for Library of Congress.

Softcover reprint of the hardcover 1st edition 1990

2127/3130-543210 – Printed on acid-free paper

List of Contributors

J. F. BATTEY, Laboratory of Neurochemistry, NINDS, NIH, Bldg. 36, Rm. 4D20, 9000 Rockville Pike, Bethesda, MD 20892, USA

B. BEUTLER, The Howard Hughes Medical Institute and The Department of Internal Medicine of the University of Texas Southwestern Medical Center, 5323 Harry Hines Boulevard, Room Y5-210, Dallas, TX 75235-9050, USA

L. BONEWALD, Division of Endocrinology and Metabolism, University of Texas, Health Science Center, 7703 Floyd Curl Drive, San Antonio, TX 78284, USA

R. L. CATE, Department of Molecular Biology, Biogen Inc., 14 Cambridge Center, Cambridge, MA 02142, USA

M. V. CHAO, Department of Cell Biology and Anatomy Division of Hematology/Oncology, Cornell University Medical College, 1300 York Ave., New York, NY 10021, USA

P. K. DONAHOE, Pediatric Surgical Research Laboratory, Massachusetts General Hospital, and Department of Surgery, Harvard Medical School, Warren Building, 32 Fruit Street, Boston, MA 02114, USA

K. ELGJO, Institute of Pathology, Rikshospitalet, University of Oslo, 0027 Oslo 1, Norway

J. FOLKMAN, Departments of Surgery and Anatomy and Cellular Biology, Children's Hospital and Harvard Medical School, 300 Longwood Ave., Boston, MA 02115, USA

T. GRAF, European Molecular Biology Laboratory, Postfach 102209, Meyerhofstr. 1, 6900 Heidelberg, FRG

R. GROSSE, Central Institute of Molecular Biology, Academy of Sciences of the German Democratic Republic, Robert-Rössle-Str. 10, DDR-1115 Berlin

M. E. GURNEY, Department of Microbiology-Immunology and Department of Cell, Molecular and Structural Biology, Northwestern University, 303 East Chicago Ave., Chicago, IL 60611, USA

V. K. M. Han, The Lawson Research Institute, St. Joseph's Health Centre, 268 Grosvenor Street, London, Ontario, Canada N6A 4V2

C.-H. Heldin, Ludwig Institute for Cancer Research, Biomedical Center, Uppsala Branch, Box 595, 751 23 Uppsala, Sweden

A. Hsueh, Department of Reproductive Medicine, University of California, La Jolla, CA 92093, USA

M. Klagsbrun, Enders 10, Departments of Biological Chemistry and Surgery, Children's Hospital and Harvard Medical School, 300 Longwood Ave., Boston, MA 02115, USA

O. D. Laerum, Department of Pathology, University of Bergen, The Gade Institute, Haukeland Hospital, 5016 Bergen, Norway

P. Langen, Central Institute of Molecular Biology, Academy of Sciences of the German Democratic Republic, Robert-Rössle-Str. 10, DDR-1115 Berlin

A.-M. Lebacq-Verheyden, Unité de Génétique Cellulaire, UCL-74.59, Institute of Cellular and Molecular Pathology, 74 ave Hippocrate, 1200 Brussels, Belgium

D. C. Lee, Department of Microbiology and Immunology, and the Lineberger Cancer Research Center, School of Medicine, University of North Carolina at Chapel Hill, Chapel Hill, NC 27599-7295, USA

A. Leutz, European Molecular Biology Laboratory, Postfach 102209, Meyerhofstr. 1, 6900 Heidelberg, FRG

L. A. Liotta, Laboratory of Pathology, National Cancer Institute, National Institutes of Health, Building 10, Room 2A33, Bethesda, MD 20892, USA

D. T. MacLaughlin, Pediatric Surgical Research Laboratory, Massachusetts General Hospital, and Department of Surgery, Warren Building, Room 1133, 32 Fruit Street, Boston, MA 02114, USA

G. R. Martin, National Institute on Aging, Gerontology Research Center, Francis Scott Key Medical Center, Baltimore, MD 21224, USA

K. Miyazono, Ludwig Institute for Cancer Research, Box 595, Biomedical Center, 75123 Uppsala, Sweden

D. Monard, Friedrich Miescher-Institut, Postfach 2543, 4002 Basel, Switzerland

M. A. S. Moore, James Ewing Laboratory of Developmental Hematopoiesis, Section #6136, RM 717 C Rockefeller Bldg. Memorial Sloan Kettering Cancer Center, 1275 York Avenue, New York, NY 10021, USA

D. E. Mullins, Department of Pharmacology, Schering-Plough Research, Bloomfield, NJ 07003, USA

G. R. MUNDY, Division of Endocrinology and Metabolism, University of Texas, Health Science Center, 7703 Floyd Curl Drive, San Antonio, TX 78284, USA

C. F. NATHAN, Beatrice and Samuel A. Seaver Laboratory, Division of Hematology-Oncology, Department of Medicine, Cornell University Medical College, Box 57, 1300 York Avenue, New York, NY 10021, USA

W. R. PAUKOVITS, Department of Growth Regulation, Institute of Tumor Biology, University of Vienna, Borschkegasse 8a, 1090 Vienna, Austria

W. E. PAUL, Laboratory of Immunology, National Institute of Allergy and Infectious Diseases, National Institutes of Health, Building 10, Room 11N311, Bethesda, MD 20892, USA

J. PFEILSCHIFTER, Universitätsklinik Heidelberg, Abteilung für Endokrinologie, 6900 Heidelberg, FRG

D. B. RIFKIN, Department of Cell Biology and Kaplan Cancer Center, New York University Medical Center and the Raymond and Beverly Sackler Foundation, 550 First Avenue, New York, NY 10016, USA

C. RIVIER, The Clayton Foundation Laboratories for Peptide Biology, The Salk Institute for Biological Studies, P.O. Box 85800, San Diego, CA 92138, USA

A. C. SANK, Laboratory of Developmental Biology and Anomalies, National Institute of Dental Research, National Institutes of Health, Building 30, Room 416, Bethesda, MD 20892, USA

E. A. SAUSVILLE, Departments of Medicine and Pharmacology, Georgetown University School of Medicine, Washington, DC 20007, USA

E. SCHIFFMANN, Department of Health & Human Services, Public Health Services, National Institutes of Health, Building 10, Room 2A33, Bethesda, MD 20892, USA

M. L. STRACKE, Laboratory of Pathology, National Cancer Institute, National Institutes of Health, Building 10, Room 2A33, Bethesda, MD 20892, USA

J. TREPEL, Medicine Branch, National Cancer Institute, National Institutes of Health, Building 10, Room 12N230, 9000 Rockville Pike, Bethesda, MD 20892, USA

W. VALE, The Clayton Foundation Laboratories for Peptide Biology, The Salk Institute for Biological Studies, P.O. Box 85800, San Diego, CA 92138, USA

J. VILČEK, Department of Microbiology, New York University Medical Center, 550 First Avenue, New York, NY 10016, USA

E. S. Vitetta, Department of Microbiology, University of Texas Southwestern Medical Center at Dallas, 5323 Harry Hines Blvd., Dallas, TX 75235, USA

S. M. Wahl, Laboratory of Immunology, National Institute of Dental Research, National Institutes of Health, Building 30, Room 326, Bethesda, MD 20892, USA

H. L. Wong, Laboratory of Immunology, National Institute of Dental Research, National Institutes of Health, Building 30, Room 334, Bethesda, MD 20892, USA

J. Yu, Department of Basic and Clinical Research, Scripps Clinic and Research Foundation, La Jolla, CA 92037, USA

Preface

This two-volume treatise, the collected effort of more than 50 authors, represents the first comprehensive survey of the chemistry and biology of the set of molecules known as peptide growth factors. Although there have been many symposia on this topic, and numerous publications of reviews dealing with selected subsets of growth factors, the entired field has never been covered in a single treatise. It is essential to do this at the present time, as the number of journal articles on peptide growth factors now makes it almost impossible for any one person to stay informed on this subject by reading the primary literature. At the same time it is becoming increasingly apparent that these substances are of universal importance in biology and medicine and that the original classification of these molecules, based on the laboratory setting of their discovery, as "growth factors," "lymphokines," "cytokines," or "colony-stimulating factors," was quite artifactual; they are in fact the basis of a common language for intercellular communication. As a set they affect essentially every cell in the body, and in this regard they provide the basis to develop a unified science of cell biology, germane to all of biomedical research.

This treatise is divided into four main sections. After three introductory chapters, its principal focus is the detailed description of each of the major peptide growth factors in 26 individual chapters. These chapters provide essential information on the primary structure, gene structure, gene regulation, cell surface receptors, biological activity, and potential therapeutic applications of each growth factor (to the extent that these are known). The last two sections of these volumes deal with the coordinate actions of sets of growth factors, since it is clear that to understand their physiology, one must consider the interactions between these peptides. There are six chapters on specific cells and tissues, including bone marrow, brain, bone and cartilage, lymphocytes, macrophages, and connective tissue. The final six chapters consider the role of growth factors in controlling fundamental processes that pertain to many different cells and tissues, such as proteolysis, inflammation and repair, angiogenesis, and embryogenesis.

The editors accept full responsibility for the table of contents, and apologize for any significant omissions. We have deliberately excluded classic peptide hormones whose actions are principally endocrine. We have not included many peptides whose biological activcities may have been described, but which have not yet been purified to homogeneity and sequenced. Essentially all of the molecules included in this treatise act at specific cell surface receptors, although we have included one peptide (glia-derived neurotrophic factor) which is a protease inhibitor. Because of the importance of growth-inhibitory actions of peptides, we have included chapters on two new sets of inhibitors (mammary-derived growth inhibitor and pentapeptide growth inhibitors) for which there is only very preliminary knowledge of their structure, function, and mechanism of action.

The present era in research on peptide growth factors is indeed an exciting one. The biological activities resulting from the actions of these substances have been known for a very long time, as described in John Hunter's detailed descriptions of the healing of severed tendons and gunshot wounds, made over 200 years ago. Modern cell and tissue culture owes its inception to the actions of growth factors present in the embryo extracts used almost 100 years ago by the pioneers of this technique. However, it has been the recent introduction of new methods for purification of peptides, and the cloning and expression of genes with recombinant DNA technology, which have truly revolutionized this field. These methods provide the scientific basis for the present treatise and the current excitement in this field. They also have provided the practical methodology for the creation of a whole new biotechnology industry, which has made major commitments to develop peptide growth factors as clinical therapeutic agents. Applications being developed include wound healing and other aspects of soft-tissue repair, repair of bone and cartilage, immunosuppression, enhancement of immune cell function, enhancement of bone marrow function in many disease states, and prevention and treatment of many proliferative diseases, including atherosclerosis and cancer. Since many of the peptides described in these pages function as growth factors in the embryo, they raise hope that they may be used some day to arrest or reverse the ravages of aging and degenerative disease.

There is no question that the peptides decribed here are of fundamental importance for understanding the behavior of all cells, and that they will be of major importance in the practice of clinical medicine in the years to come. These volumes, then, are a celebration not only of new basic knowledge, but also of the new therapeutic potential of this entire family of molecules, which have such intense potency to make cells move, grow, divide, and differentiate. We hope that this treatise will be of value to both scientists and clinicians in their pursuit and application of new knowledge in this promising area.

We wish to express our appreciation to many individuals who have participated in this venture from its inception. We are greatly indebted to Pedro Cuatrecasas, who initially suggested the writing of this treatise and has been a constant and enthusiastic supporter. We thank all of the authors for their devoted efforts to assimilate the huge literature in their respective fields and to condense this information into readable single chapters. Our secretary, Karen Moran, has been of invaluable help with numerous aspects of the organization and publication of these volumes. Finally, we would like to express our gratitude to Doris Walker and the staff at Springer-Verlag for all of their efforts in bringing this treatise to publication.

MICHAEL B. SPORN
ANITA B. ROBERTS

Contents

Section B: Individual Growth Factors and Their Receptors (Cont'd from Part I)

CHAPTER 20

Cachectin/Tumor Necrosis Factor and Lymphotoxin

CHAPTER 21

Bombesin and Gastrin-Releasing Peptide: Neuropeptides, Secretogogues, and Growth Factors

CHAPTER 26

The Inhibin/Activin Family of Hormones and Growth Factors

CHAPTER 27

Mammary-Derived Growth Inhibitor

CHAPTER 28

Pentapeptide Growth Inhibitors

Section C: Coordinate Actions of Growth Factors in Specific Tissues or Cells

CHAPTER 29

Coordinate Actions of Hematopoietic Growth Factors in Stimulation of Bone Marrow Function

CHAPTER 30

Peptide Growth Factors and the Nervous System

CHAPTER 34

Extracellular Matrices, Cells, and Growth Factors

Section D: Processes Regulated by Growth Factors

CHAPTER 35

Induction of Proteases and Protease Inhibitors by Growth Factors

CHAPTER 36

Inflammation and Repair

H. L. WONG and S. M. WAHL. With 1 Figure 509

CHAPTER 37

Angiogenesis

M. KLAGSBRUN and J. FOLKMAN. With 1 Figure 549

Contents of Companion Volume 95, Part I

Section B: Individual Growth Factors and Their Receptors (Cont'd from Part I)

CHAPTER 19

Interferons

J. VILČEK

A. What Are Interferons?

Interferons are a heterogeneous family of multifunctional cytokines whose first demonstrated biological activity was the induction of cellular resistance to virus infection. For many years antiviral activity was the only recognized biological function of the interferons. Today we appreciate the fact that antiviral activity is a characteristic feature of the interferons, but their actions on cell growth and differentiation and their many immunoregulatory activities are probably of greater fundamental biological significance.

Two very distinct families of proteins are counted among the interferons. The IFN-α/β "superfamily" (also called type I IFN) encompasses a group of structurally related genes and proteins that are further subdivided into the subfamilies IFN-α_I, IFN-α_{II}, and IFN-β (Table 1). The second "family" consists of a single gene encoding a single protein termed IFN-γ (also called type II IFN or immune IFN). It should be made clear at the outset that IFN-γ is structurally unrelated to the members of the IFN-α/β superfamily. The reasons for discussing IFN-α/β and IFN-γ in a single chapter are largely

Table 1. Classification of the interferons

	IFN-α/β (type I IFN)	IFN-γ (type II or immune IFN)
Subfamilies	IFN-α_I($\geqq$14 subtypes in man) IFN-α_{II} IFN-β	–
Location of structural gene(s)	Chromosome 9 (human) Chromosome 4 (mouse)	Chromosome 12 (human) Chromosome 10 (mouse)
Introns in genes	None	3
Location of receptor gene	Chromosome 21 (human) Chromosome 16 (mouse)	Chromosome 6 (human) Chromosome 10 (mouse)
Major biological functions postulated	Regulation of cell growth and differentiation, induction of class I MHC antigens, promotion of embryo implantation in the uterus, antiviral action	Immunoregulation (monocyte/macrophage functions, immunoglobulin synthesis), induction of class II MHC antigens

historical. Interferon was first described by Isaacs and Lindenmann (1957) as a product of virus-infected cells capable of inducing resistance to infection with homologous or heterologous viruses. A functionally related virus-inhibitory protein (today termed IFN-γ) was described by Wheelock (1965) as an "interferon-like" substance produced by mitogen-activated T lymphocytes. For many years the only properties that made it possible to distinguish IFN-γ from the other interferons were its lack of stability at pH 2 (Wheelock 1965) and distinct antigenic specificity (Youngner and Salvin 1973). Only when the sequences of the proteins and genes of the major interferons were revealed in the early 1980s did it become clear what the relationship of the different interferons is to each other. We recognize now that IFN-γ is primarily an immunoregulatory cytokine whereas the potential actions of IFN-α/β extend to a broader variety of cells and tissues.

It is impossible to review in detail all facets of interferon structure, synthesis, and action in this chapter. The text is written more as an overview (or should we say underview?) than a complete review, and the references to literature are only representative at best, rather than comprehensive. I tried to cover in somewhat more detail areas in which there has been significant recent progress and which are relevant to the field of growth factors. For information not covered in full detail in this chapter the reader is referred to other sources: for a detailed review of the structural aspects of interferon genes and proteins see Weissmann and Weber (1986), for an authoritative coverage of the biochemical mechanisms of interferon action, Lengyel (1982), and for an excellent, detailed account of all aspects of interferon research the book by de Maeyer and de Maeyer-Guignard (1988).

B. Structure of Interferon Genes and Proteins

I. Interferon-α/β (Type I IFN)

Members of the IFN-α/β superfamily represent the classical interferons. The first clear indication of the heterogeneity of the type I interferon proteins came from studies showing that interferons derived from human leukocytes and fibroblasts are antigenically distinct (Havell et al. 1975). Eventually leukocyte and fibroblast interferons were designated IFN-α and -β, respectively (Committee on Interferon Nomenclature 1980). Most of the information on interferon structure has been derived from gene cloning studies.

All animal species examined have large IFN-α gene families, but most have only one structural gene for IFN-β. A notable exception is the ungulates, which have multiple IFN-β genes. Since all of the genes and proteins comprising this superfamily are related to each other structurally and the genes form a single cluster (located in the human species on the short arm of chromosome 9), it is generally held that this superfamily evolved from a single ancestor by gene duplication and spontaneous mutation (reviewed by Weissmann and Weber 1986). It has been estimated that the split between IFN-α and IFN-β

occurred before the emergence of vertebrates perhaps 400 million years ago, whereas the split between the IFN-α_I and IFN-α_{II} subfamilies must have occurred before the mammalian radiation, i.e., more than 85 million years ago.

1. Human IFN-α/β Genes and Proteins

a) IFN-α ("Leukocyte" IFN)

At least 24 nonallelic human IFN-α genes and pseudogenes have been identified. They can be divided into two distinct subfamilies, termed IFN-α_I and -α_{II} (WEISSMANN and WEBER 1986). The IFN-α_I subfamily comprises 14 potentially functional genes and several pseudogenes. The IFN-α_{II} subfamily is known to comprise only one functional gene and five or six nonallelic pseudogenes. IFN-α_I genes encode mature proteins consisting of 165–166 amino acids; the IFN-α_{II} gene encodes a mature protein 172 amino acids long (Fig. 1).

```
                    1S         11S        21S 1          11         21         31         41
                                          $     $                           $$  $$  $  $  $     $ $$
IFN-α Consensus     MALSFSLLMA LVVLSYKSIC SLG CDLPQTHSLG NRRTLMLLAQ MGRISPFSCL KDRHDFGFPQ EEFDGNQFQK
                                          #   # #   # #   #       # #  ##   ##  ##  #  #  #     # ##

IFN-αI1             ..SP.A...V .....C..S. ... ....E....D .......... .S....S... M......... ..........
IFN-αII1            ...L.P..A. ..MT..SPVG ... .....N.G.L S.N..V..H. .R.....L.. ...R..R... .MVK.S.L..
IFN-β               .TNKCL.QI. .LLCFSTTAL .MS YN.LGFLQRS SNFQCQK.LW QLNGRLEY.. K..MN.DI.E .IKQLQ....
                                           1          11         21         31         41         51

                    51         61         71         81         91         101        111
                            $   $  $  $     $$                   $                           $
IFN-α Consensus     AQAISVLHEM IQQTFNLFST KDSSAAWDET LLDKFYTELY QQLNDLEACV IQEVGVEETP LMNEDSILAV
                           ##   ## # ## #   ##       ##      #  ##    #     #                #

IFN-αI1             .P.......L ...I....T. .........D .....C.... .......... M..ER.G... ...A......
IFN-αII1            .HVM...... L..I.S..H. ER.....NM. ...QLH...H ...QH..T.L L.V..EG.SA GAISSPA.TL
IFN-β               ED.ALTIY.. L.NI.AI.RQ DS..TG.N.. IVENLLANV. H.I.H.KTVL EEKLEK.DFT RGKLM.S.HL
                            61         71         81         91         101        111        121

                    121        131        141        151        161        171
                      $   $  $ $    $$ $$ $  $$ $$               $
IFN-α Consensus     RKYFQRITLY LTEKKYSPCA WEVVRAEIMR SFSLSTNLQE RLRRKE
                      ##  #  # #   ### ## ##### ###  #   #       #  #

IFN-αI1             K...R..... .......... .......... .L........ ......
IFN-αII1            .R...G.RV. .K.....D.. .....M...K .LF....M.. ...S.DRDLG SS
IFN-β               KR.YG..LH. .KA.E..H.. .TI..V..L. N.YFINR.TG Y..N
                            131        141        151        161
```

Fig. 1. Amino acid sequences of representative human α/β interferons. The consensus sequence was derived from 15 nonallelic IFN-α sequences by determining for each position the amino acid occurring in the largest number of sequences. The amino acids in the IFN-α_I1, IFN-α_{II}1, and IFN-β sequences coinciding with the consensus IFN-α are represented by *dots*. Numbers preceded by *S* refer to the signal peptide. #, positions conserved in all human α interferons; $, position conserved in all human α interferons and IFN-β. Since on maturation IFN-β precursors lose 21, and IFN-α precursors 23 amino-terminal amino acids, position *n* in all IFN-α proteins is homologous to position $n+2$ in the IFN-β sequence. The one-letter code for amino acids is *A*, Ala; *R*, Arg; *N*, Asn; *D*, Asp; *C*, Cys; *Q*, Gln; *E*, Glu; *G*, Gly; *H*, His; *I*, Ile; *L*, Leu; *K*, Lys; *M*, Met; *F*, Phe; *P*, Pro; *S*, Ser; *T*, Thr; *W*, Trp; *Y*, Tyr; *V*, Val. Sequence information shown in this figure is from WEISSMANN and WEBER (1986) and TANIGUCHI et al. (1980)

All of the genes encode N-terminal secretory signal peptide presequences (generally 23 residues long) which are removed by proteolytic cleavage before the release of the mature interferon molecule from the cell. While it is clear that a high degree of homology is found among all human IFN-α genes and proteins, the IFN-α_{II} sequences have diverged significantly from the -α_I sequences, warranting their classification into a separate subfamily (CAPON et al. 1985). In fact, it has been suggested that the IFN-α_{II} subfamily be named IFN-ω (ADOLF 1987).

All human IFN-α proteins feature four conserved cysteines (positions 1, 29, 98, and 138) thought to form two intramolecular disulfide bonds (Cys1–Cys98 and Cys29–Cys138) (WETZEL 1981). However, since the first four amino acid residues can be removed without significantly reducing biological activity, it is clear that the integrity of the Cys1–Cys98 bond is not essential for the preservation of activity (WEISSMANN and WEBER 1986). Human IFN-α species lack potential N-glycosylation sites and most members of the IFN-α subfamilies in their native state are not glycosylated (PESTKA 1983). Several natural human IFN-α proteins have been purified to homogeneity. They were shown to range in their apparent molecular weights from 16000 to 21000 (RUBINSTEIN et al. 1981). The reason for these large differences in the apparent molecular weights has not been fully explained.

b) IFN-β ("Fibroblast" IFN)

A single gene for human IFN-β encodes a 166-residue-long mature protein (Fig. 1). Homology between IFN-β and members of the IFN-α_I subfamily is about 25%–30% at the amino acid level and about 45% in the coding sequences at the nucleotide level (TANIGUCHI et al. 1980). In addition, there is also extensive homology in the 5′ nucleotide flanking regions which contain transcriptional promoter and enhancer sequences, reflecting the fact that IFN-α and -β genes are often coordinately induced (DEGRAVE et al. 1981). There are three cysteine residues in IFN-β of which Cys41 and Cys141 form a disulfide bridge and Cys17 is free. Unlike the human IFN-α proteins, IFN-β has a potential N-glycosylation site at position 80, and the mature protein is known to be glycosylated (KNIGHT 1976).

c) "IFN-β_2" (Interleukin-6)

WEISSENBACH et al. (1980) reported the isolation of two cDNA clones from human fibroblasts thought to be related to IFN-β structurally and functionally, for which they proposed the name "IFN-β_2." Later it was shown that the protein encoded by the "IFN-β_2" DNA lacks a significant degree of homology with IFN-β and that it is also quite distinct functionally, e.g., it lacks antiviral activity (HAEGEMAN et al. 1986). In what turned out to be one of the unexpected twists of molecular biology, "IFN-β_2" was found to be identical in its sequence to the independently purified and cloned B-cell stimulatory factor 2 (BSF2), now known as interleukin-6. Most investigators now feel that there is no compelling reason to group this protein among the interferons. Interleukin-6 is reviewed in Part I, Chapter 14 by HIRANO and KISHIMOTO.

2. IFN-α/β Genes and Proteins of Other Animal Species

Apart from man, interferons have been studied quite extensively in the murine and bovine species. The murine IFN-α/β gene cluster has been localized to chromosome 4 (KELLEY et al. 1983). At least 12 nonallelic murine IFN-α genes have been identified so far (reviewed by WEISSMANN and WEBER 1986; DE MAEYER and DE MAEYER-GUIGNARD 1988). The structure of murine IFN-α genes and proteins is very similar to their human counterparts. Most murine IFN-α genes code for mature proteins 165–166 amino acids long. A cysteine residue is present in position 86, in addition to four cysteines found in positions virtually identical to the human IFN-α species. However, unlike human IFN-α, murine IFN-α proteins contain a typical N-glycosylation site at position 78, and most murine IFN-α species appear to be glycoproteins.

A single murine IFN-β gene encodes a 161-amino-acid-long mature protein, showing only about 48% homology with its human counterpart. (This significant structural difference explains why human and murine IFN-β show virtually no cross-species biological activity). Other unique features include the presence of three N-glycosylation sites, and of a single cysteine residue, indicating the absence of an intramolecular disulfide bond.

Four bovine IFN-α genes have been sequenced (CAPON et al. 1985). Three of these belong to the IFN-α_I subfamily and one is the prototype of IFN-α_{II} subfamily genes. Bovine IFN-α_{II} is more closely related to human IFN-α_{II} than to the bovine IFN-α_I species (showing only about 54% homology with the latter at the amino acid level). This finding shows that the IFN-α_I and -α_{II} families diverged prior to the mammalian radiation, i.e., at least 85 million years ago (WEISSMANN and WEBER 1986).

Although only a single IFN-β gene is clearly demonstrable by Southern analysis in various primates, as well as in the mouse, rabbit, cat, and lion, at least five genes related to IFN-β have been revealed in the cow, horse, pig, and blackbuck (WILSON et al. 1983; LEUNG et al. 1984). Three bovine genomic IFN-β clones were isolated and sequenced by LEUNG et al. (1984). They have three cysteine residues in positions similar to human IFN-β, but they have two potential N-glycosylation sites, compared with one in human IFN-β.

II. Interferon-γ (Type II IFN)

As mentioned earlier, there is no significant structural homology between members of the IFN-α/β superfamily and IFN-γ. Only a single IFN-γ gene has been identified in all animal species examined to date. Another important difference is that the IFN-γ gene contains three introns and its chromosomal location too is different from the IFN-α/β genes (Table 1).

The human IFN-γ gene, located on chromosome 12 (Naylor et al. 1983), encodes a 143-amino-acid-long basic protein, in addition to a 23-amino-acid signal peptide sequence which is removed before the release of the molecule (Gray et al. 1982; Rinderknecht et al. 1984). There are two N-glycosylation sites at positions 28 and 100. The fully glycosylated molecule has an apparent molecular weight of 25000 under denaturing conditions, but some naturally produced molecules are only partially glycosylated (mol. wt., 20000), and a small portion of the secreted IFN-γ completely lacks carbohydrate and shows an apparent molecular weight of 15500 (Kelker et al. 1984; Rinderknecht et al. 1984). Mature IFN-γ contains no cysteines, two cysteines once considered to form the first and third residues at the NH_2-terminus of the mature protein now are known to be removed before or at the time of release (CysS21 and CysS23; see Fig. 2). Native IFN-γ is thought to be a homodimer (Yip et al. 1982).

Murine, rat, and bovine IFN-γ genes have been cloned and sequenced (reviewed by Weissmann and Weber 1986). The overall organization of the IFN-γ genes in these species is very similar to that of the human gene. Although the coding sequences of human and mouse IFN-γ are about 65% homologous, amino acid sequence homology is only about 40% (Fig. 2). Murine IFN-γ gene has been mapped to chromosome 10 (Naylor et al. 1984).

```
S1                                  S10                                     S20         S23  1
met lys tyr thr ser tyr ile leu ala phe gln leu cys ile val leu gly ser leu gly cys tyr cys GLN ASP PRO TYR VAL LYS GLU
 *           *           *   *   *       *   *                       *       *   *   *   *
met asn ala thr his cys ile leu ala leu gln leu phe leu met ala val ser     gly cys tyr cys HIS GLY THR VAL ILE GLU SER

        10                                      20                                      30
ALA GLU ASN LEU LYS LYS TYR PHE ASN ALA GLY HIS SER ASP VAL ALA ASP ASN GLY THR LEU PHE LEU GLY ILE LEU LYS ASN TRP LYS
     *       *           *   *   *                   *   *                       *   *   *       *           *   *
LEU GLU SER LEU ASN ASN TYR PHE ASN SER SER GLY ILE ASP VAL     GLU GLU LYS SER LEU PHE LEU ASP ILE TRP ARG ASN TRP GLN

        40                                      50                                      60
GLU GLU SER ASP ARG LYS ILE MET GLN SER GLN ILE VAL SER PHE TYR PHE LYS LEU PHE LYS ASN PHE LYS ASP ASP GLN SER ILE GLN
             *       *   *       *   *   *   *       *   *   *           *   *               *   *       *       *
LYS ASP GLY ASP MET LYS ILE LEU GLN SER GLN ILE ILE SER PHE TYR LEU ARG LEU PHE GLU VAL LEU LYS ASP ASN GLN ALA ILE SER

        70                                      80                                      90
LYS SER VAL GLU THR ILE LYS GLU ASP MET ASN VAL LYS PHE PHE ASN SER ASN LYS LYS LYS ARG ASP ASP PHE GLU LYS LEU THR ASN
                     *                               *   *               *       *       *       *
ASN ASN ILE SER VAL ILE GLU SER HIS LEU ILE THR THR PHE PHE SER ASN SER LYS ALA LYS LYS ASP ALA PHE MET SER ILE ALA LYS

        100                                     110                                     120
TYR SER VAL THR ASP LEU ASN VAL GLN ARG LYS ALA ILE HIS GLU LEU ILE GLN VAL MET ALA GLU LEU SER PRO ALA ALA LYS THR GLY
         *                   *   *   *       *           *   *   *       *               *       *
PHE GLU VAL ASN ASN PRO GLN VAL GLN ARG GLN ALA PHE ASN GLU LEU ILE ARG VAL VAL HIS GLN LEU LEU PRO GLU SER SER LEU ARG

        130                                     140         143
LYS ARG LYS ARG SER GLN MET LEU PHE ARG GLY ARG ARG ALA SER GLN
 *   *   *   *   *
LYS ARG LYS ARG SER ARG CYS
                        134
```

Fig. 2. Comparison of human (*upper*) and murine (*lower*) IFN-γ amino acid sequences. *Asterisks* indicate identical sequences. Amino acids thought to be part of the signal peptide sequence are identified by the letter *S* and are shown in *lowercase*. The potential N-glycosylation sites are *underlined*. Two gaps have been introduced in the murine sequence in order to maximize homology. (From Vilček et al. 1985)

C. Interferon Induction and Production

I. Production of IFN-α/β

Many different types of cells can produce IFN-α and/or IFN-β. The predominant form of interferon produced by white blood cells (especially by monocytes/macrophages and B lymphocytes) is often IFN-α or "leukocyte" IFN. Nonhemopoietic cells often produce IFN-β ("fibroblast" IFN) as the only or predominant type of interferon. However, it is very common that a single cell population produces a mixture of several IFN-α subtypes along with IFN-β (HAVELL et al. 1978; HAYES et al. 1979; ALLEN and FANTES 1980). The proportion of interferon subtypes present can vary depending on the producing cell and conditions.

It is customary to refer to the interferons as inducible proteins. This is certainly correct and the potent inducing activity of many viruses, double-stranded RNAs, and some other agents has been extensively documented. However, there are many examples of what appears to be spontaneous secretion of small amounts of interferon by cells and tissues (SMITH and WAGNER 1967; ZAJAC et al. 1969; JARVIS and COLBY 1978). IFN-α mRNA was found in several tissues from normal humans, suggesting a low level of constitutive synthesis of IFN-α (TOVEY et al. 1987). Recently, increasing attention has been paid to the possible significance of such spontaneously produced "autocrine" interferons, and especially their possible role in the regulation of cell growth and differentiation (see Sect. E.II.4).

The most extensively studied inducers of interferon are viruses and their constituents, especially double-stranded RNA. Since the discovery that double-stranded RNA can induce interferon (FIELD et al. 1967), interferon induction by many viruses has been attributed to some form of viral double-stranded RNA (LAI and JOKLIK 1973; CLAVELL and BRATT 1971). (Double-stranded RNA also plays an important role in interferon action, because it is required for the activation of two interferon-induced enzymes; see Sect. E.I) However, not all forms of interferon induction by viruses can be attributed to double-stranded RNA. In many instances the triggering event responsible for interferon induction has not been identified, e.g., in cells infected with DNA viruses. Interferon induction in mononuclear leukocytes in some cases is triggered by viral envelope glycoproteins, rather than by a double-stranded RNA-dependent mechanism (ITO et al. 1978; CAPOBIANCHI et al. 1985).

Bacteria, mycoplasma, protozoa, and some of their constituents also can stimulate interferon production, especially in mononuclear leukocytes (reviewed by DE MAEYER and DE MAEYER-GUIGNARD 1988). Lipopolysaccharide (LPS) from gram-negative bacteria is a potent inducer of IFN-α/β in monocytes/macrophages. Many other substances, including several chemically defined low-molecular-weight compounds, have been identified as interferon inducers; they are usually active only when administered to the intact animal or, less frequently, also in cultures of some hemopoietic cells (reviewed in STEWART 1979; HO 1984). In general, very little is known about the mechanism of interferon induction by these nonviral inducers.

Of great potential interest are some recent reports of interferon induction by growth factors and cytokines. MOORE et al. (1984) described the stimulation of IFN-α/β release from cultures of murine bone marrow macrophages stimulated in vitro with M-CSF (CSF-1). In addition, the M-CSF-induced proliferative response was enhanced by the addition of antibodies to murine IFN-α/β, suggesting that the generation of interferon acted as a negative "feedback" mechanism regulating M-CSF action. In another recently described system, tumor necrosis factor (TNF) was found to stimulate IFN-β synthesis in the M1 mouse myeloid cell line, in which both TNF and IFN-β act as antiproliferative and differentiation-promoting agents (ONOZAKI et al. 1988). The authors proposed that generation of IFN-β in this system was an important differentiation-inducing signal. In addition, TNF was shown to produce a modest increase in IFN-β mRNA levels in the HEp-2 cell line (JACOBSEN et al. 1989) and in human diploid fibroblasts (REIS et al. 1989). This inducing action of TNF is thought to contribute to the antiviral actions seen with TNF in these cells.

Finally, there is some evidence that interferons themselves can promote interferon synthesis. It has been known for some time that treatment of cells with interferon can enhance ("prime") subsequent interferon induction (reviewed by STEWART 1979). Recent findings suggest that, under some conditions, interferon treatment can not only prime, but actually induce the release of small amounts of interferons (KOHASE et al. 1987; HUGHES and BARON 1987). These findings can now be better understood in the light of increasing evidence that there are many similarities in the control of interferon-inducible genes and the control of IFN-α/β expression (see Sect. E.I.3).

II. Molecular Mechanisms of IFN-α/β Induction

For many years virtually nothing was known about the molecular events governing interferon induction, but some light is now being shed on this interesting process. It has been demonstrated that the induction of the IFN-β gene by viruses or by the synthetic double-stranded RNA, poly(I)·poly(C), is due primarily to transcriptional activation (RAJ and PITHA 1983) and that the 5′ flanking regions of IFN-α and -β genes contain all information necessary for inducibility (RAGG and WEISSMANN 1983; OHNO and TANIGUCHI 1983; ZINN et al. 1983). Recent efforts are being directed mainly toward the identification of *cis*-acting enhancer elements in the 5′ flanking regions that can influence the activity of promoter regions located immediately upstream of the "cap" site. In addition, *trans*-acting factors binding to enhancer elements are also being identified.

Boundaries of the human IFN-α1 (IFN-α_I1) promoter region required for full inducibility by virus have been determined by ligating various segments of the 5′ flanking sequence of the IFN-α1 gene to a truncated rabbit β globin gene. The constructs, transfected into murine cells, were then analyzed for the induction of β globin synthesis by virus infection (RYALS et al. 1985). The results showed that a 46-bp DNA segment (between positions -109 and -64

relative to the cap site) could confer maximal inducibility. A similar type of analysis was carried out with the human IFN-β gene by two groups of investigators. Using transfection analysis in mouse C127 cells, GOODBOURN et al. (1984) determined that the 5′ boundary of the sequences required for maximal inducibility lies between −77 and −73, while FUJITA et al. (1985), who employed mouse L929 cells for transfection, located the 5′ boundary between −117 and −105. The difference is apparently due to the use of different expression vectors and different cell lines for the transfection analysis.

It has been proposed that the human IFN-β enhancer ("interferon regulatory element" or IRE) is composed of both positive and negative regulatory DNA sequences (ZINN and MANIATIS 1986; KELLER and MANIATIS 1988). In uninduced cells a repressor protein is bound to the negative regulatory domain (NRDI) at the 3′ half of the IRE region (−63 to −37) which partly overlaps with one of two positive regulatory domains, PRDI and PRDII (−77 to −64 and −66 to −55, respectively). It is believed that induction by virus or poly(I)·poly(C) leads to the release of the negative *trans*-acting factor and the binding of transcription factors to the positive control region. Three proteins binding to the IRE DNA have been identified, of which two (PRDII-BF and PRDI-BF_c) were found in nuclear extracts from both uninduced and induced human MG-63 cells and one (PRDI-BF_i) was found only in extracts from induced cells. KELLER and MANIATIS (1988) proposed that induction may involve generation of PRDI-BF_i, possibly from PRDI-BF_c.

FUJITA et al. (1987) showed that virus-induced transcriptional activation of the human IFN-β gene can be mediated by repetitive hexanucleotide motifs within the enhancer region. The consensus sequence was deduced as AA(A/G)(T/G)GA. The authors noted that similar hexameric sequences present within the genomes of IFN-α and TNF are likely to control virus inducibility of the latter genes as well. Indeed, KUHL et al. (1987) showed that tetrameric AAGTGA, and also tetrameric GAAAGT, placed upstream of a promoter conferred virus inducibility on the rabbit globin gene.

MIYAMOTO et al. (1988) have recently purified an interferon transcription factor ("IRF-1") from mouse L929 cells and cloned its cDNA. IRF-1 was identified on the basis of its ability to bind to the tandemly repeated hexanucleotide sequences present in the 5′ flanking region of the IFN-β gene which were earlier shown to confer virus inducibility. The precise relationship of IRF-1 to the factors identified by KELLER and MANIATIS (1988) on the basis of their binding to the IRE region (but not yet sequenced and cloned at the time of this writing) is not known, but MIYAMOTO et al. (1988) suggested that IRF-1 might correspond to PRDI-BF_i.

III. IFN-γ Induction

As indicated earlier in this chapter, IFN-γ is structurally unrelated to IFN-α/β. Although IFN-γ too is an inducible protein, the conditions under which it is produced are completely different from those established for IFN-α/β. Production of IFN-γ was demonstrated only in lymphocytes, i.e., T cells and

NK cells, with T lymphocytes apparently representing the major source of IFN-γ under most circumstances (reviewed by Epstein 1984; Trinchieri and Perussia 1985). Helper T cells are probably more important as a source of IFN-γ than suppressor T cells. In the murine system, the TH1 subset of T-helper cell lines, and not the TH2 subset, was found capable of producing IFN-γ (Mosmann et al. 1986).

IFN-γ is a lymphokine whose expression is induced by various treatments leading to lymphocyte activation. The physiological stimulus for T-cell activation is antigen, presented by an antigen-presenting cell in association with MHC molecules to the antigen-specific T-cell receptor/CD3 complex on T cells. Under experimental conditions antigen is often replaced with non-specific T-cell activators, e.g., phytohemagglutinin (PHA) or concanavalin A (Con A). IFN-γ is usually induced coordinately with interleukin-2 (IL-2) (Hardy et al. 1987), and IL-2, together with monocyte-derived IL-1, exert a positive regulatory influence on the generation of IFN-γ (Farrar et al. 1981; Kasahara et al. 1983; Le et al. 1986).

The induction of IFN-γ appears to be regulated at the level of transcription and initial attempts have been made to identify regulatory sequences within the IFN-γ DNA responsible for induction (reviewed by Taniguchi 1988). Stable transfection of an 8.6-kb DNA fragment containing the human IFN-γ coding sequence and its flanking regions into murine T cells led to an inducible expression of human IFN-γ, whereas transfection of the same DNA fragment into murine fibroblasts did not (Young et al. 1986). Hardy et al. (1987) identified several DNAse I hypersensitive sites, localized either within the 5′ flanking region or within the first intron of the human IFN-γ gene, that appear to be involved in regulating the expression of this gene in T cells. However, no clear understanding of the molecular mechanisms of IFN-γ induction has emerged so far.

D. Interferon Receptors

I. IFN-α/β Receptor

There is ample evidence to indicate that IFN-α/β and IFN-γ bind to two distinct, nonoverlapping receptors. Binding of ^{125}I-labeled recombinant human IFN-αA to a line of human lymphoblastoid cells was prevented in the presence of an excess of unlabeled human IFN-β, indicating that IFN-α and IFN-β bind to the same receptor (Branca and Baglioni 1981). However, IFN-γ did not compete with the binding of IFN-α. Similar results were obtained in mouse cells with murine interferons (Aguet et al. 1982). The structural gene for the human IFN-α/β receptor has been functionally "mapped" to the long arm of chromosome 21 (Epstein and Epstein 1976). Mouse-human somatic cell hybrids containing chromosome 21 as the only human chromosome are sensitive to the antiviral action of human IFN-β, but not of human IFN-γ (Radziuddin et al. 1984). Sensitivity to murine IFN-α/β correlates with the presence of murine chromosome 16, suggesting that this chromosome carries the murine IFN-α/β receptor gene (Cox et al. 1980; Lin et al. 1980). Both in man and in the mouse the locus controlling sensitivity to IFN-α/β is linked to the superoxide dismutase locus.

Neither complete purification and sequencing of the IFN-α/β receptor nor cloning of the corresponding gene have yet been reported. The major difficulty lies in the small number of binding sites which generally range from a few hundred to a few thousand per cell (reviewed by de Maeyer and de Maeyer-Guignard 1988). Preliminary evidence of the cloning of the human IFN-α/β receptor was recently presented (Shulman 1988), but the sequence of the cDNA isolated was not available at this writing. Scatchard analysis revealed the presence of a single class of high-affinity binding sites in most human or murine cells examined, with the K_d values established ranging from 1 to 5×10^{-10} M. Crosslinking of recombinant human IFN-α2 to its receptor in human lymphoblastoid cells revealed a single prominent band with an apparent molecular weight of approximately 150000 (Joshi et al. 1982). This finding suggests that the molecular weight of the receptor protein itself is approximately 130000 (assuming that IFN-α binds in monomeric form).

Many studies have analyzed events associated with IFN-α/β binding to its receptor, e.g., receptor-mediated endocytosis following ligand binding, receptor downregulation, etc. These studies have been reviewed by de Maeyer and de Maeyer-Guignard (1988).

II. IFN-γ Receptor

The existence of a specific receptor for IFN-γ, separate from the receptor for IFN-α/β, was demonstrated in numerous studies (Anderson et al. 1982; Novick et al. 1987; Molinas et al. 1987). As with the receptor for IFN-α/β, the number of specific binding sites on many different types of cells that have been examined ranges from several hundred to several thousand. Binding affinity ranged from a K_d of 10^{-9} to 5×10^{-11} M. Crosslinking experiments with radiolabeled IFN-γ revealed ligand/receptor complexes with an apparent molecular weight of approximately 80–130 kDa, both in human (Sarkar and Gupta 1984) and murine (Mariano et al. 1987) cells.

Crosslinking studies in human-mouse and human-hamster hybrid cells have localized the human IFN-γ receptor gene to the long arm of chromosome 6 (Rashidbaigi et al. 1986), in close proximity to the c-*ros* oncogene (Pfizenmaier et al. 1988). However, the presence of human chromosome 6 alone, while sufficient to confer specific binding, did not render the somatic hybrid cells sensitive to the antiviral action of human IFN-γ. (Note that human IFN-γ, because of its species specificity, does not exert antiviral activity in mouse or hamster cells.) Only when human chromosome 21 was present in the somatic hybrid cells, together with human chromosome 6, could antiviral activity be demonstrated (Jung et al. 1987). It is not yet clear what function is encoded by chromsome 21. It is possible that the IFN-γ receptor is composed of two chains, one of which (encoded by chromosome 6) is sufficient for specific binding, but the other (encoded by chromosome 21) is needed to activate the process of signal transduction. Alternatively, chromosome 21 could be encoding some other species-specific function needed to set in motion the full repertoire of cellular changes associated with IFN-γ action. It is intriguing that chromosome 21 is also believed to encode the IFN-α/β receptor gene (see above). Whether the function required for the expression of IFN-γ activity is identical to, or independent of, the gene(s) on chromosome 21 controlling the binding and actions of IFN-α/β has not been established. In any case, the requirement for a function encoded by human chromosome 21, along with chromosome 6, might explain why cells with three copies of chromosome 21 show enhanced sensitivity to the antiviral action of IFN-γ, as well as IFN-α/β (EPSTEIN 1984). A recent study suggested that a biological response to IFN-γ can be elicited in mouse cells transfected with human IFN-γ cDNA devoid of a signal peptide sequence (SANCÉAU et al. 1987).

The murine IFN-γ receptor gene (or at least its binding component) was localized to murine chromosome 10 (MARIANO et al. 1987). As in the human system, the presence of murine chromosome 10 alone was not sufficient to confer sensitivity to the antiviral action of murine IFN-γ in mouse-hamster somatic cell hybrids.

Several groups have recently succeeded in the purification of the human IFN-γ receptor. As a source served the Raji lymphoblastoid cell line (AGUET and MERLIN 1987), human fibroblasts (NOVICK et al. 1987), and human placenta (CALDERON et al. 1988). The intact isolated receptor migrated as a single species of 90 kDa. N-linked carbohydrate represents about 20% of the molecular mass. Limited proteolysis yielded a 55-kDa fragment that contains the ligand-binding site (CALDERON et al. 1988).

AGUET et al. (1988) have reported the cloning of the human IFN-γ receptor cDNA. A cDNA library from Raji cells was transfected into mouse cells in a lambda expression vector and cells expressing the human IFN-γ receptor were identified and isolated with the aid of an antiserum to the receptor. The cDNA isolated has an open reading frame encoding a 489-amino-acid protein, including a putative N-terminal hydrophobic signal peptide (Fig. 3). No significant homology to known proteins was found. Hydropathy index computation revealed a hydrophobic domain in the middle of the molecule compatible with a transmembrane-anchoring portion. The putative extracellular domain contains five potential N-glycosylation sites, with two additional ones on the putative cytoplasmic portion. Extensive glycosylation could account for the difference between the apparent size of 90 kDa of the purified natural receptor and the size of 54 kDa predicted from the amino acid sequence deduced from the cDNA. Southern blot analysis confirmed that the gene coding for this protein is located on chromosome 6. Murine L1210 cells transfected with an expression vector containing the isolated cDNA showed specific binding of human IFN-γ. However, such transfectants were not fully responsive to the biological action of human IFN-γ (induction of MHC class I antigens or of 2′-5′ oligoadenylate synthetase), in agreement with the earlier conclusion that at least two different structural genes are required for the establishment of sensitivity to the action of IFN-γ.

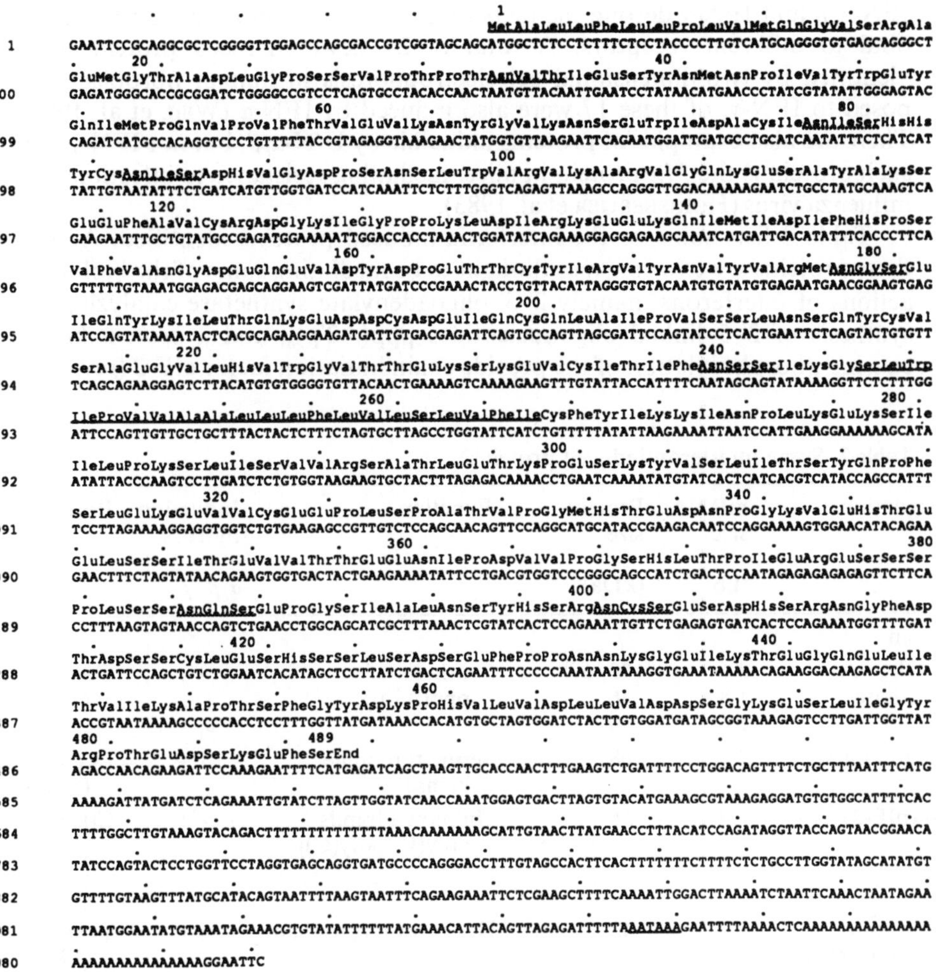

Fig. 3. The nucleotide and predicted amino acid sequences of the human IFN-γ receptor cDNA. Hydrophobic sequences predicted to form the signal peptide (1–14) and transmembrane anchoring (246–266) regions are *underlined.* Potential N-glycosylation sites are marked with *undulating lines.* (From AGUET et al. 1988, with permission from Cell Press)

E. Interferon Actions

I. Molecular Mechanisms

1. Proteins Induced by the Interferons

Interferons are potent modulators of gene expression (reviewed by REVEL and CHEBATH 1986). Many of the biological actions of the interferons can be attributed to the induction of specific proteins. Many of the proteins identified are induced by both the IFN-α/β species and IFN-γ; this overlapping pattern of gene activation explains why there are so many similarities in the actions of

IFN-α/β and IFN-γ despite the facts that they are structurally unrelated and bind to different receptors. A total of 24 newly induced proteins were identified by two-dimensional gel electrophoresis in human fibroblasts exposed to IFN-γ; of these 17 were also induced by IFN-α (WEIL et al. 1982). There are also examples of proteins inducible by IFN-α/β, but not IFN-γ; one such protein is the *Mx* protein in mice, mediating resistance to infection with influenza virus (HORISBERGER et al. 1983).

Table 2 lists some of the interferon-induced proteins identified. Among these proteins are two enzymes thought to be mediating some of the biological actions of interferons, namely 2′-5′ oligoadenylate synthetase catalyzing the synthesis of 2′-5′-linked oligonucleotides ppp(A2′p)$_n$, and a protein kinase causing phosphorylation of the peptide chain initiation factor eIF-2 (reviewed

Table 2. Some interferon-induced proteins[a]

Designation	mRNA size (kb)	Protein size (kDa)	Function	Induction by IFN α/β γ	References
Human					
2′5′ Oligo A synthase	3.6 1.8 1.6	100, 67[b] 46 40	Synthesis of ppp(A2′p)$_n$	+ > +	CHEBATH et al. (1987)
Protein kinase	2.5	68	Serine-threonine kinase	+ > +	LAURENT et al. (1985)
C56 (pIF-2)	2.0	56	Unknown [binds poly(I) · poly(C)]	+ > +	CHEBATH et al. (1983), LARNER et al. (1984)
Indoleamine 2,3-dioxygenase	2.2	42	Degrades tryptophan	+ ≪ +	YOSHIDA et al. (1981), RUBIN et al. (1988)
pIF-1	2.9	42, 58	Unknown	+ ?	LARNER et al. (1984)
1–8	0.8	?	Unknown	+ ?	FRIEDMAN et al. (1984)
6–16	1.0	13 (secreted)	Unknown	+ ?	FRIEDMAN et al. (1984)
Metallothionein-IIa	0.6	7	Metal detoxification	+ ?	FRIEDMAN et al. (1984)
HLA-A, B, C	1.8	44	Class I HLA, heavy chain	+ < +	FELLOUS et al. (1982)
B_2-microglobulin	0.9	14	Class I HLA, invariant light chain	+ < +	WALLACH et al. (1982)
HLA-DRα	1.3	34	Class II HLA heavy and light chain	+ ≪ +	REVEL and CHEBATH (1986)

Table 2 (continued)

Designation	mRNA size (kb)	Protein size (kDa)	Function	Induction by IFN α/β	γ	References
GBP	4.0	67	Guanylate binding protein	+	?	CHENG et al. (1983)
IP-10	1.5	10 (secreted)	Related to platelet factor IV	+	< +	LUSTER et al. (1985)
IP-30	1.1	30 (secreted)	Unknown (lysosomal localization)	?	+	LUSTER et al. (1988)
15-kDa protein	0.7	15	Unknown (cytoplasmic)	+	?	BLOMSTROM et al. (1986)
Murine						
Mx	3.5	72	Inhibits influenza virus	+	–	STAEHELI et al. (1986)
C202	2.0	56	Homologous to human C56	+	–	SAMANTHA et al. (1986)
TNF		17	Tumor necrosis factor/cachectin	1.7	+	COLLART et al. (1986)

[a] From REVEL and CHEBATH (1986), with modifications.
[b] Whether both the 100- and 67-kDa proteins are derived from the 3.6-kb mRNA has not been firmly established.

by LENGYEL 1982; PESTKA et al. 1987; DE MAEYER and DE MAEYER-GUIGNARD 1988). A unique feature of these two enzymes is that they are both activated by double-stranded RNA.

2′-5′ Oligoadenylate synthetase activity actually is associated with a family of functionally related proteins. They are the most extensively studied interferon-induced proteins. Three forms of mRNA (1.6, 1.8, and 3.6 kb in size) and four forms of protein (40, 46, 69, and 100 kDa) have been demonstrated in human cells (CHEBATH et al. 1987). The two smaller forms of mRNA (1.6 and 1.8 kb), derived from the same gene by differential splicing, give rise to the 40- and 46-kDa proteins, respectively. The sequence of the 3.6-kb RNA is not known, nor is it firmly established whether it gives rise to the two higher molecular weight forms of the protein. The four different forms of 2′-5′ oligoadenylate synthetase differ in their subcellular localization and in the activation requirements by double-stranded RNA.

The enzyme 2′-5′ oligoadenylate synthetase converts ATP to oligonucleotides ppp(A2′p)$_n$, where *n* ranges from 2 to about 15, with the trimer being the most abundant (HOVANESSIAN et al. 1977). The only known function of these oligoadenylates is to bind to and activate a latent RNase, termed RNase L. Trimers (and in some cases tetramers) of adenylate were the smallest form able to activate RNase L in vitro (reviewed by LENGYEL 1982).

Activation involves a direct binding of ppp(A2′p)$_n$ to the enzyme. Levels of oligoadenylate appear to be controlled by a 2′-5′ phosphodiesterase which can cleave preferentially 2′-5′ phosphodiester bonds. Although RNase L is present in cells constitutively, at least under some conditions it also appears to be induced by interferon treatment. RNase L cleaves all known forms of natural single-stranded RNA, predominantly at the 3′ side of UA, UG, and UU sequences.

The double-stranded RNA-dependent serine and threonine protein kinase induced by interferon (LEBLEU et al. 1976; ROBERTS et al. 1976; ZILBERSTEIN et al. 1976) is known to phosphorylate only two endogenous cellular proteins: a 35-kDa protein which is the α-subunit of protein synthesis initiation factor eIF2 and a 68-kDa protein (in human cells) which is the kinase itself (GALABRU and HOVANESSIAN 1985). Phosphorylation of eIF2 results in its inactivation (and inhibition of protein synthesis). In contrast, autophosphorylation of the p68 kinase apparently results in an increase of its own kinase activity. Autophosphorylation can occur only when the protein is activated by double-stranded RNA. However, heparin and some other polyanionic compounds can substitute for double-stranded RNA as activators of the p68 kinase. A 48-kDa polypeptide, until recently thought to be part of the kinase complex, now is known to be a product of proteolytic degradation of the 68-kDa protein (HOVANESSIAN and GALABRU 1987).

Induction characteristics and properties of some of the other interferon-induced proteins listed in Table 2 have been reviewed by REVEL and CHEBATH (1986).

2. Mechanisms of Gene Activation by Interferons

Very little is known about mechanisms of signal transduction mediating gene activation by the interferons. It has been reported that a very rapid and transient increase in diacylglycerol concentration occurs in human cell lines upon exposure to IFN-α or IFN-β (YAP et al. 1986). This finding suggests a role for an activation of protein kinase C in IFN-α/β action. However, no independent confirmation of these findings has been reported. There is stronger evidence for a role of protein kinase C activation in the action of IFN-γ on macrophages (HAMILTON et al. 1985; FAN et al. 1988).

Despite the paucity of information about the signal transduction pathways of interferon action, a great deal is being learned about the nuclear events mediating gene activation by the interferons. FRIEDMAN and STARK (1985) proposed that a sequence of about 30 bp, present in the 5′ flanking regions of four interferon-inducible genes (metallothionein-II, MHC class II, and two MHC class I genes), is important in the regulated transcription of interferon-responsive genes. Sequences homologous with the Friedman-Stark consensus sequence ("interferon response sequence" or IRS) were also found in the 3′ noncoding region of the murine β_2-microglobulin gene (ISRAEL et al. 1986) and in the coding sequence of the first exon in the murine 202 gene (SAMANTA et al. 1986). In addition, sequences similar to the IRS were found in the gene for a 56-kDa human interferon-inducible protein at three different locations (one on the sense strand and two on the anti-sense strand) (WATHELET et al. 1987). The IRS sequence was shown to confer inducibility by interferon to a reporter gene (SUGITA et al. 1987). However, KORBER et al. (1988) observed that the *cis*-acting sequences necessary to elicit an interferon response in a murine class I MHC gene vary among different cell lines, with some cell lines requiring a region upstream of the IRS region, in addition to the IRS. They also concluded that promoter region requirements were different for induction by IFN-α/β and IFN-γ.

LEVY et al. (1988) identified a shorter, approximately 15-nucleotide-long sequence, contained within the IRS sequence, that is present in the 5′ flanking region of 13 different interferon-responsive genes. The importance of this sequence (termed IFN-stimulated response element or ISRE) in the control of interferon inducibility was also established by deletion analysis and point mutations. These authors then identified two classes of nuclear factors that interact with the ISRE sequence. One factor (B1) was present constitutively in untreated cells and two others (B2 and B3) appeared only after the exposure of cells to IFN-α. B2, although strongly induced by IFN-α, did not appear until 2 h after treatment and its appearance was inhibited by cycloheximide, whereas the appearance of B3 was faster and not inhibited by cycloheximide. The authors proposed that B3 might act as a positive activator and B2 as a negatively acting transcription factor.

Quite similar results were reported by SHIRAYOSHI et al. (1988), who identified a sequence of only about ten base pairs, present in interferon-inducible genes, sufficient to bind three distinct nuclear factors. The proposed model of control of interferon-induced genes by these three nuclear factors is also quite similar to the general model of LEVY et al. (1988).

A sequence virtually identical to the ISRE sequence of LEVY et al. (1988) was also found in the 5′ flanking region of the interferon-inducible murine *Mx* gene by HUG et al. (1988). In addition, these authors noted that the *Mx* promoter region mediating induction by IFN-α/β contains five sequences of the type GAAANN, and that these sequences occur, in either orientation, in all interferon-inducible genes. Finally, the authors mention evidence indicating that the presence of $(GAAACT)_4$ can confer inducibility by interferon. Most interestingly, the same hexanucleotide apparently also can mediate inducibility by virus.

3. Common Mechanisms of Gene Activation by Interferons, Viruses, Double-Stranded RNA, Growth Factors, and Cytokines

Several unexpected recent observations suggest that the regulatory mechanisms for the expression of interferon-inducible genes overlap with the mechanisms regulating the expression of the IFN-α/β genes themselves. WATHELET et al. (1987) noted that two interferon-inducible genes, the 42-kDa human 2′-5′ oligoadenylate synthetase and a 56-kDa protein of unknown function, contain sequences highly homologous with the interferon response element (IRE) sequence in the 5′ flanking region of the human IFN-β gene (see Sect. C.II). Since the IRE region confers inducibility by viruses or double-stranded RNA, the authors examined inducibility of the synthetase and 56-kDa protein genes by poly(I)·poly(C) in two human cell lines. Induction was demonstrable by double-stranded RNA as well as by interferon treatment. A similar dual inducibility by IFN-α or Newcastle disease virus was demonstrated for the murine *Mx* gene by Hug et al. (1988). These authors state that tetramers of the hexanucleotide GAAACT can mediate induction by both interferon or virus. Korber et al. (1988) also observed a similarity between the interferon response sequence in the upstream region of a murine class I MHC gene and the IRE region of the IFN-β gene. They proposed that enhancement of gene expression by IFN-α/β may be controlled in part by an interferon-mediated induction of a transcription factor binding to the IRE sequence.

Evidence of an overlap between the control of interferon-inducible genes and the expression of IFN-α/β genes was also obtained in other experiments. Enoch et al. (1986) have shown that two signals are required for IFN-β induction in HeLa cells, one that is provided by virus or double-stranded RNA and one that is elicited by the treatment of cells with interferon. Working along somewhat similar lines, Tiwari et al. (1987) have analyzed the regulation of interferon-inducible genes in the HeLaM cell line that appears to be deficient in a constitutive factor involved in interferon action. The authors proposed the following model of interferon action in HeLaM cells. IFN-α produces a signal (signal 1) which enhances the synthesis of protein X (a factor present constitutively in other cell lines). IFN-α also produces signal 2 which does not require protein synthesis. Both protein X and signal 2 are needed for the induction of interferon-regulated mRNA 561, i.e., IFN-α alone can induce mRNA 561, but cycloheximide can inhibit this induction by IFN-α. (The protein coded by mRNA 561 is thought to be homologous to the 56-kDa protein listed in Table 2 under the name C56 or pIF-2.) IFN-γ, epidermal growth factor (EGF), and platelet-derived growth factor (PDGF) each could produce signal 1, but not signal 2. In contrast, two forms of double-stranded RNA could produce signal 2, but not signal 1. Double-stranded RNA, combined with IFN-γ, EGF, or PDGF, led to mRNA 561 induction.

Other interesting relationships have emerged from studies on a nuclear factor termed interferon regulatory factor-1 or IRF-1 (Miyamoto et al. 1988; see also Sect. C.II above). IRF-1 was identified on the basis of its ability to bind to upstream regulatory sequences in the human IFN-β gene, and it is very likely that IRF-1 plays an important role in the regulation of transcription of the IFN-α/β genes. Several interferon-induced genes (e.g., class I MHC genes) also contain regulatory sequences recognized by IRF-1, suggesting that they too are modulated by this factor. Furthermore, synthesis of IRF-1 itself was induced by Newcastle disease virus and the IRF-1 gene appears to contain a virus-inducible promoter. Another interesting and likely possibility mentioned by the authors is that interferons also induce IRF-1 expression, as might some growth factors and cytokines. Finally, IRF-1 might be expected to regulate the expression of its own gene.

Two cytokines, tumor necrosis factor (TNF) and interleukin-1 (IL-1), were found to have some interferon-like actions. Both TNF (KOHASE et al. 1986; MESTAN et al. 1986; REIS et al. 1988) and IL-1 (CONTENT et al. 1985; VAN DAMME et al. 1987) in some cells induce an interferon-like antiviral action. In addition, TNF and IL-1 were shown to induce the synthesis of some proteins that are also induced by the interferons (PFIZENMAIER et al. 1987; BERESINI et al. 1988; RUBIN et al. 1988). Furthermore, like interferon, TNF and IL-1 can modulate IFN-β synthesis; under appropriate conditions both cytokines can prime cells to produce increased levels of IFN-β (KOHASE et al. 1988), and in some cells TNF (ONOZAKI et al. 1988; JACOBSEN et al. 1989; REIS et al. 1989) or IL-1 (VAN DAMME et al. 1985) can actually induce IFN-β. PDGF (ZULLO et al. 1985) and nerve growth factor (SAARMA et al. 1986) too were shown to induce the 2′-5′ oligoadenylate synthetase.

These recent findings lead to several conclusions: (a) signals that activate IFN-α/β genes (e.g., double-stranded RNA, viruses) are much less selective than they were thought to be; they also lead to the activation of several other genes, including many interferon-regulated genes. (b) The actions of interferons also are not as unique as they were thought to be; several cytokines and growth factors can mimic at least some actions characteristic for the interferons. (c) Double-stranded RNA or virus infection cannot be considered physiological inducers. If interferons indeed function as regulators of cell growth and differentiation, other signals should exist for IFN-α/β induction. Cytokines, growth factors, and even interferons themselves are likely candidates for the role of physiological regulators of IFN-α/β production.

II. Spectrum of Biological Activities

1. Inhibition of Cell Growth

The most extensively studied biological activities of interferons are those related to their antiviral effects (reviewed by LENGYEL 1982; SEN 1984; DE MAEYER and DE MAEYER-GUIGNARD 1988). These studies (which will not be surveyed in this chapter) led to many important discoveries whose significance extends to other activities of the interferons. Among their many known actions, much recent attention has been focusing on the effects of the interferons on cell growth and differentiation.

In view of the enormous volume of published studies, it is impossible to include here a comprehensive review of the literature on interferons' effects on cell growth. The subject has been covered more extensively in several recent reviews (SREEVALSAN 1984; CLEMENS and MCNURLAN 1985; TAMM et al. 1987). The sensitivity of cells in culture to the antiproliferative action of interferon varies greatly. Some human B lymphoblastoid cell lines (e.g., the Daudi line) count among the most sensitive, while many other cells (either normal or transformed) are relatively resistant or completely refractory to the antiproliferative action. Individual cell lines can also differ greatly in their sensitivity to the antiproliferative action of different interferon types and subtypes. In many (but not all) cell lines IFN-γ is a more potent inhibitor of cell growth than either IFN-α or IFN-β.

The actions of interferons on various stages of the cell cycle have been extensively studied, especially in murine 3T3 fibroblast lines. In G_0-arrested cells interferon treatment can inhibit serum- or growth factor-induced entry into G_1 and transition to S phase (SOKAWA et al. 1977; BALKWILL and TAYLOR-PAPADIMITRIOU 1978). However, other stages of the cell cycle can be affected by interferons as well, and a prolongation of the G_1, S, or mitotic phases have been reported in fibroblasts and other types of cells (CLEMENS and MCNURLAN 1985). These actions can occur in the absence of a significant change in the overall rate of cellular RNA and protein synthesis. In human diploid fibroblasts the principal action is on cell division and not cell "growth" per se; treatment with IFN-β results in a prolongation of all phases of the cell cycle and a significant increase in cell volume (PFEFFER et al. 1979).

A mutually antagonistic relationship between various growth factors and interferons has been noted. The inhibitory action of IFN-α/β on DNA synthesis in 3T3 cells can be overcome by the addition of a mixture of several growth-stimulatory agents (TAYLOR-PAPADIMITRIOU et al. 1981). In addition, it has been noted that PDGF and fibroblast growth factor (FGF) under some conditions can inhibit the antiviral action of interferon (OLESZAK and INGLOT 1980). These findings, together with the fact that the antiproliferative actions of interferon are reversible and that they generally occur in the absence of cell toxicity, have led to the view that interferons function as "negative growth factors."

No satisfactory unifying hypothesis can explain all aspects of interferons' antiproliferative actions. Considerable attention has been devoted to the analysis of the effects of interferon treatment on cellular protooncogenes, especially c-*myc*. In the Daudi cell line, growth inhibition by IFN-α or -β correlates with reduced steady-state levels of c-*myc* mRNA. Two groups concluded that this action is due to a reduction in c-*myc* mRNA half-life (KNIGHT et al. 1985; DANI et al. 1985), while another group ascribed it to a reduction of c-*myc* transcription (EINAT et al. 1985). However, reduction in c-*myc* levels is often not sufficient, and sometimes not even necessary, for the establishment of the antiproliferative action of interferon. For example, TOMINAGA and LENGYEL (1985) showed that treatment of growth-arrested BALB/c 3T3 cells with IFN-α/β before their stimulation with PDGF failed to inhibit the PDGF-induced enhancement of c-*myc* mRNA levels even though the mitogenic action was inhibited. An antiproliferative action of IFN-α in the absence of a significant inhibitory action on c-*myc* mRNA levels was also seen in a subline of Daudi cells (DRON et al. 1986).

These and other seemingly contradictory findings suggest that the antiproliferative action of interferon cannot be ascribed to a single event. Moreover, the inhibitory actions apparently can take place at different stages of the cell cycle. A reduction in c-*myc* levels probably reflects an inhibitory effect on the early mitogenic signals occurring during G_0–G_1 transition or during the early G_1 phase (LIN et al. 1986). On the other hand, some other observations suggest that the inhibitory action of interferon correlates with the suppression of events occurring during the late G_1 phase. EBSWORTH et al. (1986) noted that the kinetics of interferon's inhibitory action in Swiss 3T3 cells mirrored the kinetics of the mitogenic action of tubulin-disrupting agents (colchicine or nocodazole) which are thought to act during the late G_1 phase, and they suggested that interferon may be acting via the microtubules. This idea was supported by the demonstration that microtubule-disrupting agents inhibited the antiproliferative action of interferon. Earlier it was shown that interferon treatment causes marked changes in the cytoskeleton (CHANY et al. 1980; PFEFFER et al. 1980), but whether these changes are causally related to the antiproliferative action is not known.

Another action that may be revelant to growth inhibition by interferons is the modulation of growth factor receptors. It has been shown that in some cell lines treatment with interferon can reduce the specific binding of insulin (PFEFFER et al. 1987), transferrin (BESANÇON et al. 1985), EGF (ZOON et al. 1986), or M-CSF (CHEN et al. 1986). However, with the exception of insulin binding, the relationship of these changes in receptor binding to the antiproliferative action of interferon has not been analyzed. An increase in the expression of receptors for the growth inhibitory agent tumor necrosis factor (TNF) after treatment with interferons, especially IFN-γ, has also been noted (AGGARWAL et al. 1985; RUGGIERO et al. 1986). However, this action probably does not contribute in a major way to the potentiation of TNF's cytostatic/cytotoxic action in various tumor cell lines (TSUJIMOTO et al. 1986).

Also not fully resolved is the question of which intracellular biochemical pathways activated by the interferons are important in the antiproliferative actions. An involvement of the 2′-5′ oligoadenylate synthetase/ribonuclease L pathway has been suggested because under some conditions the addition of 2′-5′ oligoadenylates to cells was found to inhibit cell proliferation and because fluctuations in the intracellular levels of 2′-5′ oligoadenylate synthetase were seen during the cell cycle (reviewed by REVEL and CHEBATH 1986; DE MAEYER and DE MAEYER-GUIGNARD 1988). However, for a number of reasons this mechanism alone cannot account for many antiproliferative effects of the interferons and, almost certainly, a number of different molecular events are involved. Another interferon-induced protein, indoleamine 2,3-dioxygenase (see Table 2), is also a candidate for the mediation of the antiproliferative action of IFN-γ (RUBIN et al. 1988).

2. Stimulation of Cell Growth

Although inhibition of cell growth is much more commonly seen with the interferons, stimulatory effects have also been reported. BRADLEY and RUSCETTI (1981) examined the effect of several partially purified natural IFN-α/β preparations on colony formation of freshly isolated human tumor cells in soft agar. Although most tumor cell populations tested showed a variable degree of inhibition, some formed an increased number of colonies in the presence of interferon (1000 U/ml). LUDWIG et al. (1983) examined the action of IFN-α (both natural and recombinant) in the clonogenic assay with cells from fresh human tumor tissues. A growth stimulation was seen in about 13% of the samples (from acute myeloid leukemia, renal cancer, breast cancer, and, less frequently, from melanomas). The fact that *E. coli*-derived recombinant IFN-α had a similar stimulatory effect as natural IFN-α makes it very unlikely that this effect was due to a contaminant, rather than IFN-α itself. A similar increase in colony formation was also seen with an established myeloma cell line, suggesting that the stimulation was not mediated indirectly through an action on other host cells contaminating the tumor cell preparations.

A stimulation of cell proliferation was also seen with IFN-α in cultures of peripheral blood leukocytes from 3 out of 29 patients with chronic B-lymphocytic leukemia (Östlund et al. 1986). Blast formation and plasmacytoid differentiation after interferon treatment was seen in cultures from 19 of 29 patients. In view of the heterogeneity of the cell populations used in these experiments, these stimulatory effects of IFN-α could have been either direct or indirect.

Increased proliferation was also seen in cultures of human synovial or skin fibroblasts treated with recombinant human IFN-γ (Brinckerhoff and Guyre 1985). The mitogenic effect was relatively modest and it could be demonstrated only after a prolonged incubation of cells with IFN-γ, suggesting that it might have been due to the induction of an autocrine growth factor by IFN-γ. Other studies showed that IFN-γ increased the proliferation of mitogen-activated normal human B cells (Kehrl et al. 1987); this action is likely to be related to the regulatory effect of IFN-γ on immunoglobulin synthesis (Snapper and Paul 1987).

3. Other Biological Activities

Table 3 lists the major biological actions observed with the interferons in cells in culture and/or in the intact organism. Some of the actions listed in Table 3 are complex and almost certainly involve multiple gene functions (e.g., activation of monocytes/macrophages), while other actions can be explained by a direct activation by the interferons of known genes (e.g., class I or II MHC antigens). The induction of 2′-5′ oligoadenylate synthetase and double-stranded RNA-dependent protein kinase are known to be important in the antiviral action against some viruses, but there are many examples of antiviral actions whose molecular mechanisms are not known (reviewed by Johnston and Torrence 1984; Sen 1984; Friedman and Pitha 1984; de Maeyer and de Maeyer-Guignard 1988).

Numerous other actions of interferon have been reported, especially in many in vitro models of differentiation (reviewed by Grossberg and Taylor 1984, Rossi 1985). One interesting action is the reversion of the transformed phenotype seen in several tumor cell lines treated with interferon in vitro (Chany et al. 1980). In a long terminal repeat (LTR)-activated Ha-*ras* oncogene-transformed mouse cell line, prolonged treatment with interferon produced revertant cells that did not form colonies in soft agar or tumors in

Table 3. Major biological actions of the interferons

Activity	Observed with	
	IFN-α/β	IFN-γ
Induction of antiviral state	+	+
Inhibition of cell growth	+	+
Stimulation of cell growth	+	+
Induction of class I MHC antigens	+	+
Induction of class II MHC antigens	±	+
Activation of monocytes/macrophages	+	+
Activation of natural killer cells	+	+
Activation of cytotoxic T cells	+	+
Stimulation of Ig synthesis in B cells	+	+
Induction of F_c receptors in monocytes	−	+
Inhibition of the growth of nonviral intracellular pathogens	−	+
Pyrogenic action	+	+

+, positive effect; ±, weak or variable effect; −, negative.

nude mice (Samid et al. 1984). In most such lines the revertant phenotype was retained on prolonged culture in the absence of interferon, even though the level of *ras* mRNA or p21 protein in the revertant lines was comparable to that in the fully transformed parent line. The suggestion has been made (Friedman 1987) that incubation with interferon may result either in a stable inactivation of a locus associated with transformation or activation of an anti-oncogene suppressor factor.

4. Possible Physiological Roles

Although high levels of interferon are produced upon some inducing treatments, e.g., virus infection (in the case of IFN-α/β) or mitogenic stimulation of T cells (for IFN-γ), circumstantial evidence suggests that the production of much lower levels of interferon in the localized environment of certain tissues can have physiological significance (Bocci 1985). In general, it is relatively easy to visualize a physiological role for IFN-γ, because it is produced by the ubiquitous T cells and NK cells and, mainly, because IFN-γ is known to exert potent regulatory effects, even at low concentrations, on monocytes/macrophages, B cells, and other cells of the immune system.

There is evidence that small amounts of IFN-α can be produced by cells and tissues (especially monocytes/macrophages) of healthy individuals (Tovey et al. 1987). It has been shown that macrophages freshly explanted from normal mice are in a state of interferon-induced resistance to virus infection (Belardelli et al. 1984), but a truly physiological role for IFN-α that is produced in healthy animals has not been revealed. Nor is it clear whether such small amounts of IFN-α are produced spontaneously or in response to some infectious agent, e.g., in response to small amounts of LPS released into the circulation by intestinal bacteria.

Ovine trophoblast protein-1 (oTP-1), a protein produced by 13- to 21-day-old preimplantation sheep embryos that is implicated in eliciting maternal recognition of pregnancy, was unexpectedly identified as a member of the IFN-α family (IMAKAWA et al. 1987; HANSEN et al. 1988). The closest homology (about 60%) was found between oTP-1 and bovine IFN-α_{II}, suggesting that oTP-1 is a member of the sheep IFN-α_{II} subfamily. Secretion of oTP-1 after day 13 of pregnancy is believed to be essential for preventing a return to ovarian cyclicity and rejection of the embryo. Production of oTP-1 mRNA by the trophobast (preplacenta) peaks sharply on day 14 of pregnancy and returns to undetectable levels by day 22. During the short period of maximal synthesis, oTP-1 is thought to be the major protein produced by the conceptus. The exact mechanism of action of oTP-1 during pregnancy is not yet known, but the recognition that it is a member of the IFN-α family should help in the elucidation of its functions. Conversely, this model may yield the first clear demonstration of a physiological function for an IFN-α protein.

There are no data available from animal experiments that would suggest a physiological role for IFN-β. However, several groups of investigators have obtained evidence suggesting that the generation of small amounts of IFN-β in various types of cells in culture can affect cell growth and differentiation. Most of the evidence for a role of endogenous ("autocrine") IFN-β is indirect and based on the demonstration that the addition of antisera to IFN-β either can enhance cell growth (CREASEY et al. 1983; KOHASE et al. 1986; RESNITZKY et al. 1986) or suppress the appearance of a function known to be upregulated by interferon, e.g., an increase in the levels of 2′-5′ oligoadenylate synthetase or MHC antigen expression (SOKAWA et al. 1981; YARDEN et al. 1984). It is not clear how such very low concentrations of autocrine IFN-β that escape demonstration in sensitive assays can have a significant effect on cell growth or 2′-5′ oligoadenylate synthetase levels. One possibility is that concentrations of IFN-β insufficient to produce a biological effect on their own become effective when some other cytokine or growth factor is present in the same culture. For example, there are many reports of a synergistic action of interferons and tumor necrosis factor on gene expression (TRINCHIERI et al. 1986; LAPIERRE et al. 1988) or inhibition of virus multiplication (WONG and GOEDDEL 1986; KOHASE et al. 1987; see also Sect. E.I.3 "Common Mechanisms of Gene Activation by Interferons, Viruses, Double-Stranded RNA, Growth Factors and Cytokines").

5. Roles in Pathophysiology and Therapeutic Applications

Production of IFN-α/β during virus infections is generally beneficial as it serves to limit the spread of virus and promote recovery (GRESSER et al. 1976). However, in mice infected with lymphocytic choriomeningitis (LCM) virus the generation of interferon is in fact responsible for the symptoms of disease (liver cell degeneration, stunted growth, and death), and administration of antiserum to murine IFN-α/β exerts a protective influence (RIVIÈRE et al. 1977). In at least one human virus infection, namely Argentine hemorrhagic fever caused by Junin virus (a member of the arenavirus family, as is murine LCM virus), high fever and mortality show a positive correlation with the presence of extremely high levels of IFN-α in the patients' serum, suggesting a pathogenetic role for IFN-α (LEVIS et al. 1984).

There is evidence suggesting a role for interferon in the pathogenesis of some autoimmune diseases, especially in systemic lupus erythematosus (SLE). The serum of many SLE patients contains levels of interferon demonstrable in conventional antiviral assays (SKURKOVICH and EREMKINA 1975; HOOKS et al. 1979), whereas serum from normal individuals rarely, if ever, contains detectable interferon levels. On the basis of its neutralization with antibodies the interferon in SLE patients was identified as IFN-α (PREBLE et al. 1982). However, unlike other known forms of IFN-α, the interferon activity in the serum of SLE patients is acid-labile, i.e., it shows a variable degree of inactivation upon exposure to pH 2. Since the interferon from SLE patients' sera has not been isolated, it is unclear whether it corresponds to any one of the known IFN-α protein(s). Although there is no direct evidence linking increased production of IFN-α to the pathogenesis of SLE, some data supporting such a link have been obtained in mouse models of SLE. These studies along with other considerations of the involvement of interferon in SLE and other autoimmune disorders have been thoughtfully and incisively analyzed by DE MAEYER and DE MAEYER-GUIGNARD (1988). Evidence for a role of IFN-γ has also been obtained, both in some human autoimmune disorders and in animal models (BILLIAU et al. 1988). Elevated IFN-α showing a characteristic acid-lability was also found in a majority of AIDS patients and in some patients with AIDS-related lymphadenopathy (DESTEFANO et al. 1982). The cellular origin of this interferon, the inducing stimulus (HIV virus itself or an autoimmune component?), and its possible role in the pathogenesis of AIDS have not been elucidated.

Ever since its discovery by ISAACS and LINDENMANN (1957), there was hope that one day interferon could be employed therapeutically. The expectations of therapeutic applicability were a powerful impetus for work on the purification, cDNA cloning, and production of interferons. In the past few years several types of interferon preparations have been licensed for clinical use. In the United States *E. coli*-derived recombinant human IFN-α2 and IFN-αA have been approved for use in the treatment of hairy cell leukemia. IFN-α2 and IFN-αA are both members of the IFN-α_{I} subfamily and they differ from each other in a single amino acid in position 23 (Arg in α2 and Lys in αA). One of the preparations has also been approved for the treatment of condylomata acuminata. Other interferon preparations too have been approved for clinical use in some countries, e.g., a natural mixture of several IFN-α subtypes produced in the Namalwa line of human lymphoblastoid cells or natural human IFN-β produced in cultured fibroblasts. The approved use of these interferon preparations in some countries includes chronic active hepatitis B, acute viral encephalitides, and nasopharyngeal carcinoma. A preparation of *E. coli*-derived recombinant human IFN-γ has been approved for therapeutic use in rheumatoid arthritis in the German Federal Republic. Approved and experimental therapeutic applications of interferons have been extensively covered in a volume devoted to this topic (FINTER and OLDHAM 1985).

Although the therapeutic achievements have so far fallen short of the most optimistic expectations of a decade ago (especially in solid tumors and common virus infections), new potential clinical applications are still being found (EZEKOWITZ et al. 1988). Most importantly, work on the interferons has served as a paradigm for the pharmacological development of other cytokines and peptide growth factors.

Acknowledgements. I thank Jian-Xin Lin for the preparation of sequence data, Peter Lengyel for critical reading of the manuscript, Rena Feinman and Tadatsugu Taniguchi for helpful information, Edward de Maeyer and Jaqueline de Maeyer-Guignard for having written their splendid book *Interferons and Other Regulatory Cytokines,* which made my job so much easier, and Ilene Toder for careful preparation of the manuscript.

References

Adolf GR (1987) Antigenic structure of human interferon ω1 (Interferon alpha II): comparison with other human interferons. J Gen Virol 68:1669–1676

Aggarwal BB, Eessalu TE, Hass PE (1985) Characterization of receptors for human tumour necrosis factor and their regulation by γ-interferon. Nature 318:665–666

Aguet M, Merlin G (1987) Purification of human γ interferon receptors by sequential affinity chromatography on immobilized monoclonal antireceptor antibodies and human γ interferon. J Exp Med 165:988–999

Aguet M, Belardelli F, Blanchard B, Marcucci F, Gresser I (1982) High affinity binding of ^{125}I-labeled mouse interferon to a specific cell surface receptor. IV. Mouse γ interferon and cholera toxin do not compete for the common receptor site of α/β interferon. Virology 117:541–544

Aguet M, Dembić Z, Merlin G (1988) Molecular cloning and expression of the human interferon-γ receptor. Cell 55:273–280

Allen G, Fantes KH (1980) A family of structural genes for human lymphoblastoid (leukocyte-type) interferon. Nature 287:408–411

Anderson P, Yip YK, Vilček J (1982) Specific binding of ^{125}I-human interferon-γ to high affinity receptors on human fibroblasts. J Biol Chem 257:11301–11304

Balkwill F, Taylor-Papadimitriou J (1978) Interferon affects both G1 and S+G2 in cells stimulated from quiescence to growth. Nature 274:798–800

Belardelli F, Vignaux F, Proietti E, Gresser I (1984) Injection of mice with antibody to interferon renders peritoneal macrophages permissive for vesicular stomatitis virus and encephalomyocarditis virus. Proc Natl Acad Sci USA 81:602–608

Beresini MH, Lempert MJ, Epstein LB (1988) Overlapping polypeptide induction in human fibroblasts in response to treatment with interferon-alpha, interferon-gamma, interleukin 1 alpha, interleukin 1 beta and tumor necrosis factor. J Immunol 140:485–493

Besançon F, Bourgeade MF, Testa U (1985) Inhibition of transferrin receptor expression by interferon-α in human lymphoblastoid cells and mitogen-induced lymphocytes. J Biol Chem 260:13074–13080

Billiau A, Heremans H, Vandekerckhove F, Dijkmans R, Sobis H, Meulepas E, Carton H (1988) Enhancement of experimental allergic encephalomyelitis in mice by antibodies against IFN-gamma. J Immunol 140:1506–1510
Blomstrom DC, Fahey D, Kutny R, Korant BD, Knight E Jr (1986) Molecular characterization of the interferon-induced 15-kDa protein. Molecular cloning and nucleotide and amino acid sequence. J Biol Chem 261:8811–8816
Bocci V (1985) The physiological interferon response. Immunol Today 6:7–9
Bradley EC, Ruscetti FW (1981) Effect of fibroblast, lymphoid and myeloid interferons on human tumor colony formation in vitro. Cancer Res 41:244–249
Branca AA, Baglioni C (1981) Evidence that types I and II interferons have different receptors. Nature 294:768–770
Brinckerhoff CE, Guyre PM (1985) Increased proliferation of human synovial fibroblasts treated with recombinant immune interferon. J Immunol 134(5):3142-6
Calderon J, Sheehan KCF, Chance C, Thomas ML, Schreiber RD (1988) Purification and characterization of the human interferon-γ receptor from placenta. Proc Natl Acad Sci USA 85:4837–4841
Capobianchi MR, Facchini J, di Marco P, Antonelli G, Dianzani F (1985) Induction of α interferon by membrane interaction between viral surface and peripheral blood mononuclear cells. Proc Soc Exp Biol Med 178:551–556
Capon DJ, Shepard HM, Goeddel DV (1985) Two distinct families of human and bovine interferon-α genes are coordinately expressed and encode functional polypeptides. Mol Cell Biol 5:768–779
Chany C, Rousset S, Bourgeade MF, Mathieu D, Gregoire A (1980) Role of receptors and the cytoskeleton in reverse transformation and steroidogenesis induced by interferon. Ann NY Acad Sci 350:254–265
Chebath J, Merlin G, Metz R, Benech P, Revel M (1983) Interferon-induced 56,000 Mr protein and its mRNA in human cells: molecular cloning and partial sequence of the cDNA. Nucleic Acids Res 11:1213–1226
Chebath J, Benech P, Hovanessian A, Galabru J, Revel M (1987) Four different forms of interferon-induced 2′,5′-oligo(a) synthetase identified by immunoblotting in human cells. J Biol Chem 262:3852–3857
Chen BDM (1986) Interferon-induced inhibition of receptor-mediated endocytosis of colony-stimulating factor (CSF-1) by murine peritoneal exudate macrophages. J Immunol 136:174–180
Cheng YSE, Colonnot RJ, Yin FH (1983) Interferon induction of fibroblast proteins with guanylate binding activity. J Biol Chem 258:7746–7750
Clavell LA, Bratt MA (1971) Relationship between the ribonucleic acid-synthesizing capacity of ultraviolet-irradiated virus and its ability to induce interferon. J Virol 8:500–508
Clemens MJ, McNurlan MA (1985) Regulation of cell proliferation and differentiation by interferons. Biochem J 226:345–360
Collart MA, Belin D, Vassali J-D, de Kossodo S, Vassali P (1986) Gamma interferon enhances macrophage transcription of the tumor necrosis factor/cachectin, interleukin 1, and urokinase genes, which are controlled by short-lived repressors. J Exp Med 164:2113–2118
Committee on Interferon Nomenclature (1980) Interferon nomenclature. Nature 286:110
Content J, De Wit L, Poupart P, Opdenakker G, Van Damme J, Billiau A (1985) Induction of a 26-kDa-protein mRNA in human cells treated with an interleukin-1-related, leukocyte-derived factor. Eur J Biochem 152:253–257
Cox DR, Epstein LB, Epstein CJ (1980) Genes coding for sensitivity to interferon (IfRec) and soluble superoxide dismutase (SOD-1) are linked in mouse and man and map to mouse chromosome 16. Proc Natl Acad Sci USA 77:2168–2172
Creasey AA, Eppstein DA, Marsh YV, Khan Z, Merigan TC (1983) Growth regulation of melanoma cells by interferon and (2′-5′) oligoadenylate synthetase. Mol Cell Biol 3:780–786

Dani C, Mechti N, Piechaczy M, Lebleu B, Jeanteur P, Blanchard JM (1985) Increased rate of degradation of *c-myc* in interferon-treated Daudi cells. Proc Natl Acad Sci USA 82:4896–4899

Degrave W, Derynck R, Tavernier J, Haegeman G, Fiers W (1981) Nucleotide sequence of the chromosomal gene for human fibroblast (β1) interferon and of the flanking regions. Gene 14:137–143

De Maeyer E, De Maeyer-Guignard J (1988) Interferons and other regulatory cytokines. Wiley, New York

DeStefano E, Friedman RM, Friedman-Kien AE, Goedert JJ, Henriksen D, Preble OT, Sonnabend JA, Vilček J (1982) Acid-labile human leukocyte interferon in homosexual men with Kaposi's sarcoma and lymphadenopathy. J Infect Dis 146:451–455

Dron M, Modjtahedi N, Brison O, Tovey MG (1986) Interferon modulation of *c-myc* expression in cloned Daudi cells: relationship to the phenotype of interferon resistance. Mol Cell Biol 6:1374–1378

Ebsworth NE, Rozengurt E, Taylor-Papadimitriou J (1986) Microtubule-disrupting agents reverse the inhibitory effect of interferon on mitogenesis in 3T3 cells. Exp Cell Res 165:255–262

Einat M, Resnitzky D, Kimchi A (1985) Close link between reduction of *c-myc* expression by interferon and G_0/G_1 arrest. Nature 313:597–600

Enoch T, Zinn K, Maniatis T (1986) Activation of the human β-interferon gene requires an interferon-inducible factor. Mol Cell Biol 6:801–810

Epstein LB (1984) The special significance of interferon-gamma. In: Vilček J, De Maeyer E (eds) Interferon, vol 2. Elsevier, Amsterdam, pp 185–220

Epstein LB, Epstein CJ (1976) Localization of the gene AVG for the antiviral expression of immune and classical interferon to the distal portion of the long arm of chromosome 21. J Infect Dis 133:A56–A62

Ezekowitz RAB, Dinauer MC, Jaffe HS, Orkin SH, Newburger PE (1988) Partial correction of the phagocyte defect in patients with X-linked chronic granulomatous disease by subcutaneous interferon gamma. N Engl J Med 319:146–151

Fan S-D, Goldberg M, Bloom BR (1988) Interferon-gamma-induced transcriptional activation is mediated by protein kinase C. Proc Natl Acad Sci USA 85:5122–5125

Farrar WL, Johnson HM, Farrar JJ (1981) Regulation of the production of immune interferon and cytotoxic T lymphocytes by interleukin 2. J Immunol 126:1120–1125

Fellous M, Nir U, Wallach D, Merlin G, Rubinstein M, Revel M (1982) Interferon-dependent induction of mRNA for the major histocompatibility antigens in human fibroblasts and lymphoblastoid cells. Proc Natl Acad Sci USA 79:3082–3086

Field AK, Tytell AA, Lampson GP, Hilleman MR (1967) Inducers of interferon and host resistance. II. Multistranded synthetic polynucleotide complexes. Proc Natl Acad Sci USA 58:1004–1010

Finter NB, Oldham RK (eds) (1985) Interferon, vol 4, In vivo and clinical studies. Elsevier, Amsterdam

Friedman RM (1987) Interferon-induced reversion in oncogene-transformed cells. In: Baron S (ed) The interferon system. University of Texas Press, Austin, pp 409–411

Friedman RM, Pitha P (1984) The effect of interferon on membrane-associated viruses. In: Friedman RM (ed) Interferon, vol 3. Elsevier, Amsterdam, pp 319–341

Friedman RL, Stark GR (1985) α-Interferon-induced transcription of HLA and metallothionein genes containing homologous upstream sequences. Nature 314: 637–639

Friedman RL, Manly SP, McMahon M, Kerr IM, Stark GR (1984) Transcriptional and posttranscriptional regulation of interferon-induced gene expression in human cells. Cell 38:745–755

Friedman-Einat M, Revel M, Kimchi A (1982) Initial characterization of a spontaneous interferon secreted during growth and differentiation of Friend erythroleukemia cells. Mol Cell Biol 2:1472–1480

Fujita T, Ohno S, Yasumitsu H, Taniguchi T (1985) Delimitation and properties of DNA sequences required for the regulated expression of human interferon-β gene. Cell 41:489–496

Fujita T, Shibuya H, Hotta H, Yamanishi K, Taniguchi T (1987) Interferon-β gene regulation: tandemly repeated sequences of a synthetic 6 bp oligomer function as a virus-inducible enhancer. Cell 49:357

Galabru J, Hovanessian AG (1985) Two interferon-induced proteins are involved in the protein kinase complex dependent on double-stranded RNA. Cell 43:685–694

Garcia-Blanco MA, Lengyel P, Morrison E, Brownlee C, Stiles CD, Rutherford M, Hannigan G, Williams BRG (1989) Regulation of 2′,5′-oligoadenylate synthetase gene expression by interferons and platelet-derived growth factor. Mol Cell Biol 9:1060–1068

Goodbourn S, Zinn K, Maniatis T (1985) Human β-interferon gene expression is regulated by an inducible enhancer element. Cell 41:509–520

Gray PW, Leung DW, Pennica D, Yelverton E, Najarian R, Simonsen C, Derynck R, Sherwood PJ, Wallace DM, Berger SL, Levinson AD, Goeddel D (1982) Expression of human immune interferon cDNA in *E. coli* and monkey cells. Nature 295:503–508

Gresser I, Tovey MG, Maury C, Bandu MT (1976) Role of interferon in the pathogenesis of virus diseases in mice as demonstrated by the use of anti-interferon serum. II. Studies with herpes simplex, Moloney sarcoma, vesicular stomatitis, Newcastle disease and influenza viruses. J Exp Med 144:1316–1323

Grossberg SE, Taylor JL (1984) Interferon effects on cell differentiation. In: Friedman RM (ed) Interferon, vol 3. Elsevier, Amsterdam, pp 299–317

Haegeman G, Content J, Volckaert G, Derynck R, Tavernier J, Fiers W (1986) Structural analysis of the sequence coding for an inducible 26-kDa protein in human fibroblasts. Eur J Biochem 159:625–632

Hamilton TA, Becton DL, Somers SD, Gray PW, Adams DO (1985) Interferon-γ modulates protein kinase C activity in murine peritoneal macrophages. J Biol Chem 260:1378–1381

Hansen TR, Imakawa K, Polites HG, Marotti KR, Anthony RV, Roberts RM (1988) Interferon RNA of embryonic origin is expressed transiently during early pregnancy in the ewe. J Biol Chem 263:12801–12804

Hardy KJ, Manger B, Newton M, Stobo JD (1987) Molecular events involved in regulating human interferon-γ gene expression during T cell activation. J Immunol 138:2353–2358

Havell EA, Berman B, Ogburn C, Berg K, Paucker K, Vilček J (1975) Two antigenically distinct species of human interferon. Proc Natl Acad Sci USA 72:2185–2189

Havell EA, Yip YK, Vilček J (1978) Characteristics of human lymphoblastoid (Namalva) interferon. J Gen Virol 38:51–60

Hayes TG, Yip YK, Vilček J (1979) Le interferon production by human fibroblasts. Virology 98:351–363

Ho M (1984) Induction and inducers of interferon. In: Billiau A (ed) Interferon 1. General and applied aspects. Elsevier, Amsterdam, pp 79–124

Hooks JJ, Moutsopoulos HM, Geis SA, Stahl NI, Decker JL, Notkins AL (1979) Immune interferon in the circulation of patients with autoimmune disease. N Engl J Med 301:5–8

Horisberger MA, Staeheli P, Haller O (1983) Interferon induces a unique protein in mouse cells bearing a gene for resistance to influenza virus. Proc Natl Acad Sci USA 80:1910–1914

Hovanessian AG, Galabru J (1987) The double-stranded RNA dependent protein kinase from human cells. In: Cantell K, Schellekens H (eds) The biology of the interferon system 1986. Nijhoff, Dordrecht, pp 73–77

Hovanessian AG, Brown RE, Kerr I (1977) Synthesis of low molecular weight inhibitor of protein synthesis with enzyme from interferon-treated cells. Nature 268:537–539

Hug H, Costas M, Staeheli P, Aebi M, Weissmann C (1988) Organization of the murine *Mx* gene and characterization of its interferon- and virus-inducible promoter. Mol Cell Biol 8:3065–3079

Hughes TK, Baron S (1987) Do interferons act singly or in combination? J Interferon Res 7:603–614

Imakawa K, Anthony RV, Kazemi M, Marotti KR, Polites HG, Roberts RM (1987) Interferon-like sequence of ovine trophoblast protein secreted by embryonic trophectoderm. Nature 330:377–379

Isaacs A, Lindenmann J (1957) Virus interference. I. The interferon. Proc R Soc Lond [Biol] 147:258–267

Israel A, Kimura A, Fournier A, Fellous M, Kourilsky P (1986) Interferon response sequence potentiates activity of an enhancer in the promoter region of a mouse H-2 gene. Nature 322:743–746

Ito Y, Nishiyama Y, Shimokata K, Nagata I, Takeyama H, Kunii A (1978) The mechanism of interferon induction in mouse spleen cells stimulated with HVJ. Virology 88:128–137

Jacobsen H, Mestan J, Mittnacht S, Dieffenbach CW (1989) Beta interferon subtype 1 induction by tumor necrosis factor. Mol Cell Biol 9:3037–3042

Jarvis AP, Colby C (1978) Murine interferon system regulation: isolation and characterization of a mutant 3T6 cell engaged in the semiconstitutive synthesis of interferon. Cell 14:355–363

Johnston MI, Torrence PF (1984) The role of interferon-induced proteins, double-stranded RNA and 2′5′-oligoadenylate in the interferon-mediated inhibition of viral translation. In: Friedman RM (ed) Interferon, vol 3. Elsevier, Amsterdam, pp 189–298

Joshi AR, Sarkar FH, Gupta SL (1982) Interferon receptors. Cross-linking of human leukocyte interferon $\alpha 2$ to its receptor on human cells. J Biol Chem 257:13884–13887

Jung V, Rashidbaigi A, Jones C, Tischfield JA, Shows TB, Pestka S (1987) Human chromosome 6 and 21 are required for sensitivity to human interferon γ. Proc Natl Acad Sci USA 84:4151–4155

Kasahara T, Hooks JJ, Dougherty SF, Oppenheim JJ (1983) Interleukin 2-mediated immune interferon (IFN-γ) production by human T cells and T cell subsets. J Immunol 130:1784–1789

Kehrl JH, Miller A, Fauci AS (1987) Effect of tumor necrosis factor alpha on mitogen-activated human B cells. J Exp Med 166:786–791

Kelker HC, Le J, Rubin BY, Yip YK, Nagler C, Vilček J (1984) Three molecular weight forms of natural human interferon-γ revealed by immunoprecipitation with monoclonal antibody. J Biol Chem 259:4301–4304

Keller AD, Maniatis T (1988) Identification of an inducible factor that binds to a positive regulatory element of the human β-interferon gene. Proc Natl Acad Sci USA 85:3309–3313

Kelley KA, Kozak CA, Dandoy F, Sor F, Skup D, Windass JD, De Maeyer-Guignard J et al. (1983) Mapping of murine interferon-α genes to chromosome 4. Gene 26:181–188

Knight E (1976) Interferon: purification and initial characterization from human diploid cells. Proc Natl Acad Sci USA 73:520–523

Knight E Jr, Anton ED, Fahey D, Friedland BK, Jonak GJ (1985) Interferon regulates *c-myc* gene expression in Daudi cells at the post-transcriptional level. Proc Natl Acad Sci USA 82:1151–1154

Kohase M, Henriksen-DeStefano D, May LT, Vilček J, Sehgal P (1986) Induction of β_2-interferon by tumor necrosis factor: a homeostatic mechanism in the control of cell proliferation. Cell 45:659–666

Kohase M, May LT, Tamm I, Vilček J, Sehgal PB (1987) A cytokine network in human diploid fibroblasts: interactions of β interferons, tumor necrosis factor, platelet-derived growth factor and interleukin-1. Mol Cell Biol 7:273–280

Kohase M, Zhang Y, Lin J-X, Yamazaki S, Sehgal PB, Vilček J (1988) Interleukin-1 can inhibit interferon-beta synthesis and its antiviral action: comparison with tumor necrosis factor. J Interferon Res 8:559–570
Korber B, Mermod N, Hood L, Stroynowski I (1988) Regulation of gene expression by interferons: control of H-2 promoter responses. Science 239:1302–1306
Kuhl D, Fuenta JDL, Chaturvedi M, Parimoo S, Ryals J, Meyer F, Weissmann C (1987) Reversible silencing of enhancers by sequences derived from the human IFN-α promoter. Cell 50:1057–1069
Lai MHT, Joklik K (1973) The induction of interferon by temperature-sensitive mutants of reovirus, UV-irradiated reovirus, and subviral reovirus particles. Virology 51:191–204
Lapierre LA, Fiers W, Pober JS (1988) Three distinct classes of regulatory cytokines control endothelial cell MHC antigen expression. J Exp Med 167:794–804
Larner AC, Jonak G, Cheng YSE, Korant B, Knight E, Darnell JR (1984) Transcriptional induction of two genes in human cells by β-interferon. Proc Natl Acad Sci USA 81:6733–6737
Laurent AG, Kurst B, Galabru J, Svab J, Hovanessian AG (1985) Monoclonal antibodies to interferon induced 68,000-Mr protein and their use for the detection of dsRNA dependent protein kinase in human cells. Proc Natl Acad Sci USA 82:4341–4345
Le J, Lin J-X, Henriksen-DeStefano D, Vilček J (1986) Bacterial lipopolysaccharide-induced interferon-gamma production: roles of interleukin 1 and interleukin 2. J Immunol 136:4525–4530
Lebleu B, Sen GC, Shaila S, Cabrer B, Lengyel P (1976) Interferon, double-stranded RNA, and protein phosphorylation. Proc Natl Acad Sci USA 73:3107–3111
Lengyel P (1982) Biochemistry of interferons and their actions. Annu Rev Biochem 51:251–282
Leung DW, Capon DJ, Goeddel DV (1984) The structure and bacterial expression of three distinct bovine interferon-β genes. Biotechnology 2:458–464
Levis SC, Saavedra MC, Ceccoli C, Falcoff E, Feuillade MR, Enria DAM, Maiztegui JI, Falcoff R (1984) Endogenous interferon in Argentine hemorraghic fever. J Infect Dis 149:428–433
Levy DE, Kessler DS, Pine R, Reich N, Darnell JE Jr (1988) Interferon-induced nuclear factors that bind a shared promoter element correlate with positive and negative transcriptional control. Genes Dev 2:383–393
Lin PF, Slate DL, Lawyer FC, Ruddle FH (1980) Assignment of the murine interferon sensitivity and cytoplasmic superoxide dismutase genes to chromosome 16. Science 209:285–287
Lin SL, Kikuchi T, Pledger WJ, Tamm I (1986) Interferon inhibits the establishment of competence in Go/S-phase transition. Science 233:356–359
Ludwig CU, Durie BG, Salmon SE, Moon TE (1983) Tumor growth stimulation in vitro by interferons. Eur J Cancer Clin Oncol 19:1625–1632
Luster AD, Unkeless JC, Ravetch JV (1985) γ-Interferon transcriptionally regulates an early-response gene containing homology to platelet proteins. Nature 315:672–676
Luster AD, Weinshank RL, Feinman R, Ravetch JV (1988) Molecular and biochemical characterization of a novel γ-interferon-inducible protein. J Biol Chem 263:12036–12043
Mariano TM, Kozak CA, Langer JA, Pestka S (1987) The mouse immune interferon receptor gene is located on chromosome 10. J Biol Chem 262:5812–5814
Mestan J, Digel W, Mittnacht S, Hillen H, Blohm D, Moller A, Jacobsen H, Kirchner H (1986) Antiviral effects of recombinant tumor necrosis factor in vitro. Nature 323:816–819
Miyamoto M, Fujita T, Kimura Y, Maruyama M, Harada H, Sudo Y, Miyata T, Taniguchi T (1988) Regulated expression of a gene encoding a nuclear factor, IRF-1, that specifically binds to IFN-β gene regulatory elements. Cell 54:903–913

Molinas FC, Wietzerbin J, Falcoff E (1987) Human platelets possess receptors for a lymphokine: demonstration of high specific receptors for Hu INF-γ. J Immunol 138:802–806

Moore RN, Larsen HS, Horohov DW, Rouse BT (1984) Endogenous regulation of macrophage proliferative expansion by colony-stimulating factor-induced interferon. Science 223:178–181

Mosmann TR, Cherwinski H, Bond MW, Giedlin MA, Coffman RL (1986) Two types of murine helper T cell clone. I. Definition according to profiles of lymphokine activities and secreted proteins. J Immunol 136:2348–2357

Naylor SL, Sakaguchi AY, Shows TB, Law ML, Goeddel DV, Gray PW (1983) Human immune interferon gene is located on chromosome 12. J Exp Med 157:1020–1027

Naylor SL, Gray PW, Lalley PA (1984) Mouse immune interferon (IFN-γ) gene is on chromosome 10. Somatic Cell Mol Genet 10:531–534

Novick D, Orchansky P, Revel M, Rubinstein M (1987) The human interferon-α receptor purification, characterization, and preparation of antibodies. J Biol Chem 262:8483–8487

Ohno S, Taniguchi T (1983) The 5′-flanking sequence of human interferon-β gene is responsible for viral induction of transcription. Nucleic Acids Res 11:5403–5412

Oleszak E, Inglot AD (1980) Platelet derived growth factor (PDGF) inhibits antiviral and anticellular action of interferon in synchronized mouse or human cells. J Interferon Res 1:37–48

Onozaki K, Urawa H, Tamatani T, Iwamura Y, Hashimoto T, Baba T, Suzuki H, Yamada M, Yamamoto S, Oppenheim JJ, Matsushima K (1988) Synergistic interactions of interleukin 1, interferon-β, and tumor necrosis factor in terminally differentiating a mouse myeloid leukemic cell line (M1). Evidence that interferon-β is an autocrine differentiating factor. J Immunol 140:112–119

Östlund L, Einhorn S, Robert K-H, Juliusson G, Biberfeld P (1986) Chronic B-lymphocytic leukemia cells proliferate and differentiate following exposure to interferon in vitro. Blood 67:152–159

Pestka S (1983) The human interferons: from protein purification and sequence to cloning and expression in bacteria: before, between and beyond. Arch Biochem Biophys 221:1–37

Pestka S, Langer JA, Zoon KC, Samuel CE (1987) Interferons and their actions. Annu Rev Biochem 56:727–777

Pfeffer LM, Murphy JS, Tamm I (1979) Interferon effects on the growth and division of human fibroblasts. Exp Cell Res 121:111–120

Pfeffer LM, Wang E, Tamm I (1980) Interferon effects on microfilament organization, cellular fibronectin distribution, and cell motility in human fibroblasts. J Cell Biol 85:9–17

Pfeffer LM, Donner DB, Tamm I (1987) Interferon-α down-regulates insulin receptors in lymphoblastoid (Daudi) cells. Relationship to inhibition of cell proliferation. J Biol Chem 262:3665–3670

Pfizenmaier K, Scheurich P, Schlüter C, Krönke M (1987) Tumor necrosis factor enhances HLA-A,B,C and HLA-DR gene expression in human tumor cells. J Immunol 138:975–980

Pfizenmaier K, Wiegmann K, Scheurich P, Krönke M, Merlin G, Aguet M, Knowles BB, Ucer U (1988) High affinity human IFN-gamma-binding capacity is encoded by a single receptor gene located in proximity to *c-ros* on human chromosome region 6q16 to 6q22. J Immunol 141:856–860

Preble OT, Black RJ, Friedman RM, Klippel JH, Vilček J (1982) Systemic lupus erythematosus: presence in human serum of an unusual acid-labile leukocyte interferon. Science 216:429–431

Ragg H, Weissmann C (1983) Not more than 117 base pairs of 5′-flanking sequence are required for inducible expression of a human IFN-α gene. Nature 303:439–442

Raj NBK, Pitha PM (1983) Two levels of regulation of β-interferon gene expression in human cells. Proc Natl Acad Sci USA 80:3923–3927
Rashidbaigi A, Langer JA, Jung V, Jones C, Morse HG, Tischfield JA, Trill JJ, Kung HF, Pestka S (1986) The gene for the human immune interferon receptor is located on chromosome 6. Proc Natl Acad Sci USA 83:384–388
Raziuddin A, Sarkar FH, Dutkowski R, Shulaman L, Ruddle FH, Gupta SL (1984) Receptors for human α and β interferon but not for γ interferon are specified by human chromosome 21. Proc Natl Acad Sci USA 81:5504–5508
Reis LFL, Le J, Hirano T, Kishimoto T, Vilček J (1988) Antiviral action of tumor necrosis factor in human fibroblasts is not mediated by B cell stimulatory factor 2/IFN-beta$_2$, and is inhibited by specific antibodies to IFN-beta. J Immunol 140:1566–1570
Reis LFL, Lee TH, Vilček J (1989) Tumor necrosis factor acts synergistically with autocrine interferon-β and increases interferon-β mRNA levels in human fibroblasts. J Biol Chem 264:16351–16354
Resnitzky D, Yarden A, Zipori D, Kimchi A (1986) Autocrine β-related interferon controls c-*myc* suppression and growth arrest during haematopoietic cell differentiation. Cell 46:31–40
Revel M, Chebath J (1986) Interferon-activated genes. Trends Biochem Sci 11:166–170
Rinderknecht E, O'Connor BH, Rodriguez H (1984) Natural human interferon-gamma. Complete amino acid sequence and determination of sites of glycosylation. J Biol Chem 259:6790–6797
Rivière Y, Gresser I, Guillon JC, Tovey MG (1977) Inhibition by anti-interferon serum of lymphocytic choriomeningitis virus disease in suckling mice. Proc Natl Acad Sci USA 74:2135–2139
Roberts WK, Hovanessian A, Brown RE, Clemens MJ, Kerr IM (1976) Interferon-mediated protein kinase and low-molecular weight inhibitor of protein synthesis. Nature 264:477–480
Rossi GB (1985) Interferons and cell differentiation. In: Gresser I (ed) Interferon 6. Academic, London, pp 31–68
Rubin BY, Anderson SL, Lunn RM, Richardson NK, Hellerman GR, Smith LJ, Old LJ (1988) Tumor necrosis factor and IFN induce a common set of proteins. J Immunol 141:1180–1184
Rubin BY, Anderson SL, Hellermann GR, Richardson NK, Lunn RM, Valinsky JE (1988) The development of antibody to the interferon-induced indoleamine 2,3-dioxygenase and the study of regulation of its synthesis. J Interferon Res 8:691–701
Rubinstein M, Levy WP, Moschera JA, Lai CY, Hershberg RD, Bartlett RT, Pestka S (1981) Human leukocyte interferon: isolation and characterization of several molecular forms. Arch Biochem Biophys 210:307–318
Ruggiero V, Tavernier J, Fiers W, Baglioni C (1986) Induction of the synthesis of tumor necrosis factor receptors by interferon-gamma. J Immunol 136:2445–2450
Ryals J, Dierks P, Ragg H, Weissmann C (1985) A 46-nucleotide promoter segment from an IFN-α gene renders an unrelated promoter inducible by virus. Cell 41:497–507
Saarma M, Toots U, Raukas E, Zhelkovsky A, Pivazian A, Neuman T (1986) Nerve growth factor induces changes in (2′-5′) oligo (A) synthetase and 2′-phosphodiesterase activities during differentiation of PC12 pheochromocytoma cells. Exp Cell Res 166:229–236
Samanta H, Engel DA, Chao HM, Thakur A, Garcia-Blanco MA, Lengyel P (1986) Interferons as gene activators. Cloning of the 5′ terminus and the control segment of an interferon activated gene. J Biol Chem 261:11849–11858
Samid D, Chang EH, Friedman RM (1984) Biochemical correlates of phenotypic reversion in interferon-treated mouse cells transformed by a human oncogene. Biochem Biophys Res Commun 119:21–28

Sancéau J, Sondermeyer P, Beranger F, Falcoff R, Vaquero C (1987) Intracellular human γ-interferon triggers an antiviral state in transformed murine L cells. Proc Natl Acad Sci USA 84:2906–2910

Sarkar FH, Gupta SL (1984) Receptors for human gamma interferon: binding and cross-linking of ^{125}I-labeled recombinant human gamma interferon to receptors on WISH cells. Proc Natl Acad Sci USA 81:5164–5168

Sen GC (1984) Biochemical pathways in interferon action. Pharmacol Ther 24:235–257

Shirayoshi Y, Burke PA, Appella E, Ozato K (1988) Interferon-induced transcription of a major histocompatibility class I gene accompanies binding of inducible nuclear factors to the interferon consensus sequence. Proc Natl Acad Sci USA 85:5884–5888

Shulman LM (1988) Molecular cloning of the human IFN-α/β receptor cDNA (Abstr). J Interferon Res 8 (Suppl):S16

Skurkovich SV, Eremkina EI (1975) The probable role of interferon in allergy. Ann Allergy 35:356–360

Smith TJ, Wagner RR (1967) Rabbit macrophage interferons. II. Some physicochemical properties and estimations of molecular weights. J Exp Med 125:579–593

Snapper CM, Paul WE (1987) Interferon-γ and B cell stimulatory factor-1 reciprocally regulate Ig isotype production. Science 236:944–947

Sokawa Y, Watanabe Y, Kawade Y (1977) Interferon suppresses the transition of quiescent 3T3 cells to a growing state. Nature 268:236–238

Sokawa Y, Nagata K, Ichikawa Y (1981) Induction and function of 2′,5′-oligoadenylate synthetase in differentiation of mouse myeloid leukemia cells. Exp Cell Res 135:191–197

Sreevalsan T (1984) Effects of interferons on cell physiology. In: Friedman RM (ed) Interferon, vol 3. Elsevier, Amsterdam, pp 343–387

Staeheli P, Haller O, Boll W, Lindenmann J, Weissmann C (1986) Mx protein: constitutive expression in 3T3 cells transformed with cloned Mx cDNA confers selective resistance to influenza virus. Cell 44:147–158

Stewart WE II (1979) The interferon system. Springer, Vienna New York

Sugita K, Miyazaki J-I, Appella E, Ozato K (1987) Interferons increase transcription of a major histocompatibility class I gene via a 5′ interferon consensus sequence. Mol Cell Biol 7:2625–2630

Tamm I, Lin SL, Pfeffer LM, Sehgal PB (1987) Interferons α and β as cellular regulatory molecules. In: Gresser I (ed) Interferon 9. Academic, New York, pp 14–74

Taniguchi T (1988) Regulation of cytokine gene expression. Annu Rev Immunol 6:439–464

Taniguchi T, Mantei N, Schwarzstein M, Nagata S, Muramatsu M, Weissmann C (1980) Human leukocyte and fibroblast interferons are structurally related. Nature 285:547–549

Taylor-Papadimitriou J, Shearer M, Rozengurt E (1981) Inhibitory effect of interferon on cellular DNA synthesis: modulation by pure mitogenic factors. J Interferon Res 1:401–410

Tiwari RK, Kusari J, Sen GC (1987) Functional equivalents of interferon-mediated signals needed for induction of an mRNA can be generated by double-stranded RNA and growth factors. EMBO J 6:3373–3378

Tominaga SI, Lengyel P (1985) β-Interferon alters the pattern of proteins secreted from quiescent and platelet-derived growth factor-treated BALB/c 3T3 cells. J Biol Chem 260:1975–1978

Tovey MG, Streuli M, Gresser I, Gugenheim J, Blanchard B, Guymarho J, Vignaux F, Gigou M (1987) Interferon messenger RNA is produced constitutively in the organs of normal individuals. Proc Natl Acad Sci USA 84:5038–5042

Trinchieri G, Perussia B (1985) Immune interferon: a pleiotropic lymphokine with multiple effects. Immunol Today 6:131–136

Trinchieri G, Kobayashi M, Rosen M, Loudon R, Murphy M, Perussia B (1986) Tumor necrosis factor and lymphotoxin induce differentiation of human myeloid cell lines in synergy with immune interferon. J Exp Med 164:1206–1225
Tsujimoto M, Feinman R, Vilček J (1986) Differential effects of type I IFN and IFN-gamma on the binding of tumor necrosis factor to receptors in two human cell lines. J Immunol 137:2272–2276
Van Damme J, Opdenakker G, Billiau A, De Somer P, De Wit L, Poupart P, Content J (1985) Stimulation of fibroblast interferon production by a 22 K protein from human leukocytes. J Gen Virol 66:693–700
Van Damme J, De Ley M, Van Snick J, Dinarello CA, Billiau A (1987) The role of interferon-beta$_1$ and the 26 kDa protein (interferon-beta$_2$) as mediators of the antiviral effect of interleukin 1 and tumor necrosis factor. J Immunol 139:1867–1872
Vilček J, Gray PW, Rinderknecht E, Sevastopoulos C (1985) Interferon γ: a lymphokine for all seasons. In: Pick E (ed) Lymphokines, vol 11. Academic, Orlando, pp 1–32
Wallach D, Fellous M, Revel M (1982) Preferential effect of γ interferon on the synthesis of HLA antigens and their mRNAs in human cells. Nature 299:833–836
Wathelet MG, Clauss IM, Nols CB, Content J, Huez GA (1987) New inducers revealed by the promoter sequence analysis of two interferon-activated human genes. Eur J Biochem 109:313–321
Weil J, Epstein CJ, Epstein LB, Sedmak JJ, Sabran JL, Grossberg SE (1982) A unique set of polypeptides is induced by γ interferon in addition to those induced in common with α and β interferon. Nature 301:437–439
Weissenbach J, Chernajovsky Y, Zeevi M, Shulman L, Soreq H, Nir U, Wallach D, Perricaudet M, Tiollais P, Revel M (1980) Two interferon mRNAs in human fibroblasts: in vitro translation and *Escherichia coli* cloning studies. Proc Natl Acad Sci USA 77:7152–7156
Weissmann C, Weber H (1986) The interferon genes. Prog Nucleic Acid Res Mol Biol 33:251–300
Wetzel R (1981) Assignment of the disulphide bonds of leukocyte interferon. Nature 289:606–607
Wheelock EF (1965) Interferon-like virus-inhibitor induced in human leukocytes by phytohemagglutinin. Science 149:310–311
Wilson V, Jeffreys AJ, Barrie PA, Boseley PG, Slocombe PM, Easton A, Burke DC (1983) A comparison of vertebrate interferon gene families detected by hybridization with human interferon DNA. J Mol Biol 166:457–475
Wong GHW, Goeddel DV (1986) Tumour necrosis factor alpha and beta inhibit virus replication and synergize with interferons. Nature 323:819–822
Yap WH, Teo TS, Tan YH (1986) An early event in the interferon-induced transmembrane signaling process. Science 234:355–358
Yarden A, Shure-Gottlieb H, Chebath J, Revel M, Kimchi A (1984) Autogenous production of interferon-beta switches on HLA genes during differentiation of histiocytic lymphoma U937 cells. EMBO J 3:969–973
Yip YK, Barrowclough B, Urban C, Vilček J (1982) Purification of two subspecies of human gamma (immune) interferon. Proc Natl Acad Sci USA 79:1820–1824
Yoshida R, Imanishi J, Oku T, Kishida T, Hayaishi O (1981) Induction of pulmonary indoleamine 2,3-dioxygenase by interferon. Proc Natl Acad Sci USA 78:129–132
Young HA, Dray JF, Farrar WL (1986) Expression of transfected human interferon-γ DNA: evidence for cell specific regulation. J Immunol 136:4700
Youngner JS, Salvin SB (1973) Production and properties of migration inhibitory factor and interferon in the circulation of mice with delayed hypersensitivity. J Immunol 111:1914–1922
Zajac BA, Henle W, Henle G (1969) Autogenous and virus induced interferons from lines of lymphoblastoid cells. Cancer Res 29:1467–1475

Zilberstein A, Federman P, Shulman L, Revel M (1976) Specific phosphorylation in vitro of a protein associated with ribosomes of interferon-treated mouse L cells. FEBS Lett 68:119–124
Zinn K, Maniatis T (1986) Detection of factors that interact with the human β-interferon regulatory region in vivo by DNAse I footprinting. Cell 45:611–618
Zinn K, DiMaio D, Maniatis T (1983) Identification of two distinct regulatory regions adjacent to the human β-interferon gene. Cell 34:865–879
Zoon KC, Karasaki Y, zur Nedden DL, Hu R, Arnheiter H (1986) Modulation of epidermal growth factor receptors by human α-interferon. Proc Natl Acad Sci USA 83:8226–8230
Zullo JN, Cochran BH, Huang AS, Stiles CD (1985) Platelet-derived growth factor and double-stranded ribonucleic acid stimulate expression of the same genes in 3T3 cells. Cell 43:793–800

CHAPTER 20

Cachectin/Tumor Necrosis Factor and Lymphotoxin

B. BEUTLER

A. Introduction

As it acts to eradicate viruses, bacteria, and parasitic pathogens, the host immune system may sometimes prove to be a liability rather than an asset. The inflammatory response to invasive organisms may injure the host as surely as the organisms themselves, and, in many instances, the immune response is itself a major pathogenetic vehicle. Thus, physicians strive to temper the immune response in their attempts to manage some infectious disease states. At present, this is achieved through the use of glucocorticosteroids and, occasionally, cytotoxic drugs.

As our understanding of the inflammatory response has grown, we have come to appreciate that many aspects of this primitive reaction to host invasion are governed by polypeptide hormones, produced by immune effector cells. Moreover, an awareness of the multifaceted nature of the immune response has developed, such that disease entities so distinct as endotoxic shock and cancer cachexia are now perceived as syndromes with a related etiologic basis. Then, too, the notion of immune system regulation of host metabolism has gained credibility.

Unlike lymphocytes, mononuclear phagocytic cells, though known to be essential participants in the immune response, lack specificity and memory. The mechanisms by which they distinguish self from nonself remain obscure. Yet it is clear that these, the most primitive of immunocytes, react vigorously to exogenous challenge in many forms. During the past decade, it became clear that macrophages work many of their effects through the elaboration of so-called "monokines." Interleukin-1(IL-1)-α and -β, and a second molecule widely known as "cachectin" or "tumor necrosis factor-α" (TNF), have been the most studied of these. Cachectin/TNF is one of the most abundant products of activated macrophages, and its production may bring about profound alterations of host physiology. The remainder of this review will be devoted to a description of cachectin/TNF as the molecule is presently understood, the story of its discovery, its actions, and its potential use as a tool in biology and medicine.

B. "Factor-Mediated" Diseases: The Hematopoietic Origin of Factors

The imperfect nature of the host immune response is frequently highlighted in the form of anaphylactic reactions, complement-mediated injury to the host, and injury caused by cell-mediated hypersensitivity. In each case, the inciting stimulus per se is relatively harmless, and an intact immune effector mechanism is required to cause injury or death.

It might therefore come as no surprise to witness injury caused by other immune factors of hematopoietic origin. Only 8 years have elapsed since the primary role of lymphoreticular cells in the mediation of host responses to bacterial endotoxin (lipopolysaccharide; LPS) was clearly established (MICHALEK et al. 1980). Radiation chimeras, produced by reconstitution of the hematopoietic progenitor pool of endotoxin-resistant mice with precursors derived from endotoxin-sensitive individuals, were shown to be normally sensitive to endotoxin; conversely, endotoxin-sensitive animals, if irradiated prior to rescue by infusion of stem cells derived from endotoxin-resistant mice, were highly resistant to challenge with endotoxin. Thus, a cell of hematopoietic origin, or its products, appeared to confer sensitivity to LPS.

The macrophage was suspected to play an important role in the mediation of endotoxicity, since "priming" stimuli capable of causing reticuloendothelial hyperplasia are known to render animals highly sensitive to the effects of LPS (VOGEL et al. 1980; HA et al. 1983; WOOD and CLARK 1984; HARANAKA et al. 1984). Moreover, endotoxin-activated macrophages were shown to produce a factor capable of killing endotoxin-resistant mice (CERAMI et al. 1985).

The link between hematopoietically derived cells and endotoxicity was thus clearly established. However, other disease states in which the action of endogenous mediators was suspected were less amenable to experimentation. The wasting diathesis so commonly observed in chronic infectious and neoplastic diseases had also long been attributed to a mediator, produced by invasive organisms, tumors, or cells of the host. Had a model of tumor-induced wasting (comparable to the murine model of endotoxin response and resistance) been available, it might have been feasible to demonstrate the importance of hematopoietic cells in the production of cachexia, and other problems related to chronic disease. However, a more direct approach to the catabolic effects of chronic disease was required.

The isolation of cachectin and TNF, and the demonstration of their identity, led to widespread appreciation of the influence exerted by the immune system over the metabolic activities of many host tissues, and to the multifaceted response that may derive from the action of a single mediator.

C. Cachectin

The origins of cachectin may be traced to the mid 1970s, and to the studies of ROUZER and CERAMI (1980), who analyzed the paradoxical hypertriglyceridemic state to which rabbits infected with *Trypanosoma brucei* organisms are subject. In the face of a profound wasting diathesis, rabbits typically show triglyceride levels that are elevated by factor of 10 during the final stages of the illness. This elevation in plasma triglyceride was shown to result from an acquired, systemic deficiency of the enzyme lipoprotein lipase (LPL), which acts to clear triglyceride from the blood. At all times during the course of the infection, parasite burden is modest, and wasting cannot be attributed to a direct competitive effect.

CERAMI et al. (1985) reasoned that the elevation of plasma triglyceride might in some way be related to the wasting diathesis observed, and that the immune system, responding to parasite invasion, might mediate the entire process. A second model system was designed to detect endogenous mediators capable of regulating lipid metabolism (KAWAKAMI and CERAMI 1981). Endotoxin-sensitive (C3H/HeN) mice, when challenged with LPS, were shown to exhibit a hypertriglyceridemic response similar to that observed in trypanosome-infected rabbits. While endotoxin-resistant (C3H/HeJ) animals did not exhibit this response following direct challenge with LPS, a hypertriglyceridemia related to LPL suppression could be elicited by infusion of postendotoxin serum derived from sensitive mice. Thus, a factor present in serum acted to suppress LPL expression by fatty tissues throughout the recipient.

Moreover, macrophages isolated from endotoxin-treated sensitive animals were shown to produce an LPL-suppressing factor identical to that measured in serum (KAWAKAMI and CERAMI 1981). The factor was also capable of suppressing LPL expression by cultured adipocytes of the 3T3-L1 cell line (KAWAKAMI et al. 1982). Termed "cachectin" because of its presumed role in the pathogenesis of cachexia, this factor was purified to homogeneity from cells of the mouse macrophage line RAW 264.7 and partially sequences (BEUTLER et al. 1985c, d). Active at picomolar concentrations, cachectin was produced in great abundance by macrophages of this line, as well as by thioglycolate-elicited peritoneal macrophages. Indeed, it comprised 1%–2% of the total secretory product of activated cells (BEUTLER et al. 1985d). Correspondingly, large quantities of the protein (milligrams per kilogram body mass) were later found to be produced in vivo after LPS challenge (BEUTLER et al. 1985b; ABE et al. 1985). However, no cachectin production was observed in unstimulated cultures (BEUTLER et al. 1985d; MAHONEY et al. 1985). A high-affinity receptor for cachectin was identified and characterized on cells of diverse lineage (BEUTLER et al. 1985d). These included the 3T3-L1 cells, as well as C2 myotubules; subsequently, a large number of other cultured cell lines, as well as freshly isolated tissues, were shown to bear the cachectin receptor in varying quantities (BEUTLER et al. 1985b).

When radioiodinated and injected into mice by an intravenous route (BEUTLER et al. 1985b), cachectin was cleared from the blood with a half-life of approximately 6–7 min. The primary mode of clearance appeared to be binding to its tissue receptor; very little was removed by a renal mechanism. The majority of an injected dose was observed to bind to the liver, gastrointestinal tract, and skin. Very little appeared to enter the brain, or to bind to fatty tissues.

The amino-terminal sequence of cachectin was soon found to be strongly homologous to the sequence determined for human TNF (BEUTLER et al. 1985c; AGGARWAL et al. 1985a; PENNICA et al. 1984). Cachectin/TNF, isolated by a very different approach, was subsequently shown to be identical to cachectin by serologic and molecular cloning studies, as well as by direct comparison of biological activities (BEUTLER et al. 1985c; PENNICA et al. 1984; CAPUT et al. 1986).

D. Tumor Necrosis Factor

The ability of host immune mechanisms to deter the growth and metastasis of malignant tumors has been much debated. While many tumors appear to provoke an inflammatory response, the capacity of the immune system to exercise "immune surveillance" has been challenged by the observation that profound immunodeficiencies of genetic origin are not associated with marked enhancement of tumor formation. However, it cannot be doubted that, under certain circumstances, immune mechanisms may dramatically arrest tumor progression. One of the most dramatic cases in point is the hemorrhagic necrosis of tumors observed in humans or animals in the course of an intercurrent bacterial infection, particularly one involving gram-negative organisms. COLEY (1893, 1906), intrigued by the hemorrhagic necrosis and subsequent involution of certain tumors observed in patients with streptococcal (erysipelas) infections, attempted to reproduce this phenomenon by injection of live or killed bacterial organisms, and filtrates derived from their culture. Some success with this form of chemotherapy was claimed; however, as a result of the toxicity inherent in administration of bacterial products to human cancer patients, the technique eventually fell into disfavor.

Other investigators, however, remained intrigued by the phenomenon of bacterially induced tumor necrosis and attempted to isolate the bacterial product responsible for it. Shear and his colleagues succeeded in purifying a factor derived from gram-negative organisms which they termed the "bacterial polysaccharide"; this substance, now known as lipopolysaccharide, was very effective at inducing tumor necrosis (SHEAR and ANDERVONT 1936; SHEAR et al. 1943a, b; HARTWELL et al. 1943; KAHLER et al. 1943; SHEAR 1944). However, it was also among the most toxic constituents present in the culture from which it was derived.

Parallel studies (ALGIRE et al. 1952) indicated that hemorrhagic necrosis might reflect generalized vascular collapse, and was histologically similar to changes observed in tumors infarcted by ischemia. The work of Coley and Shear, motivated by the wish to develop antineoplastic agents with low inherent toxicity, thus appeared to have reached an impasse.

However, later work (O'MALLEY et al. 1962) suggested that hemorrhagic necrosis might be attributable to the production of an endogenous mediator released in response to LPS, rather than to LPS itself. Shock serum derived from endotoxin-treated mice was shown to induce hemorrhagic necrosis of tumors implanted in the skin of untreated recipients. The factor responsible was not endotoxin itself, but a heat-labile material, presumed to be a protein. This observation might have been forgotten were it not for the fact that a similar experiment (CARSWELL et al. 1975) revealed much the same phenomenon; serum derived from mice treated with Bacillus Calmette-Guerin (BCG) prior to injection of endotoxin was shown to contain a necrotizing factor capable of acting upon tumors implanted in the skin of recipient mice. The same factor appeared to have a cytolytic effect on several transformed cell lines in vitro (CARSWELL et al. 1975; HELSON et al. 1975; GREEN et al. 1976; DARZYNKIEWICZ et al. 1984). Dubbed "tumor necrosis factor" (TNF), the LPS-induced mediator became the object of enormous basic and clinical interest.

Human tumor necrosis factor was purified to homogeneity, partially sequenced (AGGARWAL et al. 1985a), and cloned (PENNICA et al. 1984; WANG et al. 1985; SHIRAI et al. 1985) nearly a decade later.

E. Physical Structure of Cachectin/TNF: Homology to Lymphotoxin

Cachectin/TNF is a protein with a subunit size of 17 kDa. In man, the molecule is nonglycosylated; in certain other species (notably the mouse) glycosylation occurs, but the sugar moiety is not essential for biological activity. Each monomer consists largely of a β-pleated sheet structure (DAVIS et al. 1987), and three monomers combine noncovalently to form the active hormone (SMITH and BAGLIONI 1987; ECK et al. 1988). Cachectin/TNF may be renatured after exposure to chaotropic agents such as urea, SDS, or guanidinium hydrochloride with recovery of as much as 50% of the initial biological activity. This renaturability perhaps reflects the limited number of internal disulfide bonds (one per monomer) required for maintenance of structure.

Cachectin/TNF is initially synthesized as a prohormone (Fig. 1) containing an amino-terminal peptide of varying length, depending upon species. The propeptide segment of the molecule is highly conserved. Approximately 86% of the 79 amino acids preceding the mature hormone in the mouse are identical to the 76 amino acids preceding the mature hormone in man. By contrast, the mature hormone (156 amino acids in mouse and 157 amino acids in man) is conserved to the extent of 79% (CAPUT et al. 1986; FRANSEN et al. 1985; PENNICA et al. 1985). The function of the propeptide remains unclear. Recent data suggest that it may be essential for the secretion of cachectin/TNF, which may occur through cleavage of the molecule at the cell surface (KRIEGLER et al. 1988). While the existence of a "membrane-associated" form of cachec-

```
Human        -76  MSTESMIRDVELAEEALPKKTGGPQGSRRCLFLSLFSFLIVAGATTLFCLLHFGVIGPQR
Bovine       -77  MSTKSMIRDVELAEEVLSEKAGGPQGSRSCLCLSLFSFLLVAGATTLFCLLHFGVIGPQR
Rabbit       -79  MSTESMIRDVELAEGPLPKKAGGPQGSKRCLCLSLFSFLLVAGATTLFCLLHFRVIGPQE
Mouse        -79  MSTESMIRDVELAEEALPQKMGGFQNSRRCLCLSLFSFLLVAGATTLFCLLNFGVIGPQR
Consensus         MST SMIRDVELAE  L  K GG Q S  CL LSLFSFL VAGATTLFCLL F VIGPQ

                                     1
Human        -16  EEF-PRDLSLISPLAQA--VRSSSRTPSDKPVAHVVANPQAEGQLQWLNRRANALLANGV
Bovine       -17  EEQVPSGPSINSPLVQ--TLRSSSQASSNKPVAHVVADINSPGQLRWWDSYANALMANGV
Rabbit       -19  EEQSPNNLHLVNPVAQMVTLRSASRALSDKPLAHVVANPQVEGQLQWLSQRANALLANGM
Mouse        -19  DEKFPNGLPLISSMAQTLTLRSSSQNSSDKPVAHVVANHQVEEQLEWLSQRANALLANGM
Consensus          E  P          Q    RS S   S KP AHVVA      QL W    ANAL ANG

Human         42  ELRDNQLVVPSEGLYLIYSQVLFKGQGCPSTHVLLTHTISRIAVSYQTKVNLLSAIKSPC
Bovine        42  KLEDNQLVVPAEGLYLIYSQVLFRGQGCP-PPPVLTHTISRIAVSYQTKVNILSAIKSPC
Rabbit        42  KLTDNQLVVPADGLYLIYSQVLFSGQGCR-SYVLLTHTVSRFAVSYPNKVNLLSAIKSPC
Mouse         42  DLKDNQLVVPADGLYLVYSQVLFKGQGCP-DYVLLTHTVSRFAISYQEKVNLLSAVKSPC
Consensus          L DNQLVVP  GLYL YSQVLF GQGC      LTHT SR A SY  KVN LSA KSPC

Human        102  QRETPEGAEAKPWYEPIYLGGVFQLEKGDRLSAEINRPDYLDFAESGQVYFGIIAL    157
Bovine       101  HRETPEWAEAKPWYEPIYQGGVFQLEKGDRLSAEINLPDYLDYAESGQVYFGIIAL    156
Rabbit       101  HRETPEEAEPMAWYEPIYLGGVFQLEKGDRLSTEVNQPEYLDLAESGQVYFGIIAL    156
Mouse        101  PKDTPEGAELKPWYEPIYLGGVFQLEKGDQLSAEVNLPKYLDFAESGQVYFGVIAL    156
Consensus            TPE AE   WYEPIY GGVFQLEKGD LS E N P YLD AESGQVYPG IAL
```

Fig. 1. Amino acid sequences of human (PENNICA et al. 1984), bovine (GOEDDEL et al., unpublished), rabbit (ITOH and WALLACE 1985), and murine (PENNICA et al. 1985) TNF-α. Amino acids of the propeptide: positions -79 through -1. *Consensus* indicates residues that are identical in all four species. (From GOEDDEL et al. 1986)

tin/TNF has been reported (DECKER et al. 1987; KRIEGLER et al. 1988), the physiologic significance of this molecule remains to be determined. It would appear that the prohormone is biologically inactive, and its ability to form trimers has yet to be established. It is also possible that the propeptide may have biological activities of its own, distinct from those of the mature polypeptide.

Several regions of the cachectin/TNF molecule are highly conserved among species. One such region spans amino acids 115 through 130; a second area that is generally conserved is near the carboxy terminus. Indeed, it has been reported that deletion of the C-terminal amino acid (an invariant leucine) from cachectin/TNF ablates biological activity. These regions are also conserved in lymphotoxin (for review see PAUL and RUDDLE 1988), a second polypeptide hormone derived from a closely linked gene, which displays approximately 30% homology to cachectin/TNF at the amino acid level.

The cachectin/TNF and lymphotoxin genes (Fig. 2) are separated by only about 1100 base pairs. Lymphotoxin lies 5′ to cachectin/TNF, and both genes reside within the major histocompatibility complex (chromosome 6 in man, and chromosome 17 in the mouse) (NEDWIN et al. 1985; SPIES et al. 1986; NEDOSPASOV et al. 1986; SEMON et al. 1987). Within the MHC complex of the mouse, cachectin/TNF and lymphotoxin are located approximately 70 kilobases proximal to the *D*-locus (MULLER et al. 1987).

Despite its relatively limited sequence homology to cachectin/TNF, lymphotoxin fulfills a spectrum of biological activities that are, both in vivo

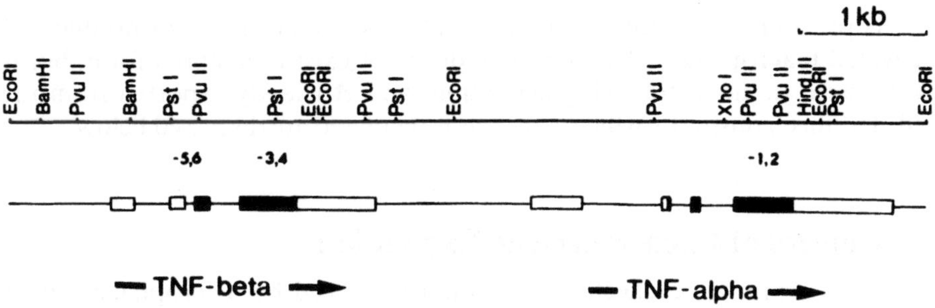

Fig. 2. The human TNF locus. TNF-α and TNF-β genes lie within the major histocompatibility complex, and are tightly linked. Each gene contains four exons. Regions coding for the mature protein are indicated by *solid bars*. (From NEDOSPASOV et al. 1986)

and in vitro, virtually identical to those of cachectin/TNF. Initially identified as a presumptive mediator of cell-mediated hypersensitivity (RUDDLE and WAKSMAN 1967, 1968a, b, c), lymphotoxin was shown some time ago to cause hemorrhagic necrosis of tumors in vivo, and to bring about the lysis of certain transformed cell lines. The two hormones share a common cell surface receptor, binding to it with comparable affinity (AGGARWAL et al. 1985b). However, the tissue of origin and stimuli required for elaboration of the two hormones are quite different.

Crystals of both cachectin/TNF and lymphotoxin have been produced and preliminary structural data from work performed on the former have recently been published (HAKOSHIMA and TOMITA 1988; ECK et al. 1988). It is reasonable to expect that high-resolution models of both molecules will be available in the near future.

F. Cachectin/TNF and Lymphotoxin: Production Sources, Kinetics, and Stimuli

It was once believed that cachectin/TNF was strictly a product of mononuclear phagocytic cells, whereas lymphotoxin was derived only from lymphocytes. This initial picture has been somewhat complicated by the observation that lymphocytes are also capable of producing cachectin/TNF when exposed to such stimuli as phorbol ester, in conjunction with a calcium ionophore (e.g., A23187) (CUTURI et al. 1987). To date, no reports of lymphotoxin production by macrophages have been published. Cachectin/TNF is also reportedly produced (albeit in very small quantities) by natural killer (NK) cells (DEGLIANTONI et al. 1985; PETERS et al. 1986). Finally, certain transformed cell lines, rendered resistant to lymphotoxin by continuous exposure to the hormone, are said to produce small amounts of cachectin/TNF spontaneously (RUBIN et al. 1986), and certain tumors derived from human sources are also reported to be capable of spontaneous TNF production (CORDINGLEY et al. 1988).

As previously mentioned, cachectin/TNF is one of the major products of activated macrophages. The hormone does not exist in a stored form, but is synthesized de novo following activation, and efficiently exported from the cell. The mechanisms by which synthesis is regulated are discussed below.

G. Control of Cachectin Gene Expression

So far as is known, cachectin/TNF is not produced by macrophages under normal circumstances. Indeed, as described below, its production and release may have deleterious effects upon the organism, and, therefore, its synthesis must be tightly governed. At the same time, cachectin/TNF is known to be produced in great abundance by cells following induction with LPS, or other invasive stimuli. It is also produced over a very short period: circulating cachectin/TNF has been measured within 15 min of LPS challenge (BEUTLER et al. 1985b).

Cachectin/TNF biosynthesis is controlled at multiple levels (BEUTLER et al. 1986; SARIBAN et al. 1988; COLLART et al. 1986; GIFFORD and LOHMANN-MATTHES 1986). In response to LPS, cachectin/TNF gene transcription (which is detectable in resting macrophages) is accelerated approximately threefold; however, cachectin/TNF mRNA levels rise by a factor of 50–100, and cachectin/TNF protein secretion, which is undetectable in quiescent cell cultures, rises by a factor of about 10000. Under some circumstances, cachectin/TNF mRNA may be detected within cells in the absence of hormone production or release. This is most readily observed in macrophages derived from C3H/HeJ (endotoxin-unresponsive) mice, or in dexamethasone-treated macrophages that are subsequently stimulated with LPS. To a large extent, therefore, cachectin/TNF biosynthesis appears to be controlled at a translational level.

The posttranscriptional control of cachectin/TNF gene expression seems to be a function of sequences that reside within the 3′-untranslated segment of the molecule. CAPUT et al. (1986) noted that the 3′-untranslated region of cachectin/TNF mRNA contains a long UA-exclusive region bearing overlapping and repeating octameric elements (UUAUUUAU). This sequence was found to be conserved in toto between human and murine forms of the cDNA. Moreover, many inflammatory cytokine mRNAs, as well as mRNAs encoding certain protooncogenes, were found to have similar 3′-untranslated sequences.

It was soon determined that one such sequence, derived from the human GM-CSF gene, when inserted into a rabbit β-globin gene in the 3′-untranslated region, would cause marked instability of the modified mRNA (SHAW and KAMEN 1986). Subsequently, the existence of a selective nucleolytic activity capable of hydrolyzing mRNA bearing multiple copies of the cachectin/TNF UUAUUUAU sequence was demonstrated (BEUTLER et al. 1988).

Recently, it has become clear that the instability of mRNA molecules bearing UUAUUUAU-rich sequences is a specific example of a more general principle. It was noted that the dinucleotide UpA is unusually susceptible to hydrolysis by cytoplasmic ribonucleases when present in a single-stranded form (BEUTLER et al. 1988). The marked susceptibility of UpA to ribonucleases has led to a scarcity of UpA in most mRNA molecules, save those which, like cachectin/TNF mRNA, conserve the dinucleotide to encourage instability (BEUTLER et al. 1988). The propensity for strand breakage at UpA may, indeed, have been responsible for the adoption of UAA and UAG as the two universal "stop" codons, for which no anticodon exists, since in primordial biosystems UpA may have been entirely absent from RNA (BEUTLER et al. 1988). Beyond its ability to confer instability, the UUAUUAU-rich sequence of GM-CSF appeared to confer superinducibility (SHAW and KAMEN 1986). Moreover, the UUAUUAU-rich sequence of interferon-β, when linked to a reporter gene, suppresses translation in *Xenopus* oocytes, and in reticulocyte lysates, but not in wheat germ translation systems (KRUYS et al. 1987, 1988). Suppression of translation, in these instances, does not appear related to instability. These observations would suggest that specific proteins within the cell, in addition to ribonuclease, must interact with such UA-rich sequences, perhaps altering the access of the mRNA to translation. Further work will be required to characterize such factors, and to understand how their activity is governed.

The effect of glucocorticoid hormones on cachectin/TNF mRNA translation, and on cachectin/TNF gene transcription, bears special mention, since this action of corticosteroids probably explains their marked protective effect against lipopolysaccharide toxicity. Glucocorticoids are only capable of inhibiting cachectin/TNF biosynthesis if administered prior to LPS; their effect is a preemptive one (BEUTLER et al. 1986). If administered following LPS challenge, cachectin/TNF biosynthesis proceeds unimpeded. This fact may explain the limited utility of glucocorticoids in the treatment of septic shock as it occurs in a clinical setting.

H. Cachectin/TNF Receptor and Postreceptor Mechanisms

After cachectin/TNF is secreted, it interacts with a plasma membrane receptor that is widely represented on mammalian cells and tissues (BEUTLER et al. 1985b; SUGARMAN et al. 1985). As a direct result of hormone-receptor interaction, one of several biological effects may be triggered. Approximately one-third of transformed cell lines surveyed (SUGARMAN et al. 1985) were killed or growth-inhibited by cachectin/TNF. On the other hand, certain fibroblast lines (SUGARMAN et al. 1985; VILCEK et al. 1986) are actually growth-stimulated by cachectin/TNF. Some of the effects of cachectin/TNF are manifested rapidly (GAMBLE et al. 1985; TRACEY et al. 1986a), whereas most take hours to develop. The nature of the cachectin/TNF receptor, from which all of these cellular effects take origin, remains quite poorly understood, as does the nature of the signal evoked upon binding of the hormone.

Recent studies suggest that the intact cachectin/TNF receptor is a protein with a molecular weight of approximately 300000 (SMITH et al. 1986), possibly consisting of dissimilar subunits (CREASEY et al. 1987). The binding subunit of the receptor, as established by crosslinking studies, has a molecular size of approximately 75 kDa (KULL et al. 1985; CREASEY et al. 1987; SCHEURICH et al. 1986; ISRAEL et al. 1986; YOSHIE et al. 1986).

A glycosyl moiety is suggested by experiments in which certain lectins have been shown to prevent the biological action of cachectin/TNF on its target cells, without impeding hormone binding (AGGARWAL et al. 1986). Moreover, a dissociation between binding and exertion of biological effect has been suggested by studies in which cachectin/TNF derived from different species has been shown to bind to a given species of receptor with similar affinity, yet trigger an effect (cytolysis) of different magnitude (SMITH et al. 1986).

When cachectin/TNF binds to its receptor, it does so with an affinity constant of approximately 3×10^9 (BEUTLER et al. 1985d). Variable numbers of cachectin/TNF receptors are observed on different tissues; however, receptor number and affinity do not correlate with cytotoxic effect. Considerable effort has been devoted to the elucidation of the postreceptor mechanism of action of cachectin/TNF. Some time ago, it was noted that lymphotoxin-mediated cell injury was associated with fragmentation of genomic DNA, occurring well before cell lysis (SCHMID et al. 1986, 1987). This finding prompted speculation that DNA fragmentation might play an essential role in the destruction of target tumor cells. More recently, it was shown that agents inhibiting ADP-ribosyltransferase activity are capable of blocking the cytotoxic effect of cachectin/TNF, at least in some target cells (AGARWAL et al. 1988). This would suggest that ADP ribosylation, induced by DNA fragmentation, might eventuate destruction of the cell. However, the proximal message leading to these events remains a mystery.

J. Biological Effects of Cachectin/TNF and Lymphotoxin: In Vivo and In Vitro

With the availability of recombinant cytokines, the ability of cachectin/TNF and lymphotoxin to modulate cellular function, both in vivo and in vitro, has been intensively studied. The action of these hormones has been examined both in isolation and in concert with related proinflammatory polypeptides, including interferon-γ, interleukin-1, and interleukin-2. A large number of differentiated cell lines have been studied, as have a wide variety of physiologic parameters. Despite specific examples to the contrary (JACOB and MCDEVITT 1988), it may reasonably be claimed that the principal actions of cachectin/TNF and lymphotoxin are catabolic and inflammatory; in many instances, the effects of the hormone on isolated tissues and cells reflect the responses of the entire organism. Below, we summarize the biological actions of cachectin/TNF as established in a number of experimental systems dealing with specific cell and tissue types. The physiologic responses to cachectin/TNF, as observed in living animals following administration of the hormone, are covered subsequently.

I. Adipose Tissue

Cachectin/TNF, by that name, was first identified as a hormone capable of modulating the metabolic activities of adipocytes (ROUZER and CERAMI 1980; KAWAKAMI and CERAMI 1981; KAWAKAMI et al. 1982; PEKALA et al. 1983; HOTEZ et al. 1984; TORTI et al. 1985; MAHONEY et al. 1985; BEUTLER et al. 1985c, d). It was noted to cause suppression of lipoprotein lipase, acetyl coenzyme A (CoA) carboxylase, and fatty acid synthetase when applied as a crude factor, and to activate the release of glycerol from differentiated fat cells, presumably indicating an effect on the hormone-sensitive lipase. Subsequently, work with purified recombinant cachectin/TNF confirmed that the hormone did, indeed, mediate most of these changes (ENERBACK et al. 1988; PATTON et al. 1986, 1987; KAWAKAMI et al. 1987; SEMB et al. 1987; ZECHNER et al. 1988). Cachectin/TNF was also shown to inhibit the morphological differentiation of adipocytes (TORTI et al. 1985) and, when chronically administered to these cells, to induce morphological dedifferentiation. It may be supposed that these effects on lipid metabolism could serve an adaptive function in an acute, self-limited illness, by favoring the mobilization of energy stores for use in the immune response.

II. Muscle

Befitting its role as a presumptive mediator of wasting in chronic disease, cachectin/TNF was regarded as a potential mediator of protein catabolism in muscle cells, and as a mediator of the diminished transmembrane potential frequently observed in shock. Indeed, cachectin/TNF has been shown to lower transmembrane potential in isolated skeletal muscle preparations (TRACEY et al. 1986a). While it may also be involved in the breakdown of skeletal muscle protein, no direct effect of this type has been demonstrated (GOLDBERG et al. 1988; KETTELHUT and GOLDBERG 1988). Recent work has demonstrated, however, that TNF modulates LPL in cardiac muscle (HÜLSMANN and DUBELAAR 1988; HÜLSMANN et al. 1988). Moreover, at high concentrations, cachectin/TNF has been shown to stimulate glucose uptake by L-6 muscle cells in vitro; the significance of this insulin-like effect (LEE et al. 1987) remains to be determined. By lowering transmembrane potential, cachectin/TNF may possibly act to encourage the influx of sodium and water into the "third space" as is often observed clinically in septic shock.

III. Liver

The toxic effects of cachectin/TNF on hepatocytes have only been studied recently. However, for some time it has been apparent that galactosamine hepatotoxicity is a function of lymphoreticular cells in mice (CHOJKIER and FIERER 1985). Hence, a factor of hematopoietic origin must ultimately mediate the necrotizing effect of this compound. Recently, GHEZZI et al. (1986) have shown that recombinant TNF depresses cytochrome P450-dependent drug metabolism in mice. Many aspects of the acute-phase response, including diminished production of albumin and enhanced production of fibrinogen, α1-acid glycoprotein, and cysteine protease inhibitor, seem to depend upon the action of TNF and IL-1, acting in concert (Koj et al. 1987; Neta et al. 1987; Sipe et al. 1987; Perlmutter et al. 1986). Total hepatic protein synthesis is suppressed by cachectin/TNF (Moldawer et al. 1988), and in rats chronic administration of cachectin/TNF has been shown to cause liver injury that is discernable at a light microscopic level (Gaskill 1988).

As a repository of Kupffer cells, the liver is the first reticuloendothelial organ to be exposed to blood-draining enteric structures, and its endotoxin burden, the capacity of the liver to produce TNF, and its function as a locally active hormone remain to be determined (Decker et al. 1987).

IV. Gastrointestinal Tract

A large fraction (approximately 10%) of an injected dose of cachectin/TNF quickly localizes in the gastrointestinal tract (Beutler et al. 1985b). As noted below, systemic administration of cachectin/TNF to experimental animals leads to bowel necrosis, concentrated in the region of the cecum (Tracey et al. 1986; Patton et al. 1987). This effect of the hormone may trigger the release of additional LPS from the bowel, leading to "irreversible endotoxin shock" (Lillehei and Maclean 1958). It is believed that the necrotizing effect of cachectin/TNF may be mediated by platelet-activating factor (Sun and Hsueh 1988), which the hormone is known to induce. Ultimately, the cause of necrosis may be vascular, as witnessed in the case of susceptible tumors. Cachectin/TNF appears to delay gastric emptying (Patton et al. 1987) and may, in part, produce anorexia through its effects on gastrointestinal structures.

V. Central Nervous System

While cachectin/TNF crosses the blood-brain barrier in vanishing small amounts (Beutler et al. 1985b), it is known that central production of the hormone may occur (Mustafa 1989; Leist et al. 1988). High levels of cachectin/TNF have been measured in the cerebrospinal fluid of patients with bacterial, but not viral, meningitis (Leist et al. 1988), and hormone concentrations may prove to be of prognostic importance. Once within the CNS, cachectin/TNF is capable of inducing a pyrogenic response (Dinarello et al. 1986; Kettelhut and Goldberg 1988), apparently through a direct effect on hypothalamic neurons (which produce enhance quantities of PGE-2 in response to cachectin/TNF) and through induction of IL-1 release. Cachectin/TNF also causes anorexia, which may, in part, have a central etiology (Wei et al. 1987; Michie et al. 1987; Plata-Salamàn et al. 1988; Moldawer et al. 1988). Cachectin/TNF has also been shown to stimulate thermogenesis by a central effect (Rothwell 1988).

Recently, it has been reported that α-melanocyte-stimulating hormone is capable of blocking some of the central effects of both cachectin/TNF and IL-1, including the thermoregulatory effect (ROBERTSON et al. 1988). In the wake of reports that IL-1 stimulates adrenocorticotropic hormone secretion in a negative feedback loop (BESEDOVSKY et al. 1986), related observations have been made with cachectin/TNF (MCCANN et al. 1989), which also appears to influence the secretion of prolactin (MCCANN et al. 1989).

Considerable speculation has been devoted to the notion that cachectin/TNF, or the related hormone lymphotoxin, may be involved in chronic inflammatory central nervous disorders, such as multiple sclerosis. In experimental allergic encephalomyelitis, T-lymphocyte clones capable of producing lymphotoxin also appear to be capable of causing the most severe neurologic lesions (BROOME-POWELL, personal communication). Further work will be required to establish a cause-and-effect relationship, and to understand how glial injury is actually mediated.

VI. Adrenal

Cachectin/TNF has been noted, in vivo, to cause adrenal hemorrhage (TRACEY et al. 1986b). It has also been noted that adrenalectomy sensitizes mice to the lethal effect of both IL-1 and cachectin/TNF (BERTINI et al. 1988). Since the inhibitory effect of glucocorticoid hormones on production of cachectin/TNF is well established (BEUTLER et al. 1986), it would seem likely that tonic production of glucocorticoid hormones may exert a physiologic restraint both on the production of cachectin/TNF and on its action in vivo. Cortisol levels are markedly increased following infusion of cachectin/TNF in toxic doses (TRACEY et al. 1987a); this may reflect an indirect response to stress or a direct effect of the hormone either at a central level or upon cells of the adrenal gland itself.

VII. Skin

A remarkable fraction (approximately 30%) of an intravenously administered dose of cachectin/TNF may be recovered from skin (BEUTLER et al. 1985b). Interestingly, the phenomenon of hemorrhagic necrosis is also principally observed in tumors with a dermal blood supply. It would seem that the dermal vasculature, which in any case is subject to a high degree of autoregulation, is quite responsive to the effects of cachectin/TNF. Both cachectin/TNF and lymphotoxin have been shown to induce dermal inflammation (SHARPE et al. 1987; JEFFES et al. 1987; AVERBOOK et al. 1987; CYBULSKY et al. 1988). In some measure cachectin/TNF, together with IL-1, may contribute to the localized Shwartzman phenomenon, and to its generalized counterpart. Human dermal fibroblasts (DAYER et al. 1985) have been shown to secrete collagenase and PGE-2 in response to cachectin/TNF. Other fibroblast-derived cells (FS4 cells) are known to undergo a proliferative response following contact with the hormone. In addition, macrophages are stimulated to produce cachectin/TNF by contact with advanced glycosylation end products that characteristically accumulate in mesenchymal tissues (VLASSARA et al. 1988). Thus, several inflammatory and regenerative processes may depend upon the action of the hormone.

VIII. Bone and Cartilage

A variety of inflammatory joint diseases might potentially depend upon the elaboration of cachectin/TNF, IL-1, or related hormones. Cachectin/TNF has been shown to stimulate the resorption of proteoglycan in cartilage (SAKLATVALA 1986) and to stimulate bone resorption (BERTOLINI et al. 1986). The production of new proteoglycan and bone are inhibited in the respective assay systems. Cachectin/TNF and lymphotoxin appear to synergize with IL-1 in their ability to prompt bone resorption (STASHENKO et al. 1987) although the latter hormone, acting in isolation, is a far more potent activator of osteoclasts than either of the former. Osteoclast-like cells of the line C3-E1 (SATO et al. 1987) are stimulated by cachectin/TNF to produce macrophage colony-stimulating factor (M-CSF) and PGE-2. The role played by these secondary mediators in bone resorption remains to be fully quantitated. Transforming growth factor-β, which antagonizes the effects of cachectin/TNF in many other systems, also seems to have the ability to stimulate bone resorption and PGE-2 production in cultured mouse calvarium (TASHJIAN et al. 1985).

The clinical significance of lymphotoxin as an osteoclast activator has been best demonstrated by the studies of GARRETT et al. (1987), who have shown that human myeloma cells may produce lymphotoxin in an autonomous manner, and that monoclonal antibody directed against lymphotoxin inhibits the bone-resorbing potential of myeloma cells in vitro. Thus, at least some of the destructive effects of myeloma may be directly attributable to the elaboration of lymphotoxin.

IX. Vascular Endothelium

The shock-promoting effects of cachectin/TNF, and its ability to induce hemorrhagic necrosis of certain tumors in vivo, appear to depend largely upon its vascular effects. Cachectin/TNF decreases endothelial cell expression of thrombomodulin, and causes the elaboration of a procoagulant activity (STERN and NAWROTH 1986) as well as the release of IL-1 from endothelial cells in vitro (NAWROTH et al. 1986a; LIBBY et al. 1986). The latter hormone exerts effects on endothelial cells substantially similar to those of cachectin/TNF itself (POHLMAN et al. 1986; POBER et al. 1986a, 1987; NAWROTH et al. 1986; LEUNG et al. 1986; SCHLEEF et al. 1988; COLOTTA et al. 1988). Cachectin/TNF also stimulates adhesion of neutrophils to vascular endothelial cells through effects on each of the cellular participants (GAMBLE et al. 1985;

BROUDY et al. 1987; POHLMAN et al. 1986). This effect of the hormone may relate to the initial phase of neutropenia observed following LPS administration, and to the margination and transudation of neutrophils that occurs during local inflammatory processes. Among its other effects on vascular endothelial cells, cachectin/TNF is growth inhibitory (SATO et al. 1986; SCHWEIGERER et al. 1987) and causes morphological changes (SATO et al. 1986; STOLPEN et al. 1986) and reorganization of vascular endothelial monolayers (STOLPEN et al. 1986) such that the cells flatten, overlap, rearrange actin filaments, and lose stainable fibronectin when treated with the hormone. Synergy between interferon-γ and cachectin/TNF in causing these changes has been noted (STOLPEN et al. 1986). Cachectin/TNF modulates the expression of various surface antigens on endothelial cells (POHLMAN et al. 1986; POBER et al. 1986a, b; COLLINS et al. 1986). It also induces endothelial phospholipase A2 activity (BOMALASKI et al. 1987). Tissue-type plasminogen activator is suppressed, and type-1 plasminogen activator inhibitor is induced in endothelial cells following exposure to cachectin/TNF (SCHLEEF et al. 1988).

Interleukin-1 and cachectin/TNF have both been reported to sensitize vascular endothelial cells to lysis by circulating antibodies obtained from patients with Kawasaki syndrome. Conceivably, this might predispose to the vascular complications associated with this disease (LEUNG et al. 1986).

X. Hematopoietic Elements

1. Neutrophils

No less important than the effect of cachectin/TNF on endothelial cells is its effect on neutrophils. Cachectin/TNF prompts neutrophil adhesion by a direct effect (GAMBLE et al. 1985; SHALABY et al. 1985; KLEBANOFF et al. 1986) that is maximal within 5 min of contact with the hormone, and does not appear to require protein synthesis. This effect, in vivo, may lead to their margination, transudation from the intravascular space into the tissue, and subsequent tissue injury. The importance of neutrophils in the pathogenesis of endotoxic shock is widely accepted (PINGLETON et al. 1975; BRIGHAM and MEYRICK 1984), and considerable significance may be attached to the fact that cachectin/TNF appears to be a major mediator of neutrophil adhesion and activation.

In addition to enhancing neutrophil adhesion to endothelial surfaces, cachectin/TNF stimulates enhanced phagocytosis of latex beads (SHALABY et al. 1985), enhanced production of superoxide anion (TSUJIMOTO et al. 1986), release of lysozyme and hydrogen peroxide (KLEBANOFF et al. 1986), and degranulation (KLEBANOFF et al. 1986). Neutrophils also show enhanced microbicidal activity when stimulated by cachectin/TNF (DJEU et al. 1986).

Cachectin/TNF has recently been shown to modulate the neutrophil response to F-Met-Leu-Phe receptor activation (ATKINSON et al. 1988; FERRANTE et al. 1988). In addition, cachectin/TNF increases the expression of complement receptor on human neutrophils, and may thereby contribute to localization within an inflammatory focus (BERGER et al. 1988). By some accounts (MING et al. 1987; CYBULSKY et al. 1988) cachectin/TNF appears to trigger the emigration of neutrophils into skin, perhaps by a direct chemotactic effect. The chemotactic activity of cachectin/TNF has been disputed by other workers (SHALABY et al. 1987), who nonetheless note that cachectin/TNF appears to inhibit neutrophil migration.

2. Eosinophils

Cachectin/TNF has been shown to enhance the toxicity of eosinophils to schistosomula in vitro (SILBERSTEIN and DAVID 1986); it would appear that cachectin/TNF is the principal factor produced by U-937 cells responsible for this bioactivity (SILBERSTEIN et al. 1987). Recombinant granulocyte-macrophage colony-stimulating factor (GM-CSF) also enhances the cytotoxicity of eosinophils for schistosomula (SILBERSTEIN et al. 1986); the mechanism by which the cytotoxic responses augmented remains to be determined, as does the significance of the phenomenon in vivo. However, these observations support the notion that cachectin/TNF may have a beneficial role to play in chronic parasitic infections (TAVERNE et al. 1984; PLAYFAIR et al. 1984; BATE et al. 1988).

3. Monocyte/Macrophages

Studies by a large number of investigators have led to the conclusion that the macrophage, in one or more of its many forms, is the principal source of cachectin/TNF (MATTHEWS 1978, 1981 a, b; CLARK 1978; MANNEL et al. 1980; SATOMI et al. 1981; ZACHARCHUK et al. 1983; FISCH and GIFFORD 1983; WOZENCRAFT et al. 1984; WOOD and CLARK 1984; WATANABE et al. 1984; ITOH et al. 1984; BLOKSMA et al. 1984; CHEN et al. 1985; WARREN and RALPH 1986; GIFFORD and LOHMANN-MATTHES 1986; KORNBLUTH and EDGINGTON 1986) produced in response to many types of invasive stimuli, LPS being the most potent of these. Less work, to date, has concerned the effects of cachectin/TNF on monocyte/macrophages themselves, although it is clear that these cells bear receptors for cachectin/TNF (IMAMURA et al. 1987) and internalize cachectin/TNF once it is bound to them. Cachectin/TNF has been reported to exert a chemotactic effect on monocytes, just as it does on endothelial cells (MING et al. 1987). Macrophage activation appears to cause a marked enhancement of LPL secretion by macrophages; however, this enhancement does not appear to be mediated by cachectin/TNF (BEHR and KRAEMER 1986). Cachectin/TNF enhances the production of GM-CSF by a variety of cell types, and thus may affect macrophage function indirectly, if not directly (MUNKER et al. 1986). Cachectin/TNF seems to induce the differentiation of certain myeloid cell lines in vitro (TAKEDA et al. 1986) and has been shown to stimulate the release of IL-1 and PGE_2 production from resting macrophages (BACHWICH et al. 1986). Like eosinophils, macrophages are capable of killing schistosomula when activated by cachectin/TNF or lymphotoxin, and are synergistically activated to achieve this effect by coincubation with interferon-γ (ESPARZA et al. 1987). Cachectin/TNF, alone or in combination with IL-2, has been shown to activate macrophages, allowing them to kill *Mycobacterium avium* organisms (BERMUDEZ and YOUNG 1988).

To date, no clear data have emerged supporting the concept that cachectin/TNF is capable of inducing cachectin/TNF biosynthesis by primary macrophages. The function of cachectin/TNF as it concerns macrophages themselves is still rather poorly understood, and much additional work remains to be done before this issue can be fully resolved.

4. Lymphocytes

Over a decade ago, prior to the cloning of cachectin/TNF, the ability of postendotoxin serum to promote antigen-directed B-cell activation was reported, and presumptively ascribed to the action of TNF (HOFFMANN et al. 1977). Moreover, crude preparations of cachectin/TNF were reported to direct the maturation of thymocytes in vitro (ABBOTT et al. 1981). Crude preparations of cachectin/TNF were also noted to be cytotoxic to human lymphocytes (UMEDA et al. 1983) as was lymphotoxin (CONTA et al. 1985; RUDDLE 1985).

More recently, the effects of cachectin/TNF on T and B lymphocytes have been studied in greater detail. Cachectin/TNF induces the expression of additional cachectin/TNF receptors on primary cultures of T lymphocytes (SCHEURICH et al. 1987) and also increases the expression of HLA-DR antigen and high-affinity IL-2 receptor. Thus, cachectin/TNF-treated T cells show an enhanced proliferative response to IL-2. Cachectin/TNF also enhances IL-2-dependent production of interferon-γ. Other workers (PLELA and KORN 1987) have suggested that lymphocyte-fibroblast adhesion is induced by interferon-γ and cachectin/TNF. It has been reported that B-cell proliferation and differentiation, occurring in response to the B-cell activator pokeweed mitogen, is inhibited by cachectin/TNF (KASHIWA et al. 1987). Finally, cachectin/TNF, lymphotoxin, and various interferons have been shown to enhance the T-cell-mediated response to antigenic challenge through an effect on T cells (TROPPMAIR et al. 1988).

The significance of these observations which involve in vitro studies remains to be demonstrated at the level of the organism. However, the importance of cachectin/TNF as a modulator of B-cell and T-cell function and as an autocrine factor capable of influencing the progression of certain hematopoietic malignancies (CORDINGLEY et al. 1988) would appear likely. Very recent work has implicated cachectin/TNF and lymphotoxin as essential participants in mixed lymphocyte reactions (SHALABY et al. 1988). These data, as well as other studies pointing to the involvement of cachectin/TNF as a reactant in the early phase of graft-versus-host disease (PIGUET et al. 1987), suggest that the hormone does, indeed, affect T-cell function so as to produce important systemic consequences.

K. Gross Physiologic and Pathologic Consequences of Cachectin/TNF Production or Administration

The importance of cachectin/TNF as a mediator of shock was suggested long before this monokine was purified or cloned (CLARK 1978, 1982a, b; CLARK et al. 1981). At the same time, there existed a strong suspicion that the same hormone must confirm a protective effect (CLARK 1979; PARANT et al. 1980; TAVERNE et al. 1981, 1982, 1984; HAIDARIS et al. 1983; WOZENCRAFT et al. 1984; PLAYFAIR et al. 1984). Such speculations could be subjected to test only after the purification of the hormone was achieved. The identity of cachectin/TNF (BEUTLER et al. 1985c) underscored the possibility that a single polypeptide hormone might fulfil both beneficial and deleterious functions, and might mediate many, if not all, of the biological effects of lipopolysaccharide. Passive immunization studies (BEUTLER et al. 1985a) revealed that the elaboration of cachectin/TNF was, indeed, an important element in the pathogenesis of endotoxicity. Subsequent work (TRACEY et al. 1987b; MATHISON et al. 1988) in which monoclonal reagents were employed, and in which other species were examined, has supported this view.

Cachectin/TNF has been shown to produce a shock syndrome when administered in isolation to rats and dogs (TRACEY et al. 1986b, 1987a). The hormone induces hypotension, tachypnea, metabolic acidosis, an initial phase of hyperglycemia followed by hypoglycemia, hemoconcentration, and multiple end-organ damage very similar to that seen in endotoxin-poisoned animals. Such changes, witnessed following acute administration of cachectin/TNF, were also reported by other workers (MANNEL et al. 1987; LEHMANN et al. 1987; KETTELHUT et al. 1987; PATTON et al. 1987; BAUSS et al. 1987). Lethality resulting from bolus administration of cachectin/TNF was, in the short run, usually attributable to respiratory arrest, which occurred as a result of acute interstitial pneumonitis (TRACEY et al. 1986). When smaller doses of the hormone were administered, cecal necrosis was often observed, occasionally with perforation of the gastrointestinal tract, leading to acute intraabdominal infection.

Interestingly, chronic administration of the hormone in smaller doses led to a syndrome entirely distinct from that observed following injection of large quantities of cachectin/TNF. Oliff et al. inoculated nude mice with neoplastic (Chinese hamster ovary) cells that had been genetically modified, causing them continuously to secrete cachectin/TNF (OLIFF et al. 1987). The resulting tumors in these animals were small and nonmetastatic; however, the chronic elaboration of cachectin/TNF was associated with a profound wasting diathesis, confirming the original supposition that chronic exposure to the hormone might lead to cachexia (CERAMI et al. 1985; BEUTLER et al. 1985d). The pathogenesis of wasting was not entirely clarified in these studies; however, it appeared that anorexia was largely responsible for the weight loss observed.

Continuous infusion of cachectin/TNF has also been shown to cause hepatic toxicity (GASKILL 1988) and to suppress food intake, hepatic protein synthesis (MOLDAWER et al. 1988; PLATA-SALAMÀN et al. 1988; WEI et al. 1987; PERLMUTTER et al. 1986), and anemia (TRACEY et al. 1988) in animals.

Some effects of cachectin/TNF appear to be mediated centrally, as noted above. Appetite suppression may be among these. In addition, cachectin/TNF has well-defined effects on thermogenesis (COOMBES et al. 1987; ROTHWELL 1988) that are both central and peripheral, and that may contribute to the dissipation of energy observed during acute and chronic infectious disorders.

Supporting the notion that cachectin/TNF is an important mediator of endotoxicity is the finding that agents capable of sensitizing animals to the lethal effect of LPS (e.g., galactosamine and actinomycin) also appear to sensitize to cachectin/TNF (LEHMANN et al. 1987; WALLACH et al. 1988); moreover, adrenalectomy, which greatly diminishes the mean lethal dose of LPS in many species, also predisposes to the toxicity of cachectin/TNF (BERTINI et al. 1988). In each of these studies, IL-1 (like TNF) was shown to be lethal to sensitized animals. This would suggest that, despite their utilization of a different plasma membrane receptor, both factors may evoke similar postreceptor responses within sensitive cell populations.

L. Disease States Associated with Elevated Levels of Cachectin/TNF

Since it was demonstrated that cachectin/TNF could play an important role in the pathogenesis of endotoxic shock (BEUTLER et al. 1985a; TRACEY et al. 1986b) and cachexia (OLIFF et al. 1987), a number of attempts have been made to measure the hormone in biological fluids derived from human and animal subjects suffering from a variety of infectious and neoplastic diseases. Conflicting results have arisen in these studies, in large part because of the novelty of the available assays. Cachectin/TNF was demonstrated in serum derived from patients with proven meningococcal septicemia, and correlated high levels with a negative outcome (WAAGE et al. 1987). Induction of cachectin/TNF was also shown to occur in mouse lung tissue during experimental infection with *Legionella pneumophila* (BLANCHARD et al. 1987); however, these authors concluded (BLANCHARD et al. 1988) that the hormone exerts a protective effect in this disease. Cachectin/TNF was measured in a variety of parasitic infections (SCUDERI et al. 1986) in humans; however, in this study, a high percentage of normal controls also displayed elevated hormone concentrations in their serum. Human patients with acquired immunodeficiency syndrome (LAHDEVIRTA et al. 1988) also display elevated serum concentrations of cachectin/TNF; while levels are markedly correlated with the stage of the disease, it remains to be determined whether the inducing stimulus is the human immunodeficiency virus itself or the infectious complications that characterize AIDS. It has been suggested that cerebral malaria in man might depend, in large part, upon the elaboration of cachectin/TNF (CLARK et al. 1981; CLARK 1982a). Recent experimental work (CLARK et al. 1987; GRAU et al. 1987) has supported the view that cachectin/TNF is an essential mediator of murine cerebral malaria. Separate studies (PIGUET et al. 1987) indicate that cachectin/TNF is involved in the mediation of graft-versus-host disease in mice. The potent abortifacient effect of LPS caused some suspicion that cachectin/TNF might cause premature delivery, as witnessed during intrauterine infection. Recent work suggests that macrophage-like cells of the placental decidua are capable of producing abundant quantities of cachectin/TNF, and that cachectin/TNF inhibits the growth of cells of the amnion, possibly leading to rupture of the fetal membranes (CASEY et al. 1989). Indeed, high levels of cachectin/TNF can be detected in amniotic fluid derived from patients with intrauterine infections (CASEY, personal communication).

Cachectin/TNF has been assayed in serum derived from a large number of human cancer patients, and found to be at detectable levels in a substantial number of cases (BALKWILL et al. 1987). Conflicting data have been reported by a second laboratory (SOCHER et al. 1988), in which cachectin/TNF was absent from the serum of 19 individuals with cachexia related to cancer. It remains to be fully established whether the hormone actually participates in most cases of unexplained wasting, whether it is produced intermittently, or whether undetectable hormone concentrations are adequate to cause wasting if chronically present.

M. Cachectin/TNF and Its Clinical Applications: To Be or Not To Be

In large part, the successful treatment of disease consists of the restoration of homeostasis. In the realm of endocrinology, this implies the judicious administration or antagonism of specific hormones. An insulin-dependent diabetic may be effectively treated by insulin replacement therapy; a patient with Graves' disease may be helped by pharmacological measures that inhibit thyroxin release. Difficulties arise when diseases are treated by administration of hormones that were never deficient to begin with.

The use of cytokines as antineoplastic agents provides a case in point. The rational basis for administering immunomodulatory agents to cancer patients is quite slender. Moreover, many cytokines, like cachectin/TNF, have markedly toxic effects that may prove dose-limiting long before a therapeutic level can be reached.

At present, our understanding of the role for which cachectin/TNF evolved is too narrow to allow a precise assessment of conditions under which administration or removal of the hormone would prove to be of benefit. Already, it is clear that the hormone can kill when overproduced; it is equally clear that it did not evolve to produce a lethal effect.

Like the inflammatory response of which it is a part, cachectin/TNF is a mixed blessing. Perhaps in understanding its physiological function, we may come to comprehend the benefits and liabilities of inflammation as a whole, and gain insight into the circumstances under which "cachectin/TNF deficiency" and "cachectin/TNF excess" obtain. Only then will a rational and specific therapeutic approach to diseases that entail the action of this hormone be feasible.

References

Abbott J, Doyle PJ, Ngiam K, Olson CL (1981) Ontogeny of murine T lymphocytes. I. Maturation of thymocytes induced in vitro by tumor necrosis factor-positive serum (TNF+)1,2. Cell Immunol 57:237–250

Abe S, Gatanaga T, Yamazaki M, Soma G, Mizuno D (1985) Purification of rabbit tumor necrosis factor. FEBS Lett 180:203–206

Agarwal S, Drysdale B-E, Shin HS (1988) Tumor necrosis factor-mediated cytotoxicity involves ADP-ribosylation. J Immunol 140:4187–4192

Aggarwal BB, Kohr WJ, Hass PE, Moffat B, Spencer SA, Henzel WJ, Bringman TS, Nedwin GE, Goeddel DV, Harkins RN (1985a) Human tumor necrosis factor. Production, purification, and characterization. J Biol Chem 260:2345–2354

Aggarwal BB, Eessalu TE, Hass PE (1985b) Characterization of receptors for human tumour necrosis factor and their regulation by gamma-interferon. Nature 318:665–667

Aggarwal BB, Traquina PR, Eessalu TE (1986) Modulation of receptors and cytotoxic response of tumor necrosis factor-alpha by various lectins. J Biol Chem 261:13652–13656

Algire GH, Legallais FY, Anderson BF (1952) Vascular reactions of normal and malignant tissues in vivo. V. The role of hypotension in the action of a bacterial polysaccharide on tumors. JNCI 12:1279–1295

Atkinson YH, Marasco WA, Lopez AF, Vadas MA (1988) Recombinant human tumor necrosis factor-alpha: regulation of *n*-formylmethionylleucylphenylalanine receptor affinity and function on human neutrophils. J Clin Invest 81:759–765

Averbook B, Ulich T, Jeffes E, Yamamoto R, Chow G, Masunaka I, Granger G (1987) Human alpha lymphotoxin and TNF induce different types of inflammatory responses in normal tissue. Fed Proc 46(3):562

Bachwich PR, Chensue SW, Larrick JW, Kunkel SL (1986) Tumor necrosis factor stimulates interleukin-1 and prostaglandin E2 production in resting macrophages. Biochem Biophys Res Commun 136:94–101

Balkwill F, Burke F, Talbot D, Tavernier J, Osborne R, Naylor S, Durbin H, Fiers W (1987) Evidence for tumour necrosis factor/cachectin production in cancer. Lancet 2(8570):1229–1232

Bate CAW, Taverne J, Playfair JHL (1988) Malarial parasites induce TNF production by macrophages. Immunology 64:227–231

Bauss F, Droge W, Mannel DN (1987) Tumor necrosis factor mediates endotoxic effects in mice. Infect Immun 55:1622–1625

Behr SR, Kraemer FB (1986) Effects of activation on lipoprotein lipase secretion by macrophages. Evidence for autoregulation. J Exp Med 164:1362–1367

Berger M, Wetzler EM, Wallis RS (1988) Tumor necrosis factor is the major monocyte product that increases complement receptor expression on mature human neutrophils. Blood 71:151–158

Bermudez LEM, Young LS (1988) Tumor necrosis factor, alone or in combination with IL-2, but not IFN-gamma, is associated with macrophage killing of *Mycobacterium avium* complex. J Immunol 140:3006–3013

Bertini R, Bianchi M, Ghezzi P (1988) Adrenalectomy sensitizes mice to the lethal effects of interleukin 1 and tumor necrosis factor. J Exp Med 167:1708–1712

Bertolini DR, Nedwin G, Bringman T, Smith D, Mundy GR (1986) Stimulation of bone resorption and inhibition of bone formation in vitro by human tumour necrosis factor. Nature 319:516–518

Besedovsky H, del Rey A, Sorkin E, Dinarello CA (1986) Immunoregulatory feedback between interleukin-1 and glucocorticoid hormones. Science 233:652–654
Beutler B, Milsark IW, Cerami A (1985a) Passive immunization against cachectin/tumor necrosis factor (TNF) protects mice from the lethal effect of endotoxin. Science 229:869–871
Beutler B, Milsark IW, Cerami A (1985b) Cachectin/tumor necrosis factor: production, distribution, and metabolic fate in vivo. J Immunol 135:3972–3977
Beutler B, Greenwald D, Hulmes JD, Chang M, Pan Y-CE, Mathison J, Ulevitch R, Cerami A (1985c) Identity of tumour necrosis factor and the macrophage-secreted factor cachectin. Nature 316:552–554
Beutler B, Mahoney J, Le Trang N, Pekala P, Cerami A (1985d) Purification of cachectin, a lipoprotein lipase-suppressing hormone secreted by endotoxin-induced RAW 264.7 cells. J Exp Med 161:984–995
Beutler B, Krochin N, Milsark IW, Luedke C, Cerami A (1986) Control of cachectin (tumor necrosis factor) synthesis: mechanisms of endotoxin resistance. Science 232:977–980
Beutler B, Thompson P, Keyes J, Hagerty K, Crawford D (1988) Assay of a ribonuclease that preferentially hydrolyses mRNAs containing cytokine-derived UA-rich instability sequences. Biochem Biophys Res Commun 152:973–980
Beutler E, Gelbart T, Han J, Koziol JA, Beutler B (1988) Evolution of the genome and the genetic code: selection at the dinucleotide level by methylation and polyribonucleotide cleavage. Proc Natl Acad Sci USA 86:192–196
Blanchard DK, Djeu JY, Klein TW, Friedman H, Stewart WE II (1987) The induction of tumor necrosis factor (TNF) in murine lung tissue during infection with *Legionella pneumophila:* a potential protective role of TNF (Abstr). Lymphokine Res 6:1421
Blanchard DK, Djeu JY, Klein TW, Friedman H, Stewart WE II (1988) Protective effects of tumor necrosis factor in experimental *Legionella pneumophila* infections of mice via activation of PMN function. J Leukocyte Biol 43:429–435
Bloksma N, Hofhuis FM, Willers JM (1984) Role of mononuclear phagocyte function in endotoxin-induced tumor necrosis. Eur J Cancer Clin Oncol 20:397–403
Bomalaski JS, Chen M-J, Clark MA (1987) Induction of phospholipase A2 activity and synthesis of a phospholipase A2 activating protein (PLAP) by tumor necrosis factor. Arthritis Rheum 30:S28
Brigham KL, Meyrick B (1984) Interactions of granulocytes with the lungs. Circ Res 54:623–635
Broudy VC, Harlan JM, Adamson JW (1987) Disparate effects of tumor necrosis factor-alpha/cachectin and tumor necrosis factor-beta/lymphotoxin on hematopoietic growth factor production and neutrophil adhesion molecule expression by cultured human endothelial cells. J Immunol 138:4298–4302
Caput D, Beutler B, Hartog K, Brown-Shimer S, Cerami A (1986) Identification of a common nucleotide sequence in the 3′-untranslated region of mRNA molecules specifying inflammatory mediators. Proc Natl Acad Sci USA 83:1670–1674
Carswell EA, Old LJ, Kassel RL, Green S, Fiore N, Williamson B (1975) An endotoxin-induced serum factor that causes necrosis of tumors. Proc Natl Acad Sci USA 72:3666–3670
Casey ML, Beutler B, MacDonald PC (1989) Cachectin/tumor necrosis factor-alpha action in human amnion and decidua: potential role in infection-associated preterm labor. J Clin Invest 83:430–436
Cerami A, Ikeda Y, Le Trang N, Hotez PJ, Beutler B (1985) Weight loss associated with an endotoxin-induced mediator from peritoneal macrophages: the role of cachectin (tumor necrosis factor). Immunol Lett 11:173–177
Chen AR, McKinnon KP, Koren HS (1985) Lipopolysaccharide (LPS) stimulates fresh human monocytes to lyse actinomycin D-treated WEHI-164 target cells via increased secretion of a monokine similar to tumor necrosis factor. J Immunol 135:3978–3987

Chojkier M, Fierer J (1985) D-Galactosamine hepatotoxicity is associated with endotoxin sensitivity and mediated by lymphoreticular cells in mice. Gastroenterology 88:115–121
Clark IA (1978) Does endotoxin cause both the disease and parasite death in acute malaria and babesiosis? Lancet 2:75–77
Clark IA (1979) Resistance to *Babesia* spp. and *Plasmodium* sp. in mice pretreated with an extract of *Coxiella burnetii*. Infect Immun 24:319–325
Clark IA (1982a) Suggested importance of monokines in pathophysiology of endotoxin shock and malaria. Klin Wochenschr 60:756–758
Clark IA (1982b) Correlation between susceptibility to malaria and babesis parasites and the endotoxicity. Trans R Soc Trop Med Hyg 76:4–7
Clark IA, Virelizier JL, Carswell EA, Wood PR (1981) Possible importance of macrophage-derived mediators in acute malaria. Infect Immun 32:1058–1066
Clark IA, Cowden WB, Butcher GA, Hunt NH (1987) Possible roles of tumor necrosis factor in the pathology of malaria. Am J Pathol 129:192–199
Coley WB (1893) The treatment of malignant tumors by repeated inoculations of erysipelas; with a report of ten original cases. Am J Med Sci 105:487–511
Coley WB (1906) Late results of the treatment of inoperable sarcoma by the mixed toxins of erysipelas and *Bacillus prodigiosus*. Am J Med Sci 131:375–430
Collart MA, Berlin D, Vassalli JD, DeKossodo S, Vassalli P (1986) Gamma interferon enhances macrophage transcription of the tumor necrosis factor/cachectin, interleukin 1, and urokinase genes, which are controlled by short-lived repressors. J Exp Med 164:2113–2118
Collins T, Lapierre LA, Fiers W, Strominger JL, Pober JS (1986) Recombinant human tumor necrosis factor increases mRNA levels and surface expression of HLA-A,B antigens in vascular endothelial cells and dermal fibroblasts in vitro. Proc Natl Acad Sci 83:446–450
Colotta F, Lampugnani MG, Polentarutti N, Dejana E, Mantovani A (1988) Interleukin-1 induces *c-fos* protooncogene expression in cultured human endothelial cells. Biochem Biophys Res Commun 152:1104–1110
Conta BS, Powell MB, Ruddle NH (1985) Activation of LyT-1+ and LyT-2+ T cell cloned lines: stimulation of proliferation, lymphokine production, and self-destruction. J Immunol 134:2185–2190
Coombes RC, Rothwell NJ, Shah P, Stock MJ (1987) Changes in thermogenesis and brown fat activity in response to tumour necrosis factor in the rat. Biosci Rep 7:791–799
Cordingley FT, Hoffbrand AV, Heslop HE, Turner M, Bianchi A, Reittie JE, Vyakarnam A, Meager A, Brenner MK (1988) Tumour necrosis factor as an autocrine tumour growth factor for chronic B-cell malignancies. Lancet 1:969–971
Creasey AA, Yamamoto R, Vitt CR (1987) A high molecular weight component of the human tumor necrosis factor receptor is associated with cytotoxicity. Proc Natl Acad Sci USA 84:3293–3297
Cuturi MC, Murphy M, Costa-Giomi MP, Weinmann R, Perussia B, Trinchieri G (1987) Independent regulation of tumor necrosis factor and lymphotoxin production by human peripheral blood lymphocytes. J Exp Med 165:1581–1594
Cybulsky MI, McComb DJ, Movat HZ (1988) Neutrophil leukocyte emigration induced by endotoxin. J Immunol 140:3144–3149
Darzynkiewicz Z, Williamson B, Carswell EA, Old LJ (1984) Cell cycle-specific effects of tumor necrosis factor. Cancer Res 44:83–90
Davis JM, Narachi MA, Alton NK, Arakawa T (1987) Structure of human tumor necrosis factor alpha derived from recombinant DNA. Biochemistry 26:1322–1326
Dayer J-M, Beutler B, Cerami A (1985) Cachectin/tumor necrosis factor (TNF) stimulates collagenase and PGE2 production by human synovial cells and dermal fibroblasts. J Exp Med 162:2163–2168

Decker T, Lohmann-Matthes M-L, Gifford GE (1987) Cell-associated tumor necrosis factor (TNF) as a killing mechanism of activated cytotoxic macrophages. J Immunol 138:957–962

Degliantoni G, Murphy M, Kobayashi M, Francis MK, Perussia B, Trinchieri G (1985) Natural killer (NK) cell-derived hematopoietic colony-inhibiting activity and NK cytotoxic factor. Relationship with tumor necrosis factor and synergism with immune interferon. J Exp Med 162:1512–1530

Dinarello CA, Cannon JG, Wolff SM, Bernheim HA, Beutler B, Cerami A, Palladino MA, O'Connor JV (1986) Tumor necrosis factor (cachectin) is an endogenous pyrogen and induces production of interleukin-1. J Exp Med 163:1433–1450

Djeu JY, Blanchard DK, Halkias D, Friedman H (1986) Growth inhibition of *Candida albicans* by human polymorphonuclear neutrophils: activation by interferon-gamma and tumor necrosis factor. J Immunol 137:2980–2984

Eck MJ, Beutler B, Kuo G, Merryweather JP, Sprang SR (1988) Crystallization of trimeric recombinant human tumor necrosis factor (cachectin). J Biol Chem 263:12816–12819

Enerback S, Semb H, Tavernier J, Bjursell G, Olivecrona T (1988) Tissue-specific regulation of guinea pig lipoprotein lipase; effects of nutritional state and of tumor necrosis factor on mRNA levels in adipose tissue, heart and liver. Gene 64: 97–106

Esparza I, Mannel D, Ruppel A, Falk W, Krammer PH (1987) Interferon-gamma (IFN-gamma) and lymphotoxin (LT) or tumor necrosis factor (TNF) synergize to activate macrophages for tumoricidal and schistosomulicidal functions. Lymphokine Res 6:1715

Ferrante A, Nandoskar M, Walz A, Goh DHB, Kowanko IC (1988) Effects of tumour necrosis factor alpha and interleukin-1 alpha and beta on human neutrophil migration, respiratory burst and degranulation. Int Arch Allergy Appl Immunol 86:82–91

Fisch H, Gifford GE (1983) In vitro production of rabbit macrophage tumor cell cytotoxin. Int J Cancer 32:105–112

Fransen L, Muller R, Marmenout A, Tavernier J, van der Heyden J, Kawashima E, Chollet A, Tizard R, Van Heuverswyn H, Van Vliet A, Ruysschaert M-R, Fiers W (1985) Molecular cloning of mouse tumour necrosis factor cDNA and its eukaryotic expression. Nucleic Acids Res 13:4417–4429

Gamble JR, Harlan JM, Klebanoff SJ, Lopez AF, Vadas MA (1985) Stimulation of the adherence of neutrophils to umbilical vein endothelium by human recombinant tumor necrosis factor. Proc Natl Acad Sci USA 82:8667–8671

Garrett R, Durie BGM, Nedwin GE, Gillespie A, Bringman T, Sabatini M, Bertolini DR, Mundy GR (1987) Production of lymphotoxin, a bone-resorbing cytokine, by cultured human myeloma cells. N Engl J Med 317:526–532

Gaskill HV III (1988) Continuous infusion of tumor necrosis factor: mechanisms of toxicity in the rat. J Surg Res 44:664–671

Ghezzi P, Saccardo B, Bianchi M (1986) Recombinant tumor necrosis factor depresses cytochrome P450-dependent microsomal drug metabolism in mice. Biochem Biophys Res Commun 136:316–321

Gifford GE, Lohmann-Matthes ML (1986) Requirement for the continual presence of lipopolysaccharide for production of tumor necrosis factor by thioglycollate-induced peritoneral murine macrophages. Int J Cancer 38:135–137

Goldberg AL, Kettelhut IC, Furuno K, Fagan JM, Baracos V (1988) Activation of protein breakdown and prostaglandin E_2 production in rat skeletal muscle in fever is signaled by a macrophage product distinct from interleukin 1 or other known monokines. J Clin Invest 81:1378–1383

Grau GE, Fajardo LF, Piguet P-F, Allet B, Lambert P-H, Vassalli P (1987) Tumor necrosis factor (cachectin) as an essential mediator in murine cerebral malaria. Science 237:1210–1212

Green S, Dobrjansky A, Carswell EA, Kassel RL, Old LJ, Fiore N, Schwartz MK (1976) Partial purification of a serum factor that causes necrosis of tumors. Proc Natl Acad Sci USA 73:381–385

Ha DK, Gardner ID, Lawton JW (1983) Characterization of macrophage function in *Mycobacterium lepraemurium*-infected mice: sensitivity of mice to endotoxin and release of mediators and lysosomal enzymes after endotoxin treatment. Parasite Immunol 5:513–526

Haidaris CG, Haynes JD, Meltzer MS, Allison AC (1983) Serum containing tumor necrosis factor is cytotoxic for the human malaria parasite *Plasmodium falciparum*. Infect Immun 42:385–393

Hakoshima T, Tomita K-I (1988) Crystallization and preliminary x-ray investigation reveals that tumor necrosis factor is a compact trimer furnished with 3-fold symmetry. J Mol Biol 201:455–457

Haranaka K, Satomi N, Sakurai A, Haranaka R (1984) Role of first stimulating agents in the production of tumor necrosis factor. Cancer Immunol Immunother 18:87–90

Hartwell JL, Shear MJ, Adams JR Jr (1943) Chemical treatment of tumors. VII. Nature of the hemorrhage-producing fraction from *Serratia marcescens* (*Bacillus prodigiosus*) culture filtrate. JNCI 4:107–122

Helson L, Green S, Carswell E, Old LJ (1975) Effect of tumour necrosis factor on cultured human melanoma cells. Nature 258:731–732

Hoffmann MK, Green S, Old LJ, Oettgen HF (1976) Serum containing endotoxin-induced tumour necrosis factor substitutes for helper T cells. Nature 263:416–417

Hoffmann MK, Galanos C, Koenig S, Oettgen HF (1977) B-cell activation by lipopolysaccharide. Distinct pathways for induction of mitosis and antibody production. J Exp Med 146:1640–1647

Hotez PJ, Le Trang N, Fairlamb AH, Cerami A (1984) Lipoprotein lipase suppression in 3T3-L1 cells by a haematoprotozoan-induced mediator from peritoneal exudate cells. Parasite Immunol 6:203–209

Hülsmann WC, Dubelaar M-L (1988) Effects of tumor necrosis factor (TNF) on lipolytic activities of rat heart. Mol Cell Biochem 79:147–151

Hülsmann WC, Dubelaar M-L, de Wit LEA, Persoon NLM (1988) Cardiac lipoprotein lipase: effects of lipopolysaccharide and tumor necrosis factor. Mol Cell Biochem 79:137–145

Imamura K, Spriggs D, Kufe D (1987) Expression of tumor necrosis factor receptors on human monocytes and internalization of receptor bound ligand. J Immunol 139:2989–2992

Israel S, Hahn T, Holtmann H, Wallach D (1986) Binding of human TNF-alpha to high-affinity cell surface receptors: effect of IFN. Immunol Lett 12:217–224

Itoh A, Iizuka K, Natori S (1984) Induction of a TNF-like factor by murine macrophage-like cell line J774.1 on treatment with *Sarcophaga* lectin. FEBS Lett 175:59–62

Itoh H, Wallace RB (1985) A novel human physiologically active polypeptide. European patent application no 84105149.3

Jacob CO, McDevitt HO (1988) Tumour necrosis factor-alpha in murine autoimmune "lupus" nephritis. Nature 331:356–358

Jeffes EWB, Averbook B, Ulich T, Yamamoto R, Chow G, Masunaka I, Granger G (1987) Human alpha lymphotoxin and tumor necrosis factor induce inflammatory responses in normal tissue. Clin Res 35:692A

Kahler H, Shear MJ, Hartwell JL (1943) Chemical treatment of tumors. VIII. Ultracentrifugal and electrophoretic analysis of the hemorrhage-producing fraction from *Serratia marcescens* (*Bacillus prodigiosus*) culture filtrate. JNCI 4:123–129

Kashiwa H, Wright SC, Bonavida B (1987) Regulation of B cell maturation and differentiation. I. Suppression of pokeweed mitogen-induced B cell differentiation by tumor necrosis factor (TNF). J Immunol 138:1383–1390

Kawakami M, Cerami A (1981) Studies of endotoxin-induced decrease in lipoprotein lipase activity. J Exp Med 154:631–639

Kawakami M, Pekala PH, Lane MD, Cerami A (1982) Lipoprotein lipase suppression in 3T3-L1 cells by an endotoxin-induced mediator from exudate cells. Proc Natl Acad Sci USA 79:912–916

Kawakami M, Murase T, Ogawa H, Ishibashi S, Mori N, Takaku F, Shibata S (1987) Human recombinant TNF suppresses lipoprotein lipase activity and stimulates lipolysis in 3T3-L1 cells. J Biochem 101:331–338

Kettelhut IC, Goldberg AL (1988) Tumor necrosis factor can induce fever in rats without activating protein breakdown in muscle or lipolysis in adipose tissue. J Clin Invest 81:1384–1389

Kettelhut IC, Fiers W, Goldberg AL (1987) The toxic effects of tumor necrosis factor in vivo and their prevention by cyclooxygenase inhibitors. Proc Natl Acad Sci USA 84:4273–4277

Klebanoff SJ, Vadas MA, Harlan JM, Sparks LH, Gamble JR, Agosti JM, Waltersdorph AM (1986) Stimulation of neutrophils by tumor necrosis factor. J Immunol 136:4220–4225

Koj A, Kurdowska A, Magielska-Zero D, Rokita H, Sipe JD, Dayer JM, Demczuk S, Gauldie J (1987) Limited effects of recombinant human and murine interleukin 1 and tumour necrosis factor on production of acute phase proteins by cultured rat hepatocytes. Biochem Int 14:553–560

Kornbluth RS, Edgington TS (1986) Tumor necrosis factor production by human monocytes is a regulated event: induction of TNF-alpha-mediated cellular cytotoxicity by endotoxin. J Immunol 137:2585–2591

Kriegler M, Perez C, DeFay K, Albert I, Lu SD (1988) A novel form of TNF/cachectin is a cell surface cytotoxic transmembrane protein: ramifications for the complex physiology of TNF. Cell 53:45–53

Kruys VI, Wathelet M, Poupart P, Contreras R, Fiers W, Content J, Huez G (1987) The 3′ untranslated region of the human interferon-beta mRNA has an inhibitory effect on translation. Proc Natl Acad Sci USA 84:6030–6034

Kruys VI, Wathelet MG, Huez GA (1988) Identification of a translation inhibitory element (TIE) in the 3′ untranslated region of the human interferon-beta mRNA. Gene 72:191–200

Kull FC Jr, Jacobs S, Cuatrecasas P (1985) Cellular receptor for ^{125}I-labeled tumor necrosis factor: specific binding, affinity labeling, and relationship to sensitivity. Proc Natl Acad Sci USA 82:5756–5760

Lahdevirta J, Maury CPJ, Teppo A-M, Repo H (1988) Raised circulating cachectin/tumor necrosis factor in patients with the acquired immunodeficiency syndrome. Am J Med 86:289–291

Lee MD, Zentella A, Pekala PH, Cerami A (1987) Effect of endotoxin-induced monokines on glucose metabolism in the muscle cell line L6. Proc Natl Acad Sci USA 84:2590–2594

Lehmann V, Freudenberg MA, Galanos C (1987) Lethal toxicity of lipopolysaccharide and tumor necrosis factor in normal and *d*-galactosamine-treated mice. J Exp Med 165:657–663

Leist TP, Frei K, Kam-Hansen S, Zinkernagel RM, Fontana A (1988) Tumor necrosis factor α in cerebrospinal fluid during bacterial, but not viral, meningitis. Evaluation in murine model infections and in patients. J Exp Med 167:1743–1748

Leung DY, Geha RS, Newburger JW, Burns JC, Fiers W, Lapierre LA, Pober JS (1986) Two monokines, interleukin 1 and tumor necrosis factor, render cultured vascular endothelial cells susceptible to lysis by antibodies circulating during Kawasaki syndrome. J Exp Med 164:1958–1972

Libby P, Ordovas JM, Auger KR, Robbins AH, Birinyi LK, Dinarello CA (1986) Endotoxin and tumor necrosis factor induce interleukin-1 gene expression in adult human vascular endothelial cells. Am J Pathol 124:179–185

Lillehei RC, Maclean LD (1958) The intestinal factor in irreversible endotoxin shock. Ann Surg 148:513–525

Mahoney JR Jr, Beutler BA, Le Trang N, Vine W, Ikeda Y, Kawakami M, Cerami A (1985) Lipopolysaccharide-treated RAW 264.7 cells produce a mediator which inhibits lipoprotein lipase in 3T3-L1 cells. J Immunol 134:1673–1675
Mannel DN, Moore RN, Mergenhagen SE (1980) Macrophages as a source of tumoricidal activity (tumor-necrotizing factor). Infect Immun 30:523–530
Mannel DN, Falk W, Northoff H (1987) Endotoxic activities of tumor necrosis factor independent of IL1 secretion by macrophages/monocytes. Lymphokine Res 6:151–159
Mathison JC, Wolfson E, Ulevitch RJ (1988) Participation of tumor necrosis factor in the mediation of gram negative bacterial lipopolysaccharide-induced injury in rabbits. J Clin Invest 81:1925–1937
Matthews N (1978) Tumour-necrosis factor from the rabbit. II. Production by monocytes. Br J Cancer 38:310–315
Matthews N (1981 a) Tumour-necrosis factor from the rabbit. V. Synthesis in vitro by mononuclear phagocytes from various tissues of normal and BCG-injected rabbits. Br J Cancer 44:418–424
Matthews N (1981 b) Production of an anti-tumour cytotoxin by human monocytes. Immunology 44:135–142
McCann SM, Rettori V, Milenkovic L, Jurcovicova J, Snyder G, Beutler B (1989) Role of interleukin 1 and cachectin in control of anterior pituitary hormone release. In: Perez-Polo JR (ed) Neural control of reproductive function. Liss, New York (in press)
Michalek SM, Moore RN, McGhee JR, Rosenstreich DL, Mergenhagen SE (1980) The primary role of lymphoreticular cells in the mediation of host responses to bacterial endotoxin. J Infect Dis 141:55–63
Michie HR, Spriggs DR, Rounds J, Wilmore DW (1987) Does cachectin cause cachexia? Surg Forum 38:38–40
Ming WJ, Bersani L, Mantovani A (1987) Tumor necrosis factor is chemotactic for monocytes and polymorphonuclear leukocytes. J Immunol 138:1469–1474
Moldawer LL, Andersson C, Gelin J, Lundholm KG (1988) Regulation of food intake and hepatic protein synthesis by recombinant-derived cytokines. Am J Physiol 254:G450–G456
Muller U, Jongeneel CV, Nedospasov SA, Lindahl KF, Steinmetz M (1987) Tumor necrosis factor and lymphotoxin genes map close to H-2D in the mouse major histocompatibility complex. Nature 325:265–267
Munker R, Gasson J, Ogawa M, Koeffler HP (1986) Recombinant human TNF induces production of granulocyte-monocyte colony-stimulating factor. Nature 323:79–82
Mustafa M (1989) Measurement of CSF cachectin (TNF-alpha) activity in experimental *Haemophilus influenzae* type b (Hib) meningitis (Abstr). Proc Int Conf Antimicrob Actions Chemother 28 (in press)
Nawroth P, Bank I, Handley D, Cassimeris J, Chess L, Stern D (1986 a) Tumor necrosis factor/cachectin interacts with endothelial cell receptors to induce release of interleukin 1. J Exp Med 163:1363–1375
Nawroth PP, Handley DA, Esmon CT, Stern DM (1986 b) Interleukin 1 induces endothelial cell procoagulant while suppressing cell-surface anticoagulant activity. Proc Natl Acad Sci USA 83:3460–3464
Nedospasov SA, Hirt B, Shakhov AN, Dobrynin VN, Kawashima E, Accolla RS, Jongeneel CV (1986) The genes for tumor necrosis factor (TNF-alpha) and lymphotoxin (TNF-beta) are tandemly arranged on chromosome 17 of the mouse. Nucleic Acids Res 14:7713–7725
Nedwin GE, Naylor SL, Sakaguchi AY, Smith D, Jarrett-Nedwin J, Pennica D, Goeddel DV, Gray PW (1985) Human lymphotoxin and tumor necrosis factor genes: structure homology and chromosomal localization. Nucleic Acids Res 13:6361–6373

Neta R, Vogel SN, Sipe JD, Oppenheim JJ, Giclas PC, Douches SD (1987) Comparison of the in vivo effects of rIL-1 and rTNF in radioprotection, induction of CSF and of acute phase reactants. Fed Proc 46:1200

Oliff A, Defeo-Jones D, Boyer M, Martinez D, Kiefer D, Vuocolo G, Wolfe A, Socher SH (1987) Tumors secreting human TNF/cachectin induce cachexia in mice. Cell 50:555–563

O'Malley WE, Achinstein B, Shear MJ (1962) Action of bacterial polysaccharide on tumors. II. Damage of sarcoma 37 by serum of mice treated with *Serratia marcescens* polysaccharide, and induced tolerance. JNCI 29:1169–1175

Parant MA, Parant FJ, Chedid LA (1980) Enhancement of resistance to infection by endotoxin-induced serum factor from *Mycobacterium bovis* BCG-infected mice. Infect Immun 28:654–659

Patton JS, Shepard HM, Wilking H, Lewis G, Aggarwal BB, Eessalu TE, Gavin LA, Grunfeld C (1986) Interferons and tumor necrosis factors have similar catabolic effects on 3T3 L1 cells. Proc Natl Acad Sci USA 83:8313–8317

Patton JS, Peters PM, McCabe J, Crase D, Hansen S, Chen AB, Liggitt D (1987) Development of partial tolerance to the gastrointestinal effects of high doses of recombinant tumor necrosis factor-alpha in rodents. J Clin Invest 80:1587–1596

Paul NL, Ruddle NH (1988) Lymphotoxin. Annu Rev Immunol 6:407–438

Pekala PH, Kawakami M, Angus CW, Lane MD, Cerami A (1983) Selective inhibition of synthesis of enzymes for de novo fatty acid biosynthesis by an endotoxin-induced mediator from exudate cells. Proc Natl Acad Sci USA 80:2743–2747

Pennica D, Nedwin GE, Hayflick JS, Seeburg PH, Derynck R, Palladino MA, Kohr WJ, Aggarwal BB, Goeddel DV (1984) Human tumor necrosis factor: precursor structure, expression and homology to lymphotoxin. Nature 312:724–729

Pennica D, Hayflick JS, Bringman TS, Palladino MA, Goeddel DV (1985) Cloning and expression in *Escherichia coli* of the cDNA for murine tumor necrosis factor. Proc Natl Acad Sci USA 82:6060–6064

Perlmutter DH, Dinarello CA, Punsal PI, Colten HR (1986) Cachectin/tumor necrosis factor regulates hepatic acute-phase gene expression. J Clin Invest 78:1349–1354

Peters PM, Ortaldo JR, Shalaby MR, Svedersky LP, Nedwin GE, Bringman TS, Hass PE, Aggarwal BB, Herberman RB, Goeddel DV, Palladino MA Jr (1986) Natural killer-sensitive targets stimulate production of TNF-alpha but not TNF-beta (lymphotoxin) by highly purified human peripheral blood large granular lymphocytes. J Immunol 137:2592–2598

Piguet PF, Grau G, Allet B, Vassalli P (1987) Tumor necrosis factor (TNF) is an important mediator of the mortality and morbidity induced by the graft-versus-host reaction (GVHR) (Abstr). Immunobiology 175:27

Pingleton WW, Coalson JJ, Guenter CA (1975) Significance of leukocytes in endotoxic shock. Exp Mol Pathol 22:183–194

Plata-Salamán CR, Oomura Y, Kai Y (1988) Tumor necrosis factor and interleukin-1β: suppression of food intake by direct action in the central nervous system. Brain Res 448:106–114

Playfair JHL, Taverne J, Matthews N (1984) What is tumour necrosis factor really for. Immunol Today 5:165–166

Plela TH, Korn JH (1987) Lymphocyte-fibroblast adhesion induced by interferon gamma and tumor necrosis factor. Arthritis Rheum 30:S90

Pober JS, Gimbrone MA Jr, Lapierre LA, Mendrick DL, Fiers W, Rothlein R, Springer TA (1986a) Overlapping patterns of activation of human endothelial cells by interleukin 1, tumor necrosis factor, and immune interferon. J Immunol 137:1893–1896

Pober JS, Bevilacqua MP, Mendrick DL, Lapierre LA, Fiers W, Gimbrone MA Jr (1986b) Two distinct monokines, interleukin 1 and tumor necrosis factor, each independently induce biosynthesis and transient expression of the same antigen on the surface of cultured human vascular endothelial cells. J Immunol 136:1680–1687

Pober JS, Lapierre LA, Stolpen AH, Brock TA, Springer TA, Fiers W, Bevilacqua MP, Mendrick DL, Gimbrone MA Jr (1987) Activation of cultured human endothelial cells by recombinant lymphotoxin: comparison with tumor necrosis factor and interleukin 1 species. J Immunol 138:3319–3324

Pohlman TH, Stanness KA, Beatty PG, Ochs HD, Harlan JM (1986) An endothelial cell surface factor(s) induced in vitro by lipopolysaccharide, interleukin 1, and tumor necrosis factor-alpha increases neutrophil adherence by a CDw18-dependent mechanism. J Immunol 136:4548–4553

Robertson B, Dostal K, Daynes RA (1988) Neuropeptide regulation of inflammatory and immunologic responses: the capacity of α-melanocyte-stimulating hormone to inhibit tumor necrosis factor and IL-1-inducible biologic responses. J Immunol 140:4300–4307

Rothwell NJ (1988) Central effects of TNF-alpha on thermogenesis and fever in the rat. Bioscience Rep 8:345–352

Rouzer CA, Cerami A (1980) Hypertriglyceridemia associated with *Trypanosoma brucei brucei* infection in rabbits: role of defective triglyceride removal. Mol Biochem Parasitol 2:31–38

Rubin BY, Anderson SL, Sullivan SA, Williamson BD, Carswell EA, Old LJ (1986) Nonhematopoietic cells selected for resistance to tumor necrosis factor produce tumor necrosis factor. J Exp Med 164:1350–1355

Ruddle NH (1986) Lymphotoxin production in AIDS. Immunol Today 7:8–9

Ruddle NH, Waksman BH (1967) Cytotoxic effect of lymphocyte-antigen interaction in delayed hypersensitivity. Science 157:1060–1062

Ruddle NH, Waksman BH (1968 a) Cytotoxicity mediated by soluble antigen and lymphocytes in delayed hypersensitivity. III. Analysis of mechanism. J Exp Med 128:1267–1279

Ruddle NH, Waksman BH (1968 b) Cytotoxicity mediated by soluble antigen and lymphocytes in delayed hypersensitivity. I. Characterization of the phenomenon. J Exp Med 128:1237–1254

Ruddle NH, Waksman BH (1968 c) Cytotoxicity mediated by soluble antigen and lymphocytes in delayed hypersensitivity. II. Correlation of the in vitro response with skin reactivity. J Exp Med 128:1255–1265

Saklatvala J (1986) Tumour necrosis factor alpha stimulates resorption and inhibits synthesis of proteoglycan in cartilage. Nature 322:547–549

Sariban E, Imamura K, Luebbers R, Kufe D (1988) Transcriptional and posttranscriptional regulation of tumor necrosis factor gene expression in human monocytes. J Clin Invest 81:1506–1510

Sato K, Kasono K, Fujii Y, Kawakami M, Tsushima T, Shizume K (1987) Tumor necrosis factor type alpha (cachectin) stimulates mouse osteoblast-like cells (MC3T3-E1) to produce macrophage-colony stimulating activity and prostaglandin E2. Biochem Biophys Res Commun 145:323–329

Sato N, Goto T, Haranaka K, Satomi N, Nariuchi H, Mano-Hirano Y, Sawasaki Y (1986) Actions of tumor necrosis factor on cultured vascular endothelial cells: morphologic modulation, growth inhibition, and cytotoxicity. JNCI 76:1113–1121

Satomi N, Haranaka K, Kunii O (1981) Research on the production site of tumor necrosis factor (TNF). Jpn J Exp Med 51:317–322

Scheurich P, Ucer U, Kronke M, Pfizenmaier K (1986) Quantification and characterization of high-affinity membrane receptors for tumor necrosis factor on human leukemic cell lines. Int J Cancer 38:127–133

Scheurich P, Thoma B, Ucer U, Pfizenmaier K (1987) Immunoregulatory activity of recombinant human tumor necrosis factor (TNF)-alpha: induction of TNF receptors on human T cells and TNF-alpha-mediated enhancement of T cell responses. J Immunol 138:1786–1790

Schleef RR, Bevilacqua MP, Sawdey M, Gimbrone MA Jr, Loskutoff DJ (1988) Cytokine activation of vascular endothelium. Effects on tissue-type plasminogen activator and type 1 plasminogen activator inhibitor. J Biol Chem 263:5797–5803

Schmid DS, Tite JP, Ruddle NH (1986) DNA fragmentation: manifestation of target cell destruction mediated by cytotoxic T-cell lines, lymphotoxin-secreting helper T-cell clones, and cell-free lymphotoxin-containing supernatant. Proc Natl Acad Sci USA 83:1881–1885

Schmid DS, Hornung R, McGrath KM, Paul N, Ruddle NH (1987) Target cell DNA fragmentation is mediated by lymphotoxin and tumor necrosis factor. Lymphokine Res 6:195–202

Schweigerer L, Malerstein B, Gospodarowicz D (1987) Tumor necrosis factor inhibits the proliferation of cultured capillary endothelial cells. Biochem Biophys Res Commun 143:997–1004

Scuderi P, Lam KS, Ryan KJ, Petersen E, Salmon SE, Sterling KE, Finley PR, Ray CG, Slymen DJ (1986) Raised serum levels of tumour necrosis factor in parasitic infections. Lancet 2:1364–1365

Semb H, Peterson J, Tavernier J, Olivecrona T (1987) Multiple effects of tumor necrosis factor on lipoprotein lipase in vivo. J Biol Chem 262:8390–8394

Semon D, Kawashima E, Jongeneel CV, Shakhov AN, Nedospasov SA (1987) Nucleotide sequence of the murine TNF locus, including the TNF-alpha (tumor necrosis factor) and TNF-beta (lymphotoxin) genes. Nucleic Acids Res 15:9083–9084

Shalaby MR, Aggarwal BB, Rinderknecht E, Svedersky LP, Finkle BS, Palladino MA Jr (1985) Activation of human polymorphonuclear neutrophil functions by interferon-gamma and tumor necrosis factors. J Immunol 135:2069–2073

Shalaby MR, Palladino MA Jr, Hirabayashi SE, Eessalu TE, Lewis GD, Shepard HM, Aggarwal BB (1987) Receptor binding and activation of polymorphonuclear neutrophils by tumor necrosis factor-alpha. J Leukocyte Biol 41:196–204

Shalaby MR, Espevik T, Rice GC, Ammann AJ, Figari IS, Ranges GE, Palladino MA Jr (1988) The involvement of human tumor necrosis factors-α and -β in the mixed lymphocyte reaction. J Immunol 141:499–503

Sharpe RJ, Margolis RJ, Amento EP, Granstein RD (1987) Induction of dermal acute inflammation by tumor necrosis factor. Clin Res 35:716A

Shaw G, Kamen R (1986) A conserved AU sequence from the 3′ untranslated region of GM-CSF mRNA mediates selective mRNA degradation. Cell 46:659–667

Shear MJ (1944) Chemical treatment of tumors. IX. Reactions of mice with primary subcutaneous tumors to injection of a hemorrhage-producing bacterial polysaccharide. JNCI 4:461–476

Shear MJ, Andervont HB (1936) Chemical treatment of tumors. III. Separation of hemorrhage-producing fraction of *B. coli* filtrate. Proc Soc Exp Biol Med 34:323–325

Shear MJ, Turner FC, Perrault A, Shovelton J (1943a) Chemical treatment of tumors. V. Isolation of the hemorrhage-producing fraction from *Serratia marcescens* (*Baccillus prodigiosus*) culture filtrate. JNCI 4:81–97

Shear MJ, Perrault A, Adams JR Jr (1943b) Chemical treatment of tumors. VI. Method employed in determining the potency of hemorrhage-producing bacterial preparations. JNCI 4:99–105

Shirai T, Yamaguchi H, Ito H, Todd CW, Wallace RB (1985) Cloning and expression in *Escherichia coli* of the gene for human tumour necrosis factor. Nature 313:803–806

Silberstein DS, David JR (1986) Tumor necrosis factor enhances eosinophil toxicity to *Schistosoma mansoni* larvae. Proc Natl Acad Sci USA 83:1055–1059

Silberstein DS, Owen WF, Gasson JC, DiPersio JF, Golde DW, Bina JC, Soberman R, Austen KF, David JR (1986) Enhancement of human eosinophil cytotoxicity and leukotriene synthesis by biosynthetic (recombinant) granulocyte-macrophage colony-stimulating factor. J Immunol 137:3290–3294

Silberstein DS, Dessein AJ, Elsas PP, Fontaine B, David JR (1987) Characterization of a factor from the U937 cell line that enhances the toxicity of human eosinophils to *Schistosoma mansoni* larvae. J Immunol 138:3042–3050

Sipe JD, Vogel SN, Douches S, Neta R (1987) Tumor necrosis factor/cachectin is a less potent inducer of serum amyloid A synthesis than interleukin 1. Lymphokine Res 6:93–101

Smith RA, Baglioni C (1987) The active form of tumor necrosis factor is a trimer. J Biol Chem 262:6951–6954

Smith RA, Kirstein M, Fiers W, Baglioni C (1986) Species specificity of human and murine tumor necrosis factor. J Biol Chem 261:14871–14874

Socher SH, Martinez D, Craig JB, Kuhn JG, Oliff A (1988) Tumor necrosis factor not detectable in patients with clinical cancer cachexia. JNCI 80:595–598

Spies T, Morton CC, Nedospasov SA, Fiers W, Pious D, Strominger JL (1986) Genes for the tumor necrosis factors alpha and beta are linked to the human major histocompatibility complex. Proc Natl Acad Sci USA 83:8699–8702

Stashenko P, Dewhirst FE, Peros WJ, Kent RL, Ago JM (1987) Synergistic interactions between interleukin 1, tumor necrosis factor, and lymphotoxin in bone resorption. J Immunol 138:1464–1468

Stern DM, Nawroth PP (1986) Modulation of endothelial hemostatic properties by tumor necrosis factor. J Exp Med 163:740–745

Stolpen AH, Guinan EC, Fiers W, Pober JS (1986) Recombinant tumor necrosis factor and immune interferon act singly and in combination to reorganize human vascular endothelial cell monolayers. Am J Pathol 123:16–24

Sugarman BJ, Aggarwal BB, Hass PE, Figari IS, Paladino MA Jr, Shepard HM (1985) Recombinant human tumor necrosis factor-alpha: effects on proliferation of normal and transformed cells in vitro. Science 230:943–945

Sun X-M, Hsueh W (1988) Bowel necrosis induced by tumor necrosis factor in rats is mediated by platelet-activating factor. J Clin Invest 81:1328–1331

Takeda K, Iwamoto S, Sugimoto H, Takuma T, Kawatani N, Noda M, Masaki A, Morise H, Arimura H, Konno K (1986) Identity of differentiation inducing factor and tumour necrosis factor. Nature 323:338–340

Tashjian AH, Voelkel EF, Lazzaro M, Singer FR, Roberts AB, Derynck R, Winkler ME, Levine L (1985) Alpha and beta human transforming growth factors stimulate prostaglandin production and bone resorption in cultured mouse calvaria. Proc Natl Acad Sci USA 82:4535–4538

Taverne J, Dockrell HM, Playfair JH (1981) Endotoxin-induced serum factor kills malarial parasites in vitro. Infect Immun 33:83–89

Taverne J, Depledge P, Playfair JH (1982) Differential sensitivity in vivo of lethal and nonlethal malarial parasites to endotoxin-induced serum factor. Infect Immun 37:927–934

Taverne J, Matthews N, Depledge P, Playfair JH (1984) Malarial parasites and tumour cells are killed by the same component of tumour necrosis serum. Clin Exp Immunol 57:293–300

Torti FM, Dieckmann B, Beutler B, Cerami A, Ringold GM (1985) A macrophage factor inhibits adipocyte gene expression: an in vitro model of cachexia. Science 229:867–869

Tracey KJ, Lowry S, Beutler B, Cerami A, Albert J, Shires GT (1986a) Cachectin/tumor necrosis factor mediates changes in skeletal muscle transmembrane potential. J Exp Med 164:1368–1373

Tracey KJ, Beutler B, Lowry SF, Merryweather J, Wolpe S, Milsark IW, Hariri RJ, Fahey TJ III, Zentella A, Albert JD, Shires GT, Cerami A (1986b) Shock and tissue injury induced by recombinant human cachectin. Science 234:470–474

Tracey KJ, Lowry SF, Fahey TJ III, Albert JD, Fong Y, Hesse D, Beutler B, Manogue KR, Calvano S, Wei H, Cerami A, Shires GT (1987a) Cachectin/tumor necrosis factor induces lethal shock and stress hormone responses in the dog. Surg Gynecol Obstet 164:415–422

Tracey KJ, Fong Y, Hesse DG, Manogue KR, Lee AT, Kuo GC, Lowry SF, Cerami A (1987b) Anti-cachectin/TNF monoclonal antibodies prevent septic shock during lethal bacteraemia. Nature 330:662–666

Tracey KJ, Wei H, Manogue KR, Fong Y, Hesse DG, Nguyen HT, Kuo GC, Beutler B, Cotran RS, Cerami A, Lowry SF (1988) Cachectin/tumor necrosis factor induces cachexia, anemia, and inflammation. J Exp Med 167:1211–1227

Troppmair J, Auböck J, Niederwieser D, Schönitzer D, Huber C (1988) Interferons (IFNs) and tumor necrosis factors (TNFs) in T cell-mediated immune responses against alloantigens. I. Influence on the activation of resting and antigen-primed T cells. Immunobiology 176:236–254

Tsujimoto M, Yokota S, Vilcek J, Weissmann G (1986) Tumor necrosis factor provokes superoxide anion generation from neutrophils. Biochem Biophys Res Commun 137:1094–1100

Umeda T, Hara T, Niijima T (1983) Cytotoxic effect of tumor necrosis factor on human lymphocytes and specific binding of the factor to the target cells. Cell Mol Biol 29:349–352

Vilcek J, Palombella VJ, Henriksen-Destefano D, Swenson C, Feinman R, Hirai M, Tsujimoto M (1986) Fibroblast growth enhancing activity of tumor necrosis factor and its relationship to other polypeptide growth factors. J Exp Med 163: 632–643

Vlassara H, Brownlee M, Manogue KR, Dinarello CA, Pasagian A (1988) Cachectin/TNF and IL-1 induced by glucose-modified proteins: role in normal tissue remodeling. Science 240:1546–1548

Vogel SN, Moore RN, Sipe JD, Rosenstreich DL (1980) BCG-induced enhancement of endotoxin sensitivity in C3H/HeJ mice. I. In vivo studies. J Immunol 124:2004–2009

Waage A, Halstensen A, Espevik T (1987) Association between tumour necrosis factor in serum and fatal outcome in patients with meningococcal disease. Lancet 1(8529):355–357

Wallach D, Holtmann H, Engelmann H, Nophar Y (1988) Sensitization and desensitization to lethal effects of tumor necrosis factor and Il-1. J Immunol 140:2994–2999

Wang AM, Creasey AA, Ladner MB, Lin LS, Strickler J, van Arsdell JN, Yamamoto R, Mark DF (1985) Molecular cloning of the complementary DNA for human tumor necrosis factor. Science 228:149–154

Warren MK, Ralph P (1986) Macrophage growth factor CSF-1 stimulates human monocyte production of interferon, tumor necrosis factor, and colony stimulating activity. J Immunol 137:2281–2285

Watanabe N, Sone H, Neda H, Niitsu Y, Urushizaki I (1984) Mechanisms of production of tumor necrosis factor (TNF): reconstitution experiment with nude mice. Gan To Kagaku Ryoho 11:1284–1289

Wei H, Tracey K, Manogue K, Nguyen H, Fong Y, Hesse D, Beutler B, Solomon R, Cerami A, Lowry S (1987) Cachectin mediates suppressed food intake and anemia during chronic administration. Fed Proc 46:1338

Wood PR, Clark IA (1984) Macrophages from *Babesia* and malaria infected mice are primed for monokine release. Parasite Immunol 6:309–317

Wozencraft AO, Dockrell HM, Taverne J, Targett GAT, Playfair JHL (1984) Killing of human malaria parasites by macrophage secretory products. Infect Immun 43:664–669

Yoshie O, Tada K, Ishida N (1986) Binding and crosslinking of 125I-labeled recombinant human tumor necrosis factor to cell surface receptors. J Biochem 100:531–541

Zacharchuk CM, Drysdale B-E, Mayer MM, Shin HS (1983) Macrophage-mediated cytotoxicity: role of a soluble macrophage cytotoxic factor similar to lymphotoxin and tumor necrosis factor. Proc Natl Acad Sci USA 80:6341–6345

Zechner R, Newman TC, Sherry B, Cerami A, Breslow JL (1988) Recombinant human cachectin/tumor necrosis factor but not interleukin-1α downregulates lipoprotein lipase gene expression at the transcriptional level in mouse 3T3-L1 adipocytes. Mol Cell Biol 8:2394–2401

CHAPTER 21

Bombesin and Gastrin-Releasing Peptide: Neuropeptides, Secretogogues, and Growth Factors

A.-M. Lebacq-Verheyden, J. Trepel, E. A. Sausville, and J. F. Battey

A. Introduction

In the early 1960s, V. Erspamer and colleagues observed that crude methanol extracts of amphibian skin exerted a wide range of pharmacological actions on vascular and extravascular smooth muscle, on external and internal secretions, and on renal circulation and function. This observation prompted the isolation of numerous bioactive peptides that were remarkably abundant in amphibian skin. The knowledge gained from their structure and function often helped to identify homologous peptides in mammalian tissues, where they existed only in minute amounts. Structurally related peptides had the same pharmacological profile and were grouped into families. The major families included the tachykinins, the caeruleins, and the bombesins (Erspamer and Melchiorri 1973). Peptides of these families were originally identified as digestive hormones and as central modulators of metabolism and behavior; more recently, they became also known as growth factors. The present review deals with bombesin and with its mammalian homolog, the gastrin-releasing peptide (GRP).

B. Structure and Cellular Localization of the Peptides

I. Bombesin-Related Peptides

The 14-amino-acid peptide bombesin was isolated in 1971 from the skin of the European frog *Bombina bomina,* using a smooth muscle contraction assay (Anastasi et al. 1971, 1972). As many as 13 bombesin-related peptides were isolated from different amphibian species and grouped into three subfamilies according to their C-terminal tripeptide (Erspamer et al. 1988). Bombesin and alytesin have a C-terminal His-Leu-Met-NH_2; the ranatensins and litorin have a C-terminal His-Phe-Met-NH_2; and the phyllolitorins have Ser-(Leu or Phe)-Met-NH_2. In 1978, a structurally related peptide was isolated from porcine gastric tissue, using the rise of plasma immunoreactive gastrin as a bioassay. This 27-amino-acid peptide was therefore termed the gastrin-releasing peptide (GRP) (McDonald et al. 1978, 1979). Closely related GRP molecules were subsequently isolated from the chicken proventriculus (McDonald et al.

1980), canine intestine (REEVE et al. 1983), rat intestine (REEVE et al., unpublished), and a human neuroendocrine tumor (ORLOFF et al. 1984). A mammalian homolog of ranatensin, named neuromedin B, was isolated from porcine spinal cord (MINAMINO et al. 1983, 1985), as well as a shorter form of GRP referred to as neuromedin C (MINAMINO et al. 1984). No mammalian homolog of the phyllolitorins has been formally identified, but preliminary results suggest the existence of a phyllolitorin-related peptide in human leukemic lymphoblasts (ERSPAMER et al. 1988).

II. Structure of Bombesin and GRP

The structure of bombesin and GRP is shown in Fig. 1. The conservation of amino acids is remarkable in the C-terminal halves of these molecules and decreases in their N-terminal regions, where amino acid substitutions and insertions are observed. It is noteworthy that an identical α-amidated C-terminal heptapeptide is shared by amphibian, avian, and mammalian bombesin-related peptides. Structure-activity studies on smooth muscle preparations established that the C-terminal heptapeptide was the minimal structure associated with biological activity, and that the C-terminal nonapeptide of bombesin was as potent as the natural peptide (BROCCARDO et al. 1975). The same structural results were observed when parallel binding studies and biological assays were performed on rat brain membranes (MOODY et al. 1978, 1982), guinea pig pancreatic acinar cells (JENSEN et al. 1978), a rat pituitary cell line (WESTENDORF and SCHONBRUNN 1982, 1983), a hamster pancreatic islet cell line (SWOPE and SCHONBRUNN 1984, 1987), and a mouse embryo fibroblast cell line (ROZENGURT and SINNETT-SMITH 1983; ZACHARY and ROZENGURT 1985a, b). These observations located the biologically active domain and the receptor-binding domain of bombesin and GRP to their well-conserved C-terminal nonapeptide.

III. Molecular Forms of GRP

Radioimmunoassays with C-terminal antisera detected GRP-like peptides in tissue extracts from a variety of mammalian species. Several molecular forms, closely related to GRP_{1-27} and to its cleavage products GRP_{14-27} and/or GRP_{18-27}, were present in varying proportions in most tissues investigated. The presence of the C-terminal decapeptide of GRP in mammalian tissues was definitively confirmed by the isolation from canine intestine (REEVE et al. 1983) and rat spinal cord (MINAMINO et al. 1984) of a peptide shown by amino acid sequencing to be GRP_{18-27}.

The tissues shown to contain GRP_{1-27} and at least one of its C-terminal fragments included gastrointestinal tissues (BROWN et al. 1978; DOCKRAY et al. 1979; WALSH et al. 1979; ERSPAMER et al. 1979; HUTCHISON et al. 1981; YANAIHARA et al. 1981; PRICE et al. 1984; GREELEY et al. 1986a; HOLST et al. 1987a), brain (VILLARREAL and BROWN 1978; WALSH et al. 1979; MOODY and PERT 1979; MOODY et al. 1981a; YANAIHARA et al. 1981; ROTH et al. 1983a; PANULA et al. 1984; GHATEI et al. 1984a; KITA et al. 1986), spinal cord (MOODY et al. 1981c; YAKSH et al. 1988), sensory ganglia (PANULA et al. 1983),

HUMAN GRP:	Val	Pro	Leu	Pro	Ala	Gly	Gly	Gly	Thr	Val	Leu	Thr	Lys	Met	Tyr	Pro	Arg	Gly	Asn	His	Trp	Ala	Val	Gly	His	Leu	Met-NH$_2$
PORCINE GRP:	Ala	-	Val	Ser	Val	-	-	-	-	-	-	Ala	-	-	-	-	-	-	-	-	-	-	-	-	-	-	- NH$_2$
RAT GRP:	Ala	-	Val	Ser [Thr Gly]	-	-	-	-	-	-	-	Ala	-	-	-	-	-	-	Ser	-	-	-	-	-	-	-	- NH$_2$
CANINE GRP:	Ala	-	Val	-	Gly	-	Gln	-	-	-	-	Asp	-	-	-	-	-	-	-	-	-	-	-	-	-	-	- NH$_2$
AVIAN GRP:	Ala	-	-	Gln	Pro	-	-	Ser	Pro	Ala	-	-	-	Ile	-	-	-	-	Ser	-	-	-	-	-	-	-	- NH$_2$
FROG BOMBESIN:														Glu	Gln	Arg	Leu	-	-	Gln	-	-	-	-	-	-	- NH$_2$

Fig. 1. Comparison of GRP peptide sequences with bombesin. The amino acid sequences of GRP from five different species (human, SPINDEL et al. 1984; porcine, McDONALD et al. 1979; rat, LEBACQ-VERHEYDEN et al. 1988b; canine, REEVE et al. 1983; avian, McDONALD et al. 1980; frog bombesin, ANASTASI et al. 1972) are shown, and compared with the frog peptide bombesin. Conserved residues are indicated by a *dash;* note that rat GRP is 29 amino acids long due to insertion of a Thr-Gly in the amino-terminal half of the peptide. The carboxyl-terminal heptapeptide is absolutely conserved, consistent with its identification as the biologically active domain

sympathetic ganglia (SCHULTZBERG 1983), pituitary gland (MAJOR et al. 1983), pancreas (GHATEI et al. 1984b; KNUHTSEN et al. 1987a), adrenals (LEMAIRE et al. 1986), and genitourinary tract (GHATEI et al. 1985a). GRP-like peptides were hardly detectable in the normal adult respiratory tract (GHATEI et al. 1982b), but were clearly present in the same molecular forms in human fetal lung (PRICE et al. 1983; GHATEI et al. 1983; YAMAGUCHI et al. 1983; YOSHIZAKI et al. 1984), and in neuroendocrine tumors arising in the lung (WOOD et al. 1981; YAMAGUCHI et al. 1983; TAMAI et al. 1983; ROTH et al. 1983b; PRICE et al. 1985; YOSHIZAKI et al. 1986) and thyroid (MATSUBAYASHI et al. 1984; YAMAGUCHI et al. 1984; GHATEI et al. 1985b). GRP_{1-27} and its C-terminal fragment(s) were also present in small cell lung cancer cell lines (ERISMAN et al. 1982), in their culture media, and in the plasma of a patient with small cell lung carcinoma (SORENSON et al. 1982). However, rat plasma (BROWN et al. 1978), rat spinal cord (PANULA et al. 1983), human cerebrospinal (YAMADA et al. 1981), and a small cell lung carcinoma cell line (MOODY et al. 1981b, 1983b) were reported to contain only the smaller GRP_{18-27} or GRP_{14-27}.

IV. Cellular Localization of GRP

In normal adult mammalian tissues, the GRP-like immunoreactivity detected by immunohistochemistry in various organs was located to neuronal bodies, nerve fibers, and nerve terminals, and very rarely to diffuse neuroendocrine cells (DOCKRAY et al. 1979). In human developing tissues, however, GRP-like immunoreactivity was observed in pulmonary (WHARTON et al. 1978) and thyroid (SUNDAY et al. 1988b) neuroendocrine cells. Neuroendocrine tumors which probably derive from these cells in the adult also contained GRP-like peptides (WOOD et al. 1981; KAMEYA et al. 1983a; reviewed in SUNDAY et al. 1988a).

1. Neuronal GRP

GRP-like immunoreactive (GRP-li) nerve fibers were widely distributed in the digestive tract and in the pancreas, whereas neuronal bodies were restricted to the submucosal plexus, myenteric plexus, and intrapancreatic ganglia (DOCKRAY et al. 1979; YANAIHARA et al. 1981; BUFFA et al. 1982; HARRISON and WAHUTA 1982; FERRI et al. 1983; MOGHIMZADEH et al. 1983; IWANAGA 1983; COSTA et al. 1984; EKBLAD et al. 1984; LEANDER et al. 1984). GRP-li fibers projecting from neurons in the myenteric plexus were found in the celiac and mesenteric ganglia (KONDO et al. 1983; SCHULTZBERG and DALSGAARD 1983; DALSGAARD et al. 1983; COSTA et al. 1984; HAMAJI et al. 1987). GRP-li fibers and rare GRP-li cell bodies were associated with the genitourinary tract (STJERNQUIST et al. 1983, 1986; GHATEI et al. 1985a) and GRP-li nerve fibers were found in the skin (O'SHAUGHNESSY et al. 1983). GRP-li fibers and occasional neuronal bodies were also located in the hypogastric ganglia (HELÉN et al. 1984) and in dorsal root ganglia (PANULA et al. 1983); only GRP-li fibers were present in the spinal cord, mainly in the dorsal horns (PANULA et al. 1982; FUXE et al. 1983; MASSARI et al. 1983; O'DONOHUE et al. 1984; YAKSH et al. 1988). Finally, numerous GRP-li nerve fibers and nerve terminals were scattered throughout the brain, whereas neuronal bodies were only found in a few hypothalamic and brain stem nuclei, including the suprachiasmatic nucleus, paraventricular nucleus, the nucleus of the solitary tract, the dorsal parabrachial nucleus, and the substantia gelatinosa of the trigeminal complex (ROTH et al. 1982; PANULA et al. 1982, 1984; reviewed in PANULA 1986).

2. Neuroendocrine GRP

Nonmammalian species including amphibia (LECHAGO et al. 1978; EL-SALHY et al. 1981), birds (TIMSON et al. 1979; VAILLANT et al. 1979), and some fishes (HOLMGREN et al. 1982) have GRP-li neuroendocrine cells rather than GRP-li nerve fibers in their digestive tissues. An evolutionary transition appears to be represented by an elasmobranch fish, which has both GRP-li neuroendocrine cells and nerve fibers, whereas a holostean fish has only GRP-li nerve fibers (HOLMGREN and NILSSON 1983a, b). In normal adult mammals, there are very few diffuse neuroendocrine cells shown to contain GRP-like immunoreactivity. However, some developing tissues from humans and primates contain numerous GRP-li neuroendocrine cells. These cells were located in fetal and neonatal lung (WHARTON et al. 1978; CUTZ et al. 1981; JOHNSON et al. 1982; YAMAGUCHI et al. 1983; TSUTSUMI et al. 1983; IWANAGA 1983; GHATEI et al. 1983; STAHLMAN et al. 1985; DAYER et al. 1985) and in neonatal thyroid (SUNDAY et al. 1988b). Neuroendocrine tumors containing GRP-like peptides included pulmonary carcinoids, small cell lung carcinomas (SCLCs), and occasional lung adenocarcinomas with neuroendocrine differentiation features (YAMAGUCHI et al. 1983; YANG et al. 1983; TSUTSUMI et al. 1983; TAMAI et al. 1983; KAMEYA et al. 1983b; BOSTWICK et al. 1984; SAID et al. 1985; LEWIN et al. 1985). They also included medullary thyroid carcinomas (KAMEYA et al. 1983a; MATSUBAYASHI et al. 1984; YAMAGUCHI et al. 1984; GHATEI et al. 1985b) and neuroendocrine tumors from the colon, skin, and other sites (CHEJFEC et al. 1985). The presence of GRP-like immunoreactivity in tumor cells appears to be restricted to tumors of neuroendocrine lineage; however, not all neuroendocrine tumors show GRP-like immunoreactivity.

C. Molecular Genetics of the Prepro-GRP Gene

I. The Human Prepro-GRP Gene

1. Structure

Detailed analysis of the gene encoding the gastrin-releasing peptide precursor has been reported for both the human and the rat. The human prepro-GRP gene was present in a single copy per haploid genome (SAUSVILLE et al. 1986; SPINDEL et al. 1987b), consisted of three exons (SPINDEL et al. 1987b), and mapped to chromosome 18 (NAYLOR et al. 1987), at band 18q21 (LEBACQ-VERHEYDEN et al. 1987). Analysis of cDNA clones derived from a pulmonary carcinoid tumor (SPINDEL et al. 1984) and from a small cell lung carcinoma (SCLC) cell line (SAUSVILLE et al. 1986) showed that the prohormone coding domain consisted of a 23-amino-acid signal sequence, followed by the 27-amino-acid GRP, a glycine amidation donor, a dibasic amino acid cleavage site, and a 95-amino-acid extension peptide (Fig. 2). Three different prepro-GRP mRNAs were initially found in tumor cells expressing the prepro-GRP gene (SAUSVILLE et al. 1986; SPINDEL et al. 1986). These three forms of mRNA were identical over regions encoding the signal sequence, GRP, and the 68 N-terminal amino acids of the extension peptide; they differed only in the region encoding the remaining C-terminal amino acids of the extension peptide. This heterogeneity was generated by alternative splicing of the primary transcripts at the second and third exon splice junction (SAUSVILLE et al. 1986). So far the

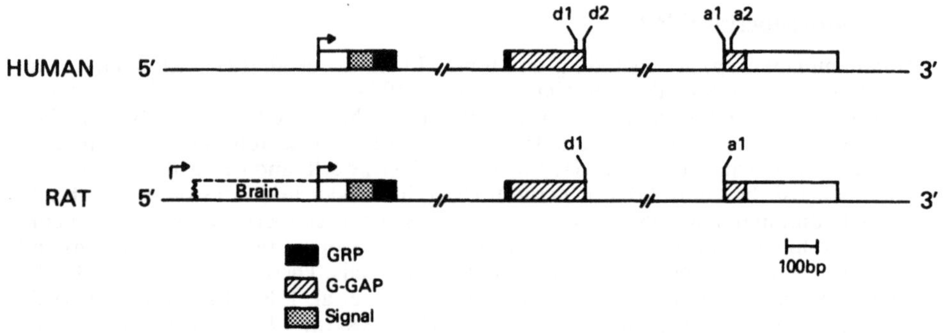

Fig. 2. Comparison of the human and rat prepro-*GRP* genes. The three-exon structure of the human (SPINDEL et al. 1987b) and rat (LEBACQ-VERHEYDEN et al. 1988b) *GRP* genes is shown. The human gene is remarkable for the alternate splice donor (d_1 and d_2) and acceptor (a_1 and a_2) sites flanking the second intron, resulting in three mRNA isoforms. The rat gene does not show this diversity in removal of intron two. Two mRNA forms are observed for the rat prepro-GRP gene which originate from two initiation sites. The upstream initiation site is only active in the rat CNS. The *arrows* indicate initiation sites, the *dashed box* indicates a domain found only in the brain-specific mRNA form, the *dotted box* designates the signal sequence, the *filled box* the GRP peptide domain, and the *hatched box* shows the pro-GRP extension peptide domain (G-GAP). The two *white boxes* are 5' and 3' untranslated sequences

three forms of mRNA were present in the same molar ratio in all normal and malignant cells known to express the prepro-GRP gene (SAUSVILLE et al. 1986; SPINDEL et al. 1987b; CUTTITTA et al. 1988). Therefore, the alternative processing of the human prepro-GRP gene does not show cell-type specific regulation.

2. Expression

Earlier biochemical and immunohistochemical studies established that the human developing lung was an abundant source of GRP-like peptides (Sect. B.III; TRACK and CUTZ 1982) and that they were localized to pulmonary neuroendocrine cells (Sect. B.IV.2). Northern blot analysis confirmed that human fetal lung expressed the prepro-GRP gene (SPINDEL et al. 1987b; BATTEY et al. 1988). A developmental study using in situ hybridization analysis localized mRNA to pulmonary neuroendocrine cells as early as 8 weeks of gestation; their numbers reached a peak around midgestation (18–23 weeks), decreased during late gestation and after birth, and were further reduced in normal adult lung (SPINDEL et al. 1987a). A similar developmental study showed that neonatal thyroid contained numerous prepro-GRP-expressing neuroendocrine cells; their numbers declined during the postnatal period and became undetectable in normal adult thyroid (SUNDAY et al. 1988b). These studies suggested that GRP might be important in controlling normal cell growth during human lung and human thyroid development. Northern blot analysis (SPINDEL et al. 1984, 1987b; SAUSVILLE et al. 1986) and in situ hybridization studies (SUZUKI et al. 1987; SUNDAY et al. 1988b; HAMID et al. 1989) also confirmed the expression of the prepro-GRP gene in neuroendocrine tumors derived from the lung and thyroid.

3. Regulation

Detectable levels of steady-state prepro-GRP mRNA were observed only in a limited subset of tissues and cell lines examined to date (SAUSVILLE et al. 1986). Therefore, transcription of the human prepro-GRP gene appeared to be under strict cell type regulation, with gene expression absent in most cultured cell lines examined. Nuclear run-on analysis comparing a series of small cell carcinoma cell lines that expressed varying amounts of prepro-GRP mRNA showed that the principal level of control was at initiation of primary transcription (MARKOWITZ et al. 1988). In addition, a cluster of DNAse hypersensitivity sites, located within several kilobases of the promoter, was invariably found in small cell carcinoma cell lines expressing prepro-GRP mRNA, and was absent in their nonexpressing counterparts (MARKOWITZ et al. 1988). DNAse hypersensitivity sites may result from an altered distribution of DNA-binding proteins that modulate recognition of the prepro-GRP promoter by RNA polymerase II. Further studies will be needed to determine the precise location and importance of these candidate *cis*-binding sites and *trans*-binding proteins that appear to regulate transcription of the prepro-GRP gene.

II. Rat Prepro-GRP Gene

1. Structure

Comparison between the human and rat prepro-GRP genes showed a similar overall organization (Fig. 2). However, several differences were also noted. The three-exon rat prepro-GRP gene encoded a 23-amino-acid signal peptide, followed by a 29-(rather than 27-)amino-acid GRP, a glycine amidation donor, a dibasic amino acid cleavage site, and a 90-amino-acid extension peptide (LEBACQ-VERHEYDEN et al. 1988b). The predicted amino acid sequence of rat pro-GRP extension peptide showed two regions of considerable homology with human pro-GRP; they were rat pro-GRP_{46-71}, with 23/26 amino acids identical to the N-terminal region common to the three isoforms of human pro-GRP, and rat pro-$GRP_{102-124}$, with 20/23 amino acids identical to the C-terminal region specific for the third isoform of human pro-GRP. In contrast to human prepro-GRP, there was no alternative processing of rat prepro-GRP at the second and third exon splice junction; the rat prepro-GRP mRNA was homologous to the third form of human prepro-GRP mRNA, and there was no heterogeneity in mRNAs over this region. However, heterogeneity was found at the 5′ end of the transcripts (LEBACQ-VERHEYDEN et al. 1988b).

Analysis of rat prepro-GRP mRNAs derived from brain and gastrointestinal tissues indicated at least two promoter regions for transcription initiation. The 3′ promoter was located in a position analogous to the one mapped previously in the human prepro-GRP gene (SPINDEL et al. 1987b; MARKOWITZ et al. 1988); it encoded a 1.1-kb mRNA which initiated from a single site, and was active in both brain and gut. The 5′ promoter encoded a 1.5-kb mRNA; it was heterogeneous in its precise site of initiation, and appeared to be active only in the brain and spinal cord (LEBACQ-VERHEYDEN et al. 1988b). The additional 400 bases of 5′ sequences in the 1.5-kb transcripts contained no long open reading frames, did not appear to alter the structure of the translated prohormone, and were not contained in an additional 1.8-kb transcript recently identified in stomach and colon (LEBACQ-VERHEYDEN and BATTEY, unpublished data.

2. Expression

Northern blot analysis detected prepro-GRP transcripts in a limited number of adult rat tissues including brain, spinal cord, stomach, intestine, and testis (LEBACQ-VERHEYDEN et al. 1988b; LEBACQ-VERHEYDEN and BATTEY, unpublished data). In situ hybridization studies were performed on rat brain with cRNA probes that would detect only the 1.5-kb mRNA or both 1.1-kb and 1.5-kb forms (ZOELLER et al. 1989). Brain regions showing the highest levels of prepro-GRP mRNA, namely the cingulate cortex, subiculum and Ammon's horn in the hippocampus, initiated transcription from both promoters. Additional brain regions expressing prepro-GRP mRNA apparently initiated from the 3′ promoter alone included the frontoparietal cortex, striate cortex, amygdala, suprachiasmatic nucleus, interpeduncular nucleus, and the nucleus of the solitary tract. The above distribution of prepro-GRP-expressing neurons in rat brain confirmed and extended previous immunohistochemical studies (ROTH et al. 1982; PANULA et al. 1982, 1984). In situ hybridization appeared to be a more sensitive method for detecting GRP gene products, particularly in cortical neuronal cell bodies. However, the failure to detect GRP gene products in the paraventricular nucleus of the hypothalamus indicated that these well-established GRP-like immunoreactive neurons may contain another bombesin-related peptide. It was also noted that the distribution of prepro-GRP-expressing neurons in rat brain overlapped the distribution of bombesin-binding sites (WOLF et al. 1983; WOLF and MOODY 1985; ZARBIN et al. 1985), suggesting that in these areas GRP might function as a central neuromodulator or neurotransmitter.

III. Human Pro-GRP-Derived Peptides

1. Posttranslational Processing

An α-amidated C-terminal methionine residue was essential for the binding of bombesin and GRP to high-affinity cell surface receptors and to initiation of a biological response (MOODY et al. 1978). This methionine residue is located 27 amino acids from the N-terminal end in human pro-GRP; it is adjacent to a glycine residue that serves as an amidation donor, and to a pair of basic amino acids that separates GRP from the C-terminal extension peptide of pro-GRP. Given this prohormone structure, it is apparent that both specific cleavage and α-amidation activities are needed after prohormone translation for the generation of biologically active GRP. In addition, the human-rat cross-species conservation of peptide domains in the C-terminal extension peptide raised the possibility that other biologically active peptides might be encoded in pro-GRP. Presumably, these peptides would also be released by posttranslational cleavage and modification events.

Analysis of the posttranslational processing of human pro-GRP in different cell types revealed considerable variation in the pattern and efficiency of prohormone processing. Murine Swiss 3T3 fibroblasts transfected with a constitutively expressed human prepro-GRP gene construct expressed abundant levels of mRNA and secreted intact pro-GRP, with no evidence for any posttranslational processing event other than the cotranslational removal of the signal peptide from prepro-GRP (LEBACQ-VERHEYDEN et al. 1988c). A moth ovary cell line expressing high levels of human pro-GRP after infection with a recombinant baculovirus cleaved the prohormone at four specific sites. Three of the four cleavage sites were located in the C-terminal extension peptide domain, producing a family of pro-GRP-derived peptides. One cleavage occurred between the methionine residue at the C-terminal end of GRP_{1-27} and its glycine amidation donor, generating a nonamidated GRP_{1-27} peptide which is expected to be several orders of magnitude less potent as a ligand than its amidated counterpart (LEBACQ-VERHEYDEN et al. 1988a). Since neither Swiss 3T3 nor moth ovary cells normally express pro-GRP, these cells may not be expected faithfully to perform the necessary posttranslational modifications. In fact, cell-type variation of provasopressin-neurophysin II processing was also observed in heterologous cell lines expressing a transfected metallothionein-vasopressin fusion gene (CWIKEL and HABENER 1987). Posttranslational processing was also studied in a human SCLC line that expressed its endogenous prepro-GRP gene. This cell line secreted GRP_{1-27} that was cleaved and amidated appropriately. In addition, a number of novel peptides were generated by additional specific cleavages in the C-terminal prohormone domain (LEBACQ-VERHEYDEN et al. 1988a). Recently, this analysis was extended to another SCLC cell line that also expressed its endogenous prepro-GRP gene; this study showed that there was considerable variation in the patterns of posttranslational cleavage of pro-GRP when the two SCLC lines were compared.

2. Expression

Isoforms I/II of the extension peptides of human pro-GRP were localized by immunohistochemistry to fetal pulmonary neuroendocrine cells and to a small cell carcinoma cell line; in both locations, they were shown to be coexpressed with GRP (CUTTITTA et al. 1988). The extension peptides were also found in neuroendocrine tumors derived from the lung (HAMID et al. 1987) and from other sites (SPRINGALL et al. 1986); in these studies, they were often expressed in the absence of detectable GRP-like immunoreactivity, which might reflect a failure of GRP processing and was associated with a poorer prognosis (HAMID et al. 1987).

D. Pharmacological Effects of Bombesin and GRP

Exogenous bombesin and GRP exert a wide range of pharmacological effects when injected into a number of mammalian species including man. They influence mainly digestive functions, metabolism, and behavior, as extensively documented over the past 10 or 15 years. More recently, these peptides were shown to have trophic effects on the stomach and pancreas, and they were found to promote growth in vitro in several culture systems.

I. Effects Unrelated to Growth

At the cellular level, bombesin and GRP promote the release of circulating and local hormones, as well as classical and peptidergic neurotransmitters (see below). As a result, their effects are widespread in the organism, complex to analyze, and variable from species to species. The pharmacological effects of bombesin-related peptides on exocrine and endocrine secretions, smooth muscle, metabolism, and behavior are summarized in Tables 1–4, together with the references pertinent to their discovery and characterization in individual species (see Appendix). Whenever appropriate, reference will be made to these tables rather than to individual publications in the next three sections.

1. In Vivo Effects

The in vivo effects of bombesin and GRP were highly complex, due to the interplay of direct and indirect mechanisms of action both in peripheral organs and in the CNS. Specific pharmacological effects were elicited by systemic injection of the peptides; in addition, the same, opposite, or unrelated effects were observed after their administration into the brain or spinal cord. For example, intravenous injection of bombesin or GRP stimulated gastric acid secretion, an effect which was secondary to the release of gastrin (reviewed in WALSH 1988); in contrast, intracerebral or intraspinal injection of the peptides produced a strong inhibition of basal and stimulated gastric acid secretion (reviewed in TACHÉ 1988) (Table 1). Both peripheral and central administration of bombesin-related peptides delayed gastric emptying, and their central injection further delayed intestinal transit; however, these in vivo inhibitory effects contrasted sharply with the stimulatory effects on gastrointestinal smooth muscle observed in vitro (Table 2). The effects on metabolism (Table 3), as well as some effects on behavior including compulsive scratching, grooming, biting, and general arousal (Table 4), were only observed after central administration of the peptides. In contrast, other behavioral effects in-

Table 1. Pharmacological effects of bombesin and GRP on exocrine and endocrine secretions

	In vivo effects		In vitro effects	
	Peripheral[a]	Central[b]	Organs[c]	Cells[d]
Gastric secretions				
Acid	↑[1,2,3]	↓[4,5]		
Mucus	→[6]	↑[6]		
Gastrin	↑[1,2,3,7,8,9]; →[10]	↑[4,5]	↑[11,12]	↑[13]
Somatostatin	↑[10,14]		↑[11,12]	
Gastric inhibitory peptide	↑[15,16]; →[10]			
Intestinal secretions				
Cholecystokinin	↑[17,18,19]; →[10]			
Enteroglucagon	↑[19,20,21]; →[10]			
Motitilin	↑[22,23]			
Neurotensin	↑[24,25]; →[26]			
Vasoactive intestinal polypeptide	↑[23]			
Pancreatic secretions				
Amylase	↑[17,27]	↓[28]	↑[29]	↑[30,31]
Pancreatic polypeptide	↑[10,32,33]		↑[34]; →[35]	
Insulin	↑[9,10,36,37,38]	↓[39,40]	↑[41,42,43]	↑[44]; ↓[45]
Glucagon	↑[9,10,36,37,38]	↑[39,40]	↑[41,43]	
Adrenal secretions				
Epinephrine	→[39,40]	↑[39,40]		
Corticosterone	→[39]; ↑[46]	→[46]		→[46]
Pituitary secretions				
Prolactin	↑[47,48]; →[49,50]; ↓[51]	↑[47]; ↓[49,51]		↑[52]; →[47,53]
Growth hormone	↑[47]; →[48,50,54]	↑[47]; →[51]; ↓[55]		↑[53,56]; →[57]
Thyroid-stimulating hormone	→[50]; ↓[58]			
Luteinizing hormone	→[50]; ↑[59]	→[60]		
Other secretions				
Calcitonin	↑[61]			
Parathyroid hormone	↓[61]; →[62]			
Renin-angiotensin	↑[63]			

Stimulation (↑), inhibition (↓), or no effect (→).
[a] After systemic injection.
[b] After intracerebral or intraspinal injection.
[c] On isolated perfused organs or on organ fragments.
[d] On isolated cells or cell lines.
1–63, groups of references pertinent to the discovery and characterization of pharmacological effects in individual species (see Appendix).

Table 2. Pharmacological effects of bombesin and GRP on smooth muscle

	In vivo effects		In vitro effects	
	Peripheral[a]	Central[b]	Organs[c]	Cells[d]
Stomach				
Motility corpus	↓[1,2]; ↑[3,4]		↑[1,2,3]	
Motility antrum	↑[1,2,3,4,5]	↑[6]	↑[1,7,8,9]	
Electric activity	△[10]			
Gastric emptying	↓[11,12]; →[13]	↓[14]		
Small intestine				
Motility	↓[1,4]	↑[15]	↑[1,16,17,18]	↑[19]
Electric activity	△[10,20]			
Transit		↓[14]		
Large intestine				
Motility			↑[1,21,22,23]	
Electric activity	→[10]			
Transit	→[1]	↑[13]		
Motility of:				
Gallbladder	↑[24]		↑[25]	
Urinary bladder	↑[22,24]; →[26]		↑[22,24]; →[26]	
Uterus			↑[22,27]; →[28]	
Seminal vesicles			↑[29]	
Respiratory tract	↑[30]			
Cardiovascular system				
Blood pressure	↑[31]; →[32]	↑[32]		
Heart rate	→[32]	↓[32]		

Stimulation (↑), inhibition (↓), modification (△), or no effect (→).
[a] After systemic injection.
[b] After intracerebral injection.
[c] On isolated organs or muscle strips.
[d] On isolated muscle cells.
1–32, groups of references pertinent to the discovery and characterization of pharmacological effects in individual species (see Appendix).

cluding reduction in food intake and retention of memory were elicited chiefly by peripheral administration (Table 4). Microinjections into brain nuclei further delineated some central sites of action of bombesin-related peptides (Tables 3, 4), and surgical manipulations established whether the adrenals, the pituitary, and/or the vagus nerve were required for the mediation of these effects (Table 5). Further work will be needed to elucidate the diverse mechanisms by which centrally administered bombesin-related peptides influence digestive functions, metabolism, and behavior.

Table 3. Pharmacological effects of bombesin and GRP on metabolism

	Peripheral effects[a]	Central effects when given	
		icis or icv[b]	In brain nucleus[c]
At 4° C			
Rectal temperature	→[1]	↓[2,3]	↓[4,5] [POA]; →[5] [PAG]
Oxygen consumption		↓[6]	↓[4] [POA]
Heat loss		→[7]	
Heart rate		↓[8]	
Plasma glucose		↑[9]	
At 20°–22° C			
Rectal temperature		↓[10,11]; →[12,13]	↓[14] [PAG]
Oxygen consumption		↑[15]; →[16]	
Heat loss		↑[17]	
Heart rate	→[18]	↓[18]	
Blood pressure	→[18]	↑[18]	
Minute ventilation		↑[19,20]	↑[21] [NAm]
Plasma glucose	→[22]	↑[23]	↑[24] [DHA]; ↑[25] [VMH, LHA]
At 20°–22° C, in food-deprived animal			
Rectal temperature		↓[26]	↓[27] [SNi]; ↓[27] [POA]

Stimulation (↑), inhibition (↓), or no effect (→).
[a] After systemic injection.
[b] After intracisternal (icis) or intracerebroventricular (icv) injection.
[c] After brain parenchymal microinjection.
POA, preoptic area, anterior hypothalamus; PAG, periaqueductal gray; NAm, nucleus ambiguus; DHA, dorsal hypothalamic area; VMH, ventromedial hypothalamus; LHA, lateral hypothalamic area; SNi, substantia nigra; 1–27, groups of references pertinent to the discovery and characterization of pharmacological effects in individual species (see Appendix).

2. In Vitro Effects on Isolated Organs

In vitro effects on isolated organs and organ fragments were still complex, due to the combined effects of the peptides on multiple cell types. For example, exogenous bombesin induced the release of both gastrin and somatostatin in the isolated, vascularly perfused rat stomach; however, pharmacological analysis and immunoneutralization showed that somatostatin partially inhibited the bombesin-induced release of gastrin in a paracrine way (Table 1). Interesting observations were made on the indirect effects of bombesin-related peptides on smooth muscle. In guinea pig antrum longitudinal muscle strips, bombesin stimulated directly the frequency of spontaneous contractions, but was shown to increase muscle tone indirectly, through the release of [^{3}H]acetylcholine from intramural neurons (Kantoh et al. 1985). Bombesin also stimulated in-

Table 4. Pharmacological effects of bombesin and GRP on behavior

	Peripheral effects[a]	Central effects when given		
		icis or icv[b]	In brain nucleus[c]	ith[d]
Food intake[P]	↓[1,2,3,4,5]	↓[6,7,8,9]	↓[10] [SNi], ↓[11] [LHA] ↓[12] [most], ↓[13] [SOL]	
Scratching/grooming	→[14]	↑[15,16,17]	↑[18] [PAG], ↑[12] [all]	
Biting/scratching	→[19,20]			↑[19,20]
Perception of pain		→[21]	↓[22] [PAG]	
Locomotor activity		↑[23]	↑[24] [POA], ↑[25] [NAc, VTA]	
Sleep		↓[26]	↓[12] [all]	
Memory[P]	↑[27]	↑[27]		

Stimulation (↑), inhibition (↓), or no effect (→).
[a] After systemic injection.
[b] After intracisternal (icis) or intracerebroventricular (icv) injection.
[c] After brain parenchymal microinjection.
[d] After intrathecal injection (ith).
SNi, substantia nigra; LHA, lateral hypothalamic area; "most", ventromedial hypothalamic area, paraventricular nucleus, dorsomedial nucleus, lateral hypothalamic area, amygdala, periaqueductal gray; SOL, nucleus tractus solitarius area; POA, preoptic area, anterior hypothalamus; NAc, nucleus accumbens; VTA, ventral tegmental area; "all," "most" and posterior hypothalamus, septum, reticular formation, ventral tegmental area; [P], behavioral effect observed chiefly after peripheral injection; 1–27, groups of references pertinent to the discovery and characterization of pharmacological effects in individual species (see Appendix).

Table 5. Peripheral pathways mediating central effects of bombesin and GRP

Effects on	Adrenals	Pituitary	Vagus
Digestive functions			
Gastric acid inhibition	–[1]	–[2]	–[1]
Gastric mucus stimulation	+[3]		
Gastrointestinal transit delay	+[4]	+[4]	+[5]
Metabolism			
Hypothermia	–[6,13]	–[7]	
Bradycardia	–[8]		
Blood pressure rise	+[8]		
Ventilation stimulation			+[9]
Hyperglycemia	+[6,10]	–[10]	
Behavior			
Anorexia	–[11,12]		–[12]
Locomotor hyperactivity	+[13]		
Compulsive scratching	–[14]	–[14]	
Memory retention			+[15]

Effect dependent (+) or independent (–) upon the presence of the organ; 1–15, references (see Appendix).

directly the contraction of the canine muscularis mucosae of the colon, apparently by releasing substance P from peptidergic intramural neurons (ANGEL et al. 1984).

3. Direct Effects and Cellular Distribution of Receptors

In vitro effects on isolated cell populations or cell lines usually indicate a direct action of peptides, mediated through binding to specific cell surface receptors. Bombesin-related peptides were shown to act directly upon and/or to bind to antral gastrin cells, exocrine pancreatic acinar cells, endocrine pancreatic islet cells, anterior pituitary cells (Table 1), intestinal smooth muscle cells (Table 2), and human pancreatic membranes (SCEMAMA et al. 1986). In addition, specific binding sites were located by autoradiography to endocrine cells in the antral mucosa, to circular and longitudinal smooth muscle layers in varying segments of the gastrointestinal tract, and to the entire myenteric plexus (VIGNA et al. 1987; MORAN et al. 1988). The latter binding sites account for the neuronal modulation exerted by exogenous bombesin-related peptides on gastrointestinal functions.

4. Induced Release of Endogenous GRP

A direct action of bombesin on GRP neurons was demonstrated on the isolated, hemisected toad spinal cord (PHILLIS and KIRKPATRICK 1979), as well as by iontophoretic application of the peptide to single cortical neurons in the anesthetized rat (PHILLIS and LIMACHER 1974). More recently, iontophoretic application of bombesin to hypothalamic neurons in the preoptic area was shown to inhibit the firing rates of cold-sensitive neurons and to stimulate the firing rates of heat-sensitive neurons; these findings were in accordance with the hypothermia and heat losses induced by centrally administered bombesin (LIN and LIN 1986). Specific bombesin/GRP-binding sites were located by radioreceptor assay (MOODY et al. 1978, 1980) and autoradiography (WOLF et al. 1983; WOLF and MOODY 1985; GILLATI et al. 1984; ZARBIN et al. 1985) to various regions of the brain including the olfactory bulb and tubercle, nucleus accumbens, suprachiasmatic and paraventricular nucleus of the hypothalamus, central medial thalamic nucleus, medial amygdaloid nucleus, hippocampus, dentate gyrus, subiculum, nucleus of the solitary tract, and substantia gelatinosa. Binding sites were also located on the dorsal horns of the spinal cord (O'DONOHUE et al. 1984).

From these observations, it is clear that bombesin-related peptides can act upon and/or bind to neurons that are widespread in the peripheral and CNSs. Furthermore, the colocalization of GRP-like immunoreactive nerve terminals and bombesin/GRP receptors suggested that the endogenous peptides might be released under specific physiologic conditions and function in situ as neurotransmitters or neuromodulators. Immunoneutralization studies suggested that endogenous hypothalamic GRP-like peptides were involved in the inhibition of growth hormone and prolactin release (KENTROTI et al. 1988b). The potassium-stimulated release of GRP-like peptides in a calcium-dependent manner from rat brain synaptosomes, hypothalamic slices (MOODY et al. 1980), and spinal cord slices (MOODY et al. 1981 c) further supported the notion that they were of physiologic relevance in the CNS. Similarly, indirect immunoneutralization studies and direct radioimmunoassays documented the release of GRP-like peptides from the rat stomach (DU VAL et al. 1971; SCHUSDZIARRA et al. 1983, 1984, 1986; SCHUBERT et al. 1985; SHORT et al. 1985; JAIN et al. 1985; WOLFE et al. 1987), pig stomach (KNUHTSEN et al. 1984; HOLST et al. 1987 a, b), and pig pancreas (KNUHTSEN et al. 1985), consistent with involvement of this peptide in the neural regulation of gastrin release, somatostatin release, amylase release, and possibly insulin and glucagon release.

II. Effects on Growth

1. In Vitro Studies

a) Swiss 3T3 Embryo Fibroblasts

The initial landmark demonstration that the bombesin family of peptides could function as mitogens was in studies performed on Swiss 3T3 murine embryonal fibroblasts. ROZENGURT and SINNETT-SMITH (1983) showed that either bombesin or GRP could stimulate cell division and DNA synthesis in quiescent cells. The magnitude of the effect was dose dependent; it was markedly potentiated by the addition of insulin, platelet-derived growth factor, or fibroblast growth factor, which lowered the bombesin dose required to achieve a half-maximal response as well as increased the magnitude of the maximal response.

In subsequent studies, the bombesin-dependent mitogenic stimulation was directly associated with the presence of a single class of high-affinity ($K_d \approx 1$ nM) cell surface receptors which bound either bombesin or GRP. Structure-activity studies indicated that both receptor-binding and mitogenic stimulation required a peptide ligand containing an amidated C-terminal domain either identical with or very similar to the conserved heptapeptide defining the bombesin peptide family (ZACHARY and ROZENGURT 1985). To obtain full mitogenic activity, the last eight amino acids of GRP, namely GRP_{20-27}, were needed. Deletion of either His20 or Trp21 resulted in significant loss of potency in both receptor-binding and thymidine uptake assays (HEIMBROOK et al. 1988). A bombesin receptor antagonist, the (D-Arg1, D-Pro2, D-Trp7,9, Leu11)-substance P analog previously shown to be an inhibitor of bombesin-binding and bombesin-stimulated amylase secretion (JENSEN et al. 1984), specifically inhibited bombesin binding and mitogenic stimulation of Swiss 3T3 cells (ZACHARY and ROZENGURT 1985 a, 1986; CORPS et al. 1985).

b) Chick Embryo Otic Vesicle

Given the numerous and varied tissue types which contain neural and neuroendocrine cells expressing GRP, it was interesting to consider the possibility that GRP might activate growth of specific target cells in embryonic developing tissues. This hypothesis was examined in part in cultured chick embryo otic vesicles. Cultured otic vesicles maintained a normal pattern of cell growth and differentiation for at least 48 h when grown in medium supplemented with 20% fetal calf serum. Growth and development were reversibly arrested by serum deprivation. After serum deprivation, bombesin at 100 n*M* concentration reactivated cell division and development in growth-arrested vesicles (REPRESA et al. 1988). The effect was dose dependent, saturable, and potentiated by insulin, which was ineffective by itself. This intriguing observation establishes that bombesin can modulate growth in embryonic developing tissues. Whether or not bombesin modulation of growth is important in vivo during chick embryo otic vesicle development will require additional investigation, in particular the demonstration of biologically active bombesin-like peptides in otic vesicles during normal development at this stage.

c) Normal Human Bronchial Epithelial Cells

In normal human bronchial epithelial cells, either bombesin or the C-terminal half of GRP, $GRP_{14\text{-}27}$ increased the clonal growth rate and colony-forming efficiency by about twofold when the cells were grown in serum-free medium (WILLEY et al. 1984). The maximal effect observed was at peptide concentrations of approximately 100 n*M*. This observation was of particular interest since increases in GRP immunoreactivity and pulmonary endocrine cell hyperplasia accompanied disease states where the lung was responding to injury, such as bronchopulmonary dysplasia (JOHNSON et al. 1982) and bronchiestasis (TSUTSUMI et al. 1983, reviewed by SUNDAY et al. 1988 a).

d) Small Cell Lung Carcinoma Cell Lines

Biochemical studies of human SCLC cell lines often detected intracellular bombesin-like peptides (Sect. B.III; CARNEY et al. 1985), subsequently shown to be *GRP* gene products (SAUSVILLE et al. 1986). In addition, high-affinity receptors for bombesin/GRP were found on the cell surface of a number of SCLC cell lines (MOODY et al. 1983 a). Given these facts, it was of interest to examine whether or not autocrine GRP growth stimulation might be important in augmenting the growth of these cell lines.

Two SCLC cell lines showed elevated levels of thymidine incorporation when maintained in medium containing micromolar concentrations of $GRP_{1\text{-}27}$ or $GRP_{14\text{-}27}$; $GRP_{1\text{-}16}$ had no effect. In addition, the mitogenic effect was observed only on the two SCLC cell lines, and not on either an adenocarcinoma or a squamous cell carcinoma cell line (WEBER et al. 1985). CARNEY et al. (1987) studied the effects of bombesin and gastrin-releasing peptide on the soft agar cloning efficiency of a panel of human lung carcinoma cell lines. Nine of ten SCLC cell lines showed a greater than sevenfold increase in soft agar cloning efficiency when 50 n*M* bombesin was added to the medium. The clonal growth-promoting effect was only observed when the cells were maintained and grown in a defined, serum-free medium. The bombesin analog des-Leu13-Met14-bombesin shown previously to be inactive as a ligand for bombesin receptors (RIVIER and BROWN 1978) had no growth-promoting effect. Nine cell lines derived from other lung tumor histologies showed no bombesin-dependent colony-forming stimulation.

A monoclonal antibody binding to the C-terminal heptapeptide of either bombesin or GRP was used as a bombesin antagonist, competing with receptor for endogenously secreted GRP (CUTTITTA et al. 1985). Soft agar cloning efficiency of two SCLC cell lines decreased significantly when the anti-bombesin monoclonal antibody was included in the serum-free medium at about 20 n*M* concentration. The decrease in cloning efficiency was abolished by the addition of bombesin at 50 n*M* concentration along with the antibody; in addition, an isotype-matched indifferent antibody added at 20 n*M* concentration had no effect on soft agar cloning, indicating that the effect on growth was specific. Nude mouse heterotransplants were established using one of the two SCLC cell lines studies for clonal growth. Intraperitoneal administration of about 600 pmol (100 μg) antibombesin monoclonal antibody three times a week suppressed the in vivo growth of the tumors for 4 weeks, while the indifferent antibody had no discernable effect on tumor formation. In another study, a peptide with bombesin receptor antagonist activity, [D-Arg1, D-Phe5, D-Trp-7–9, Leu11] substance P, was also shown to be effective in inhibiting the mass culture growth of cultured SCLC cells maintained in serum-free medium. Half maximal effects were observed at 24 μ*M* concentrations of the antagonist (WOLL and ROZENGURT 1988). TREPEL et al. (1988b) showed that a more potent bombesin receptor antagonist peptide, [Leu13ψ-CH_2NH-Leu14]bombesin, inhibited cultured SCLC colony formation by about twofold in a soft agar clonogenic assay, when the antagonist was present at 100-, 500- or 1000-n*M* concentrations. Taken together, these results are consistent with the view that, under certain circumstances, SCLC cells are dependent on endogenously secreted GRP for optimal growth. It is of interest that, in the small cell lung carcinoma system, clonogenic growth in soft agar is more sensitive to stimulation by bombesin than mass culture growth. This raises the possibility that bombesin action in this system is most important for optimal growth in the solid matrix. Recent studies highlighting other somatic genetic changes found in this tumor, including chromosomal deletion and alteration in the structure and regulation of protooncogenes, underscore the fact that GRP and autocrine growth factors are only one aspect of a complicated cascade of mechanisms leading to the deregulated growth of SCLC cells (MINNA et al. 1987).

2. In Vivo Studies

a) Gastrin Cell Hyperplasia

A measurable and reproducible increase in the number of gastrin cells in the antral mucosa of the rat stomach was noted following chronic administration of bombesin (LEHY et al. 1983). These authors also observed a 45% increase in the number of actively dividing G cells, as measured by [^{3}H]thymidine incorporation. No change in the neighboring somatostatin cells was seen. It should be noted that there was no direct evidence bearing on the issue of whether the hyperplasia was a direct or indirect effect of bombesin on G cells.

b) Pancreatic Growth

Chronic treatment of rats for 5 days with bombesin induced a dose-dependent pancreatic cell hypertrophy (LHOSTE et al. 1985a). The growth was characterized by increase in weight, protein, RNA, and enzyme content, but not DNA content. After a longer period of treatment (15 days), increase in pancreatic DNA content was also observed, consistent with pancreatic hypertrophy and hyperplasia (LHOSTE et al. 1985b). Both bombesin and GRP appeared to be equipotent in their trophic effect on rat pancreas (DAMGÉ et al. 1988). Whether or not GRP is important in regulating pancreatic acinar cell growth or G-cell growth during normal development is an interesting issue awaiting future investigations.

E. Cellular Responses to Bombesin and GRP

I. Introduction to Bombesin-Mediated Signal Transduction

Hormones and neurotransmitters affect cellular metabolism by activating signal transduction pathways usually characteristic of each ligand, utilizing a specific cell-surface receptor. Four general types of signal transduction mechanisms may be distinguished: phospholipase activation, calcium flux, tyrosine kinase activation, and cyclic nucleotide accumulation. As illustrated schematically in Fig. 3, and as will be discussed in detail here, bombesin-related peptides as exemplified by mammalian GRP have major effects on the first three of these mechanisms in a number of pertinent systems, and in selected instances may also perturb cyclic nucleotide metabolism as well.

II. Bombesin Binding to Cells/Membranes: Definition of the Bombesin Receptor

Guinea pig pancreatic acinar cells specifically bound ^{125}I[Tyr4]-bombesin with a K_d in the nanomolar range (JENSEN et al. 1978). Likewise, specific binding to rat brain membranes was characterized by similar affinities, and this system allowed definition of the importance of the C-terminal segment of the peptide in mediating specific binding (MOODY et al. 1978). This observation is in accord with the preservation of this portion of the peptide through phylogeny. The binding of bombesin peptides to rat pituitary cells was not affected by thyrotropin-releasing hormone, epidermal growth factor, or somatostatin (WESTENDORF and SCHONBRUNN 1983).

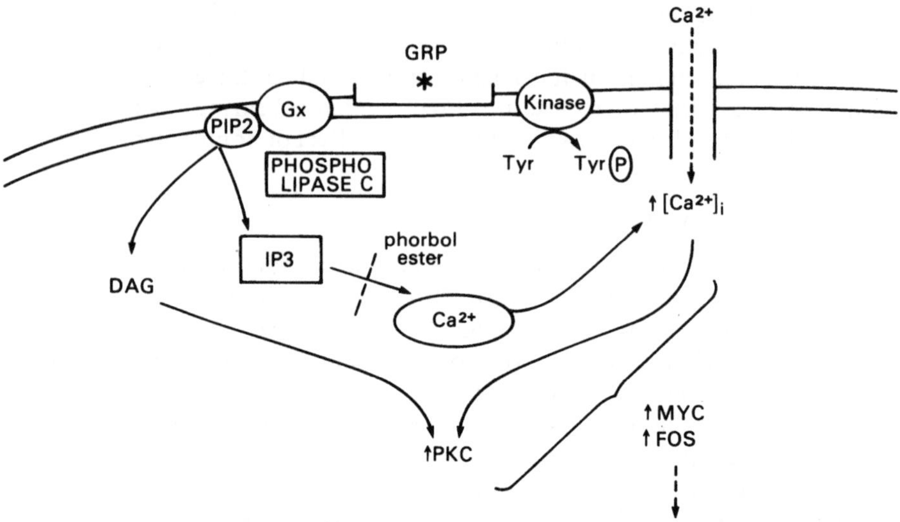

Fig. 3. Signal transduction elicited by GRP. Binding of GRP to its receptor causes activation of phospholipase C, increased intracellular Ca^{2+}, and activation of a tyrosine kinase and protein kinase C. These stimuli increase transcription of c-*myc* and c-*fos*, presumably leading to DNA synthesis and cell division

The binding of bombesin congeners to SCLC cell lines has revealed a K_d in the same nanomolar range as nonneoplastic sources, with, however, less than 2000 sites/cell in most examples studied. There was little correlation between the level of endogenously produced bombesin-like immunoreactivity and the degree of receptor binding. High-affinity binding was restricted to SCLC as opposed to non-SCLC-derived lung tumor cell lines. The binding was trypsin sensitive, occurred at 4° or 37° C, and was ethylene glycol tetraacetic acid (EGTA) insensitive. Using affinity chromatographic techniques, a 78-kDa cellular protein was associated with bombesin binding in SCLC cell lines (MOODY et al. 1985).

The molecular nature of bombesin-binding cell surface molecules was studied in Swiss 3T3 cells, which express in selected examples >100000 receptors/cell (ZACHARY and ROZENGURT 1985b). Affinity crosslinking studies showed that a ligand-binding component of the bombesin receptor expressed in these cells is an M_r 65000–85000 surface protein(s) with a pI of 6.0–6.5. The ligand-crosslinked complex bound to wheat germ agglutinin agarose at 4° C, and was reduced by incubation with an *N*-glycanase to an $M_r = 45000$ product, suggesting that the bombesin-binding molecule is heavily glycosylated (ZACHARY and ROZENGURT 1987a; KRIS et al. 1987).

As the physiologic concomitants of bombesin action are complex, it is possible that the bombesin receptor has the capacity to associate or influence a number of cellular enzymatic activities. Thus, the future definition of the molecular structure of the bombesin-binding molecules revealed by crosslinking experiments will allow the generation of reagents to characterize these interactions in detail.

III. Desensitization/Internalization of the Receptor

In rat pituitary cells at 37° C, bombesin peptides were internalized and degraded by a chloroquine-sensitive process (WESTENDORF and SCHONBRUNN 1983). In guinea pig pancreatic acinar cells, ^{125}I[Tyr4]-bombesin binding reached a steady state by 15 min. Preincubation of the acinar cells with unlabeled bombesin, followed by washing, followed then by exposure to labeled peptide, revealed a decrease in specific binding (PANDOL et al. 1982). This implies that the act of bombesin binding desensitizes the receptor to subsequent binding in a way that differs from simple competition with unlabeled peptide. This may be of physiologic significance, as desensitization of amylase release by bombesin in these cells also occurs as the period of preincubation with bombesin is lengthened. Desensitization of amylase secretion is not affected by chloroquine. Recovery of sensitivity occurred over a period of approximately 90 min, but not at 4° C. Of great interest is the observation that dissociation of labeled bombesin requires greater than 90 min, but reversal of desensitization of stimulated amylase secretion is complete by the end of the 2- to 3-min washing period.

Swiss 3T3 cells also internalize and degrade labeled GRP. Chloroquine increases cell-associated label, without effects on the binding of GRP, but with a decrease in the rate of GRP degradation. In this system, after degradation of ^{125}I-GRP, there is rebinding of as much ^{125}I-GRP as was observed prior to GRP treatment, and this was correlated with the lack of effect of GRP treatment on ^{125}I-GRP crosslinking to the 75000- to 85000-M_r species described above (ZACHARY and ROZENGURT 1987a, b). It is difficult to define precisely a molecular mechanism for these effects at this time. Undoubtedly greater insight will be afforded after the availability of nonligand-dependent receptor-specific probes. It is apparent that the traffic of the receptor subsequent to binding of ligand may be different in different receptor-bearing cells.

IV. Phospholipase Activation

Numerous ligands, neurotransmitters, and classical endocrine hormones have as an initial aspect of their effects in responsive cells an increase in intracellular Ca^{2+} concentration. The realization that these same ligands were those which could evoke apparent turnover of phospholipid constituents, prominently phosphatidylinositol (PI), in the cell membrane led to the series of experiments which demonstrated that release of intracellular calcium in response to ligand occurs via the action of inositol-(1,4,5)-trisphosphate (I(1,4,5)P_3). This metabolite is proposed to be produced by a hormone receptor-coupled phospholipase C, acting on the membrane lipid phosphatidylinositol-4,5-bisphosphate $(PIP)_2$ (MICHELL et al. 1981; BERRIDGE

1987). Thus, a major focus in defining elements of the biochemical effects evoked by agonists is to understand the molecular basis for agonist-induced phosphatidyl inositol (PI) turnover.

Investigation of the effect of bombesin, vasopressin, platelet-derived growth factor (PDGF), and tumor-promoting phorbol esters to decrease the binding of ^{125}I-EGF to Swiss 3T3 cells led to the observation that bombesin was a potent stimulus for PI turnover in these cells (BROWN et al. 1984). A more detailed analysis of this effect revealed that bombesin incubation with Swiss 3T3 cells led to the accumulation of $I(1,4,5)P_3$ within 10 s of stimulation. Mono- and diphosphates as well as inositol-(1,3,4)-trisphosphate ($I(1,3,4)P_3$) and inositol tetrakisphosphate (IP_4) increased more slowly, but in a sustained fashion. Insulin greatly increased the efficiency of the bombesin-mediated increases in labeled inositol monophosphate accumulation, with less marked but evident effects on the accumulation of the di- and triphosphates as well (HESLOP et al. 1986). Bombesin-induced PI turnover in Swiss 3T3 cells was inhibited after treatment of cells with active tumor-promoting phorbol esters such as tetradecanoyl-phorbol acetate (TPA). Pretreatment of the cells for long periods with TPA, known to lead to decreased levels of protein kinase C, eliminated the ability of short-term treatment with TPA to abrogate the bombesin-induced response (K. D. BROWN et al. 1987).

In SCLC, bombesin congeners evoked PI hydrolysis with increases in $I(1,4,5)P_3$ to a maximal extent within 10 s. This was followed by a decline in this metabolite by 30 s after stimulation, at which time $I(1,3,4)P_3$ levels increased. These cells also had a substantial accumulation of an inositol tetrakisphosphate (IP_4) species which did not increase after short periods of stimulation, but was increased by long exposures to the agonist. Both the short-term increases in $I(1,4,5)P_3$ and the long-term accumulation of IP and IP_4 were sensitive to inhibition by TPA (TREPEL et al. 1988a).

Thus, two types of responses involving PI metabolites may be distinguished in bombesin-responsive cells. There is a very early increase in metabolites attributable to activation of a phospholipase C with PIP_2 as substrate. Over longer periods, there is an increase in inositol monophosphates attributable either to catabolism of the higher-order inositol phosphomonoesters or to action of a less tightly receptor-coupled phospholipase C on PI itself, to account for inositol 1-monophosphate (IP_1), or phosphatidylinositol-4-phosphate (PIP), to account for inositol-1,4-bisphosphate ($I(1,4)P_2$). Whether all PI metabolites derive from one pathway, through agonist-induced PIP_2 hydrolysis using a receptor-coupled phospholipase C, or whether other phospholipases-C or other lipid precursors participate in this response, is currently not clear, particularly at late times after hormone stimulation.

The early hydrolysis of PIP_2 metabolites correlated in secretory tissues with secretion of the target hormone. In pituitary GH_4C_1 cells, bombesin (which stimulates prolactin secretion by 4 s) and TRH increased IP_3 within 2 s of stimulation, and IP_1 did not increase until later (BJORO et al. 1987; PACHLER et al. 1988). In insulinoma cells, bombesin increased IP_3 (and IP_2) prior to the burst of hormone secretion evoked in this cell line, whereas stimulation of IP_1

was evident after the secretory burst but persisted for a longer period. In addition, production of diacylglycerol occurred in synchrony with the burst of insulin release (SWOPE and SCHONBRUNN 1988).

Bombesin also stimulated uptake of ^{32}P into phosphatidylcholine (PC), as well as release of labeled choline and phosphocholine in Swiss 3T3 cells. Pretreatment for prolonged periods with TPA abrogated this effect (MUIR and MURRAY 1987). Whether this results from the action of a receptor-coupled PC-specific phospholipase C or phospholipase D remains to be established, but the participation of PC in bombesin-evoked responses could provide another source of diacylglycerol, proposed to be an important second messenger in agonist action (BERRIDGE 1987).

V. Guanine Nucleotide-Binding Protein/Bombesin Receptor Interaction

Guanine nucleotide-binding proteins (G proteins) were originally recognized as important in mediating receptor-effector coupling in studies of the adenylate cyclase and retinal cyclic GMP phosphodiesterase systems (GILMAN 1987; CASEY and GILMAN 1988). Recently, interest has been intense in two other areas of potential involvement, including the coupling of receptors to phospholipase activation (FAIN et al. 1988) and modulation of ion channel activity (BROWN and BIRNBAUMER 1988). Cholera and pertussis toxins have been used as probes of potential G-protein involvement, as these bacterial toxins influence activity of known G proteins (GILMAN 1987).

Great interest was stimulated by the observation that bombesin-evoked increases in c-*myc* and DNA synthesis in Swiss 3T3 cells were sensitive to low concentrations (5 ng/ml) and brief periods of treatment with pertussis toxin (LETTERIO et al. 1986), without effect on the mitogenic responses observed with PDGF or phorbol dibutyrate. This raised the possibility that a G-protein substrate for pertussis toxin action could be responsible for transducing the bombesin-evoked mitogenic signal. However, subsequent experiments revealed that, while pertussis toxin did block bombesin-induced mitogenesis, pertussis toxin did not inhibit bombesin binding, PI turnover, or Ca^{2+} mobilization evoked by bombesin, or activation of protein kinase C in bombesin-treated cells. Moreover, the inhibitory effect of pertussis toxin on the mitogenic response to bombesin was reversed by treatment with insulin (ZACHARY et al. 1987; TAYLOR et al. 1988).

In GH_4C_1 rat pituitary cells, rat insulinoma cells, Swiss 3T3 cells, and rat brain tissue, nonhydrolyzable GTP analogs inhibited agonist binding to membrane fraction bombesin receptors, although the magnitude of this effect varied in the three cell types. In a bombesin receptor with a remarkably low K_d of 24 p*M*, guanine nucleotides decreased receptor affinity without decreasing the receptor number, and increased the rate of ligand dissociation. Pertussis toxin at 100 ng/ml did not affect binding of bombesin to Swiss 3T3 membranes or inhibit rapid increases in Ca^{2+} produced by bombesin. Cholera toxin pretreatment (500 ng/ml for 18–20 h) did not affect the ability of bombesin agonists to bind to their receptor and did not influence the inhibitory effect of nonhydrolyzable guanine nucleotide analogs on bombesin agonist binding (FISCHER and SCHONBRUNN 1988). Although cholera toxin pretreatment did not affect binding of bombesin to GH_4C_1, insulinoma, or Swiss 3T3 membranes, bombesin-induced increases in Ca^{2+} and PI turnover were inhibited by prolonged cholera toxin pretreatment in SCLC (TREPEL et al. 1988a). This effect correlated with the loss of susceptibility of G_s in cholera toxin-treated cells to ADP ribosylation by exogenous toxin after preparation of membrane fractions.

Taken together, there is evidence for a G protein which influences bombesin binding in several systems but which is itself evidently not subject to inhibition by pertussis toxin. While there is no evidence for an effect of cholera toxin on this putative G protein in a way that would affect bombesin binding, it is possible that cholera toxin in SCLC may affect events in bombesin-evoked signal transduction not directly coupled to effects on agonist binding.

Evidence linking bombesin-induced signal transduction to a *ras* gene product was provided by experiments in which NIH3T3 cells transfected with a mouse mammary tumor virus promoter (MMTV; inducible by steroids)-directed N-*ras* protooncogene (i.e., germ line N-*ras*) demonstrated responsiveness to bombesin agonists (WAKELAM et al. 1986). This responsiveness was manifest by increases in PI turnover and thymidine incorporation subsequent to addition of bombesin congeners. N-*ras* also increased responsiveness to bradykinin, but not EGF. It was proposed that the N-*ras* protooncogene was in fact the putative coupling factor between the receptor and the phospholipase. Subsequent experiments have revealed that the effect of bombesin to increase PI turnover in N-*ras*-expressing cells is apparent only at low-density culture conditions, and that this effect may relate to desensitization at high cell density of the bombesin-evoked response by the presence of other as yet undefined autocrine growth factors (WAKELAM 1988).

In contrast to these results obtained with the germline N-*ras* protooncogene, Ca^{2+} mobilization and PI turnover responses to bombesin were lost in activated-*ras* transformed NIH3T3 cells (ALONSO et al. 1988), as would be expected if transformation were associated with uncoupling of a growth-factor-specific control. However, this effect extended to a number of membrane-localized oncogenes, including *src, mos, raf, met,* and *trk.*

Thus, at this time it is difficult to link by a specific mechanism a *ras* gene product to the putative bombesin receptor-related G protein. In part this may extend from the lack of reagents directed at components of the receptor apparatus that are independent of bombesin binding. Also, inferences have been made from biologic experiments such as the effect of pertussis toxin on bombesin-induced mitogenesis which may relate more to the biologic features of toxin interaction with specific cell types than with constant components of the bombesin receptor mechanism present in different cells. Nonetheless, it is clear that definition of the molecular basis for the effect of guanine nucleotides on bombesin binding will be of great import in completely characterizing the effect of this hormone on its effector cells.

VI. Ion Fluxes

Bombesin caused efflux of $^{45}Ca^{2+}$ from prelabeled pancreatic islets (JENSEN et al. 1978; DESCHODT-LANCKMAN et al. 1976; MAY et al. 1978). This was associated with transient increases in cyclic GMP, but not short-term increases in cyclic AMP (DORFLINGER and SCHONBRUNN 1983). However, long-term treatment with bombesin of some cells, notably GH_4C_1 cells, did show an increase in cyclic AMP (BJORO et al. 1987). More recent studies in Swiss 3T3 cells have confirmed that bombesin increases Ca^{2+} efflux (MENDOZA et al. 1986) and in addition documented effects of bombesin on other ionic constituents of these cells. Specifically, there was stimulation of the ouabain-sensitive Rb^+ uptake, as well as an increase in ^{22}Na uptake which correlated with an increase in the intracellular pH. These data were interpreted to indicate that bombesin treatment of these cells resulted in activation of the Na^+/H^+ exchanger. Bombesin stimulation of Ca^{2+} efflux occurred in Na^+-free medium. In Swiss 3T3 cells, bombesin stimulated not only an increase in $^{45}Ca^{2+}$ efflux, but also increased the influx of $^{45}Ca^{2+}$ as well as influx of 3′-*O*-methyl glucose. These changes correlated with an increase in PIP_2 hydrolysis and a sustained increase in 1,2-diacylglycerol levels (TAKUWA et al. 1987).

The mechanism of bombesin-stimulated effects on Ca^{2+} is under active investigation. Chelation of external Ca^{2+} by EGTA had no effect on the increase in intracellular Ca^{2+} concentration ($[Ca^{2+}]_i$) in Swiss 3T3 cells and in one report which studied SCLC (MENDOZA et al. 1986; MOODY et al. 1988). Thus, these data imply a significant portion of the increased $[Ca^{2+}]_i$ derived from internal stores. A series of studies in SCLC (HEIKKILA et al. 1987; SAUSVILLE et al. 1988) confirmed that while a substantial proportion of the increase in $[Ca^{2+}]_i$ did derive from internal stores, there was a consistent effect of EGTA to reduce the magnitude of the increase in $[Ca^{2+}]_i$ in normal, depolarizing, or Na^+-depleted medium, and these data were interpreted to indicate that a portion of the increased $[Ca^{2+}]_i$ after bombesin treatment may derive also from extracellular sources. In addition, the latter studies demonstrated that the increase in $[Ca^{2+}]_i$ observed after bombesin treatment was Na^+ independent and was not affected by depolarization of the cells with high K^+ or gramicidin.

In insulinoma cells treated with bombesin, there is a very rapid increase in $[Ca^{2+}]_i$, the peak concentration of which is unaffected by coincubation with EGTA. There is an EGTA-sensitive component of the time-dependent increase in $[Ca^{2+}]_i$ in those cells. This may reflect depletion of internal stores by EGTA (SWOPE and SCHONBRUNN 1988), but may also reflect the prevention of bombesin-induced Ca^{2+} influx occurring at a later time after ligand addition than is the maximal time of release of Ca^{2+} from internal stores.

In Swiss 3T3 cells, the release of Ca^{2+} from internal stores occurs almost instantaneously after ligand addition, and is a feature of bombesin's action in Swiss 3T3 cells which is shared with vasopressin. This is in contrast to the action of the platelet-derived growth factor in these cells, where there is a considerably greater lag period observed prior to the increase in $[Ca^{2+}]_i$ (LOPEZ-RIVAS et al. 1987).

The bombesin-dependent hydrolysis of PIP_2 results in an increase in diacylglycerol levels which increases the activity of protein kinase C in some cell types. In bovine aortic cells, activated protein kinase C has been suggested to lead ultimately to a decrease in calcium channel activity. Bombesin does promote a downmodulation of Ca^{2+}-channel activity in these cells (GALIZZI et al. 1987). Concurrent studies have also demonstrated that protein kinase C activation with TPA decreased bombesin-mediated $^{45}Ca^{2+}$ efflux in Swiss 3T3 cells (K. D. BROWN et al. 1987).

Thus a great many studies agree that bombesin is a powerful stimulus to increase $[Ca^{2+}]_i$ in a variety of cell types. What is not clear is the source of the increased Ca^{2+}. This issue is of some interest because of the recent description (MELDOLESI and POZZAN 1987) of ligand-coupled or ligand-operated Ca^{2+} channels whose regulation may be an important aspect of signal transduction.

VII. Protein Phosphorylation

Initial efforts to examine activation of protein kinases by bombesin congeners focused on the activation of protein kinase C. Using the state of phosphorylation of an 80-kDa protein which had previously been shown to be a substrate for a protein kinase activated by phorbol esters, there was clear evidence in Swiss 3T3 cells of bombesin-stimulated 80-kDa protein phosphorylation (ZACHARY et al. 1986). The phosphorylation state of this protein is very labile, with a half-time for dephosphorylation of this species of less than 2 min after removal of bombesin (RODRIGUEZ-PENA et al. 1986).

Subsequent experiments have additionally examined the occurrence of tyrosine phosphorylation. In contrast to PDGF, which caused an increase in total phosphotyrosine in Swiss 3T3 cells, bombesin and GRP did not significantly increase total phosphotyrosine levels in treated Swiss 3T3 cells (ISACKE et al. 1986). Using a specific antiphosphotyrosine antibody, however, it was possible to demonstrate the early appearance of a 115-kDa protein phosphorylated on tyrosine in Swiss 3T3 cells shortly after bombesin treatment (CIRILLO et al. 1986). This species appeared after less than 2 min of treatment and was maximal after 30 min. The 115-kDa protein could be solubilized from the membrane and immunoprecipitated, and then was observed to incorporate additional phosphate from γ-[^{32}P]-ATP in a bombesin-dependent fashion. ^{125}I-labeled bombesin congeners were specifically immunoprecipitated with the 115-kDa species by antiphosphotyrosine antibodies.

Further studies have examined the 115-kDa species in SCLC (GAUDINO et al. 1988). This species was observed in the basal, non-bombesin-stimulated state in the "classic" bombesin-producing cell line NCI-H128, but not in a "variant" non-bombesin-producing cell line. In accord with this observation, extracts of NCI-H128 were immunoprecipitated with antiphosphotyrosine antibodies, and both 115-kDa and 70-kDa proteins were observed to be substrates for a phosphorylation reaction that again was not stimulated by exogenous bombesin. Bombesin-Sepharose also specifically bound a protein of

115 kDa which is phosphorylated on tyrosine from detergent extracts of NCI-H128.

Taken together, these results suggest that bombesin congeners in both SCLC and Swiss 3T3 cells do not cause large increases in intracellular phosphotyrosine. However, there is clearly evidence for the specific phosphorylation of a limited number of species, one of which, the 115-kDa protein, has features suggesting that it is intimately associated with if not indeed an actual part of a bombesin-receptor complex. This implies therefore the existence of at least two components for such a complex: one a 75- to 85-kDa species which is heavily glycosylated to account for the results of receptor affinity crosslinking experiments described above, as well as the 115-kDa species. However, it again must be emphasized that until specific probes are available for these elements the nature of the bombesin receptor remains largely conjectural.

VIII. Bombesin Receptor Antagonists

Due to the extensive pharmacologic effects of bombesin congeners, potent and specific antagonists are of general interest. A goal motivate their development would be to modulate hormonal secretion and tumor cell growth. A serendipitous observation which was characterized in detail was that the substance P analog [D-Arg1,D-Pro2,D-Trp7,9,Leu11]-substance P acted as an effective competitive antagonist for bombesin-evoked amylase secretion with a calculated K_d of 2.9 μM and IC_{50} for amylase release of 7.2 μM (JENSEN et al. 1984). Interestingly, this antagonist was found also to inhibit vasopressin-induced DNA synthesis in Swiss 3T3 cells (CORPS et al. 1985).

[D-Arg1,D-Pro2,D-Trp7,9,Leu11]-substance P was then used in a variety of studies to demonstrate that the response in question was bombesin specific. For example, this antagonist inhibited bombesin-induced 80-kDa protein phosphorylation (ZACHARY et al. 1986), phosphatidylcholine turnover (MUIR and MURRAY 1987), c-*fos* and c-*myc* transcription (ROZENGURT and SINNETT-SMITH 1987), as well as bombesin- or GRP-induced but not cholera toxin-, prostaglandin-, or 8-Br-cAMP-induced mitogenesis in Swiss 3T3 cells (ZACHARY and ROZENGURT 1985a). Other substance P analogs have been found to have anti-bombesin-receptor activity. [D-Arg1,D-Phe5,D-Trp7,9,Leu11]substance P was recently observed to be fivefold more potent than [D-Arg1,D-Pro2,D-Trp7,9,Leu11]substance P in inhibition of ^{125}I-GRP binding, reduction of affinity crosslinking of the 75- to 85-kDa cell-surface-associated GRP-binding protein species, and in reducing the increase in $[Ca^{2+}]_i$ evoked in these cells by bombesin. [D-Arg1,D-Trp7,9,Leu11]substance P (spantide) showed no antagonist activity in these assays. Both [D-Arg1,D-Pro2,D-Trp7,9,Leu11]- and [D-Arg1,D-Phe5,D-Trp7,9,Leu11]-substance P inhibited the growth of SCLC, although doses of 150 μM were required (WOLL and ROZENGURT 1988). The specificity of this effect for a bombesin-mediated pathway has been challenged (LAYTON et al. 1988). Moreover, both of these substance P antagonists acted to inhibit vasopressin- as well as bombesin-induced signals in Swiss 3T3 cells.

Bombesin-based structural analogs as potential bombesin receptor antagonists were initially explored with the development of [D-Phe12]-bombesin (HEINZ-ERIAN et al. 1972). This compound clearly inhibited bombesin-induced amylase release from pancreatic cells with an IC_{50} of 5 μM versus an agonist concentration of 0.2 nM. The calculated K_d was 4–5 μM, and therefore the antagonist showed 2000-fold lower affinity for receptor than bombesin itself. Recent developments in peptide synthesis technoloy have allowed the synthesis of the reduced peptide bond antagonist [Leu13-4-CH_2NH-Leu14]-bombesin. This compound demonstrated a 100-fold improvement in binding affinity as compared with prior bombesin receptor antagonists, inhibiting the bombesin-induced proliferation of Swiss 3T3 cells with an IC_{50} of 18 nM, without any inhibition of substance P binding at concentrations up to 10 mM (COY et al. 1988). [Leu13-″-CH_2NH-Leu14]-bombesin has recently been shown to block binding of labeled GRP congeners to Swiss 3T3 cells (WOLL et al. 1988). Under conditions where the peptide can inhibit clonal growth of an SCLC cell line, it also inhibited ^{125}I-Tyr4-bombesin binding, I(1,4,5)P_3 generation, and increases in $[Ca^{2+}]_i$ (TREPEL et al. 1988b).

IX. Consequences of Bombesin-Evoked Second Messenger Production

1. Secretion

Bombesin is a potent pancreatic secretogogue, stimulating amylase release from the normal pancreas (JENSEN et al. 1978; DESCHODT-LANCKMAN et al. 1976; MAY et al. 1978), and insulin from insulinoma cells (SWOPE and SCHONBRUNN 1988). The mechanism of secretion in the latter case has been described as consisting of a "burst" phase associated with an increase in diacylglycerol followed by a sustained phase of insulin secretion. The response of the cells to added ionomycin plus active phorbols mimicked the bombesin-induced response, implying that, while acting as a secretogogue, bombesin utilizes both protein kinase C-related and Ca^{2+}-related mechanisms.

Pituitary preparations of various types are responsive to bombesin with elaboration of anterior pituitary hormones. Secretion of both prolactin and growth hormone is stimulated in steroid-primed rats administered intravenous bombesin (WESTENDORF and SCHONBRUNN 1982). Interestingly, estrogen- or insulin-primed cells responded better to bombesin than nonprimed cells. Bombesin increased both vasoactive-intestinal polypeptide and epidermal growth factor-induced secretion, but not that due to thyrotropin-releasing hormone. This suggests that both bombesin and thyrotropin-releasing hormone may stimulate prolactin release through a common mechanism. In GH_4C_1 cells, the effective concentration for release of 50% of maximal prolactin response is approximately 2 nM (BJORO et al. 1987).

2. Receptor Transmodulation

In Swiss 3T3 cells, bombesin changed the epidermal growth factor receptor from a high-affinity to a low-affinity state, without effect on the maximal amount of EGF binding. This occurred at 37° but not at 4° C (BROWN et al. 1984; KORC et al. 1984). Both PDGF and vasopressin also caused receptor transmodulation. All three of these agonists had in common an ability to increase protein kinase C activity, and it was therefore proposed that this effect on EGF binding was mediated indirectly through protein kinase C. Subsequent experiments are in accord with this point of view (ZACHARY et al. 1986; ZACHARY and ROZENGURT 1985b), as there is a prompt decline in EGF affinity after addition of active phorbol esters, with only a delayed and attenuated decrement in EGF affinity caused by the addition of calcium ionophores.

3. Protooncogene Expression

A bombesin-specific early increase of steady-state c-*fos* mRNA (induced by 60, declining by 90 min) followed temporally by an increase in c-*myc* mRNA (persistent through 120 min) was observed in Swiss 3T3 cells to be distinctive from Ki-*ras* or Ha-*ras,* whose steady-state levels did not change after bombesin stimulation (ROZENGURT and SINNETT-SMITH 1987; PALUMBO et al. 1986; BRAVO et al. 1987). Of interest, in the presence of insulin, bombesin stimulated DNA synthesis at subnanomolar concentrations, but had only a small effect on c-*fos* and c-*myc* levels. This result implies that there can be dissociation between protooncogene induction and DNA synthesis. This is supported by the observation that the induction of c-*fos* and c-*myc* by bombesin is not augmented by insulin cotreatment, whereas the stimulation of DNA synthesis is synergistically affected (BRAVO et al. 1987). Also, downregulation of protein kinase C by long exposure of the cells to phorbol esters resulted in decreased c-*fos* and c-*myc* induction by bombesin (ROZENGURT and SINNETT-SMITH 1987; PALUMBO et al. 1986; BRAVO et al. 1987). Calcium ionophore had by itself a minimal effect on the expression of these protooncogenes, but markedly potentiated the effect of phorbol dibutyrate to increase c-*fos* and c-*myc* (ROZENGURT and SINNETT-SMITH 1987). The induction of c-*fos* and c-*myc* by EGF, PDGF, and serum was independent of depletion of protein kinase C (BRAVO et al. 1987; MCCAFFREY et al. 1987).

Thus it is clear that a very early response to bombesin in the Swiss 3T3 system is induction of c-*fos* and c-*myc* through a pathway which is responsive to alterations of both protein kinase C and Ca^{2+}, although there appears to be synergy in the action of Ca^{2+} and activated protein kinase C. In SCLC, a correlation exists between biochemical responsiveness of cell cultures and constitutive expression of GRP and *myc*-gene family members. Specifically, bombesin-responsive cell lines in general tend to express both GRP itself and L-*myc,* without constitutive high-level expression of c-*myc* (SAUSVILLE et al. 1988). These correspond in a general way to the "classic" SCLC cell lines with typical neuroendocrine features (CARNEY et al. 1985). In contrast, bombesin-nonresponsive cells tend to have low levels of endogenous GRP and prominent c-*myc* expression, corresponding to "variant" SCLCs which have lost many typical neuroendocrine features. Thus it is possible that in SCLC the response to bombesin-like peptides may reflect the pattern of steady-state *myc*-family gene expression.

4. DNA Synthesis

In Swiss 3T3 cells, insulin clearly potentiates the effect of bombesin to increase [^{3}H]thymidine incorporation. Insulin alone is without effect, whereas, importantly, bombesin by itself does provoke DNA synthesis and is in that sense a complete mitogen, although it is greatly potentiated by insulin (HESLOP et al. 1986; ROZENGURT and SINNETT-SMITH 1983). Vasopressin and phorbol esters did not enhance the bombesin-related effect, implying that a similar pathway may be activated by these agents. However, in contrast to the effect on DNA synthesis of bombesin and insulin, low concentrations of serum were used in the demonstration of effects of bombesin and insulin to increase actual cell number. This raises the possibility that the major effect of bombesin and insulin may be specifically related to entry into S-phase, rather than complete progression through the cell cycle. The structure-activity relationship of bombesin-binding inhibitors corresponded with the structure-activity relationship for bombesin-stimulated DNA synthesis (ZACHARY and ROZENGURT 1985a).

An open question is the basis for the synergy of bombesin or GRP and insulin in causing Swiss 3T3 DNA synthesis. An important result is the observation that pertussis-toxin inhibition of bombesin-induced DNA synthesis is reversed by insulin (ZACHARY et al. 1987) even while bombesin-stimulated increases in $[Ca^{2+}]_i$ and phospholipase activity are not affected by pertussis toxin. Thus, the basis for insulin-mediated increases in bombesin-mediated DNA synthesis could be a process sensitive to pertussis toxin yet distinct from the early bombesin-mediated events at the cell membrane.

Further evidence that early mobilization of Ca^{2+} could be separated from events leading to DNA synthesis was provided by the experiments of HENDLEY and MAMRACK (1988), who observed in normal human fibroblasts that both fetal bovine serum and thrombin stimulated PI turnover, DNA synthesis, and cytoplasmic alkalinization. In contrast bombesin did stimulate PI turnover but did not affect either cytoplasmic pH or induce DNA synthesis. In this regard, MATUOKA et al. (1988) demonstrated that an antibody to phosphatidylinositol bisphosphate (PIP_2) could inhibit mitogenesis by bombesin (20%–30% decrease) or PDGF (50% decrease) but not EGF, fibroblast growth factor (FGF), insulin, or serum.

Therefore a major theme that will continue to be explored is to what extent early events evoked by bombesin such as calcium flux, PI turnover, tyrosine phosphorylation, and effects on membrane activities such as the Na^+/H^+ transporter are necessary or sufficient for DNA synthesis to occur. It is entirely possible that the most relevant signal transductions in causing DNA synthesis remain to be defined.

In SCLC, bombesin congeners have been reported to increase [^{3}H]thymidine incorporation and cell number (WEBER et al. 1985), although the generality of this finding with respect to SCLC grown in liquid culture has been questioned (LAYTON et al. 1988). However, it is important to remember that, since SCLCs in many cases actually produce the peptide as well, defining a basal state with which to compare stimulated DNA synthesis or growth could be difficult. Furthermore, it is entirely possible that an important role for bombesin peptides in SCLC could be the secretogogue function of the peptide, as opposed to direct effects on DNA synthesis per se. These effects may require a matrix-dependent assay system for efficient detection.

F. Conclusions

Bombesin and its mammalian homolog GRP are an evolutionarily conserved mechanism for transmitting various signals between a bombesin-releasing cell and receptor-bearing effector cells. This mechanism is used to modulate neurotransmission, stimulate secretion of other biologically active peptides, and promote cell growth. The consequences of receptor-ligand interaction in different types of target cells varies, despite striking similarities in the molecular mechanisms of signal transduction. The ligand-dependant response elicited depends to a large extent on pre-established properties of the ligand-binding cell. It is therefore a context-dependent response, whose nature depends on the milieu of the stimulating and responding cell. At this time, the molecular mechanisms allowing the target cell to choose among the various responses to bombesin remain obscure. Further definition of this process is critical to a more complete understanding of the function of these peptides in physiologic and pathologic contexts.

Acknowledgements. We would like to thank Gena Parris for her expert assistance in the preparation of this manuscript, and John Minna for his encouragement and support.

Appendix

References for Table 1

1 (dog), BERTACCINI et al. (1973, 1974a); HIRSCHOWITZ and GIBSON (1978); MIYATA et al. (1980); MODLIN et al. (1981a); MATERIA et al. (1981); HIRSCHOWITZ and MOLINA (1983); LAMBERT et al. (1984); BUNNETT et al. (1985); *2* (man), DELLE FAVE et al. (1980); VARNER et al. (1981); WALSH et al. (1981); KNIGGE et al. (1984); *3* (cat), VAGNE et al. (1982, 1987); *4* (rat), TACHÉ et al. (1980b, 1981); TACHÉ and COLLU (1982); DUBRASQUET et al. (1982); GUGLIETTA et al. (1985); *5* (dog), PAPPAS et al. (1985); LENZ et al. (1986); *6* (rat), TACHÉ (1982); *7* (dog), TAYLOR et al. (1979); MODLIN et al. (1980b, 1981b); MCDONALD et al. (1981, 1983); SINGER et al. (1981); VAYSSE et al. (1981); INOUE et al. (1983); KLEIBEUKER et al. (1985); MUKAI et al. (1987); *8* (man), BASSO et al.

(1975b, 1976); GHATEI et al. (1982a); FLETCHER et al. (1983); WOOD et al. (1983); ANNIBALE et al. (1985); DE JONG et al. (1987a); *9* (rat), GREELEY and THOMPSON (1984); NAMBA et al. (1984); GREELEY and THOMPSON (1984); NAMBA et al. (1984); GREELEY et al. (1986b); *10* (calf), BLOOM et al. (1983); *11* (rat), CHIBA et al. (1980); DU VAL et al. (1981); MARTINDALE et al. (1982a); RICHELSEN et al. (1983); GUO et al. (1987); *12* (pig), HOLST et al. (1987a, b); *13* (dog), GIRAUD et al. (1987); SUGANO et al. (1987); *14* (dog), SCHUSDZIARRA et al. (1980); DE GRAEF and WOUSSEN-COLLE (1985); *15* (dog), MCDONALD et al. (1981, 1983); *16* (rat), GREELEY et al. (1986b); *17* (dog), ERSPAMER et al. (1974); KONTUREK et al. (1976); MIYATA et al. (1980); INOUE et al. (1983); DE JONG et al. (1987b); *18* (man), GHATEI et al. (1982a); JANSEN and LAMERS (1983); DE JONG et al. (1987a); *19* (rat), NAMBA et al. (1984); *20* (dog), MATSUYAMA et al. (1980); MCDONALD et al. (1981, 1983); *21* (man), GHATEI et al. (1982a); BRUZZONE et al. (1983); KNIGGE et al. (1984); *22* (dog), POITRAS et al. (1983); *23* (man), GHATEI et al. (1982a); *24* (man), GHATEI et al. (1982a); WOOD et al. (1983); *25* (rat), ROKAEUS et al. (1982); *26* (man), FLETCHER et al. (1983); KNIGGE et al. (1984); *27* (man), BASSO et al. (1975a); *28* (rat), DUBRASQUET et al. (1982); *29* (pig), KNUHTSEN et al. (1985); *30* (rat, mouse), DESCHODT-LANKMAN et al. (1976); IWATSUKI and PETERSEN (1978); IWAMOTO et al. (1983); HOWARD et al. (1985); LOGSDON et al. (1987); *31* (guinea pig), JENSEN et al. (1978); MAY et al. (1978); UHLEMANN et al. (1979); LEE et al. (1980); *32* (dog), TAYLOR et al. (1979); MODLIN et al. (1980a, 1981a); MCDONALD et al. (1981, 1983); MATERIA et al. (1981); SINGER et al. (1981); INOUE et al. (1983); *33* (man), DE MAGISTRIS et al. (1981); GHATEI et al. (1982a); FLETCHER et al. (1983); KNIGGE et al. (1984); DE JONG et al. (1987a); *34* (pig), KNUHTSEN et al. (1987b); *35* (dog), ADRIAN et al. (1978); *36* (dog), KANETO et al. (1978); SCHUSDZIARRA et al. (1980); MATSUYAMA et al. (1980); VAYSSE et al. (1981); MCDONALD et al. (1981, 1983); *37* (man), FALLUCCA et al. (1977); GHATEI et al. (1982a); WOOD et al. (1983); BRUZZONE et al. (1983); KNIGGE et al. (1984); *38* (mouse and rat), PETTERSSON and AHRÉN (1987, 1988); *39* (rat), BROWN et al. (1977c, 1979); *40* (dog), BROWN (1983); *41* (dog), IPP and UNGER (1979); HERMANSEN (1980); *42* (dog), MARTINDALE et al. (1982b); *43* (pig), KNUHTSEN et al. (1987b); *44* (hamster), SWOPE and SCHONBRUNN (1984); *45* (rat), TAMINATO et al. (1978); *46* (dog and rat), THOMAS and SANDER (1985); SANDER and PORTER (1988); *47* (rat) RIVIER et al. (1978); *48* (man), PONTIROLI et al. (1980); *49* (rat), KARASHIMA et al. (1984); *50* (man), MORLEY et al. (1980); GHATEI et al. (1982a); *51* (rat), TACHÉ et al. (1979a); MATSUSHITA et al. (1983); *52* (rat), WESTENDORF and SCHONBRUNN (1982); KENTROTI and MCCANN (1985); BJORO et al. (1987); *53* (calf), BICKNELL and CHAPMAN (1983); *54* (rat), GÜLLNER et al. (1982); KARASHIMA et al. (1984); *55* (rat), ABE et al. (1981); KARASHIMA et al. (1984); KABAYAMA et al. (1984); KENTROTI and MCCANN (1985); KENTROTI et al. (1988a); *56* (rat), WESTENDORF and SCHONBRUNN (1982); *57* (rat), RIVIER et al. (1978); KABAYAMA et al. (1984); KENTROTI et al. (1988a); *58* (rat), GÜLLNER et al. (1982); MITSUMA et al. (1985); *59* (rat), GÜLLNER et al. (1982); *60* (rat), TACHÉ et al. (1979a); *61* (man), GHATEI et al. (1982a); *62* (man), PONTIROLI et al. (1980); MORLEY et al. (1980); *63* (dog), ERSPAMER et al. (1973).

References for Table 2

1 (man), BERTACCINI et al. (1974b); BERTACCINI and IMPICCIATORE (1975); *2* (dog), BERTACCINI and IMPICCIATORE (1975); *3* (rat), BERTACCINI and IMPICCIATORE (1975); *4* (dog), FOX and MCDONALD (1984); *5* (cat), VAGNE et al. (1982, 1987); *6* (rat), SPENCER and TALMAN (1987); *7* (dog), MAYER et al. (1982, 1986); *8* (rat), ENDEAN et al. (1975); GIRARD et al. (1984); *9* (guinea pig), KANTOH et al. (1985); *10* (dog), CAPRILLI et al. (1975); *11* (man), SCARPIGNATO et al. (1981); *12* (rat), SCARPIGNATO and BERTACCINI (1981); *13* (rat), PORRECA and BURKS (1983); *14* (rat), PORRECA and BURKS (1983); GMEREK et al. (1983); GMEREK and COWAN (1984); KOSLO et al. (1986); *15* (rat), FULGINITI et al. (1984); *16* (dog), HIRNING and BURKS (1984); *17* (rat, guinea pig), ERSPAMER et al. (1972a); *18* (cat), ERSPAMER et al. (1972a); ENDEAN et al. (1975); BROCCARDO et al.

(1975); Mazzanti et al. (1982); Erspamer et al. (1988); *19* (man), Micheletti et al. (1988); *20* (rat), Al-Saffar (1984); *21* (dog), Angel et al. (1984); *22* (rat), see *17; 23* (guinea pig), see *18;* Leander et al. (1984); *24* (guinea pig), see *17; 25* (guinea pig), Mazzanti et al. (1982); *26* (dog, monkey), Erspamer et al. (1972a); *27* (rat), Stjernquist et al. (1986); *28* (cat, guinea pig, hamster, rabbit), Erspamer et al. (1972a); *29* (guinea pig), Stjernquist et al. (1983); *30* (guinea pig), Impicciatore and Bertaccini (1973); *31* (varia), Erspamer et al. (1972b); *34* (rat), Fisher and Brown (1984); Fisher et al. (1985).

References for Table 3

1 (rat), Brown et al. (1977a); *2* (rat), Brown et al. (1977a, b, 1980, 1987); Rivier and Brown (1978); Loosen et al. (1978); Moody et al. (1978, 1982); Taché et al. (1979b, 1980a); Wunder et al. (1980); Francesconi and Mager (1981); Märki et al. (1981); Avery et al. (1981); Brown (1982); Lin and Lin (1986); *3* (mouse), Mason et al. (1980); *4* (rat), Wunder et al. (1980); *5* (rat), Pittman et al. (1980); *6* (rat), Wunder et al. (1980); Brown (1982); Lin and Lin (1986); *7* (rat), Francesconi and Mager (1981); Lin and Lin (1986); *8* (rat), Fisher and Brown (1984); *9* (rat), Brown et al. (1977c); *10* (rat), Brown et al. (1977c); Taché et al. (1980a); Francesconi and Mager (1981); Avery et al. (1981); Morley et al. (1982); Girard et al. (1983); Rasler (1983); Lin and Lin (1986); Chiu et al. (1987); *11* (mouse), Nemeroff et al. (1979); Mason et al. (1980); *12* (rat), Brown et al. (1977a); Pert et al. (1980); *13* (rabbit, pig), Lipton and Glynn (1980); Parrot and Baldwin (1982); *14* (rat), Pert et al. (1980); *15* (rat), Lin and Lin (1986); *16* (rat), Brown (1982); *17* (rat), Taché et al. (1980a); Francesconi and Mager (1981); Lin and Lin (1986); *18* (rat), Fisher and Brown (1984); Fisher et al. (1985); *19* (rat), Niewoehner et al. (1983); Hedner et al. (1985); *20* (cat), Holtman et al. (1983); *21* (rat), Hedner et al. (1985); *22* (rat), Brown et al. (1977c); Namba et al. (1984); *23* (rat), Brown et al. (1977c, 1979, 1980); Märki et al. (1981); Gunion et al. (1984); *24* (rat), Brown (1983); *25* (rat), Iguchi et al. (1984); *26* (rat), Avery and Calisher (1982); Babcock and Wunder (1984); *27* (rat), Calisher and Avery (1984).

References for Table 4

1 (rat), Gibbs et al. (1979, 1981); Morley and Levine (1981); Stein and Woods (1982); Kulkosky et al. (1982a, b); Stuckey et al. (1985); Wager-Srdar et al. (1986); Geary et al. (1986); *2* (mouse), McLaughlin and Baile (1981); Taylor and Garcia (1985); *3* (monkey), Woods et al. (1983); Figlewicz et al. (1985); *4* (wolf pups), Morley et al. (1986); *5* (hamster), Bartness et al. (1986); *6* (rat), Gibbs et al. (1981); Morley and Levine (1981); Avery and Calisher (1982); Kulkosky et al. (1982a, b); *7* (mice), McLaughlin and Baile (1981); *8* (pig), Parrott and Baldwin (1982); *9* (monkey), Figlewicz et al. (1986); *10* (rat), Calisher and Avery (1984); *11* (rat), Stuckey and Gibbs (1982); *12* (rat), Kyrkouli et al. (1987); *13* (rat), de Beaurepaire and Suaudeau (1988); *14* (rat), Gmerek and Cowan (1983); Kulkosky et al. (1982b); *15* (rat), Brown et al. (1977a); Gibbs et al. (1981); Meisenberg (1982); Kulkosky et al. (1982a, b); Girard et al. (1983); Gmerek and Cowan (1983); Merali et al. (1983); Crawley and Moody (1983); Schulz et al. (1984); Rasler (1984); Cowan et al. (1985); van Wimersma Greidanus et al. (1985a, b); Negri (1986); *16* (mouse), Katz (1980); Meisenberg (1982); Cowan et al. (1985); Meisenberg and Simmons (1986); *17* (guinea pig, rabbit, monkey), Cowan et al. (1985); *18* (rat), Gmerek and Cowan (1983); *19* (rat), Gmerek et al. (1983); *20* (mouse), O'Donohue et al. (1984); Bishop et al. (1986); *21* (mouse), Nemeroff et al. (1979); *22* (rat), Pert et al. (1980); *23* (rat), Pert et al. (1980); Merali et al. (1983); Schulz et al. (1984); *24* (rat), Hawkins and Avery (1983); *25* (rat), Schulz et al. (1984); *26* (rat), Schulz et al. (1984); Rasler (1984); *27* (mouse), Flood and Morley (1988).

References for Table 5

1, Taché et al. (1980b); *2,* Taché and Collu (1982); *3,* Taché (1982); *4,* Gmerek and Cowan (1984); *5,* Porreca and Burks (1983); *6,* Brown et al. (1977c); *7,* Rasler (1983); *8,* Fisher et al. (1985); Fisher and Brown (1984); *9,* Hedner et al. (1985); *10,* Brown et al. (1979); *11,* Morley and Levine (1981); *12,* Gibbs et al. (1981); *13,* Hawkins and Avery (1983); *14,* Gmerek and Cowan (1983); *15,* Flood and Morley (1988).

References

Abe H, Chihara K, Minamitani N, Iwasaki J, Chiba T, Matsukura S, Fujita T (1981) Stimulation by bombesin of immunoreactive somatostatin release into rat hypophysial portal blood. Endocrinology 109:229–234

Adrian TE, Bloom SR, Hermansen K, Iversen J (1978) Pancreatic polypeptide, glucagon and insulin secretion from the isolated perfused canine pancreas. Diabetologia 14:413–417

Alonso T, Morgan, Marvizon JC, Zarbl H, Santos E (1988) Malignant transformation by *ras* and other oncogenes produces common alteration in inositol phospholipid signalling pathways. Proc Natl Acad Sci USA 85:4271–4275

Al-Saffar A (1984) Somatostatin inhibits bombesin-induced effects on migrating myoelectric complexes in the small intestine of the rat. Regul Pept 9:11–19

Anastasi A, Erspamer V, Bucci M (1971) Isolation and structure of bombesin and alytesin, two analogous active peptides from the skin of the European amphibians *Bombina* and *Alytes*. Experientia 27:166–167

Anastasi A, Erspamer V, Bucci M (1972) Isolation and amino acid sequences of alytesin and bombesin, two analogous active tetradecapeptides from the skin of European discoglossid frogs. Arch Biochem Biophys 148:443–446

Angel F, Go VLW, Szurszewski JH (1984) Innervation of the muscularis mucosae of canine proximal colon. J Physiol (Lond) 357:93–108

Annibale B, Corleto V, Severi C, de Magistris L, de Toma G, Delle Fave G (1985) Evidence that bombesin releases extragastric gastrin in man. Regul Pept 11:43–49

Avery DD, Calisher SB (1982) The effects of injections of bombesin into the cerebral ventricles on food intake and body temperature in food-deprived rats. Neuropharmacology 21:1059–1063

Avery DD, Hawkins MF, Wunder BA (1981) The effects of injections of bombesin into the cerebral ventricles on behavioral thermoregulation. Neuropharmacology 20:23–27

Babcock AM, Wunder BA (1984) Effects of bombesin on body temperature and oxygen consumption in food-deprived rats. Neuropharmacology 23:1357–1358

Bartness TJ, Morley JE, Levine AS (1986) Photoperiod-peptide interactions in the energy intake of Siberian hamsters. Peptide 7:1079–1085

Basso N, Giri S, Improta G, Lezoche E, Melchiorri P, Percoco M, Speranza V (1975a) External pancreatic secretion after bombesin infusion in man. Gut 16:994–998

Basso N, Lezoche E, Materia A, Giri S, Speranza V (1975b) Effect of bombesin on extragastric gastrin in man. Am J Dig Dis 20:923–927

Basso N, Giri S, Lezoche N, Materia A, Melchiorri P, Speranza V (1976) Effects of secretin, glucagon and duodenal acidification on bombesin-induced hypergastrinemia in man. Am J Gastroenterol 66:448–451

Battey JF, Lebacq-Verheyden AM, Krystal G, Markowitz SD, Sartor O, Way J (1988) Regulation of the expression of the human preprogastrin-releasing peptide gene and post-translational processing of its gene product. Ann NY Acad Sci 547:30–40

Berridge MJ (1987) Inositol trisphosphate and diacylglycerol: two interacting second messengers. Annu Rev Biochem 56:159–193

Bertaccini G, Impicciatore M (1975) Action of bombesin on the motility of the stomach. Naunyn Schmiedebergs Arch Pharmacol 289:149–156

Bertaccini G, Erspamer V, Impicciatore M (1973) The actions of bombesin on gastric secretion of the dog and the rat. Br J Pharmacol 49:437–444

Bertaccini G, Erspamer V, Melchiorri P, Sopranzi N (1974a) Gastrin release by bombesin in the dog. Br J Pharmacol 52:219–225

Bertaccini G, Impicciatore M, Molina E, Zappia L (1974b) Action of bombesin on human gastrointestinal motility. Rend Gastroenterol 6:45–51

Bicknell RJ, Chapman C (1983) Bombesin stimulates growth hormone secretion from cultured bovine pituitary cells. Neuroendocrinology 36:33–38

Bishop JF, Moody TW, O'Donohue TL (1986) Peptide transmitters of primary sensory neurons: similar actions of tachykinins and bombesin-like peptides. Peptides 7:835–842

Bjoro T, Torjesen PA, Ostberg BC, Sand O, Iversen JG, Gautvik KM, Haug E (1987) Bombesin stimulates prolactin secretion from cultured rat pituitary tumour cells (GH_4C_1) via activation of phospholipase C. Regul Pept 19:169–182

Bloom SR, Edwards AV, Ghatei MA (1983) Endocrine responses to exogenous bombesin and gastrin releasing peptide in conscious calves. J Physiol (Lond) 344:37–48

Bostwick DG, Roth KA, Evans CJ, Barchas JD, Bensch KG (1984) Gastrin-releasing peptide, a mammalian analog of bombesin, is present in human neuroendocrine lung tumors. Am J Pathol 117:195–200

Bravo R, MacDonald-Bravo H, Muller R, Hubsch D, Almendral JM (1987) Bombesin induced c-*fos* and c-*myc* expression in quiescent 3T3 cells. Exp Cell Res 170:103–115

Broccardo M, Erspamer GF, Melchiorri P, Negri L, de Castiglione R (1975) Relative potency of bombesin-like peptides. Br J Pharmacol 55:221–227

Brown AM, Birnbaumer L (1988) Direct G-protein gating of ion channels. Am J Physiol 254:H401–H410

Brown KD, Blay J, Irvine RF, Heslop JP, Berridge MJ (1984) Reduction of epidermal growth factor receptor affinity by heterologous ligands: evidence for a mechanism involving the breakdown of phosphoinositide and the activation of protein kinase C. Biochem Biophys Res Commun 123:377–384

Brown KD, Blakely DM, Hamon MH, Stuart-Laurie M, Corps AN (1987) Protein kinase C-mediated negative-feed back inhibition of unstimulated and bombesin-stimulated polyphosphoinositide hydrolysis in Swiss-mouse 3T3 cells. Biochem J 245:631–639

Brown MR (1982) Bombesin and somatostatin related peptides: effects on oxygen consumption. Brain Res 242:243–246

Brown MR (1983) Central nervous system sites of action of bombesin and somatostatin to influence plasma epinephrine levels. Brain Res 276:253–257

Brown MR, Rivier J, Vale W (1977a) Bombesin: potent effects on thermoregulation in the rat. Science 196:998–1000

Brown MR, Rivier J, Vale W (1977b) Actions of bombesin, thyrotropin releasing factor, prostaglandin E2 and naloxone on thermoregulation in the rat. Life Sci 20:1681–1688

Brown MR, Rivier J, Vale W (1977c) Bombesin affects the central nervous system to produce hyperglycemia in rats. Life Sci 21:1729–1734

Brown MR, Allen R, Villarreal J, Rivier J, Vale W (1978) Bombesin-like activity: radioimmunologic assessment in biological tissues. Life Sci 23:2721–2728

Brown MR, Taché Y, Fischer D (1979) Central nervous system action of bombesin: mechanism to induce hyperglycemia. Endocrinology 105:660–665

Brown MR, Märki W, Rivier J (1980) Is gastrin releasing peptide mammalian bombesin? Life Sci 27:125–128

Brown MR, Allen R, Fisher L (1987) Bombesin alters the sympathetic nervous system response to cold exposure. Brain Res 400:35–39

Bruzzone R, Tamburrano G, Lala A, Mauceri M, Annibale B, Severi C, de Magistris L, Leonetti F, Delle Fave G (1983) Effect of bombesin on plasma insulin, pancreatic glucagon and gut glucagon in man. J Clin Endocrinol 56:643–647

Buffa R, Solovieva I, Fiocca R, Giorgino S, Rindi G, Solcia E, Mochizuchi T, Yanaihara C, Yanaihara N (1982) Localization of bombesin and GRP (gastrin releasing peptide) sequences in gut nerves or endocrine cells. Histochemistry 76:457–467

Bunnett NW, Clark B, Debas HT, del Milton RC, Kovacs TOG, Orloff MS, Pappas TN, Reeve JR Jr, Rivier JE, Walsh JH (1985) Canine bombesin-like gastrin releasing peptides stimulate gastrin release and acid secretion in the dog. J Physiol (Lond) 365:121–130

Calisher SB, Avery DD (1984) Injections of bombesin into the substantia nigra produce hypothermia and hypophagia in food-reprived rats. Neuropharmacology 23:1201–1206

Caprilli R, Melchiorri P, Improta G, Vernia P, Frieri G (1975) Effects of bombesin and bombesin-like peptides on gastrointestinal myoelectric activity. Gastroenterology 68:1228–1235

Carney DN, Gazdar AF, Bepler G, Guccion JG, Marangos PJ, Moody TW, Zweig MH, Minna JD (1985) Establishment and identification of small cell lung cancer cell lines having classic and variant features. Cancer Res 45:2913–2923

Carney DN, Cuttitta F, Moody TW, Minna JD (1987) Selective stimulation of small cell lung cancer clonal growth by bombesin and gastrin-releasing peptide. Cancer Res 47:821–825

Casey PJ, Gilman AG (1988) G-protein involvement in receptor-effector coupling. J Biol Chem 263:2577–2580

Chejfec G, Lee I, Warren WH, Gould VE (1985) Bombesin in human neuroendocrine (NE) neoplasms. Peptides [Suppl 3]6:107–112

Chiba T, Taminato T, Kadowaki S, Inoue Y, Mori K, Seino Y, Abe H, Chihara K, Matsukura S, Fujita T, Goto Y (1980) Effects of various gastrointestinal peptides on gastric somatostatin release. Endocrinology 106:145–149

Chiu WT, Lin LS, Shih CJ, Lin MT (1987) Bombesin-induced hypothermia: possible involvement of cholinergic and dopaminergic receptors in the rat hypothalamus. Exp Neurol 95:368–377

Cirillo DM, Gaudino G, Naldini L, Comoglio PM (1986) Receptor for bombesin with associated tyrosine kinase activity. Mol Cell Biol 6:4641–4649

Corps AN, Rees LH, Brown KD (1985) A peptide that inhibits the mitogenic stimulation of Swiss 3T3 cells by bombesin or vasopressin. Biochem J 231:781–784

Costa M, Furness J, Yanaihara N, Yanaihara C, Moody TW (1984) Distribution and projections of neurons with immunoreactivity for both gastrin-releasing peptide and bombesin in the guinea pig small intestine. Cell Tissue Res 235:285–293

Cowan A, Khunawat P, Zu Zhu X, Gmerek DE (1985) Effects of bombesin on behavior. Life Sci 37:135–145

Coy DH, Heinz-Erian P, Jiang Y, Sasaki Y, Taylor J, Moreau J-P, Wolfrey WT, Gardner JD, Jensen RT (1988) Probing peptide backbone function in bombesin. A reduced peptide bond analog with potent and specific receptor activity. J Biol Chem 263:5056–5060

Crawley JN, Moody TW (1983) Anxiolytics block excessive grooming behavior induced by $ACTH_{1-24}$ and bombesin. Brain Res Bull 10:399–401

Cuttitta F, Carney DN, Mulshine J, Moody TW, Fedorko J, Fischler A, Minna JD (1985) Bombesin-like peptides can function as autocrine growth factors in human small-cell lung cancer. Nature 316:823–826

Cuttitta F, Fedorko J, Gu J, Lebacq-Verheyden AM, Linnoila RI, Battey JF (1988) Gastrin-releasing peptide gene-associated peptides are expressed in normal human fetal lung and small cell lung cancer: a novel peptide family found in man. J Clin Endocrinol Metab 67:576–583

Cutz E, Chan W, Track NS (1981) Bombesin, calcitonin and leu-enkephalin immunoreactivity in endocrine cells of human lung. Experientia 37:765–767

Cwikel BJ, Habener JF (1987) Provasopressin-neurophysin II processing is cell-specific in heterologous cell lines expressing a metallothionein-vasopressin fusion gene. J Biol Chem 262:14235–14240

Dalsgaard CJ, Hokfelt T, Schultzberg M, Lundberg JM, Terenius L, Dockray GJ, Goldstein M (1983) Origin of peptide-containing fibers in the inferior mesenteric ganglion of the guinea pig: immunohistochemical studies with antisera to substance P, enkephalin, vasoactive intestinal polypeptide, cholecystokinin and bombesin. Neuroscience 9:191–211

Damgé C, Hajri A, Lhoste E, Aprahamian M (1988) Comparative effect of chronic bombesin, gastrin-releasing peptide and caerulein on the rat pancreas. Regul Pept 20:141–150

Dayer AM, de Mey J, Will JA (1985) Localization of somatostatin-, bombesin- and serotonin-like immunoreactivity in the lung of the fetal rhesus monkey. Cell Tissue Res 239:621–625

De Beaurepaire R, Suaudeau C (1988) Anorectic effect of calcitonin, neurotensin and bombesin infused in the area of the rostral part of the nucleus of the tractus solitarius in the rat. Peptides 9:729–733

De Graef J, Woussen-Colle MC (1985) Effects of sham feeding, bethanechol, and bombesin on somatostatin release in dogs. Am J Physiol 248:G1–G7

De Jong AJL, Klamer M, Jansen JBMJ, Lamers CBHW (1987 a) Effect of atropine and somatostatin on bombesin-stimulated plasma gastrin, cholecystokinin and pancreatic polypeptide in man. Regul Pept 17:285–293

De Jong AJL, Singer MV, Lamers CBH (1987 b) Effect of pancreatic polypeptide antiserum on bombesin-stimulated pancreatic exocrine secretion in dogs. Peptides 8:973–976

Delle Fave G, Kohn A, de Magistris L, Mancuso M, Sparvoli C (1980) Effect of bombesin-stimulated gastrin on gastric acid secretion in man. Life Sci 27:993–999

De Magistris L, Delle Fave G, Kohn A, Schwartz TW (1981) Differential stimulation of pancreatic-polypeptide and gastrin secretion by bombesin in man. Life Sci 28:2617–2621

Deschodt-Lanckman M, Robberecht P, de Neef P, Lammens M, Christophe J (1976) In vitro action of bombesin and bombesin-like peptides on amylase secretion, calcium efflux, and adenylate cyclase activity in the rat pancreas. A comparison with other secretagogues. J Clin Invest 58:891–898

Dockray GJ, Vaillant C, Walsh JH (1979) The neuronal origin of bombesin-like immunoreactivity in the rat gastrointestinal tract. Neuroscience 4:1561–1568

Dorflinger LJ, Schonbrunn A (1983) Somatostatin inhibits vasoactive intestinal peptide stimulated cyclic adenosine monophosphate accumulation in GH pituitary cells. Endocrinology 113:1541–1550

Dubrasquet M, Roze C, Ling N, Florencio H (1982) Inhibition of gastric and pancreatic secretions by cerebroventricular injections of gastrin-releasing peptide and bombesin in rats. Regul Pept 3:105–112

Du Val JW, Saffouri B, Weir GC, Walsh JH, Arimura A, Makhlouf GM (1981) Stimulation of gastrin and somatostatin secretion from the isolated rat stomach by bombesin. Am J Physiol 241:G242–G247

Ekblad E, Ekman R, Håkanson R, Sundler F (1984) GRP neurones in the rat small intestine issue long anal projections. Regul Pept 9:279–287

El-Salhy M, Grimelius L, Wilander E, Abu-Sinna G, Lundqvist G (1981) Histological and immunohistochemical studies of the endocrine cells of the gastrointestinal mucosa of the toad (*Bufo regularis*). Histochemistry 71:53–65

Endean R, Erspamer V, Erspamer GF, Improta G, Melchiorri P, Negri L, Sopranzi N (1975) Parallel bioassay of bombesin and litorin, a bombesin-like peptide from the skin of *Litoria aurea*. Br J Pharmacol 55:213–219

Erisman MD, Linnoila RI, Hernandez O, DiAugustine RP, Lazarus LH (1982) Human lung small-cell carcinoma contains bombesin. Proc Natl Acad Sci USA 79:2379–2383

Erspamer V, Melchiorri P (1973) Active polypeptides of the amphibian skin and their synthetic analogues. Pure Appl Chem 35:463–494

Erspamer V, Erspamer GF, Inselvini M, Negri L (1972 a) Occurrence of bombesin and alytesin in extracts of the skin of three European discoglossid frogs and

pharmacological actions of bombesin on extravascular smooth muscle. Br J Pharmacol 45:333–348
Erspamer V, Melchiorri P, Sopranzi N (1972 b) The action of bombesin on the systemic arterial blood pressure of some experimental animals. Br J Pharmacol 45:442–450
Erspamer V, Melchiorri P, Sopranzi N (1973) The action of bombesin on the kidney of the anesthetized dog. Br J Pharmacol 48:438–455
Erspamer V, Improta G, Melchiorri P, Sopranzi N (1974) Evidence of cholecystokinin release by bombesin in the dog. Br J Pharmacol 52:227–232
Erspamer V, Erspamer GF, Melchiorri P, Negri L (1979) Occurrence and polymorphism of bombesin-like peptides in the gastrointestinal tract of birds and mammals. Gut 20:1047–1056
Erspamer GF, Severini C, Erspamer V, Melchiorri P, delle Fave G, Nakajima T (1988) Parallel bioassay of 27 bombesin-like peptides on 9 smooth muscle preparations. Structure-activity relationships and bombesin receptor subtypes. Regul Pept 21:1–11
Fain JN, Wallace MA, Wojcikiewicz RJH (1988) Evidence for involvement of guanine nucleotide binding regulatory proteins in the activation of phospholipases by hormones. FASEB J 2:2569–2574
Fallucca F, Delle Fave GF, Gambardella S, Mirabella C, de Magistris L, Carratu R (1977) Glucagon secretion induced by bombesin in man. Lancet 2:609–610
Ferri GL, Adrian TE, Ghatei MA, O'Shaughnessy DJ, Probert L, Lee YC, Buchan AMJ, Polak JM, Bloom SR (1983) Tissue localization and relative distribution of regulatory peptides in separated layers from the human bowel. Gastroenterology 84:777–786
Figlewicz DP, Stein LJ, Woods SC, Porte D Jr (1985) Acute and chronic gastrin-releasing peptide decreases food intake in baboons. Am J Physiol 248:R578–583
Figlewicz DP, Sipols A, Porte D Jr, Woods SC (1986) Intraventricular bombesin can decrease single meal size in the baboon. Brain Res Bull 17:535–537
Fischer JB, Schonbrunn A (1988) The bombesin receptor is coupled to a guanine nucleotide-binding protein which is insensitive to pertussis and cholera toxins. J Biol Chem 263:2808–2816
Fisher LA, Brown MR (1984) Bombesin-induced stimulation of cardiac parasympathetic innervation. Regul Pept 8:335–343
Fisher LA, Cave CR, Brown MR (1985) Central nervous system cardiovascular effects of bombesin in conscious rats. Am J Physiol 248:H425–H431
Fletcher DR, Shulkes A, Bladin PHD, Hardy KJ (1983) The effect of atropine on bombesin and gastrin releasing peptide stimulated gastrin, pancreatic polypeptide and neurotensin release in man. Regul Pept 7:31–40
Flood JF, Morley JE (1988) Effects of bombesin and gastrin-releasing peptide on memory processing. Brain Res 460:314–322
Fox JET, McDonald TJ (1984) Motor effects of gastrin releasing peptide (GRP) and bombesin in the canine stomach and small intestine. Life Sci 35:1667–1673
Francesconi R, Mager M (1981) Thermoregulatory effects of centrally administered bombesin, bradykinin and methionine-enkephalin. Brain Res Bull 7:63–68
Fulginiti JT, Porreca F, Burks TF (1984) Centrally administered bombesin stimulates intestinal motility. Proc West Pharmacol Soc 27:141–142
Fuxe K, Agnati LF, McDonald T, Locatelli V, Hokfelt T, Dalsgaard CJ, Battistini N, Yanaihara N, Mutt V, Cuello AG (1983) Immunohistochemical indications of gastrin-releasing peptide-bombesin-like immunoreactivity in the nervous system of the rat. Codistribution with substance P-like immunoreactive nerve terminal systems and coexistence with substance P-like immunoreactivity in dorsal root ganglion cell bodies. Neurosci Lett 37:17–22
Galizzi JP, Qar J, Fosset M, van Renteryhens C, Lazdunski M (1987) Regulation of calcium channels in aortic muscle cells by protein kinase C activators (diacylglycerol and phorbol esters) and by peptides (vasopressin and bombesin) that stimulate phosphoinositide breakdown. J Biol Chem 262:6947–6950

Gaudino G, Cirillo D, Naldini L, Rossino P, Comoglio PM (1988) Activation of the protein-tyrosine kinase associated with the bombesin receptor complex in small cell lung carcinomas. Proc Natl Acad Sci USA 85:2166–2170
Geary N, Smith GP, Gibbs J (1986) Pancreatic glucagon and bombesin inhibit meal size in ventromedial hypothalamus-lesioned rats. Regul Pept 15:261–268
Ghatei MA, Jung RT, Stevenson JC, Hillyard CJ, Adrian TE, Lee YC, Christofides ND, Sarson DL, Mashiter K, MacIntyre I, Bloom SR (1982a) Bombesin: action on gut hormones and calcium in man. J Clin Endocrinol Metab 54:980–985
Ghatei MA, Sheppard MN, O'Shaughnessy DJ, Adrian TE, McGregor GP, Polak JM, Bloom SR (1982b) Regulatory peptides in the mammalian respiratory tract. Endocrinology 111:1248–1254
Ghatei MA, Sheppard MN, Henzen-Logman S, Blank MA, Polak JM, Bloom SR (1983) Bombesin and vasoactive intestinal polypeptide in the developing lung: marked changes in acute respiratory distress syndrome. J Clin Endocrinol Metab 57:1226–1232
Ghatei MA, Bloom SR, Langevin H, McGregor GP, Lee YC, Adrian TE, O'Shaughnessy DJ, Blank MA, Uttenthal LO (1984a) Regional distribution of bombesin and seven other regulatory peptides in the human brain. Brain Res 293:101–109
Ghatei MA, George SK, Major JH, Carlei F, Polak JM, Bloom SR (1984b) Bombesin-like immunoreactivity in the pancreas of man and other mammalian species. Experientia 40:884–886
Ghatei MA, Gu J, Allen JM, Polak JM, Bloom SR (1985a) Bombesin-like immunoreactivity in female rat genito-urinary tract. Neurosci Lett 54:13–19
Ghatei MA, Springall DR, Nicholl CG, Polak JM, Bloom SR (1985b) Gastrin-releasing peptide-like immunoreactivity in medullary thyroid carcinoma. Am J Clin Pathol 84:581–586
Gibbs J, Fauser DJ, Rowe EA, Rolls BJ, Rolls ET, Maddison SP (1979) Bombesin suppresses feeding in rats. Nature 282:208–210
Gibbs J, Kulkosky PJ, Smith GP (1981) Effects of peripheral and central bombesin on feeding behavior of rats. Peptides [Suppl 2]2:179–183
Gillati M, Moody T (1984) The development of rat brain bombesin-like peptides and their receptors. Brain Res 15:286–289
Gilman AG (1987) G proteins: Transducers of receptor-generated signals. Annu Rev Biochem 56:615–649
Girard F, Aubé C, St-Pierre S, Jolicoeur FB (1983) Structure-activity studies on neurobehavioral effects of bombesin (BB) and gastrin-releasing peptide (GRP). Neuropeptides 3:443–452
Girard F, Bachelard H, St-Pierre S, Rioux F (1984) The contractile effect of bombesin, gastrin releasing peptide and various fragments in the rat stomach strip. Eur J Pharmacol 102:489–497
Giraud AS, Soll AH, Cuttitta F, Walsh JH (1987) Bombesin stimulation of gastrin release from canine gastrin cells in primary culture. Am J Physiol 252:G413–G420
Gmerek DE, Cowan A (1983) Studies on bombesin-induced grooming in rats. Peptides 4:907–913
Gmerek DE, Cowan A (1984) Pituitary-adrenal mediation of bombesin-induced inhibition of gastrointestinal transit in rats. Regul Pept 9:299–304
Gmerek DE, Cowan A, Vaught JL (1983) Intrathecal bombesin in rats: effects on behaviour and gastrointestinal transit. Eur J Pharmacol 94:141–143
Greeley GH Jr, Thompson JC (1984) Insulinotropic and gastrin-releasing action of gastrin-releasing peptide (GRP). Regul Pept 8:97–103
Greeley GH Jr, Partin M, Spannagel A, Dinh T, Hill FLC, Trowbridge J, Salter M, Chuo HF, Thompson JC (1986a) Distribution of bombesin-like peptides in the alimentary canal of several vertebrate species. Regul Pept 16:169–181
Greeley GH Jr, Spannagel A, Hill FLC, Thompson JC (1986b) Comparison of the actions of bombesin, gastrin-releasing peptide-27, neuromedin B, and gastrin-

releasing peptide-10 in causing release of gastrin and gastric inhibitory peptide in rat. Proc Soc Exp Biol Med 183:136–139
Guglietta A, Strunk CL, Irons BJ, Lazarus LH (1985) Central neuromodulation of gastric acid secretion by bombesin-like peptides. Peptides [Suppl 3]6:75–81
Güllner HG, Owen WW, Yajima H (1982) Effect of porcine gastrin releasing peptide on anterior pituitary hormone release. Biochem Biophys Res Commun 106:831–835
Gunion MW, Grijalva CV, Taché Y, Novin D (1984) Lateral hypothalamic lesions or transections block bombesin hyperglycemia in rats. Brain Res 299:239–246
Guo YS, Mok L, Cooper CW, Greeley GH Jr, Thompson JC, Singh P (1987) Effect of gastrin-releasing peptide analogues on gastrin and somatostatin release from isolated rat stomach. Am J Physiol 253:G206–G210
Hamaji M, Kawai Y, Kawashima Y, Tohyama M (1987) Projections of bombesin-like immunoreactive fibers from the rat stomach to the celiac ganglion revealed by a double-labeling technique. Brain Res 416:192–194
Hamid QA, Addis BJ, Springall DR, Ibrahim NBN, Ghatei MA, Bloom SR, Polak JM (1987) Expression of the C-terminal peptide of human pro-bombesin in 361 lung endocrine tumours, a reliable marker and possible prognostic indicator for small cell carcinoma. Virchows Arch 411:185–192
Hamid QA, Bishop AE, Springall DR, Adams C, Giaid A, Denny P, Ghatei M, Legon S, Cuttitta F, Rode J, Spindel E, Bloom SR, Polak JM (1989) Detection of human probombesin mRNA in neuroendocrine (small cell) carcinoma of the lung. In situ hybridization with cRNA probe. Cancer 63:266–271
Harrison FA, Wahuta EM (1982) The presence of bombesin-like and substance P-like immunoreactivity in the ovine digestive tract. J Physiol (Lond) 322:55–56P
Hawkins MF, Avery DD (1983) Effects of centrally-administered bombesin and adrenalectomy on behavioral thermoregulation and locomotor activity. Neuropharmacology 22:1249–1255
Hedner J, Mueller RA, Hedner T, McCown TJ, Breese GR (1985) A centrally elicited respiratory stimulant effect by bombesin in the rat. Eur J Pharmacol 115:21–29
Heikkila R, Trepel JB, Cuttitta F, Neckers LM, Sausville EA (1987) Bombesin-related peptides induce calcium mobilization in a subset of human small cell lung cancer cell lines. J Biol Chem 262:16456–16460
Heimbrook DC, Boyer ME, Garsky VM, Balishin NL, Kiefer DM, Oliff A, Riemen MW (1988) Minimal ligand analysis of gastrin releasing peptide-receptor binding and mitogenesis. J Biol Chem 263:7016–7019
Heinz-Erian P, Coy DH, Tamura M, Jones SW, Gardner JD, Jensen RT (1987) [D-Phe^{12}]bombesin analogues: a new class of bombesin receptor antagonists. Am J Physiol 252:G439–G442
Helén P, Panula P, Yang HYT, Rapoport SI (1984) Bombesin/gastrin-releasing peptide (GRP)- and Met^5-enkephalin-Arg^6-Gly^7-Leu^8-like immunoreactivities in small intensely fluorescent (SIF) cells and nerve fibers of rat sympathetic ganglia. J Histochem Cytochem 32:1131–1138
Hendley B, Mamrack MD (1988) Differential response of normal human fibroblasts to bombesin versus thrombin. J Cell Physiol 136:486–492
Hermansen K (1980) Effects of substance P and other peptides on the release of somatostatin, insulin and glucagon in vitro. Endocrinology 107:256–261
Heslop JP, Blakely DM, Brown KD, Irvine RF, Berridge MJ (1986) Effects of bombesin and insulin on inositol (1,4,5)trisphosphate and inositol (1,3,4) trisphosphate formation in Swiss 3T3 cells. Cell 47:703–709
Hirning LD, Burks TF (1984) Bombesin stimulates motility of the canine isolated small intestine by a neurogenic mechanism. Proc West Pharmacol Soc 27:403–405
Hirschowitz BI, Gibson RG (1978) Stimulation of gastrin release and gastric secretion: effect of bombesin and a nonapeptide in fistula dogs with and without fundic vagotomy. Digestion 18:227–239
Hirschowitz BI, Molina E (1983) Relation of gastric acid and pepsin secretion to serum gastrin levels in dogs given bombesin and gastrin-17. Am J Physiol 244:G546–G551

Holmgren S, Nilsson S (1983a) Bombesin-, gastrin/CCK-, 5-hydroxytryptamine-, neurotensin-, somatostatin-, and VIP-like immunoreactivity and catecholamine fluorescence in the gut of the elasmobranch, *Squalus acanthias*. Cell Tissue Res 234:595–618

Holmgren S, Nilsson S (1983b) VIP-, bombesin- and neurotensin-like immunoreactivity in neurons of the gut of the holostean fish *Lepisosteus platyrhincus*. Acta Zool (Stockh) 64:25–32

Holmgren S, Vaillant C, Dimaline R (1982) VIP-, substance P-, gastrin/CCK-, bombesin-, somatostatin- and glucagon-like immunoreactivities in the gut of the rainbow trout *Salmo gairdneri*. Cell Tissue Res 223:141–153

Holst JJ, Knuhtsen S, Orskov C, Skak-Nielsen T, Poulsen SS, Jensen SL, Nielsen OV (1987a) GRP nerves in pig antrum: role of GRP in vagal control of gastrin secretion. Am J Physiol 253:G643–G649

Holst JJ, Knuhtsen S, Orskov C, Skak-Nielsen T, Poulsen SS, Nielsen OV (1987b) GRP-producing nerves control antral somatostatin and gastrin secretion in pigs. Am J Physiol 253:G767–G774

Holtman JR Jr, Jensen RT, Buller A, Hamosh P, Taveira da Silva AM, Gillis RA (1983) Central respiratory stimulant effect of bombesin in the cat. Eur J Pharmacol 90:449–451

Howard JM, Jensen RT, Gardner JD (1985) Bombesin-induced residual stimulation of amylase release from mouse pancreatic acini. Am J Physiol 248:G196–G199

Hutchison JB, Dimaline R, Dockray GJ (1981) Neuropeptides in the gut: quantification and characterization of cholecystokinin octapeptide-, bombesin- and vasoactive intestinal polypeptide-like immunoreactivities in the myenteric plexus of the guinea pig small intestine. Peptides 2:23–30

Iguchi A, Matsunaga H, Nomura T, Gotoh M, Sakamoto N (1984) Glucoregulatory effects of intrahypothalamic injections of bombesin and other peptides. Endocrinology 114:2242–2246

Impicciatore M, Bertaccini G (1973) The bronchoconstrictor action of the tetradecapeptide bombesin in the guinea-pig. J Pharm Pharmacol 25:872–875

Inoue K, McKay D, Yajima H, Rayford PL (1983) Effect of synthetic porcine gastrin-releasing peptide on plasma levels of immunoreactive cholecystokinin pancreatic polypeptide and gastrin in dogs. Peptides 4:153–157

Ipp E, Unger RH (1979) Bombesin stimulates the release of insulin and glucagon, but not pancreatic somatostatin, from the isolated perfused dog pancreas. Endocrinol Res Commun 6:37–42

Isacke CM, Meisenhelder J, Brown KD, Gould KL, Gould SJ, Hunter T (1986) Early phosphorylation events following the treatment of Swiss 3T3 cells with bombesin and the mammalian bombesin-related peptide, gastrin releasing peptide. EMBO J 5:2889–2898

Iwamoto Y, Nakamura R, Akanuma Y (1983) Effects of porcine gastrin-releasing peptide on amylase release, 2-deoxyglucose uptake, and α-aminoisobutyric acid uptake in mouse pancreatic acini. Endocrinology 113:2106–2112

Iwanaga T (1983) Gastrin-releasing peptide (GRP)/bombesin-like immunoreactivity in the neurons and paraneurons of the gut and lung. Biomed Res 4:93–104

Iwatsuki N, Petersen OH (1978) In vitro action of bombesin on amylase secretion, membrane potential, and membrane resistance in rat and mouse pancreatic acinar cells. A comparison with other secretagogues. J Clin Invest 61:41–46

Jain DK, Wolfe MM, McGuigan JE (1985) Functional and anatomical relationships between antral gastrin cells and gastrin-releasing peptide neurons. Histochemistry 82:463–467

Jansen JBMJ, Lamers CBHW (1983) Molecular forms of cholecystokinin in human plasma during infusion of bombesin. Life Sci 33:2197–2205

Jensen RT, Moody T, Pert C, Rivier JE, Gardner JD (1978) Interaction of bombesin and litorin with specific membrane receptors on pancreatic acinar cells. Proc Natl Acad Sci USA 75:6139–6143

Jensen RT, Jones SW, Folkers K, Gardner JD (1984) A synthetic peptide that is a bombesin receptor antagonist. Nature 309:61–63

Johnson DE, Lock JE, Elde RP, Thompson TR (1982) Pulmonary neuroendocrine cells in hyaline membrane disease and bronchopulmonary dysplasia. Pediatr Res 16:446–454

Kabayama Y, Kato Y, Shimatsu A, Ohta H, Yanaihara N, Imura H (1984) Inhibition by gastrin-releasing peptide of growth hormone (GH) secretion induced by human pancreatic GH-releasing factor in rats. Endocrinology 115:649–653

Kameya T, Bessho T, Tsumuraya M, Yamaguchi K, Abe K, Shimosato Y, Yanaihara N (1983a) Production of gastrin releasing peptide by medullary carcinoma of the thyroid. An immunohistochemical study. Virchows Arch 401:99–108

Kameya T, Shimosato Y, Kodama T, Tsumuraya M, Koide T, Yamaguchi K, Abe K (1983b) Peptide hormone production by adenocarcinomas of the lung: its morphologic basis and histogenetic considerations. Virchows Arch 400:245–257

Kaneto A, Kaneko T, Nakaya S, Kajinuma H, Kosaka K (1978) Effect of bombesin infused intrapancreatically on glucagon and insulin secretion. Metabolism 27:549–553

Kantoh M, Takahashi T, Yamamura T, Ishikawa Y, Utsunomiya J (1985) Bombesin evoked acetylcholine release from the guinea pig antrum. Life Sci 36:2445–2452

Karashima T, Okajima T, Kato KI, Ibayashi H (1984) Suppressive effects of cholecystokinin and bombesin on growth hormone and prolactin secretion in urethane anesthetized rats. Endocrinol Jpn 31:539–547

Katz R (1980) Grooming elicited by intracerebroventricular bombesin and eledoisin in the mouse. Neuropharmacology 19:143–146

Kentroti S, McCann SM (1985) The effect of gastrin-releasing peptide on growth hormone secretion in the rat. Endocrinology 117:1363–1367

Kentroti S, Aguila MC, McCann SM (1988a) The inhibition of growth hormone release by gastrin-releasing peptide involves somatostatin release. Endocrinology 122:2407–2411

Kentroti S, Dees WL, McCann SM (1988b) Evidence for a physiological role of hypothalamic gastrin-releasing peptide to suppress growth hormone and prolactin release in the rat. Proc Natl Acad Sci USA 85:953–957

Kita T, Chihara K, Abe H, Minamitani N, Kaji H, Kodama H, Chiba T, Fujita T, Yanaihara N (1986) Regional distribution of gastrin-releasing peptide- and somatostatin-like immunoreactivity in the rabbit hypothalamus. Brain Res 398:18–22

Kleibeuker JH, Kauffman GL Jr, Walsh JH (1985) Intravenous histamine reduces bombesin-stimulated gastrin release in dogs. Regul Pept 11:209–215

Knigge U, Holst JJ, Knuhtsen S, Petersen B, Krarup T, Holst-Pedersen J, Christiansen PM (1984) Gastrin-releasing peptide: pharmacokinetics and effects on gastro-entero-pancreatic hormones and gastric secretion in normal men. J Clin Endocrinol Metab 59:310–315

Knuhtsen S, Holst JJ, Knigge U, Olesen M, Nielsen OV (1984) Radioimmunoassay pharmacokinetics, and neuronal release of gastrin-releasing peptide in anesthetized pigs. Gastroenterology 87:372–378

Knuhtsen S, Holst JJ, Jensen SL, Knigge U, Nielsen OV (1985) Gastrin-releasing peptide: effect on exocrine secretion and release from isolated perfused porcine pancreas. Am J Physiol 248:G281–G286

Knuhtsen S, Holst JJ, Baldissera FGA, Skak-Nielsen T, Poulsen SS, Jensen SL, Nielsen OV (1987a) Gastrin-releasing peptide in the porcine pancreas. Gastroenterology 92:1153–1158

Knuhtsen S, Holst JJ, Schwartz TW, Jensen SL, Nielsen OV (1987b) The effect of gastrin-releasing peptide on the endocrine pancreas. Regul Pept 17:269–276

Kondo H, Iwanaga T, Yanaihara N (1983) On the occurrence of gastrin releasing peptide (GRP)-like immunoreactive nerve fibers in the celiac ganglion of rats. Brain Res 289:326–329

Konturek SJ, Krol R, Tasler J (1976) Effect of bombesin and related peptides on the release and action of intestinal hormones on pancreatic secretion. J Physiol (Lond) 257:663–672

Korc M, Matrisian LM, Magun BE (1984) Cytosolic calcium regulates epidermal growth factor endocytosis in rat pancreas and cultured fibroblasts. Proc Natl Acad Sci USA 81:461–465

Koslo RJ, Gmerek DE, Cowan A, Porreca F (1986) Intrathecal bombesin-induced inhibition of gastrointestinal transit: requirement for an intact pituitary-adrenal axis. Regul Pept 14:237–242

Kris RM, Hazan R, Villines J, Moody TW, Schlessinger J (1987) Identification of the bombesin receptor on murine and human cells by cross linking experiments. J Biol Chem 262:11215–11220

Kulkosky PJ, Gibbs J, Smith GP (1982a) Behavioral effects of bombesin administration in rats. Physiol Behav 28:505–512

Kulkosky PJ, Gibbs J, Smith GP (1982b) Feeding suppression and grooming repeatedly elicited by intraventricular bombesin. Brain Res 242:194–196

Kyrkouli SE, Stanley BG, Leibowitz SF (1987) Bombesin-induced anorexia: sites of action in the rat brain. Peptides 8:237–241

Lambert JR, Hansky J, Soveny C, Hunt P (1984) Comparative effects of bombesin and porcine gastrin-releasing peptide in the dog. Dig Dis Sci 29:1036–1040

Layton JE, Scanlon DB, Soveny C, Morstyn G (1988) Effects of bombesin antagonists on the growth of small cell lung cancer cells in vitro. Cancer Res 48:4783–4789

Leander S, Ekman R, Uddman R, Sundler F, Håkanson R (1984) Neuronal cholecystokinin, gastrin-releasing peptide, neurotensin, and β-endorphin in the intestine of the guinea pig. Distribution and possible motor functions. Cell Tissue Res 235:521–531

Lebacq-Verheyden AM, Bertness V, Kirsch I, Hollis GF, McBride OW, Battey J (1987) Human gastrin-releasing peptide gene maps to chromosome band 18q21. Somatic Cell Mol Genet 13:81–86

Lebacq-Verheyden AM, Kasprzyk PG, Raum MG, van Wyke Coelingh KA, Lebacq JA, Battey JF (1988a) Posttranslational processing of endogenous and of baculovirus-expressed human gastrin-releasing peptide precursor. Mol Cell Biol 8:3129–3135

Lebacq-Verheyden AM, Krystal G, Sartor O, Way J, Battey JF (1988b) The rat prepro gastrin releasing peptide gene is transcribed from two initiation sites in the brain. Mol Endocrinol 2:556–563

Lebacq-Verheyden AM, Segal S, Cuttitta F, Battey JF (1988c) Swiss 3T3 mouse embryo fibroblasts transfected with a human prepro-GRP gene synthesize and secrete pro-GRP rather than GRP. J Cell Biochem 36:237–248

Lebacq-Verheyden AM, Kasprzyk PG, Battey JF (1989) A study of post-translational processing of the human progastrin-releasing peptide in two small cell carcinoma cell lines. J Clin Endocrinol Metab (in press)

Lechago J, Holmquist AL, Rosenquist GL, Walsh JH (1978) Localization of bombesin-like peptides in frog gastric mucosa. Gen Comp Endocrinol 36:553–558

Lee PC, Jensen RT, Gardner JD (1980) Bombesin-induced desensitization of enzyme secretion in dispersed acini from guinea pig pancreas. Am J Physiol 238:G213–G218

Lehy T, Accary JP, Labeille D, Dubrasquet M (1983) Chronic administration of bombesin stimulates antral gastrin cell proliferation in the rat. Gastroenterology 84:914–919

Lemaire S, Chouinard L, Mercier P, Day R (1986) Bombesin-like immunoreactivity in bovine adrenal medulla. Regul Pept 13:133–146

Lenz HJ, Klapdor R, Hester SE, Webb VJ, Galyean RF, Rivier JE, Brown MR (1986) Inhibition of gastric acid secretion by brain peptides in the dog. Role of the autonomic nervous system and gastrin. Gastroenterology 91:905–912

Letterio JJ, Coughlin SR, Williams LT (1986) Pertussis toxin-sensitive pathway in the stimulation of c-*myc* expression and DNA synthesis by bombesin. Science 234:1117–1119

Lewin KJ, Layfield L, Cheng L (1985) Disseminated bombesin-producing carcinoid tumor of pulmonary origin. Am J Surg Pathol 9:129–134

Lhoste E, Aprahamian M, Pousse A, Hoeltzel A, Stock-Damgé C (1985a) Combined effect of chronic bombesin and secretin or cholecystokinin on the rat pancreas. Peptides [Suppl 3]6:83–87

Lhoste E, Aprahamian M, Pousse A, Hoeltzel A, Stock-Damgé C (1985b) Trophic effect of bombesin on the rat pancreas: is it mediated by the release of gastrin or cholecystokinin? Peptides [Suppl 3]6:89–97

Lin KS, Lin MT (1986) Effects of bombesin on thermoregulatory responses and hypothalamic neuronal activities in the rat. Am J Physiol 251:R303–R309

Lipton JM, Glynn JR (1980) Central administration of peptides alters thermoregulation in the rabbit. Peptides 1:15–18

Logsdon CD, Zhang J, Guthrie J, Vigna S, Williams JA (1987) Bombesin binding and biological effects on pancreatic acinar AR42J cells. Biochem Biophys Res Commun 144:463–468

Loosen PT, Nemeroff CB, Bissette G, Burnett GB, Prange AJ Jr, Lipton MA (1978) Neurotensin-induced hypothermia in the rat: structure-activity studies. Neuropharmacology 17:109–113

Lopez-Rivas A, Mendoza SA, Nanberg E, Sinnett-Smith J, Rozengurt E (1987) Ca^{2+}-mobilizing action of platelet-derived growth factor differs from those of bombesin and vasopressin in swiss 3T3 mouse cells. Proc Natl Acad Sci USA 84:5768–5772

Major J, Ghatei MA, Bloom SR (1983) Bombesin-like immunoreactivity in the pituitary gland. Experientia 39:1158–1159

Märki W, Brown M, Rivier JE (1981) Bombesin analogs: effects on thermoregulation and glucose metabolism. Peptides [Suppl 2]2:169–177

Markowitz S, Krystal G, Lebacq-Verheyden AM, Way J, Sausville EA, Battey J (1988) Transcriptional activation and DNAse I hypersensitive sites are associated with selective expression of the gastrin-releasing peptide gene. J Clin Invest 82:808–815

Martindale R, Kauffmann GL, Levin S, Walsh JH, Yamada T (1982a) Differential regulation of gastrin and somatostatin secretion from isolated perfused rat stomachs. Gastroenterology 83:240–244

Martindale R, Levin S, Alfin-Slater R (1982b) Effects of caerulein and bombesin on insulin and glucagon secretion from the isolated, perfused rat pancreas. Regul Pept 3:313–324

Mason GA, Nemeroff CB, Luttinger D, Hatley OL, Prange AJ Jr (1980) Neurotensin and bombesin: differential effects on body temperature of mice after intracisternal administration. Regul Pept 1:53–60

Massari VJ, Tizabi Y, Park CH, Moody TW, Helke CJ, O'Donohue TL (1983) Distribution and origin of bombesin, substance P and somatostatin in cat spinal cord. Peptides 4:673–681

Materia A, Modlin IM, Albert D, Sank A, Crochelt RF, Jaffe BM (1981) The effect of somatostatin and 16,16-dimethyl-prostaglandin E2 on bombesin-stimulated canine gastric acid, plasma gastrin and pancreatic polypeptide secretion. Regul Pept 1:297–305

Matsubayashi S, Yanaihara C, Ohkubo M, Fukata S, Hayashi Y, Tamai H, Nakagawa T, Miyauchi A, Kuma K, Abe K, Suzuki T, Yanaihara N (1984) Gastrin-releasing peptide immunoreactivity in medullary thyroid carcinoma. Cancer 53:2472–2477

Matsushita N, Kato Y, Katakami H, Shimatsu A, Yanaihara N, Imura H (1983) Inhibition of prolactin secretion by gastrin releasing peptide (GRP) in the rat. Proc Soc Exp Biol Med 172:118–121

Matsuyama T, Namba M, Nonaka K, Tarui S, Tanaka R, Shima K (1980) Decrease in blood glucose and release of gut glucagon-like immunoreactive materials by bombesin infusion in the dog. Endocrinol Jpn 1:115–119

Matuoka K, Fukami K, Nakanishi O, Kawai S, Takenawa T (1988) Mitogenesis in response to PDGF and bombesin is abolished by microinjection of antibody to PIP_2. Science 239:640–643
May RJ, Conlon TP, Erspamer V, Gardner JD (1978) Action of peptides isolated from amphibian skin on pancreatic acinar cells. Am J Physiol 235:E112–E118
Mayer EA, Elashoff J, Walsh JH (1982) Characterization of bombesin effects on canine gastric muscle. Am J Physiol 243:G141–G147
Mayer EA, Reeve JR Jr, Khawaja S, Chew P, Elashoff J, Clark B, Walsh JH (1986) Potency of natural and synthetic canine gastrin-releasing decapeptide on canine antral muscle. Am J Physiol 250:G581–G587
Mazzanti G, Erspamer GF, Piccinelli D (1982) Relative potencies of porcine bombesin-like heptacosapeptide (PB-27), amphibian bombesin (B-14) and litorin, and bombesin C-terminal nonapeptide (B-9) on in vitro and in vivo smooth muscle preparations. J Pharm Pharmacol 34:120–121
McCaffrey P, Ran W, Campisi J, Rich Rosner M (1987) Two independent growth factor-generated signals regulate c-*fos* and c-*myc* RNA levels in Swiss 3T3 cells. J Biol Chem 262:1442–1445
McDonald TJ, Nilsson G, Vagne M, Ghatei M, Bloom SR, Mutt V (1978) A gastrin releasing peptide from the porcine non-antral gastric tissue. Gut 19:767–774
McDonald TJ, Jornvall H, Nilsson G, Vagne M, Ghatei M, Bloom SR, Mutt V (1979) Characterization of a gastrin releasing peptide from porcine non-antral gastric tissue. Biochem Biophys Res Commun 90:227–233
McDonald TJ, Jornvall H, Ghatei M, Bloom SR, Mutt V (1980) Characterization of an avian gastric (proventricular) peptide having sequence homology with the porcine gastrin-releasing peptide and the amphibian peptides bombesin and alytesin. FEBS Lett 122:45–48
McDonald TJ, Ghatei MA, Bloom SR, Track NS, Radziuk J, Dupre J, Mutt V (1981) A qualitative comparison of canine plasma gastroenteropancreatic hormone responses to bombesin and the porcine gastrin-releasing peptide (GRP). Regul Pept 2:293–304
McDonald TJ, Ghatei MA, Bloom SR, Adrian TE, Mochizuki T, Yanaihara C, Yanaihara N (1983) Dose-response comparisons of canine plasma gastroenteropancreatic hormone responses to bombesin and the porcine gastrin-releasing peptide. Regul Pept 5:125–137
McLaughlin CL, Baile CA (1981) Obese mice and the satiety effects of cholecystokinin, bombesin and pancreatic polypeptide. Physiol Behav 26:433–437
Meisenberg G (1982) Short-term behavioural effects of neurohypophyseal hormones: pharmacological characteristics. Neuropharmacology 21:309–316
Meisenberg G, Simmons WH (1986) Behavioral alterations induced by substance P, bombesin and related peptides in mice. Peptides 7:557–561
Meldolesi J, Pozzan T (1987) Pathways of Ca^{+2} influx at the plasma membrane: voltage-, receptor-, and second messenger-operated channels. Exp Cell Res 171:217–283
Mendoza SA, Schneider JA, Lopez-Rivas A, Sinnett-Smith JW, Rozengurt E (1986) Early events elicited by bombesin and structurally related peptides in quiescent Swiss 3T3 cells. II. Changes in Na^+ and Ca^{+2} fluxes, Na/K pump activity, and intracellular pH. J Cell Biol 102:2223–2233
Merali Z, Johnston S, Zalcman S (1983) Bombesin-induced behavioural changes: antagonism by neuroleptics. Peptides 4:693–697
Micheletti R, Grider JR, Makhlouf GM (1988) Identification of bombesin receptors on isolated muscle cells from human intestine. Regul Pept 21:219–226
Michell RH, Kirk CJ, Jones LM, Downes CP, Creba J (1981) The stimulation of inositol lipid metabolism that accompanies calcium mobilization in stimulated cells: defined characteristics and unanswered questions. Philos Trans R Soc Lond [Biol] 296:123–137

Minamino N, Kangawa K, Matsuo H (1983) Neuromedin B: a novel bombesin-like peptide identified in porcine spinal cord. Biochem Biophys Res Commun 114:541–548

Minamino N, Kangawa K, Matsuo H (1984) Neuromedin C: a bombesin-like peptide identified in porcine spinal cord. Biochem Biophys Res Commun 119:14–20

Minamino N, Sudoh T, Kangawa K, Matsuo H (1985) Neuromedin B-32 and B-30: two "big" neuromedin B identified in porcine brain and spinal cord. Biochem Biophys Res Commun 130:685–691

Minna JD, Battey JF, Brooks BJ, Cuttitta F, Gazdar AF, Johnson BE, Ihde DC, Lebacq-Verheyden AM, Mulshine J, Nau MM, Oie HK, Sausville EA, Seifter E, Vinocour M (1987) Molecular genetic analysis reveals chromosomal deletion, gene amplification, and autocrine growth factor production in the pathogenesis of human lung cancer. Cold Spring Harbor Symp Quant Biol 51:843–853

Mitsuma T, Nogimori T, Chaya M (1985) Bombesin inhibits thyrotropin secretion in rats. Acta Endocrinol (Copenh) 108:79–84

Miyata M, Rayford PL, Thompson JC (1980) Hormonal (gastrin, secretin, cholecystokinin) and secretory effects of bombesin and duodenal acidification in dogs. Surgery 87:209–215

Modlin IM, Lamers CB, Jaffe BM (1980a) Evidence for cholinergic dependence of pancreatic polypeptide release by bombesin – a possible application. Surgery 88:75–85

Modlin IM, Lamers C, Walsh JH (1980b) Mechanisms of gastrin release by bombesin and food. J Surg Res 28:539–546

Modlin IM, Lamers CBH, Walsh JH (1981a) Stimulation of canine pancreatic polypeptide, gastrin, and gastric acid secretion by ranatensin, litorin, bombesin nonapeptide and substance P. Regul Pept 1:279–288

Modlin IM, Lamers CBH, Walsh JH, Jaffe BM (1981b) Bombesin: a vagally independent stimulator of gastrin release. Am J Surg 141:98–104

Moghimzadeh E, Ekman R, Håkanson R, Yanaihara N, Sundler F (1983) Neuronal gastrin-releasing peptide in the mammalian gut and pancreas. Neuroscience 10:553–563

Moody TW, Pert CB (1979) Bombesin-like peptides in rat brain: quantitation and biochemical characterization. Biochem Biophys Res Commun 90:7–14

Moody TW, Pert CB, Rivier J, Brown MR (1978) Bombesin: specific binding to rat brain membranes. Proc Natl Acad Sci USA 75:5372–5376

Moody TW, Thoa NB, O'Donohue TL, Pert CB (1980) Bombesin-like peptides in rat brain: localization in synaptosomes and release from hypothalamic slices. Life Sci 26:1707–1712

Moody TW, O'Donohue TL, Jacobowitz DM (1981a) Biochemical localization and characterization of bombesin-like peptides in discrete regions of rat brain. Peptides 2:75–79

Moody TW, Pert CB, Gazdar AF, Carney DN, Minna JD (1981b) High levels of intracellular bombesin characterize human small-cell lung carcinoma. Science 214:1246–1248

Moody TW, Thoa NB, O'Donohue TL, Jacobowitz D (1981c) Bombesin-like peptides in the rat spinal cord: biochemical characterization, localization and mechanism of release. Life Sci 29:2273–2279

Moody TW, Crawley JN, Jensen RT (1982) Pharmacology and neurochemistry of bombesin-like peptides. Peptides 3:559–563

Moody TW, Bertness V, Carney DN (1983a) Bombesin-like peptides and receptors in human tumor cell lines. Peptides 4:683–686

Moody TW, Russel EK, O'Donohue TL, Linden CD, Gazdar AF (1983b) Bombesin-like peptides in small cell lung cancer: biochemical characterization and secretion from a cell line. Life Sci 32:487–493

Moody TW, Carney DN, Cuttitta F, Quattrochi K, Minna JD (1985) High affinity receptors for bombesin/GRP-like peptides on human small cell lung cancer. Life Sci 37:105–113

Moody TW, Murphy A, Mahmoud S, Fiskum G (1988) Bombesin-like peptides elevate cytosolic calcium in small cell lung cancer cells. Biochem Biophys Res Commun 147:189–195

Moran TH, Moody TW, Hostetler AM, Robinson PH, Goldrich M, McHugh PR (1988) Distribution of bombesin binding sites in the rat gastrointestinal tract. Peptides 9:643–649

Morley JE, Levine AS (1981) Bombesin inhibits stress-induced eating. Pharmacol Biochem Behav 14:149–151

Morley JE, Varner AA, Modlin IM, Carlson HE, Braunstein GD, Walsh JH, Hershman JM (1980) Failure of bombesin to alter anterior pituitary hormone secretion in man. Clin Endocrinol 13:369–373

Morley JE, Levine AS, Oken MM, Grace M, Kneip J (1982) Neuropeptides and thermoregulation: the interactions of bombesin, neurotensin, TRH, somatostatin, naloxone and prostaglandins. Peptides 3:1–6

Morley JE, Levine AS, Hertel H, Tandeski T, Seal US (1986) The effect of peripheral administration of peptides on food intake, glucose and insulin in wolf pups. Peptides 7:969–972

Muir JG, Murray AW (1987) Bombesin and phorbol esters stimulate phosphatidylcholine hydrolysis by phospholipase C: evidence for a role of protein kinase C. J Cell Physiol 130:382–391

Mukai H, Kawai K, Suzuki Y, Yamashita K, Munekata E (1987) Stimulation of dog gastropancreatic hormone release by neuromedin B and its analogues. Am J Physiol 252:E765–E771

Namba M, Ghatei MA, Adrian TE, Bacarese-Hamilton AJ, Mulderry PK, Bloom SR (1984) Effect of neuromedin B on gut hormone secretion in the rat. Biomed Res 5:229–234

Naylor SL, Sakaguchi AY, Spindel E, Chin WW (1987) Human gastrin-releasing peptide gene is located on chromosome 18. Somatic Cell Mol Genet 13:87–91

Negri L (1986) Satiety and scratching; effects of bombesin-like peptides. Eur J Pharmacol 132:207–212

Nemeroff CB, Osbahr AJ III, Manberg PJ, Ervin GN, Prange AJ Jr (1979) Alterations in nociception and body temperature after intracisternal administration of neurotensin, β-endorphin, other endogenous peptides and morphine. Proc Natl Acad Sci USA 76:5368–5371

Niewoehner DE, Levine AS, Morley JE (1983) Central effects of neuropeptides in ventilation in the rat. Peptides 4:277–281

O'Donohue TL, Massari VJ, Pazoles CJ, Chronwall BM, Shults CW, Quirion R, Chase TN, Moody TW (1984) A role for bombesin in sensory processing in the spinal cord. J Neurosci 4:2956–2962

Orloff MS, Reeve JR Jr, Miller Ben-Avram C, Shively JE, Walsh JH (1984) Isolation and sequence analysis of human bombesin-like peptides. Peptides 5:865–870

O'Shaughnessy DJ, McGregor GP, Ghatei MA, Blank MA, Springall DR, Gu J, Polak JM, Bloom SR (1983) Distribution of bombesin, somatostatin, substance-P and vasoactive intestinal polypeptide in feline and porcine skin. Life Sci 32:2827–2836

Pachler JA, Law GJ, Dannies PS (1988) Bombesin stimulates inositol polyphosphate production in GH_4C_1 pituitary tumor cells: comparison with TRH. Biochem Biophys Res Commun 154:654–659

Palumbo AP, Rossino P, Comoglio PM (1986) Bombesin stimulation of c-*fos* and c-*myc* gene expression in cultures of Swiss 3T3 cells. Exp Cell Res 167:276–280

Pandol SJ, Jensen RT, Gardner JD (1982) Mechanism of [Tyr^4]bombesin-induced desensitization in dispersed acini from guinea pig pancreas. J Biol Chem 257:12024–12029

Panula PA (1986) Histochemistry and function of bombesin-like peptides. Med Biol 64:177–192

Panula PA, Yang HYT, Costa E (1982) Neuronal location of bombesin-like immunoreactivity in the central nervous system of the rat. Regul Pept 4:275–283

Panula PA, Hadjiconstantinou M, Yang HYT, Costa E (1983) Immunohistochemical localization of bombesin/gastrin-releasing peptide and substance P in primary sensory neurons. J Neurosci 3:2021–2029
Panula PA, Yang HYT, Costa E (1984) Comparative distribution of bombesin/GRP and substance-P-like immunoreactivities in rat hypothalamus. J Comp Neurol 224:606–617
Pappas T, Hamel D, Debas H, Walsh JH, Taché Y (1985) Cerebroventricular bombesin inhibits gastric acid secretion in dogs. Gastroenterology 89:43–48
Parrott RF, Baldwin BA (1982) Centrally-administered bombesin produces effects unlike short-term satiety in operant feeding pigs. Physiol Behav 28:521–524
Pert A, Moody TW, Pert CB, de Wald LA, Rivier J (1980) Bombesin: receptor distribution in brain and effects on nociception and locomotor activity. Brain Res 193:209–220
Pettersson M, Ahrén B (1987) Gastrin releasing peptide (GRP): effects on basal and stimulated insulin and glucagon secretion in the mouse. Peptides 8:55–60
Pettersson M, Ahrén B (1988) Insulin and glucagon secretion in the rat: effects of gastrin releasing peptide. Neuropeptides 12:159–163
Phillis JW, Kirkpatrick JR (1979) Actions of various gastrointestinal peptides on the isolated amphibian spinal cord. Can J Physiol Pharmacol 57:887–899
Phillis JW, Limacher JJ (1974) Excitation of cerebral cortical neurons by various polypeptides. Exp Neurol 43:414–423
Pittman QJ, Taché Y, Brown MR (1980) Bombesin acts in preoptic area to produce hypothermia in rats. Life Sci 26:725–730
Poitras P, Tassé D, Laprise P (1983) Stimulation of motilin release by bombesin in dogs. Am J Physiol 245:G249–G256
Pontiroli AE, Alberetto M, Restelli L, Facchinetti A (1980) Effects of bombesin and ceruletide on prolactin, growth hormone, luteinizing hormone, and parathyroid hormone release in normal human males. J Clin Endocrinol Metab 51:1303–1305
Porreca F, Burks TF (1983) Centrally administered bombesin affects gastric emptying and small and large bowel transit in the rat. Gastroenterology 85:313–317
Price J, Penman E, Bourne GL, Rees LH (1983) Characterisation of bombesin-like immunoreactivity in human fetal lung. Regul Pept 7:315–322
Price J, Penman E, Wass JAH, Rees LH (1984) Bombesin-like immunoreactivity in human gastrointestinal tract. Regul Pept 9:1–10
Price J, Nieuwenhuijzen Kruseman AC, Doniach I, Howlett TA, Besser GM, Rees LH (1985) Bombesin-like peptides in human endocrine tumors: quantitation, biochemical characterization and secretion. J Clin Endocrinol Metab 60:1097–1103
Rasler FE (1983) Bombesin produces hypothermia in hypophysectomized rats. Life Sci 32:2503–2507
Rasler FE (1984) Behavioral and electrophysiological manifestations of bombesin: excessive grooming and elimination of sleep. Brain Res 321:187–191
Reeve JR Jr, Walsh JH, Chew P, Clark B, Hawke D, Shively JE (1983) Amino acid sequences of three bombesin-like peptides from canine intestine extracts. J Biol Chem 258:5582–5588
Represa JJ, Miner C, Barbosa E, Giraldez F (1988) Bombesin and other growth factors activate cell proliferation in chick embryo otic vesicles in culture. Development 102:87–96
Richelsen B, Rehfeld JF, Larsson LI (1983) Antral gland cell column: a method for studying release of gastric hormones. Am J Physiol 245:G463–G469
Rivier C, Rivier JE, Vale W (1978) The effect of bombesin and related peptides on prolactin and growth hormone secretion in the rat. Endocrinology 102:519–522
Rivier JE, Brown MR (1978) Bombesin, bombesin analogues and related peptides: effects on thermoregulation. Biochemistry 17:1766–1771
Rodriguez-Pena A, Zachary I, Rozengurt E (1986) Rapid dephosphorylation of a MR 80,000 protein, a specific substrate of protein kinase C upon removal of phorbol esters, bombesin, and vasopressin. Biochem Biophys Res Commun 140:379–385

Rokaeus A, Yanaihara N, McDonald TJ (1982) Increased concentration of neurotensinlike immunoreactivity (NTLI) in rat plasma after administration of bombesin and bombesin-related peptides (porcine and chicken gastrin-releasing peptides). Acta Physiol Scand 114:605–610

Roth KA, Weber E, Barchas JD (1982) Distribution of gastrin releasing peptide-bombesin-like immunostaining in rat brain. Brain Res 251:277–282

Roth KA, Evans CJ, Lorenz RG, Weber E, Barchas JD, Chang JK (1983a) Identification of gastrin releasing peptide-related substances in guinea pig and rat brain. Biochem Biophys Res Commun 112:528–536

Roth KA, Evans CJ, Weber E, Barchas JD, Bostwick DG, Bensch KG (1983b) Gastrin-releasing peptide-related peptides in a human malignant lung carcinoid tumor. Cancer Res 43:5411–5415

Rozengurt E, Sinnett-Smith J (1983) Bombesin stimulation of DNA synthesis and cell division in cultures of Swiss 3T3 cells. Proc Natl Acad Sci USA 80:2936–2940

Rozengurt E, Sinnett-Smith JW (1987) Bombesin induction of c-*fos* and c-*myc* protooncogenes in Swiss 3T3 cells: significance for the mitogenic response. J Cell Phys 131:218–225

Said JW, Vimadalal S, Nash G, Shintaku IP, Heusser RC, Sassoon AF, LLoyd RV (1985) Immunoreactive neuron-specific enolase, bombesin, and chromogranin as markers for neuroendocrine lung tumors. Hum Pathol 16:236–240

Sander LD, Porter JR (1988) Influence of bombesin, CCK, secretin and CRF on corticosterone concentration in the rat. Peptides 9:113–117

Sausville EA, Lebacq-Verheyden AM, Spindel ER, Cuttitta F, Gazdar AF, Battey JF (1986) Expression of the gastrin-releasing peptide gene in human small cell lung cancer: evidence for alternative processing resulting in three distinct mRNAs. J Biol Chem 261:2451–2457

Sausville EA, Moyer JD, Heikkila R, Neckers LM, Trepel JB (1988) A correlation of bombesin-responsiveness with *myc*-family gene expression in small cell lung carcinoma cell lines. Ann NY Acad Sci 547:310–321

Scarpignato C, Bertaccini G (1981) Bombesin delays gastric emptying in the rat. Digestion 21:104–106

Scarpignato C, Micali B, Vitulo F, Zimbaro G, Bertaccini G (1981) The effect of bombesin on gastric emptying of solids in man. Peptides [Suppl 2]2:199–203

Scemama JL, Zahidi A, Fourmy D, Fagot-Revurat P, Vaysse N, Pradayrol L, Ribet A (1986) Interactions of [^{125}I]-Tyr^{4}-bombesin with specific receptors on normal human pancreatic membranes. Regul Pept 13:125–132

Schubert ML, Saffouri B, Walsh JH, Makhlouf GM (1985) Inhibition of neurally mediated gastrin secretion by bombesin antiserum. Am J Physiol 248:G456–G462

Schultzberg M (1983) Bombesin-like immunoreactivity in sympathetic ganglia. Neuroscience 8:363–374

Schultzberg M, Dalsgaard CJ (1983) Enteric origin of bombesin immunoreactive fibers in the rat coeliac-superior mesenteric ganglion. Brain Res 269:190–195

Schulz DW, Kalivas PW, Nemeroff CB, Prange AJ Jr (1984) Bombesin-induced locomotor hyperactivity: evaluation of the involvement of the mesolimbic dopamine system. Brain Res 304:377–382

Schusdziarra V, Rouiller D, Harris V, Pfeiffer EF, Unger RH (1980) Effect of bombesin upon plasma somatostatin-like immunoreactivity, insulin, and glucagon in normal and chemically sympathectomized dogs. Regul Pept 1:89–96

Schusdziarra V, Bender H, Pfeiffer EF (1983) Release of bombesin-like immunoreactivity from the isolated perfused rat stomach. Regul Pept 7:21–29

Schusdziarra V, Bender H, Pfeffer A, Pfeiffer EF (1984) Modulation of acetylcholine-induced secretion of gastric bombesin-like immunoreactivity by cholinergic and histamine H_2-receptors, somatostatin and intragastric pH. Regul Pept 8:189–198

Schusdziarra V, Schmid R, Bender H, Schusdziarra M, Rivier J, Vale W, Classen M (1986) Effect of vasoactive intestinal peptide, peptide histidine isoleucine and growth hormone-releasing factor-40 on bombesin-like immunoreactivity, somatostatin and gastrin release from the perfused rat stomach. Peptides 7:127–133

Short GM, Reel GM, Doyle JW, Wolfe MM (1985) Effect of GRP on β-adrenergic stimulated gastrin and somatostatin release in the isolated rat stomach. Am J Physiol 249:G197–G202

Singer V, Niebel W, Lamers C, Becker S, Vesper J, Hartmann W, Diemel J, Goebell H (1981) Effects of truncal vagotomy and antrectomy on bombesin-stimulated pancreatic secretion, release of gastrin and pancreatic polypeptide in the anesthetized dog. Dig Dis Sci 26:871–877

Sorenson GD, Bloom SR, Ghatei MA, del Prete SA, Cate CC, Pettengill OS (1982) Bombesin production by human small cell carcinoma of the lung. Regul Pept 4:59–66

Spencer SE, Talman WT (1987) Centrally administered bombesin modulates gastric motility. Peptides 8:887–891

Spindel ER, Chin WW, Price J, Rees LH, Besser GM, Habener JF (1984) Cloning and characterization of cDNAs encoding human gastrin-releasing peptide. Proc Natl Acad Sci USA 81:5699–5703

Spindel ER, Zilberberg MD, Habener JF, Chin WW (1986) Two prohormones for gastrin-releasing peptide are encoded by two mRNAs differeing by 19 nucleotides. Proc Natl Acad Sci USA 83:19–23

Spindel ER, Sunday ME, Hofler H, Wolfe HJ, Habener JF, Chin WW (1987a) Transient elevation of messenger RNA encoding gastrin-releasing peptide, a putative pulmonary growth factor in human fetal lung. J Clin Invest 80:1172–1179

Spindel ER, Zilberberg MD, Chin WW (1987b) Analysis of the gene and multiple messenger ribonucleic acids (mRNAs) encoding human gastrin-releasing peptide: alternate RNA splicing occurs in neural and endocrine tissue. Mol Endocrinol 1:224–232

Springall DR, Ibrahim NBN, Rode J, Sharpe MS, Bloom SR, Polak JM (1986) Endocrine differentiation of extra-pulmonary small cell carcinoma demonstrated by immunohistochemistry using antibodies to PGP 9.5, neuron-specific enolase and the C-flanking peptide of human pro-bombesin. J Pathol 150:151–162

Stahlman MT, Kasselberg AG, Orth DN, Gray ME (1985) Ontogeny of neuroendocrine cells in human fetal lung. II. An immunohistochemical study. Lab Invest 52:52–60

Stein LJ, Woods SC (1982) Gastrin-releasing peptide reduces meal size in rats. Peptides 3:833–835

Stjernquist M, Håkanson R, Leander S, Owman C, Sundler F, Uddman R (1983) Immunohistochemical localization of substance P, vasoactive intestinal polypeptide and gastrin-releasing peptide in vas deferens and seminal vesicle, and the effect of these and eight other neuropeptides on resting tension and neurally evoked contractile activity. Regul Pept 7:67–86

Stjernquist M, Ekblad E, Owman C, Sundler F (1986) Neuronal localization and motor effects of gastrin-releasing peptide (GRP) in rat uterus. Regul Pept 13:197–205

Stuckey JA, Gibbs J (1982) Lateral hypothalamic injection of bombesin decreases food intake in rats. Brain Res Bull 8:617–621

Stuckey JA, Gibbs J, Smith GP (1985) Neural disconnection of gut from brain blocks bombesin-induced satiety. Peptides 6:1249–1252

Sugano K, Park J, Soll AH, Yamada T (1987) Stimulation of gastrin release by bombesin and canine gastrin releasing peptides. Studies with isolated canine G cells in primary culture. J Clin Invest 79:935–942

Sunday ME, Kaplan LM, Motoyama E, Chin WW, Spindel ER (1988a) Biology of disease. Gastrin-releasing peptide (mammalian bombesin) gene expression in health and disease. Lab Invest 59:5–24

Sunday ME, Wolfe HJ, Roos BA, Chin WW, Spindel ER (1988b) Gastrin-releasing peptide gene expression in developing, hyperplastic, and neoplastic human thyroid C-cells. Endocrinology 122:1551–1558

Suzuki M, Yamaguchi K, Abe K, Adachi N, Nagasaki K, Asanuma F, Adachi I, Kimura S, Terada M, Taya Y, Matsuzaki J, Miki K (1987) Detection of gastrin-releasing peptide mRNA in small cell lung carcinomas and medullary thyroid carcinomas using synthetic oligodeoxyribonucleotide probes. Jpn J Clin Oncol 17:157–163

Swope SL, Schonbrunn A (1984) Bombesin stimulates insulin secretion by a pancreatic islet cell line. Proc Natl Acad Sci USA 81:1822–1826

Swope SL, Schonbrunn A (1987) Characterization of ligand binding and processing by bombesin receptors in an insulin-secreting cell line. Biochem J 247:731–738

Swope SL, Schonbrunn A (1988) The biphasic stimulation of insulin secretion by bombesin involves both cytosolic free calcium and protein kinase C. Biochem J 253:193–202

Taché Y (1982) Bombesin: central nervous system action to increase gastric mucus in rats. Gastroenterology 83:75–80

Taché Y (1988) CNS peptides and regulation of gastric acid secretion. Annu Rev Physiol 50:19–39

Taché Y, Collu R (1982) CNS mediated inhibition of gastric secretion by bombesin: independence from interaction with brain catecholamingeric, and serotoninergic pathways and pituitary hormones. Regul Pept 3:51–59

Taché Y, Brown M, Collu R (1979a) Effects of neuropeptides on adenohypophyseal hormone response to acute stress in male rats. Endocrinology 105:220–224

Taché Y, Simard P, Collu R (1979b) Prevention by bombesin of cold-restraint stress induced hemorrhagic lesions in rats. Life Sci 24:1719–1726

Taché Y, Pittman Q, Brown M (1980a) Bombesin-induced poikilothermy in rats. Brain Res 188:525–530

Taché Y, Vale W, Rivier J, Brown M (1980b) Brain regulation of gastric secretion: influence of neuropeptides. Proc Natl Acad Sci USA 77:5515–5519

Taché Y, Märki W, Rivier J, Vale W, Brown M (1981) Central nervous system inhibition of gastric secretion in the rat by gastrin-releasing peptide, a mammalian bombesin. Gastroenterology 81:298–302

Takuwa N, Takuwa Y, Bollag WE, Rasmussen H (1987) The effects of bombesin on polyphosphoinositide and calcium metabolism in Swiss 3T3 cells. J Biol Chem 262:182–188

Tamai S, Kameya T, Yamaguchi K, Yanai N, Abe K, Yanaihara N, Yamazaki H, Kageyama K (1983) Peripheral lung carcinoid tumor producing predominantly gastrin-releasing peptide (GRP). Morphologic and hormonal studies. Cancer 52:273–281

Taminato T, Seino Y, Goto Y, Matsukura S, Imura H, Sakura N, Yanaihara N (1978) Bombesin inhibits insulin release from isolated pancreatic islets of rats in vitro. Endocrinol Jpn 25:305–307

Taylor CW, Blakely DM, Corps AN, Berridge MJ, Brown KD (1988) Effects of pertussis toxin on growth factor-stimulated inositol phosphate formation and DNA synthesis in Swiss 3T3 cells. Biochem J 249:917–920

Taylor IL, Garcia R (1985) Effects of pancreatic polypeptide, caerulein and bombesin on satiety in obese mice. Am J Physiol 248:G277–G280

Taylor IL, Walsh JH, Carter D, Wood J, Grossman MI (1979) Effects of atropine and bethanechol on bombesin-stimulated release of pancreatic polypeptide and gastrin in dog. Gastroenterology 77:714–718

Thomas RM, Sander LD (1985) Influence of CCK and bombesin on ACTH and cortisol secretion in the conscious dog. Peptides 6:703–707

Timson CM, Polak JM, Wharton J, Ghatei MA, Bloom SR, Usellini L, Capella C, Solcia E, Brown MR, Pearse AGE (1979) Bombesin-like immunoreactivity in the avian gut and its localisation to a distinct cell type. Histochemistry 61:213–221

Track NS, Cutz E (1982) Bombesin-like immunoreactivity in developing human lung. Life Sci 30:1553–1556

Trepel JB, Moyer JD, Heikkila R, Sausville EA (1988a) Modulation of bombesin-induced phosphatidylinositol hydrolysis in a small cell lung cancer cell line. Biochem J 255:403–410

Trepel JB, Moyer JD, Cuttitta F, Frucht H, Coy DH, Natale RB, Mulshine JL, Jensen RT, Sausville EA (1988b) A novel bombesin receptor antagonist inhibits autocrine signals in a small cell lung carcinoma cell line. Biochem Biophys Res Commun 156:1383–1389

Tsutsumi Y, Osamura RY, Watanabe K, Yanaihara N (1983) Immunohistochemical studies on gastrin-releasing peptide- and adrenocorticotropic hormone-containing cells in the human lung. Lab Invest 48:623–632

Uhlemann ER, Rottman AJ, Gardner JD (1979) Actions of peptides isolated from amphibian skin on amylase release from dispersed pancreatic acini. Am J Physiol 236:E571–E576

Vagne M, Gelin ML, McDonald TJ, Chayvialle JA, Minaire Y (1982) Effect of bombesin on gastric secretion and motility in the cat. Digestion 24:5–13

Vagne M, Collinet M, Cuber JC, Bernard C, Chayvialle JA, McDonald TJ, Mutt V (1987) Effect of porcine gastrin releasing peptide on gastric secretion and motility and the release of hormonal peptides in conscious cats. Peptides 8:423–430

Vaillant C, Dockray GJ, Walsh JH (1979) The avian proventriculus is an abundant source of endocrine cells with bombesin-like immunoreactivity. Histochemistry 64:307–314

Van Wimersma Greidanus TJB, Donker DK, van Zinnicq Bergmann FFM, Bekenkamp R, Maigret C, Spruijt B (1985a) Comparison between excessive grooming induced by bombesin or by ACTH: the differential elements of grooming and development of tolerance. Peptides 6:369–372

Van Wimersma Greidanus TJB, Donker DK, Walhof R, van Grafhorst JCA, de Vries N, van Schaik SJ, Maigret C, Spruijt BM, Colbern DL (1985b) The effects of neurotensin, naloxone and haloperidol on elements of excessive grooming behavior induced by bombesin. Peptides 6:1179–1183

Varner AA, Modlin IM, Walsh JH (1981) High potency of bombesin for stimulation of human gastrin release and gastric acid secretion. Regul Pept 1:289–296

Vaysse N, Pradayrol G, Chayvialle JA, Pignal F, Esteve JP, Susini C, Descos F, Ribet A (1981) Effects of somatostatin-14 and somatostatin-28 on bombesin-stimulated release of gastrin, insulin and glucagon in the dog. Endocrinology 108:1843–1847

Vigna SR, Mantyh CR, Giraud AS, Soll AH, Walsh JH, Mantyh PW (1987) Localization of specific binding sites for bombesin in the canine gastrointestinal tract. Gastroenterology 93:1287–1295

Villarreal JA, Brown MR (1978) Bombesin-like peptide in hypothalamus: chemical and immunological characterization. Life Sci 23:2729–2734

Wager-Srdar SA, Morley JE, Levine AS (1986) The effect of cholecystokinin, bombesin and calcitonin on food intake in virgin, lactating and postweaning female rats. Peptides 7:729–734

Wakelam MJO (1988) Inhibition of the amplified bombesin-stimulated inositol phosphate response in N-*ras* transformed cells by high density culturing. FEBS Lett 228:182–186

Wakelam MJO, Davies SA, Houslay MD, McKay I, Marshall CJ, Hall A (1986) Normal p21 N-*ras* couples bombesin and other growth factor receptors to inositol phosphate production. Nature 323:173–176

Walsh JH (1988) Peptides as regulators of gastric acid secretion. Annu Rev Physiol 50:41–63

Walsh JH, Wong HC, Dockray GJ (1979) Bombesin-like peptides in mammals. Fed Proc 38:2315–2319

Walsh JH, Maxwell V, Ferrari J, Varner AA (1981) Bombesin stimulates human gastric function by gastrin-dependent and independent mechanisms. Peptides [Suppl 2]2:193–198

Weber S, Zuckerman JE, Bostwick DG, Bensch KG, Sikic BI, Raffin TA (1985) Gastrin-releasing peptide is a selective mitogen for small cell lung carcinoma in vitro. J Clin Invest 75:306–309

Westendorf JM, Schonbrunn A (1982) Bombesin stimulates prolactin and growth hormone release by pituitary cells in culture. Endocrinology 110:352–358

Westendorf JM, Schonbrunn A (1983) Characterization of bombesin receptors in a rat pituitary cell line. J Biol Chem 258:7527–7535

Wharton J, Polak JM, Bloom SR, Ghatei MA, Solcia E, Brown MR, Pearse AGE (1978) Bombesin-like immunoreactivity in the lung. Nature 273:769–770

Willey JC, Lechner JF, Harris CC (1984) Bombesin and the C-terminal tetradecapeptide of gastrin-releasing peptide are growth factors for normal human bronchial epithelial cells. Exp Cell Res 153:245–248

Wolf SS, Moody TW (1985) Receptors for GRP-bombesin-like peptides in the rat forebrain. Peptides [Suppl 1]6:111–114

Wolf SS, Moody TW, O'Donohue TL, Zarbin MA, Kuhar MJ (1983) Autoradiographic visualization of rat brain binding sites for bombesin-like peptides. Eur J Pharmacol 87:163–164

Wolfe MM, Short GM, McGuigan JE (1987) β-adrenergic stimulation of gastrin release mediated by gastrin-releasing peptide in rat antral mucosa. Regul Pept 17:133–142

Woll PJ, Rozengurt E (1988) [D-Arg^1, D-Phe^5, D-Trp^{7,9}, Leu^{11}] substance P, a potent bombesin antagonist in murine Swiss 3T3 cells, inhibits the growth of human small cell lung cancer cells in vitro. Proc Natl Acad Sci USA 85:1859–1863

Woll PJ, Coy DH, Rozengurt E (1988) [Leu^{13}-ψ-(CH_2NH)Leu^{14}]bombesin is a specific bombesin antagonist in Swiss 3T3 cells. Biochem Biophys Res Commun 155:359–365

Wood SM, Wood JR, Ghatei MA, Lee YC, O'Shaughnessy D, Bloom SR (1981) Bombesin, somatostatin and neurotensin-like immunoreactivity in bronchial carcinoma. J Clin Endocrinol Metab 53:1310–1312

Wood SM, Jung RT, Webster JD, Ghatei MA, Adrian TE, Yanaihara N, Yanaihara C, Bloom SR (1983) The effect of the mammalian neuropeptide, gastrin-releasing peptide (GRP), on gastrointestinal and pancreatic hormone secretion in man. Clin Sci 65:365–371

Woods SC, Stein LJ, Figlewicz PD, Porte D Jr (1983) Bombesin stimulates insulin secretion and reduces food intake in the baboon. Peptides 4:687–697

Wunder BA, Hawkins MF, Avery DD, Swan H (1980) The effects of bombesin injected into the anterior and posterior hypothalamus on body temperature and oxygen consumption. Neuropharmacology 19:1095–1097

Yaksh TL, Michener SR, Bailey JE, Harty GJ, Lucas DL, Nelson DK, Roddy DR, Go VLW (1988) Survey of distribution of substance P, vasoactive intestinal polypeptide, cholecystokinin, neurotensin, Met-enkephalin, bombesin and PHI in the spinal cord of cat, dog, sloth and monkey. Peptides 9:357–372

Yamada T, Takami MS, Gerner RH (1981) Bombesin-like immunoreactivity in human cerebrospinal fluid. Brain Res 223:214–217

Yamaguchi K, Abe K, Kameya T, Adachi I, Taguchi S, Otsubo K, Yanaihara N (1983) Production and molecular size heterogeneity of immunoreactive gastrin-releasing peptide in fetal and adult lungs and primary lung tumors. Cancer Res 43:3932–3939

Yamaguchi K, Abe K, Adachi I, Suzuki M, Kimura S, Kameya T, Yanaihara N (1984) Concomitant production of immunoreactive gastrin-releasing peptide and calcitonin in medullary carcinoma of the thyroid. Metabolism 33:724–727

Yanaihara N, Yanaihara C, Mochizuki T, Iwahara K, Fujita T, Iwanage T (1981) Immunoreactive GRP. Peptides [Suppl 2]:2:185–191

Yang K, Ulich T, Taylor I, Cheng L, Lewin KJ (1983) Pulmonary carcinoids. Immunohistochemical demonstration of brain-gut peptides. Cancer 52:819–823

Yoshizaki K, de Bock V, Solomon S (1984) Origin of bombesin-like peptides in human fetal lung. Life Sci 34:835–843

Yoshizaki K, de Bock V, Takai I, Wang NS, Solomon S (1986) Bombesin-like peptides in human small cell carcinoma of the lung. Regul Pept 14:11–20

Zachary I, Rozengurt E (1985 a) High-affinity receptors for peptides of the bombesin family in Swiss 3T3 cells. Proc Natl Acad Sci USA 82:7616–7620

Zachary I, Rozengurt E (1985 b) Modulation of the epidermal growth factor receptor by mitogenic ligands: effects of bombesin and role of protein kinase C. Cancer Surv 4:729–765

Zachary I, Rozengurt E (1986) A substance P antagonist also inhibits specific binding and mitogenic effects of vasopressin and bombesin-related peptides in Swiss 3T3 cells. Biochem Biophys Res Commun 137:135–141

Zachary I, Rozengurt E (1987 a) Identification of a receptor for peptides of the bombesin family in Swiss 3T3 cells by affinity cross linking. J Biol Chem 262:3947–3950

Zachary I, Rozengurt E (1987 b) Internalization and degradation of peptides of the bombesin family in Swiss 3T3 cells occurs without ligand-induced receptor down regulation. EMBO J 6:2233–2239

Zachary I, Sinnett-Smith JW, Rozengurt E (1986) Early events elicited by bombesin and structurally related peptides in quiescent Swiss 3T3 cells. I. Activation of protein kinase C and inhibition of epidermal growth factor binding. J Cell Biol 102:2211–2222

Zachary I, Millar J, Nanberg E, Higgins T, Rozengurt E (1987) Inhibition of bombesin-induced mitogenesis by pertussis toxin: dissociation from the phospholipase C pathway. Biochem Biophys Res Commun 146:456–463

Zarbin MA, Kuhar MJ, O'Donohue TL, Wolf SS, Moody TW (1985) Autoradiographic localization of [^{125}I-Tyr^4] bombesin-binding sites in rat brain. J Neurosci 5:429–437

Zoeller RT, Lebacq-Verheyden AM, Battey JF (1989) Distribution of two distinct messenger ribonuclear acids encoding gastrin-releasing peptide in rat brain. Peptides 10:415–422

CHAPTER 22

Platelet-Derived Endothelial Cell Growth Factor

K. MIYAZONO and C.-H. HELDIN

A. Introduction

Angiogenesis, the formation of new blood vessels, is an important process in various normal and pathological conditions, such as embryogenesis, tumor growth, and wound healing, and is composed of three different stages, i.e., migration, proliferation, and maturation of endothelial cells (FOLKMAN 1984).

The best-characterized endothelial cell mitogens are fibroblast growth factors (FGFs), which are a family of related polypeptides with high affinities for heparin (GOSPODAROWICZ et al. 1986; LOBB et al. 1986). FGFs were originally isolated from neural tissues, but have been found also in other cell types, including macrophages (BAIRD et al. 1985).

Platelets are a rich source of growth regulatory proteins; the best characterized of these are platelet-derived growth factor (PDGF) (HELDIN et al. 1985; ROSS et al. 1986) and transforming growth factor-β (TGF-β) (SPORN et al. 1987). PDGF stimulates the growth of vascular smooth muscle cells and fibroblasts, but not of vascular endothelial cells. TGF-β has a growth inhibitory effect on most cell types, including endothelial cells. In spite of these findings, platelets have been shown to play an important role in the growth and metabolism of vascular endothelial cells. MACA et al. (1977) found that intact platelets enhance DNA synthesis and replication of cultured endothelial cells, and that this growth-promoting effect is limited to platelets and does not occur in erythrocytes or leukocytes. Growth-promoting activity for endothelial cells, which could be separated from the fibroblast growth-promoting activity by gel chromatography, was recently identified in lysates of human platelets (KING and BUCHWALD 1984; MIYAZONO et al. 1985). The structural and functional characteristics of this factor, in combination with the fact that FGFs have not been found in platelets, suggested that the endothelial cell growth-promoting activity resided in a novel growth factor; this factor was later denoted platelet-derived endothelial cell growth factor (PD-ECGF).

The focus of the present review is this novel factor, which was recently purified and cloned; PD-ECGF is the major mitogen for endothelial cells in platelet lysates, and has biological and biochemical properties that are distinct from those of other known mitogens.

B. Purification and Biochemical Characterization of PD-ECGF

I. Purification of PD-ECGF

During the purification of PD-ECGF, stimulation of [^{3}H]thymidine incorporation into porcine aortic endothelial cells was used to assay the activity (MIYAZONO et al. 1985).

A procedure for the purification of PD-ECGF in small quantities was described in 1987 (MIYAZONO et al. 1987). Recently, the purification procedure has been improved, and now allows the purification of 30–40 μg PD-ECGF/batch (MIYAZONO and HELDIN 1989).

As starting material, a side fraction from the purification of PDGF from human platelets was used. Platelet lysate corresponding to 800–1000 liters human blood was first chromatographed on a CM-Sephadex column. Since PDGF is a cationic protein, it bound to this column, whereas PD-ECGF was recovered in the flow-through fraction. This fraction was then further purified by chromatography on QAE-Sephadex, ammonium sulfate precipitation, and chromatography on DEAE-Sepharose, high-performance hydroxylapatite, and alkyl-Superose (Fig. 1). Since PD-ECGF is destroyed at extreme pH values, all procedures were performed at neutral pH. To improve the

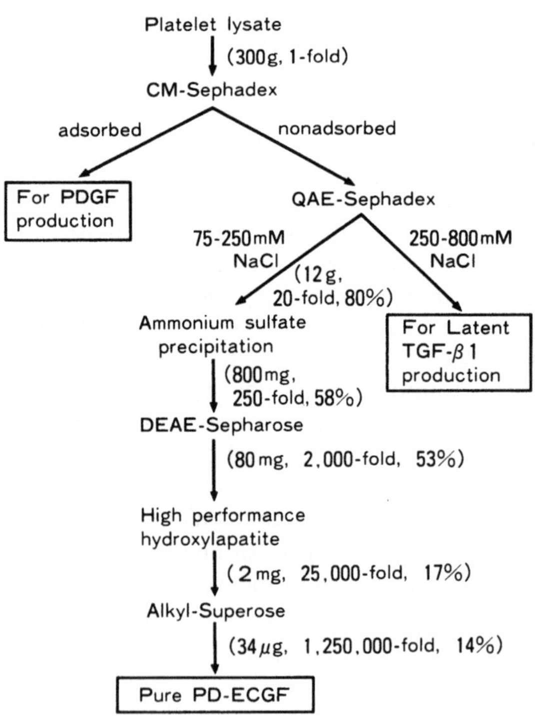

Fig. 1. Purification of PD-ECGF. The amounts of protein, purification fold, and activity yield are shown in parentheses

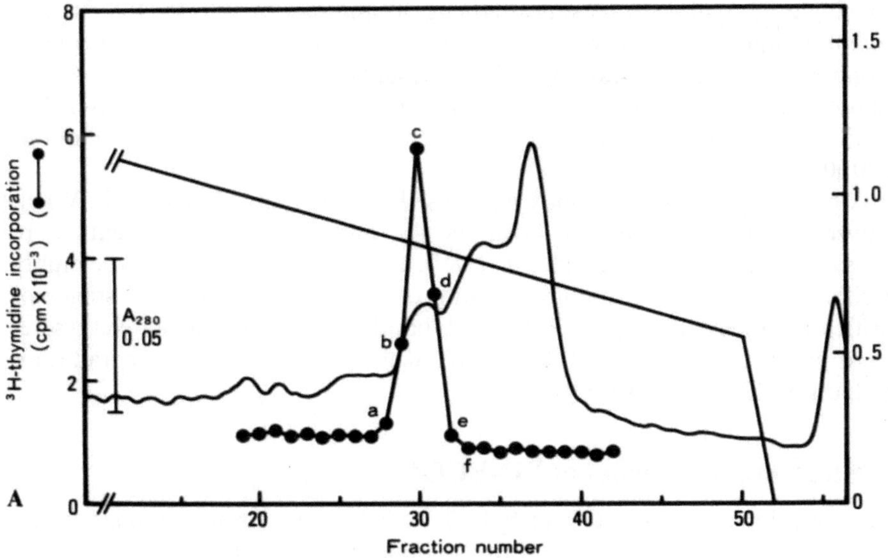

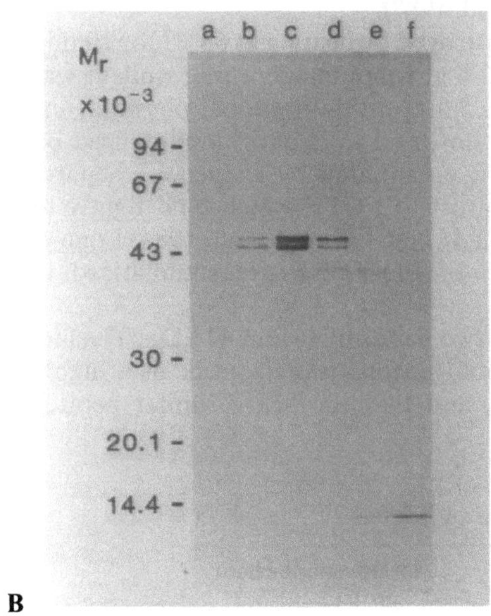

Fig. 2 A, B. Alkyl-Superose chromatography of PD-ECGF. **A** Alkyl-Superose chromatography of the material obtained by high-performance hydroxylapatite chromatography. **B** Analysis by SDS-gel electrophoresis and silver staining of the fractions from the alkyl-Superose chromatography

recovery, the purification was designed to minimize dialyses and concentrations between the steps. The two final chromatographic steps on hydroxylapatite and alkyl-Superose HPLC-grade columns were crucial in the purification because of their very high resolving capacities.

Alkyl-Superose chromatography takes advantage of hydrophobic interactions between proteins and the column matrix. In this step PD-ECGF eluted as a single peak in a decreasing gradient of ammonium sulfate (Fig. 2). The protein composition of the purified material was analyzed by SDS-gel electrophoresis and silver staining. The two bands of M_r 46000 and 44000 coeluted with the mitogenic activity and represent PD-ECGF (see below). The purification procedure yielded an overall increase in specific activity of about 1200000-fold at a recovery of 14%.

A polyclonal rabbit antibody was raised against native PD-ECGF. When analyzed by immunoblotting, both the 46- and 44-kDa components were recognized (MIYAZONO and HELDIN 1989). The antiserum seemed to be highly specific for PD-ECGF; immunoblotting of material from various stages of purification revealed only the PD-ECGF bands except in crude platelet lysate, where some components of lower M_r which may represent degradation products of PD-ECGF were seen.

II. Structural Properties of PD-ECGF

The migration of PD-ECGF in SDS-gel electrophoresis was similar under reducing and nonreducing conditions, indicating that it is a single-chain polypeptide. Analysis by chromatofocusing revealed that the isoelectric point of PD-ECGF is about 4.8 (MIYAZONO et al. 1987).

The bioactivity of PD-ECGF was destroyed by heating at 65° C for 10 min or by exposure to pH 4.5 or lower for 2 h at room temperature, while it was resistant to reducing conditions, such as 5 m*M* dithiothreitol at physiological pH and ionic strength. PD-ECGF did not bind to heparin immobilized on Sepharose, and the bioactivity was not potentiated by exogenous heparin (MIYAZONO et al. 1987). This is in contrast to FGFs, which bind tightly to heparin-Sepharose and elute only at 1–2 *M* NaCl (SHING et al. 1984; LOBB et al. 1986). The biochemical properties of PD-ECGF are summarized in Table 1.

The purified PD-ECGF occurred as two bands of 46 and 44 kDa. The fact that both bands eluted as a single, homogeneous protein peak in a highly resolving reverse-phase HPLC system, and that they gave similar peptide

Table 1. Biochemical properties of PD-ECGF

Molecular weight	45000, single-chain protein
Isoelectric point	4.8
Stability – Activity destroyed by	Acid (below pH 4.5) Heat (65° C, 30 min) Trypsin
– Activity resistant to	Dithiothreitol
No affinity for	Heparin Concanavalin A Wheat germ agglutinin

maps after staphylococcal V8 protease digestion, indicated that both components represent PD-ECGF (MIYAZONO and HELDIN 1989). The microheterogeneity is probably due to partial proteolysis of PD-ECGF.

When native PD-ECGF was treated with trypsin, the bioactivity was destroyed, accompanied by a conversion of the 44- to 46-kDa bands to components of 37–39 kDa, which were rather resistant to further degradation. This observation indicates that PD-ECGF contains a trypsin-resistant domain, but that the intact molecule is needed for its bioactivity (MIYAZONO and HELDIN 1989).

C. Primary Sequence of PD-ECGF

In order to clone the cDNA for PD-ECGF, immunoblotting experiments using the antiserum against PD-ECGF were performed to find an appropriate source of mRNA for a cDNA library. Although PD-ECGF is present in human platelets, none of the hemopoietic cell lines investigated produced appreciable quantities of PD-ECGF (USUKI et al., unpublished observation). In contrast, extracts from human term placenta contained large amounts of PD-ECGF. Thus, a human placental cDNA library was constructed in λgt10 and screened by oligonucleotide probes deduced from the amino acid sequence that was obtained by sequencing of peptides of PD-ECGF. A clone with a 1.8-kilobase (kb) insert was identified and found to represent a full-length cDNA clone for PD-ECGF (ISHIKAWA et al. 1989).

PD-ECGF consists of 482 amino acids as deduced from the sequence of the cDNA clone; almost 400 of these were verified by protein sequencing

```
         ↓
MAALMTPGTGAPPAPGDFSGEGSQGLPDPSPEPKQLPELIRMKRDGGRLSEADIRGFVAA    60

  #
VVNGSSQGAQIGAMLMAIRLRGMDLEETSVLTQALAQSGQQLEWPEAWRQQLVDKHSTGG   120

               * *
VGDKVSLVLAPALAACGCKVPMISGRGLGHTGGTLDKLESIPGFNVIQSPEQMQVLLDQA   180

 **
GCCIVGQSEQLVPADGILYAARDVTATVDSLPLITASILSKKLVEGLSALVVDVKFGGAA   240

                                       *             *
VFPNQEQARELAKTLVGVGASLGLRVAAALTAMDKPLGRCVGHALEVEEALLCMDGAGPP   300

DLRDLVTTLGGALLWLSGHAGTQAQGAARVAAALDDGSALGRFERMLAAQGVDPGLARAL   360

*
CSGSPAERRQLLPRAREQEELLAPADGTVELVRALPLALVLHELGAGRSRAGEPLRLGVG   420

                                                  @
AELLVDVGQRLRRGTPWLRVHRDGPALSGPQSRALQEALVLSDRAPFAAPSPFAELVLPPQQ 482
```

Fig. 3. Amino acid sequence of PD-ECGF. An amino-terminal processing site is indicated by an *arrow*. Cysteine residues (*), a potential N-glycosylation site (#) and a possible site of polymorphism @ are indicated

(Fig. 3). The sequence is unique; no significant homology to other proteins was found. Short stretches of internal repeats were, however, noted. The amino-terminal sequence of PD-ECGF starts at position 10; proteolytic processing probably occurs at this position after synthesis. The predicted M_r of the mature protein is about 48 000, which is close to the estimate by SDS-gel electrophoresis and silver staining of purified PD-ECGF (Fig. 2). PD-ECGF contains only one tyrosine residue, which explains why it has been difficult to radiolabel it by the chloramine-T method (MIYAZONO and HELDIN 1989). There is one potential N-glycosylation site, located at position 63. However, PD-ECGF is most likely not glycosylated, since it does not bind to concanavalin A-Sepharose or wheat germ agglutinin-Sepharose (our unpublished data). There are seven cysteine residues in the molecule, indicating that at least one of them has a free SH-group.

In analogy with acidic (JAYE et al. 1986) and basic (ABRAHAM et al. 1986a, b) FGF, PD-ECGF does not have an amino-terminal hydrophobic signal sequence. This indicates that these endothelial cell mitogens are not secreted by the common mechanism. The fact that these factors act on responder cells via cell surface receptors indicates that some mechanism for release from the producer cell must prevail; how this occurs, however, remains to be elucidated.

The PD-ECGF cDNA clone was expressed in NIH 3T3 cells under the control of Moloney leukemia virus long-terminal repeats. A transfected cell line was selected that produced PD-ECGF. A lysate of these cells contained a 45-kDa component when analyzed by immunoblotting using the PD-ECGF antiserum, and stimulated the [^{3}H]thymidine incorporation into porcine aortic endothelial cells (ISHIKAWA et al. 1989). Consistent with the lack of hydrophobic leader sequence, no PD-ECGF was found in the conditioned medium of the transfected cell line.

D. Biological Activities of PD-ECGF

I. In Vitro Effects of PD-ECGF

Since endothelial cells from large vessels and capillaries have different properties (ZETTER 1984), it was important to test the effect of PD-ECGF on different types of endothelial cells. As shown in Fig. 4, pure PD-ECGF stimulated the growth of endothelial cells derived from porcine aorta, murine lung capillaries, as well as human umbilical vein. Maximal stimulation was achieved at about 16 ng/ml (350 p*M*). However, PD-ECGF had no effect on the growth of different types of fibroblasts (MIYAZONO et al. 1987), in contrast to FGFs that stimulate growth of both fibroblasts and endothelial cells.

When added to the cell culture medium, the PD-ECGF antiserum neutralized the mitogenic activity of native PD-ECGF. Furthermore, the antiserum completely inhibited the endothelial cell mitogenic activity of a platelet lysate,

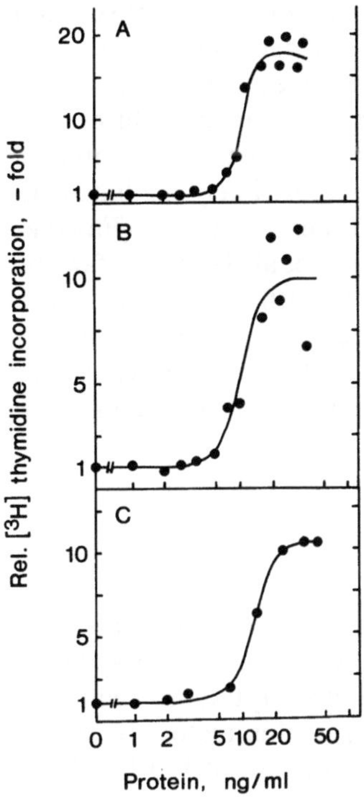

Fig. 4 A–C. Growth-promoting activity of PD-ECGF on various kinds of endothelial cells. Different concentrations of pure PD-ECGF were added to endothelial cells from (**A**) porcine aorta, (**B**) murine lung capillary (LE-II cells; gift from T Maciag), and (**C**) human umbilical vein

indicating that PD-ECGF is the principal mitogen for these cells in platelets (MIYAZONO and HELDIN 1988).

The growth-promoting activity of PD-ECGF could also be completely neutralized by suramin (our unpublished observation). Suramin is a low-M_r cyclic compound that antagonizes the effect of PDGF, EGF, and TGF-β (BETSHOLTZ et al. 1986; COFFEY et al. 1987). Thus suramin seems to be able to interfere with the interactions between a broad range of growth regulatory proteins and their receptors.

PD-ECGF was also shown to stimulate endothelial cell chemotaxis in vitro (ISHIKAWA et al. 1989), another crucial step in the process of angiogenesis. Half-maximal stimulation was obtained at ab out 1 ng/ml pure PD-ECGF and the chemotactic activity was specific for endothelial cells; PD-ECGF did not induce the migration of vascular smooth muscle cells (ISHIKAWA et al. 1989).

II. In Vivo Effects of PD-ECGF

The in vitro effects of PD-ECGF described above revealed that PD-ECGF stimulates both growth and chemotaxis of endothelial cells. Interestingly, PD-ECGF was also found to induce angiogenesis in vivo in the chick chorioallantoic membrane assay; the PD-ECGF antiserum inhibited this response (ISHIKAWA et al. 1989). These biological properties of PD-ECGF are thus similar to those of acidic and basic FGF, which stimulate chemotaxis and proliferation of endothelial cells in vitro and induce angiogenesis in vivo.

Several angiogenic factors have been described (FOLKMAN and KLAGSBRUN 1987). The mechanism of action of these factors may differ. For example, tumor necrosis factor-α and TGF-β inhibit growth of endothelial cells in vitro, and thus their angiogenic activity is most likely mediated indirectly, probably by induction of synthesis of other molecules (FRÀTER-SCHRÖDER et al. 1987; LEIBOVICH et al. 1987; ROBERTS et al. 1986). So far, the only angiogenic factors found that also stimulate endothelial cell growth in vitro are FGFs and PD-ECGF. However, the target-cell specificity of FGFs are rather broad, whereas PD-ECGF seems to be specific for endothelial cells. Thus, PD-ECGF represents a novel type of angiogenic factor that seems to act specifically on endothelial cells.

E. Conclusion

PD-ECGF is a novel 45-kDa angiogenesis factor. The availability of fairly large quantities of pure factor as well as of a specific antiserum and a cDNA clone for PD-ECGF will now make it possible to address questions related to the mechanism of action and the in vivo function of this factor. It will be particularly interesting to explore the possible involvement of PD-ECGF in normal and pathological angiogenesis.

References

Abraham JA, Mergia A, Whang JL, Tumolo A, Friedman J, Hjerrild KA, Gospodarowicz D, Fiddes JC (1986a) Nucleotide sequence of a bovine clone encoding the angiogenic protein, basic fibroblast growth factor. Science 233:545–548

Abraham JA, Whang JL, Tumolo A, Mergia A, Friedman J, Gospodarowicz D, Fiddes JC (1986b) Human basic fibroblast growth factor: nucleotide sequence and genomic organization. EMBO J 5:2523–2528

Baird A, Mormède P, Böhlen P (1985) Immunoreactive fibroblast growth factor in cells of peritoneal exudate suggest its identity with macrophage-derived growth factor. Biochem Biophys Res Commun 126:358–364

Betsholtz C, Johnsson A, Heldin C-H, Westermark B (1986) Efficient reversion of siman sarcoma virus transformation and inhibition of growth factor-induced mitogenesis by suramin. Proc Natl Acad Sci USA 83:6440–6444

Coffey RJ, Leof EB, Shipley GD, Moses HL (1987) Suramin inhibition of growth factor receptor binding and mitogenicity in AKR-2B cells. J Cell Physiol 132:143–148

Folkman J (1984) Angiogenesis. In: Jaffe EA (ed) Biology of endothelial cells. Nijhoff, Boston, pp 412–428

Folkman J, Klagsbrun M (1987) Angiogenic factors. Science 235:442–447

Fràter-Schröder M, Risau W, Hallman R, Gautschi P, Böhlen P (1987) Tumor necrosis factor type α, a potent inhibitor of endothelial cell growth in vitro, is angiogenic in vivo. Proc Natl Acad Sci USA 84:5277–5281

Gospodarowicz D, Neufeld G, Schweigerer L (1986) Fibroblast growth factor. Mol Cell Endocrinol 46:187–204

Heldin C-H, Wasteson Å, Westermark B (1985) Platelet-derived growth factor. Mol Cell Endocrinol 39:169–187

Ishikawa F, Miyazono K, Hellman U, Drexler H, Wernstedt C, Hagiwara K, Usuki K, Takaku F, Risau W, Heldin C-H (1989) Identification of angiogenic activity and the cloning and expression of platelet-derived endothelial cell growth factor. Nature 338:557–562

Jaye M, Howk R, Burgess W, Ricca GA, Chiu I-M, Ravera MW, O'Brien SJ, Modi WS, Maciag T, Drohan WN (1986) Human endothelial cell growth factor: cloning, nucleotide sequence, and chromosome localization. Science 233:541–545

King GL, Buchwald S (1984) Characterization and partial purification of an endothelial cell growth factor from human platelets. J Clin Invest 73:392–396

Leibovich SJ, Polverini PJ, Shepard HM, Wiseman DM, Shively V, Nuseir N (1987) Macrophage-induced angiogenesis is mediated by tumor necrosis factor-α. Nature 329:630–632

Lobb RR, Harper JW, Fett JW (1986) Purification of heparin-binding growth factor. Anal Biochem 154:1–14

Maca RD, Fry GL, Hoak JC, Loh PT (1977) The effects of intact platelets on cultured human endothelial cells. Thromb Res 11:715–727

Miyazono K, Heldin C-H (1989) High yield purification of platelet-derived endothelial cell growth factor: structural characterization and establishment of a specific antiserum. Biochemistry 28:1704–1710

Miyazono K, Okabe T, Ishibashi S, Urabe A, Takaku F (1985) A platelet factor stimulating the proliferation of vascular endothelial cells: partial purification and characterization. Exp Cell Res 159:487–494

Miyazono K, Okabe T, Urabe A, Takaku F, Heldin C-H (1987) Purification and properties of an endothelial cell growth factor from human platelets. J Biol Chem 262:4098–4103

Roberts AB, Sporn MB, Assoian RK, Smith JM, Roche NS, Wakefield LM, Heine UI, Liotta LA, Falanga V, Kehrl JH, Fauci AS (1986) Transforming growth factor type-β: rapid induction of fibrosis and angiogenesis in vivo and stimulation of collagen formation in vitro. Proc Natl Acad Sci USA 83:4167–4171

Ross R, Raines EW, Bowen-Pope DF (1986) The biology of platelet-derived growth factor. Cell 46:155–159

Shing Y, Folkman J, Sullivan R, Butterfield C, Murray J, Klagsbrun M (1984) Heparin affinity: purification of a tumor-derived capillary endothelial cell growth factor. Science 223:1296–1299

Sporn MB, Roberts AB, Wakefield LM, de Crombrugghe B (1987) Some recent advances in the chemistry and biology of transforming growth factor-beta. J Cell Biol 105:1039–1045

Zetter BR (1984) Culture of capillary endothelial cells. In: Jaffe EA (ed) Biology of endothelial cells. Nijhoff, Boston, pp 14–26

CHAPTER 23

Nerve Growth Factor

M. V. CHAO

A. Introduction

The discovery of nerve growth factor (NGF) nearly 4 decades ago and the research that followed have had a profound impact upon our understanding of neurotrophic factors and neuronal survival and development. The requirement of NGF by selective neuronal populations implies that many other neurotrophic factors are functioning in the nervous system. Yet NGF still remains as the only polypeptide factor that has been shown to be physiologically required in vivo (LEVI-MONTALCINI and ANGELETTI 1963; LEVI-MONTALCINI 1987). The fortuitious discovery that large quantities of NGF are produced in the submaxillary gland (COHEN 1960) provided a source for purification and a means to establish its biological significance.

The genes encoding NGF and its receptor have been recently isolated and used to measure the pattern of expression in neuronal and nonneuronal tissues. In many ways these findings have revealed new insights into the distribution of NGF and its receptor, altered previous notions about the sites and modes of action of NGF, and suggested new approaches to investigate further the broad actions that NGF possesses.

This chapter will review the molecular structure of the genes encoding the NGF molecule and its receptor and studies concerning NGF's mechanism of action. Attention will be given to the identification and isolation of a number of cellular genes known to be activated by NGF treatment; possible second messengers that serve to transmit NGF's effects; and the role of the receptor in signal transduction. Due to the wide variety of specific responses to NGF, multiple signals and pathways are probably utilized. A common requirement for the biological effects of NGF is interaction with a specific cell surface receptor. The definition of functional NGF receptors will ultimately clarify the mechanism by which NGF exerts its many effects upon responsive cells.

B. Nerve Growth Factor Gene Structure

I. Nerve Growth Factor Protein Complex

Nerve growth factor is found in the mouse submaxillary gland as a 7S complex composed a dimer of three subunits α, β, and γ (BRADSHAW 1978; GREENE and Shooter 1980). The biological activities of NGF are entirely contained within the β-subunit of 118 amino acids, which exists as a 2.5S form. The β-subunit is produced from a precursor molecule which is processed by proteolytic cleavage at both the N and C termini. Both the α- and γ-chains are members of the kallikrein family of serine proteases; however, only the γ-subunit contains enzymatic activity. These α- and γ-chains are believed to protect the β-subunit and the γ-subunit is thought to be involved in processing the β-subunit precursor.

II. Gene Structure

Molecular clones for the β-subunit of NGF were initially isolated by the use of oligonucleotides made to the determined amino acid sequence (Angeletti and Bradshaw 1971). These oligonucleotides were used to identify cDNA clones from a library of mouse submaxillary gland (Scott et al. 1983). A genomic clone encoding human NGF was isolated using a murine cDNA clone (Ullrich et al. 1983). The gene encoding the human β-NGF has been localized to chromosome 1 (p22.1) by somatic cell hybridization (Francke et al. 1983; Zabel et al. 1985). The mouse homolog has been identified in chromosome 3, along with the gene encoding epidermal growth factor (Zabel et al. 1985).

The entire mouse β-NGF gene spans nearly 45 kilobases (kb) and is represented in Fig. 1. A large intron of 32 kb makes up a large portion of the gene. At least four exons are present, the most 3′ of which (exon IV) contains the β-NGF sequence. A preliminary comparison of the murine and human genes indicates that the two genes are organized with similar intron/exon structure (Selby et al. 1987a). The coding region of the gene predicts that NGF is made as a precursor molecule of 327 amino acids.

A single polyadenylation site is present approximately 130 nucleotides into the 3′ untranslated region. A remarkable conservation of sequence has been noticed in the 3′ untranslated region of diverse species from avian to mammalian. This sequence is characterized by an abundance of AU residues and contains a consensus sequence ATTTA. This conserved 3′ untranslated sequence has been shown to be responsible for imparting mRNA instability for a number of genes including interleukin-1, colony-stimulating factor, tumor necrosis factor, and protooncogenes such as *fos, sis, myc,* and *myb* (Shaw and Kamen 1986).

Two major NGF transcripts have been detected which differ significantly in tissue distribution and size. In the mouse placenta and submaxillary gland,

Fig. 1. Genomic structure for the mouse β-NGF gene. Exons are shown *in solid boxes,* with the coding regions of the β-subunit at the 3′ end of exon IV

the major mRNA species is a longer transcript (*A*), whereas in all other tissues, including muscle and brain, a shorter transcript (*B*) is preferentially expressed over the *A* transcript (EDWARDS et al. 1986). These two transcripts are the product of alternative splicing of exon II, resulting in the absence of the second exon in the short transcript. Significantly, the ratio of the *A* and *B* transcripts is the same in all the tissues examined, suggesting the initiation of NGF transcription is regulated. Transcript *B* is generally present at least three times the level of transcript *A*, in those tissues where it predominates. At least two other minor transcripts have been mapped, both of which are missing sequences in the second exon. These minor transcripts are present at a very low level, considerably less than the *A* transcript (SELBY et al. 1987a). All of these transcripts are derived from a single copy *NGF* gene.

The major consequence of differential splicing of NGF mRNAs is that the N-terminal sequence of the precursor form of NGF is altered. No differences exist in the coding sequence for the β-NGF. Transcript *A* can produce a precursor protein of 34 kDa and transcript *B* generates a smaller precursor of 27 kDa. A hydrophobic sequence is present in the second exon and it has been hypothesized that this hydrophobic segment may serve as a transmembrane-spanning domain for the longer precursor. Alternatively, the presence of a hydrophobic stretch of amino acids may serve as a secretory signal peptide in the shorter precursor (EDWARDS et al. 1986). Therefore, differential splicing may dictate the cellular localization of *NGF*. The precursor protein for the β-subunit for NGF is not known to encode any other functional protein. The α- and γ-subunits are encoded by separate genes, and are not part of the prepro-β-NGF protein.

III. Nerve Growth Factor Gene Promoter

Due to the large size of the NGF gene and the small size of the 5′ exons, the precise initiation sites of transcription have been difficult to detect. However, primer extension and S1 nuclease protection experiments have defined the start point of transcription for the NGF gene in a variety of tissues, including brain, submaxillary gland, placenta, heart, kidney, and intestine. Two promoter sequences have been detected, one for *A* and *B* transcripts, and a second, which probably accounts for the minor *C* transcript (SELBY et al. 1987a).

The major promoter sequence is rich in GC residues, but contains two AT-rich sequences, TAAATA and TTAAA, which are within 50 bp from the start site (Fig. 2). This promoter sequence has been shown to be transcriptionally active in fibroblasts using the human growth hormone gene as a reporter gene (ZHENG and HEINRICH 1988). A second promoter has been found upstream of the sequence shown in Fig. 2. It does not contain any TATA consensus sequences, but two CCAAT elements have been found within 200 bp upstream of exon IA. This second promoter site is believed to account for the initiation of the minor C transcript.

```
GCTATGTCCCATCAACTCGGGAGCTATCCATCCCTTGTCCCCAGGACCCTTACAACCCGGACCCC
   -220                   -200                  -180

TGGGTCTAGTCACAGCAGGTGCGGGCTGGGATTGGAGTTGGCCAGAGAGGGAGGGGTCGGGTGAG
-160                -140                 -120                 -100

TGGGGGGCAGGATTTGGAGAGGGTGTGACGAGCCTGGAGGAGGGGC[TAAATA]CAGTCAGGAAGCC
   -90       -80        -70        -60        -50        -40

TG[TTAAA]GAAGCTCTGTGCTCCAGCACGGC[AGAGAGCGCCTGGAGCCGGAGGGGAGCGCATCG]
     -20        -10
```

Fig. 2. Promoter sequence of the mouse NGF gene. The promoter for the most abundant transcripts (*A* and *B*) is shown. The bracketed sequence is the IB exon (see Fig. 1). This sequence was derived from SELBY et al. (1987) and ZHENG and HEINRICH (1988)

IV. Amino Acid Sequence

The majority of biological studies have been conducted with NGF isolated from the mouse submaxillary gland. Mouse NGF has been used to obtain biological responses in chicken, rat, human, bovine tissues, and cells. Since the responses have been detected over the same concentration ranges, it is clear that NGF must be evolutionarily highly conserved in structure and function.

This prediction has been borne out by the molecular cloning of chick (MEIER et al. 1986; GOEDERT 1986; EBENDAL et al. 1986; WION et al. 1986), rat (WHITTEMORE et al. 1988), bovine (MEIER et al. 1986), cobra (SELBY et al. 1987b), and guinea pig β-NGF genes (SCHWARZ et al. 1988) molecules. Figure 3 indicates the high degree of similarity between the coding regions of β-NGF. A large number of structural features are highly conserved. These include the six cysteine residues, potential N-linked glycosylation sites, histidine and tryptophan residues essential for activity, and the dibasic amino acids involved in the proteolytic cleavage of the prepro-β-NGF molecule.

It is known that anti-mouse NGF antibodies do not exhibit the same effects on the biological activity of NGF in other species, such as chick (LEVI-MONTALCINI and ANGELETTI 1968). This result is surprising in light of the extraordinary conservation of sequence. The few differences in amino acid sequence most likely account for the poor immunological crossreactivity between different species of NGF (THOENEN and BARDE 1980). These changes in sequence are found in primarily hydrophilic regions, between amino acids 60–65 and 90–95. Since hydrophilic residues are generally believed to contain antigenic determinants, these differences may explain why antibodies made against such a highly conserved protein do not crossreact.

A single hydrophilic region in the vicinity of amino acid 33 is conserved in every species of NGF examined. Anti-peptide antibodies to this conserved region were found to be most effective in recognizing native NGF (Ebendal et al. 1989). It is believed this region is involved in the biological activity of NGF

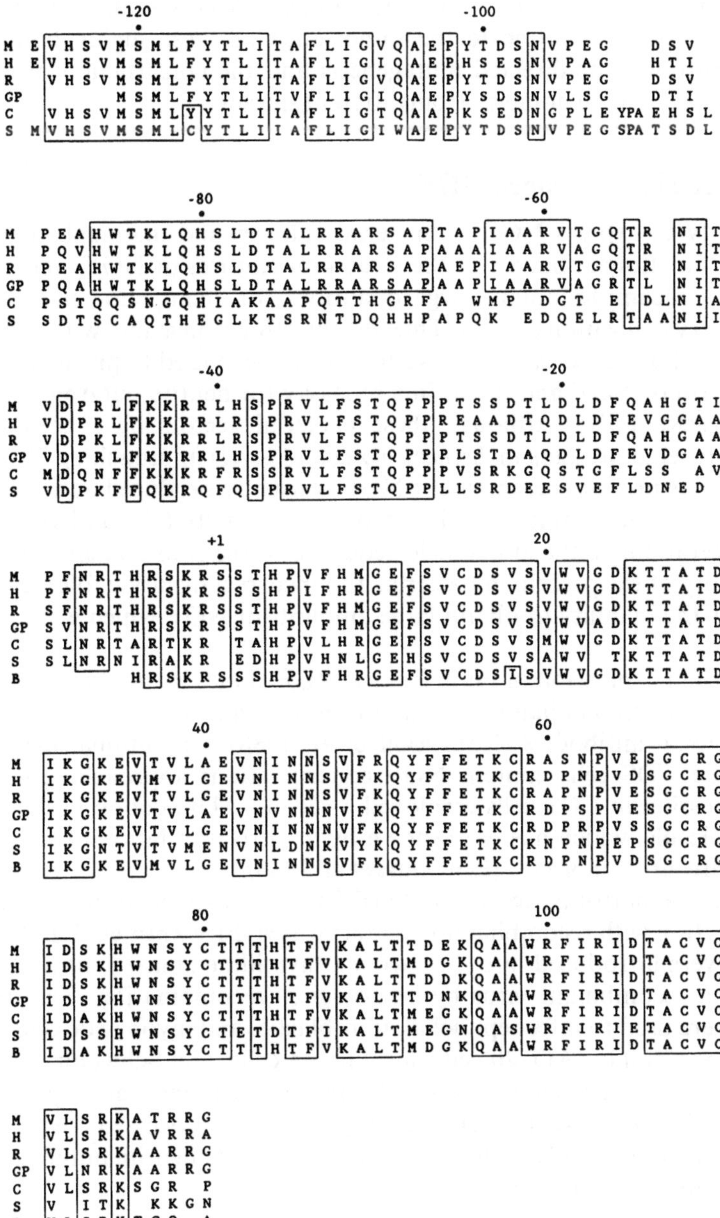

Fig. 3. Amino acid sequence of the *β*-precursor protein. Amino acid residues extending from −125 to +120 are shown. Conserved blocks of sequence homology are *boxed.* The sequence for the mouse (*M*), human (*H*), rat (*R*), guinea pig (*GP*), chicken (*C*), and snake (*S*) are represented. The N-terminus of the mature *β*-subunit is designated *+1*

and perhaps the binding of NGF to its receptor (MEIER et al. 1986). It should be feasible now to delineate the importance of these highly homologous regions through site-directed mutagenesis and the construction of chimeric NGF molecules with domains from different species.

V. Expression of Cloned NGF

The ability to express NGF in bacterial cultures and mammalian cells will permit a functional analysis of the structural features of NGF, and will also provide a means to obtain large quantities for therapeutic purposes and eventual crystallization studies. Thus far the efforts have met with mixed success. Several bacterial expression systems have been used to produce a recombinant form of NGF and tested in a bioassay for neurite outgrowth in PC12 cells (GREENE 1977). Although these bacterial vectors have been used to produce foreign proteins in *Escherichia coli* at levels approaching 20% of the total cellular protein, recombinant NGF is usually produced at a lower level (0.1%–1%), and exhibits only low biological activity (HU and NEET 1988). Slight alterations in the amino acid sequence of the recombinant NGF used, the stability of the protein in bacteria, and incorrect disulfide folding may account for the inability to obtain large amounts of NGF which is biologically active.

Transient expression of rat and chick NGF clones in monkey kidney COS cells generates recombinant NGF that is biologically active and is recognized by polyclonal antibodies (HALLBOOK et al. 1988). Conditional media from transfected COS cells contain NGF concentrations approaching 40 ng/ml. The expression in COS cells employed a second AUG codon in the prepro-NGF molecule, which has been shown by in vitro translation to produce the 27-kDa precursor molecule (EDWARDS et al. 1986).

In an alternative approach, β-NGF has been expressed in cultured cells after infection with recombinant vaccinia viruses (EDWARDS et al. 1988). NGF cDNAs were introduced into the vaccinia thymidine kinase gene and used to generate a recombinant virus by homologous recombination. The resulting viruses were used to infect epithelial, fibroblastic, and myoblastic cells, as well as anterior pituitary AtT-20 cells, insulinoma cells, and Schwann cells. NGF was present in the supernatants of infected cell lines at approximately 25 ng/ml. The expressed NGF was appropriately processed, secreted, and shown to be biologically active by the stimulation of neurite outgrowth from chicken dorsal root ganglion cells.

VI. The α- and γ-Subunits

The α- and γ-proteins that constitute the 7S complex of NGF are representatives of a family of serine proteases which give rise to bioactive peptides such as epidermal growth factor (EGF) and renin. Both of these subunits bind specifically to β-NGF but at different sites. The genes encoding the mouse α- and γ-subunits are linked within 10 kb (B. A. EVANS and RICHARDS 1985). The

```
              -20                                             +1
α   M W F L I L F L A L S L G G I D A A P P V Q S     Q V         D C E   N S Q P W
γ                                                     I V G G F K C E K N S Q P W
K   M W F L I L F L A L S L G R N D A A P P V Q S X X R V V G G Y N C E M N S Q P W

                        20                                      40
α   H V A V Y R F N K Y Q C G G V L L D R N W V L T A A H C Y N D K Y Q V W L G K N
γ   H V A V Y R Y T Q Y L C G G V L L D P N W V L T A A H C Y D D N Y K V W L G K N
K   Q V A V Y Y F G E Y L C G G V L I D P S W V I T A A H C A T D N Y Q V W L G R N

                        60                                      80
α   N F L E D E P S D Q H R L V S K A I P H P D F N M S L L N E H T P Q P E D D Y S
γ   N L F K D E P S A Q H R F V S K A I P H P G F N M S L M R         F L E Y D Y S
K   N L Y E D E P F A Q H R L V S Q S F P H P G F N Q D L I W N H T R Q P G D D Y S

                        100                                     120
α   N D L M L L R L S K P A D I T D V V K P I T L P T E E P K L G S T C L A S G W G
γ   N D L M L L R L S K P A D I T D T V K P I T L P T E E P K L G S T C L A S G W G
K   N D L M L L H L S Q P A D I T D G V K V I D L P I E E P K V G S T C L A S G W G

                        140                                     160
α   S T T P I K F K Y P D D L Q C V N L K L L P N E D C D K A H K M K V T D A M L C
γ   S I T P T K F Q F T D D L Y C V N L K L L P N E D C A K A H I E K V T D A M L C
K   S I T P D G L E L S K K L Q C V N I D L L S N E K C E A H K E E V T T D L M L C

                        180                                     200
α   A G E M D G G S Y T C E H D S G G P L I C D G I L Q G I T S W G P E P C G E P T
γ   A G E M D G G K D T C K G D S G G P L I C D G V L Q G I T S W G H T P C G E P D
K   A G E M D G G K D T C K G D S G G P L I C N G V L Q G I T S W G F N P C G E P K

                        220
α   E P S V Y T K L I K F S S W I R E T M A N N P
γ   M P G V Y T K L N K F T S W I K D T M A K N P
K   K P G I Y T K L I K F T P W I K E V M K E N P
```

Fig. 4. Structural comparison of the α- and γ-subunit protein (Isaakson et al. 1984). The sequence of the rat pancreatic kallikrein (*K*) is shown (Swift et al. 1982). The regions of strongest homology are *boxed*

gene structures of the α- and γ-subunits of NGF suggest tat the two genes may be duplications of an ancestral kallikrein-like gene.

The α- (Isackson and Bradshaw 1984) and γ- (Ullrich et al. 1984a) subunits contains nearly 80% similarity in amino acid sequence (Fig. 4). The γ-subunit has arginine enteropeptidase activity and potentially could cleave the prepro-β-NGF molecule at two dibasic amino acid sites. A number of distinctive features in the structure of the α-protein, however, account for the lack of enzymatic activity. They include a substitution of a glutamine for an arginine at the signal peptide cleavage site, a deletion of four crucial residues at the N-terminus, and several substitutions that disrupt the active site (Isackson and Bradshaw 1984; Isackson et al. 1984). Although there is strong homology of the α-protein to the kallikrein family, its functional role in the biosynthesis and storage of the β-NGF protein is unknown.

C. In Vivo Expression of NGF

Nerve growth factor is generally found at much lower levels in target tissues than in the most abundant source, the submaxillary gland. Using the β-NGF cDNA in a hybridization assay, mRNA levels were quantitated in a variety of rat tissues (HEUMANN et al. 1984; SHELTON and REICHARDT 1984). These measurements indicated that the density of sympathetic innervation has been directly correlated with the levels of NGF mRNA. Similar results were also achieved by an immunoassay for NGF protein (KORSCHING and THOENEN 1983). For example, heart atrium and iris displayed more NGF expression than lung or kidney. It must be emphasized that the levels of NGF mRNA in peripheral organs were quite low. The methods were highly sensitive, since the levels of detection of NGF mRNA approached 8 fg and levels of NGF protein were in the 0.2-pg range.

The ability to measure exceedingly low levels of NGF and its mRNA allowed for the detection of NGF in several brain regions, including the hippocampus, cortex, septum, and olfactory bulb (KORSCHING et al. 1985; SHELTON and REICHARDT 1986). Interestingly, expression of NGF was localized specifically to hippocampal neurons (AYER-LELIEVRE et al. 1988a). The presence of NGF in these regions has raised the possibility of a physiological function of NGF in the CNS (WHITTEMORE and SEIGER 1987; THOENEN et al. 1987). The induction of choline acetyltransferase activity by NGF in basal forebrain cholinergic neurons (GNAHN et al. 1983) and the ability of ^{125}I-NGF to be specifically transported from the hippocampus and cortex to the basal forebrain (SEILER and SCHWAB 1984) have demonstrated that NGF acts on selective neuronal populations in the brain in a manner similar to the periphery.

The expression of both NGF and its receptor are activated upon sciatic nerve transection (TANIUCHI et al. 1986a; HEUMANN et al. 1987). NGF and the receptor for NGF are produced by nonneuronal Schwann cells distal to the lesion. The biological significance of the high levels of NGF and NGF receptors observed in Schwann cells is unclear at the present time. NGF may be influencing the production of the certain components of the extracellular matrix (THOENEN et al. 1987). Alternatively, NGF may be sequestered by NGF receptors present on Schwann cells for use by the regenerating axon (E. M. JOHNSON et al. 1988).

The in vivo expression of receptor mRNA can also be correlated with the requirement for NGF by responsive tissues. The classic target cell populations for NGF have been sympathetic and sensory neurons. In the sympathetic ganglia, which require NGF for survival and function throughout the entire life span, NGF receptor mRNA levels are found to increase significantly. In contrast, dorsal root ganglia sensory neurons that require NGF for survival only during early development show significant decreases in receptor mRNA after birth (BUCK et al. 1987). These observations indicate that the appearance and disappearance of NGF receptors in neuronal tissues may be a significant parameter that determines the ability of NGF to influence selective populations. Since NGF appears to be synthesized at a low, constant level in many tissues, regulation of receptor affinity and gene expression may determine which cells respond to NGF.

D. Mechanism of Signal Transduction

Despite the extensive characterization of the trophic and differentiative properties of NGF (THOENEN and BARDE 1980; GREENE and SHOOTER 1980), the molecular signals that are transduced by NGF remain poorly understood. NGF is known to be secreted by target tissues, is bound by a high-affinity cell surface receptor, and is internalized by endocytosis (LEVI et al. 1980). The internalized NGF and its receptor are then retrogradely transported to the perikaryon (HENDRY et al. 1974; E. M. JOHNSON et al. 1978). The intracellular signals generated by the binding and internalization of NGF are probably generated in this pathway.

Although several reports support the hypothesis that NGF itself or the NGF/receptor complex may be involved in nuclear binding (ANDRES et al. 1977; YANKNER and SHOOTER 1980), NGF does not appear to be directly responsible for transmission of the intracellular signal. Microinjection of NGF or antibodies to NGF does not have any effect upon neurite formation in PC12 cells (HEUMANN et al. 1981; SEELEY et al. 1983). Although these are negative results, the apparent lack of response by NGF placed intracellularly in PC12 cells implies that other second messengers are operational.

I. Second Messengers

Neurite outgrowth has provided a convenient assay by which to test potential second messengers for NGF. Perhaps the strongest candidate has been cAMP, which elicits some of the actions of NGF, including process formation and phosphorylation of cellular substrates (HALEGOUA and PATRICK 1980). Elevated levels of cAMP have been proposed to be the second messenger responsible for nerve fiber growth (SCHUBERT et al. 1978). However, it is now apparent that cAMP is neither necessary nor sufficient for many NGF-stimualted activities (RICHTER-LANDSBERG and JASTORFF 1986; VAN BUSKIRK et al. 1985). This conclusion is based upon the differential effects of cAMP analogs and NGF on the extent of neurite formation and phosphorylation, and the ability of PC12 cells defective in the cAMP-dependent protein kinase to respond to NGF. Involvement of cAMP in NGF-induced neurite outgrowth has been further ruled out by the use of a competitive cAMP antagonist, which blocks the neurotrophic effects of cAMP analogs, but not of NGF (RYDEL and GREENE 1988). Moreover, there appears to be no effect of NGF on adenylate cyclase activity (RACE and WAGNER 1985), suggesting that the regulation of the cAMP-dependent protein kinase by NGF may not involve conventional coupling of the NGF receptor to the cyclase enzyme.

The turnover of phosphoinositides has been implicated in the transduction of signals in a wide variety of receptor systems (NISHIZUKA 1984). The production of inositol triphosphate increases intracellular calcium and 1,2 diacylglycerol, activating protein kinase C. In fact, changes in intracellular Ca^{2+} have been correlated with NGF responses (SCHUBERT et al. 1978). But no significant effects on calcium fluxes by NGF in PC12 cells have been detected (LANDRETH et al. 1980). It nevertheless is believed that a certain threshold of calcium must be available for the phosphorylation of proteins carried out by calcium-dependent kinases (HAMA et al. 1986; CREMINS et al. 1986).

The question of phosphoinositide turnover in response to NGF binding has been examined. Phosphoinositide (PI) intermediates have been detected in rat superior cervical ganglia (LAKSHMANAN 1978) and in PC12 cells (TRAYNOR et al. 1982) and accumulation of phosphoinositide was found in PC12 cells 15 s after NGF treatment (CONTRERAS and GUROFF 1986). These results are complemented by other reports that the effects of NGF on PI turnover are primarily to potentiate the effects of agonists that stimulate phospholipase C (VAN CALKER and HEUMANN 1987; VOLONTE et al. 1988). Although these data implicate PI turnover in NGF responses, no definitive evidence has been provided to show directly that there is a functional involvement.

The production of arachidonate from phospholipid pools provides another potential second messenger. Increased metabolism of arachidonate has been observed in NGF-treated PC12 cells (DEGEORGE et al. 1988), suggesting that the metabolic products of arachidonate (such as prostaglandin or thromboxane) may be responsible for neurite outgrowth. Another potential intermediate from the release of arachidonate is diacylglycerol, a direct activator of protein kinase C. The production of diacylglycerol has also been detected in PC12 cells from the hydrolysis of glycosylphosphatidylinositol (CHAN et al. 1989). Abundant data indicate that protein phosphorylation induced by NGF involves both cAMP-dependent kinase (HALEGOUA and PATRICK 1980; ALBERT et al. 1984; YU et al. 1980) and calcium, phospholipid-dependent kinases (MCTIGUE et al. 1985; HAMA et al. 1986; CREMINS et al. 1986). Overlap exists between the phosphorylation patterns induced by NGF and protein kinase C-activating phorbol esters (HAMA et al. 1986; BLENIS and ERIKSON 1986; CREMINS et al. 1986). The potential significance of protein kinase C activation in NGF action is dramatically illustrated by the effect of inhibitors of protein kinase C such as K252a (KOIZUMI et al. 1988) and sphingosine (HALL et al. 1988) on NGF-induced neurite outgrowth. These inhibitors specifically and selectively reduce the extent of process formation induced by NGF.

To summarize, the precise role and identity of second messengers systems for NGF are not clear, but pathways with multiple second messengers are probably required. One of the major obstacles in elucidating a signal transduction mechanism for NGF is the enormous number of cellular events that are affected. Some events are specific for NGF-induced differentiation, while many others are in common with other growth factors. Furthermore, many responses occur at widely differing time intervals. NGF-specific responses must require interactions between different signal transduction mechanisms at different times and at different metabolic levels.

II. Role of Oncogenes

Aside from the neurite outgrowth, one of the hallmark effects of NGF upon PC12 cells has been growth arrest. Since a number of cellular oncogenes influence cell proliferation in response to growth factors and can change the pattern of cellular differentiation, oncogene products have been introduced into PC12 cells to assess the signaling pathways which control NGF-induced differentiation and growth control.

The c-*myc* gene product and the adenovirus E1A-transforming protein are nuclear proteins that allow primary cells to grow indefinitely as established cells. When introduced into PC12 cells, c-*myc* and E1A diminish the ability of PC12 cells to respond (MARUYAMA et al. 1987). The expression of either c-*myc* or E1A in PC12 cells leads to increased DNA synthesis and the lack of neurite outgrowth from NGF treatment. These cells display the same number of NGF receptors with affinities similar to the parental PC12 cells, implying that the effects of these transforming proteins occur at a step after receptor binding. It is interesting that c-*myc* expression has such a distinct effect upon PC12 cells, since c-*myc* expression is increased transiently upon NGF treatment (see below).

Quite opposite effects in PC12 cells are observed with the introduction of transforming *ras* protein (BAR-SAGI and FERAMISCO 1985), the N-*ras* gene (GUERRERO et al. 1986), or the Kirsten and Harvey murine sarcoma virus (NODA et al. 1985). PC12 cells expressing the oncogenic form of *ras* displayed morphological changes and arrest of DNA synthesis similar to NGF treatment. Introduction of normal cellular *ras* protein does not influence PC12 cells. These effects were observed soon after infection with the virus or introduction of the activated *ras* protein. The implication from these observations is that *ras* function is a necessary step in the mediation of NGF effects in PC12 cells. This conclusion is strongly supported by experiments in which monoclonal antibodies to the *ras* protein can block the effects of NGF in PC12 cells (HAGAG et al. 1986). The antibodies were directly microinjected into fused PC12 cells and specifically blocked neurite process formation. These results have been extended in primary cultures of chicken sensory neurons (BORASIO et al. 1989). Oncogenic forms of *ras* protein microinjected into embryonic neurons supported neurite outgrowth and cell survival. *Ras* proteins are known to be closely related to G proteins and possess GTPase activity. Since G proteins function to transduce signals through ligand-receptor interactions, NGF may utilize this signaling pathway to exert its cellular effects.

Significantly, infection of PC12 cells with a *sarc*-containing retrovirus also leads to similar effects to *ras*. Expression of v-*src* in PC12 cells causes morphological differentiation and the lack of [^{3}H]thymidine incorporation (ALEMA et al. 1985). Whether the effects by v-*src* on PC12 cells are transmitted through the same pathway as NGF or *ras* is not clear. The use of phosphotyrosine antibodies has indicated that NGF stimulates the phosphorylation of several proteins on tyrosine residues (MAHER 1988). The finding that v-*src* and *ras* expression elicit the same kind of responses as NGF implies that these proteins mimic steps in the pathway of neuronal differentiation by NGF. But since v-*src* and *ras* gene products have quite distinct biochemical properties, it is difficult to reconcile the mechanism of NGF action with the known properties of both of these proteins. A more reasonable conclusion is that, since NGF exerts so many different effects, several distinct and disparate pathways are utilized. It may therefore be unrealistic to assign a single second messenger and signal transduction mechanism to accommodate all the potential functions of NGF.

III. Genes Induced by NGF

Another approach to defining the steps involved in NGF action has been to identify specific genes that are regulated by NGF. The expression of specific genes during the process of differentiation is likely to be sequentially regulated. Altered expression of specific genes has been inferred by the appearance of new proteins or new RNAs after treatment by NGF. Therefore, identification of NGF-inducible genes has been accomplished by differential screening of cDNA libraries with enriched RNA populations isolated from induced cells.

PC12 cells have been particularly useful in the isolation of cDNAs which are inducible by NGF. In response to NGF, PC12 cells stop proliferating and are converted to cells resembling a sympathetic neuron (GREENE and TISCHLER 1976). In contrast to sympathetic neurons, PC12 cells do not depend upon NGF for survival.

The biological actions of NGF in PC12 cells have been divided into rapid, short-term or delayed, long-term responses (GREENE 1984). Early responses occurring within seconds to several hours of the interaction of NGF with responsive cells include regulation of the uptake of nutrients and anabolic precursors such as amino acids and sugars (MCGUIRE and GREENE 1979) and many phosphorylation events such as those occurring on the ribosomal protein S6 and tyrosine hydroxylase (HALEGOUA and PATRICK 1980). Examples of delayed responses, requiring a period of hours or days, include initiation of neurite outgrowth (GREENE and TISCHLER 1976) and activation of biosynthetic enzymes such as ornithine decarboxylase (MACDONNELL et al. 1977) and tyrosine hydroxylase (THOENEN et al. 1971).

1. Early Response Genes

Initial approaches to identification of inducible genes were to examine protooncogene expression in PC12 cells. The earliest molecular events that were activated by NGF included c-*fos* and c-*myc* protooncogenes, as well as actin and ornithine decarboxylase genes (GREENBERG et al. 1985). These genes were all activated by an increase in the rate of transcription within 30 min of NGF treatment. The expression of the c-*fos* gene, in particular, occurs very rapidly and does not depend upon ongoing protein synthesis. Protein synthesis inhibitors such as cycloheximide will cause a superinduction of c-*fos*

(CURRAN and MORGAN 1985; MILBRANDT 1986), presumably by inhibiting the synthesis of regulatory molecules that either decrease transcription or degrade RNA.

The c-*fos* gene is activated in a wide number of mammalian cell types by many stimuli, including serum, phorbol esters, agents that increase intracellular calcium and cAMP, and other growth factors such as insulin, fibroblast growth factor, and epidermal growth factor (GREENBERG et al. 1985). Indeed, EGF stimulates the transcription of the c-*myc*, c-*fos*, ornithine decarboxylase, and actin genes in PC12 cells. Since EGF does not induce differentiation of PC12 cells, the induction of this class of genes does not appear to be sufficient for differentiation of PC12 cells by NGF.

It is interesting to note that dexamethasone, which causes the differentiation of PC12 cells into chromaffin-like cells, does not stimulate induction of the c-*fos* gene (KRUIJER et al. 1985). Again, this implies that while *fos* expression is not sufficient for NGF action, the c-*fos* protein may be necessary for NGF responses.

Several other genes that are activated rapidly and transiently by NGF have been identified. Some of these cDNAs have also been identified by differential screening of 3T3 fibroblasts induced with platelet-derived growth factor, serum (LAU and NATHANS 1987), and tumor promoters (KUJUBU et al. 1987). These genes were identified in cells undergoing proliferation, in which induction is thought to occur as cells proceed from G_0 to G_1.

Novel NGF-inducible genes have been identified by carrying out subtractive hybridization on differentiating PC12 cells at different times of NGF treatment. For example, a putative DNA-binding protein called *NGFI-A* (MILBRANDT 1987) or early growth response (*Egr-1*) gene (SUKHATME et al. 1988) is detected within 15 min of NGF treatment. The *Egr-1* gene is a murine counterpart of the rodent NGFI-A gene. This gene is stimulated in not only PC12 cells by NGF but also following mitogenic stimulation in fibroblasts, epithelial cells, and lymphocytes. Remarkably, the kinetics of induction are similar to that of c-*fos*. The protein contains three DNA-binding zinc finger domains similar to other regulatory proteins known to interact specifically with DNA such as 5S ribosomal TFIIIA factor and the *Drosophila* Kruppel protein. The presence of these DNA-binding elements implies that this protein may function as a specific transcription factor.

The distinctive time course of expression of these genes has led to the terms "primary" response or "immediate early" genes (LAU and NATHANS 1987). *NGFI-B* has been shown to be related to a steroid receptor (MILBRANDT 1988) by virtue of its strong homology to glucocorticoid receptors. The ligand for this novel steroid receptor has not been identified, but by analogy to the role of glucocorticoid receptors as a transcription factor (R. EVANS 1988) it is probable that the *NGFI-B* gene product acts transcriptionally upon other cellular genes. If so, the *NGFI-A* and *NGFI-B* products are among the first regulatory molecules that initiate the series of events which eventually lead to differentiation by NGF.

2. Later Responses

A number of gene products are induced in PC12 cells at later times by NGF. A highly induced clone, VGF8a, displayed the highest steady-state level of mRNA after 5 h of NGF treatment (LEVI et al. 1985), and maximum induction of the neural specific gene, SCG10, occurs after 24–28 h of NGF treatment (STEIN et al. 1988). The proteins encoded by SCG10 and VGF8a are not related to any known proteins.

Using differential screening of cDNA libraries made from PC12 cells treated for 2 weeks, 10 unique cDNAs were isolated from approximately 13000 clones (LEONARD et al. 1987). Among the proteins identified are tyrosine hydroxylase, thymosin β_4, and a novel intermediate filament gene. In contrast to the immediate early genes induced by NGF, many of these cDNAs are not induced by other growth factors, tumor promoters. Other distinctive gene products which are activated by NGF by increases in RNA levels are calcium-binding proteins (MASIAKOWSKI and SHOOTER 1988), neural adhesion molecule (PRENTICE et al. 1987), neurofilament proteins (LINDENBAUM et al. 1988), and Thy-1 (DICKSON et al. 1986). In addition to many enzymatic activities that are stimulated by NGF (GREENE and SHOOTER 1980), the genes encoding the Alzheimer's β-amyloid protein and the prion protein are also up regulated by NGF treatment (MOBLEY et al. 1988). These observations indicate that NGF is capable of modulating the activities of a wide variety of proteins by changing the transcription pattern of target genes. The rapidly induced gene products probably have a direct influence upon these transcriptional responses.

E. Receptor for NGF

All of the diverse responses of NGF are initiated by an interaction with a specific cell surface receptor. The cell surface receptor for NGF has been detected on a wide variety of both neuronal and nonneuronal derivatives of the neural crest including sympathetic neurons (FRAZIER et al. 1974; MASSAGUE et al. 1981), Schwann cells (TANIUCHI et al. 1986a; DISTEFANO and JOHNSON 1988a), neurofibromas (ROSS et al. 1984), and cell lines of neural crest origin, specifically melanoma (FABRICANT et al. 1977), neuroblastoma (SONNENFELD and ISHII 1982; MARCHETTI and PEREZ-POLO 1987), and pheochromocytoma cells (LANDRETH and SHOOTER 1980; GROB et al. 1983; HOSANG and SHOOTER 1985; GREEN and GREENE 1986).

Monoclonal antibodies generated against the NGF receptor have provided a primary means to study the distribution and biochemical properties of the receptor. The ME20.4 antibody was originally detected by its ability to block NGF binding to receptors found in human melanoma cells (ROSS et al. 1984) and has been used in immunocytochemical studies (HEFTI et al. 1986; SPRINGER et al. 1987; SCHATTEMAN et al. 1988; GARIN CHESA et al. 1988). The 192-IgG monoclonal antibody raised against the rat PC12 receptor enhances NGF binding (CHANDLER et al. 1984) and is most likely to recognize a different site on the receptor. These two monoclonal antibodies are species specific.

I. Biochemical Analysis

The identification of the receptor protein has been primarily accomplished by three different methods: (a) affinity crosslinking of the receptor to labeled NGF, (b) immunoprecipitation of the receptor using monoclonal antibodies, or (c) affinity chromatography with NGF-Sepharose or with monoclonal antibodies.

Using affinity crosslinking, the major radioiodinated complex isolated from neurons, pheochromocytoma, neuroblastoma and melanoma cell lines has an apparent molecular weight of 100 kDa. Subtraction of the molecular weight of the β-dimer of NGF from the crosslinked complex gives an apparent molecular weight for the receptor of 75 kDa.

Immunoprecipitation with monoclonal antibodies directed against the receptor yields a similar protein of 75 kDa. Antibodies specific for the human (ME20.4) and rat (IgG-192) NGF receptor both immunoprecipitate a similarly sized 75- to 80-kDa protein (GROB et al. 1985; TANIUCHI et al. 1986a, b, c; GREEN and GREENE 1986; MARANO et al. 1987). However, the monoclonal antibodies directed against the human and rat NGF receptors are species specific and do not display any crossreactivity.

Affinity chromatography with NGF-Sepharose has also been used to isolate a 75- to 80-kDa protein from rabbit sympathetic neurons (KOUCHALAKOS and BRADSHAW 1986) and human melanoma cells (PUMA et al. 1983; MARANO et al. 1987). Further characterization of the receptor using metabolic labeling and pulse-chase experiments has demonstrated that the NGF receptor from human melanoma cells (GROB et al. 1985) is a glycosylated protein which is processed from a 59-kDa precursor. Both tunicamycin-sensitive and -insensitive processing was observed, indicating that the glycosylated protein contains both N-linked and O-linked sugars.

Two other molecular weight species can be detected in variable amounts depending on the methods of isolation and cell source. A 200-kDa protein is found in most receptor preparations (PUMA et al. 1983; GROB et al. 1983; TANIUCHI et al. 1986a, b, c; MARANO et al. 1987) and has been shown to be a homodimer of the 75-kDa species by peptide mapping (BUXSER et al. 1985). A second species of 135–158 kDa can be identified by crosslinking ^{125}I-NGF with the receptor using the heterobifunctional and relatively lipophilic crosslinking reagent hydroxysuccinimidyl-4-azido benzoate (HSAB) (MASSAGUE et al. 1981; HOSANG and SHOOTER 1985; GREEN and GREENE 1986). This protein is crosslinked only on cells which express high as well as low-affinity receptor (PC12 cells and neurons). However, using a different crosslinking reagent ethyldimethylaminopropyl-carbodimide (EDC) to crosslink the NGF receptor from SY5Y neuroblastoma cells expressing only the high-affinity form of the receptor yielded a lower molecular weight species of 110 kDa for the receptor-NGF complex or a calculated 84 kDa for the receptor alone (GREEN and GREENE 1986).

A smaller species of the receptor is released in the media of cultured cells and is also found in serum and urine. This form of the receptor was detected by binding of ^{125}I-NGF followed by immunoprecipitation with the 192-IgG antibody (DISTEFANO and JOHNSON 1988 b). From the size of the protein (50000 Da) it appears to represent the extracellular portion of the NGF receptor (see below). This truncated form of the NGF receptor is found most abundantly in cultured Schwann cells, and appears to be generated by a proteolytic event at the cell surface. The presence of the truncated form of the NGF receptor may explain the difficulty in raising antibodies against the receptor due to the immunological tolerance of animals toward the receptor.

II. Cloning of the NGF Receptor Gene

DNA-mediated gene transfer has been used as a functional assay for the expression and molecular cloning of the NGF receptor (CHAO et al. 1986; RADEKE et al. 1987). This method circumvented the requirement for mRNA enrichment or for protein purification and thus allowed for the isolation of a gene whose product is expressed in relatively low levels.

Mouse fibroblast cells (Ltk^-) were exposed to high molecular weight DNA with the purified herpes virus thymidine kinase gene in a calcium phosphate precipitate. After 2 weeks in HAT selection, tk^+ colonies were screened for NGF receptor expression by an in situ rosette assay employing the ME20.4 monoclonal antibody (CHAO et al. 1986) or by a cell sorter using the 192-IgG monoclonal antibody (RADEKE et al. 1987).

The human NGF receptor gene was rescued from a transfected mouse L-cell line using human Alu repetitive sequences. The rat NGF receptor gene was isolated following cDNA subtractive hybridization. Poly $(A)^+$ RNA isolated from the L-cell transfectant was converted to ^{32}P-cDNA; the common cDNA sequences were removed by hybridization with poly $(A)^+$ RNA from Ltk^- cells.

The 3.8-kb mRNA for the NGF receptor contains a large 3′ untranslated region of nearly 2 kb. This sequence is uninterrupted and is contained in the largest receptor exon (2.3 kb) together with sequences representing the intracellular domain of the receptor. The significance of such a long 3′ untranslated region is unknown, but lengthy noncoding sequences have been detected in the mRNAs of many genes encoding cell surface molecules. The receptor mRNA contains a single polyadenylation sequence, ATTAAA, nine bases upstream from a poly(A) tail. No other polyadenylation sites have been uncovered for the receptor mRNA.

The validity of the receptor sequence deduced from cDNA cloning was verified by gas phase sequencing of NGF receptor protein purified from the melanoma A875 cell line (MARANO et al. 1987); by cDNA expression in mammalian cells (D. JOHNSON et al. 1986; RADEKE et al. 1987); and by expression in *Xenopus laevis* oocytes by microinjection of receptor RNA produced in vitro (SEHGAL et al. 1988 a).

The full-length cDNA predicts a signal sequence of 28 amino acids; an extracellular domain containing four 40-amino-acid repeats with 6 cysteine residues at conserved positions, followed by a region rich in threonine and serine residues; a single transmembrane domain and a 155-amino-acid

```
                                    +1                  20                  40
Human  MGAGATGRAMDGPRLLLLLLLLGVSLGGAKEACPTGLYTHSGECCKACNLGEGVAQPCGANQTVCEPCL
Rat    -R---ACS---RL------I----S-----T-S------------------------------------

                    60                    80                    100
DSVTFSDVVSATEPCKPCTECVGLQSMSAPCVEADDAVCRCAYGYYQDETTGRCEACRVCEAGSGLVFSCQ
-N--------------------L--------------------------E--H----S---V---------

       120                 140                 160                 180
DKQNTVCEECPDGTYSDEANHVDPCLPCTVCEDTERQLRECTRWADAECEEIPGRWITRSTPPEGSDSTAP
-----------E------------------------------P--------------P-------------

                  200                 220                 240
STQEPEAPPEQDLIASTVAGVVTTVMGSSQPVVTRGTTDNLIPVYCSILAAVVVGLVAYIAFKRWNSCKQN
------V------VP----DM--------------------------------------------------

      260                 280                 300                 320
KQGANSRPVNQTPPPEGEKLHSDSGISVDSQSLHDQQPHTQTASGQALKGDGGLYSSLPPAKREEVEKLLN
--------------------------------------T--------------N-----LT----------

           340                 360                 380
GSAGDTWRHLAGELGYQPEHIDSFTHEACPVRALLASWATQDSATLDALLAALRRIQRADLVESLCSESTATSPV
-xxx----------------------------------GA--------------------I--------------
```

Fig. 5. Amino acid sequence of the NGF receptor. A comparison of the human (D. Johnson et al. 1986) and rat (Radeke et al. 1987) receptor sequences is displayed. Identical amino acids are represented by a *dash*. The *boxed* segment represents the transmembrane domain. The NH_2-terminal amino acid sequence has been verified by a gas-phase sequence analysis of the NGF receptor purified from human melanoma cells (Marano et al. 1987)

cytoplasmic domain. The amino acid sequence of the human and rat NGF receptors is displayed in Fig. 5. Of 28 cysteine residues in the predicted amino acid sequence, 24 are found within the first 160 residues of the receptor. Alignment of the cysteine residues indicates that the 40-amino-acid repeats contain considerable amino acid similarities in residues other than cysteines, leading to the hypothesis that these repeats are evolutionarily related. Similar cysteine-rich repeats have been detected in the extracellular domains of a number of proteins including cell surface proteins such as the EGF receptor (Ullrich et al. 1984b), the low-density lipoprotein (LDL) receptor (Yamamoto et al. 1984), the neurogenic *notch* gene (Wharton et al. 1985), and the *neu* oncogene (Bargmann et al. 1986). No similarity exists in the spacing of the cysteine residues in the NGF receptor with other known cysteine-rich proteins. However, it is likely since the cysteine repeats constitute 70% of the extracellular portion of the receptor that this region constitutes part or all of the binding site for NGF.

The extracellular sequence is also distinguished by a highly negative charge. The bulk of the negative charge is contributed by the cysteine-rich domain in which the net charge is −24. No charged amino acids are found in residues 223–244, which have all the characteristics of a transmembrane domain. This hydrophobic segment is followed by several basic amino acids which commonly mark the cytoplasmic border of the membrane-spanning region. Receptors for asialoglycoprotein, transferrin, EGF, interleukin-1, interleukin-2, interleukin-6, and platelet-derived growth factor (PDGF) also contain a single transmembrane domain. The receptors for EGF and LDL contain cysteine-rich regions in the N terminus which face the exterior of the cell (YAMAMOTO et al. 1984; ULLRICH et al. 1984b). Hence it is likely that the NGF receptor is oriented with the amino-terminal region outside the cell, and the carboxy terminus in the cytoplasm.

Though the deduced amino acid sequence of the human NGF receptor reveals structural features in common with other receptors whose structures have been elucidated, the NGF receptor lacks significant amino acid similarities with any other known receptor or protein. In particular, the primary sequence of the intracellular domains does not display any similarities to known growth factors, oncogenes, or the tyrosine kinase family of genes. This lack of homology to known tyrosine kinases agrees with the absence of tyrosine kinase activity by the NGF receptor from human melanoma (GROB et al. 1985) and from PC12 cells and rat superior cervical ganglia (TANIUCHI et al. 1986b), in spite of reports that tyrosine phosphorylation has been observed in NGF-treated PC12 cells (MAHER 1988). The lack of similarity between the NGF receptor and the tyrosine kinase family of receptors is not unexpected since the primary target cells influenced by NGF do not undergo cell division in response to the hormone. The absence of such activity for the NGF receptor suggests that the mechanism of action of NGF differs significantly from other peptide growth factors such as PDGF, FGF, EGF, CSF-1, and insulin.

III. Features of the NGF Receptor Gene

The structural features of the NGF receptor are reflected in the organization of the gene (Fig. 6). The isolation of overlapping bacteriophages spanning over 50 kb allowed for the characterization of the boundaries of the human NGF receptor gene (SEHGAL et al. 1988b). Analysis of the splice junction sites for the six exons of the receptor gene reveals that structurally distinct domains of the receptor are in separate exons. For example, the second and third exons encode the N-terminal amino acid sequence of the mature receptor protein and a discrete domain of 161 amino acids, of which 24 are cysteine residues. It is of interest that the splice site for the third exon is located precisely after the 24th cysteine residue. Similarly, the transmembrane region found in the fourth exon contains an intron-exon boundary in the stop transfer sequence directly following the hydrophobic-containing amino acids. Such precise divisions of the structural features of the receptor molecule are supportive of a functional role of these domains.

In situ hybridization of somatic cell lines has established that the human gene is located on the long arm of chromosome 17 (HUEBNER et al. 1986). The monoclonal antibody against the human receptor was independently used to assign the receptor gene to chromosome 17 (RETTIG et al. 1986). Polymorphic

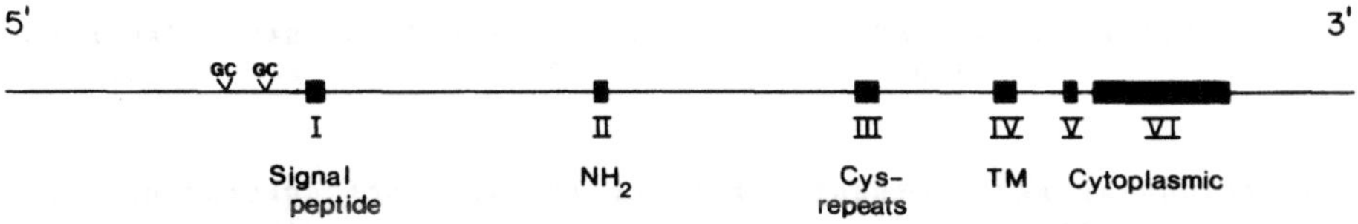

Fig. 6. Genomic map of the human NGF receptor gene. The NGF receptor gene is shown in a 5′ to 3′ orientation with exons represented as *boxes*. A single polyadenylation site exists (D. JOHNSON et al. 1986). The promoter sequence is designated by GC sequences. The precise intron-exon organization of the NGF receptor gene was defined by DNA sequence analysis (SEHGAL et al. 1988b). The human NGF receptor gene spans nearly 25 kb

analyses using genomic probes for the human receptor have ruled out any linkage to familial dysautonomia (BREAKEFIELD et al. 1986) and also von Recklinghausen neurofibromatosis, though the gene defect is also on chromosome 17 (SEIZINGER et al. 1987).

The initiation site for the NGF receptor gene in human A875 melanoma cells and in rat PC12 cells was localized by primer extension and S1 nuclease protection experiments (SEHGAL et al. 1988b). For both cell lines the same start site for transcription was found. Isolation and sequencing of 5′ human and rat genomic sequences was carried out by hybridization of 5′ cDNA fragments.

The sequence of the receptor promoter reveals a high G+C content but lacks consensus "TATA" and "CAAT" sequences (Fig. 7). Several conserved GGGCGG sequences are present within 200 nucleotides of the initiator ATG codon. These GC-rich sequences represent potential binding sites for the transcription factor Sp1 (DYNAN and TJIAN 1983). Several eukaryotic genes have promoters which are rich in GC content but lack the characteristic TATA and CAAT boxes. Most of these genes encode enzymes with housekeeping functions, such as hypoxanthine phosphoribosyltransferase, HMG CoA reductase, and 3-phosphoglycerate kinase. The expression of these genes is at a low constitutive level in diverse tissues (MELTON et al. 1984; REYNOLDS et al. 1984; SINGER-SAM et al. 1984). However, the promoters of several cellular growth control genes such as c-Harvey *ras* (ISHII et al. 1985a), c-Kirsten *ras* (HOFFMAN et al. 1987), the EGF receptor (ISHII et al. 1985b), and the insulin receptor (ARAKI et al. 1987) also do not contain the typical TATA and CAAT transcriptional consensus sequence, but have multiple GC box motifs.

The promoter for the NGF receptor therefore resembles those genes which are constitutively expressed or involved in growth regulation. An unusual feature of this promoter is that it appears to direct transcription primarily from a major site of initiation instead of using multiple sites of initiation as occurs with many constitutively expressed genes. The single initiation site was observed in both human melanoma cells and PC12 cells (SEHGAL et al. 1988b). The sequences which direct transcription from this initiation site are not defined, but do not involve the consensus TATA sequence. Neuronal-specific

```
GCAGCCGGAGAGAGGAACAGGAACCGCAGTGGACGCCCCTGGTCCCCGGGACGACGCGCAGTAGCGC
                    -250                  -230                  -210

CGGGAACTGGGTACCAGGCGGGTGGGTGAGAGGCTCTAAGGGACAAGGCAGGGAGAAGCGCACGGGT
             -190                 -170                  -150

GCGGGGAACCAGCCCTCCCTTTGCCTCTGCTTCCCACCCCGAGGCGGCAGGGCGGGCGGGCAGGTTC
   -130                -110                 -90                   -70

CGGGGGTGGCGGGCTGGCGGGGCGGAGGCGGGGGCCGACAGCTGGCTTCACCAGCCTTCCGCCGTAG
                  -50                 -30                  -10
```

Fig. 7. Promoter sequence for the human NGF receptor gene. The promoter is mapped by primer extension and S1 nuclease protection analysis (SEHGAL et al. 1988b). The *boxed* segments represent potential binding sites for the Sp1 transcription factor

expression and developmental regulation of the receptor gene are most likely due to other discrete regulatory elements.

The constitutive nature of the receptor promoter sequence may be a partial explanation for why this tissue-specific gene is expressed efficiently in a variety of nonneuronal cells after genomic gene transfer. Significantly, the genes for the EGF receptor, insulin receptor, and c-*ras* share promoter characteristics with the NGF receptor gene. Since these gene products are in some way involved with growth control and their promoter elements are similar, regulation of these genes may share common mechanisms.

IV. Kinetic Forms of the NGF Receptor

Equilibrium binding of ^{125}I-NGF to cells reveals two distinct affinity states for the NGF receptor (SUTTER et al. 1979; LANDRETH and SHOOTER 1980; SCHECHTER and BOTHWELL 1981). In most responsive cells, such as neurons and PC12 cells, approximately 10%–15% of the receptors display high-affinity binding with a K_d of 10^{-11} *M*, with the remainder of the receptors possessing a K_d of 10^{-9} *M*. The difference in equilibrium binding is accounted by a 100-fold difference in the rate of dissociation of NGF, thereby accounting for the designation of high- and low-affinity receptors as "slow" and "fast" receptors, respectively.

Several lines of evidence suggest that the high-affinity receptor is not only internalized upon NGF binding, but is responsible for the actions of NGF. First, the SY5Y human neuroblastoma cell line contains a small number of high-affinity sites and no low-affinity sites (SONNENFELD and ISHII 1982), yet remains responsive to NGF. Secondly, the concentration of NGF necessary to elicit responses in PC12 cells (GREENE 1977) and in sympathetic and sensory ganglion neurons (FRAZIER et al. 1974; HILL and HENDRY 1976) is consistent with mediation by the high-affinity receptor. However, the strongest evidence for a biological role for the high-affinity receptor comes from mutant PC12 cell lines which lack the high-affinity receptor but which retain low-affinity sites (GREEN et al. 1986). These mutant cells are unable to respond to NGF as determined by the inability of the cells to extend neurites in the presence of NGF and to fail to induce ornithine decarboxylase or c-*fos* and c-*myc* protooncogene transcription (GREENBERG et al. 1985). Furthermore, internalization of the low-affinity receptors is not observed, suggesting that high-affinity receptor binding and internalization is necessary for NGF action in normal PC12 cells. These dramatic differences in the actions of the two kinetic forms of the NGF receptor strongly imply that the molecular events which mediate the change in receptor affinity are involved with the signal transduction mechanism of NGF.

V. Expression of Cloned NGF Receptors

Introduction of cDNA clones encoding the NGF receptor into fibroblasts and melanoma cells generates cells which express only the low-affinity form of the receptor (CHAO et al. 1986; D. JOHNSON et al. 1986; RADEKE et al. 1987). As many as 500000 receptors/cell can be observed after transfection; however, none of these lines display any detectable responses with NGF treatment. The lack of responses to NGF in cells expressing abundant NGF receptors suggests that many cells do not possess the proper cellular environment to respond to NGF.

The introduction of a full-length human receptor cDNA in a variant PC12 cell line (NR18 cells) resulted in the generation of both high- and low-affinity receptors with similar affinities to the kinetic classes found in sympathetic neurons and PC12 cells (HEMPSTEAD et al. 1989). The NR18 cell line was chosen due to the lack of endogenous NGF receptors and the inability to respond to NGF (BOTHWELL et al. 1980). The appearance of high-affinity human NGF receptors in this cell line was correlated with functional responses, the induction of c-*fos* transcription by NGF. These results indicate that a cloned human receptor cDNA can give rise to functional NGF receptors in appropriate cells. The detection of only one mRNA species in cell lines that display only low-affinity receptors (melanoma and transfected fibroblasts) and cell lines that have predominantly high-affinity receptors (SY-5Y neuroblastoma) also demonstrates that one gene gives rise to a single mRNA species, specifying both kinetic forms of the NGF receptor.

The lack of appropriate responses to NGF in many cells expressing abundant NGF receptors strongly suggests that most cells do not possess either the correct cellular environment or the appropriate signaling machinery. The lack of a response can also be correlated with the absence of the high-affinity form of the receptor, which mediates many of the biological responses of NGF (GREEN et al. 1986). Hence, clarification of the biochemical nature of the high-affinity form is crucial in understanding the mechanisms by which the receptor protein transmits its intracellular signal after binding to NGF. Differential splicing does not appear to give rise to the two kinetic forms of the receptor and, furthermore, no other evidence has been found for another gene related to the receptor that binds NGF (HEMPSTEAD et al. 1988). Although posttranslational modifications may account for a difference of affinity, there is presently no indication that a posttranslational event influences ligand affinity for NGF.

The most plausible explanation for the difference in receptor ligand affinity for the NGF receptor is that a distinct regulatory protein interacts with the receptor and dictates the affinity of the receptor for its ligand. A number of different observations makes this a tenable hypothesis. First, several groups, using affinity crosslinking agents to label NGF receptors, have described crosslinked species larger than the 100-kDa receptor-NGF complex which can be isolated from cells expressing the high-affinity form of the receptor (MASSAGUE et al. 1981; HOSANG and SHOOTER 1985). Second, examination of the human (D. JOHNSON et al. 1986) and rat (RADEKE et al. 1987) NGF receptor sequences indicates an extremely conserved region of amino acid sequence in the transmembrane region. This domain is nearly 90 amino acids in length and extends 50 amino acids into the cytoplasmic domain (Fig. 5). No differences exist in any of the amino acids between rat and human NGF receptors in this transmembrane region. The isolation and analysis of the chicken NGF receptor gene (ERNFORS et al. 1988; LARGE et al. 1989) confirms that this region is extraordinarily conserved between species. It is attractive to speculate that this particular domain of the receptor is so highly conserved because of important interactions with other proteins closely associated with the plasma membrane.

Thus, the two kinetic forms of the NGF receptor appear to be encoded by the same protein (GREEN and GREENE 1986), which is the product of a single gene. Several possibilities remain for an associated protein which is capable of modulating the affinity state of the NGF receptor. Whether NGF-mediated events are dependent upon interactions of the receptor with *ras* or *sarc* proteins has not yet been determined. The structure of the receptor protein does not contain any sequences which interact with known G proteins. Resolution of the molecular basis of the two kinetic forms of the NGF receptor will undoubtedly begin to unravel the complex series of events that result from NGF binding to responsive cells.

F. Conclusions

The genes encoding NGF and its receptor are both present in the mammalian genome as a single copy. The NGF and receptor proteins contain a number of structural features which most undoubtedly carry out crucial functions. These include cysteine residues which are conserved in all the NGF and receptor molecules analyzed to date, the binding domains, and the transmembrane and intracellular domains of the receptor. By site-directed mutagenesis and expression of mutant molecules, many of these sites of interaction with NGF and its receptor and the functional consequences will be defined. Though closely related genes have not been detected, it is tempting to speculate that other neurotrophic molecules and their receptors may have common features with NGF.

Although the actions of NGF were originally thought to be restricted to the sympathetic and neural crest derived sensory neurons, it is clear that effects of NGF extend to the CNS and beyond. The effects of NGF on cholinergic neurons in the basal forebrain are well established and provide impetus to investigate what neurotrophic effects NGF has in other distinctive brain regions such as the cerebellum, where NGF receptors have been detected early in development. The ability of NGF to prevent retrograde neuronal cell death in the basal forebrain area and partially to improve spatial memory in aged rats (WILLIAMS et al. 1986; FISCHER et al. 1987) has suggested that abnormal functioning of neurotrophic factors such as NGF may contribute to neuronal degeneration in neuropathologies such as Alzheimer's disease. The biochemical and molecular definition of the mechanisms of action of NGF will ultimately lead to an understanding of the changes that occur in neurodegenerative disease.

Furthermore, the detection of NGF and NGF receptors in a wide number of nonneuronal regions such as testis (AYER-LELIEVRE et al. 1988 b), muscle (RAIVICH et al. 1985), and nonneuronal tumor cells (GARIN CHESA et al. 1988) implies that the actions of NGF are not restricted only to cells derived from the neural crest. Nor are the potential biological effects confined to the traditional functions ascribed to this molecule. For example, NGF has distinctive proliferative effects upon mast cells (ALOE and LEVI-MONTALCINI 1977), binds to T-lymphocyte populations (THORPE et al. 1987), and influences differentiation of eosinophils and basophils (MATSUDA et al. 1988). Hence the activities of NGF appear to extend to the immune system. It is likely as the molecular and immunological probes for NGF and the receptor are further applied that NGF will be found to possess even more powers than previously thought.

References

Albert KA, Helmer-Matyjek E, Nairn AG, Mukller TH, Haycock JW, Greene LA, Goldstein M, Greengard P (1984) Calcium/phospholipid-dependent protein kinase (protein kinase C) phosphorylates and activates tyrosine hydroxylase. Proc Natl Acad Sci USA 81:7713–7717

Alema S, Casalbore P, Agostini E, Tato F (1985) Differentiation of PC12 pheochromocytoma cells induced by v-*src* oncogene. Nature 316:557–559

Aloe L, Levi-Montalcini R (1977) Mast cells increase in tissues of neonatal rats injected with the nerve growth factor. Brain Res 133:356–366

Andres RY, Jeng I, Bradshaw RA (1977) Nerve growth factor receptors: identification of distinct classes in plasma membranes and nuclei of embryonic dorsal root neurons. Proc Natl Acad Sci USA 74:2785–2789

Angeletti RH, Bradshaw RA (1971) Nerve growth factor from mouse submaxillary gland: amino acid sequence. Proc Natl Acad Sci USA 68:2417–2420

Araki E, Shimada F, Uzawa H, Mori M, Ebina Y (1987) Characterization of the promoter region of the human insulin receptor gene. J Biol Chem 262:16186–16191

Ayer-LeLievre C, Olson L, Ebendal T, Seiger A, Persson H (1988 a) Expression of the β-nerve growth factor gene in hippocampal neurons. Science 240:1339–1341

Ayer-LeLievre C, Olson L, Ebendal T, Hallbook F, Persson H (1988 b) Nerve growth factor mRNA and protein in the testis and epididymis of mouse and rat. Proc Natl Acad Sci USA 85:2628–2632
Bargmann CI, Hung MC, Weinberg RA (1986) The *neu* oncogene encodes an epidermal growth factor receptor-related protein. Nature 319:226–230
Bar-Sagi D, Feramisco JR (1985) Microinjection of the *ras* oncogene protein into PC12 cells induces morphological differentiation. Cell 42:841–848
Blenis J, Erikson RL (1986) Regulation of protein kinase activities in PC12 pheochromocytoma cells. EMBO J 5:3441–3447
Borasio GD, John J, Wittinghofer A, Barde Y-A, Sendtner M, Heumann R (1989) ras p21 protein promotes survival and fiber outgrowth of cultured embryonic neurons. Neuron 2:1087–1096
Bothwell MA, Schechter AL, Vaughn KM (1980) Clonal variants of PC12 pheochromocytoma cells with altered response to nerve growth factor. Cell 21:857–866
Bradshaw RA (1978) Nerve growth factor. Annu Rev Biochem 47:191–216
Breakefield XO, Ozelius L, Bothwell MA, Chao MV, Axelrod F, Kramer PL, Kidd KK, Lanahan AA, Johnson DE, Ross AH, Gusella JF (1986) DNA polymorphism for the nerve growth factor receptor gene excludes its role in familial dysautonomia. Mol Biol Med 3:483–494
Buck CR, Martinez H, Black IB, Chao MV (1987) Developmentally regulated expression of the nerve growth factor receptor gene in the periphery and brain. Proc Natl Acad Sci USA 84:3060–3063
Buxser S, Puma P, Johnson GL (1985) Properties of the nerve growth factor receptor. J Biol Chem 260:1917–1926
Chan BL, Chao MV, Saltiel AR (1989) Nerve growth factor stimulates the hydrolysis of glycosyl-phosphatidylinositol in PC12 cells: a novel mechanism of protein kinase C regulation. Proc Natl Acad Sci USA 86:1756–1760
Chandler CE, Parsons LM, Hosang M, Shooter EM (1984) A monoclonal antibody modulates the interaction of nerve growth factor with PC12 cells. J Biol Chem 259:6882–6889
Chao MV, Bothwell MA, Ross AH, Koprowski H, Lanahan A, Buck CR, Sehgal A (1986) Gene transfer and molecular cloning of the human NGF receptor. Science 232:418–421
Cohen S (1960) Purification of a nerve growth promoting protein from mouse salivary gland and its neurocytotoxic antiserum. Proc Natl Acad Sci USA 46:302–311
Contreras ML, Guroff G (1986) Calcium-dependent nerve growth factor-stimulated hydrolysis of phosphoinositides in PC12 cells. J Neurochem 48:1466–1472
Cremins J, Wagner JA, Halegoua S (1986) Nerve growth factor action is mediated by cyclic AMP- and Ca^{2+} phospholipid-dependent protein kinases. J Cell Biol 103:887–893
Curran T, Morgan JI (1985) Superinduction of c-*fos* by nerve growth factor in the presence of peripherally active benzodiazepines. Science 229:1265–1268
DeGeorge JJ, Walenga R, Carbonetto S (1988) Nerve growth factor rapidly stimulates arachidonate metabolism in PC12 cells: potential involvement in nerve fiber growth. J Neurosci Res 21:323–332
Dickson G, Prentice H, Julien J-P, Ferrari G, Lean A, Walsh FS (1986) Nerve growth factor activates Thy-1 and neurofilament gene transcription in rat PC12 cells. EMBO J 5:3449–3453
DiStefano PS, Johnson EM (1988 a) Nerve growth factor receptors on cultured rat Schwann cells. J Neurosci 8:231–241
DiStefano PS, Johnson EM (1988 b) Identification of a truncated form of the nerve growth factor receptor. Proc Natl Acad Sci USA 85:270–274
Dynan WS, Tjian R (1983) The promoter-specific transcription factor Sp1 binds to upstream sequences in the SV40 early promoter. Cell 35:79–87
Ebendal T, Larhammar D, Persson H (1986) Structure and expression of the chicken β nerve growth factor gene. EMBO J 5:1483–1487

Ebendal T, Persson H, Larhammar D, Lundstromer K, Olson L (1989) Characterization of antibodies to synthetic nerve growth factor (NGF) and proNGF peptides. J Neurosci Res 22:223–240
Edwards RE, Selby M, Rutter WJ (1986) Differential RNA splicing predicts two distinct nerve growth factor precursors. Nature 319:784–787
Edwards RE, Selby MJ, Mobley WC, Weinrich SL, Hruby DE, Rutter WJ (1988) Processing and secretion of nerve growth factor: expression in mammalian cells with a vaccinia virus vector. Mol Cell Biol 8:2456–2464
Ernfors P, Hallbook F, Ebendal T, Shooter EM, Radeke MJ, Misho TP, Persson H (1988) Developmental and regional expression of β-nerve growth factor receptor messenger RNA in the chick and rat. Neuron 1:983–996
Evans BA, Richards RI (1985) Genes for the α and γ subunits of mouse nerve growth factor are contiguous. EMBO J 4:133–138
Evans R (1988) The steroid and thyroid hormone receptor superfamily. Science 240:889–895
Fabricant RF, DeLarco JE, Todaro GJ (1977) Nerve growth factor receptors on human melanoma cells in culture. Proc Natl Acad Sci USA 74:565–569
Fischer W, Wictorin K, Bjorklund A, Williams LR, Varon S, Gage FH (1987) Amelioration of cholinergic neuron atrophy and spatial memory impairment in aged rats by nerve growth factor. Nature 329:65–68
Francke U, Martinville B, Coussens L, Ullrich A (1983) The human gene for the β subunit of nerve growth factor is located on the proximal short arm of chromosome 1. Science 222:1248–1251
Frazier WA, Boyd LF, Bradshaw RA (1974) Properties of the specific binding of ^{125}I nerve growth factor to responsive peripheral neurons. J Biol Chem 249:5513–5519
Garin Chesa P, Rettig WJ, Thomson TM, Old LJ, Melamed MR (1988) Immunohistochemical analysis of nerve growth factor expression in normal and malignant human tissues. J Histochem Cytochem 36:383–389
Gnahn H, Hefti F, Heumann R, Schwab ME, Thoenen H (1983) NGF-mediated increase of choline acetyltransferase (ChAT) in the neonatal rat forebrain: evidence for a physiological role of NGF in the brain? Dev Brain Res 9:45–52
Goedert M (1986) Molecular cloning of the chicken nerve growth factor gene: mRNA distribution in developing and adult tissues. Biochem Biophys Res Commun 141:1116–1122
Green SH, Greene LA (1986) A single $M_r = 103{,}000$ ^{125}I-β-nerve growth factor-affinity-labeled species represents both the low and high affinity forms of the nerve growth factor receptor. J Biol Chem 261:15316–15326
Green SH, Rydel RE, Connolly JL, Greene LA (1986) PC12 cell mutants that possess low- but not high affinity nerve growth factor receptors neither respond to nor internalize nerve growth factor. J Cell Biol 102:830–843
Greenberg ME, Greene LA, Ziff EB (1985) Nerve growth factor and epidermal growth factor induce rapid transient changes in proto-oncogene transcription in PC12 cells. J Biol Chem 260:14101–14110
Greene LA (1977) A quantitative bioassay for nerve growth factor (NGF) activity employing a clonal pheochromocytoma cell line. Brain Res 133:350–353
Greene LA (1984) The importance of both early and delayed responses in the biological actions of nerve growth factor. Trends Neurosci 7:91–94
Greene LA, Shooter EM (1980) The nerve growth factor: biochemistry, synthesis, and mechanism of action. Annu Rev Neurosci 3:353–402
Greene LA, Tischler AS (1976) Establishment of a noradrenergic clonal line of rat adrenal pheochromocytoma cells which respond to nerve growth factor. Proc Natl Acad Sci USA 73:2424–2428
Grob PM, Berlot CH, Bothwell MA (1983) Affinity labeling and partial purification of nerve growth factor receptors from rat pheochromocytoma and human melanoma cells. Proc Natl Acad Sci USA 80:6819–6823
Grob PM, Ross AH, Koprowski H, Bothwell MA (1985) Characterization of the human melanoma nerve growth factor receptor. J Biol Chem 260:8044–8049

Guerrero I, Wong H, Pellicer A, Burstein DE (1986) Activated N-*ras* gene induces neuronal differentiation of PC12 rat pheochromocytoma cells. J Cell Physiol 129:71–76

Hagag N, Halegoua S, Viola M (1986) Inhibition of growth factor-induced differentiation of PC12 cells by microinjection of antibody to *ras* p21. Nature 319:680–682

Halegoua S, Patrick J (1980) Nerve growth factor mediates phosphorylation of specific proteins. Cell 22:571–581

Hall FL, Fernyhough P, Ishii DN, Vulliet PR (1988) Suppression of nerve growth factor-directed neurite outgrowth in PC12 cells by sphingosine, an inhibitor of protein kinase C. J Biol Chem 263:4460–4466

Hallbook F, Ebendal T, Persson H (1988) Production and characterization of biologically active recombinant beta nerve growth factor. Mol Cell Biol 8:452–456

Hama T, Huang KP, Guroff G (1986) Protein kinase C as a component of a nerve growth factor-sensitive phosphorylation system in PC12 cells. Proc Natl Acad Sci USA 83:2353–2357

Hefti F, Hartikka J, Salvatierra A, Weiner WJ, Mash D (1986) Localization of nerve growth factor receptors on cholinergic neurons of the human basal forebrain. Neurosci Lett 69:37–41

Hempstead BL, Patil N, Olson K, Chao M (1988) Molecular analysis of the nerve growth factor receptor. Cold Spring Harbor Symp Quant Biol 53:477–485

Hempstead BL, Schleifer LS, Chao MV (1989) Expression of functional nerve growth factor receptors after gene transfer. Science 243:373–375

Hendry IA, Stockel K, Thoenen H, Iversen LL (1974) The retrograde axonal transport of nerve growth factor. Brain Res 68:103–121

Heumann R, Schwab M, Thoenen H (1981) A second messenger required for nerve growth factor biological activity? Nature 292:838–840

Heumann R, Korsching S, Scott J, Thoenen H (1984) Relationship between levels of nerve growth factor (NGF) and its messenger RNA in sympathetic ganglia and peripheral target tissues. EMBO J 3:3183–3189

Heumann R, Korsching S, Bandtlow C, Thoenen H (1987) Changes of nerve growth factor synthesis in non-neuronal cells in response to sciatic nerve transection. J Cell Biol 104:1623–1631

Hill CE, Hendry IA (1976) Differences in sensitivity to nerve growth factor of axon formation and tyrosine hydroxylase induction in cultured sympathetic neurons. Neuroscience 1:489–496

Hoffman EK, Trusko SP, Freeman N, George DL (1987) Structural and functional characterization of the promoter region of the mouse c-Ki-*ras* gene. Mol Cell Biol 7:2592–2596

Hosang M, Shooter EM (1985) Molecular characteristics of nerve growth factor receptors on PC12 cells. J Biol Chem 260:655–662

Hu GL, Neet KE (1988) Expression of the cDNA for mouse β-nerve growth factor protein in *Escherichia coli*. Gene (in press)

Huebner K, Isobe M, Chao M, Bothwell M, Ross AH, Finan J, Hoxie JA, Sehgal A, Buck CR, Lanahan A, Nowell PC, Koprowski H, Croce CM (1986) The nerve growth factor receptor gene is at human chromosome region 17q12-17q22, distal to the chromosome 17 breakpoint in acute leukemias. Proc Natl Acad Sci USA 83:1403–1407

Isackson PJ, Bradshaw RA (1984) The α-subunit of mouse 7S nerve growth factor is an inactive serine protease. J Biol Chem 259:5380–5383

Isackson PJ, Ullrich A, Bradshaw RA (1984) Mouse 7S nerve growth factor: complete sequence of a cDNA coding for the α-subunit precursor and its relationship to serine proteases. Biochemistry 23:5997–6002

Ishii S, Merlino GT, Pastan I (1985 a) Promoter region of the human Harvey *ras* proto-oncogene: similarity to the EGF receptor proto-oncogene promoter. Science 230:1378–1381

Ishii S, Xu YH, Stratton BA, Roe GT, Merlino GT, Pastan I (1985b) Characterization and sequence of the promoter region of the human epidermal growth factor receptor gene. Proc Natl Acad Sci USA 82:4902–4904

Johnson D, Lanahan A, Buck CR, Sehgal A, Morgan C, Mercer E, Bothwell M, Chao M (1986) Expression and structure of the human NGF receptor. Cell 47:545–554

Johnson EM, Andres RY, Bradshaw RA (1978) Characterization of the retrograde transport of nerve growth factor (NGF) using high specifity activity (^{125})NGF. Brain Res 150:319–331

Johnson EM, Taniuchi M, DiStefano PS (1988) Expression and possible function of nerve growth factor receptors on Schwann cells. Trends Neurosci 11:299–304

Koizumi S, Contreras ML, Matsuda Y, Hama T, Lazarovici P, Guroff G (1988) K-252a: a specific inhibitor of the action of nerve growth factor on PC12 cells. J Neurosci 8:715–721

Korsching S, Thoenen H (1983) Nerve growth factor in sympathetic ganglia and corresponding target organs of the rat: correlation with density of sympathetic innervation. Proc Natl Acad Sci USA 80:3513–3516

Korsching S, Auburger G, Heumann R, Scott J, Thoenen H (1985) Levels of nerve growth factor and its mRNA in the central nervous system of the rat correlate with cholinergic innervation. EMBO J 4:1389–1393

Kouchalakos RN, Bradshaw RA (1986) Nerve growth factor receptor from rabbit sympathetic ganglia membranes. J Biol Chem 261:16054–16059

Kruijer W, Schubert D, Verma IM (1985) Induction of the proto-oncogene *fos* by nerve growth factor. Proc Natl Acad Sci USA 82:7330–7334

Kujubu DA, Lim RW, Varnum BC, Herschman HR (1987) Induction of transiently expressed genes in PC-12 pheochromocytoma cells. Oncogene 1:257–262

Lakshmanan J (1978) Nerve growth factor induced turnover of phosphatidylinositol in rat superior cervical ganglia. Biochem Biophys Res Commun 82:767–775

Landreth GE, Shooter EM (1980) Nerve growth factor receptors on PC12 cells: ligand-induced conversion from low- to high-affinity states. Proc Natl Acad Sci USA 77:4751–4755

Landreth GE, Cohen P, Shooter EM (1980) Ca^{2+} transmembrane fluxes and nerve growth factor action on a clonal cell line of rat pheochromocytoma. Nature 283:202–204

Large TH, Weskamp G, Helder JC, Radeke MJ, Misko TP, Shooter EM, Reichardt LF (1989) Structure and developmental expression of the nerve growth factor receptor in the chicken central nervous system. Neuron 2:1123–1134

Lau LF, Nathans D (1987) Expression of a set of growth-related immediate early genes in BALB/c 3T3 cells: coordinate regulation with c-*fos* or c-*myc*. Proc Natl Acad Sci USA 84:1182–1186

Leonard DGB, Ziff EB, Greene LA (1987) Identification and characterization of mRNAs regulated by nerve growth factor in PC12 cells. Mol Cell Biol 7:3156–3167

Levi A, Shechter Y, Neufeld E, Schlessinger J (1980) Mobility, clustering, and transport of nerve growth factor in embryonal sensory cells and in a sympathetic neuronal cell line. Proc Natl Acad Sci USA 77:3469–3473

Levi A, Eldridge JD, Paterson BM (1985) Molecular cloning of a gene sequence regulated by nerve growth factor. Science 229:393–395

Levi-Montalcini R (1987) The nerve growth factor: thirty-five years later. Science 237:1154–1164

Levi-Montalcini R, Angeletti PU (1963) Essential role of the nerve growth factor in the survival and maintenance of dissociated sensory and sympathetic nerve cells in vitro. Dev Biol 7:655–659

Levi-Montalcini R, Angeletti PU (1968) Nerve growth factor. Physiol Rev 48:534–569

Lindenbaum MH, Carbonetto S, Grosveld F, Flavell D, Mushynski WE (1988) Transcriptional and post-transcriptional effects of nerve growth factor on expression of the three neurofilament subunits in PC-12 cells. J Biol Chem 263:5662–5667

MacDonnell PG, Nagaich K, Lakshmanan J, Guroff G (1977) Nerve growth factor increases activity of ornithine decarboxylase in superior cervical ganglion of young rats. Proc Natl Acad Sci USA 74:4681–4685

Maher PA (1988) Nerve growth factor induces protein-tyrosine phosphorylation. Proc Natl Acad Sci USA 85:6788–6791

Marano N, Dietzschold B, Earley JJ, Schatteman G, Grob P, Ross AH, Bothwell M, Koprowski H (1987) Purification and amino terminal sequencing of human melanoma nerve growth factor receptor. J Neurochem 48:225–232

Marchetti D, Perez-Polo JR (1987) Nerve growth factor receptors in human neuroblastoma cells. J Neurochem 49:475–486

Maruyama K, Schiavi SC, Huse W, Johnson GL, Ruley HE (1987) *myc* and E1A oncogenes alter the responses of PC12 cells to nerve growth factor and block differentiation. Oncogene 1:361–367

Masiakowski P, Shooter EM (1988) Nerve growth factor induces the genes for two proteins related to a family of calcium-binding proteins in PC12 cells. Proc Natl Acad Sci USA 85:1277–1281

Massague J, Guillette BJ, Czech MP, Morgan CJ, Bradshaw RA (1981) Identification of a nerve growth factor receptor protein in sympathetic ganglia membranes by affinity labeling. J Biol Chem 256:9419–9424

Matsuda H, Coughlin MD, Bienenstock J, Denburg JA (1988) Nerve growth factor promotes human hemopoietic colony growth and differentiation. Proc Natl Acad Sci USA 85:6508–6512

McGuire JC, Greene LA (1979) Nerve growth factor stimulation of specific protein synthesis by rat PC12 pheochromocytoma cells. J Biol Chem 254:3362–3367

McTigue M, Cremins J, Haleguoa S (1985) Nerve growth factor and other agents mediate phosphorylation and activation of tyrosine hydroxylase. J Biol Chem 260:9047–9056

Meier R, Becker-Andre M, Gotz R, Heumann R, Shaw A, Thoenen H (1986) Molecular cloning of bovine and chick nerve growth factor (NGF): delineation of conserved and unconserved domains and their relationship to the biological activity and antigenicity of NGF. EMBO J 5:1489–1493

Melton DW, Konecki DS, Brennand J, Caskey CT (1984) Structure, expression and mutation of the hypoxanthine phosphoribosyltransferase gene. Proc Natl Acad Sci USA 81:2147–2151

Milbrandt J (1986) Nerve growth factor rapidly induces c-*fos* mRNA in PC12 rat pheochromocytoma cells. Proc Natl Acad Sci USA 83:4789–4793

Milbrandt J (1987) A nerve growth factor-induced gene encodes a possible transcriptional regulatory factor. Science 238:797–799

Milbrandt J (1988) Nerve growth factor induces a gene homologous to the glucocorticoid receptor gene. Neuron 1:183–188

Mobley WC, Neve RL, Prusiner SB, McKinley MP (1988) Nerve growth factor increases mRNA levels for prion protein and β-amyloid proteins precursor in developing hamster brain. Proc Natl Acad Sci USA 85:9811–9815

Nishizuka Y (1984) Turnover of inositol phospholipids and signal transduction. Science 225:1365–1370

Noda M, Ko M, Ogura A, Liu DG, Amano T, Takano T, Ikawa Y (1985) Sarcoma viruses carrying *ras* oncogenes induce differentiation-associated properties in a neuronal cell line. Nature 318:73–75

Prentice HM, Moore SE, Dickson JG, Doherty P, Walsh FS (1987) Nerve growth factor-induced changes in neural cell adhesion molecule (N-CAM) in PC12 cells. EMBO J 6:1859–1863

Puma P, Buxser SE, Watson L, Kelleher DJ, Johnson GL (1983) Purification of the receptor for nerve growth factor from A875 cells by affinity chromatography. J Biol Chem 258:3370–3375

Race HM, Wagner JA (1985) Nerve growth factor potentiates but does not activate adenylate cyclase in PC12 cells. J Neurochem 44:1588–1592

Radeke MJ, Misko TP, Hsu C, Herzenberg LA, Shooter EM (1987) Gene transfer and molecular cloning of the rat nerve growth factor receptor. Nature 325:593–597

Raivich G, Zimmermann A, Sutter A (1985) The spatial and temporal pattern of β-NGF receptor expression in the developing chick embryo. EMBO J 4:637–644

Rettig WJ, Thomson TM, Spengler BA, Biedler JL, Old LJ (1986) Assignment of human nerve growth factor receptor gene to chromosome 17 and regulation of receptor expression in somatic cell hybrids. Somatic Cell Mol Genet 12:441–447

Reynolds GA, Basu SK, Osborne TF, Chin DJ, Gil G, Brown MS, Goldstein JL, Luskey KL (1984) HMG CoA reductase: a negatively regulated gene with unusual promoter and 5′ untranslated regions. Cell 38:275–286

Richter-Landsberg C, Jastorff B (1986) The role of cAMP in nerve growth factor-promoted neurite outgrowth in PC12 cells. J Cell Biol 102:821–826

Ross AH, Grob P, Bothwell MA, Elder DE, Ernst CS, Marano N, Ghrist BFD, Slemp CC, Herlyn M, Atkinson B, Koprowski H (1984) Characterization of nerve growth factor receptor in neural crest tumors using monoclonal antibodies. Proc Natl Acad Sci USA 81:6681–6685

Rydel R, Greene LA (1988) cAMP analogs promote survival and neurite outgrowth in cultures of rat sympathetic and sensory neurons independently of nerve growth factor. Proc Natl Acad Sci USA 85:1257–1261

Schatteman GC, Gibbs L, Lanahan AA, Claude P, Bothwell M (1988) Expression of NGF receptor in the developing and adult primate central nervous system. J Neurosci 8:860–873

Schechter AL, Bothwell MA (1981) Nerve growth factor receptors on PC12 cells: evidence for two receptor classes with differing cytoskeletal association. Cell 24:867–874

Schubert D, LaCorbiere M, Whitlock C, Stallcup W (1978) Alterations in the surface properties of cells responsive to nerve growth factor. Nature 273:718–723

Schwarz MA, Fisher D, Bradshaw RA, Isackson PJ (1988) Isolation and sequence of a cDNA clone of β-nerve growth factor from the guinea pig prostate gland. J Neurochem 52:1203–1209

Scott J, Selby M, Urdea M, Quiroga M, Bell GI, Rutter WJ (1983) Isolation and nucleotide sequence of a cDNA encoding the precursor of mouse nerve growth factor. Nature 302:538–540

Seeley PJ, Keith CH, Shelanski ML, Greene LA (1983) Pressure microinjection of nerve growth factor and anti-nerve growth factor into the nucleus and cytoplasm: lack of effects on neurite outgrowth from pheochromocytoma cells. J Neurosci 3:1488–1494

Sehgal A, Wall D, Chao M (1988 a) Efficient processing and expression of human nerve growth factor receptors in *Xenopus laevis* oocytes: effects on maturation. Mol Cell Biol 8:2242–2246

Sehgal A, Patil N, Chao MV (1988 b) A constitutive promoter directs expression of the nerve growth factor receptor gene. Mol Cell Biol 8:3160–3167

Seiler M, Schwab ME (1984) Specific retrograde transport of nerve growth factor (NGF) from neocortex to nucleus basalis in the rat. Brain Res 300:33–39

Seizinger B et al. (1987) Genetic linkage of von Recklinghausen neurofibromatosis to the nerve growth factor receptor gene. Cell 49:589–594

Selby MJ, Edwards R, Sharp F, Rutter WJ (1987 a) The mouse nerve growth factor gene: structure and expression. Mol Cell Biol 7:3057–3064

Selby MJ, Edwards RH, Rutter WJ (1987 b) Cobra nerve growth factor: structure and evolutionary comparison. J Neurosci Res 18:293–298

Shaw G, Kamen R (1986) A conserved AU sequence from the 3′ untranslated region of GM-CSF mRNA mediates selective mRNA degradation. Cell 46:659–667

Shelton DL, Reichardt LF (1984) Expression of the nerve growth factor gene correlates with the density of sympathetic innervation in effector organs. Proc Natl Acad Sci USA 81:7951–7955

Shelton DL, Reichardt LF (1986) Studies on the expression of the β nerve growth factor (*NGF*) gene in the central nervous system: level and regional distribution of NGF mRNA suggest that NGF functions as a trophic factor for several distinct populations of neurons. Proc Natl Acad Sci USA 83:2714–2718
Singer-Sam J, Keith DH, Tani K, Simmer RL, Shively L, Lindsay S, Yoshida A, Riggs AD (1984) Sequence of the promoter region of the gene for human X-linked 3-phosphoglycerate kinase. Gene 32:409–417
Sonnenfeld KH, Ishii DN (1982) Nerve growth factor effects and receptors in cultured human neuroblastoma cell lines. J Neurosci Res 8:375–391
Springer JE, Koh S, Tyrien MW, Loy R (1987) Basal forebrain magnocellular neurons stain for nerve growth factor receptor: correlation with cholinergic cell bodies and effects of axotomy. J Neurosci Res 17:111–118
Stein R, Orit S, Anderson DJ (1988) The induction of a neural-specific gene, SCG10, by nerve growth factor in PC12 cells is transcriptional, protein synthesis dependent, and glucocorticoid inhibitable. Dev Biol 127:316–325
Sukhatme VP, Cao X, Chang LC, Tsai-Morris C-H, Stamenkovich D, Ferreira PCP, Cohen DR, Edwards SA, Shows TB, Curran T, LeBeau MM, Adamson ED (1988) A zinc finger-encoding gene coregulated with c-*fos* during growth and differentiation, and after cellular depolarization. Cell 53:37–43
Sutter A, Riopelle RJ, Harris-Warrick RM, Shooter EM (1979) NGF receptors: characterization of two distinct classes of binding sites on chick embryo sensory ganglia cells. J Biol Chem 254:5972–5982
Swift GH, Dagorn JC, Ashley PL, Cummings SW, MacDonald RJ (1982) Rat pancreatic kallikrein mRNA: nucleotide sequence and amino acid sequence of the encoded preproenzyme. Proc Natl Acad Sci USA 79:7263–7267
Taniuchi M, Clark HB, Johnson EM (1986a) Induction of nerve growth factor receptor in Schwann cells after axotomy. Proc Natl Acad Sci USA 83:4094–4098
Taniuchi M, Schweitzer JB, Johnson EM (1986b) Phosphorylation of nerve growth factor receptor proteins in sympathetic neurons and PC12 cells. J Biol Chem 261:13342–13349
Taniuchi M, Schweitzer JB, Johnson EM (1986c) Nerve growth factor receptor molecules in rat brain. Proc Natl Acad Sci USA 83:1950–1954
Thoenen H, Barde Y-A (1980) Physiology of nerve growth factor. Physiol Rev 60:1284–1335
Thoenen H, Angeletti PU, Levi-Montalcini R, Kettler R (1971) Selective induction of tyrosine hydroxylase and dopamine β hydroxylase in rat superior cervical ganglia by nerve growth factor. Proc Natl Acad Sci USA 68:1598–1602
Thoenen H, Bandtlow C, Heumann R (1987) The physiological function of nerve growth factor in the central nervous system: comparison with the periphery. Rev Physiol Biochem Pharmacol 109:145–178
Thorpe LW, Stach RW, Hashim GA, Marchetti D, Perez-Polo JR (1987) Receptors for nerve growth factor on spleen mononuclear cells. J Neurosci Res 17:128–134
Traynor AE, Schubert D, Allen WR (1982) Alterations of lipid metabolism in response to nerve growth factor. J Neurochem 39:1677–1683
Ullrich A, Gray A, Berman C, Dull TJ (1983) Human beta nerve growth factor sequence highly homologous to that of mouse. Nature 303:821–825
Ullrich A, Gray A, Wood WI, Hayflick J, Seeburg PH (1984a) Isolation of a cDNA clone coding for the γ-subunit of mouse nerve growth factor using a high-stringency selection procedure. DNA 3:387–391
Ullrich A, Coussens L, Hayflick JS, Dull TJ, Gray A, Tam AW, Lee J, Yarden Y, Libermann TA, Schlessinger J, Downward J, Mayes ELV, Whittle N, Waterfield MD, Seeburg PH (1984b) Human epidermal growth factor receptor cDNA sequences and aberrant expression of the amplified gene in A431 epidermoid carcinoma cells. Nature 313:756–761
Van Buskirk R, Corcoran T, Wagner JA (1985) Clonal variants of PC12 pheochromocytoma cells with defects in cAMP-dependent protein kinases induce

ornithine decarboxylase in response to nerve growth factor but not to adenosine agonists. Mol Cell Biol 5:1984–1992
Van Calker D, Heumann R (1987) Nerve growth factor potentiates the agonist-stimulated accumulation of inositol phosphates in PC-12 pheochromocytoma cells. Eur J Pharmacol 135:259–260
Volonte C, Parries GS, Racker E (1988) Stimulation of inositol incorporation into lipids of PC12 cells by nerve growth factor and bradykinin. J Neurochem 51:1156–1162
Wharton KA, Johansen KM, Xu T, Artavanis-Tsakonas S (1985) Nucleotide sequences from neurogenic locus notch implies a gene product that shares homology with proteins containing EGF-like repeats. Cell 43:567–581
Whittemore SR, Seiger A (1987) The expression, localization and functional significance of β-nerve growth factor in the central nervous system. Brain Res Rev 12:439–464
Whittemore SR, Friedman PL, Larhammar D, Persson H, Gonzalez-Carvajal M, Hotels VR (1988) Rat β-nerve growth factor sequence and site of synthesis in the adult hippocampus. J Neurosci Res 20:403–410
Williams LR, Varon S, Peterson GM, Wictorin K, Fischer W, Bjorklund A, Gage FH (1986) Continuous infusion of nerve growth factor prevents basal forebrain neuronal death after fimbria fornix transection. Proc Natl Acad Sci USA 83:9231–9235
Wion D, Perret C, Frechin N, Keller A, Behar G, Brachet P, Auffray C (1986) Molecular cloning of the avian β-nerve growth factor gene: transcription in brain. FEBS Lett 203:82–86
Yamamoto T, Davis CG, Brown MS, Schneider WJ, Casey ML, Goldstein JL, Russell DW (1984) The human LDL receptor: a cysteine-rich protein with multiple Alu sequences in its mRNA. Cell 39:27–38
Yankner BA, Shooter EM (1980) Nerve growth factor in the nucleus: interaction with receptors on the nuclear membrane. Proc Natl Acad Sci USA 76:1269–1273
Yu MW, Tolson NW, Guroff G (1980) Increased phosphorylation of specific nuclear proteins in superior cervical ganglia and PC12 cells in response to nerve growth factor. J Biol Chem 255:10481–10492
Zabel BU, Eddy RL, Lalley PA, Scott J, Bell GI, Shows TB (1985) Chromosomal locations of the human and mouse genes for precursors of epidermal growth factor and the β subunit of nerve growth factor. Proc Natl Acad Sci USA 82:469–473
Zheng M, Heinrich G (1988) Structural and functional analysis of the promoter region of the nerve growth factor gene. Mol Brain Res 3:133–140

CHAPTER 24

A Glia-Derived Nexin Acting as a Neurite-Promoting Factor

D. MONARD

A. Introduction

Glial cells are thought to derive, as do neurons, from the neuroepithelial cells of the neural tube. Astrocytes, oligodendrocytes, Schwann cells, or microglia are the major types of glial cells, and it has long been recognized that they have a role in the development of the neuronal cells. They are also believed to support some functions of the fully differentiated neurons.

Established glioma cell lines and glia primary cultures have made it possible to identify glia-derived molecules able to affect survival, neurite outgrowth, and neurotransmitter synthesis in cultured neuronal cells. Some of these "glial factors" influencing neuronal cells have now been purified to homogeneity (BARDE et al. 1982; BARBIN et al. 1984; FUKADA 1985). Knowledge about their primary sequence will certainly be a prerequisite for the study of their mode of action and in vivo relevance.

One of the macromolecules secreted by rat glioma cells promotes neurite outgrowth in mouse neuroblastoma cells (MONARD et al. 1973). This neurite-promoting activity is also detected in the medium conditioned by rat brain primary cultures established after a critical developmental stage (SCHUERCH-RATHGEB and MONARD 1978). This chapter describes the properties and features of this glia-derived neurite-promoting factor.

B. A Glia-Derived Neurite-Promoting Factor Acting as a Protease Inhibitor

I. Biochemical Properties

This glia-derived neurite-promoting factor is distinct from the well-established nerve growth factor (MONARD et al. 1975). Experiments aimed at purification of this protein revealed that a strong serine protease inhibitor activity was always associated with the neurite-promoting activity. Nanograms of thrombin were also able to antagonize the neurite-promoting activity, and hirudin, a potent thrombin inhibitor, mimicked the activity of the glia-derived factor in neuroblastoma cells (MONARD et al. 1983). Finally, a 43-kDa protein which binds to heparin Sepharose and to Affi-gel blue Sepharose was purified to homogeneity. It was established that this glia-derived protein is both a neurite-promoting factor and a potent serine protease inhibitor (GUENTHER et al. 1985). The 43 kDa glia-derived protein forms SDS-resistant complexes with urokinase, plasminogen activator, thrombin, or trypsin. The dissociation constants of the equilibrium complexes of the factor with trypsin, urokinase, and thrombin were 17, 280, and 18 p*M*, respectively. The glia-derived protein inactivates thrombin about 200-fold faster than antithrombin III. Kinetic experiments have also indicated that the rate at which the glia-derived inhibitor reacts with thrombin increases by over 40-fold in the presence of heparin. Heparin also decreases the dissociation of the complex with thrombin by over 80-fold to 0.3 p*M* (STONE et al. 1987). Different heparin types, fractionated on the basis of their affinity for antithrombin III, give an optimal rate of α-thrombin inhibition. At optimal heparin concentrations, the rate of inactivation is 0.5–1.2 n*M*/s, which suggests that under these conditions the interaction is diffusion controlled (WALLACE et al. 1989).

II. Molecular Cloning

Since *Xenopus* oocytes released the serine protease inhibitory activity following injection of mRNA from rat glioma cells, a corresponding cDNA library was constructed and screened by hybridization-selected translation using specific antibodies to the purified 43 kDa protein. The sequencing of the cDNA clones obtained made it possible to deduce the amino acid sequence of the neurite-promoting factor released by rat glioma cells. Since the amino-terminal residue was blocked and resistant to Edmann degradation, the purified protein was submitted to limited trypsinization, and some of the tryptic peptides were purified by high-pressure liquid chromatography. The sequence of these peptides matched some fragments of the amino acid sequence deduced from the cDNA, establishing that the cDNA cloned was in fact derived from the mRNA coding for the purified 43 kDa protein (SOMMER et al. al. 1987). Its correct identity having been established, the available cDNA probe was used to screen a cDNA library corresponding to the mRNAs from a human glioma cell line. The sequencing of the human cDNA clones obtained led to the amino acid sequence of the corresponding human protein (GLOOR et al. 1986).

III. Characteristics of the Primary Structures

All of the cDNAs analyzed start with a sequence coding for a 19-amino-acid signal peptide, confirming that the proteins are secreted by the glial cells. There is an 83% homology between the rat and human protein sequences. The sequences obtained show regions of homology with known protease inhibitors: endothelial plasminogen activator inhibitor (NY et al. 1986; PANNEKOEK et al. 1986), antithrombin III (BOCK et al. 1982), and α_1-protease inhibitor (KURACHI et al. 1981). Best alignment studies revealed that the Arg345 and the Ser346 can be considered the reactive center of this glia-derived protease inhibitor. With this as given, the P_{17} residue is a glutamic acid and the P_{69} residue a lysine. This is of interest, since the formation of a salt bridge

Table 1. Sequence properties of glia-derived nexins

	Rat	Human α	Human β
Residues	378	378	379
Signal peptide (residues)	19	19	19
Residue 310	Arg	Arg	Thr–Gly
Molecular weight	41 700	41 919	41 921
Glycosylation sites (Asn–X–Thr/Ser)		99–101	99–101
	140–142	140–142	140–142
	364–366	364–366	365–367
Cys residues	117	117	117
	131	131	131
	360	209	209

between the Glu342 and the Lys290 (therefore also at a 52-residue equidistance) has been reported to stabilize the three-dimensional structure of α_1-proteinase inhibitor (LOEBERMANN et al. 1984). These sequence properties in fact fulfil the criteria proposed by CARELL and TRAVIS (1985) for definition of the serpin (*ser*ine *p*rotease *in*hibitors) superfamily. In addition the first 18 amino acids of the human protein were found to be identical to the partial amino terminal sequence known at the time for protein nexin I (SCOTT et al. 1985). These sequence data disclosed two important facts: (a) the glia-derived proteins belong to the serpins and (b) the nexins belong to the serpins. The glia-derived neurite-promoting factor can thus be termed glia-derived nexin (GDN).

Some of the properties of rat and human GDN are summarized in Table 1. In contrast to rat GDN, two distinct cDNA subclones have been identified for human GDN. These clones are identical except that, in the β-form, the two residues Thr310 and Gly311 replace the residue Arg310. This reveals the presence of two corresponding mRNAs in the human glioma cells (SOMMER et al. 1987). It is worth noting that two of the three cysteine residues are conserved at exactly the same positions (117 and 131, respectively) in rat and human GDN. It remains to be seen if a disulfide bridge between these residues is of importance for the inhibitory and the neurite-promoting activity of the protein. The putative glycosylation sites (Asn-X-Thr/Ser) conserved in the rat and human sequences are located at the same positions (364-366; 365-367 for the β-form). Again, further studies should establish if glycosylation at these residues can influence the biological properties of GDN.

Since the rate of reaction with thrombin is increased more than 40-fold in the presence of heparin, the heparin-binding site of GDN represents another domain of interest. In antithrombin III, which shows a 41% overall homology with GDN (SOMMER et al. 1987), the residues Lys107, Lys125, and Lys136 are situated within the heparin-binding site of the inhibitor (CHANG 1989). A best alignment comparison reveals that, in GDN, lysinyl residues are also found at the positions corresponding to Lys125, Lys133, and Lys136 of antithrombin III. This homology is quite appealing but the heparin site of GDN still remains to be unequivocally established.

These data have recently been supported by results demonstrating that human protease nexin I is identical to human GDN at the level of the primary structure (McGROGAN et al. 1988). The two forms of mRNA are also found in the human fibroblasts where Northern hybridization indicates that these cells produce about twice as much α-form as β-form. Transfection with the cDNA coding for either of the two forms revealed that both forms have the same protease inhibitor characteristics (McGROGAN et al. 1988).

IV. Biological Effects

As stated above, rat GDN was purified as a protein-promoting neurite outgrowth in the NB_2a clone of mouse neuroblastoma cells. The serine protease inhibitor activity of GDN having been established, other serine protease inhibitors were tested for their neurite-promoting activity in this in vitro assay. Only hirudin (the most potent thrombin inhibitor known before the characterization of GDN) and synthetic tripeptides with high antithrombin specificity can promote neurite outgrowth at concentrations similar to those of GDN. Increasing amounts of thrombin gradually antagonize the neurite-promoting effect of GDN (MONARD et al. 1983) or cause retraction of the neurites induced by GDN in neuroblastoma cells (GURWITZ and CUNNINGHAM 1988). These data indicate that protease inhibitory activity is of crucial importance for the biological activity of GDN.

GDN also promotes neurite outgrowth in cultured neurons of the chick dorsal root ganglion (ZURN et al. 1988). The GDN effect is dose dependent and is detected when the culture conditions (polyornithine coating, low cell density, serum-free medium) do not allow optimal neurite outgrowth. For example, GDN potentiates neurite outgrowth only marginally when these neurons are seeded on one of their optimal substrates, laminin. In the initial phase of neurite outgrowth, addition of GDN increases both the percentage of neuron-bearing neurites and the mean neurite length. It does not, however, support the survival of those neurons known to require the presence of nerve growth factor (NGF) (LEVI-MONTALCINI 1966). It is most interesting that, in the presence of NGF, GDN is still able, under certain conditions, to potentiate the rate of neurite outgrowth. These data suggest a permissive function for GDN and illustrate that this glia-derived protease inhibitor could assume importance primarily when the environmental conditions do not permit an optimal rate of neuritic growth (MONARD 1988). It is also noteworthy that hirudin and the synthetic tripeptides able to mimic the neurite-promoting effect of GDN in neuroblastoma cells did not influence the rate of neurite outgrowth in these primary neurons.

At a specific developmental stage of the cerebellum, the immature granule cells migrate from the external granular layer through the molecular layer and the Purkinje cell layer to their final location in the internal granular layer where they further differentiate. An increased plasminogen activator activity is associated with those migrating neuroblasts. This increased cell-associated proteolytic activity is thought to be required for the phases of de-adhesion involved in the dynamics of cellular migration (KRYSTOSEK and SEEDS 1981 a). [^{3}H]Thymidine labeling of the proliferative neuroblasts of the extragranular layer allows monitoring of the granule cell migration in explants from early postnatal mouse cerebellum cultured in serum-free medium. GDN causes an inhibition of this migration of up to 50% (LINDNER et al. 1986). Other serine protease inhibitors (aprotinin, hirudin, soybean trypsin inhibitor, leupeptin, 6-aminocaproic acid, and the synthetic antithrombin D-Phe-Pro-ArgCH$_2$Cl) do not interfere with the migration of granule cells even when tested at concentrations known to elicit their full protease inhibitory effect. Addition of thrombin (10–15 ng/ml) did not alter the pattern of migration in control cultures but did suppress this inhibitory effect of GDN, probably through the formation of the SDS-resistant complex mentioned above. The results indicated that a protein released by glial cells can modulate the migration of neuroblasts. Through this ability to influence the location of the final position of the still differentiating neurons, glial cells may exert a key influence at early stages on the formation of neuronal connectivity.

V. Localization of Glia-Derived Nexin

Northern blot analysis indicated that GDN mRNA is predominantly detected in the nervous system of the rat and that the extent of the expression seems to depend on the developmental stage (GLOOR et al. 1986). Thus, if GDN is of some relevance for neurite outgrowth in vivo, it should be found at specific times in certain nervous system structures. The antibodies used to perform the hybridization-selected translation were inappropriate for immunocytochemistry. GDN was therefore briefly treated with 2% paraformaldehyde before the immunization of rabbits. The anti-GDN immunoglobulins were purified from the antisera. Their specificity was tested on immunoblots of medium conditioned by C6 rat glioma cells and of crude homogenate of the rat olfactory bulb. In both cases, only the expected 43 kDa band, comigrating with the purified GDN, was detected. The first immunocytochemical stains revealed high concentrations of the antigen in the olfactory system of the adult rat (REINHARD et al. 1988). GDN immunoreactivity is localized both at the olfactory epithelium and at the olfactory bulb. At the level of the submucosa, there is a similarity between the anti-GDN, antifibronectin, and anti-laminin patterns, suggesting that in this structure GDN could be synthesized by the cells that ensheath the olfactory axons and secreted, together with fibronectin and laminin, to provide a suitable environment for the elongation of newly generated olfactory axons. Northern blot analysis showed the presence of GDN mRNA in both the olfactory epithelium and the olfactory bulb, indicating that the GDN staining in the submucosa and in the olfactory nerve layer is not solely attributable to the axonal transport of GDN to the CNS. In the rat olfactory bulb, the amount of GDN mRNA does not significantly change during postnatal development and is similar to the level found in the adult. Experiments in progress are aimed at the precise localization and identification of the cells synthesizing GDN. Already at this stage, localization of GDN in the olfactory system is quite intriguing. Under physiological conditions, the olfactory system is the only structure in the mammalian CNS where neurogenesis and axogenesis continue throughout life. A continuous degeneration of the short-lived receptor cells takes place, followed by a constant replacement of neurons by neurogenic stem cells in the base of the epithelium (GRAZIADEI 1973). The developing neuroblast must send its newly formed axon into the olfactory bulb. Moreover, this is the only sensory pathway which can be totally interrupted and nevertheless reconstitute functional connections with the CNS. This regeneration potential requires protection of the living neurons from proteolytic degradation and an environment that promotes neurite outgrowth. High GDN levels in structures where neuronal degeneration and regeneration are continuously taking place does not prove, but does strongly support, an in vivo relevance for this glia-derived neurite-promoting factor.

VI. Glia-Derived Nexin, a Representative of a New Family of Neurite-Promoting Factors?

Identification of the purified glia-derived neurite-promoting factor as a strong serine protease inhibitor was obviously an unexpected finding (MONARD et al. 1983). The molecular cloning identifying this glial factor as the first member of the nexin family fully characterized at the level of the primary sequence has confirmed these properties. Moreover, as already stated, recent sequencing data have established that protease nexin I (PN-I), released by human foreskin fibroblasts, is identical, at least at the primary structure level, to human GDN (MCGROGAN et al. 1988). Anti-PN-I antibodies crossreact with rat GDN (ROSENBLATT et al. 1987), and anti-GDN antibodies with purified PN-I. Purified PN-I can also promote neurite outgrowth in mouse neuroblastoma cells (unpublished results). Purified PN-I and GDN do not show the same molecular weight when analyzed on the same SDS-polyacrylamide gel. It is possible that this discrepancy is due to a different type of glycosylation. Further experiments should establish if the nature of the posttranslational modifications is significant for the mode of action in the nervous system. Cardiac myocytes and fibroblasts also release an inhibitor of neuronal plasminogen activator which potentiates neurite outgrowth from sympathetic neurons (PITTMAN and PATTERSON 1987). Two molecular-weight forms are detected in the medium conditioned by heart cells. One form is associated with heparin and cannot be stimulated upon addition of the proteoglycan. The other form can bind to heparin, which potentiates its ability to form an SDS-resistant complex with [^{125}I]urokinase. This inhibitor shows a molecular weight of about 50 kDa in the presence of SDS and is detected at about 2000 kDa under native conditions. This heart cell-derived inhibitor is more hydrophobic than PN-I but shares the same properties.

VII. Mode of Action of GDN?

The nexins (from the Latin *nexus,* to bind or link) were first described as cell-secreted proteins able selectively to link to serine proteases in the extracellular environment (BAKER et al. 1980). They were first considered to be "receptors" mediating cellular binding, internalization, and degradation of the proteases in the vicinity of the cell. In addition to protease nexin I, two other protease nexins, designated PN-II and PN-III, have been characterized on the basis of their protease specificity (KNAUER et al. 1983). Diisopropylphosphate treatment of the protease abolishes the linkage to the nexins. Thus, the complex formation of each of these nexins with their respective proteases requires a functional catalytic site of the enzyme. PN-I was shown to bind the protease and promote the association of thrombin and urokinase with the cell. PN-II preferentially associates with the EGF-binding protein (KNAUER and CUNNINGHAM 1982) and PN-III with the γ-subunit of the 7S nerve growth factor complex (KNAUER et al. 1982), two proteins with arginylesteropeptidase activity. The nexins have been reviewed (KNAUER and CUNNINGHAM 1984; BAKER et al. 1986). Some of their properties are obviously of importance in the study of the mode of action of GDN as neurite-promoting factor.

The mode of action of GDN must also be considered in relation to the increased proteolytic activities detected at the level of the growth cone (KRYSTOSEK and SEEDS 1981 b; PITTMAN 1985). Such proteolytic activities are thought to cause the transitory interactions of the microspikes (or filopodia) with their environment, thus triggering their motility. The proteases associated with or released by the growth cone could be involved in the degradation of some components of the extracellular matrix (ECM) supporting neurite outgrowth, thus creating nonpermissive zones where further adhesion would become weaker or unlikely. An excess of proteolytic activity would lead to a hypermotility of the growth cone or even become deleterious to net neuritic extension. In fact, addition of nanograms of thrombin interferes with neurite outgrowth (MONARD et al. 1983) or causes neurite retraction in mouse neuroblastoma cells (GURWITZ and CUNNINGHAM 1988). It is conceivable that the presence of GDN in the vicinity of motile growth cone structures would lead to the formation of stable complexes with some of the growth cone-associated (or -released) proteases. Such complex formation would be strongly potentiated in spots where heparin sulfate proteoglycan colocalizes with GDN. Such localized decrease in proteolytic activity at discrete zones located just at the origin of the growth cones or in the environment explored by the microspikes would reduce the de-adhesion phases characterizing microspike activity and establish new zones of preferential adhesion. Such mechanisms could trigger the immobilization of some microspikes which seem to promote net neuritic extension by dragging the growth cone forward (BRAY and CHAPMAN 1985).

Today, only fragmentary information supports this hypothetical model. Elastase cleaves GDN between the P_1 and the P_2 residues. The two fragments generated still attach together to heparin-Sepharose. The exact location of the cleavage has been confirmed by sequencing data following their resolution on HPLC chromatography. Elastase-treated GDN has lost both the protease inhibitor and the neurite-promoting activity (NICK et al. submitted). This, together with the fact that hirudin and synthetic tripeptides with antithrombin specificity mimic the effect of GDN in neuroblastoma cells (MONARD et al. 1983) but not in primary sympathetic neurons (ZURN et al. 1988), indicates that the inhibition of proteolytic activity is necessary but by itself not sufficient to explain the mode of action of GDN. These data suggest that the formation of a stable high-molecular-weight complex implies conformational changes leading to the exposition of new epitopes contributing to a better anchorage of the neuritic membrane with its environment. The molecular and immunological tools available today should make it possible to define the functional domains of GDN.

VIII. In Vivo Relevance of the Balance Between Proteases and Protease Inhibitors for Neurite Outgrowth?

The in vivo relevance of GDN remains to be demonstrated. Attention has recently been drawn to the role of proteases following lesion or in degenerative processes in the nervous system due to the discovery of a serine protease inhibitor sequence in the cDNA coding for one form of the precursor of the β-amyloid protein characteristic of the Alzheimer's plaques (PONTE et al. 1988; TANZI et al. 1988; KITAGUCHI et al. 1988). It is speculated that the form without this protease inhibitory domain would be much more sensitive to proteolytic degradation and might, at least partially, explain the formation of the plaques. The presence of α_1-antichymotrypsin in the amyloid deposits (ABRAHAM et al. 1988) further indicates that a lack of balance between proteases and their inhibitors triggers the ongoing degeneration. An excess of plaminogen activator activity has also been detected in Wallerian degeneration (BIGNAMI et al. 1982). Preliminary experiments indicate that GDN or GDN-like proteins able to complex [^{125}I]urokinase are detected following lesion in the peripheral nervous system where regeneration is possible (PATTERSON 1985). Recombinant cell-derived protease inhibitors such as GDN would be very important tools for assessing their implication in some of these in vivo paradigms.

C. Conclusion

The identification of a glia-derived neurite-promoting factor as a potent serine protease inhibitor of the nexin family suggests that the protease/protease inhibitor balance can modulate the rate of neurite outgrowth. One has to expect that distinct proteases and their specific inhibitors could be involved in this regulation. Neurite outgrowth is a complex cellular phenomenon. The protease/protease inhibitor balance is only one of the biochemical events involved and is not necessarily always rate limiting. The molecular and immunological tools now available are making it possible to define both the mode of action and the in vivo relevance of the GDN molecule with respect to this chain of events.

References

Abraham CR, Selkoe DJ, Potter H (1988) Immunochemical identification of the serine protease inhibitor α_1-antichymotrypsin in the brain amyloid deposits of Alzheimer's disease. Cell 52:487–501

Baker JB, Low DA, Simmer RL, Cunningham DD (1980) Protease-nexin: a cellular component that links thrombin and plasminogen activator and mediates their binding to cells. Cell 21:37–45

Baker JB, Knauer DJ, Cunningham DD (1986) Protease nexins: secreted protease inhibitors that regulate protease activity at or near the cell surface. In: Cohn PM (ed) The receptors, vol 3. Academic, New York, p 153

Barbin G, Manthorpe M, Varon S (1984) Purification of the chick eye ciliary neurotrophic factor. J Neurochem 43:1468–1478

Barde YA, Edgar D, Thoenen H (1982) Purification of a new neurotrophic factor from mammalian brain. EMBO J 5:549–553

Bignami A, Cella G, Chi NH (1982) Plasminogen activators in rat neural tissues during development and in Wallerian degeneration. Acta Neuropathol (Berl) 58:224–228

Bock SC, Wion KL, Vehar GA, Lawn RM (1982) Cloning and expression of the cDNA for human antithrombin III. Nucleic Acids Res 10:8113–8125

Bray D, Chapman K (1985) Analysis of microspike movements on the neuronal growth cone. J Neurosci 5:3204–3213

Carrell R, Travis J (1985) α_1-Antitrypsin and the serpins: variation and countervariation. Trends Biochem Sci 10:20–24

Chang JY (1989) Binding of heparin to human antithrombin III activates selective modification at Lys-236. Lys-107, Lys-125 and Lys-136 are situated within the heparin binding site of antithrombin III. J Biol Chem 264:3111–3115

Fukada K (1985) Purification and partial characterization of a cholinergic neuronal differentiation factor. Proc Natl Acad Sci USA 82:8795–8799

Gloor S, Odink K, Guenther J, Nick H, Monard D (1986) A glia-derived neurite promoting factor with protease inhibitory activity belongs to the protease nexins. Cell 47:687–693

Graziadei PPC (1973) Cell dynamics in the olfactory mucosa. Tissue Cell 5:113–131

Guenther J, Nick H, Monard D (1985) A glia-derived neurite-promoting factor with protease inhibitory activity. EMBO J 4:1963–1966

Gurwitz D, Cunningham DD (1988) Thrombin modulates and reverses neuroblastoma neurite outgrowth. Proc Natl Acad Sci USA 85:3440–3444

Kitaguchi N, Takahashi Y, Tokushima Y, Shiojiri S, Ito H (1988) Novel precursor of Alzheimer's disease amyloid protein shows protease inhibitory activity. Nature 331:530–532

Knauer DJ, Cunningham DD (1982) Epidermal growth factor carrier protein binds to cells via a complex with released carrier protein nexin. Proc Natl Acad Sci USA 79:2310–2314

Knauer DJ, Cunningham DD (1984) Protease nexins: cell-secreted proteins which regulate extracellular serine proteases. Trends Biochem Sci 9:231–233

Knauer DJ, Scaparro KM, Cunningham DD (1982) The γ subunit of 7S nerve growth factor binds to cells via complexes formed with two cell-secreted nexins. J Biol Chem 257:15098–15104

Knauer DJ, Thompson JA, Cunningham DD (1983) Protease nexins: cell-secreted proteins that mediate the binding, internalization, and degradation of regulatory serine proteases. J Cell Physiol 117:385–396

Krystosek A, Seeds NW (1981 a) Plasminogen activator secretion by granule neurons in cultures of developing cerebellum. Proc Natl Acad Sci USA 78:7810–7814
Krystosek A, Seeds NW (1981 b) Plasminogen activator release at the neuronal growth cone. Science 213:1532–1534
Kurachi K, Chandra T, Friezner-Degen SJ, White TT, Marchioro TL, Woo SLC, Davie EW (1981) Cloning and sequence of cDNA coding for α_1-antitrypsin. Proc Natl Acad Sci USA 78:6826–6830
Levi-Montalcini R (1966) The nerve growth factor: its mode of action on sensory and sympathetic nerve cells. Harvey Lect 60:217–259
Lindner J, Guenther J, Nick H, Zinser G, Antonicek H, Schachner M, Monard D (1986) Modulation of granule cell migration by a glia-derived protein. Proc Natl Acad Sci USA 83:4568–4571
Loebermann H, Tokuoka R, Deisenhofer J, Huber R (1984) Human α_1-proteinase inhibitor. J Mol Biol 177:531–556
McGrogan M, Kennedy J, Li MP, Hsu C, Scott RW, Simonsen CC, Baker JB (1988) Molecular cloning and expression of two forms of human protease nexin I. Biotechnology 6:172–177
Monard D (1988) Cell-derived proteases and protease inhibitors as regulators of neurite outgrowth. Trends Neurosci 11:541–544
Monard D, Solomon F, Rentsch M, Gysin R (1973) Glia-induced morphological differentiation in neuroblastoma cells. Proc Natl Acad Sci USA 70:1894–1897
Monard D, Stockel K, Goodman R, Thoenen H (1975) Distinction between nerve growth factor and glial factor. Nature 258:444–445
Monard D, Niday E, Limat A, Solomon F (1983) Inhibition of protease activity can lead to neurite extension in neuroblastoma cells. Prog Brain Res 58:359–364
Ny T, Sawdey M, Lawrence D, Millan JL, Loskutoff DJ (1986) Cloning and sequence of a cDNA coding for the human β-migrating endothelial-cell-type plasminogen activator inhibitor. Proc Natl Acad Sci USA 83:6776–6780
Pannekoek H, Veerman H, Lambers H, Diergaarde P, Verweij CL, Van Zonneveld AJ, Van Mourik JA (1986) Endothelial plasminogen activator inhibitor (PAI): a new member of the serpin gene family. EMBO 5:2539–2544
Patterson PH (1985) On the role of proteases, their inhibitors and the extracellular matrix in promoting neurite outgrowth. J Physiol (Paris) 80:207–211
Pittman RN (1985) Release of plasminogen activator and a calcium-dependent metalloprotease from cultured sympathetic and sensory neurons. Dev Biol 110:91–101
Pittman RN, Patterson PH (1987) Characterization of an inhibitor of neuronal plasminogen activator released by heart cells. J Neurosci 7:2664–2673
Ponte P, Gonzalez-De Whitt P, Schilling J, Miller J, Hsu D, Greenberg B, Davis K, Wallace W, Lieberburg I, Fuller F, Cordell B (1988) A new A4 amyloid mRNA contains a domain homologous to serine proteinase inhibitors. Nature 331:525–527
Reinhard E, Meier R, Halfter W, Rovelli G, Monard D (1988) Detectioin of glia-derived nexin in the olfactory system of the rat. Neuron 1:387–394
Rosenblatt DE, Cotman CW, Nieto-Sampedro M, Rowe JW, Knauer DJ (1987) Identification of a protease inhibitor produced by astrocytes that is structurally and functionally homologous to human protease nexin I. Brain Res 415:40–48
Schuerch-Rathgeb Y, Monard D (1978) Brain development influences the appearance of glial factor-like activity in rat brain primary cultures. Nature 273:308–309
Scott RW, Bergman BL, Bajpai A, Hersh RT, Rodriguez H, Jones BJ, Barreda C, Watts S, Baker JB (1985) Protease nexin: properties and a modified purification procedure. J Biol Chem 260:7029–7034
Sommer J, Gloor SM, Rovelli GF, Hofsteenge J, Nick H, Meier R, Monard D (1987) cDNA sequence coding for a rat glia-derived nexin and its homology to members of the serpin superfamily. Biochemistry 26:6407–6410
Stone S, Nick H, Hofsteenge J, Monard D (1987) Glial-derived neurite-promoting factor is a slow-binding inhibitor of trypsin, thrombin, and urokinase. Arch Biochem Biophys 252:237–244

Tanzi RE, McClatchey AI, Lamperti ED, Villa-Komaroff L, Gusella JF, Neve RL (1988) Protease inhibitor domain encoded by an amyloid protein precursor mRNA associated with Alzheimer's disease. Nature 331:528–530
Wallace A, Rovelli G, Hofsteenge J, Stone SR (1989) Effect of heparin on the glia-derived nexin-thrombin interaction. Biochem J 257:191–196
Zurn AD, Nick H, Monard D (1988) A glia-derived nexin promotes neurite outgrowth in cultured chick sympathetic neurons. Dev Neurosci 10:17–24

CHAPTER 25

Mullerian-Inhibiting Substance

R. L. CATE, P. K. DONAHOE, and D. T. MACLAUGHLIN

A. Introduction

During the development of the male and female reproductive tracts, the Mullerian duct gives rise to the uterus, Fallopian tubes, and part of the vagina and the Wolffian duct gives rise to the epididymis, vas deferens, and seminal vesicles. The two ducts develop in the early embryonic stages of both sexes. Regression of one or the other duct occurs as a consequence of gonadal differentiation. In the female, Wolffian duct regression occurs passively due to the lack of testosterone, while, in the male, regression of the Mullerian duct is an active process controlled by a testicular factor called Mullerian-inhibiting substance (MIS) (JOST 1946a, b, 1947a, b). A critical step in characterizing MIS was the development of an organ culture assay where urogenital ridges from 14½-day-old fetal rats were excised and cultured in vitro and then evaluated histologically for regression of the duct (PICON 1969). Based on this assay, MIS was identified as a protein produced by Sertoli cells of the fetal (BLANCHARD and JOSSO 1974) and newborn testis (DONAHOE et al. 1977a) and subsequently purified to near homogeneity (PICARD and JOSSO 1984; BUDZIK et al. 1985). The purified protein, a 140-kDa glycoprotein, is composed of two identical 70-kDa subunits.

Using sequence information derived from bovine MIS and antibodies raised against bovine MIS, bovine cDNAs for MIS have been isolated (CATE et al. 1986a; PICARD et al. 1986a). The human gene for MIS has also been isolated and expressed in Chinese hamster ovary (CHO) cells (CATE et al. 1986a, b). The recombinant protein is biologically active in the organ culture assay with a potency similar to that of the natural product. In addition to allowing the production of human MIS, the expression of MIS in CHO cells has provided a valuable method for studying its biosynthesis, which has led to the discovery that MIS undergoes proteolytic processing (PEPINSKY et al. 1988).

The determination of the primary amino acid sequences of bovine and human MIS has revealed that MIS is a member of a rather large family of proteins involved in growth regulation and differentiation. The family has been named after transforming growth factor type β (TGF-β), a very potent inhibitor of growth or function of a wide variety of cell types (MASSAGUE 1987; SPORN et al. 1987). Members of this family have now been found in *Xenopus* and *Drosophila,* indicating the presence of an ancestral gene prior to the divergence of these species. While MIS has clearly evolved as a protein with very specialized activities distinct from those of TGF-β, the similarities in their structures and biochemical properties have allowed some important insights into the regulation and function of MIS.

Expression of MIS is not confined solely to immature Sertoli cells of the fetal testis. Although MIS expression is highest during the period when the Mullerian duct regresses (TRAN et al. 1987), expression persists after birth for a short time at a fairly high level (DONAHOE et al. 1977a), and is still produced in the adult testis at a low level (JOSSO et al. 1979). It is also produced by the granulosa cells of the late developing and adult ovary (HUTSON et al. 1981; VIGIER et al. 1984; TAKAHASHI et al. 1986a). Granulosa cells are believed to be derived from the same progenitor cell as Sertoli cells, and share many similar structural and functional characteristics. With the MIS cDNA and gene now available, investigations have started to determine the mechanisms that are important in the specific expression of MIS in Sertoli and granulosa cells, and the nucleotide sequences near the MIS gene that control this specificity.

The expression of MIS in the adult ovary and in the testis after the Mullerian duct has regressed indicates that there may be additional roles for MIS. Potential roles have been considered for MIS in gonadal differentiation (VIGIER et al. 1987), testicular descent (HUTSON and DONAHOE 1986), and the regulation of meiosis (TAKAHASHI et al. 1986b). In addition, there is evidence that MIS may inhibit the growth of transformed cells (DONAHOE et al. 1979). It has been accepted that regression of the Mullerian duct and these other potential effects of MIS are mediated by a receptor. However, at this time very little is known about this molecule.

The main objectives of this presentation are to review the recent developments in the molecular biology of MIS and to relate them to previous biochemical and developmental observations. Although information has been included on chicken MIS, most of this presentation will focus on MIS in the mammal. For more information on MIS action and Mullerian duct regression in the chicken, the reader is directed to the review by HUTSON et al. (1983). Also, recent reviews by JOSSO and PICARD (1986) and DONAHOE et al. (1987) can be consulted for more details on the purification and immunobiology of MIS. Although MIS is known by many names including anti-Mullerian hormone (AMH), Mullerian inhibitor, and Mullerian inhibiting factor (MIF), the name MIS will be used throughout this chapter, regardless of the name used by the authors.

B. Structure of MIS

I. Bovine and Chicken MIS Proteins

Initial characterization and purification of MIS was achieved using the organ culture assay developed by PICON (1969), in which regression of the Mullerian duct is measured. In this assay, the 14½-day-old fetal rat urogenital ridge is used as the target organ, and MIS-induced changes in the Mullerian duct are

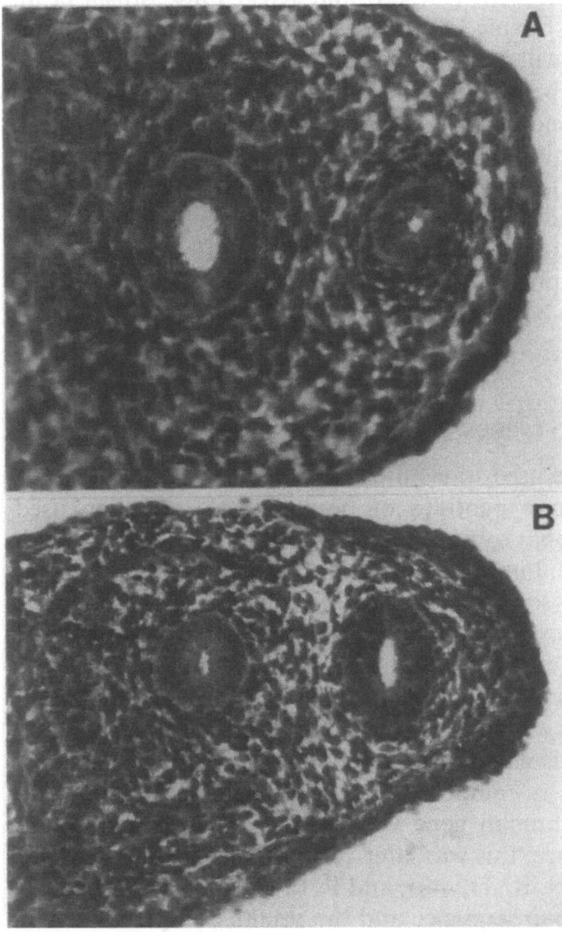

Fig. 1 A, B. Regression of the rat Mullerian duct by human recombinant MIS. **A** shows partial regression of the Mullerian duct produced by human MIS, while **B** shows the results of an organ culture assay performed in the absence of MIS. The Wolffian duct is on the *left* and the Mullerian duct is on the *right*

followed histologically. MIS from mammalian or chick testis and ovary is active in this assay; an example showing partial regression produced by human recombinant MIS is shown in Fig. 1. The main features of MIS-induced regression are the dissolution of the basement membrane of the Mullerian duct epithelial cells, the dissociation of the epithelial cells, and the condensation of mesenchyme around the regressing duct. Although the assay is very specific for MIS, it has low sensitivity and an imprecise histological end point. A number of immunological assays have been developed which have been described in detail in recent reviews (DONAHOE et al. 1987; JOSSO 1986).

Various approaches have been used to purify MIS from fetal and neonatal bovine testis (see above reviews). Biochemical analysis of purified bovine MIS showed that it is a 140-kDa disulfide-linked dimer composed of two identical 70- to 74-kDa subunits (PICARD and JOSSO 1984, BUDZIK et al. 1983, 1985). Under reducing conditions, a minor species of 56 kDa has been observed in preparations of purified bovine MIS (PICARD and JOSSO 1984; SHIMA et al. 1984), suggesting that a site in bovine MIS is sensitive to proteolysis. The significance of this observation will be discussed in the section on proteolytic processing of human MIS (see below). Bovine MIS contains a high proportion of hydrophobic amino acids and both N- and O-linked sugars (PICARD et al. 1986b). The level of glycosylation has been estimated at 8.3% (BUDZIK et al. 1980), 11% (SWANN et al. 1979), and 13.5% (PICARD et al. 1986b).

MIS has also been purified from 8-week-old chicken testis (TENG et al. 1987). Under reducing conditions, chicken MIS has a molecular weight of 74 kDa. The amino acid composition indicated that it is a relatively acidic protein with a ratio of acidic to basic amino acids of 1.93 and a PI of 6.1 (TENG et al. 1987).

II. Bovine and Human MIS Genes

Two different approaches were used to isolate cDNAs for bovine MIS. CATE et al. (1986a) sequenced tryptic peptides of bovine MIS, and then used degenerate oligonucleotide probes to screen a λgt10 cDNA library made with RNA of newborn bovine testis. PICARD et al. (1986a) used a rabbit antiserum raised against bovine MIS to screen a λgt11 cDNA library also made with RNA of newborn bovine testis. The complete mRNA sequence (1816 nucleotides) of bovine MIS was determined from an analysis of various cDNAs and a bovine genomic clone (CATE et al. 1986a).

Using the bovine cDNA as a hybridization probe, a cosmid clone containing the human MIS gene was isolated (CATE et al. 1986a). Sequence analysis showed that the human gene is composed of five exons. At that time, the intron-exon structure of the human gene was predicted by comparing the human with the bovine sequence; this was later verified by sequencing human MIS cDNA clones (C. HESSION, R. TIZARD, and R. L. CATE, unpublished observations). Based on the human sequence and the results of a primer extension and S1 nuclease assay of the bovine and human MIS mRNAs (CATE et al. 1986a), a schematic picture of human MIS has been composed that is shown in Fig. 2.

The 5′ untranslated regions of the bovine and human MIS mRNAs are 10 nucleotides in length, which is relatively short when compared with the typical 5′ untranslated region of 40 nucleotides (KOZAK 1983), and probably precludes the possibility of these sequences being involved in translational control. When compared with the bovine sequence, the 3′ untranslated region of the human mRNA contains a 30-nucleotide insertion, two nucleotides downstream of the TGA stop codon. At the nucleotide level, the two genes share 74%, 90%, 71%, 74%, and 84% similarity for the five exons, respectively, while the two promoters share 67% similarity. The human MIS gene has been mapped to the short arm of chromosome 19 (JOSSO 1986).

Bovine and human MIS are synthesized as precursors of 575 and 560 amino acids, respectively (Fig. 3). The N-termini of the mature proteins were determined by sequencing MIS secreted from either newborn bovine Sertoli

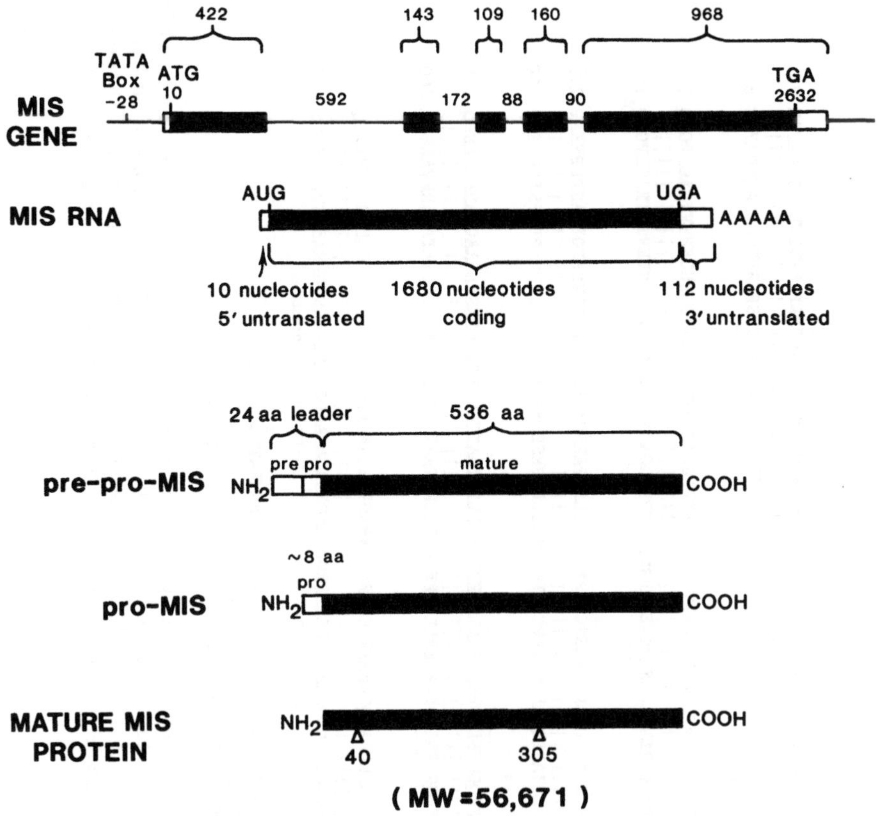

Fig. 2. Schematic diagram showing the human MIS gene, mRNA, and protein

cells (CATE et al. 1986a) or CHO cells transfected with the human MIS gene (PEPINSKY et al. 1988). Both precursors contain leaders of 24 amino acids, which can be divided into two parts; the first 16 or 18 amino acids constitute the signal sequence involved in transport across the endoplasmic reticulum, while the remaining 6 or 8 amino acids constitute a putative prosequence. The removal of the bovine prosequence requires a cleavage between a proline and an arginine, which is not known to occur for signal sequence cleavages. This implies the existence of a unique processing event, although the significance of this putative prosequence is not clear since the mature protein may require further proteolytic processing to be active (see below).

Overall, the bovine and human proteins share 78% identity. All 12 cysteines are conserved between the human and bovine proteins, as are the two potential N-glycosylation sites. The N terminus shows the weakest homology (62% over the first 110 amino acids), including two deletions of 9 and 6 amino acids in human MIS. The C terminus shows the strongest homology, with 108 of the last 112 amino acids conserved. It is this region that shares similarity with TGF-β and inhibin. The locations of the four introns within the human protein sequence are also indicated in Fig. 3.

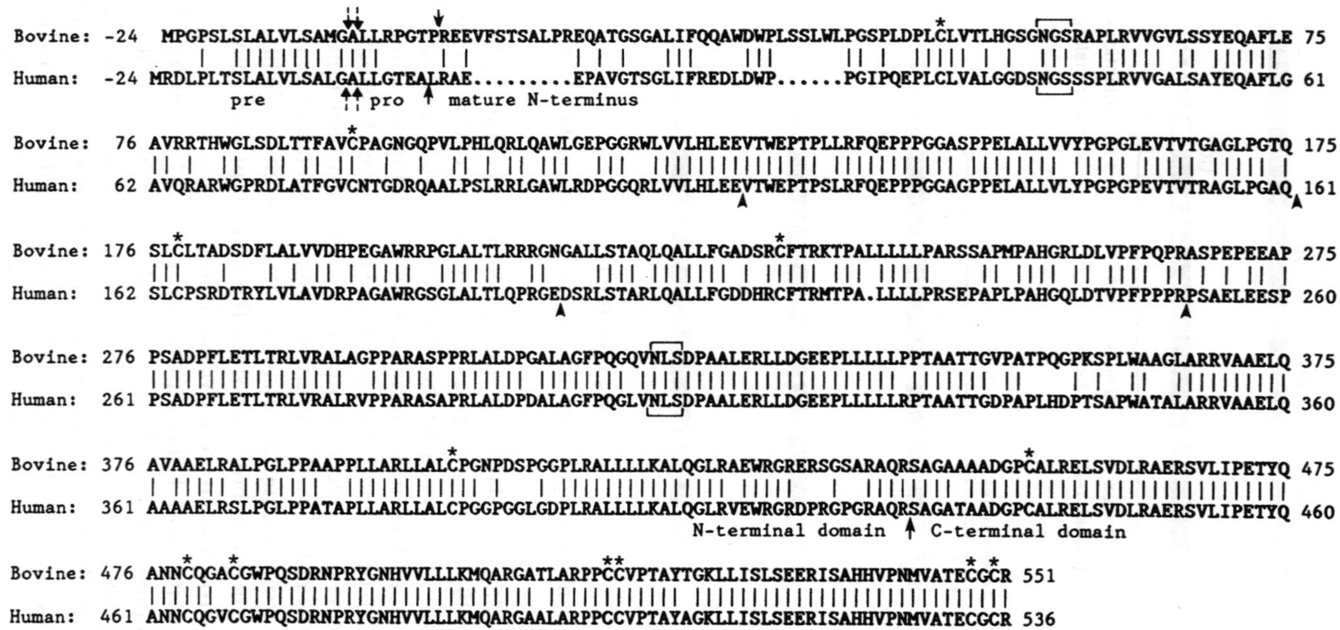

Fig. 3. Amino acid sequences of bovine and human MIS. The *broken arrows* indicate the possible cleavage sites for the signal sequence, while the adjacent *solid arrows* indicate the cleavages that generate the mature proteins. The *solid arrow* between Arg427 and Ser428 of the human sequence indicates the cleavage involved in the proteolytic processing of human MIS to produce the N- and C-terminal fragments. The potential N-glycosylation sites are marked with *brackets,* cysteines are denoted with *asterisks,* and the locations of the four introns within the human protein sequence are indicated by *arrowheads*

III. Biosynthesis of Human MIS in CHO Cells

In addition to the production of human MIS, the expression of the human gene in animal cells has provided a method for studying the biosynthesis of MIS, including glycosylation, secretion, and, as was ultimately discovered, proteolytic processing. To express the human MIS gene in animal cells, a 4.5-kilobase (kb) *Alf*II fragment containing the entire MIS gene was inserted into an expression vector carrying the SV40 early promoter. This plasmid (pBG311.hmis) was introduced along with plasmid pSV2DHFR, which contains the mouse dihydrofolate reductase (DHFR) cDNA, into Chinese hamster ovary (CHO) cells deficient in dihydrofolate reductase (CHASIN and URLAUB 1980). Ultimately, cell lines were obtained that produced on the order of 0.1 mg MIS/liter medium per day (CATE et al. 1986a, b). Cell lines expressing much higher levels of MIS (3 mg/liter per day) were obtained when pSV2DHFR was replaced with pAdD26 (KAUFMAN and SHARP 1982), which also contains the mouse DHFR cDNA but lacks an enhancer. Since this plasmid cannot transform CHO cells efficiently by itself, this approach assured that the MIS and DHFR sequences would integrate together and that the MIS gene copy number could be increased with methotrexate (WALLEN et al. 1989).

MIS was purified from the conditioned medium of one of these cell lines by a combination of ion-exchange, lentil-lectin, and immunoaffinity chromatography (WALLEN et al. 1989). The purified human protein is a disulfide-linked dimer of 140 kDa, and the sequence of the mature N terminus is Leu-Arg-Ala-Glu-Glu (Fig. 3) (PEPINSKY et al. 1988). The recombinant human protein is active in the organ culture assay, with 2.5–5 μg/ml being the minimal concentration needed for complete regression of the Mullerian duct. Bovine MIS purified from fetal testis has a very similar concentration profile in the organ culture assay (N. JOSSO, personal communication).

Analysis of [^{3}H]glucosamine and [^{3}H]leucine-labeled proteins in supernatants from CHO cells that produce MIS has demonstrated that the secreted MIS contains both N-linked and O-linked sugars (E. G. NINFA and R. L. CATE, unpublished observations). Treatment of the cells with tunicamycin, which inhibits N-linked glycosylation, resulted in the secretion of MIS with a molecular weight of about 60–62 kDa, after reduction. This decrease in size is consistent with both potential N-linked glycosylation sites (Fig. 3) being used. Since sialidase reduces the 60- to 62-kDa species to an even lower molecular weight (58–60 kDA), it has been concluded that the recombinant protein also contains O-linked sugars (KINGSLEY et al. 1986). The significance of the glycosylation is unknown, since removal of the N-linked sugars with endoglycosidase F does not affect the activity of human MIS in the organ culture assay (K. L. RAMESH, P. K. DONAHOE, and R. L. CATE, unpublished observations). A similar result was obtained with bovine MIS using tunicamycin (JOSSO et al. 1980).

The rate of secretion of MIS from CHO cells is very slow (R. L. CATE, unpublished observations), compared with other secretory proteins such as transferrin (LODISH et al. 1983). In a typical pulse-chase experiment with [^{35}S]cysteine, 50% of the MIS is still in the cell after a 5-h chase with cold cysteine. All of the intracellular MIS appears to be in the endoplasmic reticulum (ER), since it is sensitive to endoglycosidase H, which removes simple sugars. The presence of simple sugars is usually a good indication that the protein is in the ER. Secretion of MIS from bovine Sertoli cells is also very slow, requiring approximately 48 h (VIGIER et al. 1985); immunohistochemistry has shown that most of the MIS is in the ER (TRAN and JOSSO 1982; HAYASHI et al. 1984; TRAN et al. 1987). Together, these observations indicate that one of the steps in the biosynthesis of MIS that occurs in the ER is slow.

Recent studies suggest that proteins will not mature from the ER to the Golgi unless they have achieved a proper conformation (LODISH 1988; GETHING et al. 1986). Although most of the intracellular MIS appears to be dimerized (R. L. CATE, unpublished observations), its continued retention in the ER suggests that it has not yet folded into a proper conformation. When a construct encoding the N-terminal domain of MIS was expressed in CHO cells, the N-terminal domain was secreted as a dimer at a considerably faster rate than the full-length 140-kDa dimer (A. REZAIE and R. L. CATE, unpublished observations). These results suggest that folding of the C-terminal domain may be rate limiting. The rate of secretion of the full-length molecule was not affected by tunicamycin, indicating that N-linked glycosylation has no affect on the rate of secretion of MIS (R. L. CATE, unpublished observations).

C. MIS as a Member of the TGF-β Family

I. Structural Properties of the Family

The striking conservation of amino acids at the C terminus of bovine and human MIS immediately distinguished this region as an important domain. A search of the data base showed that this domain shared considerable homology with two other proteins, inhibin (MASON et al. 1985) and transforming growth factor-β (TGF-β) (DERYNCK et al. 1985). Within the past 2 years, five other proteins have been shown to be related to this group of proteins, which has become known as the TGF-β family (MASSAGUE 1987). All of these proteins are involved in developmental processes such as embryogenesis and tissue repair and act as potent regulators of cell growth and differentiation. The presence of MIS in this family has provided important insights into its regulation and function. The following section will review some of these shared properties, as well as differences that distinguish MIS from other members of the family.

Transforming growth factor-β, a 25-kDa homodimer, was originally isolated as a growth factor produced by transformed cells and was subsequently shown to be a potent growth inhibitor of a variety of cell types (SPORN et al. 1987). Three forms of TGF-β have now been identified in humans, TGF-β1

(DERYNCK et al. 1985), TGF-β2 (DE MARTIN et al. 1987), and TGF-β3 (TEN Dijke et al. 1988). TGF-β1 has been shown to affect tissue repair, wound healing, bone formation, immune suppression, and embryonic development (SPORN et al. 1987). Although in a number of assays TGF-β1 and TGF-β2 are functionally indistinguishable, certain activities appear to be unique to one form or the other (OHTA et al. 1987; ROSA et al. 1988).

Inhibin is a heterodimer ($\alpha\beta$) that inhibits follicle-stimulating hormone (FSH) secretion, and is produced in Sertoli and granulosa cells, the same cells that produce MIS (MASON et al. 1985). The β-chain exists in two forms, β_A and β_B, producing $\alpha\beta_A$ and $\alpha\beta_B$ heterodimers; however, biological differences are unknown. Dimers ($\beta_A\beta_A$ and $\beta_B\beta_B$) composed of β-chains (called activins or FSH-releasing proteins) have also been isolated; they promote the release of FSH (LING et al. 1986; VALE et al. 1986) and may have a role in erythropoiesis (YU et al. 1987). Other members of the TGF-β gene family have been discovered in *Xenopus* and *Drosophila,* but the gene products have not been characterized. The product of the *Vg1* gene of *Xenopus* appears to be involved in mesoderm formation during development (WEEKS and MELTON 1987), while the product(s) of the decapentaplegic (*dpp*) complex of *Drosophila* is associated with dorsal-ventral determination during embryogenesis (PADGETT et al. 1987).

The inhibin α- and β-chains, the TFG-β chains, and MIS are synthesized as large precursors that form disulfide-linked dimers and are subsequently secreted (Fig. 4) (MASSAGUE 1987). The dimerized precursors of the TGF-βs, inhibins, and activins are cleaved to form C-terminal bioactive dimers and N-terminal dimers which possess no known function. Cleavage of the products of the *Vg1* and *dpp* genes has also been inferred. All of these proteins except MIS contain a dibasic amino acid sequence (sometimes five basic amino acids) at the junction of the N- and C-terminal domains. Because MIS lacks this sequence and the full-length molecule is active in the organ culture assay, it has been thought that MIS is an exception and is not cleaved.

The similarity between the proteins of the TGF-β family resides in the C-terminal domain of the precursors, and is in the range of 20%–40% (PADGETT et al. 1987). Seven cysteines are conserved in this region and significant homology is found around these cysteine residues. The similarity is higher when comparing the C-terminal domains of the two inhibin β-chains with each other (70%), or the three TGF-β chains (71%–79%). Across

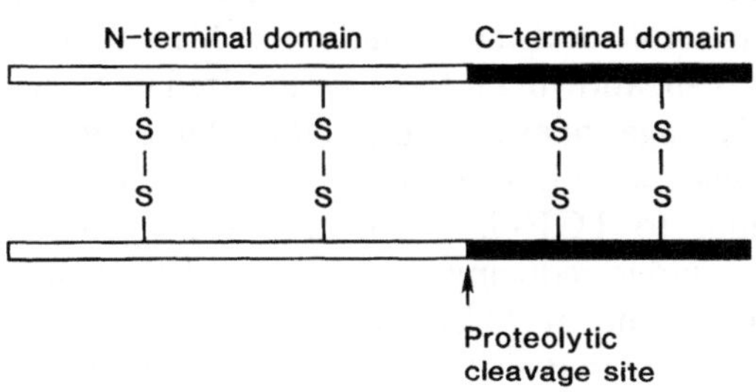

Fig. 4. Generic structure of the TGF-β family proteins

species, the C-terminal domain of the inhibin chains, TGF-β1, and MIS is highly conserved (>96%).

There is no apparent similarity between the N-terminal domains of the inhibin chains, the TGF-β chains, MIS, or the predicted *Vg1* and *dpp* gene products. The N-terminal domains of the two porcine inhibin β-chains share 50% homology (MASON et al. 1985), and the N-terminal domains of the three human TGF-β-chains share 33%–45% homology (TEN DIJKE et al. 1988). Across species, the N-terminal domain in each of these polypeptides is less conserved than the C-terminal domain. The N-terminal domains of human and murine TGF-β1 share 99% homology over the first 88 amino acids and 77% for the remainder of the domain (DERYNCK et al. 1986), while the N-terminal domains of human and bovine MIS share only 62% homology over the first 112 amino acids, but contain two regions that show a marked degree of homology (Fig. 3). No function is known for the N-terminal domain of any of the TGF-β members. BRUNNER et al. (1988) have recently demonstrated that the N-terminal domain of TGF-β1 is phosphorylated, possibly on a carbohydrate moiety.

II. Proteolytic Processing of Human MIS

Proteolytic processing is required to generate active TGF-β1 and appears to involve two steps (PIRCHER et al. 1986; LAWRENCE et al. 1985). The first step is the cleavage of the two chains of the precursor, producing a noncovalent complex of the N- and C-terminal domains, while the second step involves the dissociation of the complex with the release of the bioactive C-terminal dimer. Although in vitro, the latter step can be accomplished by using acid, urea, or heat (LAWRENCE et al. 1985; SPORN et al. 1987), it is not clear how this step is accomplished in vivo. LYONS et al. (1988) have suggested that proteolysis may release TGF-β from the latent noncovalent complex in vivo. Two other groups have described a glycoprotein (160 kDa) associated with TGF-β1 (MIYAZONO et al. 1988; WAKEFIELD et al. 1988) that may be involved in latency. However, GENTRY et al. (1987) have shown that when TGF-β1 is overexpressed in CHO cells it is secreted in a latent form without another component, indicating that an additional protein is not required for latency.

In contrast to TGF-β, it has been thought that MIS does not require proteolytic processing for activity. The full-length molecule is active in the organ culture assay, and MIS lacks a dibasic amino acid sequence at the junction of the N- and C-terminal domains. Since the receptor distribution for MIS is thought to be limited, it is possible that the regulation of MIS activity may not be as critical as it is for TGF-β, allowing it to be synthesized in an active form. However, during the early stages of evaluating CHO cell lines which expressed the human MIS gene, there were indications that MIS might undergo proteolytic processing similar to TGF-β. When analyzed by SDS-polyacrylamide gel electrophoresis under reducing conditions, MIS always appeared as a doublet with a major band of 70 kDa and a minor band of 55–57 kDa (CATE et al. 1986b), indicating that a site in human MIS was sensitive to a protease in the conditioned medium. A similar electrophoretic pattern was observed with preparations of bovine MIS (Picard and Josso 1984; Shima et al. 1984).

When human MIS was purified to homogeneity from the conditioned medium of CHO cells, a third species of 12.5 kDa was observed in addition to the 57-kDa and the major 70-kDa species (Pepinsky et al. 1988). N-terminal sequence analysis revealed that the 70- and 57-kDa species had the same sequence (Leu-Arg-Ala-Glu-Glu), while the 12.5-kDa species had the sequence Ser-Ala-Gly-Ala-Thr, indicating that 10%–20% of the MIS was being cleaved between Arg427 and Ser428 to generate the N-terminal 57-kDa species and the C-terminal 12.5-kDa species (Fig. 3, 5). The site of this cleavage fits the consensus sequence for monobasic cleavage sites described by Benoit et al. (1987). Furthermore, when MIS was analyzed under nonreducing conditions, the 12.5-kDa species migrated as a 25-kDa dimer, indicating that the 140-kDa MIS dimer was being cleaved on both chains to generate a TGF-β-like fragment. Like TGF-β, the N- and C-terminal domains were still associated in a noncovalent complex, which could be dissociated by acid or boiling (Fig. 5).

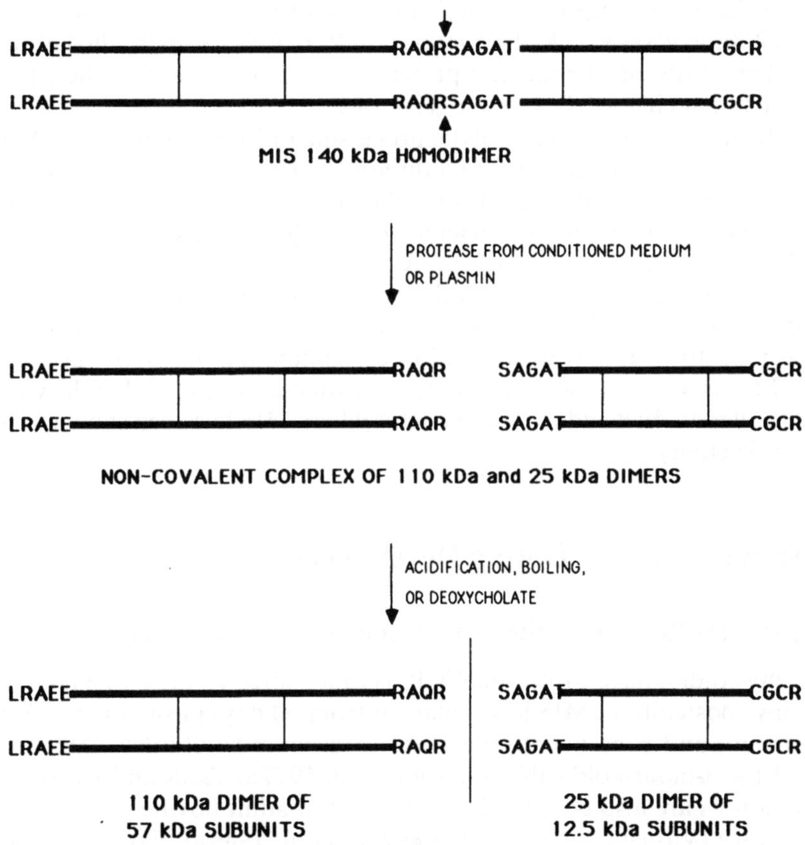

Fig. 5. Schematic for processing of human MIS

In addition to the biochemical similarity between MIS and TGF-β, there is biological evidence which is consistent with MIS undergoing proteolytic processing. Plasmin completely cleaves MIS between Arg427 and Ser428, to generate the noncovalent complex, which retains all activity in the organ culture assay (PEPINSKY et al. 1988). However unlike TGF-β, when acid or boiling was used to dissociate the complex, all activity was lost. Activity of the plasmin-cleaved MIS was retained when the complex was dissociated with 1% deoxycholate, indicating that the N- and/or C-terminal domains of MIS may be heat and acid sensitive (PEPINSKY et al. 1988).

Preliminary results suggest that both the N- and C-terminal domains generated by processing are necessary for regression of the Mullerian duct. C-terminal dimer has been prepared by cleaving MIS with plasmin, breaking the noncovalent complex with 1 *M* acetic acid, and purifying the TGF-β-like fragment by gel filtration (PEPINSKY et al. 1988). N-terminal dimer was purified from CHO cells expressing a construct that encodes the N-terminal domain of MIS (A. REZAIE and R. L. CATE, unpublished observations). When the N-terminal and C-terminal dimers were assayed individually in the organ culture assay, they were inactive. However, when they were incubated together with the 14½-day fetal urogenital ridge, regression of the Mullerian duct was observed (R. B. PEPINSKY, P. K. DONAHOE, and R. L. CATE, unpublished observations). Thus MIS may be the first protein in the TGF-β family, where the N-terminal domain has a defined biological role.

To determine whether proteolytic processing of MIS is obligatory, Arg427 in the cleavage site was changed to a threonine. This altered form of MIS was secreted from CHO cells as a 140-kDa dimer. The purified protein could not be cleaved by plasmin, and was inactive in the organ culture assay (A. REZAIE, R. B. PEPINSKY, P. K. DONAHOE, and R. L. CATE, unpublished observations). This result provides compelling evidence that full-length MIS is latent (as is the precursor of TGF-β) and that proteolytic processing is necessary for generating active MIS. A major goal in understanding the mechanism of action of MIS is to determine how this conversion is accomplished in the vicinity of the Mullerian duct and in other systems where MIS has been shown to have biological activity.

D. MIS Expression During Development

I. Upstream Regions of the Bovine and Human MIS Genes

Ontogeny studies have shown that MIS is first expressed in bovine fetal testis at 43 days postcoitum. MIS levels plateau from 50 days postcoitum to 8 days postpartum, and subsequently decrease to become barely detectable in the testis of the 3-month-old calf (DONAHOE et al. 1977a; TRAN and JOSSO 1982). Low amounts are also detected in maturing and adult bovine granulosa cells of the ovary (VIGIER et al. 1984; TAKAHASHI et al. 1986a). Expression of the bovine MIS mRNA (Fig. 6) correlates perfectly with this expression profile of

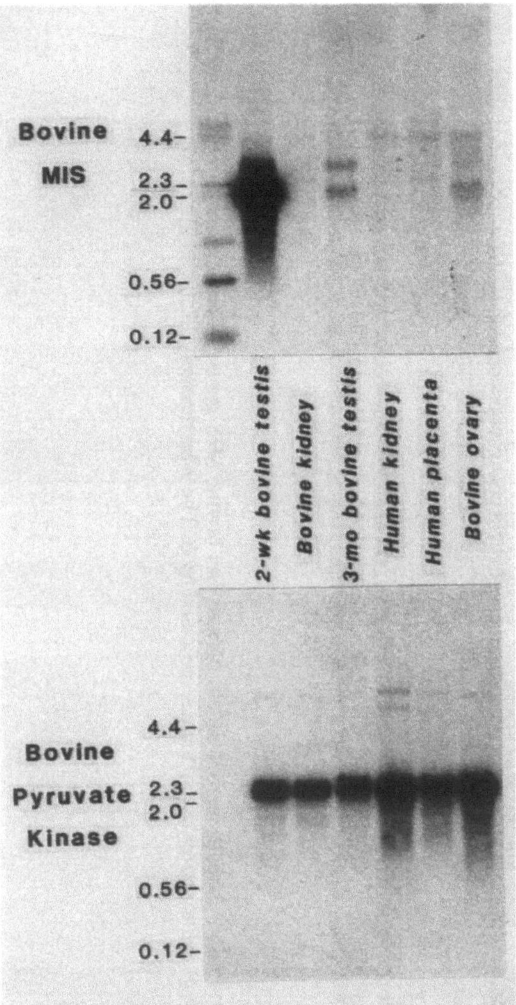

Fig. 6. Northern analysis of bovine MIS mRNA. Two identical filters containing RNAs from the indicated tissues were hybridized with probes for bovine MIS (*top*) or bovine pyruvate kinase (*bottom*), which is expressed in all tissues

the protein (PICARD et al. 1986a; CATE et al. 1986a). Neither the protein nor the mRNA for MIS have been detected in cells other than Sertoli or granulosa cells, indicating that the cell-specific expression of MIS is achieved at the level of transcription.

Transcriptional regulation of the MIS gene is probably mediated by factors that bind to specific nucleotide sequences (enhancers) near the 5′ or 3′ end of the gene and modulate transcription. Since the immature and mature Sertoli cells and the granulosa cell may respond to different extracellular signals, regulation of MIS transcription within these cells may involve different transcriptional factors induced through different signal transduction pathways. In this section, regulatory regions of the MIS gene that may interact

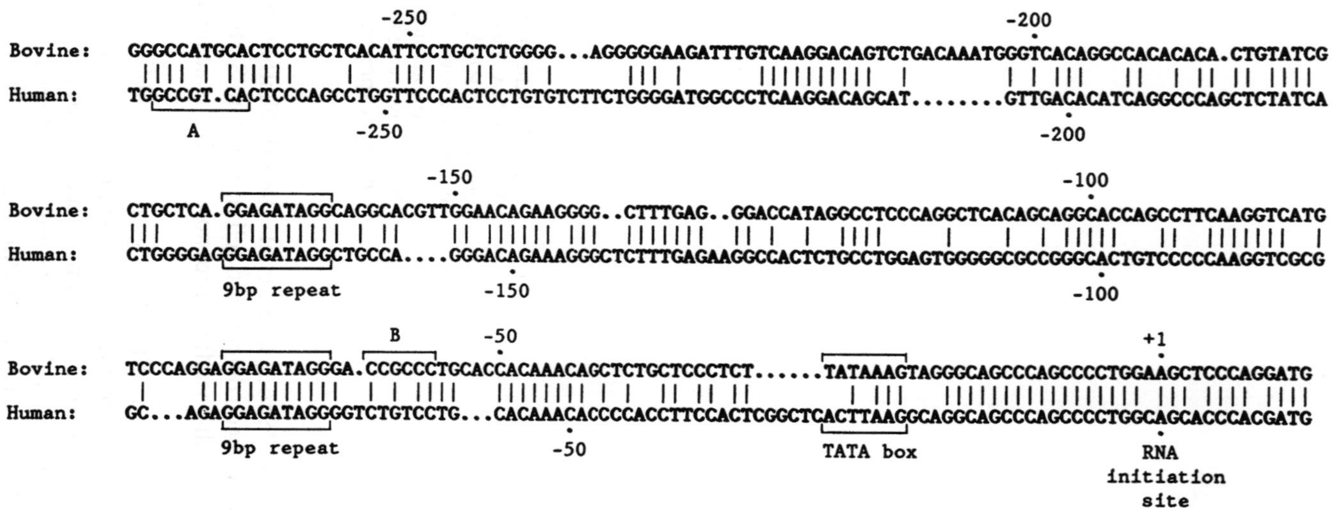

Fig. 7. Nucleotide sequences of the promoter and upstream regions of the human and bovine MIS genes. A sequence with one mismatch to the consensus CRE sequence is indicated by *A*, while a potential SP1 site is indicated by *B*. The 9-bp repeat is discussed in the text

with transcriptional factors will be examined. This will be followed by a review of MIS expression in the testis and ovary, focusing on both the regulation and function of MIS in these tissues.

The promoters and upstream regions of the bovine and human MIS genes are shown in Fig. 7, aligned at their cap sites. Overall, the nucleotide sequences in this region share 67% homology, although gaps have been inserted to better align the sequences. The bovine promoter contains an almost perfect TATA box, while the human TATA box differs significantly from the consensus sequence TATA(A/T)A(A/T) (BREATHNACH and CHAMBON 1981). It is known that the TATA box functions primarily to ensure that transcripts are accurately initiated (JONES et al. 1988). Although both promoters are active in vitro with HeLa cell extracts, the human promoter is less precise, with 50% of the transcripts starting upstream of the cap site. Heterogeneity of mRNA initiation was also observed in vivo, with longer RNAs being detected in human neonatal testis (E. G. NINFA, P. K. DONAHOE, and R. L. CATE, unpublished observations).

A number of consensus nucleotide sequences have been identified that interact with known transcriptional factors or have been shown to mediate the hormonal induction of second messengers (JONES et al. 1988). The cyclic AMP responsive element (CRE) (G/T)(A/T)CGTCA mediates the cyclic AMP induction of many eukaryotic genes (MONTMINY et al. 1986). Since MIS expression may be regulated by gonadotropins via a cyclic AMP pathway, it was of interest to see if this element was present in the upstream region of the MIS genes. The exact CRE sequence is not present in either gene, but a sequence with one mismatch is found near the human promoter (Fig. 7). This sequence is not conserved in the bovine promoter, so its significance in the absence of binding studies is unclear.

The TPA (12-O-tetradecanoylphorbol-13-acetate)-responsive element TGACTCA (RAUSCHER et al. 1988), which is the binding site for AP-1 and may mediate the effects of protein kinase C, is also not found in this region of the two genes, or further upstream in the human gene. An SP1-binding site CCGCCC (KADONAGA et al. 1987) is present in the bovine gene (Fig. 7) and further upstream in the human gene. However, functional SP1 sites are usually found in association with AP-1 sites or in multiple copies, and due to the high GC content of the MIS gene the CCGCCC sequence is found quite often in the coding sequences of both genes. Again, it is not clear whether these sequences are important in transcriptional control.

Although the upstream regions of the bovine and human MIS genes do not contain elements that have been established as enhancers, a 9-bp sequence (GGAGATAGG) is present that may be important in MIS regulation. It occurs twice in the upstream regions of both genes at similar locations (Fig. 7), but is not found in any other regions of the two genes. This is the first example where this 9-bp nucleotide sequence has been implicated in transcriptional regulation, but confirmation that the sequence is important in the tissue-specific expression of MIS awaits the outcome of DNA-binding studies. In addition, the expression of hybrid genes consisting of potential regulatory regions of the MIS gene and reporter genes (i.e., CAT or human growth hormone) in Sertoli cells or in transgenic mice should help identify the sequences that confer tissue specificity.

II. Expression of MIS in the Testis

During fetal development of the rat, pre-Sertoli cells are the first recognizable cell type to differentiate in the testis (JOST 1972; MAGRE and JOST 1980). At day 13 of development, these cells enlarge acquiring vesicles of the rough endoplasmic reticulum, aggregate with each other, and encompass the germ cells to form seminiferous cords. The expression of MIS in these immature Sertoli cells precedes the production of testosterone by Leydig cells, which differentiate outside the cords (JOSSO et al. 1977). Expression of MIS may be regulated (directly or indirectly) by testis-determining factor (TDF) encoded by the Y-chromosome and presumed to be responsible for all aspects of sexual dimorphism.

A gene has recently been cloned from the sex-determining region of the human Y-chromosome which is believed to be TDF (PAGE et al. 1987). This gene encodes a protein with a tandem array of cysteine- and histidine-rich finger domains similar to those in frog transcription factor IIIA and human transcription factor SP1. By analogy, it has been accepted that TDF is also a transcription factor that binds to DNA (or RNA) in a sequence-specific manner, thereby regulating the one or many genes involved in the pathway of gonadal differentiation (PAGE et al. 1987). When the purified TDF protein is available, it should be possible to determine whether TDF can bind to nucleotide sequences near the MIS gene and regulate transcription, or induce the synthesis of other transcriptional factors that regulate MIS transcription.

In addition to causing regression of the Mullerian duct during fetal development, MIS may have a role in normal testicular differentiation. VIGIER et al. (1987) have found that MIS causes a freemartin effect (LILLIE 1916) in fetal rat ovaries, which is characterized by germ cell depletion, the formation of structures resembling seminiferous cords, and the appearance of cells that resemble Sertoli cells. A scenario is possible in which the expression of TDF in Sertoli cells turns on the expression of MIS, which then acts in an autocrine or paracrine mode to turn on more genes involved in testicular differentiation. MCLAREN (1984) has observed that XX/XY chimeras tend to develop into normal males, suggesting that a critical proportion of XY cells can induce XX cells to participate in the formation of the testis, possibly by producing a secreted factor such as MIS. Evidence against this hypothesis comes from experiments in which a female rabbit was immunized against MIS (TRAN et al. 1986). Although this led to persistence of the Mullerian ducts in male offspring, there was no impairment of testicular differentiation. However, TRAN et al. (1986) have observed that the timing of the immunization was geared to affect Mullerian development and may not have been adequate to affect other MIS actions.

Another role has been considered for MIS during fetal development, in the regulation of male germ cell meiosis. In normal mouse development, XX germ cells in the ovary enter the prophase of meiosis well before birth, while XY germ cells enter a state of mitotic arrest. Mitotic proliferation resumes in the immediate postnatal period, but the first meiotic stages do not begin until 8–10 days after birth (McLaren 1981). Jost (1972) suggested that a Sertoli cell factor might prevent germ cells from entering into meiotic prophase. The temporal expression of MIS in fetal testis is consistent with MIS being this factor, but evidence exists to the contrary. Germ cells were found to enter meiosis in cultured mouse genital ridges from male embryos (10½ days postcoitum), despite the appearance of pre-Sertoli cells, although cords did not develop (McLaren 1985). Agelopoulou et al. (1984) have shown that pre-Sertoli cells in cultured rat gonads produce MIS, even when cord formation is prevented. Together, these results indicate that the presence of pre-Sertoli cells which are capable of producing MIS is not sufficient to inhibit the entry of germ cells into meiosis.

After birth, the level of MIS produced by the Sertoli cell falls precipitously. This decrease in MIS production coincides with the differentiation of the immature Sertoli cell to the mature Sertoli cell (Tran et al. 1981), characterized by a cessation of mitotic activity (although cells continue to increase in size), and an increase in tight junctional complexes between Sertoli cells, resulting in the creation of the blood-testis barrier. This compartmentalization provides a specific microenvironment for germ cell development; thus Sertoli cells are thought to regulate germ cell development (Bellve 1979). Although low, production of MIS by the mature Sertoli cell persists and can be detected in rete testis fluid (Josso et al. 1979; Vigier et al. 1983). This suggests that MIS may have a role in spermatogenesis, although there is no evidence available at this time to support this hypothesis.

Gonadotropins and steroids have been considered the most likely candidates to be involved in the regulation of MIS in immature and mature Sertoli cells. FSH and testosterone have both been shown to increase the production of another Sertoli cell protein, androgen-binding protein (Louis and Fritz 1979). Bercu et al. (1978, 1979) have demonstrated that administration of an antibody against luteinizing-hormone-releasing hormone to pregnant rats leads to an elevation of MIS levels in the male offspring. FSH administered to the male neonates reduced the MIS levels to normal. Thus, in vivo, FSH may downregulate the expression of MIS in the newborn testis.

In contrast, MIS expression in cultured newborn bovine Sertoli cells was not affected by FSH or testosterone (Vigier et al. 1985; LaQuaglia et al. 1986), and in cultured 20-week fetal human testis cells FSH had no effect on MIS mRNA levels (Voutilainen and Miller 1987). Thus, in vitro, FSH does not appear to affect the expression of MIS in Sertoli cells. However, it is possible that Sertoli cells grown in tissue culture lose the ability to modulate MIS activity, especially in view of the fact that MIS expression is gradually lost after Sertoli cells are removed from the testis and placed into tissue culture

(VIGIER et al. 1985; LAQUAGLIA et al. 1986). HADLEY et al. (1985) have shown that Sertoli cells grown in a reconstituted basement membrane gel form cords and provide an environment permissive for germ cell differentiation. Such an experimental approach may permit a more accurate assessment of the factors that affect MIS expression.

III. Expression of MIS in the Ovary

In 1981 it was discovered that the chick ovary produces biologically active MIS (HUTSON et al. 1981). Subsequently, MIS was detected in bovine follicular fluid by bioassay (VIGIER et al. 1984) and radioimmune assay (VIGIER et al. 1984; NECKLAWS et al. 1986) and its site of production was localized to the granulosa cells of antral follicles by immunohistochemistry (TAKAHASHI et al. 1986a). Granulosa cells share many structural and functional characteristics with Sertoli cells. Believed to be derived from a common progenitor cell, both cell types form a diffusion barrier shielding germ cells from the somatic environment, and release their secretion products into a fluid which bathes germ cells during normal maturation.

Similar activity profiles in both the bioassay and radioimmune assay indicate that the ovarian protein is identical to the testicular protein; however, biochemical analysis of ovarian MIS has been difficult due to the low level of production in the ovary. JOSSO (1986) has estimated that on a weight basis the ovary produces less than 0.1% of the MIS produced in fetal or neonatal testis. Although the exact role of MIS in the ovary is unknown, ontogeny studies have identified the stages of follicular development where MIS is expressed. This in turn has indicated possible roles for MIS, including the regulation of meiosis, for which some experimental evidence has been obtained.

The development of the follicle in the mammalian ovary is a complex process in which accessory cells (i.e., granulosa and theca) respond to polypeptide hormones (gonadotropins) to control the maturation of the oocyte and its release (ovulation). Primary oocytes in the newborn female are surrounded by a single layer of granulosa cells and are arrested in prophase of meiotic division I. Some primary follicles grow in response to FSH, with granulosa cells proliferating and producing estrogen. A portion of these growing follicles develop a cavity which accumulates fluid, and are termed antral follicles. A surge of luteinizing hormone (LH) during each ovarian cycle activates selected antral follicles to complete development. The enclosed primary oocyte matures to complete meiotic division I as the stimulated follicle enlarges and ruptures at the surface of the ovary, releasing the secondary oocyte. Under the influence of LH, the mature follicle (without the oocyte) is transformed into the corpus luteum, which produces progesterone, necessary to prepare the uterine lining to receive an embryo. Other growing follicles in the ovary that do not reach a critical maturity by this time become atretic and disappear.

The ontogeny of MIS expression during follicular development has been studied in the cow, sheep, and rat. MIS was detected in the granulosa cells of preantral and antral follicles of bovine ovary over an age span of 1 day to 5 years (TAKAHASHI et al. 1986a), but not in primary and growing follicles or the corpus luteum. Immunocytochemical studies of the ovary of the developing ewe indicated that a low level of MIS is produced at birth by granulosa cells of antral follicles, while a much higher level is produced in adult antral follicles (BEZARD et al. 1987). In the adult follicles, most of the MIS was located in the innermost granulosa cell layer, close to the oocyte and lining the antral cavity. In the rat ovary, MIS was detected in granulosa cells of preantral and antral follicles during every stage of the estrus cycle, but not in atretic follicles or the corpus luteum (UENO et al. 1989a). Again, the highest level of MIS was found in the granulosa cells that contact the oocyte and those that border the antral cavity.

A dramatic change in MIS staining was observed during proestrus in the granulosa cells of rat preovulatory follicles (UENO et al. 1989b). Early in the day of estrus, staining of MIS within these follicles was intense. Subsequently, there was a considerable decrease in MIS staining that correlated with the rise in serum progesterone which follows the LH surge. During this time, the cumulus cells became dispersed and separated from the oocyte and the more peripheral granulosa cell layers, while the oocyte underwent the first meiotic division marked by the disappearance of the germinal vesicle.

The transient expression of MIS during follicular development suggests that it may have a physiological role in the ovary, perhaps in regulating oogenesis. TAKAHASHI et al. (1986b) have shown that MIS may be involved in preventing the oocyte from completing the first meiotic division. Highly purified bovine MIS inhibited the resumption of meiosis (i.e., germinal vesicle breakdown) in both denuded and cumulus-enclosed rat oocytes in vitro. This effect was dose dependent, reversible, cyclic AMP independent, and not influenced by gonadotropins or steroids. Partially purified human recombinant MIS also inhibited germinal vesicle breakdown in rat oocytes. The inhibitory effect was lost when the human MIS was purified to homogeneity by immunoaffinity chromatography, but could be restored by addition of Nonidet P-40 (10^{-7} g/ml) (UENO et al. 1988). The inhibitory effect of human MIS (produced in the presence of Nonidet P-40) could be blocked by a polyclonal antibody raised against human MIS, and by EGF, which inhibits MIS action in the organ culture assay.

In contrast to the above results, TSAFRIRI et al. (1988) did not observe inhibition of germinal vesicle breakdown in rat oocytes using immunoaffinity-purified bovine MIS. This negative result may be due to the fact that the oocytes were harvested from large follicles of human chorionic gonadotropin(hCG)-stimulated rats; such oocytes apparently do not respond to MIS (S. UENO and P. K. DONAHOE, unpublished observations). If MIS is involved in inhibiting meiosis of the oocyte in antral follicles, one must still account for how oocytes in primary and atretic follicles are held in meiotic prophase, since granulosa cells in these follicles do not make MIS.

The ontogeny studies indicate that both the stage of follicle development and the position of the granulosa cell within the follicle are important factors in the regulation of MIS expression in the ovary. MIS levels are highest in the granulosa cells lining the antrum and those closest to the oocyte (the cumulus oophorus), which are the only granulosa cells in the follicle that proliferate (HIRSHFIELD and MIDGLEY 1984). This stem cell population is incapable of undergoing terminal differentiation under FSH stimulation. In contrast, the peripheral granulosa cells located near the basal lamina do not express MIS and undergo terminal cytodifferentiation when challenged by FSH, characterized by acquisition of LH/hCG receptor, steroidogenic enzymes, and lipid droplets (ERICKSON et al. 1985). Thus, in both the testis and ovary, MIS expression is highest in immature cells that are capable of proliferating (BEZARD et al. 1987). Since the granulosa cells that produce MIS are not sensitive to FSH, this gonadotropin does not appear to be involved in regulating MIS expression. In vitro, FSH has no affect on the level of MIS mRNA in cultured human adult granulosa cells (VOUTILAINEN and MILLER 1987).

E. Mechanism of Action

I. MIS Receptor and Mullerian Duct Regression

The action of MIS is probably mediated by a receptor on target cells. Potential target cells include the epithelial and mesenchymal cells of the Mullerian duct in the fetus, and possibly pre-Sertoli cells of the immature testis and germ cells of the adult testis and ovary. Binding of MIS to its receptor presumably activates certain signal transduction pathways within these target cells. Subsequently new genetic programs may be initiated, which ultimately lead to Mullerian duct regression, Sertoli cell differentiation (if MIS acts as an autocrine factor for testicular differentiation), and germ cell maturation (if MIS is an affector of spermatogenesis and oogenesis).

To date, the MIS receptor has resisted all attempts at characterization. A requirement for proteolytic processing may account for part of the difficulty in detecting the receptor, since the full-length MIS molecule may not bind to its receptor as is apparently the case for latent TGF-β (WAKEFIELD et al. (1987). There is a possibility that two MIS receptors exist, one for the C-terminal fragment and one for the N-terminal fragment, since both appear to be necessary for regression of the Mullerian duct. Because of the lack of progress in the biochemical characterization of the MIS receptor, the present discussion will concentrate on the morphological changes that occur during Mullerian duct regression. The spatial and temporal nature of these changes has provided insight into the potential early changes in gene activity affected by MIS and the identity of the target cells in which these genes are modulated.

Dissolution of the Mullerian duct basement membrane is a major early event in ductal regression (TRELSTAD et al. 1982). Basement membrane constituents, including laminin, heparin sulfate proteoglycan, and type IV collagen disappear progressively as the duct regresses (IKAWA et al. 1984a). It is known that the disruption of the basement membrane may cause topographic and functional changes in the basal surface of epithelial cells and affect the organization of the cytoskeleton (SUGRUE and HAY 1981). Such changes in the membrane and cytoskeleton of epithelial cells in the Mullerian duct may lead to the dissociation of these cells observed during regression. In addition to the loss of the basement membrane constituents, mesenchymal matrix components including fibronectin (IKAWA et al. 1984a) and hyaluronate (HAYASHI et al. 1982) are lost from the mesenchyme immediately around the duct. Loss of periductal fibronectin may cause a change in attachment and motility of both epithelial and mesenchymal cells, contributing to epithelial cell dissociation and leading to mesenchymal condensation. Thus the dissolution of the matrix may be responsible for the major morphological changes observed during regression.

These observations suggest that MIS might interact with either the epithelial or mesenchymal cell to either induce the synthesis of enzymes that degrade the matrix or decrease the production of matrix components. Regression of the chick Mullerian duct in vitro can be inhibited by ultrasound irradiation and X-ray irradiation, and by raising the incubation temperature to 41 °C; the effects of these physical manipulations have been attributed to inactivation of proteolytic enzymes (LUTZ and LUTZ-OSTERTAG 1956; SALZGEBER 1961). Human MIS has been tested on a number of cell lines to determine whether it can affect the synthesis of extracellular matrix components. However, no changes were observed in the levels of fibronectin, collagen type IV, and plasminogen activator inhibitor type I mRNAs, in A431, HT1080, and SKOV-3 cells (A. REZAIE and R. L. CATE, unpublished observations). It is important to note that these experiments were performed with full-length MIS which may be inactive if proteolytic processing is required to generate the active form of MIS that can bind to the receptor.

There is evidence that the mesenchyme is an important target of MIS action. TAGUCHI et al. (1984) have shown that, when human male reproductive tracts are transplanted and grown in athymic mice, the mesenchyme (after exposure to endogenous MIS) fails to develop into the anlage of endometrial stroma and myometrium. PARANKO and VIRTANEN (1986) have argued that the mesenchyme is the primary target, since during regression the epithelial cells appear to continue their program of epithelial differentiation as evidenced by an increase in cytokeratin content. A model in which the mesenchyme responds to MIS and then controls the fate of the epithelial cells of the duct is consistent with the model of mesenchyme-regulated epithelial cell differentiation that has been demonstrated for androgen action (KRATOCHWIL and SCHWARTZ 1976; CUNHA et al. 1981).

In the chick, there is evidence that the epithelial cell is the primary target of MIS action. TENG and WANG (1987) have detected binding of MIS to the Mullerian duct epithelial cell of the 7½-day male chick embryo. Binding of MIS was prevented by prenatal treatment with diethylstilbestrol (DES), which is known to cause retention of the chick Mullerian duct (WOLFF 1939). Since DES does not inhibit secretion of MIS (HUTSON et al. 1982), it has been concluded that estrogens act at the Mullerian duct to suppress regression

(MACLAUGHLIN et al. 1983). These results suggest that the epithelial cells of the chick Mullerian duct possess the receptor for MIS, which can be downregulated by estrogen.

II. Modulators of MIS Action and Mullerian Duct Regression

Early studies on the mechanism of action of MIS focused on factors that could affect MIS activity in the organ culture assay. An examination of MIS buffer constituents resulted in the serendipitous discovery that EDTA could cause apparent regression of the Mullerian duct (BUDZIK et al. 1982; SCHWARTZ et al. 1984). Subsequently it was found that Zn^{2+} could reverse this EDTA effect, and also inhibit the action of MIS (BUDZIK et al. 1982). Since preincubation with EDTA or Zn^{2+} neither activated nor inactivated MIS, it was concluded that the divalent cation effect was on the tissue rather than the MIS molecule. These observations led to a proposal that MIS action may be mediated by an extracellular enzyme that is inhibited by Zn^{2+}. Chelation of Zn^{2+} with EDTA therefore would result in activation of the enzyme and Mullerian duct regression.

BRAUTIGAN et al. (1981) had reported the existence of a membrane-bound phosphotyrosyl phosphatase which is inhibited by Zn^{2+} and activated by EDTA. Experiments were performed to ascertain whether the action of MIS is mediated by a similar phosphatase. It was found that a weak stimulator of this phosphatase, sodium fluoride, could cause Mullerian duct regression, while an inhibitor of the phosphatase, sodium vanadate, inhibited MIS-induced regression (HUTSON et al. 1984). Thus MIS action was mimicked by agents that stimulate the phosphatase, and inhibited by agents that inhibit the phosphatase, consistent with a phosphatase mediating MIS action. Because this potential mechanism would presumably involve a decrease in tyrosine phosphorylation of membrane proteins, factors that stimulate tyrosine phosphorylation were tested to see if they might block MIS action. Epidermal growth factor (EGF) specifically inhibited MIS-induced regression in the presence of manganese ion, while insulin, platelet-derived growth factor, and nerve growth factor had no effect (HUTSON et al. 1984).

Since EGF appeared to block MIS action, experiments were performed to determine if MIS could block the action of EGF, by inhibiting autophosphorylation of the EGF receptor. In both whole cells and isolated A431 cell membranes, bovine MIS inhibited EGF-stimulated receptor phosphorylation (COUGHLIN et al. 1987). This effect was not due to MIS acting as a phosphatase and hydrolyzing ATP, or due to MIS affecting the number of EGF-binding sites. Partially purified human recombinant MIS also inhibited EGF-stimulated receptor phosphorylation; this inhibition was blocked by a monoclonal antibody against human MIS (CIGARROA et al. 1989). However, when purified to homogeneity by immunoaffinity chromatography, human recombinant MIS failed to demonstrate this effect. It is possible that the chaotropic buffer used to release MIS from the immunoaffinity matrix may destroy this particular activity without altering biological activity, or that a necessary cofactor is removed during purification. Also, since MIS may require proteolytic processing to generate an active molecule, the active form may be present during the initial stages of purification and lost after the immunoaffinity step. The latter possibility is being investigated by testing the N- and C-terminal fragments of MIS in the EGF receptor assay.

Much of the previous work stemmed from the initial observation that EDTA could cause regression of the Mullerian duct. A closer examination revealed that EDTA caused only partial regression, with the main feature being breakdown of the basement membrane; there was no loss of fibronectin and mesenchymal condensation and epithelial cell migration did not occur (SCHWARTZ et al. 1984; DONAHOE et al. 1984). Subsequently, it was discovered that nucleotide pyrophosphatase (NPPase) may be involved in these other aspects of MIS-induced regression. The effects of NPPase on regression were studied after the enzyme was detected around the regressing Mullerian duct following exposure to both MIS and testosterone (FALLAT et al. 1983, 1984). The NPPase effects on the duct included hyperplasia of the epithelial cells and early mesenchymal condensation. However, the enzyme had no effect on the basement membrane. Thus the effects produced by EDTA and NPPase appear to discriminate distinct events that occur during Mullerian duct regression.

Regression of the Mullerian duct is also affected by steroids and the second messenger cyclic AMP. In vitro, testosterone (10^{-7} *M*) has been shown to augment the action of bovine MIS on the rat Mullerian duct, while dihydrotestosterone and estradiol have no effect on MIS activity (IKAWA et al. 1982). In vivo, DES prevents regression of the Mullerian duct in the chick (WOLFF 1939) and in the mouse (NEWBOLD et al. 1984). Dibutyryl cyclic AMP and phosphodiesterase inhibitors which increase intracellular levels of cyclic AMP have also been shown to inhibit MIS action on the rat Mullerian duct (PICON 1976; IKAWA et al. 1984b). The inhibitory effect of DES and dibutyryl cyclic AMP on Mullerian duct regression occurs at the target organ, and is not due to suppression of MIS production in the testis (HUTSON et al. 1982; IKAWA et al. 1984b). Steroids and cyclic AMP may act by altering the state of differentiation of the epithelial or mesenchymal cells, thereby affecting their sensitivity to the action of MIS.

F. Potential Activities of MIS

I. Descent of the Testis

Descent of the testis can be separated into two phases: the first part consists of transabdominal movement of the testis from the posterior abdominal wall to the inguinal region and the second phase occurs when the testis descends through the inguinal canal to the scrotum. Although it had been generally believed that androgens mediate this two-stage process, in recent years evidence has accumulated that androgens alone are not sufficient to cause complete testicular descent, and may only be important in the last phase of descent (for a review, see HUTSON and DONAHOE 1986). Antiandrogens, such as cyproterone acetate, fail to block descent in the fetus, and studies of children and mice with androgen resistance also fail to show an effect of androgens in the transabdominal phase of descent.

The inability of androgens to influence the initial phase of descent had led to the suggestion that it may be controlled by an alternative hormone such as MIS. There are three lines of evidence in support of this hypothesis. (1) MIS levels in infants with undescended testes are less than normal (DONAHOE et al. 1977b). (2) In individuals with retained Mullerian duct syndrome in which MIS and/or the MIS receptor is apparently defective, 70% of the males have completely undescended testes, with the testes occupying the positions of normal ovaries (HUTSON and DONAHOE 1986). (3) Fetal mice exposed to estrogen have undescended testes and retained Mullerian ducts; estrogen apparently does not suppress androgen secretion in this setting, but has been shown to inhibit MIS action (NEWBOLD et al. 1984). The gubernaculum which undergoes an outgrowth reaction during transabdominal testicular descent has been suggested as a potential target for MIS (HUTSON and DONAHOE 1986).

Most of the evidence cited above is circumstantial; direct evidence that MIS is involved in testicular descent has not yet been obtained. Josso et al. (1983) have concluded that the high incidence of undescended testes in patients with retained Mullerian ducts is due to a mechanical impediment to descent. There is some experimental evidence suggesting that MIS is not involved in descent of the testis. When a female rabbit was immunized against MIS, this led to persistence of the Mullerian ducts in male offspring, but testicular descent was not affected (TRAN et al. 1986).

II. Fetal Lung Development

Human male neonates have an increased incidence of respiratory distress syndrome, which is characterized by a deficiency of pulmonary surfactant. Dihydrotestosterone has been shown to inhibit surfactant production, both in vitro and in vivo (NIELSEN et al. 1982; NIELSEN 1985). CATLIN et al. (1988) have investigated whether MIS can affect the production of the major phospholipid of surfactant, disaturated phosphatidylcholine (DSPC), during lung development. Female fetal rat lungs (17½ days) produced less DSPC after a 5-day incubation with either testis or bovine MIS than did lungs cocultured with ovary. A decrease in DSPC production was also observed with picomolar concentrations of human recombinant MIS. When human MIS was injected subcutaneously into 19-day-old fetuses, a significant reduction in the levels of DSPC in the lungs was observed just prior to birth (E. CATLIN, S. M. POWELL, and P. K. DONAHOE, unpublished observations).

III. Antiproliferative Effects of MIS

Since MIS is involved in the growth regulation of normal cells during development, a considerable effort has been made to determine whether MIS can inhibit the growth of transformed cells. In particular, tumors of tissues that derive from the Mullerian duct (uterus, Fallopian tubes, and vagina) and may express the receptor for MIS have been considered potential targets for MIS action (DONAHOE et al. 1979). Ovarian tumors also have been considered likely targets since both the ovarian and Mullerian duct epithelium derive from the coelomic epithelium of the urogenital ridge. In fact epithelial carcinomas of the ovary which represent 96% of ovarian cancers in humans resemble histologically Mullerian duct tissues (SCULLY 1970). A correlation exists between the onset of ovarian carcinoma in the postmenopausal ovary and a decrease in MIS production, suggesting that MIS prevents malignant transformation in the normal female ovary.

In early studies, the effects of bovine MIS at various stages of purity were determined on human ovarian and endometrial cell lines in a number of cell growth assays. Partially purified bovine MIS inhibited the growth of a human ovarian cell line HOC-21, in a monolayer microcytotoxicity assay (DONAHOE et al. 1979), in a soft agar colony inhibition assay (FULLER et al. 1982), and in nude mice (DONAHOE et al. 1981). Inhibition of growth was not seen in monolayer cultures of the human endometrial cell line HEC-1 (ROSENWAKS et al. 1984), but was observed in nude mice (FULLER et al. 1984). Highly purified fractions of bovine MIS also inhibited colony growth of a large number of primary ovarian and endometrial cancers derived from patients (FULLER et al. 1985).

In a recent study, the effects of highly purified recombinant human MIS were determined on 11 ovarian, 6 endometrial, and 2 nongynecological human tumor cell lines. The human MIS had no effect on proliferation of these cell lines in five independent assays (WALLEN et al. 1989). Forty-three primary human tumor explants were also examined in human tumor colony-forming assays, gel-supported primary culture assays, and subrenal capsule assays. Human MIS significantly inhibited the growth of five of these tumors including four ovarian tumors and one small cell lung cancer explant. The overall response rate for ovarian cancer was 15.4% (WALLEN et al. 1989). The concentration of MIS (10 μg/ml) needed to produce this inhibition was high, which may be due to inefficient proteolytic processing required to generate the active form(s) of MIS. In future experiments it will be important to assess the effects of the N- and C-terminal fragments of MIS not only on primary tumors, but also on the tumor cell lines that did not respond to the full-length human MIS molecule.

G. Summary

Significant progress has been made over the past 40 years in the characterization of MIS, since Jost first suggested its existence. In the past 3 years, advances in our knowledge have come fast, as the information gained from isolating the cDNA and gene for MIS has been assimilated and used to probe further the mechanisms involved in the regulation and function of MIS. In particular, the discovery that MIS is a member of the TGF-β gene family has permitted a large step forward in understanding the structure of the MIS molecule.

The major challenges of the future are to define the extra-Mullerian activities of MIS in the fetal testis and in the adult testis and ovary, and to characterize the receptor on target cells that mediates the action of MIS. Introduction of the gene into mice should provide experimental models for investigating the function of MIS in the testis and ovary throughout development. Although still a difficult task, the identification and isolation of the MIS receptor may now be possible, given our new perspective on the MIS molecule, which may require proteolytic processing before it can bind to the receptor.

References

Agelopoulou R, Magre S, Patsavoudi E, Jost A (1984) Initial phases of the rat testis differentiation in vitro. J Embryol Exp Morphol 83:15–31

Bellve AR (1979) The molecular biology of mammalian spermatogenesis. Oxf Rev Reprod Biol 1:159–261

Benoit R, Ling N, Esch F (1987) A new prosomatostatin-derived peptide reveals a pattern for prohormone cleavage at monobasic sites. Science 238:1126–1129

Bercu BB, Morikawa Y, Jackson IMD, Donahoe PK (1978) Increased secretion of Mullerian inhibiting substance after immunological blockade of endogenous luteinizing hormone releasing hormone in the rat. Pediatr Res 12:139–142

Bercu BB, Morikawa Y, Jackson IMD, Donahoe PK (1979) Inhibition of Mullerian inhibiting substance secretion by FSH. Pediatr Res 13:246–249

Bezard J, Vigier B, Tran D, Mauleon P, Josso N (1987) Immunocytochemical study of anti-Mullerian hormone in sheep ovarian follicles during fetal and post-natal development. J Reprod Fertil 80:509–516

Blanchard M, Josso N (1974) Source of anti-Mullerian hormone synthesized by the fetal testis: Mullerian inhibiting activity of fetal bovine Sertoli cells in tissue culture. Pediatr Res 8:968–971

Brautigan DL, Bornstein P, Gallis B (1981) Phosphotyrosyl-protein phosphatase: specific inhibition by zinc ion. J Biol Chem 256:6519

Breathnach R, Chambon P (1981) Organization and expression of eucaryotic split genes coding for proteins. Annu Rev Biochem 50:349–383

Brunner AM, Gentry LE, Cooper JA, Purchio AF (1988) Recombinant type 1 transforming growth factor β precursor produced in Chinese hamster ovary cells is glycosylated and phosphorylated. Mol Cell Biol 8:2229–2232

Budzik GP, Swann DA, Hayashi A, Donahoe PK (1980) Enhanced purification of Mullerian inhibiting substance by lectin affinity chromatography. Cell 21:909–915

Budzik GP, Hutson JM, Ikawa H, Donahoe PK (1982) The role of zinc in Mullerian duct regression. Endocrinology 110:1521–1525

Budzik GP, Powell SM, Kamagata S, Donahoe PK (1983) Mullerian inhibiting substance fractionation by dye affinity chromatography. Cell 34:307–314

Budzik GP, Donahoe PK, Hutson JM (1985) A possible purification of Mullerian inhibiting substance and a model for its mechanism of action. In: Lash JW, Saxen L (eds) Developmental Mechanisms: Normal and Abnormal. Alan R. Liss, Inc., New York, pp 207–223

Cate RL, Mattaliano RJ, Hession C, Tizard R, Farber NM, Cheung A, Ninfa EG, Frey AZ, Gash DJ, Chow EP, Fisher RA, Bertonis JM, Torres G, Wallner BP, Ramachandran KL, Ragin RC, Manganaro TF, MacLaughlin DT, Donahoe PK (1986a) Isolation of the bovine and human genes for Mullerian inhibiting substance and expression of the human gene in animal cells. Cell 45:685–698

Cate RL, Ninfa EG, Pratt DJ, MacLaughlin DT, Donahoe PK (1986b) Development of Mullerian inhibiting substance as an anti-cancer drug. Cold Spring Harbor Symp Quant Biol 51:641–647

Catlin E, Manganaro T, Donahoe PK (1988) Mullerian inhibiting sustance depresses accumulation in vitro of disaturated phosphatidylcholine in fetal rat lung. Am J Obstet Gynecol 159:1299–1303

Chasin L, Urlaub G (1980) Isolation of Chinese hamster cell mutants deficient in dihydrofolate reductase activity. Proc Natl Acad Sci USA 77:4216–4220

Cigarroa F, Coughlin JP, Donahoe PK, White MF, Uitvlugt N, MacLaughlin DT (1989) Recombinant Mullerian inhibiting substance inhibits epidermal growth factor receptor tyrosine kinase. Growth Factors 1:179–191

Coughlin JP, Donahoe PK, Budzik GP, MacLaughlin DT (1987) Mullerian inhibiting substance blocks autophosphorylation of the EGF receptor by inhibiting tyrosine kinase. Mol Cell Endocrinol 49:75–86

Cunha GR, Shannon JM, Neubauer BL, Sawyer LM, Fujii H, Taguchi O, Chung LWK (1981) Mesenchymal-epithelial interactions in sex differentiation. Hum Genet 58:68–77

De Martin R, Haendler B, Hofer-Warbinek R, Gaugitsch H, Wrann M, Schlusener H, Seifert JM, Bodmer S, Fontana A, Hofer E (1987) Complimentary DNA for human glioblastoma-derived T cell suppressor factor, a novel member of the transforming growth factor-β gene family. EMBO J 6:3673–3677

Derynck R, Jarrett JA, Chen EY, Eaton DH, Bell JR, Assoian RK, Roberts AB, Sporn MB, Goeddel DV (1985) Human transforming growth factor-β complementary DNA sequence and expression in normal and transformed cells. Nature 316:701–705

Derynck R, Jarrett JA, Chen EY, Goeddel DV (1986) The murine transforming growth factor-β precursor. J Biol Chem 261:4377–4379

Donahoe PK, Ito Y, Price JM, Hendren WH III (1977a) Mullerian inhibiting substance activity in bovine fetal, newborn and prepubertal testes. Biol Reprod 16:238–243

Donahoe PK, Ito Y, Morikawa Y, Hendren WH (1977b) Mullerian inhibiting substance in human testes after birth. J Pediatr Surg 12:323–330

Donahoe PK, Swann DA, Hayashi A, Sullivan MD (1979) Mullerian duct regression in the embryo correlated with cytotoxic activity against human ovarian cancer. Science 205:913–915

Donahoe PK, Fuller AF Jr, Scully RE, Guy SR, Budzik GP (1981) Mullerian inhibiting substance inhibits growth of a human ovarian cancer in nude mice. Ann Surg 194:472–480

Donahoe PK, Budzik GP, Trelstad RL, Schwartz BR, Fallat ME, Hutson JM (1984) Molecular dissection of Mullerian duct regression. In: Trelstad RL (ed) The role of extracellular matrix in development. Liss, New York, pp 573–595

Donahoe PK, Cate RL, MacLaughlin DT, Epstein J, Fuller AF, Takahashi M, Coughlin JP, Ninfa EG, Taylor LA (1987) Mullerian inhibiting substance: gene structure and mechanism of action of a fetal regressor. Recent Prog Horm Res 43:431–467

Erickson GF, Hofeditz C, Unger M, Allen WR, Dulbecco R (1985) A monoclonal antibody to a mammary cell line recognizes two distinct subtypes of ovarian granulosa cells. Endocrinology 117:1490–1499

Fallat ME, Hutson JM, Budzik GP, Donahoe PK (1983) The role of nucleotide pyrophosphatase in Mullerian duct regression. Dev Biol 100:358–364

Fallat ME, Hutson JM, Budzik GP, Donahoe PK (1984) Androgen stimulation of nucleotide pyrophosphatase during Mullerian duct regression. Endocrinology 114:1592–1598

Fuller AF Jr, Guy SR, Budzik GP, Donahoe PK (1982) Mullerian inhibiting substance inhibits colony growth of a human ovarian carcinoma cell line. J Clin Endocrinol Metab 54:1051–1055

Fuller AF Jr, Budzik GP, Krane IM, Donahoe PK (1984) Mullerian inhibiting substance inhibition of a human endometrial carcinoma cell line xenografted in nude mice. Gynecol Oncol 17:124

Fuller AF Jr, Krane IM, Budzik GP, Donahoe PK (1985) Mullerian inhibiting substance reduction of colony growth of human gynecologic cancers in a stem cell assay. Gynecol Oncol 22:135–148

Gentry LE, Webb NR, Lim GJ, Brunner AM, Ranchalis JE, Twardzik DR, Lioubin MN, Marquardt H, Purchio AF (1987) Type 1 transforming growth factor beta: amplified expression and secretion of mature and precursor polypeptides in Chinese hamster ovary cells. Mol Cell Biol 7:3418–3427

Gething M-J, McCammon K, Sambrook J (1986) Expression of wild-type and mutant forms of influenza hemagglutinin: the role of folding in intracellular transport. Cell 46:939–950

Hadley MA, Byers SW, Suarez-Quian CA, Kleinman HK, Dym M (1985) Extracellular matrix regulates Sertoli cell differentiation, testicular cord formation, and germ cell development. J Cell Biol 101:1511–1522

Hayashi A, Donahoe PK, Budzik GP, Trelstad RL (1982) Periductal and matrix glycosaminoglycans in rat Mullerian duct development and regression. Dev Biol 92:16–26

Hayashi M, Shima H, Hayashi K, Trelstad R, Donahoe PK (1984) Immunocytochemical localization of Mullerian inhibiting substance in the endoplasmic reticulum and the Golgi apparatus in the Sertoli cells of the neonatal calf testis using a monoclonal antibody. J Histochem Cytochem 32:649

Hirshfield AN, Midgley AR (1984) Morphometric analysis of follicular development in the rat. Biol Reprod 19:597–605

Hutson JM, Donahoe PK (1986) The hormonal control of testicular descent. Endocr Rev 7:270–283

Hutson J, Ikawa H, Donahoe PK (1981) The ontogeny of Mullerian inhibiting substance in the gonads of the chicken. J Pediatr Surg 16:822–827

Hutson JM, Ikawa H, Donahoe PK (1982) Estrogen inhibition of Mullerian inhibiting substance in the chick embryo. J Pediatr Surg 17:953–959

Hutson JM, MacLaughlin DT, Ikawa H, Budzik GP, Donahoe PK (1983) Regression of the Mullerian ducts during sexual differentiation in the chick embryo: a reappraisal. Int Rev Physiol 27:177–224

Hutson JM, Fallat ME, Kamagata S, Donahoe PK, Budzik GP (1984) Phosphorylation events during Mullerian duct regression. Science 223:586–589

Ikawa H, Hutson JM, Budzik GP, MacLaughlin DT, Donahoe PK (1982) Steroid enhancement of Mullerian duct regression. J Pediatr Surg 17:453–458

Ikawa H, Trelstad RL, Hutson JM, Manganaro TF, Donahoe PK (1984a) Changing patterns of fibronectin, laminin, type IV collagen, and a basement membrane proteoglycan during rat Mullerian duct regression. Dev Biol 102:260–263

Ikawa H, Hutson JM, Budzik GP, Donahoe PK (1984b) Cyclic adenosine 3′,5′-monophosphate modulation of Mullerian duct regression. Endocrinology 114:1686–1691

Jones NC, Rigby PWJ, Ziff EB (1988) *Trans*-acting protein factors and the regulation of eukaryotic transcription: lessons from studies on DNA tumor viruses. Genes Dev 2:267–281

Josso N (1986) Anti-Mullerian hormone: new perspectives for a sexist molecule. Endocr Rev 7:421–433

Josso N, Picard JY (1986) Anti-mullerian hormone. Physiological Reviews 66:1038–1090

Josso N, Picard JY, Tran D (1977) The anti-Mullerian hormone. Recent Prog Horm Res 33:117–167

Josso N, Picard JY, Dacheux JL, Courot M (1979) Detection of anti-Mullerian activity in boar rete testis fluid. J Reprod Fertil 57:397–400

Josso N, Picard J-Y, Tran D (1980) A new testicular glycoprotein: anti-Mullerian hormone. In: Steinberger A, Steinberger E (eds) Testicular development, structure and function. Raven, New York, pp 21–31
Josso N, Fekete C, Cachin O, Nezelof C, Rappaport R (1983) Persistence of Mullerian ducts in male pseudohermaphroditism, and its relationship to cryptorchidism. Clin Endocrinol (Oxf) 19:247
Jost A (1946a) Sur la differenciation sexuelle de l'embryon de lapin. Experiences de parabiose. C R Seances Acad Sci Ser III Sci Vie 140:463–464
Jost A (1946b) Sur la differenciation sexuelle de l'embryon de lapin remarques au sujet des certaines operations chirurgical sur l'embryon. C R Seances Acad Sci Ser III Sci Vie 140:461
Jost A (1947a) Recherches sur la differenciation sexuelle de l'embryon de lapin. Arch Anat Microsc Morphol Exp 36:271–315
Jost A (1947b) Sur les derives mulleriens d'embryons de lapin des deux sexes castres a 21 jours. C R Seances Acad Sci Ser III Sci Vie 141:135–136
Jost A (1972) Donnees preliminaires sur les stades initiaux de la differenciation du testicule chez le rat. Arch Anat Microsc Morphol Exp 61:415–438
Kadonaga JT, Carner KR, Masiarz FR, Tjian R (1987) Isolation of cDNA encoding transcription factor Sp1 and functional analysis of the DNA binding domain. Cell 51:1079–1090
Kaufman RJ, Sharp PA (1982) Construction of a modular dihydrofolate reductase cDNA gene: analysis of signals utilized for efficient expression. Mol Cell Biol 2:1304–1319
Kingsley DM, Kozarsky KF, Segal M, Krieger M (1986) Three types of low density lipoprotein receptor-deficient mutant have pleiotropic defects in the synthesis of N-linked, O-linked and lipid-linked carbohydrate chains. J Cell Biol 102:1576–1585
Kozak M (1983) Comparison of initiation of protein synthesis in procaryotes, eucaryotes and organelles. Microbiol Rev 47:1–45
Kratochwil K, Schwartz P (1976) Tissue interaction in androgen response of embryonic mammary rudiment of mouse: identification of target tissue for testosterone. Proc Natl Acad Sci USA 73:4041–4044
LaQuaglia M, Shima H, Hudson P, Takahashi M, Donahoe PK (1986) Sertoli cell production of Mullerian inhibiting substance in vitro. J Urol 136:219–224
Lawrence DA, Pircher R, Jullien P (1985) Conversion of a high molecular weight latent β-TGF from chicken embryo fibroblasts into a low molecular weight active β-TGF under acidic conditions. Biochem Biophys Res Commun 133:1026–1034
Lillie F (1916) Theory of the freemartin. Science 43:611–613
Ling N, Ying S-Y, Ueno N, Shimasaki S, Esch F, Hotta M, Guillemin R (1986) Pituitary FSH is released by a heterodimer of the beta-subunits from the two forms of inhibin. Nature 321:779–782
Lodish HF (1988) Transport of secretory and membrane glycoproteins from the rough endoplasmic reticulum to the golgi. J Biol Chem 263:2107–2110
Lodish HF, Kong N, Snider M, Strous GJAM (1983) Hepatoma secretory proteins migrate from rough endoplasmic reticulum to golgi at characteristic rates. Nature 304:80–83
Louis BG, Fritz IB (1979) Follicle-stimulating hormone and testosterone independently increase the production of androgen-binding protein by Sertoli cells in culture. Endocrinology 104:454–461
Lutz H, Lutz-Ostertag Y (1956) Action des ultra-sons sur l'enzyme proteolytique des canaux de Muller de poulet male. C R Seances Acad Sci Ser III Sci Vie 150:913
Lyons RM, Keski-Oja J, Moses HL (1988) Proteolytic activation of latent transforming growth factor-β from fibroblast-conditioned medium. J Cell Biol 106:1659–1665
MacLaughlin DT, Hutson JM, Donahoe PK (1983) Specific estradiol binding in embryonic Mullerian ducts: a potential modulator of regression in the male and female chick. Endocrinology 113:141–145

Magre S, Jost A (1980) The initial phases of testicular organogenesis in the rat. An electron microscopy study. Arch Anat Microsc Morphol Exp 69:297–318
Mason AJ, Hayflick JS, Ling N, Esch F, Ueno N, Ying S, Guillemin R, Niall H, Seeburg PH (1985) Complementary DNA sequences of ovarian follicular fluid inhibin show precursor structure and homology with transforming growth factor-β. Nature 318:659–663
Massague J (1987) The TGF-β family of growth and differentiation factors. Cell 49:437–738
McLaren A (1981) The fate of germ cells in the testis of fetal sex-reversed mice. J Reprod Fertil 61:461–467
McLaren A (1984) Chimeras and sexual differentiation. In: Le Douarin N, McLaren A (eds) Chimeras in developmental biology. Academic, London, pp 381–400
McLaren A (1985) Relation of germ cell sex to gonadal differentiation. In: Halvorson HO, Monroy A (eds) The origin and evolution of sex. Liss, New York, pp 289–300
Miyazono K, Hellman U, Wernstedt C, Heldin C-H (1988) Latent high molecular weight complexes of transforming growth factor β1. J Biol Chem 263:6407–6415
Montminy MR, Sevarino KA, Wagner JA, Mandel G, Goodman RH (1986) Identification of a cyclic-AMP-responsive element within the rat somatostatin gene. 83:6682–6686
Necklaws EC, LaQuaglia MP, MacLaughlin DT, Hudson PL, Mudgett-Hunter M, Donahoe PK (1986) Detection of Mullerian inhibiting substance in biological samples by a solid phase sandwich radioimmunoassay. Endocrinology 118:791–796
Newbold RR, Suzuki Y, McLachlan JA (1984) Mullerian duct maintenance in heterotypic organ culture after in vivo exposure to diethylstilbestrol. Endocrinology 115:1863–1868
Nielsen HC (1985) Androgen receptors influence the production of pulmonary surfactant in the testicular feminization mouse fetus. J Clin Invest 76:177
Nielsen HC, Zinman HM, Torday JS (1982) Dihydrotestosterone inhibits fetal rabbit pulmonary surfactant production. J Clin Invest 69:661
Ohta M, Greenberger JS, Anklesaria P, Bassols A, Massague J (1987) Two forms of transforming growth factor-β distinguished by multipotential haematopoietic progenitor cells. Nature 329:539
Padgett RW, St Johnston RD, Gelbart WM (1987) A transcript from a *Drosophila* pattern gene predicts a protein homologous to the transforming growth factor-β family. Nature 325:81–84
Page DC, Mosher R, Simpson EM, Fisher EMC, Mardon G, Pollack J, McGillivray B, de la Chapelle A, Brown LG (1987) The sex-determining region of the human Y chromosome encodes a finger protein. Cell 51:1091–1104
Paranko J, Virtanen I (1986) Epithelial and mesenchymal cell differentiation in the fetal rat genital ducts: changes in the expression of cytokeratin and vimentin type of intermediate filaments and desmosomal plaque proteins. Dev Biol 117:135–145
Pepinsky RB, Sinclair LK, Chow E P-C, Mattaliano RJ, Manganaro TF, Donahoe PK, Cate RL (1988) Proteolytic processing of Mullerian inhibiting substance produces a TGF-β like fragment. J Biol Chem 263:18961–18964
Picard JY, Josso N (1984) Purification of testicular anti-Mullerian hormone allowing direct visualization of the pure glycoprotein and determination of yield and purification factor. Mol Cell Endocrinol 34:23
Picard JY, Benarous R, Guerrier D, Josso N, Kahn A (1986a) Cloning and expression of cDNA for anti-Mullerian hormone. Proc Natl Acad Sci USA 83:5464–6468
Picard JY, Goulut C, Bourrillon R, Josso N (1986b) Biochemical analysis of bovine testicular anti-Mullerian hormone. FEBS Lett 195:73–76
Picon R (1969) Action du testicule foetal sur le development in vitro des canaux de Muller chez le rat. Arch Anat Microsc Morphol Exp 58:1–19
Picon R (1976) Testicular inhibition of fetal Mullerian ducts in vitro: effect of dibutyryl cyclic AMP. Mol Cell Endocrinol 4:35

Pircher R, Jullien P, Lawrence DA (1986) β-Transforming growth factor is stored in human blood platelets as a latent high molecular weight complex. Biochem Biophys Res Commun 136:30–37
Rauscher FJ III, Sambucetti LC, Curran T, Distel RJ, Spiegelman BM (1988) Common DNA binding site for Fos protein complexes and transcription factor AP-1. Cell 52:471–480
Rosa F, Roberts AB, Danielpour D, Dart LL, Sporn MB, Dawid IB (1988) Mesoderm induction in amphibians: the role of TGF-β2-like factors. Science 239:783–785
Rosenwaks Z, Liu HC, Picard J-Y, Josso N (1984) Anti-Mullerian hormone is not cytotoxic to human endometrial cancer in tissue culture. J Clin Endocrinol Metab 59:166–169
Salzgeber B (1961) Evolution des gonades et des conduits genitaux de l'embryon de poulet soumis en culture a l'action des rayons x. Bull Biol Fr Belg 44:645
Schwartz BR, Trelstad RL, Hutson JM, Ikawa H, Donahoe PK (1984) Zinc chelation and Mullerian duct regression. J Exp Pathol 1:143–156
Scully RE (1970) Recent progress in ovarian cancer. Hum Pathol 1:73
Shima H, Donahoe PK, Budzik GP, Kamagata S, Hudson P, Mudgett-Hunter M (1984) Production of monoclonal antibodies for affinity purification of bovine Mullerian inhibiting substance activity. Hybridoma 3:201–214
Sporn MB, Roberts AB, Wakefield LM, de Crombrugghe B (1987) Some recent advances in the chemistry and biology of transforming growth factor-β. J Cell Biol 105:1039–1045
Sugrue SP, Hay ED (1981) Response of basal epithelial cell surface and cytoskeleton to solubilized extracellular matrix molecules. J Cell Biol 91:45–54
Swann DA, Donahoe PK, Ito Y, Morikawa Y, Hendren WH (1979) Extraction of Mullerian inhibiting substance from newborn calf testis. Dev Biol 69:73–84
Taguchi O, Cunha GR, Lawrence WD, Stanley JR (1984) Timing and irreversibility of Mullerian duct inhibition in the embryonic reproductive tract of the human male. Dev Biol 106:394–398
Takahashi M, Hayashi M, Manganaro TF, Donahoe PK (1986a) The ontogeny of Mullerian inhibiting substance in granulosa cells of the bovine ovarian follicle. Biol Reprod 35:447–453
Takahashi M, Koide SS, Donahoe PK (1986b) Mullerian inhibiting substance as oocyte meiosis inhibitor. Mol Cell Endocrinol 47:225–234
ten Dijke P, Hansen P, Iwata KK, Pieler C, Foulkes JG (1988) Identification of another member of the transforming growth factor type β gene family. Proc Natl Acad Sci USA 85:4715–4719
Teng CS, Wang JJ (1987) Characterization of the Mullerian inhibiting substance receptor: specific binding on the membrane of Mullerian duct cell by immunobiochemical technique. J Cell Biol 105:313
Teng CS, Wang JJ, Teng JIN (1987) Purification of chicken testicular Mullerian inhibiting substance by ion exchange and high-performance liquid chromatography. Dev Biol 123:245–254
Tran D, Josso N (1982) Localization of anti-Mullerian hormone in the rough endoplasmic reticulum of the developing bovine Sertoli cell using immunocytochemistry with a monoclonal antibody. Endocrinology 82:1562
Tran D, Meusy-Dessolle N, Josso N (1981) Waning of anti-Mullerian activity: an early sign of Sertoli cell maturation in the developing pig. Biol Reprod 24:923–931
Tran D, Picard JY, Vigier B, Berger R, Josso N (1986) Persistence of Mullerian ducts in male rabbits passively immunized against bovine anti-Mullerian hormone during fetal life. Dev Biol 116:160–167
Tran D, Picard JY, Campargue J, Josso N (1987) Immunocytochemical detection of anti-Mullerian hormone in Sertoli cells of various mammalian species including human. J Histochem Cytochem 35:733–743

Trelstad RL, Hayashi A, Hayashi K, Donahoe PK (1982) The epithelial-mesenchymal interface of the male rat Mullerian duct: loss of basement membrane integrity and ductal regression. Dev Biol 92:27–40

Tsafriri A, Picard J-Y, Josso N (1988) Immunopurified anti-Mullerian hormone does not inhibit spontaneous resumption of meiosis in vitro of rat oocytes. Biol Reprod 38:481–486

Ueno S, Manganaro TF, Donahoe PK (1988) Human recombinant Mullerian inhibiting substance inhibition of rat oocyte meiosis is reversed by epidermal growth factor in vitro. Endocrinology 123:1652–1659

Ueno S, Takahashi M, Manganaro TF, Ragin RC, Donahoe PK (1989 a) Cellular localization of Mullerian inhibiting substance in the developing rat ovary. Endocrinology 124:1000–1006

Ueno S, Kuroda T, MacLaughlin DT, Ragin RC, Manganaro TF, Donahoe PK (1989 b) Mullerian inhibiting substance during the estrous cycle of the adult rat ovary. Endocrinology 125:1060–1066

Vale W, Rivier J, Vaughan J, McClintock R, Corrigan A, Woo W, Karr D, Spiess J (1986) Purification and characterization of an FSH releasing protein from porcine ovarian follicular fluid. Nature 321:776–779

Vigier B, Tran D, Du Mesnil du Buisson F, Heyman Y, Josso N (1983) Use of monoclonal antibody techniques to study the ontogeny of bovine anti-Mullerian hormone. J Reprod Fertil 69:207

Vigier B, Picard J-Y, Tran D, Legeai L, Josso N (1984) Production of anti-Mullerian hormone: another homology between Sertoli and granulosa cells. Endocrinology 114:1315–1320

Vigier B, Picard J-Y, Campargue J, Forest MG, Heyman Y, Josso N (1985) Secretion of anti-Mullerian hormone by immature bovine Sertoli cells in primary culture, studied by a competition-type radioimmunoassay: lack of modulation by either FSH or testosterone. Mol Cell Endocrinol 43:141–150

Vigier B, Watrin F, Magre S, Tran D, Josso N (1987) Purified bovine AMH induces a characteristic freemartin effect in fetal rat prospective ovaries exposed to it in vitro. Development 100:43–55

Voutilainen R, Miller WL (1987) Human Mullerian inhibitory factor mRNA is hormonally regulated in the fetal testis and in adult granulosa cells. J Mol Endocrinol 1:604

Wakefield LM, Smith DM, Masui T, Harris CC, Sporn MB (1987) Distribution and modulation of the cellular receptor for transforming growth factor-beta. J Cell Biol 105:965–975

Wakefield LM, Smith DM, Flanders KC, Sporn MB (1988) Latent transforming growth factor-β from human platelets: a high molecular weight complex containing precursor sequences. J Biol Chem 263:7646–7654

Wallen JW, Cate RL, Kiefer DM, Riemen MW, Donahoe PK, Martinez D, Hoffman RM, Von Hoff DD, Pepinsky B, Oliff A (1989) Minimal antiproliferative effect of recombinant Mullerian inhibiting substance on gynecologic tumor cell lines and tumor explants. Cancer Res 49:2005–2011

Weeks DL, Melton DA (1987) A maternal mRNA localized to the vegetal hemisphere in *Xenopus* eggs codes for a growth factor related to TGF-β. Cell 51:861–867

Wolf E (1939) L'action due diethylstilboestrol sur les organes genitaux de l'embryon de Poulet. C R Seances Acad Si Ser III Sci Vie 208:1532–1534

Yu J, Shao L-E, Lemas V, Yu AL, Vaughan J, Rivier J, Vale W (1987) Importance of FSH-releasing protein and inhibin in erythrodifferentiation. Nature 330:765–767

CHAPTER 26

The Inhibin/Activin Family of Hormones and Growth Factors

W. VALE, A. HSUEH, C. RIVIER, and J. YU

Inhibins and activins were initially recognized as gonadal protein hormones which modulate follicle-stimulating hormone (FSH) production by the anterior pituitary gland. Since the recent chemical characterization of these proteins as hetero- and homodimers related to the TGF-β family, it has become clear that they are broadly distributed and have actions on multiple tissues where they may play a variety of types of roles as hormonal, paracrine, and autocrine regulators of cellular function and proliferation.

A. Chemical Characterization of Inhibins and Activins

I. Inhibin

The anterior pituitary secretes two gonadotropic hormones, luteinizing hormone (LH) and FSH, which stimulate separate cell types within the ovary and testis (FEVOLD et al. 1931; SAIRAM and PAPKOFF 1974). Most gonadotrophs produce both LH and FSH although subtypes containing only one hormone might also exist. The major acute stimulator of gonadotroph functions is gonadotropin-releasing hormone (GnRH), a decapeptide secreted episodically by hypothalamic neurons into the portal vessels supplying the anterior pituitary gland. The gonads produce substances which feedback and influence gonadotropin secretion both by acting directly at the level of the pituitary and by modulating hypothalamic GnRH production. These substances include sex steroids as well as the protein hormones that are the subject of this review.

In 1923, MOTTRAM and CRAMER reported that radiation-induced damage to the seminiferous tubules of male rats caused hypertrophy of the pituitary gland and postulated that these germinal elements produce an internal secretion that influences pituitary cells. Two groups (MARTINS and ROCHA 1931; MCCULLAGH 1932) subsequently administered aqueous extracts of the testis to castrated rats and observed prevention of the appearance of the enlarged "castration cells" normally seen in the pituitary glands of gonadectomized rats. McCullagh named the water-soluble testicular principle responsible for this activity "inhibin." Organic extracts of testis had different biological effects which were likely due to androgens. Although McCullagh continued to promulgate the inhibin concept (McCULLAGH and WALSH 1934; McCULLAGH and SCHNEIDER 1940), several other workers were unable to confirm his results and remained skeptical of the inhibin concept (NELSON and GALLAGHER 1935; VIDGOFF and VEHRS 1940; RUBIN 1941).

Clinical observations correlating the degree of damage to spermatogenesis in infertile men with bioassayable FSH rather than LH levels (KLINEFELTER et al. 1942; HOWARD et al. 1950; McCULLAGH and SCHAFFENBURG 1952) supported the concept that inhibin selectively suppresses FSH secretion. The critical examination of this hypothesis was made possible by the eventual isolation and development of specific radioimmunoassays for FSH and LH (PARKOFF 1974; PIERLE and PARSONS 1981). Additional clinical studies along with many experiments in laboratory animals supported the notion that Sertoli cells in the testis and granulosa cells in the ovary were the sources of inhibin (review in DEJONG 1988).

Fluids and extracts from male and female gonads of several species were shown selectively to inhibit FSH production in a variety of in vivo and in vitro assays (DEJONG and SHARPE 1976; SCHWARTZ and CHANNING 1977; CHANNING et al. 1985; BAKER et al. 1976; ERICKSON and HSUEH 1978; STEINBERGER and STEINBERGER 1981; DEJONG et al. 1979; FRANCHIMONT et al. 1979). These biological effects were not eliminated by removal of sex steroids from the inhibin preparations nor could the actions of the preparations be reproduced by administration of male or female sex steroids in any combination. In vivo and in vitro bioassays for inhibin were developed and used to examine the levels of inhibin-like activity in plasma or gonadal fluids under a variety of circumstances including gonadectomy, germinal element damage, gonadotropin and sex steroid administration, stages of the estrous or menstrual cycle, and maturation. Of the various assays (HUDSON et al. 1979), the one based upon the ability of inhibin to suppress basal FSH secretion by cultured anterior pituitary cells (ERICKSON and HSUEH 1978) was employed by all groups who successfully isolated the molecules now referred to as inhibin or activin. This method, a modification of VALE et al. (1972, 1975), was the basis for the purification of ovine somatostatin (BRAZEAU et al. 1973), ovine and rat corticotropin releasing factor (CRF) (VALE et al. 1981; SPIESS et al. 1983), and human and rat growth hormone releasing factor (J. RIVIER et al. 1982; GUILLEMIN et al. 1982; SPIESS et al. 1983).

In 1985, four groups reported the isolation from ovarian follicular fluids of disulfide-linked heterodimeric proteins which had potent activity to suppress FSH secretion by cultured rat anterior pituitary cells. Some N-terminal sequence analyses of each chain were performed by each laboratory. From bovine follicular fluid, ROBERTSON et al. (1985) purified a dimer of $M_r \sim 58\,000$ comprising two disulfide-linked subunits of $M_r \sim 44\,000$ and $\sim 14\,000$ referred to now by convention as the α- and β-chains respectivels. MIYAMOTO et al. (1985), LING et al. (1985), and J. RIVIER et al. (1985) purified dimers of $M_r \sim 32\,000$ from porcine follicular fluid comprising an $M_r \sim 18\,000$ α-subunit and an $M_r \sim 14\,000$ β-subunit. Subsequently, ROBERTSON et al. (1986a) purified an $M_r \sim 32\,000$ inhibin from another fraction of bovine follicular fluid and reported that they could convert the $M_r \sim 58000$ protein to $M_r \sim 32000$ by incubation in peripheral plasma. While LEVERSHA et al. (1987) described the purification of an $M_r \sim 65000$ dimeric inhibin from ovine follicular fluid, our group (BARDIN et al. 1987; VAUGHAN et al. 1989) purified an $M_r \sim 32000$ dimeric inhibin from ovine rete testis fluid. This latter finding constituted the first isolation of inhibin from a male source; partial sequence analyses of each chain supported the hypothesis that male and female inhibins have similar primary structures.

Synthetic oligonucleotide probes were designed based upon the N-terminal protein sequences of each inhibin subunit and were used to screen porcine ovarian cDNA libraries. DNA sequences of clones selected by the probes allowed the deduction of the complete protein sequences of the precursors for the porcine α-subunits (MASON et al. 1985; MAYO et al. 1986) and two distinct β-subunits, βA and βB (MASON et al. 1985). The dimer, αβA, corresponds to inhibin A and $\alpha\beta_A$ to inhibin B isolated by LING et al. (1985).

FORAGE et al. (1986) characterized bovine α- and βA-subunit precursors but did not identify a βB RNA in that species. Using porcine probes, human inhibin α-subunit clones were identified and characterized in ovarian (MASON et al. 1986; STEWART et al. 1986) and placental (MAYO et al. 1986) libraries and human βA (MASON et al. 1986; STEWART et al. 1986) and βB (MASON et al. 1986) subunit clones in ovarian libraries. ESCH et al. (1987a) and WOODRUFF et al. (1987) cloned rat α-, βA-, and βB-subunit precursors and deduced their protein sequences.

The precursors for each subunit are derived from separate RNAs which are encoded by separate genes (STEWART et al. 1986; MASON et al. 1985). According to MASON (1988), each gene comprises two exons separated by large introns of 2.1, 10, and 2.8 kb for the α, βA, and βB genes respectively. The inhibin subunit genes are chromosomally dispersed; the human α gene is on the distal portion of the long arm of chromosome 2, the βA gene is on chromosome 7 while the βB gene is located near the centromere on the short arm of chromosome 2 (MASON 1988).

The three human prohormones illustrated in Fig. 1 contain several clusters of multiple basic residues which serve as potential processing cleavage sites as well as potential Asn-linked glycosylation sites. The α-, βA-, and βB-subunits isolated as parts of the biologically active $M_r \sim 32000$ inhibin dimers are derived from the C-terminal region of each precursor and are peptides of 134, 116, and 115 amino acids respectively (Fig. 2A, B). The human $M_r \sim 18000$ α-subunit contains seven cysteine residues and two potential glycosylation sites. The $M_r \sim 44000$ α-chain identified by ROBERTSON et al. (1985) in purified bovine inhibin is an N-terminally extended version of the $M_r \sim 18000$ subunit. The β-subunits have nine cysteine residues and no consensus Asn-linked glycosylation sites. Within each species the βA- and βB-subunits are closely related, showing ~85% identity with each other and some similarity to the α-subunit. Porcine, bovine, human, and murine inhibin α-subunits are ~70% identical to one another whereas each β-subunit is virtually (>98%) identical between the species characterized. Each subunit possesses a large number of

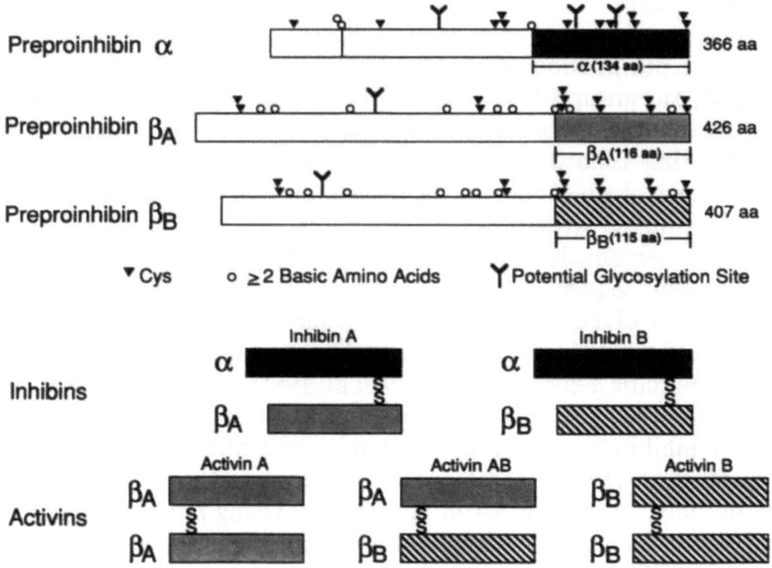

Fig. 1. Schematic of the precursors of the human inhibin α-, βA-, and βB-subunits. Mature portions of each subunit as they appear in the putative $M_r \sim 32000$ inhibin A (αβA) and inhibin B (αβB) are *shaded*

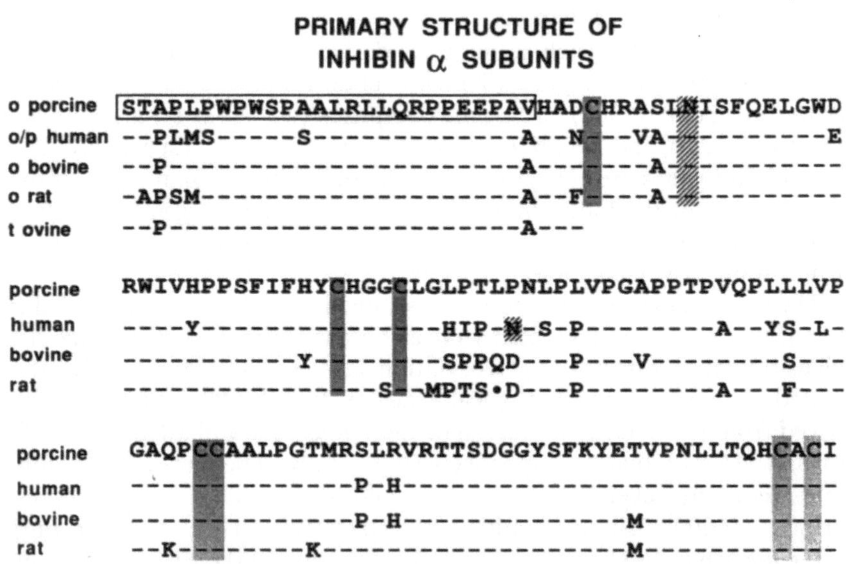

Fig. 2a

Fig. 2. a Peptide sequences of mature α-subunits from several species and tissues. Identity with porcine inhibin α(—); potential Asn-linked glycosylation sites (*crosshatched* areas); cysteine residues (*shaded areas*); peptide *enclosed by box* was synthesized for use as immunogen (VAUGHAN et al. 1989). **b** Peptide sequences of mature βA- and βB-subunits from several species and tissues. Symbols as in **a. a, b** courtesy of Academic Press

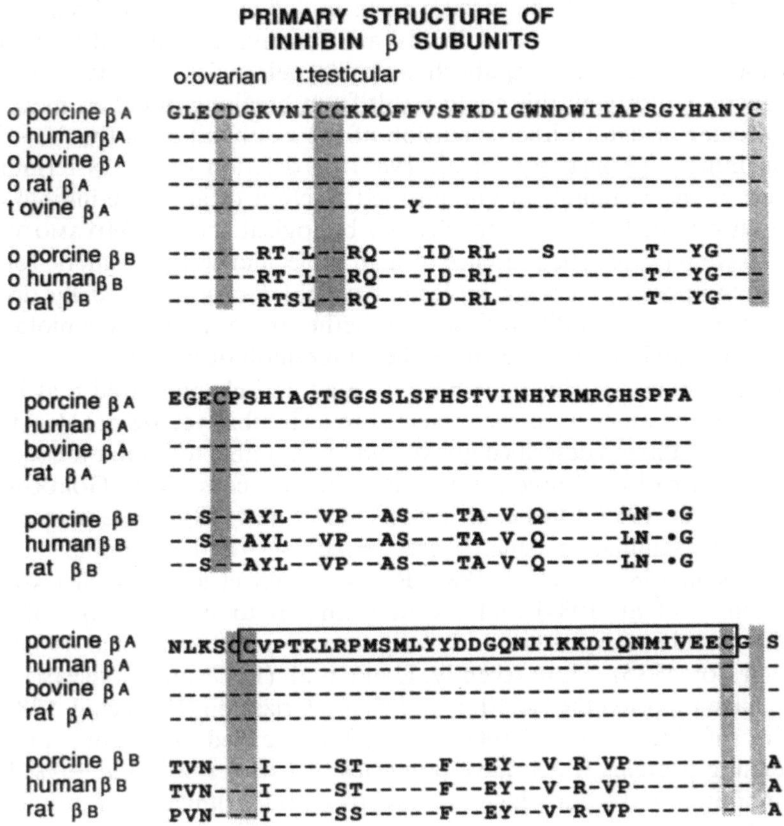

Fig. 2b

cysteine residues which participate in intramolecular and at least one intermolecular disulfide bridge.

The β-subunits and to a lesser extent the α-subunit have corresponding patterns of cysteine distributions and some additional sequence similarities not only to each other but to members of the transforming growth factor-β (TGF-β) family. This family includes TGF-β (Derynck et al. 1985; Sporn and Roberts 1988; Dijke et al. 1988), Mullerian duct inhibiting substance (MIS) (Cate et al. 1986), which causes regression of the Mullerian duct during development of the male, the fly decapentaplegic gene complex, which plays a role in embryonic dorsal-ventral determination and segementation (Padgett et al. 1987), and the *Vg*1 gene in *Xenopus* (Weeks and Melton 1987; Tannahil et al. 1988), involved in the development of primative mesoderm, and cartilage-inducing factor B (Seyedin et al. 1987).

Follicular fluids and rete testis fluids contain a variety of biologically and/or immunologically active inhibin-like species. We and others have identified free α-chain monomers and $\alpha\beta$-dimers comprising different-sized α-subunits resulting from variable glycosylation, incomplete precursor processing, or degradation of the mature form and β-chains of different lengths. Purified α-monomers in our hands are inactive on the pituitary (J.M. VAUGHAN and W. VALE, unpublished results). Biological activity is destroyed by reduction, although this could result from breaking of either inter- or intramolecular bridges. Either extension of the α-subunit as in the $M_r \sim 58000$ bovine inhibin (ROBERTSON et al. 1985) or deletion of the N-terminal 15 residues of the α-subunit in one purified fraction of an ovine inhibin dimer (VAUGHAN et al. 1989) does not destroy biological activity. MIYAMOTO et al. (1986) have reported the presence of high molecular weight trimeric forms of inhibin composed of a high molecular weight β-chains linked by disulfide bonds to $\alpha\beta$-dimers of different sizes. Whether these species were biologically active could not be determined from these immunoblot studies.

Several other proteins have been isolated and characterized that have at one time or the other have been referred to as inhibin yet are not structurally related to the dimers described above and are not able to inhibit the secretion of FSH in vitro by cultured rat anterior pituitary cells (W.L. GORDON et al. 1987; LIU et al. 1985; LING et al. 1985; VALE et al. 1988). These proteins, isolated from seminal plasma, were termed α- and β-inhibin (LI and RAMASHARMA 1987; LI et al. 1985; RAMASHARMA et al. 1984; SEIDAH et al. 1984; SHETH et al. 1984) and are now known to be fragments of larger proteins found in the prostate and seminal vesicles (AKIYAMA et al. 1985; LILJA and JEPPSSON 1985). More recently, UENO et al. (1987) and ROBERTSON et al. (1987) have reported the isolation and characterization (ESCH et al. 1987b) of proteins of $M_r \sim 31000$, 35000, and 39000, called follistatins or FSH-suppressing proteins which have potencies of between 5% and 30% that of pure inhibin in vitro but which are not structurally similar to the dimeric inhibins. By convention (BURGER and IGARASHI 1988), the term inhibin is now reserved for the dimeric inhibins and this review will deal exclusively with them.

II. Activin

While purifying inhibin from porcine follicular fluid, we noted (J. RIVIER et al. 1985) fractions that could stimulate rather than inhibit FSH secretion by cultured anterior pituitary cells. After preliminary studies indicating that this was a protein which released FSH selectively, we purified this FSH-releasing protein (FRP) by several fast protein liquid chromatographic and HPLC steps. On SDS-PAGE, purified FRP exhibited a major band of $M_r \sim 28000$ which following reduction converted to a single species of $M_r \sim 15000$. Sequence analysis of FRP yielded a single peptide sequence that was identical to the N-terminal 38 amino acids of inhibin βA. Additional peptides corresponding to residues 70–85, 88–102, and 103–116 of inhibin βA were assigned by sequence analyses of proteolytically cleaved purified fragments of FRP. Thus we established that 83 of 116 residues including the amino and carboxy termini were the same in FRP and inhibin-βA and we then proposed that FRP is a homodimer of inhibin-βA (VALE et al. 1986). Concurrently, LING et al.

(1986a) identified another gonadal FSH-releasing protein as the heterodimer, inhibin βAβB, which they named activin. Subsequently they confirmed our characterization of the βAβA homodimer (LING et al. 1986b), which they named homoactivin-A. Although not yet observed in natural sources, both groups predicted the existence of a βBβB homodimer that would stimulate FSH secretion; recently, MASON (1988) reported that recombinantly expressed inhibin βBβB was biologically active. By general agreement (BURGER and IGARASHI 1988), inhibin-βAβA is now called activin-A; inhibin βAβB is referred to as activin-AB; inhibin-βBβB is termed activin-B.

In retrospect it is not surprising to find homodimeric forms of inhibin subunits. In addition to activin-A, both MIS (CATE et al. 1986) and TGF-βs are homodimers. Activins and TGF-β are in fact reported to have similar actions on the secretion of pituitary FSH by YING et al. (1986b), who initially considered that the FSH-releasing activity of their follicular fluid fractions was due to a molecule closely related if not identical to TGF-β. However, in an extensive series of in vitro experiments conducted under a variety of conditions, we have been unable to confirm the observation that TGF-β stimulates FSH secretion (VALE et al. 1988; CAMPEN and VALE 1988a). As is described in later sections, TGF-β and activin-A do share similar actions although with different potencies on some other cell types, and on others activin-A functions as a growth/differentiation factor.

B. Actions of Inhibin and Activin on the Anterior Pituitary

Highly purified inhibins are potent inhibitors of the basal release of FSH by cultured anterior pituitary cells exhibiting EC_{50} values of between 3 and 60 p*M* depending on source of inhibin and conditions such as duration of assay, serum concentration, and presence of other factors. A several-hour latent period is required before the effects of inhibin can be detected and maximal effects are not reached until 8–24 h after initiation of treatment. After several days exposure to inhibin, the amounts of FSH in both the media as well as in the cells is reduced; this suppression of total FSH content of the culture suggests that inhibin blocks FSH biosynthesis. Consistently, in collaboration with W. Chen's group we have shown that pure inhibin lowers the level of FSH β-mRNA in pituitary cell cultures (CARROLL et al. 1989); these results are in agreement with those of MILLER (1988), who used a partially purified inhibin preparation. By contrast, inhibin has little effect on the basal secretion of LH and only slightly suppresses cellular LH content and does not lower LH-β mRNA levels (MILLER 1988; CARROLL et al. 1989). In cells stimulated by GnRH or by other secretagogues such as phorbol myristate acetate, highly purified inhibin suppresses the secretion of both LH and FSH in our studies (CAMPEN and VALE 1988a) and those of FUKUDA et al. (1987) but not of ROBERTSON et al. (1986b); subsequently, however, this group found that stimulated LH section was indeed suppressed by inhibin (BURGER 1988). Several workers had previously noted inhibition of GnRH-mediated LH secretion in vitro by crude inhibin preparations (MASSICOTTE et al. 1984a, b; SCOTT and BURGER 1981; DEJONG 1979; FRANCHIMONT et al. 1979). Despite these in vitro results, however, modest amounts of inhibin preparations are highly selective in acutely lowering FSH and not LH secretion rates in vivo, even though in the models chosen LH secretion is GnRH dependent. The reasons for the discrepancies between inhibin effects on spontaneous versus stimulated or in vivo versus in vitro LH secretion are not clear at this time.

Highly purified activin-A (inhibin-βAβA) is a potent and selective FSH secretogogue in vitro, stimulating release of FSH by cultured rat pituitary cells with an EC_{50} of $\sim$25 p*M*. The effects of activin can be distinguished from those of GnRH in several ways: GnRH acts immediately to increase FSH release whereas activin exhibits a long latency in vitro. GnRH releases LH while activin by itself has no effect on LH secretion. The action of GnRH is completely blocked by a GnRH antagonist which has no influence on the response to activin, indicating that its effects are not mediated by the GnRH receptor. Pituitary cells treated with GnRH become rapidly desensitized within hours whereas activin continues to stimulate FSH production for days with no attenuation. Prolonged exposure of pituitary cells to GnRH results in depletion of cellular FSH stores, whereas activin, while stimulating secretion to the same extent, increases amounts of stored FSH (VALE et al. 1986). Activin treatment elevates levels of FSH β-mRNA (MILLER 1988; CARROLL et al. 1989), thus supporting the notion that activin stimulates FSH biosynthesis.

Inhibin lowers the spontaneous production of FSH and obscures the effects of lower concentrations of activin while higher concentrations of activin overcome the effects of inhibin, resulting in a net stimulation of FSH secretion. The highest levels of activin completely abolish any effects of inhibin. Although the latter observation suggests that the inhibition could involve competition at a common site perhaps a plasma membrane receptor, we would still have to explain the effects of inhibin in the absence of added activin as well as the converse. As described below, our finding of inhibin α- and β-subunit mRNAs in the pituitary suggests that pituitary cells could make $\beta\beta$- or $\alpha\beta$-dimers constitutively and raises the possibility that effects seen in response to the addition of one dimer could plausibly reflect antagonism of the other. In point of fact, activin and inhibin appear functionally to antagonize one another in several other systems that we have explored. Although there are no effects of activin alone on the secretion of LH, activin does antagonize the inhibitory effects of inhibin on GnRH-stimulated LH secretion (VALE et al. 1988).

Ovarian follicular fluids have net inhibitory effects on FSH secretion. Because of the presence of a relative abundance of $\alpha\beta$-versus $\beta\beta$-dimers which are not separated by simple gel permeation methods, the existence of gonadal FSH-releasing proteins (activins) had been overlooked in the past. Previous discussions of an FSH-releasing factor (FRF) have assumed it to be present in the hypothalamus-median eminence, and partially purified preparations of FRF have been reported to be smaller and more rapidly acting than activin (MIZUNUMA et al. 1983; IGARASHI and MCCANN 1964; SCHALLY et al. 1966).

The physiological significance of activin in the control of FSH secretion is not established; we do not know whether the pituitary is exposed to biologically meaningful levels of activin from either a gonadal or pituitary source.

In our original report of the structure and biological activities of the FSH-releasing inhibin βAβA dimer (activin), we noted its inhibitory effects on spontaneous ACTH and GH secretion by cultured rat anterior pituitary cells. Activin is a very potent inhibitor of both spontaneous as well as GRF-mediated GH secretion (VALE et al. 1988; CORRIGAN et al. 1988) and GH biosynthesis (BILLESTRUP et al. 1988) as monitored either by rates of incorporation of labeled amino acids into GH or by GH mRNA levels. We reported earlier (BILLESTRUP et al. 1986) that GRF stimulates proliferation of somatotrophs in primary culture and we have found that activin prevents this effect of GRF on somatotroph cell division. The effects of cotreatment of cells with both inhibin and activin result in a shift to the right of the activin dose response curve. Whether activin and/or inhibin are involved in GH regulation in vivo has yet to be determined.

C. Development of Antisera Toward Inhibin Subunits

In order to develop antibodies directed toward inhibin we have generated antibodies against conjugated synthetic inhibin subunit fragments. This approach has been used successfully to develop antisera to a variety of larger proteins (LERNER 1982). The absence of information regarding the position of intra- and intermolecular disulfide bonds complicates the design of synthetic peptides for immunization. We have made immunogenic fragments corresponding mainly to regions lacking cysteine residues to avoid generation of antibodies against epitopes that do not exist in the intact hormone. When the region of interest lacked a tyrosine (for iodinating or coupling) at the desired terminus, a tyrosine connected by a spacer glycine was added. We have obtained useful antibodies toward the cysteine-free N-terminal region of porcine inhibin-α using porcine inhibin-α(1-25)-Gly-Tyr conjugated to human α-globulins. CUEVAS et al. (1987) have also raised antisera toward the N-terminal region of the inhibin-α subunit. For the cysteine-rich β-subunit, we have made cyclic peptides that we hoped would mimic a loop in the native protein. The best antibodies toward β-subunits were raised in rabbits immunized with conjugated cyclicAc-inhibin βA(83-113)-NH_2 and cyclicAc-inhibin βB(80-112)-NH_2 (VAUGHAN et al. 1989). With the availability of recombinant activin-A from A. Mason, Genetech, it should be possible to develop improved antisera toward the βA-subunit. The antisera now available directed toward the inhibin α- and β-subunits recognize proteins in bovine, human, ovine, rat, rhesus monkey, and human gonadal fluids and have been used for radioimmunoassays, immunoblots, immunoprecipitatioin, affinity chromatography, and passive immunization. Other groups have obtained inhibin antisera by use of native inhibin or its subunits as immunogens

(MCLACHLAN et al. 1986b; MIYAMOTO et al. 1986). Using newly developed radioimmunoassays for inhibin, studies of inhibin levels in human subjects under various conditions are now underway and have been recently reviewed (DEKRETSER et al. 1988).

D. Gonadal Production of Inhibin

The gonadal origins of inhibin, assumed since its recognition, have been affirmed by a variety of studies (review, CHANNING et al. 1985). Bioassays, immunocytochemical, and in situ hybridization methods have indicated that inhibin and its subunits are produced in granulosa cells of the ovary and Sertoli cells of the testis in vivo and in vitro (A. STEINBERGER 1979; ERICKSON and HSUEH 1978; CUEVAS et al. 1987; BARDIN et al. 1987; BICSAK et al. 1986). Recently, inhibin subunits have been localized in gonadal cells other than granulosa or Sertoli cells. MEUNIER et al. (1989) have reported inhibin α mRNA and immunoreactivity to be in theca interna and interstitial gland cells; LEE et al. (1989) and our group (ROBERTS et al. 1989) find inhibin α and β mRNAs and immunoreactivities in the Leydig cells.

I. Granulosa Cells

Following earlier studies on inhibin bioactivity secreted by granulosa cells (ERICKSON and HSUEH 1978), we have further studied the regulation of inhibin production by rat granulosa cells using a radioimmunoassay designed to detect the N-terminal portion of the porcine inhibin α-subunit (BICSAK et al. 1986). We were able to demonstrate that FSH stimulates secretion of immunoreactive inhibin. Gel permeation chromatography combined with an in vitro anterior pituitary cell inhibin bioassay revealed that the cells secrete bioactive inhibin with an $M_r \sim 32000$, as well as an $M_r \sim 50000$ protein which is immunochemically related to inhibin α but biologically inactive.

Subsequent in vitro labeling with [^{35}S]cysteine followed by immunoprecipitation with anti-inhibin α antibody showed that the $M_r \sim 32000$ species is a dimer of inhibin α- and β-subunits, while the $M_r \sim 50000$ species is probably an unprocessed monomeric precursor of the inhibin α-chain. The observation of free high molecular weight inhibin α-chain is also consistent with the fact that levels of α-subunit mRNA in the ovary are always much higher than those of β-subunit mRNA (DAVIS et al. 1986; MASON et al. 1985; MAYO et al. 1986). We were unable to detect any high molecular weight dimeric inhibin, composed of an unprocessed α- and fully processed β-subunit, as is found in bovine follicular fluid (ROBERTSON et al. 1985). This may be due to species differences.

In addition to FSH, it appears that a number of hormones are able to modulate inhibin production by granulosa cells. After induction of LH and prolactin receptors by treatment of the cells with FSH, we found that LH, but not prolactin, augmented inhibin production (BICSAK et al. 1986). The stimulatory effect of LH/hCG on inhibin production has also been confirmed using a different radioimmunoassay (RIA) (ZHANG et al. 1988). Also of particular interest was our finding that IGF-I and vasoactive intestinal peptide (VIP) both stimulated inhibin production by cultured rat granulosa cells. Both IGF-I and VIP have been shown to be present in the ovary, and may therefore play a paracrine role in the regulation of inhibin production. Our observation that IGF-I increases granulosa cell inhibin production was confirmed by a subsequent study by ZHANG et al. (1987) in which bioactive inhibin was measured in the conditioned medium from granulosa cells incubated with either FSH or IGF-I.

By contrast to gonadotropins, IGF-I, and VIP, treatment of the cells with either gonadotropin-releasing hormone (GnRH) or EGF inhibited FSH-stimulated inhibin production. Both EGF and GnRH have been shown to inhibit granulosa cell aromatase activity and LH receptor induction. Furthermore, recent evidence suggests that secretion of immunoreactive inhibin in hypophysectomized rats is stimulated by estrogen but inhibited by GnRH through a direct ovarian site of action (C. RIVIER and VALE 1989b). Because GnRH-like peptides may exist in the ovary (R. F. ATEN et al. 1986), the putative peptide could therefore exert an intraovarian effect on inhibin production.

II. Sertoli Cells

We have recently used the same inhibin RIA which was employed to study granulosa cell inhibin to investigate the production of this hormone by Sertoli cells (BICSAK et al. 1987). We employed serum-free cultures of immature rat Sertoli cells and found that FSH elicited a dose-dependent stimulation of inhibin production, while hCG and prolactin were ineffective. Inclusion of a phosphodiesterase inhibitor enhanced the FSH-stimulated inhibin production, consistent with a cAMP-mediated pathway for FSH action. The fact that forskolin, cholera toxin, and dibutyryl cAMP could all increase inhibin biosynthesis by Sertoli cells confirmed cAMP as the second messenger for FSH-stimulated inhibin production. Unlike the case of the granulosa cells, inhibin production by Sertoli cells was not enhanced by the addition of androgens or estradiol. In addition, treatment with EGF stimulates basal inhibin secretion and its effects are additive with FSH (MORRIS et al. 1988; BARDIN et al. 1987).

E. Intragonadal Actions of Inhibin and Activin

In addition to their primary roles at the pituitary level to regulate FSH secretion, inhibin and activin have also been shown to act directly at the gonadal level. These proteins may be paracrine or autocrine mediators within both the ovary and the testis.

I. Paracrine Regulation

Being a secretory product of ovarian granulosa and testicular Sertoli cells, inhibin has been shown to modulate LH-induced androgen production by both ovarian and testis cells (HSUEH et al. 1987). Using an ovarian theca-interstitial cell preparation and theca cell cultures, we have shown that the LH-stimulated androstenedione production was further augmented by inhibin in a dose-dependent manner. In contrast, activin treatment suppressed LH action. Because testicular Leydig cells are homologous to ovarian theca interna cells, the paracrine actions of inhibin and activin were also tested in these cells. In interstitial cells prepared from immature hypophysectomized rats and testis cells from neonatal rats, inhibin treatment was shown to augment LH-induced testosterone biosynthesis. Likewise, activin treatment decreased LH action. The modulatory effect of inhibin on androgen production by Leydig cells has also been found by MORRIS et al. (1988).

These data indicate that the inhibin-related gene products synthesized by Sertoli and granulosa cells may form heterodimers or homodimers to serve as intragonadal paracrine signals in the modulation of LH-stimulated androgen biosynthesis. Thus, testis and ovarian inhibin not only suppresses pituitary FSH release, but also enhances LH-regulated Leydig and theca cell production of androgens, which exert negative feedback at the pituitary LH-producing cells. In addition to the two long-loop feedback axes between pituitary (LH- and FSH-secreting) gonadotrophs and gonadal (androgen- and inhibin-producing) cells, inhibin may provide the crosscommunication between the two loops.

II. Autocrine Regulation

In addition to its potential paracrine role, it appears that inhibin and related proteins may also act directly at granulosa cells, the cell of their synthesis. A report by YING et al. (1986a) showed that, while TGF-β enhanced FSH-stimulated aromatase activity, purified inhibin was inhibitory in that respect. In contrast, a subsequent study by HUTCHINSON et al. (1987) suggested that while TGF-β did stimulate FSH-enhanced aromatose activity, as well as progesterone production, inhibin itself had no effect on either parameter. In addition, it seemed that activin was similar to TGF-β in its ability to stimulate aromatases, but differed in that it was inhibitory with respect to progesterone production. The reason for the discrepancy between these two findings is unclear. Our own unpublished studies using purified ovine inhibin and activin suggest that inhibin suppresses FSH-stimulated aromatase activity, while activin and TGF-β (LAPOLT et al. 1989) are both stimulatory in this respect. In some clonal gonadal cells lines (GONZALEZ-MANCHON and VALE 1989), we also find similar actions of TGF-β and activin-A.

Although there are still no data to demonstrate activin production by the granulosa cells, several additional studies provided interesting findings suggesting an autocrine role of activin in the regulation of granulosa cell differentiation. Using iodinated activin (also known as erythroid differentiation factor or EDF) as the radioligand, specific activin receptors were identified in the granulosa cells (SUGINO et al. 1988 b). The concentration of activin receptors was increased by FSH treatment. In addition, specific binding to the activin receptor was not displaced by inhibin or TGF-β (SUGINO et al. 1988 b). In cultured granulosa cells, treatment with activin was shown to increase FSH receptor content (HASEGAWA et al. 1988). These investigators also reported that activin augmented the FSH stimulation of progesterone production and LH receptor induction. In the absence of FSH, activin treatment alone stimulates granulosa cell inhibin secretion (SUGINO et al. 1988 b) and inhibin α-subunit mRNA levels (LAPOLT et al. 1989). It is, therefore, likely that the specific activin receptors in the granulosa cells may mediate the observed effects of activin on granulosa cell differentiation. Because granulosa cells express the inhibin β-subunit genes necessary for activin expression (MEUNIER et al. 1988 a), activin may be synthesized by these cells during selective stages of differentiation. The secreted protein may then act through specific granulosa cell receptors to exert autocrine functions.

F. Role of Inhibin in Regulation of FSH Secretion In Vivo

I. Female Rat

In the female rat, ovarian follicular development shows a rapid increase during the first weeks of life (OJEDA et al. 1980; PEDERSEN 1969; UILENBROEK et al. 1976). Because inhibin is synthesized and released by granulosa cells (BICSAK et al. 1986, 1988; ZHANG et al. 1988; ZHIWEN et al. 1987), it was not surprising to observe that plasma radioimmunoassayable inhibin levels (C. RIVIER and VALE 1987), as well as ovarian inhibin content (SANDER et al. 1985, 1986), exhibited a sharp rise during early sexual development. From day 17 of age, this increase is accompanied by a marked decline in FSH release. Following the establishment of regular 4-day cycles in the adult animal, inhibin secretion continues to change as a function of folliculogenesis (HASEGAWA et al. 1987; C. RIVIER and VALE 1989 b). We have observed in particular that inhibin secretion remains low during diestrus-1 and -2, starts to increase sharply during the late afternoon of proestrus, and stays elevated until midnight. Plasma inhibin then shows an abrupt decline, reaching a nadir between 2 and 4 a.m. on estrus. These cycle-related fluctuations in plasma radioimmunoassayable inhibin agree with prior measurements in the rat using bioassays (DEPAOLO et al. 1979; FUJII et al. 1983).

In addition to demonstrating changes in the circulating concentrations of inhibin during the estrous cycle, we (MEUNIER et al. 1989) as well as others (WOODRUFF et al. 1988) have examined the levels of inhibin mRNAs by in situ hybridization in the ovary; furthermore we have correlated these results with immunohistochemical detection of the proteins. Our studies have shown that after increasing in the follicle as it develops, inhibin-α reaches its highest level in the granulosa cells of mature follicles a few hours after the primary gonadotropin proestrus surge while at the same time the β-subunits decrease dramatically (Meunier et al. 1989). Immediately before ovulation, the α-subunit sharply declines in the preovulatory follicles and remains in the granulosa cells from nonovulatory follicles. Additionally, inhibin α-subunit mRNA and protein is also found in rat luteal tissue (Meunier et al. 1989). Furthermore, because inhibin bioactivity (Tsonis et al. 1988) and immunoreactivity (McLachlan et al. 1987) increase during the luteal phases of the sheep and the human cycle, it is probable that the corpus luteum of several species secretes inhibin.

In considering the mechanisms which regulate the changes in the pattern of secretion of gonadotropins and inhibin during the estrous cycle, one must ask whether inhibin regulates the surge(s) of FSH and, reciprocally, whether FSH modulates the changes in inhibin release measured during proestrus and estrus. Because GnRH antagonists or pentobarbital (which block GnRH secretion; Daane and Parlow 1971) interfere with the primary (proestrus) but not with the secondary estrus FSH surge (Condon et al. 1984; Hasegawa et al. 1981), the secondary surge is presently believed to be independent of endogenous GnRH. By contrast, several investigators have reported that the administration of porcine follicular fluid abolished the estrous FSH rise (Hoak and Schwartz 1980; Hoffmann et al. 1979; Schwartz and Channing 1977; Rush et al. 1981). In order to investigate the dynamic relationship between radioimmunoassayable inhibin and FSH secretion, we (C. Rivier et al. 1986, 1988 b) and others (Robertson et al. 1988) first studied the effect of FSH or FSH-like molecules on inhibin secretion. The injection of pregnant mare serum gonadotropin (PMSG), a molecule with FSH-like activity (Gospodarowicz 1972), into both immature or mature female rats, causes a rapid increase in inhibin release, an observation which agrees with previous reports using bioassays to measure FSH levels (Lee et al. 1981). A similar stimulatory effect of FSH on inhibin release has been observed by other investigators in the rat (Robertson et al. 1988), in women undergoing ovulation induction following gonadotropin treatment (McLachlan et al. 1986 a), and in cultured granulosa cells (Bicsak et al. 1986; Zhang et al. 1988; Zhiwen et al. 1987). Conversely, removal of FSH by hypophysectomy causes a rapid disappearance of inhibin from the plasma (Kaneko et al. 1987). Finally, it should be noted that PMSG increases inhibin mRNA levels in the ovary (Davis et al. 1986, 1988; Meunier et al. 1988 a), while hypophysectomy decreases it (Kaneko et al. 1987).

As mentioned above, a vast body of evidence had suggested an inverse relationship between serum FSH and inhibin (Kimura et al. 1983). This relationship can be demonstrated, in particular, in the PMSG-treated intact rat, in which the increase in the circulating levels of inhibin is followed by a sharp drop in FSH values (C. Rivier et al. 1987 a, c). Similarly, during the estrous cycle, there appears to be a functional correlation between the secretion of FSH and inhibin. In particular, the decrease in the levels of circulating inhibin and in the ratio of inhibin α- to β-subunit expression (perhaps reflecting a decrease in the ratio of inhibin to activin produced), which we observed in the late part of proestrus (MEUNIER et al. 1989), probably contributes to the secondary FSH surge of the early morning of estrus. Reciprocally, the observation that the sequential addition of PMSG, then hCG, to the medium of cultured granulosa cells caused a stimulation followed by a decrease in inhibin secretion (ZHANG et al. 1988) suggested a possible role of the rise in gonadotropins during diestrus-2 and early proestrus, in modulating plasma inhibin in late proestrus. In conclusion, therefore, presently available evidence strongly suggests that there is a finely tuned process by which FSH and LH regulates inhibin secretion during the estrous cycle, and that, reciprocally, inhibin acts as a negative feedback signal to modulate the activity of the pituitary gonadotropes.

While all of the above studies are important for our understanding of the role of gonadotropins in regulating inhibin secretion, we have also considered the possibility that other secretagogues known to affect gonadal function, such as sex steroids and GnRH, might alter inhibin release. It is generally postulated that the concentration of GnRH in peripheral blood is too low to allow a direct effect of the hypothalamic decapeptide on the gonads (SHARPE et al. 1981). However, because ovarian GnRH receptors with specificity and affinity properties indistinguishable from those of pituitary GnRH receptors have been demonstrated (MAGOFFIN and ERICKSON 1982; PIEPER et al. 1981; REEVES et al. 1980), several investigators had suggested that a GnRH-like material was acting in the ovary (BIRNBAUMER et al. 1985). In order to investigate the possibility that sex steroids and/or GnRH might act directly at the level of the ovary to modulate inhibin secretion, we administered estradiol benzoate, PMSG, and GnRH analogs to immature hypophysectomized female rats. We observed that estrogens markedly increased plasma inhibin levels, an effect which was inhibited by a GnRH agonist and augmented by a GnRH antagonist (RIVIER et al. 1989). We therefore concluded that sex steroids and an endogenous GnRH-like peptide (HSUEH and ERICKSON 1979; ATEN et al. 1986) exert a direct effect at the ovarian level, and play a physiological paracrine role in modulating inhibin secretion.

The basic concept which led to McCullagh's hypothesis of the existence of inhibin (MCCULLAGH 1932) holds that inhibin specifically interferes with FSH secretion. As discussed elsewhere in this chapter, the availability of highly purified inhibin has allowed several investigators to demonstrate the inhibitory action of this molecule on FSH secretion (YING et al. 1987; VALE et al. 1988) and pituitary FSH β-mRNA levels (MERCER et al. 1987). The hypothesis that inhibin exerts a tonic inhibitory role on FSH secretion decrees that removal of endogenous inhibin results in an increase in FSH release. Indeed, we (C. RIVIER and VALE 1979, 1987; RIVIER et al. 1986, 1987a, b) and others (CULLER and NEGRO-VILAR 1988) have observed that the i.v. injection of an antiserum directed against the N-terminus portion of the α-chain of inhibin (VAUGHAN et al. 1989) markedly augments FSH, but not LH, secretion in females 20 days old and older. No such effect is observed in younger animals (C. RIVIER and VALE 1987; CULLER and NEGRO-VILAR 1988), suggesting that inhibin does not play a physiological role in regulating FSH release by very young female rats. By contrast, in the adult cycling female rats, immunoneutralization of endogenous inhibin at any stage of the cycle results in increased plasma FSH levels (C. RIVIER et al. 1987a, c). It therefore appears that, while the control of FSH regulation is known to depend upon GnRH and sex steroids (MCCANN 1974), inhibin also plays an important function in modulating pituitary function throughout most of the life span of the female rat.

In the adult female rat, removal of inhibin by the i.v. injection of anti-inhibin serum on diestrus-1 causes a marked rise in plasma FSH levels and increase in the number of tubal ova shed during the subsequent estrus (C. RIVIER and VALE 1989a). Similarly, active immunization against inhibin (BINDON et al. 1986; CUMMINS et al. 1986; HENDERSON et al. 1984) or against the α-subunit of inhibin produced by recombinant DNA techniques (FORAGE et al. 1987) also leads to increased ovulation rates in ewes. In order to assess whether the primary mechanisms through which inhibin immunoneutralization increased the number of maturing follicles was mediated through elevated FSH secretion, we infused ovine FSH to rats during diestrus-1. Because this treatment markedly increased the number of eggs present in the tubes on the next day of estrus (C. RIVIER and VALE 1989a), we concluded that elevated levels of circulating FSH indeed represent an important mediator of the augmented ovulation rate of animals injected with anti-inhibin antibodies.

II. Male Rat

The roles of GnRH, sex steroids, and various neurotransmitters such as opiates and catecholamines in regulating sexual maturation in the male rat are well documented (BHANOT and WILKINSON 1983; ODELL and SWERDLOFF 1976; DUNCAN et al. 1983; PIACSEK and GOODSPEED 1978). By contrast, a potential role of inhibin has long remained controversial (JONES et al. 1985). On the one hand, studies showing that castration caused a more rapid rise in the plasma FSH than LH levels of young rats, and that this increase could not be blocked by exogenous androgens (HERMANS et al. 1980), had suggested a role of inhibin in mediated FSH secretion in the immature male. On the other hand, studies using the injection of steroid-free follicular fluid, which contains inhibin (CHANNING et al. 1985), to mature male rats, have not conclusively shown that this treatment could selectively decrease FSH secretion (SUMMERVILLE and SCHWARTZ 1981). Indeed, the observation that cultured Sertoli cells from young rats secrete measurably larger amounts of inhibin than those of older animals had suggested that the role of inhibin might be confined to the early stages of sexual development (ULTEE-VANGESSEL and DEJONG 1987).

In the male, various techniques using bioassays (A. STEINBERGER and STEINBERGER 1976; LEGAC and DEKRETZER 1982; AU et al. 1986; ULTEE-VANGESSEL et al. 1986), radioimmunoassays (BICSAK et al. 1987; TOEBOSCH et al. 1988), immunohistochemical localization (CUEVAS et al. 1987; C. RIVIER et al. 1988a), and in situ hybridization (BARDIN et al. 1987; C. RIVIER et al.

1988a; TOEBOSCH et al. 1988; MEUNIER et al. 1988b) have indicated that inhibin was mainly produced by the Sertoli cells of the testes. Because the activity of the Sertoli cell shows age-related changes (A. STEINBERGER and STEINBERGER 1976; DEJONG et al. 1979; VERHOEVEN and FRANCHIMONT 1983; ULTEE-VANGESSEL and DEJONG 1987), it was anticipated that inhibin secretion would also vary with the stages of sexual maturation. Indeed, in agreement with previous reports (AU et al. 1986) we observed that, following consistently high levels between day 8 and 18, plasma immunoreactive inhibin then progressively declined (C. RIVIER et al. 1988a). Immunoreactive testicular inhibin also declined with age when expressed on a wet weight basis. Furthermore, immunohistochemical techniques showed that inhibin-α was present in the cytoplasm of the Sertoli cells of very young male rats (8–15 days of age), then decreased as a function of the appearance of spermatocytes in the tubules (C. RIVIER et al. 1988a). In agreement, the expression of testicular inhibin subunit RNAs, as measured by S1 nuclease analysis, also declines during sexual maturation (RIVIER et al. 1988a; MEUNIER et al. 1988b). Although ESCH et al. (1987a) have reported that the inhibin βA mRNA is missing in the male rat, we have shown that the levels of message encoding this subunit are easily detectable in the testes of immature rats (MEUNIER et al. 1988b). With maturation, concentrations of all mRNAs decline but those of βA fall more rapidly than those of α or βB.

As mentioned above, the role of inhibin in modulating FSH in the male secretion has remained controversial (JONES et al. 1985). The immunoneutralization of endogenous inhibin causes a marked increase in the plasma FSH levels of male rats aged 10–24 days, but not in older animals (CULLER and NEGRO-VILAR 1988; C. RIVIER et al. 1988a). This has suggested that inhibin does not represent an essential modulator of FSH secretion in older rats, a hypothesis supported by the maintenance of relatively high plasma FSH levels observed in the adult male rat (ODELL and SWERDLOFF 1975). However, inhibin and activin exert effects on testicular cells and inhibin/activin subunits are localized in the pituitary (V. ROBERTS et al. 1988) and the brain (SAWCHENKO et al. 1988); thus inhibin and activin may play a variety of local roles within the male rat.

G. Tissue Expression of Inhibin Subunits

The anatomic distribution of inhibin subunits has been explored by a variety of immunologic and molecular techniques. We have employed a specific and sensitive S1-nuclease analysis method to measure the levels of inhibin-subunit mRNA levels (MEUNIER et al. 1988b) and have localized the subunit proteins by immunohistochemical methods with an array of antisera (VAUGHAN et al. 1989; SAWCHENKO et al. 1988; VALE et al. 1988). In the ovary, inhibin-α mRNA is much more abundant than either βA- or βB-subunit mRNA by Northern (MASON et al. 1985) or S1 nuclease (MEUNIER et al. 1988b) analysis. The adult testis also expresses much more α- than β-mRNAs.

Inhibin subunits are also expressed in nongonadal tissues including brain, spinal cord, pituitary, adrenal, spleen, placenta, bone marrow, and kidney. At least in the sheep adrenal, inhibin is said to be regulated by ACTH and corticosteroids (CRAWFORD et al. 1987), while pituitary inhibin immunoreactivity appears to be modulated by sex steroids (V. ROBERTS et al. 1988). Indeed, we (V. ROBERTS et al. 1988) have observed that ovariectomy increased the size and number of cells immunoreactive for inhibin-α and -βB, while estrogen prevented these effects. Because the inhibin subunits are localized in the cytoplasm of FSH- and LH-immunoreactive gonadotropes (V. ROBERTS et al. 1988), and because at the present time the predominant source of circulating inhibin-related peptides is believed to be the gonads (C. RIVIER et al. 1987a; ROBERTSON et al. 1988), it is possible that pituitary inhibin may exert local, or autocrine, effects such as the modulation of the intracellular degradation of FSH and LH (FARNWORTH et al. 1988) and the decrease of pituitary GnRH receptors (G. F. WANG et al 1988). Alternatively, inhibin or activin might be secreted and act locally to modulate secretion of FSH, GH or ACTH.

In the pituitary, adrenal, spleen, kidney, bone marrow, spinal cord, and brain, α-subunit mRNAs predominate, although the ratios of α- to β-subunits vary in different brain regions (H. MEUNIER and W. VALE, unpublished results). In placenta and bone marrow, levels of β-subunit mRNAs are in excess. One current hypothesis supported by the recombinant expression experiments of MASON (1988) is that ratios of α- to β-subunits expressed determines the amounts of inhibin relative to activin produced. Accordingly, taken as a whole the gonads would produce predominantly $\alpha\beta$-dimers whereas placenta and bone marrow would make more $\beta\beta$-dimers.

H. Inhibin and Activin in the Placenta

The demonstration of the presence of mRNAs for inhibin subunits (MAYO et al. 1986; MEUNIER et al. 1988b) in the human and rat placenta is consistent with results of bioassays and radioimmunoassays (HOCHBERG et al. 1981; MCLACHLAN et al. 1987). Our immunocytochemical studies with anti-inhibin-α serum revealed numerous immunoreactive cells in the placental villi localized in the central, cytotrophoblast layer (PETRAGLIA et al. 1987).

We have studied the production and actions of immunoreactive (IR) inhibin-related proteins in primary cultures of human trophoblasts. Secretion of IR-inhibin was stimulated by the addition of human chorionic gonadotropin (presumably acting through adenylate cyclase), 8-Br-cAMP and agents such as forskolin and cholera toxin that elevate intracellular cAMP levels (PETRAGLIA et al. 1987). Inhibin and activin modulate the production of various substances made by the placenta. We have found that activin stimulates the secretion of GnRH, hCG, and progesterone by trophoblast cultures whereas inhibin can prevent all of these effects of activin (PETRAGLIA et al. 1989). The interactions observed in the trophoblast cultures reveal an array of potential paracrine/autocrine regulatory possibilities. Inhibin and activin may play important roles in the physiology of the placenta and may be useful for the manipulation of pregnancy.

I. Activin and the Control of Oxytocin Secretion

Encouraged by the finding of inhibin subunit mRNAs in the various brain regions, we are using affinity-purified antisera to map the distributions of inhibin subunits in the colchicine-treated rat brain (SAWCHENKO et al. 1988). Antisera directed toward two different regions of the inhibin βA-subunit (66–79 and 81–113) both stained a cluster of neurons focused in the caudal part of the nucleus of the solitary tract (NTS) near the spinal-medullary transition zone and a smaller group of cells in the ventrolateral medullary reticular formation. Axon and terminals exhibiting inhibin-β immunoreactivity were also visualized in untreated rats and displayed a distribution consistent with the established projections of the NTS and/or ventrolateral medulla including the paraventricular (PVN) and supraoptic (SON) and dorsomedial nuclei of the hypothalamus, the paraventricular nucleus of the thalamus, and the bed nucleus of the stria terminalis. In view of the reproductive roles of inhibin-related proteins, it is noteworthy that the inhibin-β-stained terminals were in the GnRH-rich regions of the septal and preoptic areas, and, more prominently, in the portions of the PVN and SON in which the oxytocin-expressing magnocellular neurosecretory neurons are concentrated (SAWCHENKO et al. 1988).

We are evaluating the hypothesis that activin might play a role in the regulation of oxytocin secretion under circumstances such as lactation and parturition. The bilateral infusion of only 70 fmol purified activin into the PVN of anesthetized rats was sufficient to elevate plasma oxytocin levels within 10 min while having no effect on the secretion of vasopressin in the same animals. Infusion of vehicle or inhibin had no effect. This is the most rapid effect that we have been able to observe for inhibin-related peptides and may indicate that a different cellular mode of action is involved. Also, activation of the neuronal pathways containing inhibin-β staining by electrical stimulation of the NTS in rats led to a stimulation of oxytocin secretion. Finally, bilateral infusion of anti-inhibin-β serum into the PVN attenuated both suckling-induced increases in intramammary pressure and oxytocin secretion in lactating rats (PLOTSKY et al. 1988).

The distribution of inhibin-β-cell bodies and fibers and the effects of stimulation of those pathways and of activin administration on oxytocin secretion is consistent with a role for this or a related protein in the central regulation of oxytocin secretion. The consequences on oxytocin secretion in lactating rats of passive immunization with activin-blocking antiserum suggest that these "activinergic" pathways may be important in the regulation of milk ejection.

J. Roles of Activin and Inhibin in Erythropoiesis

The isolation and characterization of erythroid differentiation factor (EDF) (ETO et al. 1987) and the realization that it was probably identical to activin-A led us (YU et al. 1987, 1988a) to explore the effects of both activin and inhibin on the induction of hemoglobin accumulation in a human erythroleukemic cell line, K562, and on the proliferation of erythroid progenitors in human bone marrow culture. It is thus proposed that these two proteins constitute a novel humoral regulatory control over erythropoiesis (YU et al. 1987).

I. Complexity of Hematopoietic Control

The production sites of hematopoietic factors and the regulatory mechanisms involved have been subjects of much interest. Erythropoietin has been known to be produced mainly in the kidney and to a lesser extent in the liver. It and other factors like M-CSF and G-CSF are detectable in normal human plasma, while IL-3 and GM-CSF are not. Therefore, IL-3 and GM-CSF appear to be tissue factors, rather than hormones in the plasma. Moreover, IL-3 and GM-CSF are produced by activated T-lymphocytes (CLINE and GOLDE 1974; NATHAN et al. 1988; MANGAN et al. 1982). GM-CSF, M-CSF, and G-CSF are produced when fibroblasts, endothelial cells, and smooth muscle cells (ZUCKERMAN et al. 1985; TSAI et al. 1986a, b) are activated by inflammatory mediators, such as IL-1 and tumor necrosis factor, which are released by monocytes after activation with endotoxin (BAGBY et al. 1986; BEVILACQUA et al. 1986; SIEFF et al. 1987; BROUDY et al. 1986). Because most of these factors are produced either by activated T cells or through stimulation of inflammatory mediators, their actual roles in providing base-line hematopoiesis are uncertain. It is likely that they are responsible for regulation of hematopoiesis in response to inflammatory stimulation.

The hematopoietic factors promote both proliferation and differentiation of hematopoietic progenitor cells. While there is difficulty in uncoupling these two processes in normal hematopoiesis, recent studies have used various myeloid leukemic cell lines to analyze the relative effects of the different hematopoietic factors separately on cell growth and cell differentiation (J. P. MOREAU et al. 1988). A differentiation factor, leukemia inhibitory factor (LIF, see below), which induces macrophage differentiation of a leukemia cell line, can also serve as a growth factor for another myeloid cell line (J. P. MOREAU et al. 1988). The distinction between growth-promoting and differentiation-inducing activities may be largely determined by the target cell types. Not only do hematopoietic factors have colony-stimulating activities toward progenitor cells, but also their activities may be directed toward a broad range of target cells, which include the more mature cells within one particular lineage. For example, GM-CSF and G-CSF promote the proliferation of immature progenitor cells; they also increase the responsiveness of mature neutrophils to bacterial peptides, as well as increase the respiratory burst of these cells (CLARK and KAMEN 1987). Thus, these hematopoietic factors have several biological activities: promote the proliferation and differentiation of progenitors and enhance the functions of more terminally differentiated cells.

Although there are specific factors which support the growth of hematopoietic cells, a large family of other inducible, secretory cytokines have now been shown to have important effects on the growth of other hematopoietic cell lineages. They include IL-1 (MOORE and WARREN 1987), IL-2 (K. A. SMITH 1980), IL-4 (YOKOTA et al. 1986), IL-5, and IL-6 (HIRANO et al. 1986), and they were originally identified as regulators of immune systems. In addition, it has also been reported that platelet-derived growth factor (DELWICHE et al. 1983) and insulin-like growth factors (BERSCH et al. 1982; KURTZ et al. 1985; DAINIAK and KRECZKO 1985) are both mitogenic for hematopoietic cells. Recently, nerve growth factor has been shown to promote the growth of human hematopoietic colonies (MATSUDA et al. 1988), while transforming growth factor-β has been found to suppress the colony formation of many early progenitors of various hematopoietic lineages (SING et al. 1988).

Conversely, recent experiments indicate that some of the classically defined hematopoietic factors also interact with normal or malignant cells of nonhematopoietic origin. The newly isolated hematopoietic regulator, LIF, is found to be a multifunctional factor with distinct activities both in hematopoiesis and in early embryonic development (WILLIAMS et al. 1988), and placenta is capable of interacting with hematopoietic factors, such as M-CSF and erythropoietin (RETTENMIER et al. 1986; MULLER et al. 1983; SAWYER et al. 1987; KOURY et al. 1988), suggesting that the classical hematopoietic factors also play a role in normal placental functions and fetal development. These studies thus lead to some questions about the specificity (or promiscuity) of a few hematopoietic factors in regard to their target cell activity and their regulatory controls and prepare us for the notion that the inhibin family of growth factors might also serve as regulators of hematopoiesis.

II. Induction of Erythroid Differentiation

EDF which is identical to activin A, was isolated by ETO et al. 1987 from a phorbol ester stimulated human monocytic leukemic cell line (THP-1) based upon the protein's ability to differentiate murine Friend cells. In response to EDF/activin, Friend cells produce hemoglobin and stop proliferating.

We (YU et al. 1987) used the human erythroleukemic cell line K562 as a model system of erythroid differentiation (HUTSON et al. 1984; LOZZIO and LOZZIO 1975; SOKOLOSKI et al. 1986; FUKUDA 1980). In addition of activin was shown to cause increasing numbers of K562 cells to produce hemoglobin whether measured by benzamidine staining or by immunohistochemical detection. Thus, activin induces the differentiation of these erythroleukemic cells. The effective dose for maximum and half-maximum induction by activin in 3-day incubations are, respectively, 10 ng/ml ($\sim$30 pM). In keeping with this high potency, K562 cells express high-affinity ($K_d \sim 130$ pM), specific, saturable activin-binding sites (CAMPEN and VALE 1987, 1988 b). SHIBATA et al. (1987) have shown that activin-A causes a rapid and transient increase in cytoplasmic-free calcium concentration even in the absence of extracellular calcium. Also, activin increases incorporation of labeled precursors into inositol trisphosphate, bisphosphate and monophosphate, and diacylglycerol.

Thus the effects of activin may be in part a consequence of calcium mobilization and C-kinase activation. Prolonged exposure of K562 cells to activin causes them to become terminally differentiated with limited proliferation (YU et al. 1987). The accumulation of hemoglobin in the activin-A-treated K562 cells represents an approximately 23-fold increase in hemoglobin content. We estimate that activin is about twice as effective on a single cell basis as are chemical inducers (such as hemin) in inducing the production of hemoglobin in the responsive K562 cells. The induction of hemoglobin accumulation in K562 cells is antagonized by coincubation with inhibin. Thus the two polypeptides have functionally opposite effects on erythrogenesis in K562 cells.

III. Potentiation of Erythroid Colony Formation

We then studied the effect of activin on human bone marrow cultures (YU et al. 1987, 1988a, b, 1989). Differentiation of hematopoietic cells is shown in Fig. 3. The multipotent stem cell is first committed to become CFU-GEMM, which is the common origin for megakaryocytes, erythroid, CFU-GM, and eosinophils. Along the erythroid lineage, BFU-E (burst-forming unit-erythroid) is the first committed erythroid progenitor, followed by CFU-E (cluster-forming unit-erythroid), proerythroblast (or cluster-forming unit), erythroblast, and then mature red cell. The addition of activin significantly potentiates the erythropoietin-mediated formation of both CFU-E and BFU-E in serum-free cultures in a dose-dependent manner (YU et al. 1987, 1988b, 1989). The persistence of the activin effect under serum-free conditions suggests that the effect of activin does not require the presence of factors in serum. As in the K562 cells, inhibin antagonizes activin on the human bone

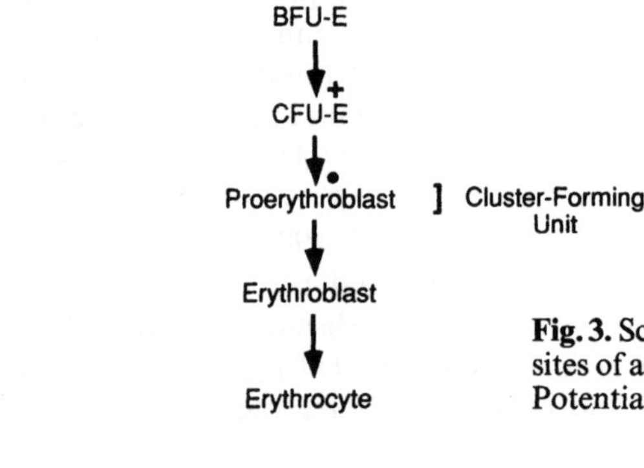

Fig. 3. Schematic of multiple sites of action of activin on erythrogenesis. Potentiation of erythropoietin (+); no effect (●)

marrow cultures by counteracting the erythropoietin-potentiating effects of activin (Yu et al. 1987). At present, it is not clear whether activin and inhibin have receptors with overlapping specificities or if they have separate receptors.

Activin increases colony formation of both CFU-E and BFU-E but, in contrast, the number of CFU-GM colonies is not affected by the addition of activin. When marrow cells are cultured for CFU-GEMM, the addition of activin does not significantly potentiate the formation of the mixed colonies and CFU-GM, but it enhances the formation of colonies with predominantly erythroid phenotypes in the culture (Yu et al. 1989). Furthermore, activin has no apparent effect on the maturation of murine megakaryocytes (identified by acetylcholinesterase measured by fluorescence, Ishibashi et al. 1987) in murine bone marrow cultures. As shown in Fig. 3, activin does not have any effect on the colony formation of CFU-GM or CFU-Meg, but it potentiates CFU-E, BFU-E, and the formation of colonies with predominant erythroid phenotypes in the cultures for CFU-GEMM. These results suggest that the effect of activin is lineage specific.

Along the erythroid lineage, activin affects BFU-E and CFU-E to almost the same degree; it does not have an effect on the proliferation of the more mature cluster-forming units (J. Yu, unpublished observations). These cluster-forming cells represent a population of erythroid cells which are more mature and about six times more sensitive than CFU-E toward erythropoietin (Monette et al. 1981). They may include pre-erythroblasts and other early erythroblasts.

Recently, many of these observations have been confirmed with the use of recombinant human activin and inhibin (Broxmeyer et al. 1988). Recombinant activin selectively enhances colony formation of erythroid progenitors in the presence of erythropoietin and recombinant inhibin decreases specifically the enhancement of erythroid colony formation by activin, but not by GM-CSF, IL-3, or IL-4. In addition, at high concentrations, activin also potentiates colony formation from CFU-GEMM. Several lines of experimental evidence demonstrate that activin can increase the proportion of the erythroid progenitors engaged in active DNA synthesis (Yu et al. 1988b, 1989). The effect of activin on DNA synthesis has a rapid onset, since it is observed after approximately 1 h incubation of marrow cells with activin (Yu et al. 1989). It has been previously shown that incubation with erythropoietin takes about 5–6 h before an increase in DNA synthesis is observed (Dessypris and Krantz 1984). Therefore, the rapidity of the activin effect on DNA synthesis cannot be easily explained on the basis of the recruitment of G_0 resting cells into cycling stage. Furthermore, it should be noted that the potentiation effect of activin on DNA synthesis is a reversible process. Nevertheless, the "rapidity" and "reversibility" of the activin-enhancing effect on DNA synthesis are reminiscent of the observations by Axelrad et al. (1987), who reported that the effects of their "negative regulatory protein" on DNA synthesis of mouse BFU-E are both rapid and reversible. Our observations are also consistent with the proposal that the proliferative state of the erythroid progenitors is not a static condition; it may be continually maintained by interactions on the surface of these progenitors with both negative and positive regulators.

The potentiation effect of activin on the colony formation of erythroid progenitors requires the cooperation of both T lymphocytes and monocytes (BROXMEYER et al. 1988; YU et al. 1989). Removal of either of these cell populations substantially diminishes the potentiation effect of activin for unfractionated marrow cells on both CFU-E and BFU-E (YU et al. 1989). Monocytes and/or T lymphocytes are known to produce various stimulatory factors after appropriate stimulation (CLARK and KAMEN 1987). However, of the currently known factors produced by these cells, none of them have specificity restricted only to erythroid lineage, as does activin. IL-3, which is produced from stimulated T lymphocytes has multilineage specificity (CLARK and KAMEN 1987). Insulin-like growth factors are known to be mitogenic for erythroid progenitors and, more recently, for CFU-GM as well (BERSCH et al. 1982; DAINIAK and KRECZKO 1985). Another erythroid potentiation activity, which was isolated by WESTBROOK et al. (1984) and apparently specific for erythroid lineage, is known to have no effect on the differentiation of K562. It is now known to be identical to the tissue inhibitor of metalloproteinase (GASSON et al. 1985; DOCHERTY et al. 1985). Therefore, in order to account for the dependency on accessory cells, it is proposed that a certain factor or factors are released from these accessory cells after stimulation with activin.

IV. Expression of Activin/Inhibin Subunits in Hematopoietic Cells

EDF was isolated from a human monocytic leukemic cell line, THP-1, and characterized to be identical to the FSH-releasing protein, activin-A (ETO et al. 1987). By S1 nuclease analysis, we (MEUNIER et al. 1988b) showed the presence of high levels of inhibin βA mRNA but were unable to detect significant α-subunit message. Subsequently, MURATA et al. (1988) confirmed the presence of inhibin βA mRNA in bone marrow. The presence of a substantial amount of βA-subunit mRNA in the bone marrow and the isolation of the βAβA homodimer (activin) from a human leukemic cell line suggests that activin could be produced in the normal bone marrow to act as a paracrine modulator. However, inhibin ($\alpha\beta$A), which requires an α-subunit, may not be formed locally, but could reach the bone marrow from other sources (YU et al. 1987).

There is now evidence that marrow stromal cells can produce hematopoietic factors and the extracellular matrix surrounding marrow stromal cells avidly binds these factors (TSAI et al. 1986a, b; R. ROBERTS et al. 1988; M.Y. GORDON et al. 1987). It is suggested that hematopoietic factors are presented to hematopoietic progenitor cells in this manner, or "direct" cell-cell transfer of hematopoietic factors may occur between stromal and progenitor cells (EAVES and EAVES 1988). Therefore, the local secretion of hematopoietic factors such as activin in the "microenvironment" of bone marrow may contribute to the regulation of hematopoiesis.

In summary, these studies lead to the speculation that these two TGF-β-like proteins inhibin and activin modulate the action of erythropoietin and that a new factor, which has yet to be isolated, is released from accessory cells after stimulation with activin and apparently is specific for erythroid lineages. Lastly, we can speculate that in some cases of anemia, where the amount of erythropoietin present in the patients is adequate, there may be a defect in the production of activin or inhibin. It is likely that these findings will have bearing on certain pathological conditions and may have some therapeutic implications in the future.

K. Conclusions

Inhibin subunits and their mRNAs are broadly distributed anatomically and the dimers have powerful activities in diverse biological systems where inhibin and activin often exhibit opposite effects (Fig. 4). The physiological roles of inhibin to regulate FSH secretion in the female rat and immature male rat are strongly supported by immunological and physiological studies involving inhibin as a feedback signal between pituitary and gonad. The significance of these dimers within the gonad, brain, pituitary, placenta, and bone marrow as endocrine, paracrine, and autocrine mediators of growth and function are compelling but have yet to be placed in physiological context. Although the panoply of functions of inhibin and activin are incompletely appreciated at this time, this family has already demonstrated a powerful mechanism for the

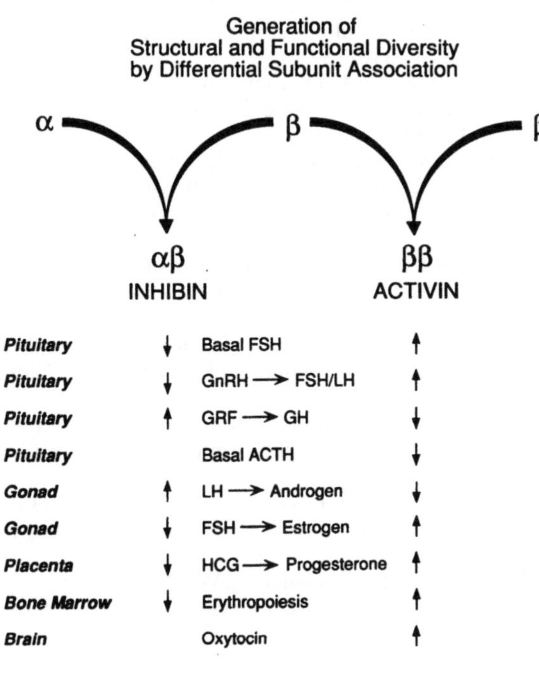

Fig. 4. Summary of biological activities of inhibin and activin on multiple tissues

generation of signal diversity whereby differential subunit association can result in the generation of dimers with opposing biological actions in multiple tissues.

Acknowledgements. The authors gratefully recognize the assistance Ms. Sandra Guerra, Ms. Bethany Coyne, Mr. David Dalton, and Ms. Joan Vaughan in the preparation of this manuscript and collaborators in our laboratories. Research was supported in part by grants DK37089, DK40218, HD13527, DK26741, and contract HD32826. Research was conducted in part by The Clayton Foundation for Research, California Division. W. Vale and C. Rivier are Clayton Foundation Investigators.

References

Akiyama K, Hoshioka Y, Schmid K, Offner GD, Troxler RF, Tsuda R, Hara M (1985) The amino acid sequence of human β-microseminoprotein. Biochim Biophys Acta 829:288–294

Aten RF, Williams AT, Behrman HR, Wolin DL (1986) Ovarian gonadotropin-releasing proteins: demonstration and characterization. Endocrinology 118:961

Au CL, Robertson DM, de Kretser DM (1986) Measurement of inhibin and an index of inhibin production by rat testes during postnatal development. Biol Reprod 35:37–43

Axelrad AA, Croizat H, Eskinazi D (1981) A washable macromolecule from Fv2rr marrow negatively regulates DNA synthesis in erythropoietic progenitor cells BFU-E. Cell 26:233–244

Axelrad AA, Croizat H, del Rizzo D, Eskinazi D, Pezzutti G, Stewart S, Van der Gaag H (1987) Properties of a protein NRP that negatively regulates DNA synthesis of the early erythropoietic progenitor cells BFU-E. In: Najman A, Guigon M, Govin N, Govin MJY (eds) The inhibitors of hematopoiesis, vol 162. Libbey, London, pp 79–92

Bagby GC Jr, Dinarello CA, Wallace P, Wagner C, Hefeneider S, McCall E (1986) Interleukin 1 stimulates granulocyte macrophage colony-stimulating activity release by vascular endothelial cells. J Clin Invest 78:1316–1323

Baker HW, Bremner WJ, Burser HG, de Kretser DM, Dulmanis A, Eddie LW, Hudson B, Keosh EJ, Lee VW, Rennie GC (1976) Testicular control of follicle-stimulating hormone secretion. Recent Prog Horm Res 32:429–476

Bardin CW, Morris PL, Chen C-L, Shaha C, Voglmayr J, Rivier J, Spiess J, Vale WW (1987) Testicular inhibin: structure and regulation by FSH, androgens and EGF. In: Burger HG et al. (eds) Proceedings of the Serono Conference on Inhibin, vol 42. Raven, New York, pp 179–190

Bersch E, Groopman JE, Golde DW (1982) Natural and biosynthetic insulin stimulates the growth of human erythroid progenitors in vitro. J Clin Endocrinol Metab 55:1209–1211

Bevilacqua MP, Schleef RR, Gimbrone MA Jr, Loskutoff DJ (1986) Regulation of the fibrinolytic system of cultured human vascular endothelium by interleukin 1. J Clin Invest 78:587–591

Bhanot R, Wilkinson M (1983) Opiatergic control of gonadotropin secretion during puberty in the rat: a neurochemical basis for the hypothalmmic "gonadostat"? Endocrinology 113:596–603

Bicsak TA, Tucker EM, Cappel S, Vaughan V, Rivier J, Vale W, Hsueh AJW (1986) Hormonal regulation of granulosa cell inhibin biosynthesis. Endocrinology 119:2711–2719

Bicsak TA, Vale W, Vaughan J, Tucker EM, Cappel S, Hsueh AJW (1987) Hormonal regulation of inhibin production by cultured Sertoli cells. Mol Cell Endocrinol 49:211–217

Bicsak TA, Cajander SB, Vale W, Hsueh AJW (1988) Inhibin: studies of stored and secreted forms by biosynthetic labeling and immunodetection in cultured rat granulosa cells. Endocrinology 122:741–748

Billestrup N, Swanson LW, Vale W (1986) Growth hormone releasing factor stimulates proliferation of somatotrophs in vitro. Proc Natl Acad Sci USA 83:6854–6857

Billestrup N, Potter E, Vale W (1988) Inhibition of growth hormone expression by FSH releasing protein (FRP) in primary pituitary cells. July 17–23, 1988, International congress of endocrinology, Kyoto (abstr)

Bindon BM, Piper LR, Cahill LP, Driancourt MA, O'Shea T (1986) Genetic and hormonal factors affecting superovulation. Theriogenology 5:53–70

Birnbaumer L, Shahabi N, Rivier J, Vale W (1985) Evidence for a physiological role of gonadotropin-releasing hormone (GnRH) or GnRH-like material in the ovary. Endocrinology 116:1367–1370

Brazeau P, Vale W, Burgus R, Ling N, Butcher M, Rivier J, Guillemin R (1973) Hypothalamic polypeptide that inhibits the secretion of immunoreactive pituitary growth hormone. Science 179:77–79

Broudy VC, Kaushansky K, Segal GM, Harlan JM, Adamson JW (1986) Tumor necrosis factor type alpha stimulates human endothelial cells to produce granulocyte/macrophage colony-stimulating factor. Proc Natl Acad Sci USA 83:7467–7471

Broxmeyer JE, Lu L, Cooper S, Schwall RH, Mason AJ, Nikolics K (1988) Selective and indirect modulation of human multipotential and erythroid hematopoietic progenitor cell proliferation by recombinant human activin and inhibin. Proc Natl Acad Sci USA 85:9052–9056

Burger HG (1988) Inhibin-regulation and mechanism of action. In: Hodgen GD, Rosenwaks Z, Spieler JM (eds) Nonsteroidal gonadal factors: physiological roles and possibilities in contraceptive development. Jones, Institute Press, Norfolk, VA, pp 137–148

Burger HG, Igarashi M (1988) Inhibin: definition and nomenclature, including related substances (letter to the editor). J Clin Endocrinol Metab 66:885–886

Campen CA, Vale W (1987) Binding of radioiodinated FSH-releasing protein (FRP) to pituitary membranes and leukemia cells. 17th Annual Meeting of the Society for Neuroscience, Nov 16–21, 1987, New Orleans, LA (pt 3), vol 13, p 1530 (abstr no 425.14)

Campen CA, Vale W (1988 a) Interaction between ovine inhibin and steroids on the release of gonadstropin: from cultured rat pituitary cells. Endocrinology 123:1320–1328

Campen CA, Vale W (1988 b) Characterizatioin of activin A binding sites on the human leukemia cell line K562. Biochem Biophys Res Commun 157:844–849

Carroll RS, Gharib SD, Corrigan AZ, Chin WW (1989) Effects of gonadal peptides on follicle-stimulating hormone (FSH) beta-subunit mRNA in cultured pituitary cells. In: 71st Annual Meeting of the Endocrine Society, June 21–24, 1989; Seattle (abstr subm)

Cate RL, Mattaliano RJ, Hession C, Tizard R, Farber NM, Cheung A, Ninfa EG, Frey AZ, Gash DJ, Chow EP (1986) Isolation of the bovine and human genes for Mullerian inhibiting substance and expression of the human gene in animal cells. Cell 45:685–698

Channing C, Gordon W, Lin W-K, Ward D (1985) Physiology and biochemistry of ovarian inhibin. Proc Soc Exp Biol Med 178:339–361

Clark SC, Kamen R (1987) The human hematopoietic colony-stimulating factors. Science 236:1229–1237

Cline MJ, Golde DW (1974) Production of colony-stimulating activity by human lymphocytes. Nature 248:703–704

Condon TP, Heber D, Stewart JM, Sawyer CH, Whitmoyer DI (1984) Differential gonadotropin secretion: blockade of periovulatory LH but not FSH secretion by a potent LHRH antagonist. Neuroendocrinology 38:357–361

Corrigan AZ, Billestrup N, Bilezikjian LM (1988) FSH releasing protein inhibits growth hormone and adrenocorticotropic hormone production by rat anterior pituitary cells. In: 70th Annual Meeting of the Endocrine Society, June 8–11, 1988, New Orleans, vol 571 (abstr no 163)
Crawford RJ, Hammond VE, Evans BA, Coghlan JP, Haralambidis J, Hudson B, Penschow JD, Richards RI, Tregear GW (1987) α-Inhibin gene expression occurs in the ovine adrenal cortex, and is regulated by adrenocorticotropin. Mol Endocrinol 1:699–706
Cuevas P, Ying SY, Ling N, Ueno N, Esch F, Guillemin R, Healy D, Ta S (1987) Immunohistochemical detection of inhibin in the gonad [published erratum appears in Biochem Biophys Res Commun (1987) 146(2):926]. Biochem Biophys Res Commun 142:23–30
Culler MD, Negro-Vilar A (1988) Passive immunoneutralization of endogenous inhibin: sex-related differences in the role of inhibin during development. Mol Cell Endocrinol 58:263–273
Cummins LJ, O'Shea T, Al-Obaidi SAR, Bindon BM, Findlay JK (1986) Increase in ovulation rate after immunization of Merino ewes with a fraction of bovine follicular fluid containing inhibin activity. J Reprod Fertil 77:365–372
Daane TA, Parlow AF (1971) Periovulatory patterns of rat serum follicle stimulating hormone and luteinizing hormone during the normal estrous cycle: effects of pentobarbital. Endocrinology 88:653–663
Dainiak N, Kreczko S (1985) Interactions of insulin, insulinlike growth factor II, and platelet-derived growth factor in erythropoietic culture. J Clin Invest 76:1237–1242
Davis SR, Dench F, Nikolaidis I, Clements JA, Forage RG, Krozowski Z, Burger HG (1986) Inhibin α-subunit gene expression in the ovaries of immature female rats is stimulated by pregnant mare serum gonadotrophin. Biochem Biophys Res Commun 138:1191–1195
Davis SR, Burger HG, Robertson DM, Farnworth PG, Carson RS, Krozowski Z (1988) Pregnant mare's serum gonadotropin stimulates inhibin subunit gene expression in the immature rat ovary: dose response characteristics and relationships to serum gonadotropins, inhibin, and ovarian steroid content. Endocrinology 123:2399–2407
De Jong FH (1979) Inhibin – fact or artifact. Mol Cell Endocrinol 13:1–10
De Jong FH, Sharpe RM (1976) Evidence for inhibin-like activity in bovine follicular fluid. Nature 263:71–71
De Jong FH, Welschen R, Hermans WP, Smith SD, van der Molen HJ (1978) Effects of testicular and ovarian inhibin-like activity, using in vitro and in vivo systems. Geilo, Norway: 125–138
De Jong FH, Welschen R, Hermans WP, Smith SD, van der Molen HJ (1979) Effects of factors from ovarian follicular fluid and Sertoli cell culture medium on in vivo and in vitro release of pituitary gonadotrophins in the rat: an evaluation of systems for the assay of inhibin. J Reprod Fertil 26:47
De Kretser DM, Robertson DM, Risbridger GP, Hedger MP, McLachlan RI, Burger HG, Findlay JK (1988) Inhibin and related peptides. Prog Endocrinol 13–23
Delwiche F, Raines E, Powell JS, Adamson JW (1983) Platelet derived growth factor (PDGF) enhances in vitro erythroid colony growth via stimulation of mesenchymal cells. Blood 62:121 a (abstr)
DePaolo LV, Shander D, Wise PM, Barraclough CA, Channing CP (1979) Identification of inhibin-like activity in ovarian venous plasma of rats during the estrous cycle. Endocrinology 105:647–654
Derynck R, Jarret JA, Chen EY, Eaton DH, Bell JR, Assoian RK, Roberts AB, Sporn MB, Goeddel DV (1985) Human transforming growth factor-beta complementary DNA sequence and expression in normal and transformed cells. Nature 316:701–705
Dessypris EN, Krantz SB (1984) Effect of pure erythropoietin on DNA-synthesis by human marrow day 15 erythroid burst forming units in short-term liquid culture. Br J Haematol 56:295–306

Dijke PT, Hansen P, Iwata KK, Pieler C, Foulkes JG (1988) Identification of another member of the transforming growth factor type beta gene family. Proc Natl Acad Sci USA 85:4715–4719

Docherty AJP, Lyons A, Smith BJ, Wright EM, Stephens PE, Harris TJR (1985) Sequence of human tissue inhibitor of metalloproteinases and its identity to erythroid-potentiating activity. Nature 318:66–69

Duncan JA, Dalkin AC, Barkan A, Regiani S, Marshall JC (1983) Gonadal regulation of pituitary gonadotropin-releasing hormone receptors during sexual maturation in the rat. Endocrinology 113:2238–2246

Eaves AC, Eaves CJ (1988) Maintenance and proliferation control of primitive hemopoietic progenitors in long-term cultures of human marrow cells. Blood Cells 14:355–368

Erickson GF, Hsueh AJW (1978) Secretion of "inhibin" by rat granulosa cells in vitro. Endocrinology 102:1275–1282

Esch FS, Shimasaki S, Cooksey K, Mercado M, Mason AJ, Ying S, Ueno N, Ling N (1987a) Complementary deoxyribonucleic acid (cDNA) cloning and DNA sequence analysis of rat ovarian inhibins. Mol Endocrinol 5:388–396

Esch FS, Shimasakki S, Mercado M, Cooksey K, Ling N, Ying S, Ueno N, Guillemin R (1987b) Structural characterization of follistatin: a novel follicle-stimulating hormone release-inhibiting polypeptide from the gonad. Mol Endocrinol 1:849–855

Eto Y, Tsuji T, Takezawa M, Takano S, Yokogawa Y, Shibai H (1987) Purification and characterization of erythroid differentiation factor (EDF) isolated from human leukemia cell line THP-1. Biochem Biophys Res Commun 142:1095–1103

Farnworth PG, Robertson DM, deKretser DM, Burger HG (1988) Effects of 31 kilodalton bovine inhibin on follicle-stimulating hormone in rat pituitary cells in vitro: actions under basal conditions. Endocrinology 122:207–213

Fevold HL, Hisaw FL, Leonard SL (1931) The gonad-stimulating and luteinizing hormones of the anterior lobe of the hypophysis. Am J Physiol 97:291–301

Forage RG, Ring JM, Brown RW, McInerney BV, Cobon GS, Gregson RP, Robertson DM, Morgan FJ, Hearn MT, Findlay JK (1986) Cloning and sequence analysis of cDNA species coding for the two subunits of inhibin from bovine follicular fluid. Proc Natl Acad Sci USA 83:3091–3095

Forage RG, Brown RW, Oliver KJ, Atrache BT, Devine PL, Hudson GC, Goss NH, Bertram KC, Tolstoshev P, Robertson DM, DeKretser DM, Doughton B, Burger HG, Findlay JK (1987) Immunizatioin against an inhibin subunit produced by recombinant DNA techniques results in increased ovulation rate in sheep. J Endocrinol 114:R1–R4

Franchimont P, Verstraelen-Proyard J, Hazee-Hagelstein MT, Renard C, Demoulin A, Bourguignon JP, Hustin J (1979) Inhibin: from concept to reality. Vitam Horm 37:243–302

Fujii T, Hoover DJ, Channing CP (1983) Changes in inhibin activity, and progesterone, oestrogen and androstenedione concentrations, in rat follicular fluid throughout the oestrous cycle. J Reprod Fertil 69:307–314

Fukuda M (1980) K562 human leukaemic cells express fetal type (i) antigen on different glycoproteins from circulating erythrocytes. Nature 285:405–407

Fukuda M, Miyamoto K, Hasegawa Y, Ibuki Y, Igarashi M (1987) Action mechanism of inhibin in vitro-cycloheximide mimics inhibin action on pituitary cells. Mol Cell Endocrinol 51:41–50

Gasson JC, Golde DW, Kaufman SE, Westbrook CA, Hewick RM, Kaufman RJ, Wong GG, Temple PA, Leary AC, Brown EL, Orr EC, Clark SC (1985) Molecular characterization and expression of the gene encoding human erythroid-potentiating activity. Nature 315:768–771

Gearing DP, Gough NM, King JA, Hiltonk DJ, Nicola AN, Simpson RJ, Nice EC, Kelson A, Metcalf D (1987) Molecular cloning and expression of cDNA encoding a murine myeloid leukaemia inhibitory factor (LIF) EMBO J 6:3995–4002

Gonzalez-Manchon C, Vale W (1989) Activin A, inhibin and TGF-beta modulate proliferation of two gonadal cell lines. Endocrinology (in press)

Gordon MY, Riley GP, Watt SM, Greaves MF (1987) Compartmentalization of a haematopoietic growth factor (GM-CSF) by glycosaminoglycans in the bone marrow microenvironment. Nature 326:403–405

Gordon WL, Liu W, Akiyama K, Tsuda R, Hara M, Schmid K, Ward DN (1987) Beta-microseminoprotein (β-MSP) is not an inhibin. Biol Reprod 36:829–835

Gospodarowicz D (1972) Purification and physicochemical properties of the pregnant mare serum gonadotropin (PMSG). Endocrinology 91:101–106

Guillemin R, Brazeau P, Bohlen P, Esch F, Ling N (1982) Growth hormone-releasing factor from a human pancreatilc tumor that caused acromegaly. Science 218:585–587

Hasegawa Y, Miyamoto K, Yazaki C, Igarashi M (1981) Regulation of the second surge of follicle-stimulating hormone; effects of antiluteinizing hormone-releasing hormone serum and pentobarbital. Endocrinology 109:130–135

Hasegawa Y, Miyamoto H, Igarashi M, Yamaka T, Sasaki I, Wamura S (1987) Changes in serum concentrations of inhibin during the estrous cycle of the rat, pig and cow. In: Ares Serono symposia, Tokyo. Raven, New York (in press)

Hasegawa Y, Miyamoto K, Abe Y, Nakamura T, Sugino H, Eto Y, Shibai H, Igarashi M (1988) Induction of follicle stimulating hormone receptor by erythroid differentiation factor on rat granulosa cell. Biochem Biophy Res Commun 156:668–674

Henderson KM, Franchimont P, Lecomte-Yerna MJ, Hudson N, Ball K (1984) Increase in ovulation rate after active immunization of sheep with inhibin partially purified from bovine follicular fluid. J Endocrinol 102:305–309

Hermans WP, vanLeeuwen CEM, Debets MHM, deJong FH (1980) Involvement of inhibin in the regulation of follicle-stimulating hormone concentrations in prepubertal and adult male and female rats. J Endocrinol 86:79–92

Hirano T, Yasukawa K, Harada H, Taga T, Watanabe Y, Matsuda T, Kashiwamura S, Nakajima K, Koyama K, Iwamatsu A, Tsunasawa S, Sakiyama F, Matsui H, Takahara Y, Taniguchi T, Kishimoto T (1986) Complementary DNA for a novel human interleukin (BSF-2) that induces B lymphocytes to produce immunoglobin. Nature 324:73–76

Hoak DC, Schwartz NB (1980) Blockade of recruitment of ovarian follicles by suppression of the secondary surge of follicle-stimulating hormone with porcine follicular fluid. Proc Natl Acad Sci USA 77:4953–4956

Hochberg ZJ, Weiss J, Richman RA (1981) Inhibin-like activity in extracts of rabbit placentae. Placenta 2:259–264

Hoffmann JC, Lorenzen JR, Weil T, Schwartz NB (1979) Selective suppression of the primary surge of follicle-stimulating hormone in the rat: further evidence for folliculostatin in porcine follicular fluid. Endocrinology 105:200–203

Howard RP, Sniffen RC, Simmons FA, Albright F (1950) Testicular deficiency: a clinical and pathologic study. J Clin Endocrinol 10:121–186

Hsueh AJW, Erickson GF (1979) Extrapituitary action of gonadotropin-releasing hormone: direct inhibition of ovarian steroidogenesis. Science 204:854–855

Hsueh AJW, Adashi EY, Jones PBC, Welsh TH Jr (1984) Hormonal regulation of the differentiation of cultured ovarian granulosa cells. Endocr Rev 5:76

Hsueh AJW, Dahl KD, Vaughan J, Tucker E, Rivier J, Bardin CW, Vale W (1987) Heterodimers and homodimers of inhibin subunits have different paracrine action in the modulation of luteinizing hormone-stimulated androgen biosynthesis. Proc Natl Acad Sci USA 84:5082–5086

Hudson B, Baker HW, Eddie LW, Higginson RE, Burger HG, deKretser DM, Dobos M, Lee VW (1979) Bioassays for inhibin: a critical review. J Reprod Fertil 26:17–29

Hutchison LA, Findlay JK, de Vos FL, Robertson DM (1987) Effects of bovine inhibin, transforming growth factor and bovine activin-A on granulosa cell differentiation. Biochem Biophys Res Commun 146:1405–1412

Hutson JM, Fallat ME, Kamagata S, Donahoe PK, Budzik GP (1984) Phosphorylation events during Mullerian duct regression. Science 223:586–589

Igarashi M, McCann SM (1964) A hypothalamic follicle stimulating hormone-releasing factor. Endocrinology 74:446–452

Ishibashi T, Koziol JA, Burstein SA (1987) Human recombinant erythropoietin promotes differentiation of murine megakaryocytes in vitro. J Clin Invest 79:286–289

Jones HM, Wood CL, Rush ME (1985) A role for inhibin in the control of follicle-stimulating hormone secretion in male rats. Life Sci 36:889–899

Kaneko H, Taya K, Sasamoto S (1987) Changes in the secretion of inhibin and steroid hormones during induced follicular atresia after hypophysectomy in the rat. Life Sci 41:1823–1830

Kimura J, Katoh M, Taya K, Sasamoto S (1983) An inverse relationship between inhibin and follicle-stimulating hormone during the period of ovulation induced by human chorionic gonadotrophin in dioestrous rats. J Endocrinol 97:313–318

Klinefelter HF Jr, Reifenstein EC Jr, Albright F (1942) Syndrome characterized by gynecomastia, aspermatogenesis without A-Leydigism, and increased excretion of follicle-stimulating hormone. J Clin Endocrinol 2:615–627

Koury MJ, Bondurant MC, Graber SE, Sawyer ST (1988) Erythropoietin messenger RNA levels in developing mice and transfer of ^{125}I-erythropoietin by the placenta. J Clin Invest 82:154–159

Kurtz A, Hartl W, Jelkmann W, Zapf I, Bauer C (1985) Activity in fetal bovine serum that stimulates erythroid colony formation in fetal mouse livers is insulinlike growth factor 1. J Clin Invest 76:1643–1648

LaPolt PS, Soto D, Su JG, Vaughan J, Vale W, Hsueh AJW (1989) Activin stimulation of inhibin secretion and alpha subunit mRNA levels in cultured granulosa cells. Mol Endocrinol (in press)

Lee VWK, McMaster J, Quigg H, Findlay J, Leversha L (1981) Ovarian and peripheral blood inhibin concentrations increase with gonadotropin treatment in immature rats. Endocrinology 108:2403–2405

Lee VWK, McMaster J, Quig H, Leversha L (1982) Ovarian and circulating inhibin levels in immature female rats treated with gonadotropin and after castration. Endocrinology 111:1849–1854

Lee W, Mason AJ, Schwall R, Szony E, Mather JP (1989) Secretion of activin by interstitial cells in the testis. Science 243:396–398

LeGac F, DeKretser DM (1982) Inhibin production by sertoli cell cultures.Mol Cell Endocrinol 28:487–498

Lerner RA (1982) Tapping the immunological repertoire to produce antibodies of predetermined specificity. Nature 299:592–596

Leversha LJ, Robertson DM, Vos FLD, Morgan FJ, Hearn MT, Wettenhall RE, Findlay JK, Burger HG, deKretser DM (1987) Isolation of inhibin from ovine follicular fluid. J Endocrinol 113:213–221

Li CH, Ramasharma K (1987) Inhibin. Annu Rev Pharmacol Toxicol 27:1–21

Li CH, Hammonds RG, Ramasharma K, Chung D (1985) Human seminal alpha inhibins: isolation, characterization, and structure. Proc Natl Acad Sci USA 82:4041–4044

Lilja H, Jeppsson J (1985) Amino acid sequence of the predominant basic protein in human seminal plasma. FEBS Lett 182:181–184

Ling N, Ying SY, Ueno N, Esch F, Denoroy L, Guillemin R (1985) Isolation and partial characterization of a Mr 32,000 protein with inhibin activity from porcine follicular fluid. Proc Natl Acad Sci USA 82:7217–7221

Ling N, Ying SY, Ueno N, Shimasaki S, Esch F, Hotta M, Guillemin R (1986a) Pituitary FSH is released by a heterodimer of the beta-subunits from the two forms of inhibin. Nature 321:779–782

Ling N, Ying SY, Ueno N, Shimasaki S, Esch F, Hotta M, Guillemin R (1986b) A homodimer of the beta-subunits of inhibin A stimulates the secretion of pituitary follicle stimulating hormone. Biochem Biophys Res Commun 138:1129–1137

Liu L, Booth J, Merriam GR, Barnes KM, Sherins RJ, Loriaux DL, Cutler GB (1985) Evidence that synthetic 31-amino acid inhibin-like peptide lacks inhibin activity. Endocr Res 11:191–197

Lorenzen JR, Schwartz NB (1979) The differential ability of porcine follicular fluid to suppress serum FSH in female rats from 6 days of age to adulthood. Adv Exp Med Biol 112:375–381

Lozzio CB, Lozzio BB (1975) Human chronic myelogenous leukemia cell-line with positive Philadelphia chromosome. Blood 45:321–334

Magoffin DA, Erickson GF (1982) Mechanisms by which GnRH inhibits androgen synthesis directly in ovarian interstitial cells. Mol Cell Endocrinol 27:191–198

Mangan KF, Chikkappa G, Bieler LZ, Scharfman WB, Parkinson DR (1982) Regulation of human blood erythroid burst-forming unit (BFU-E) proliferation by T-lymphocyte subpopulations defined by Fc receptors and monoclonal antibodies. Blood 59:990–996

Martins T, Rocha A (1931) Regulation of the hypophysis by the testicle and some problems of sexual dynamics. Endocrinology 15:421–434

Mason AJ (1988) Structure and recombinant expression of human inhibin and activin. In: Hodgen GD, Rosenwaks Z, Spieler JM (eds) Nonsteroidal gonadal factors: physiological roles and possibilities in contraceptive development. Jones, Institute Press, Norfolk VA, pp 19–29

Mason AJ, Hayflick JS, Ling N, Esch F, Ueno N, Ying SY, Guillemin R, Niall H, Seeburg PH (1985) Complementary DNA sequences of ovarian follicular fluid inhibin show precursor structure and homology with transforming growth factor-beta. Nature 318:659–663

Mason AJ, Niall HD, Seeburg PH (1986) Structure of two human ovarian inhibins. Biochem Biophys Res Commun 135:957–964

Massicotte J, Lagace L, Godbout M, Labrie F (1984a) Modulation of rat pituitary gonadotrophin secretion by porcine granulosa cell "inhibin", LH releasing hormone and sex steroids in rat anterior pituitary cells in culture. J Endocrinol 100:133–140

Massicotte J, Lagace L, Labrie F, Dorrington JH (1984b) Modulation of gonadotropin secretion by Sertoli cell inhibin, LHRH, and sex steroids. Am J Physiol 247:E495–504

Matsuda H, Coughlin MD, Bienenstock J, Denburg J (1988) Nerve growth factor promotes human hemopoietic colony growth and differentiation. Proc Natl Acad Sci USA 85:6058–6512

Mayo KE, Cerelli GM, Spiess J, Rivier J, Rosenfeld MG, Evans RM, Vale W (1986) Inhibin A-subunit cDNAs from porcine ovary and human placenta. Proc Natl Acad Sci USA 83:5849–5853

McCann SM (1974) Regulation of secretion of follicle-stimulating hormone and luteinizing hormone. In: Knobil E, Sawyer WH (eds) Handbook of physiology, sect 7, vol 4. American Physiological Society, Washington DC, pp 489–517

McCullagh D (1932) Dual endocrine activity of the testis. Science 76:19–20

McCullagh EP, Schaffenburg CA (1952) The role of the seminiferous tubules in the production of hormones. Ann NY Acad Sci 55:674–684

McCullagh DR, Schneider I (1940) The effect of nonadrenergic testis extract on the estrus cycle in rats. Endocrinology 27:899–902

McCullagh DR, Walsh EL (1934) Further studies concerning testicular function. Proc Soc Exp Biol Med 31:678–680

McLachlan RI, Robertson DM, Healy DL, deKretser DM, Burger HG (1986a) Plasma inhibin levels during gonadotropin-induced ovarian hyperstimulation for IVF: a new index of follicular function? Lancet (8492):1233–1234

McLachlan RI, Robertson DM, Burger HG, DeKretser DM (1986b) The radioimmunoassay of bovine and human follicular fluid and serum inhibin. Mol Cell Endocrinol 46:175–185

McLachlan RI, Healy DL, Robertson DM, Burger HG, deKretser DM (1987) Circulating immunoactive inhibin in the luteal phase and early gestation of women undergoing ovulation induction. Fertil Steril 48:1001–1005

Mercer JE, Clements JA, Funder JW, Clarke IJ (1987) Rapid and specific lowering of pituitary FSHβ mRNA levels by inhibin. Mol Cell Endocrinol 53:251–254

Meunier H, Rivier C, Hsueh AJW, Vale W (1988a) Distribution and regulation of inhibin subunit mRNAs. In: Proceedings of Conrad International Workshop: nonsteroidal gonadal factors: physiological roles and possibilities in contraceptive development

Meunier H, Rivier C, Evans RM, Vale W (1988b) Gonadal and extragonadal expression of inhibin alpha, beta A, and beta B subunits in various tissues predicts diverse functions. Proc Natl Acad Sci USA 85:247–251

Meunier H, Cajander SB, Roberts VJ, Rivier C, Sawchenko PE, Hsueh AJW, Vale W (1989) Rapid changes in the expression of inhibin α, βA and βB subunits in ovarian cell types during the rat estrous cycle. Mol Endocrinol 2:1352–1363

Miller WL (1988) Regulation of the subunit mRNAs of follicle-stimulating hormone by inhibin, estradiol, and progesterone. In: Hodgen GD, Rosenwaks Z, Spieler JM (eds) Nonsteroidal gonadal factors: physiological roles and possibilities in contraceptive development. Jones, Insitute Press, Norfolk VA, pp 110–124

Miyamoto K, Hasegawa Y, Fukuda M, Nomura M, Igarashi M, Kangawa K, Matsuo H (1985) Isolation of porcine follicular fluid inhibin of 32K daltons. Biochem Biophys Res Commun 129:396–403

Miyamoto K, Hasegawa Y, Fukuda M, Igarashi M (1986) Demonstration of high molecular weight forms of inhibin in bovine follicular fluid (bFF) by using monoclonal antibodies to bFF 32K inhibin. Biochem Biophys Res Commun 136:1103–1109

Mizunuma H, Sampson WK, Lumpkin MD, Moltz JH, Fawcett CP, McCann SM (1983) Purification of a bioactive FSH-releasing factor (FSHRF). Brain Res Bull 10:623–629

Monette FC, Ouellette PL, Faletra PP (1981) Characterization of murine erythroid progenitors with high erythropoietin sensitivity in vitro. Exp Hematol 9:249–256

Moore MAS, Warren DJ (1987) Surgery of interleukin 1 and granulocyte colony-stimulating factor: in vivo stimulation of stem-cell recovery and hematopoietic regeneration following 5-fluorouracil treatment of mice. Proc Natl Acad Sci USA 84:7134–7138

Moreau JP, Donaldson DD, Bennett F, Witek-Giannoti J, Clark SC, Wong GG (1988) Leukaemia inhibitory factor is identical to the myeloid growth factor human interleukin for DA cells. Nature 336:690–692

Morris PL, Vale WW, Cappel S, Bardin CW (1988) Inhibin production by primary Sertoli cell-enriched cultures: regulation by follicle-stimulating hormone, androgens, and epidermal growth factor. Endocrinology 122:717–725

Mottram JC, Cramer W (1923) Report on the general effects of exposure to radium on metabolism and tumour growth in the rat and the special effects on testis and pituitary. Q J Exp Physiol 13:209–229

Muller R, Slamon DJ, Adamson ED, Tremblay JM, Muller D, Cline MJ, Verma IM (1983) Transcription of c-*onc* genes c-*raski* and c-*fms* during mouse development. Mol Cell Biol 3:1062–1069

Murata M, Eto Y, Shibai H, Sakai M, Muramatsu M (1988) Erythroid differentiation factor is encoded by the same mRNA as that of the inhibin beta A chain. Proc Natl Acad Sci USA 85:2434–2438

Nathan DG, Chess L, Hillman DG, Clarke B, Breard J, Merler E, Houseman DE (1988) Human erythroid burst-forming unit: T-cell requirement for proliferation in vitro. J Exp Med 147:324–339

Nelson WO, Gallagher TF (1935) Studies on the anterior hypophysis. IV. The effect of male hormone preparations upon the anterior hypophysis of gonadectomized male and female rats. Anat Rec 64:129–145

Odell WD, Swerdloff RS (1975) The role of testicular sensitivity to gonadotropins in sexual maturation of the male rat. J Steroid Biochem 6:853–587
Odell WD, Swerdloff RS (1976) Etiologies of sexual maturation: a model system based on the sexually maturing rat. Recent Prog Horm Res 32:245–288
Ojeda SR, Andrews WW, Advis JP, White SS (1980) Recent advances in the endocrinology of puberty. Endocr Rev 1:228–257
Padgett RW, St Johnston RD, Gelbart WM (1987) A transcript from a *Drosophila* pattern gene predicts a protein homologous to the transforming growth factor-beta family. Nature 325:81–84
Papkoff H (1974) Chemical and biological properties of the subunits of pregnant mare serum gonadotropin. Biochem Biophys Res Commun 58:397–404
Pedersen T (1969) Kinetics of follicle growth in the ovary of the immature mouse. Acta Endocrinol (Copenh) 62:117–132
Petraglia F, Sawchenko P, Lim AT, Rivier J, Vale W (1987) Localization, secretion, and action of inhibin in human placenta. Science 237:187–189
Petraglia F, Vaughan J, Vale W (1989) Inhibin and activin modulate release of protein and steroid hormones from cultured placental cells. Proc Natl Acad Sci USA (in press)
Piacsek BE, Goodspeed MP (1978) Maturation of the pituitary-gonadal system in the male rat. J Reprod Fertil 52:29–35
Pieper DR, Richards JS, Marshall JC (1981) Ovarian gonadotropin-releasing hormone (GnRH) receptors: characterization, distribution and induction by GnRH. Endocrinology 108:1148–1155
Pierce JG, Parsons TF (1981) Glycoprotein hormones: structure and function. Annu Rev Biochem 50:465–495
Plotsky PM, Sawchenko PE, Vale W (1988) Evidence for inhibin beta-chain like-peptide mediation of suckling-induced oxytocin secretion. In: 18th Society for Neuroscience Meeting, nov 13–18, 1988, Toronto, p 627 (abstr no 256.1)
Ramasharma K, Sairam MR, Seidah NG, Chretien M, Manjunath P, Schiller PW, Yamashiro D, Li CH (1984) Isolation, structure, and synthesis of a human seminal plasma peptide with inhibin-like activity. Science 223:1199–1202
Reeves JJ, Seguin C, Lefebvre F-A, Kelly PA, Labrie F (1980) Similar luteinizing hormone-release hormone binding sites in rat anterior pituitary and ovary. Proc Natl Acad Sci USA 77:5567–5571
Rettenmier CA, Sacca R, Furman WL, Roussel MF, Holt JR, Nienhuis AW, Stanley ER, Sherr CJ (1986) Expression of the human c-*fms* protooncogene product (colony-stimulating factor-1 receptor) on pheripheral blood mononuclear cells and choriocarcinoma cell lines. J Clin Invest 77:1740–1746
Rivier C, Vale W (1979) Hormonal secretion in male rats chronically treated with [DTrp^6,Pro^9,Net]-LR. Life Sci 25:1065–1074
Rivier C, Vale W (1987) Inhibin: measurement and role in the immature female rat. Endocrinology 120:1688–1690
Rivier C, Vale W (1989a) Immunoneutralization of endogenous inhibin modifies hormone secretion and ovulation rate in the rat. Endocrinology 125:152–157
Rivier C, Vale W (1989b) Immunoreactive inhibin secretion by the hypophysectomized female rat: demonstration of the modulating effect of gonadotropin-releasing hormone (GnRH) and estrogen through a direct ovarian site of action. Endocrinology 124:195–198
Rivier C, Rivier J, Vale W (1986) Inhibin-mediated feedback control of follicle-stimulating hormone secretion in the female rat. Science 234:205–208
Rivier C, Meunier H, Vaughan J, Vale W (1987a) Role of endogenous inhibin in the female rat. In: Rolland R et al. (eds) Neuroendocrinology of reproduction. Elsevier, Amsterdam, pp 3–7
Rivier C, Rivier J, Vale W (1987b) Inhibin: measurement and role in the rat. Int J Rad Appl Instrum [B] 14:273–276

Rivier C, Vale W, Rivier J (1987 c) Studies of the inhibin family of hormones: a review. In: Girard J (ed) Hormone research. Karger, Basel, pp 104–118

Rivier C, Cajander S, Vaughan J, Hsueh AJW, Vale W (1988 a) Age-dependent changes in physiological action, content, and immunostaining of inhibin in male rats. Endocrinology 123:120–126

Rivier C, Rivier J, Vale W (1988 b) Inhibin: regulation and role in the female rat. In: Genazzani AR, Montemagno U, Nappi C, Petrablia F (eds) The brain and female reproductive function. Parthenon, New Jersey, pp 183–190

Rivier C, Roberts V, Vale W (1989) Possible role of LH and FSH in modulating inhibin secretion and expression during the estrous cycle of the rat. Endocrinology (in press)

Rivier J, Spiess J, Thorner M, Vale W (1982) Characterization of a growth hormone releasing factor from a human pancreatic tumor. Nature 300:276–278

Rivier J, Spiess J, Vale W (1983) Characterization of rat hypothalamic corticotropin-releasing factor. Proc Natl Acad Sci USA 80:4851–4855

Rivier J, Spiess J, McClintock R, Vaughan J, Vale W (1985) Purification and partial characterization of inhibin from porcine follicular fluid. Biochem Biophys Res Commun 133:120–127

Roberts R, Gallagher J, Spooncer E, Allen TB, Bloomfield F, Dexter TM (1988) Heparan sulphate bound growth factors: a mechanism for stromal cell mediated haemopoiesis. Nature 332:376–378

Roberts V, Meunier H, Vaughan J, Rivier J, Rivier C, Vale W, Sawchenko P (1988) Production and regulation of inhibin subunits in pituitary gonadotropes. Endocrinology 124:552–554

Roberts V, Meunier H, Sawchenko P, Vale W (1989) Differential production and regulation of inhibin subunits in rat testicular cell types. Endocrinology (in press)

Robertson DM, Foulds LM, Leversha L, Morgan FJ, Hearn MT, Burger HG, Wettenhall RE, deKretser DM (1985) Isolation of inhibin from bovine follicular fluid. Biochem Biophys Res Commun 126:220–226

Robertson DM, Vos FLd, Foulds LM, McLachlan RI, Burger HG, Morgan FJ, Hearn MT, deKretser DM (1986 a) Isolation of a 31 kDa form of inhibin from bovine follicular fluid. Mol Cell Endocrinol 44:271–277

Robertson DM, Giacometti MS, deKretser DM (1986 b) The effects of inhibin purified from bovine follicular fluid in several in vitro pituitary cell culture systems. Mol Cell Endocrinol 46:29–36

Robertson DM, Klein R, Vos FLd, McLachlan RI, Wettenhall REH, Hearn MTW, Burger HG, deKretser DM (1987) The isolation of polypeptides with FSH suppressing activity from bovine follicular fluid which are structurally different to inhibin. Biochem Biophys Res Commun 149:744–749

Robertson DM, Hayward S, Irby D, Jacobsen J, Clarke L, McLachlan RI, deKretser DM (1988) Radioimmunoassay of rat serum inhibin: changes after PMSG stimulation and gonadectomy. Mol Cell Endocrinol 58:1–8

Rubin D (1941) The question of an aqueous hormone from the testicle. Endocrinology 29:281–287

Rush ME, Ashiru OA, Lipner H, Williams AT, McRae C, Blake CA (1981) The actions of porcine follicular fluid and estradiol on periovulatory secretion of gonadotropic hormones in rats. Endocrinology 108:2316–2323

Sairam MR, Papkoff H (1974) Chemistry of pituitary gonadotropin. In: Handbook of physiology, vol 4, pp 111–131

Sander HJ, Meijs-Roelofs HMA, Van Leeuwen ECM, Kramer P, Van Cappellen WA (1986) Inhibin increases in the ovaries of female rats approaching first ovulation: relationships with follicle growth and serum FSH concentrations. J Endocrinol 111:159–166

Sawchenko PE, Plotsky PM, Pfeiffer SW, Cunningham ETC, Vaughan J, Rivier J, Vale W (1988) Inhibin beta in central neural pathways involved in the control of oxytocin secretion. Nature 334:615–617

Sawyer ST, Sawada K, Krantz SB (1987) Structure of the receptor for erythropoietin in murine and human erythroid cells and murine placenta. Blood 70 [Suppl 1]:184a

Schally AV, Saito T, Arimura A, Muller EE, Bowers CY, White WF (1966) Purification of follicle-stimulating hormone-releasing factor (FSH-RF) from bovine hypothalamus. Endocrinology 79:1087–1094

Schwartz NB, Channing CP (1977) Evidence for ovarian "inhibin": suppression of the secondary rise in serum follicle stimulating hormone levels in proestrous rats by injection of porcine follicular fluid. Proc Natl Acad Sci USA 12:5721–5724

Scott RS, Burger HG (1981) Mechanism of action of inhibin. Biol Reprod 24:541–550

Seidah NG, Arbatti NJ, Rochemont J, Sheth AR, Chretien M (1984) Complete amino acid sequence of human seminal plasma beta-inhibin. Prediction of post Gln-Arg cleavage as a maturation site. FEBS Lett 175:349–355

Sheth AR, Arabatti N, Carlquist M, Jornvall H (1984) Characterization of a polypeptide from human seminal plasma with inhibin (inhibition of FSH secretion) like activity. FEBS Lett 165:11–15

Shibata H, Ogata E, Etoh Y, Shibai H, Kojima I (1987) Erythroid differentiation factor stimulates hydrolysis of polyphosphoinositide in friend erythroleukemia cells. Biochem Biophys Res Commun 146:187–193

Sieff CA, Tsai S, Faller DV (1987) Interleukin 1 induces cultured human endothelial cell production of granulocyte-macrophage colony-stimulating factor. J Clin Invest 79:48–51

Sing GK, Keller JR, Ellingsworth LR, Ruscetti FW (1988) Transforming growth factor beta selectively inhibits normal and leukemic human bone marrow cell growth in vitro. Blood 72:1504–1511

Smith KA (1980) T-cell growth factor. Immunol Rev 51:337–357

Sokoloski JA, Blair OC, Sartorelli AC (1986) Alterations in glycoprotein synthesis and guanosine triphosphate levels associated with the differentiation of HL-60 leukemia cells produced by inhibitors of inosine 5′-phosphate dehydrogenase. Cancer Res 46:2314–2319

Spiess J, Rivier J, Vale W (1983) Characterization of rat hypothalamic growth hormone releasing factor. Nature 303:532

Sporn MB, Roberts AB (1988) Peptide growth factors are multifunctional. Nature 332:217–219

Steinberger A (1979) Inhibin production by Sertoli cells in culture. J Reprod Fertil 26:31–45

Steinberger A (1981) Regulation of inhibin secretion in the testis. In: Franchimont P, Channing CP (eds) Intragonadal regulation of reproduction. Academic, New York, pp 283–298

Steinberger A, Steinberger E (1971) Replication patterns of Sertoli cells in maturing testis in vivo and in organ culture. Biol Reprod 4:84–87

Steinberger A, Steinberger E (1976) Secretion of an FSH-inhibiting factor by cultured Sertoli cells. Endocrinology 99:918–921

Steinberger E, Chowdhury M (1974) Control of pituitary FSH in male rats. Acta Endocrinol (Copenh) 76:235–241

Stewart AG, Milborrow HM, Ring JM, Crowther CE, Forage RG (1986) Human inhibin genes. Genomic characterisation and sequencing. FEBS Lett 206:329–334

Sugino H, Nakamura T, Hasegawa Y, Miyamoto K, Abe Y, Igarashi M, Eto Y, Shibai H, Titani K (1988a) Erythroid differentiation factor can modulate follicular granulosa cell functions. Biochem Biophys Res Commun 153:281–288

Sugino H, Nakamura T, Hasegawa Y, Miyamoto K, Igarashi M, Eto Y, Shibai H, Titani K (1988b) Identification of a specific receptor for erythroid differentiation factor on follicular granulosa cell. J Biol Chem 263:15249–15252

Summerville JW, Schwartz NB (1981) Suppression of serum gonadotropin levels by testosterone and porcine follicular fluid in castrate male rats. Endocrinology 109:1442–1447

Tannahill D, Weeks DL, Melton DA (1988) A localized *Xenopus* maternal mRNA encodes a TGF-beta-like factor. J Cell Biochem 12A:182

Toebosch AMW, Robertson DM, Trapman J, Klaassen P, DePaus RA, DeJong FH, Grootegoed JA (1988) Effects of FSH and IGF — an immature rat Sertoli cells: inhibin α and β subunit mRNA levels and inhibin secretion. Mol Cell Endocrinol 55:101–105

Tsai S, Emerson SG, Sieff CA, Nathan DG (1986a) Isolation of a human stromal cell strain secreting hemopoietic growth factors. J Cell Biol 127:137–145

Tsai S, Sieff CA, Nathan DG (1986b) Stromal cell-associated erythropoiesis. Blood 67:1418–1426

Tsonis CG, Baird DT, Campbell BK, Leask R, Scaramuzzi RJ (1988) The sheep corpus luteum secretes inhibin. J Endocrinol 116:R3–R5

Ueno N, Lins N, Yins SY, Esch F, Shimasaki S, Guillemin R (1987) Isolation and partial characterization of follistatin: a single-chain Mr 35,000 monomeric protein that inhibits the release of follicle-stimulating hormone. Proc Natl Acad Sci USA 84:8282–8286

Uilenbroek JTJ, Wolff-Exalto EAd, Welschen R (1976) Studies on the significance of the high levels of follicle stimulating hormone for follicular development in immature rats. Ann Biol Anim Biochim Biophys 16:297–305

Ultee-van Gessel AM, deJong FH (1987) Inhibin-like activity in Sertoli cell culture media and testicular homogenates from rats of various ages. J Endocrinol 113:103–110

Ultee-van Gessel AM, Leemborg FH, deJong FH, van der Molen HJ (1986) In vitro secretion of inhibin-like activity by Sertoli cells from normal and prenatally irradiated immature rats. J Endocrinol 109:411–418

Vale V, Grant G, Amoss M, Blackwell R, Guillemin R (1972) Culture of enzymatically dispersed pituitary cells: functional validation of a method. J Clin Endocrinol Metab 91:562–572

Vale W, Brazeau P, Rivier C et al. (1975) Somatostatin. Recent Prog Horm Res 31:365–397

Vale W, Spiess J, Rivier C, Rivier J (1981) Characterization of a 41-residue ovine hypothalamic peptide that stimulates secretion of corticotropin and β-endorphin. Science 213:1394–1397

Vale W, Rivier J, Vaughan J, McClintock R, Corrigan A, Woo W, Karr D, Spiess J (1986) Purification and characterization of an FSH releasing protein from porcine ovarian follicular fluid. Nature 321:776–779

Vale W, Rivier C, Hsueh A, Campen C, Meunier H, Bicsak T, Vaughan J, Corrigan A, Bardin W, Sawchenko P, Petraglia F, Plotsky P, Spiess J, Rivier J (1988) Chemical and biological characterization of the inhibin family of proteins hormones. In: Clark JH (ed) Recent progress in hormone research, vol 44. Academic, San Diego, pp 1–34

Vaughan JM, Rivier J, Corrigan AZ, McClintock R, Campen CA, Jolley D, Vogelmayr JK, Bardin CW, Rivier C, Vale W (1989) Detection and purification of inhibin using antisera generated against synthetic peptide fragments. In: Conn PM (ed) Methods in enzymology, hormone action, pt L, vol 168: neuroendocrine peptides. Academic, Orlando, pp 588–617

Verhoeven G, Franchimont P (1983) Regulation of inhibin secretion by Sertoli cell-enriched cultures. Acta Endocrinol (Copenh) 102:136–143

Vidgoff B, Vehrs H (1940) Studies on the inhibitory hormone of the testes, IV. Effect on the pituitary, thyroid and adrenal glands of the adult male rat. Endocrinology 26:656–661

Wang C, Hsueh AJW, Erickson GF (1979) Induction of functional prolactin receptors by follicle-stimulating hormone in rat granulosa cells in vivo and in vitro. J Biol Chem 254:11330–11336

Wang QF, Farnworth PG, Findlay JK, Burger HG (1988) Effect of purified 31K bovine inhibin on the specific binding of gonadotropin-releasing hormone to rat anterior pituitary cells in culture. Endocrinology 123:2161–2166

Weeks DL, Melton DA (1987) A maternal mRNA localized to the vegetal hemisphere in *Xenopus* eggs codes for a growth factor related to TGF-beta. Cell 51:861–867

Westbrook CA, Gasson JC, Gerber SE, Selsted ME, Golde DW (1984) Purification and characterization of human T-lymphocyte-derived erythroid-potentiating activity. J Biol Chem 259:9992–9996

Williams RL, Hilton DJ, Pease S, Willson TA, Colin LS, Gearing DP, Wagner EF, Metcalf D, Nicola NA, Gough NM (1988) Myeloid leukaemia inhibitory factor maintains the developmental potential of embryonic stem cells. Nature 336:684–686

Woodruff TK, Meunier H, Jones PBC, Hsueh AJW, Mayo KE (1987) Rat inhibin: molecular cloning of α- and β-subunit complementary deoxyribonucleic acids and expression in the ovary. Mol Endocrinol 1:561–568

Woodruff TK, D'Agostino J, Schwartz NB, Mayo KE (1988) Dynamic changes in inhibin messenger rRNAs in rat ovarian follicles during the reproductive cycle. Science 239:1296–1299

Ying S-Y, Becker A, Ling N, Ueno N, Guillemin R (1986a) Inhibin and beta type transforming growth factor (TGF-β) have opposite modulating effects on the follicle stimulating hormone (FSH)-induced aromatase activity of cultured rat granulosa cells. Biochem Biophys Res Commun 136:969–975

Ying S-Y, Becker A, Baird A, Ling N, Ueno N, Esch F, Guillemin R (1986b) Type beta transforming growth factor (TGF-beta) is a potent stimulator of the basal secretion of follicle stimulating hormone (FSH) in a pituitary monolayer system. Biochem Biophys Res Commun 135:950–956

Ying S-Y, Czvik J, Becker A, Ling A, Ueno N, Guillemin R (1987) Secretion of follicle-stimulating hormone and production of inhibin are reciprocally related. Proc Natl Acad Sci USA 84:4631–4635

Yokota T, Otsuka T, Mosmann T, Banchereau J, DeFrance T, Blanchard D, De Vries JE, Lee F, Arai KI (1986) Isolation and characterization of a human interleukin cDNA clone, homologous to mouse B-cell stimulatory factor 1, that expresses B-cell and T-cell-stimulating activities. Proc Natl Acad Sci USA 83:5894–5898

Yu J, Shao L, Lemas V, Yu AL, Vaughan J, Rivier J, Vale W (1987) Importance of FSH-releasing protein and inhibin in erythrodifferentiation. Nature 330:765–767

Yu J, Shao L, Lemas V, Yu AL, Vaughan J, Rivier J, Vale W (1988a) Inhibin and FSH releasing protein modulate erythrodifferentiation. J Cell Biochem [Suppl] 12A:141 (abstr)

Yu J, Shao L, Vaughan J, Vale W, Yu AL (1988b) Characterization of the potentiation effect of activin on human erythroid colony formation in vitro. Blood 72:106a (abstr)

Yu J, Shao L-E, Vaughan J, Vale W, Yu AL (1989) Characterization of the potentiation effect of activin on human erythroid colony formation in vitro. Blood 73:952–960

Zhang Z, Carson RS, Herington AC, Lee VWK, Burger HG (1987) Folicle-stimulating hormone and somatomedin-C stimulate inhibin production by rat granulosa cells in vitro. Endocrinology 120:1633–1638

Zhang Z, Lee VWK, Carson RS, Burger HG (1988) Selective control of rat granulosa cell inhibin production by FSH and LH in vitro. Mol Cell Endocrinol 56:35–40

Zuckerman KS, Bagby GC Jr, McCall E, Sparks B, Wells J, Patel V, Goodrum D (1985) A monokine stimulates production of human erythroid burst-promoting activity by endothelial cells in vitro. J Clin Invest 75:722–725

CHAPTER 27

Mammary-Derived Growth Inhibitor

R. Grosse and P. Langen

A. Introduction

Today search for growth-regulating polypeptides is guided by a model which postulates control of cell proliferation by both positive and negative signals. Although Bullough and Lawrence (1960) and Iversen (1960) developed the chalone theory in 1960, the existence of growth inhibitors was not demonstrated until recently. This theory postulates a simple feedback mechanism: tissues produce specific compounds (chalones), which inhibit proliferation in the producing tissues. The more tissue is present, the more chalone will be produced and the less growth will proceed. Purification of growth inhibitors has progressed rather slowly. Several attempts have been undertaken to purify a chalone from the epidermis, granulocytes, lymphocytes, or liver (for reviews, see Iversen 1981; Langen 1985; Chap. 28).

The chalone theory is currently being rejuvenated within the framework of the "autocrine hypothesis" (Sporn and Roberts 1985). This concept, and the fact that transforming growth factor-β (TGF-β), the interferons, and tumor necrosis factor represent physiologically important families of growth inhibitors, have encouraged many groups to search for endogenous growth inhibitors. However, in comparison to the many growth stimulatory factors described in this book the number of growth inhibitors with a similar degree of biological characterization is rather low (Wang and Hsu 1986).

The different ways a growth inhibitory polypeptide could arrest growth, and the interaction with growth stimulatory polypeptides which may contribute to a controlled cellular proliferation, are depicted in Fig. 1. The model suggests three pathways for exerting growth inhibition by endogenous inhibitory polypeptides. These pathways have also been considered for growth stimulatory factors: a secretory pathway as discussed for many growth factors (1), a storage in some form at the plasma membrane as described for fibroblast growth factor (2) (Baird and Ling 1987), and, finally, a direct intracellular action on transcriptional events as suggested by reports indicating nuclear receptors for growth factors (3) (Rakowicz-Szulczynska et al. 1986). Growth inhibition could be brought about by an autocrine or by a paracrine mechanism. So far, evidence for either mechanism has been described in the case of TGF-β. In addition the model suggests a third, intracellular pathway implying a direct interaction of a growth inhibitory polypeptide with elements of the transcription machinery. This mode of ac-

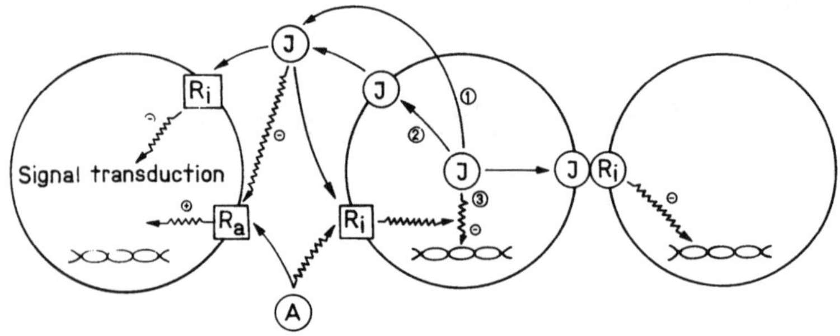

Fig. 1. Growth inhibition models. The model indicates three pathways MDGI may employ for exerting growth inhibitory activity: a secretory mechanism (*1*); transport to plasma membrane and storage as a membrane-associated protein (*2*); and direct action on the level of gene expression (*3*). *A* growth stimulary protein (EGF, insulin); *J* growth inhibitory protein; R_i J-receptor; R_a A-receptor

tion could result in an irreversible growth arrest as might occur during terminal differentiation.

In this chapter we describe a growth inhibitor, designated mammary-derived growth inhibitor (MDGI), which is present in high amounts in the lactating, terminally differentiated bovine mammary gland. Structural and biological data point to a new family of growth-regulating factors for which MDGI is the best-investigated example. The specific role we propose for MDGI is to arrest growth of mammary epithelial cells when they become committed to differentiation in the mammary gland.

B. Results

I. Purification

Purification was monitored by a 24-h in vitro proliferation assay using Ehrlich ascites mammary carcinoma cells from the stationary phase of growth in vivo (BOEHMER et al. 1985). In order to exclude enrichment of inhibitory activity due to cytotoxic agents, cells were preincubated for 4 h before being subjected to the standard in vitro assay. Cells treated this way have lost their sensitivity to inhibitory activity found in crude preparations from bovine mammary gland. They are, however, strongly inhibited by interferon-β, oxidized polyamines, or synthetic inhibitors, such as nucleoside analogs. Therefore "preincubation experiments" were performed at each step of the purification procedure to exclude cytotoxic effects. The same results were obtained if instead of "preincubated cells" Ehrlich cells were harvested from the logarithmic phase of growth in vivo. Thus some preselection of inhibitory activity toward "quiescent" cells is inherent to the assay.

Growth inhibitory activities were first detected in ascitic fluid and in extracts of lactating bovine mammary gland (LEHMANN et al. 1980). A partial enrichment of inhibitory activity was achieved by lectin-affinity chromatography of fractions from a high-speed supernatant of a homogenate of lactating gland (LEHMANN et al. 1983).

Final purification from the lactating bovine mammary gland and identification of the inhibitory component was achieved by three independent methods, namely preparative gel electrophoresis (BOEHMER et al. 1984), anion exchange chromatography (BOEHMER et al. 1985), and size exclusion high-performance liquid chromatography (HPLC) in the presence of SDS (BOEHMER et al. 1987a). For preparative gel electrophoresis 1–1.2 kg gland was homogenized, the homogenate centrifuged, the $50000 \times g$ supernatant proteins were precipitated at 60% saturation of $(NH_4)_2SO_4$ followed by a $75000 \times g$ centrifugation and ultrafiltration of the supernatant through an Amicon XM-50 membrane. The filtrate was concentrated and applied to a Sephadex G-75 column to obtain the partially purified inhibitor. This material was further analyzed by electrophoresis in 10% polyacrylamide gels with a discontinuous alkaline buffer and slicing the gels for final localization of the biological activity. Proteins were monitored by addition of tracer amounts of ^{125}I-labeled partially purified inhibitor before preparative electrophoresis. The inhibitory activity was shown to be associated with a rather diffuse band corresponding to a molecular mass of 12–14 kDa. By use of this procedure we obtained approximately 20 μg MDGI with a specific activity of 6×10^5 U/mg (1 unit = the amount of MDGI giving half-maximal inhibition in the assay) from 1 kg. In order to improve yield, the method was later modified. Instead of the Amicon XM-50 membrane strongly adsorbing the MDGI fraction, a UF 150 cellulose acetate membrane was used, and the concentrated filtrate was then applied to a 5 × 100-cm Sephadex G-50 column and the active fractions (V_e/V_0, 1.65–1.85) finally applied to a column containing DEAE-Sepharose CL-6B. The inhibitory activity was eluted with imidazole buffer containing 50 m*M* NaCl. This fraction contained one polypeptide of about 14.5 kDa as shown by SDS-PAGE under both reducing and nonreducing conditions. The polypeptide is not glycosylated.

To further substantiate the finding, obtained with both types of purification schemes, that the inhibitory activity is associated with the 14.5-kDa polypeptide, two further experiments were performed. First, specific polyclonal antisera raised against MDGI were tested for their neutralizing effect on growth inhibition by MDGI in Ehrlich ascites carcinoma cells (BOEHMER et al. 1985) and in a mammary carcinoma cell line MATU (BOEHMER et al. 1987a). Western-blotting analysis revealed that a 14.5-kDa protein comigrating with pure MDGI was the only species specifically recognized by the antibody either in MDGI preparations or in crude bovine mammary gland tissue supernatants. Additions of these antisera to cell cultures together with MDGI prevented growth inhibition. Second, a separation by size exclusion HPLC in the presence of 0.1% SDS was performed and the fractions obtained assayed for their growth inhibitory effect on Ehrlich cells. The growth inhibitory activity eluted together with the fraction consisting of the 14.5-kDa polypeptide.

The same antibodies have been used in a first attempt to obtain information on the tissue distribution of MDGI making use of an enzyme-linked immunoadsorbent assay (ELISA) (BOEHMER et al. 1985). High amounts of crossreactive antigen were found in milk fat globule membranes (MFGMs) of both bovine and human origin. A more detailed immunochemical study defined the crossreactive antigen as a 14.5-kDa polypeptide whose level increased dramatically with the onset of lactation after delivery (BRANDT et al. 1988). The 14.5-kDa antigen was purified from bovine MFGM to homogeneity by preparative SDS-PAGE and electroelution. The MFGM-derived 14.5-kDa polypeptide was found to be indistinguishable from MDGI as demonstrated by tryptic digestion and partial amino acid sequence analysis of tryptic fragments of both proteins (BRANDT et al. 1988). Thus we conclude that MDGI or a membrane-associated MDGI-form could be involved in growth regulation of mammary epithelial cells. This mode of inhibitory action of a membrane-associated inhibitor with corresponding cellular structures of neighboring cells (Fig. 1) was indicated when we showed that MFGMs contain growth inhibitory activity (HERRMANN and GROSSE 1986).

II. Amino Acid Sequence Determination and Sequence Homologies

MDGI was obtained in an N-terminally blocked form. Therefore, different cleavage methods were used to obtain fragments for sequencing as outlined in detail by BOEHMER et al. (1987a). The amino acid sequence of MDGI (Fig. 2) showed virtually no sequence homology to TGF-β or any of the structurally known interferons. Instead, a significant homology was found with intracellular proteins known to bind hydrophobic ligands, designated as fatty acid binding proteins (FABPs), and with proteins associated with the differentiated state of cells such as murine P-422, the rat cellular retinol binding protein CRBP (56% identical or conservatively exchanged residues), and the bovine cellular retinoic acid binding protein CRABP (61%) (BOEHMER et al. 1987a). The highest degree of homology of MDGI was found with members of the FABPs isolated from bovine (F. SPENER, personal communication), human (OFFNER et al. 1988), and rat heart (CLAFFEY et al. 1987; GIBSON et al. 1988) with MP2 and with P422 (Fig. 2).

FABPs are involved in lipid metabolism (BASS 1985). It has been suggested that, in heart, a heart muscle-specific FABP regulates the supply of fatty acids to the mitochondria for β-oxidation (KEMPEN et al. 1983). The mammary gland, however, is a highly lipogenic tissue and fatty acids are not likely to be a major fuel for its metabolism. More recently, two FABPs were isolated from mammary glands of the rat (P.D. JONES et al. 1988). They differ in their isoelectric point, and the major mammary form with a pI of 4.9 [designated as M-FABP(r) in Fig. 2] has been partially sequenced. The partial sequences aligned to MDGI reveal differences in eight positions. The possible existence of isoforms of FABPs in bovine heart and mammary gland could explain our observation of microheterogeneous sites at seven positions in the MDGI sequence (Fig. 2). It is possible that the inhibitory activity is exerted by only one

```
             1        10        20        30        40        50
             |         |(D)(K)   |         |     (A)  |         |
MDGI(b)      VDAFVGTWKLVSSENFDDYMKSLGVGFATRQVGNMTKPTLIISVNGDTVI
H-FABP(b)    ...........D.K..........................T..E.......
H-FABP(h)    ....L......D.K..................AS.....T..EK...ILT
M-FABP(r)                  .......         ..AS.....T..EK
H-FABP(r)    A..........D.K..................AS.....T..EK....IT
                            *  **
MP2(b)       SNK.L............E...A....L...KL..LA..RV...KK..IIT
P422(m)      C....................EV.......K.AG.A..NM.......L.T

                      60        70        80    (N) (E)90 (H)   100
                       |         |         |         |         |
MDGI(b)      IKTQSTFKNTEISFKLGVEFDETTADDRKVKSIVTLDGGKLVQVQKWNGQ
H-FABP(b)    ..........................................H.......
H-FABP(h)    L..H......................................HL...D..
M-FABP(r)      .H....  .. .Q....             .V...        ..D..
H-FABP(r)    ...H..........Q.......V.........V.........H....D..
                                       *
MP2(b)       .R.E.P...........Q..E....N..T..T...AR.S.N........N
P422(m)      .RSE...................I.........I....A.......D.K

                      110       120       130
                       |         |    (T) |
MDGI(b)      ETSLVREMVDGKLILTLTHGTAVCTRVYEKQ_
H-FABP(b)    ..........................T....A
H-FABP(h)    ..TK...LI..............S..T...EA
M-FABP(r)    ..T.T.                 S..
H-FABP(r)    ..T.T..LS...........NV.S..T...EA
                *  *        *          *  *
MP2(b)       ..TIK.KL....MVVECKMKDV....I...V
P422(m)      S.TIK.KRDGD..VVECVMKGVTS.....RA
```

Fig. 2. Comparison of the sequence of MDGI with sequences of bovine heart FABP, human heart FABP, rat mammary FABP, rat heart FABP, bovine myelin P2, and murine P422. In seven positions of MDGI, microheterogeneities were found. *Asterisks* indicate potential ligand-binding sites for MP2. *Underlined* residues represent amino acids different from H-FABP(b) even taking into account microheterogeneous sites in the MDGI sequence

of the polypeptide chains and that different forms of FABPs fulfill several unrelated functions not yet known.

In this regard, data of Bassuk et al. (1987) are to be considered identifying the liver FABP as a mitosis-associated polypeptide that is the covalent target of activated metabolites of the carcinogens *N*-2-fluorenylacetamide (2-acetylaminofluorene) in rat hepatocytes. Immunohistochemical evidence indicated a low level of FABP when growth activity of hepatocytes was low and an elevated level during mitosis in normal and regenerating liver, with a further increase during proliferation of hyperplastic and malignant hepatocytes.

BERNLOHR et al. (1984) as well as SPIEGELMAN et al. (1983) have isolated a group of cDNAs corresponding to mRNAs whose levels are increased during 3T3 preadipocyte differentiation. One of these cDNAs, pAL 422, encodes for a protein designated as P422 (BERNLOHR et al. 1984). Expression of P422 was shown to be transcriptionally activated during differentiation (BERNLOHR et al. 1985). A similar if not identical protein, aP2, was identified in an adipocyte cDNA library (K.S. COOK et al. 1985). Transcriptional activation of the aP2 gene also occurs during adipocyte differentiation. The exon structure of the aP2 gene has significant sequence homology with rat liver FABP (HUNT et al. 1986). J.S. COOK et al. (1988) reported data suggesting that the promoter region of the P422 gene contains elements mediating the response to differentiation-inducing agents such as dexamethasone or inhibitors of cAMP-dependent phosphodiesterase. DISTEL et al. (1987) identified a 18-nucleotide element in the 5′ flanking differentiation-associated sequence inhibiting the expression of the P422 gene in preadipocytes. This element seems to bind a *fos* protein. Thus, the p422 gene may interact with proteins to form nucleoprotein complexes driving either differentiation or inhibition of differentiation.

Murine P422 and MDGI share about 85% sequence homology. It is tempting to speculate that the MDGI gene contains elements important for differentiation of mammary epithelial cells. For the purified P422 protein, specific retinoic acid binding was found (MATARESE and BERNLOHR 1987). The protein also binds retinyl palmitate. However, no specific fatty acid binding was reported. Bovine myelin P2 (MP2, 78% homology with MDGI) is probably localized at the cytoplasmic side of Schwann cells and can be considered as peripheral membrane protein. It is not known whether aP2 and MP2 are identical or closely related proteins, since their sequence has not been determined within a single species. Injection of MP2 into rats induces experimental allergic neuritis, an animal model for the Guillain-Barré syndrome, a human demyelinating disease of the peripheral nerve system. The exact function of MP2 is not clear. The three-dimensional structure of MP2 has been determined (T.A. JONES et al. 1988). It was suggested that in the C-terminal half of the molecule, which is less conservative over the family of related proteins, single amino acid substitutions might be sufficient to account for the ligand-binding specificity. The arginine residues 106 and 126 were proposed to interact specifically with a fatty acid ligand because they are substituted by glutamines in CRABP and CRBP (Fig. 2). However, a vitamin A binding property of P422 would not fit into the theory because residues 106 and 126 are the same as in the FABPs. Other authors ascribed the sequence homology observed at the amino termini of FABPs, P422, and MP2, which may now be extended to MDGI, to a function of these proteins as lipid carriers. It remains to be proved whether the proteins are important carriers, or the lipids are rather stabilizing some favorable protein structure. It is also not known whether the FABPs, MP2, P422, or MDGI exert some specific function through ligand binding, as suggested for CRABP and CRBP.

As mentioned previously, CRABP and CRBP are also known to be expressed in a differentiation-dependent manner (Eriksson et al. 1986). Thus, differentiation-dependent expression seems to be a common feature of the proteins homologous to MDGI.

Another polypeptide, not listed in Fig. 2 because its structure has not yet determined, needs to be considered: the fibroblast growth regulator (FGR-s), a polypeptide growth inhibitor isolated by Wang and associates (Steck et al. 1982). In collaboration with John Wang we have examined the immunological relationship between FGR-s and MDGI (Boehmer et al. 1987b). In a series of immunodot blot and western blot assays using antibodies against MDGI it was shown that FGR-s shares common structural features with MDGI. This raises the possibility that these growth inhibitors may together define a new family of growth regulatory molecules. Furthermore, by including antibodies against H-FABP and MP2 in the study, a hierarchy of four related proteins was proposed: (a) H-FABP, (b) MDGI, (c) FRS-r, and (d) MP2. Thus, the 14.5-kDa growth inhibitor FGR-s seems to fall somewhere between MDGI and MP2.

III. Cellular Activities

1. Ehrlich Ascites Carcinoma Cells

Our first studies were directed to the detection, partial purification, and characterization of inhibitory activity starting from Ehrlich cells and ascites fluid. These initial studies were guided by the chalone concept that endogenous inhibitors are formed in the tissues on which they act. Growth of Ehrlich cells in vivo is characterized by a cell cycle of about 12 h. They reach a stationary phase at day 10–12 and may survive for another 3 weeks without killing the host. Cells from the stationary phase were taken for in vitro cultivation in the proliferation assay. Cells of that type proved to contain about 25%–35% cells in G1-phase, 35%–45% in G2-phase, and 15%–25% cells in S-phase of the cell cycle (Lehmann et al. 1980). These cells are characterized by a rather reduced metabolism, i.e., suppressed macromolecular synthesis, decreased nucleotide content, lack of oxygen, and increased glycolysis. Nevertheless, as has been demonstrated by Bichel (1971), these restrictive conditions are presumably not sufficient to explain reduced growth. The authors showed that transplantation of two different ascites tumors in one animal resulted in independent growth with a total cell number twice of what was observed in animals carrying only one tumor. The data argued against growth limitation due to depletion of nutrients and growth stimulatory peptides.

Initially, we followed Bichel's (1972) purification scheme and obtained an inhibitory fraction from the ascitic fluid of Ehrlich ascites carcinoma cells which partially blocked growth of Ehrlich cells at concentrations of 10–20 μg protein/ml. This fraction behaved similarly to MDGI in corresponding biological assays (Table 1). A dose-response curve for inhibition of prolifera-

Table 1. Action of MDGI and other growth inhibitory substances on Ehrlich ascites carcinoma cells

	Plateau in dose-response curve	Loss of inhibition after pre-incubation of cells with 4% serum	Effect on exponentially growing cells	Antagonism with	
				Insulin, EGF	2′-Deoxycytidine
MDGI	Yes	Yes	No	Yes	Yes
Interferon	No	No	NE	NE	NE
Synthetic cytostatics	No	No	Yes	No	Yes for some antimetabolites
Polyamines	No	No	Yes	NE	NE
Prostaglandin E_2	Yes	No	Yes	NE	NE
Synthetic peptide corresponding to residues 121–131 of MDGI	Yes	Partially	No	Partially	Yes

NE, not estimated.

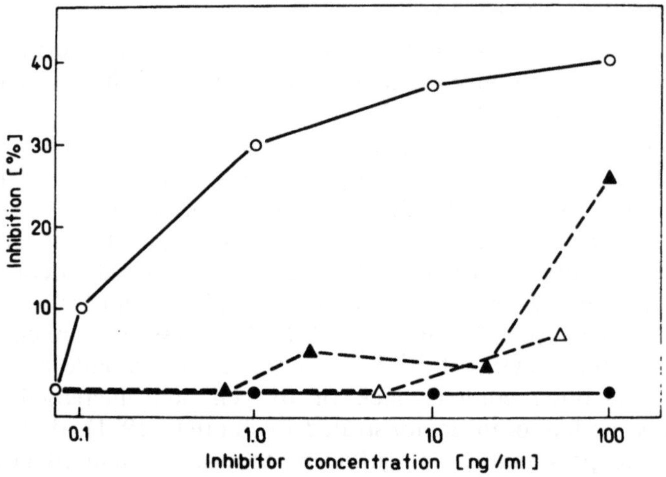

Fig. 3. Dose-response curve for inhibition of in vitro proliferation of Ehrlich ascites carcinoma cells by MDGI. Cells from the stationary growth phase in vivo were cultivated in vitro with MDGI (○–○), with MDGI plus 10 ng/ml EGF (▲–▲), or with MDGI plus 8 ng/ml insulin (△–△). In vitro preincubation of the cells for 4 h with serum before MDGI addition prevented inhibition (●–●). (BOEHMER et al. 1985)

tion of a hyperdiploid line of "stationary Ehrlich cells" by purified MDGI is given in Fig. 3. Half-maximal inhibition was obtained with 1 ng/ml (about 10^{-10} *M*). As further shown, cells became insensitive to inhibition by preincubation for 4 h with serum. Likewise, inhibition was abolished by simultaneously adding MDGI with EGF or insulin. The inhibitory activity of

MDGI is also antagonized by 10^{-5} *M* 2′-deoxycytidine. Some important properties of MDGI are summarized in Table 1 and compared with other inhibitory substances to demonstrate specific features of MDGI. Ehrlich cells were shown to possess PDGF receptors (F.-D. BOEHMER, unpublished results). An antagonistic PDGF effect was only observed if the cells were pretreated with 3–10 ng/ml PDGF 4 h before addition of MDGI. The PDGF effect then proceeded even if PDGF was washed out before MDGI was added. EGF or insulin did not prevent growth inhibition if added to Ehrlich cells during the preincubation period. In summary, our data indicate that MDGI acts by reducing the rate at which cells pass through some restriction point in G1/S. PDGF and EGF probably prevent this by interfering with MDGI action at different time points of the cell cycle. PDGF, EGF, or insulin do not influence growth of Ehrlich cells in vitro. One may speculate that the requirement for these growth factors, which has been lost during malignant transformation, is partially restored by MDGI. LEHMANN et al. (1989) have obtained evidence in favor of an MDGI-dependent retardation of growth of mammary epithelial cells during the G1/S transition. In Ehrlich cells, inhibition was lowest under conditions when G_2-phase cells were dividing as deduced from pulse cytometry. Inhibition increased with the entry of cells into G_1- and probably also into the S-phase of cell cycle. In a soft agar growth assay with MATU cells the rate of colony formation during the first 4–6 days in culture was reduced by MDGI, suggesting some influence on entry into S-phase (K. ECKERT, unpublished results). In contrast, under the same conditions TGF-*β* exerted a growth inhibition, which steadily increased, reaching a maximal value of 70% after 14 days in culture.

In parallel to MDGI, three peptides corresponding to the MDGI sequence have been synthesized and tested in Ehrlich cells (Table 2). The peptide comprising 11 amino acids of the C terminus behaved in some aspects as MDGI, i.e., its inhibitory activity was neutralized by insulin or 2′-deoxycytidine.

Table 2. Growth inhibition by MDGI-derived synthetic peptides

Peptide B[a] concentration (*M*)	Inhibition of cell proliferation (%)		
	No additions	Insulin (10^{-6} *M*)	2′-Deoxycytidine (10^{-5} *M*)
10^{-8}	17	0	0
10^{-7}	24	0	0
10^{-6}	33	27	0

[a] The following peptides were synthesized and tested: (1) EFDETTADDR, identical to residues 69–78 of MDGI; (2) TAVCTRVYEKQ, identical to residues 121–131 of MDGI; (3) TRVCTRVYELQ, two positions of peptide 2 modified. Peptides 1 and 3 were not active in the assay with Ehrlich ascites carcinoma cells.

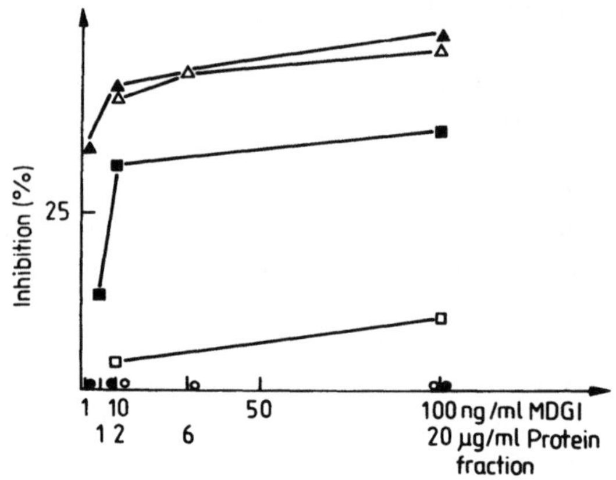

Fig. 4. Inhibition by MDGI and by protein fractions obtained from ascites fluid of proliferatioin of the wild type and a mutant strain line of Ehrlich ascites mammary carcinoma cells. The standard proliferation assay was performed as described in Sect. B.I. Wild-type cells were treated with MDGI (▲–▲), with a protein fraction enriched from the ascites fluid of the wild type (△–△), and with a protein fraction enriched from the mutant strain HD33 (□–□). The mutant strain HD33 was treated with MDGI (●–●), with the protein fraction from the ascites fluid of the wild type (○–○), and with the protein fraction from the ascites fluid of the mutant strain HD33 (■–■)

To address the question of possible effects of MDGI on protein synthesis in Ehrlich cells, cytosolic fractions were isolated from Ehrlich cells cultured in the presence of MDGI and tested for their activity to stimulate polysomal amino acid incorporation in a cell-free translation system (BIELKA et al. 1986). In parallel, MDGI was added directly to cytosolic fractions without prior interaction with Ehrlich cells. It was found that the cytosolic fraction isolated from Ehrlich cells after a 24-h in vitro incubation with MDGI suppressed protein synthesis; this did not occur with direct addition of MDGI to the cell-free translation assay. Thus, inhibition of cell proliferation seems to be related to some cytosolic activity requires signals mediated by the plasma membrane, since MDGI and its endogenous ligand do not directly affect translation.

A mutant strain HD33 of Ehrlich ascites carcinoma cells was selected by cytostatic treatment with *N*-methylcolchicamide. This strain is resistant toward colchicine treatment and does not produce glycogen. Fractions with inhibitory activities were partially enriched from the ascites fluid of HD33 and from the ascites fluid of the wild type (LEHMANN et al. 1987). These fractions and MDGI were tested for their growth inhibitory activities in mutant and wild-type Ehrlich cells (Fig. 4). The HD33 strain was completely insensitive both toward MDGI and the inhibitory activity enriched from the ascites fluid of the wild type. On the contrary, the fraction obtained from the ascites fluid of the mutant strain acted only on the mutant strain but hardly on the wild type. However, with regard to the antagonism between MDGI and growth stimulatory factors, the inhibitory fraction obtained from the HD33 strain resembled MDGI and the fraction from the wild-type ascitic fluid. The data suggest that regulation of cell proliferation by autocrine growth inhibitors can be altered by mutagen treatment.

2. Mammary Epithelial Cell Lines

The response of permanent mammary carcinoma cell lines and normal human mammary epithelial cells to MDGI was investigated (Table 3) (LEHMANN et al. submitted). The proliferation assay is based on a serum starvation to trigger cells into quiescence, followed by a restimulation with fresh medium. MDGI was present during the restimulation period for 16–20 h. Flow cytophotometric measurements with MATU, MCF-7, and normal human mammary epithelial cells proved them to be arrested in G1/G0. Obviously, synchronization of the cells in G1/G0 was a prerequisite to measure growth inhibition in the presence of MDGI. Cells not arrested by serum starvation were not responsive. This is in line with the considerations outlined previously in regard to Ehrlich cells. In Table 3 it can be seen that DNA synthesis in all but MCF-7 cells could be partially blocked by MDGI. For MATU cells the antagonistic effect of insulin was confirmed. Of special interest were the findings observed with normal human mammary epithelial cells. These cells were restimulated with 10% fetal calf serum (FCS), insulin, EGF, and pituitary extract according to STAMPFER and BARTLEY (1985). In Ehrlich cells the inhibitory effect of MDGI was compensated by EGF alone. Therefore, it is likely that normal cells may respond better to MDGI than malignant cells. In normal mammary epithelial cells inhibition was found to be correlated with

Table 3. Inhibition of DNA synthesis by MDGI in normal and malignant mammary epithelial cells

	% Restimulation			
	Control	40 ng/ml	400 ng/ml	4000 ng/ml
MATU[a]	80	70	48	37
T47D[a]	110	92	62	ND
MCF-7[a]	54	ND	64	ND
mMaCa[b]	281	213	190	122
Normal human mammary epithelial cells				
2nd passage	680	ND	680	ND
3rd passage	175	ND	136	ND
5th passage	105	ND	77	ND
6th passage	60	ND	24	
12th passage	70	ND	0	ND
14th passage	56	ND	0	ND

Serum-starved cells were restimulated in the presence of MDGI and fresh medium; DNA synthesis was determined after a 16-h restimulation period by pulse labeling with [^{3}H]TdR.

[a] Human mammary carcinoma cells lines derived from infiltrating ductal carcinomas.

[b] Derived from a spontaneous mouse adenocarcinoma.

passage number and degree of restimulation: strong inhibition was found at higher passages whose restimulation of DNA synthesis was decreased approximately ten fold compared with early passages. It is not clear whether this is due to a loss of proliferative capacity, induction of some differentiated function, or selection of responsive cell clones and how any of these events could be related to MDGI action.

IV. Biochemical and Cellular Mechanism of Action

1. Interaction with Hydrophobic Ligands

Since the proteins homologous to MDGI bind lipids, it was reasonable to assume that MDGI is also a lipid-binding protein. One might also speculate that some hydrophobic ligand is involved in MDGI-induced growth inhibition. In colloboration between Frank Boehmer of our laboratory, Morley Hollenberg (University of Calgary), and Guenter Reichmann (University of Halle, GDR) an analysis of putative endogenous MDGI ligands was performed (Table 4). The analysis revealed the presence of long-chain fatty acids, associated with MDGI prior to or during preparation, as well as the capability of MDGI to bind long-chain fatty acids in respective binding assays (BOEHMER et al. 1988). Interestingly, MDGI also carries significant amounts of covalently bound palmitic acid.

Posttranslational palmitate attachment to acyl proteins occurs via an O-ester or thiol ester bound (OLSON et al. 1985). These acyl proteins are transported specifically to cell membranes (OLSON et al. 1985; WILCOX and OLSON 1987). As noted previously, immunoreactive MDGI-related antigens have been detected in MFGM (BRANDT et al. 1988) and also in basolateral membranes prepared from lactating bovine mammary gland (E. SPITZER et al., unpublished observations). Membrane localization might have been directed in these cases by covalent attachment of fatty acids to MDGI.

As shown in Table 4, different potential MDGI ligands were investigated either by direct binding assays or by measuring the displacement of labeled fatty acid by the putative ligand. All tested ligands studied so far were found to exert only relatively weak interactions with the protein. The binding constants for the long-chain fatty acids are too low to assume depletion of some essential fatty acids from the medium by MDGI, which in turn could have brought about growth inhibition in vitro. An opposite effect was observed when [^{3}H]palmitic acid was bound to MDGI and added to MATU-cells (BOEHMER et al. 1988). The bound [^{3}H]palmitic acid was more rapidly taken up by the target cells than was free [^{3}H]palmitic acid, suggesting a lipid-carrying function of the inhibitor. Moreover, unlabeled MDGI affected the cellular uptake of the MDGI-[^{3}H]palmitic acid complex. The concentration dependence of the observed inhibition by unlabeled MDGI for the uptake of [^{3}H]palmitate-labeled MDGI suggested the presence of a saturable carrier responsible for the MDGI-facilitated uptake of [^{3}H]palmitate. This finding can be interpreted in terms of a cellular receptor for MDGI.

Table 4. Lipid binding of MDGI

	Endogenously bound lipid (%)		Displacement of MDGI-bound [^{3}H]palmitic acid		
	Extractable	Covalently bound	% of [^{3}H]-palmitic acid displaced with excess of unlabeled lipid		Binding constants (10^{-6} *M*)
			1000-fold	100-fold	
Total (nmol/nmol)	1.159±0.580	0.647±0.516			
Palmitic acid (16:0)	16.2±4.5	21.7±2.1	85	58	2.6
Stearic acid (18:0)	9.4±2.9	16.4±2.4	NE	NE	
Oleic acid (18:1)	34.9±4.2	9.4±2.4	85	88	0.9
Linoleic acid (18:2)	7.5±1.5	4.3±1.2	NE	NE	
Arachidonic acid (20:4)	1.9±0.6	4.6±1.2	82	88	1.6
Myristic acid (14:8)	1.7±0.8	2.5±0.8	66	31	NE
Caprylic acid (8:0)	ND	ND	20	6	
Heptadecanoic acid (17:0)	ND	ND	86	NE	3.3
Retinoic acid	ND	–	30	0	
Retinyl acetate	NE	–	Enhanced binding	8	
Retinol	ND	–	Enhanced binding	12.4	
Estradiol	ND	–	8	4	
Diacylglycerol	NE	–	49	22	
Phosphatidylcholine	NE	–	51	NE	
Prostaglandin E_1	ND	–	0	NE	
Prostaglandin E_2	ND	–	5	NE	
Thromboxane B_2	NE	–	58	0	
Thromboxane A_2 agonist	NE	–	64	4	

ND, not detectable; NE, not estimated.

As described before, a synthetic peptide corresponding to residues 121–131 of the MDGI sequence exhibits very similar effects to MDGI. This peptide, however, binds only negligible amounts of fatty acid. Thus, fatty acid binding might not be related to the biological activities described here. Binding of long-chain fatty acids could stabilize the MDGI structure because delipidated MDGI has a very high tendency to aggregate (Behlke et al. 1989). Aggregation of the homologous FABP from mammary tissue of lactating rat (P. D. Jones et al. 1988) has been reported. Alternatively, fatty acid binding might be related to some other function of MDGI, not related to growth inhibition.

2. Possible Role of Ribonucleotide Reductase

We have found that in Ehrlich cells 10 μ*M* 2′-deoxycytidine can overcome inhibition by MDGI; in this respect it resembles insulin or EGF (Lehmann et al. 1979). Cytidine is also effective, but uracil, uridine, and other nucleobases or nucleosides are not. Inhibition antagonized by 2′-deoxycytidine is usually ascribed to suppression of ribonucleotide reductase (RR), a key enzyme of DNA replication and cell proliferation which catalyzes the formation of 2′-deoxycytidine phosphates, which is the rate-limiting step in DNA synthesis. 2′-Deoxycytidine has been previously shown to relieve RR from inhibition by thymidine-5′-triphosphate, purine 2′-deoxyribonucleoside 5′-triphosphates, or anaerobiosis. The effect of cytidine is in accordance with the finding that the cellular content of 2′-deoxycytidine phosphates correlates directly with that of the cytidine phosphates, i.e., the enzyme works within nonsaturating substrate concentrations, and its rate can be increased by exogenous cytidine. The results are therefore also compatible with a suppression by MDGI of the metabolic pathway leading to the cytidine 5′-phosphates. These, however, affect only RR activity (and not RNA synthesis) in a way resulting in inhibition of cell proliferation. Similar effects of 2′-deoxycytidine on growth-inhibiting peptides have been described for hematopoietic cell lines (SCHUNCK et al. 1988; BHALLA et al. 1986).

C. Conclusions

We have detected a polypeptide designated MDGI involved in growth regulation of the mammary gland. Its tissue specificity remains to be elucidated. MDGI belongs to a family of structurally related polypeptides which may fulfill different functions including growth arrest during normal development of tissues and organs. MDGI may also represent a new biologically active polypeptide for the control of nonproliferative processes such as adrenoreceptor regulation.

Many important questions related to the mode of action and regulation of gene expression of MDGI remain to be answered. Although there is circumstantial evidence for both MDGI secretion and MDGI interaction with the plasma membrane, so far a receptor has not been identified. Therefore, we need to consider that MDGI might act directly intracellularly; such a mechanism could involve MDGI interaction with chromatin. Whether there is a specific MDGI ligand involved remains unclear. Further investigations may lead to more important physiological ligands than those described so far for FABPs.

The analysis of the MDGI gene will play an important role for our understanding of the biological function of MDGI. In this regard the structure of the adipocyte P422 gene has given us a first insight into the mechanism of how proliferation and differentiation may be controlled at the transcriptional level.

For MDGI, expression studies and experiments with cloned gene segments inserted into transfection vectors will help to elucidate to what extent MDGI is controlling or is being controlled by events crucial for the transition from proliferation to differentiation in the mammary gland.

Acknowledgements. We thank Drs. Frank Boehmer, Frank Vogel, and Wolfgang Zschiesche for critical reviews of this article. We are grateful to Mrs. I. Wiznerowicz and E. Hellmuth for excellent help in preparing the manuscript.

References

Baird A, Ling N (1987) Fibroblast growth factors asre present in the extracellular matrix produced by endothelial cells in vitro: implications for a role of heparinase-like enzymes in the neovascular response. Biochem Biophys Res Commun 142:428–436

Bass NM (1985) Function and regulation of hepatic and intestinal fatty acid binding proteins. Chem Phys Lipids 38:95–114

Bassuk JA, Tsichlis PN, Sorof S (1987) Liver fatty acid binding protein is the mitosis-associated polypeptide target of a carcinogen in rat hepatocytes. Proc Natl Acad Sci USA 84:7547–7551

Beheke J, Mieth M, Bochmer FD, Grosse R (1989) Hydrodynamic and circular dichroic analysis of mammary-derived growth inhibitor. Biochem Biophys Res Commun 161, 363–370

Bernlohr DA, Angus CM, Lane MD, Bolanowski MK, Kelly TK (1984) Expression of specific mRNAs during adipose differentiation: identification of an mRNA encoding a homologue of myelin P2 protein. Proc Natl Acad Sci USA 81:5468–5472

Bernlohr DA, Bolanowski MA, Kelly TJ, Lane MD (1985) Evidence for an increase in transcription of specific mRNAs during differentiation of 3T3-L1 preadipocytes. J Biol Chem 260:5563–5567

Bhalla K, Cole J, MacLaughlin W, Baker M, Arlin Z, Graham G, Grant ST (1986) Deoxycytidine stimulates the in vivo growth of normal CFU-GM and reverses the negative effects of acidic isoferritin and prostaglandin E_1. Blood 68:1136–1141

Bichel P (1971) Autoregulation of ascites tumor growth by inhibition of the G_1 and G_2 phase. Eur J Cancer 7:349–355

Bichel P (1972) Specific growth regulation in 3 ascites tumors. Eur J Cancer 8:167–173

Bielka H, Grosse R, Boehmer FD, Jughahn I, Binas B (1986) Inhibition of proliferation of Ehrlich ascites carcinoma cells is functionally correlated with reduced activity of the cytosol to stimulate protein synthesis. Biomed Biochim Acta 45:441–445

Boehmer FD, Lehmann W, Schmidt H, Langen P, Grosse R (1984) Purification of a growth inhibitor for Ehrlich ascites mammary carcinoma cells from bovine mammary gland. Exp Cell Res 150:466–476

Boehmer FD, Lehmann W, Noll F, Samtleben R, Langen P, Grosse R (1985) Specific neutralizing antiserum against a polypeptide growth inhibitor for mammary cells purified from bovine mammary gland. Biochim Biophys Acta 846:145–154

Boehmer FD, Kraft R, Otto A, Wernstedt C, Hellmann U, Kurtz A, Mueller T, Rohde K, Etzold G, Lehmann W, Langen P, Heldin CH, Grosse R (1987a) Identification of a polypeptide growth inhibitor from bovine mammary gland. J Biol Chem 262:15137–15143

Boehmer FD, Sun Q, Pepperle M, Mueller T, Eriksson U, Wang JI, Grosse R (1987b) Antibodies against mammary derived growth inhibitor (MDGI) react with a fibroblast growth inhibitor and with heart fatty acid binding protein. Biochem Biophys Res Commun 148:1425–1431

Boehmer FD, Mieth M, Reichmann G, Taube C, Grosse R, Hollenberg MD (1988) A polypeptide growth inhibitor isolated from lactating bovine mammary gland (MDGI) is a lipid carrying protein. J Cell Biochem 38:199–204

Brandt R, Pepperle M, Otto A, Kraft R, Boehmer FD, Grosse R (1988) A 13 kDa protein purified from milk fat globule membranes is closely related to a mammary-derived growth inhibitor. Biochemistry 27:1420–1425

Bullough WS, Laurence EB (1960) The control of epidermal mitotic activity in the mouse. Proc Royal Soc London Ser B 151:990–993

Claffey KP, Herrera VL, Brecher P, Riuz-Opazo N (1987) Cloning and tissue distribution of rat heart fatty acid binding protein mRNA: identical forms in heart and skeletal muscle. Biochemistry 26:7900–7904

Cook JS, Lucas JJ, Sibley E, Bolanowski MA, Christy RJ, Kelly TJ, Lane MD (1988) Expression of the differentiation-induced gene for fatty acid-binding protein is activated by glucocorticoid and cAMP. Proc Natl Acad Sci USA 85:2949–2953

Cook KS, Hunt CR, Spiegelman BM (1985) Developmentally regulated mRNAs in 3T3-adipocytes: analysis of transcriptional control. J Cell Biol 100:514–520

Distel RJ, Ro HS, Rosen BS, Groves DL, Spiegelman BM (1987) Nucleoprotein complexes that regulate gene expression in adipocyte differentiation: direct participation of c-*fos*. Cell 49:835–844

Eriksson U, Hansson E, Nilsson M, Joensson KH, Sundelin J, Peterson PA (1986) Increased levels of several retinoid binding proteins resulting from retinoic acid-induced differentiation of F9 cells. Cancer Res 46:717–722

Gibson BW, Yu Z, Aberth W, Burlingame AL, Bass NM (1988) Revision of the blocked N terminus or rat heart fatty acid binding by liquid secondary ion mass spectrometry. J BiolChem 263:4182–4185

Herrmann I, Grosse R (1986) Characterization of membrane-associated growth inhibitor activity for Ehrlich ascites mammary carcinoma cells. Biomed Biochim Acta 45:447–457

Hunt CR, Ro JH, Dobson DE, Min HY, Spiegelman BM (1986) Adipocyte P2 gene: developmental expression and homology of 5'-flanking sequences among fat cell-specific genes. Proc Natl Acad Sci USA 83:3786–3790

Iversen OH (1960) Cell metabolism in experimental skin carcinogenesis. Acta Pathol Microbiol Scand [B] 50:17–24

Iversen OH (1981) The chalones. In: Baserga R (ed) Tissue growth factors. Springer, Berlin Heidelberg New York, p 491

Jones PD, Carne A, Bass NM, Grigor MR (1988) Isolation and characterization of fatty acid binding proteins from mammary tissue of lactating rats. Biochem J 251:919–925

Jones TA, Bergfors T, Sedzik J, Unge T (1988) The three-dimensional structure of P2 myelin protein. EMBO J 7:1597–1604

Kempen HJ, Glatz J, De Lange J, Veerkamp J (1983) Concomitant increase in hepatic triacylglycerol biosynthesis and cytosolic fatty-acid-binding-protein content after feeding rats with a cholestyramine-containing diet. Biochem J 216:511–514

Langen P (1985) Chalones and other growth inhibitors. In: Torrence P (ed) Biological response modifiers. Academic, New York, pp 265–291

Lehmann W, Graetz H, Schuett M, Langen P (1979) Antagonistic effects of insulin and a negative growth regulator from ascites fluid on the growth of Ehrlich ascites carcinoma cells in vitro. Exp Cell Res 119:396–399

Lehmann W, Graetz H, Samtleben R, Schuett M, Langen P (1980) On a chalone-like factor for the Ehrlich ascites mammary carcinoma. Acta Biol Med Ger 39:93–105

Lehmann W, Samtleben R, Graetz H, Langen P (1983) Purification of a chalone-like inhibitor for Ehrlich ascites mammary carcinoma cells from bovine mammary gland. Eur J Cancer Clin Oncol 19:101–107

Lehmann W, Boehmer FD, Karsten U, Graetz H, Koberling A, Kunde D, Langen P (1987) Altered growth control by natural growth inhibitors in the mutant HD33 substrain of the Ehrlich-Lettre ascites mammary carcinome. Eur J Cancer Clin Oncol 23:69–74

Lehmann W, Widmaier R, Langen P (1989) Response of different mammary epithelial cell lines to a mammary derived growth inhibitor (MDGI) Biomed Biochim Acta 48:143–151

Matarese V, Bernlohr DA (1987) Purified 422 protein is a vitamin A binding protein. Fed Proc 46:2005

Offner GD, Brecher P, Sawlivich WB, Costello CE, Troxler RF (1988) Characterization and amino acid sequence of a fatty acid-binding protein from human heart. Biochem J 252:191–198
Olson EN, Towler DA, Glaser L (1985) Specificity of fatty acid acylation of cellular proteins. J Biol Chem 260:3784–3790
Rakowicz-Szulczynska EM, Rodeck U, Herlyn M, Koprowski H (1986) Chromatin binding of epidermal growth factor, nerve growth factor, and platelet-derived growth factor in cells bearing the appropriate surface receptor. Proc Natl Acad Sci USA 83:3728–3732
Schunk H, Schuett M, Langen P, Lord B, Laerum OD, Paukovits WR (1988) 2′-Deoxycytidine, insulin and epidermal growth factor overcome the effect of two natural inhibitors of the haemopoietic system. Acta Pathol Microbiol Immunol Scand 96 (Suppl 2):120–129
Spiegelman B, Frank M, Green H (1983) Molecular cloning of mRNA from 3T3 adipocytes. Regulation of mRNA content for glycerophosphate dehydrogenase and other differentiation-dependent proteins during adipocyte development. J Biol Chem 258:10083–10089
Sporn MB, Roberts AB (1985) Autocrine growth factors and cancer. Nature 313:745–747
Stampfer MR, Bartley JC (1985) Induction of transformation and continuous cell lines from normal human mammary epithelial cells after exposure to benzo-*a*-pyrene. Proc Natl Acad Sci USA 82:2394–2398
Steck PA, Blenis J, Voss P, Wang J (1982) Growth control in cultured 3T3 fibroblasts. II. Molecular properties of a fraction enriched in growth inhibitory activity. J Cell Biol 92:523–530
Wang JL, Hsu YM (1986) Negative regulators of cell growth. Trends Biochem Sci 11:24–26
Wilcox CA, Olson EN (1987) The majority of cellular fatty acid acylated proteins are localized to the cytoplasmic surface of the plasma membrane. Biochemistry 26:1029–1036

CHAPTER 28

Pentapeptide Growth Inhibitors

W. R. PAUKOVITS, K. ELGJO, and O. D. LAERUM

The hemoregulatory peptide and the epidermal pentapeptide are small oligopeptides (molecular weight around 500–600) involved in the regulation of proliferation in their respective tissues in vitro and in vivo. The identification of these peptides was achieved after a 20-year-long search for specific endogenous feedback inhibitors of proliferation initiated by P. WEISS and KAVANAU (1957), BULLOUGH and LAURENCE (1960), and IVERSEN (1960). Both peptides function in a similar way in their respective tissues. They probably originate from differentiated (mature) cells, i.e. polymorphonuclear neutrophil (PMN) granulocytes or mature keratinocytes, respectively, and they reduce the proliferative activity in their progenitor cells (stem cells, myeloid progenitors, or epidermal basal cells). They are similar (Fig. 1) in chemical structure, suggesting a larger family of regulatory peptides of which the hemoregulatory peptide and the epidermal pentapeptide are the first known members.

Attempts to purify the two inhibitory pentapeptides date back to the late 1960s when two of the authors (K. E. and O. D. L.) were working with cell renewal and growth regulation in hairless mouse epidermis. Later, the work came to include hematopoiesis (PAUKOVITS 1971). The attempts to purify mitosis inhibitors from the epidermis and from mature granulocytes gave no immediate results. However, in 1982 PAUKOVITS and LAERUM described the structure of a pentapeptide from mature granulocytes that reversibly inhibited myelopoiesis in vivo and in vitro. Two years later, in 1984, ELGJO and REICHELT described a pentapeptide in mouse epidermis that reversibly inhibited epidermal cell proliferation in vivo and in vitro. The two pentapeptides have many characteristics in common. The purification procedures and some tissue-related effects are, however, different. The two pentapeptides are, therefore, described separately.

HP pGlu – Glu – Asp – Cys – Lys

EPP pGlu – Glu – Asp – Ser – Gly

Fig. 1. Amino acid sequences of the hemoregulatory peptide (*HP*) and the epidermal pentapeptide (*EPP*)

A. Hemoregulatory Peptide

Proliferation of myeloid progenitor cells has been found to be directly controlled by the interaction of at least four distinct hematopoietic growth factors: interleukin-3, granulocyte CSF, granulocyte-macrophage CSF, and interleukin-6. The action of these is counterbalanced by inhibitory factors like transforming growth factors-β, interleukin-2, interleukin-6 and interferons (see the respective chapters in this book). While most of the regulatory molecules originate from tissues and cells which are not the direct progeny of their respective target cells, other experiments have revealed the existence of inhibitors which are endogenous to the particular cell lineage. It is well known that mature granulocytes inhibit the formation of granulocyte-macrophage colonies (PARAN et al. 1969; CHAN 1971; HAVEMANN et al. 1979; BJORNSON et al. 1984). This is seen in culture where bone marrow cell progenitors are present in the upper and granulocytes in the lower layer of a double-layer culture system (HINTERBERGER et al. 1977). Although a variety of causes have been discussed for this inhibition the most consistent explanation seems to be that the granulocytes produce diffusible inhibitory molecules acting on the colony-forming cells in the upper layer.

In line with this, several investigators have shown that crude or semipurified extracts of granulocytes inhibited myelopoietic cell proliferation, both in vitro and in vivo (RYTÖMAA and KIVINIEMI 1968a, b; BENESTAD et al. 1973; LAERUM and MAURER 1973; PERRINS et al. 1980; BALAZS et al. 1980; for historical review, see also IVERSEN 1976, 1981).

I. Preparation of the Hemoregulatory Peptide

1. Sources

According to experimental evidence (RYTÖMAA and KIVINIEMI 1968c; PAUKOVITS 1971, 1973; LORD et al. 1976; MAURER et al. 1976; KASTNER and MAURER 1980; BALAZS et al. 1980; PELUS et al. 1981; PAUKOVITS et al. 1983), and also in accordance with theoretical expectations for a so-called chalone hypothesis (BULLOUGH and LAURENCE 1960), cells belonging to the granulocytic lineage contain/release a low-molecular-weight inhibitor of myeloid cell production which not only inhibits the proliferative activity of myeloid cells in vitro but also in vivo. However, evidence remains inconclusive that this inhibitor mainly originates from mature granulocytes although several findings support this view (BENESTAD and HERSLETH 1984). RYTÖMAA and KIVIENIEMI (1968a, b) have observed that repeated extractions and conditioning of media by mature granulocytes lead to rapidly decreasing yields of inhibitory activity. As a consequence of these observations it is difficult to obtain good yields of the inhibitory peptide from highly purified granulocytes. Summarizing the experimental evidence it appears that (a) all cell populations containing granulocytes contain a low-molecular-weight inhibitory factor, (b) the yield of this inhibitor is dependent on the number of granulocytes present,

(c) washing of the cells decreases the yield, and (d) peripheral white blood cells are a weaker source of inhibitor than granulocytes from the bone marrow.

2. Fractionation

Dialyzates of conditioned media and *ultrafiltrates* of either conditioned media or homogenates have been used as a reliable source of the inhibitor (PAUKOVITS 1971, LAERUM and MAURER 1973; AARDAL et al. 1977; PAUKOVITS et al. 1983). The inhibitor also appeared to be soluble in polar *organic solvents* (PAUKOVITS and PAUKOVITS 1975b), which has led to the development of a procedure utilizing acetone extraction of crude conditioned media as the primary purification step (KASTNER and MAURER 1980). One of the main advantages of this method is the low solubility of most proteins and other biological molecules and inorganic salts in acetone.

Gel chromatography has been used extensively as a method for semipurification of the inhibitory peptide. It was, however, shown (PAUKOVITS et al. 1983) that dextran gels with the exception of Sephadex G-10 do not separate the hemoregulatory peptide from salts and other small molecules. Also the small pore polyacrylamide gels of the BioGel type are not applicable for desalting. The elution volume from Sephadex G-10 columns ($V_e/V_o = 1.3–1.4$) points to a molecular weight of the naturally occurring peptide below 700.

Ion Exchange Fractionation. In the naturally occurring peptide carboxyl groups outweigh amino groups by a factor of 3 (PAUKOVITS et al. 1987). Thus,

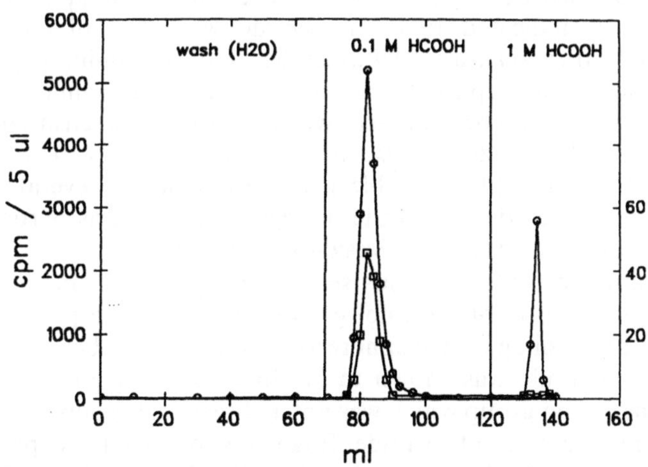

Fig. 2. Chromatography of the hemoregulatory peptide on a Dowex-1 anion exchange column. The hemoregulatory peptide binds rather strongly to the resin and can only be eluted under strongly acidic conditions (0.1 *M* formic acid). The elution was monitored by adding a small amount of reductively ^{3}H-methylated hemoregulatory peptide as a tracer to the sample and measuring the radioactivity per fraction (○) (PAUKOVITS et al. 1983). Aliquots of the fractions were tested for their inhibitory potency for CFU-GM colony formation (□)

the peptide binds to anion exchangers tightly and can be eluted only using drastic conditions. Figure 2 gives an example of the separation of peptide on a small column of the strong anion exchanger AG1 × 2. A high-resolution method for separating the peptide on mono-Q columns has recently been described by FETSCH and MAURER (1987).

Thiol-Binding Chromatography. The presence of thiol groups in the inhibitory peptide molecule (AARDAL et al. 1982; PAUKOVITS and HINTERBERGER 1978 a) provides the possibility of binding the peptide in its reduced state to polymer-bound sulfydryl reagents. A very efficient separation from non-thiol-containing contaminants can be achieved in this way. We have used the insolubilized thiopyridone derivative thiopropyl-sepharose 6B, although other materials are available which also selectively bind thiol-containing compounds (PAUKOVITS et al. 1983).

Reverse-phase HPLC of the peptide results in the expected early elution (highly polar small peptide) and is used as the final purification step and for demonstration of the purity of hemoregulatory peptide preparations (PAUKOVITS and LAERUM 1984; PAUKOVITS et al. 1986).

The *yield* of pure hemoregulatory peptide was very low irrespective of the source used. Human leukocytes, which are very convenient for "large-scale" purification, gave less than 10 μg/kg (PAUKOVITS and LAERUM 1982).

II. Structural Studies

Sequence determination of the granulocyte-associated peptide initially used information obtained during purification. The chromatographic properties mentioned above indicated that the peptide was a small (600–700 Da, PAUKOVITS and HINTERBERGER 1978 a) acidic, thiol-containing peptide . Attempts to sequence the peptide by the standard Edman technique were unsuccessful, and the reason became clear when we studied its sensitivity against a variety of proteolytic enzymes (PAUKOVITS and PAUKOVITS 1978). The peptide was found to be completely stable against the action of several proteases; the only enzyme which destroyed the biological activity was pyroglutamate aminopeptidase (PAUKOVITS and HINTERBERGER 1978 b).

The inhibitory peptide was thus sequenced by indirect methods. After [^{3}H]-carboxymethylation of the thiol group the radioactive peptide was investigated by high-voltage electrophoresis. The exact number of functional groups in the molecule was shown to be three carboxyl groups, one thiol group, and one amino group which was not a N-terminal group. Further electrophoretic investigation of peptide fragments obtained by partial acid hydrolysis then led to the determination of the distribution of these functional side-chain groups in the peptide molecule (PAUKOVITS and LAERUM 1984): N-terminal position: pyroGlu; pos. 2, carboxyl; pos. 3, carboxyl; pos. 4, thiol; pos. 5, amino plus a free C-terminal carboxyl group. The sequence consistent with these results was then deduced as:

pyroGlu-(Asp or Glu)-(Asp or Glu)-Cys-Lys-OH.

Peptides compatible with this sequence were synthesized. One of them, pyroGlu-Glu-Asp-Cys-Lys-OH (see Fig. 1), was found to possess the full range of biological properties in vitro and in vivo (PAUKOVITS and LAERUM 1982; LAERUM and PAUKOVITS 1984a). Since the formal proof was lacking that this was identical to the inhibitor(s) present in crude and semipurified granulocyte extracts, a neutral term was chosen as denominator of this class of regulators, the hemoregulatory peptide(s), together with a code indicating the number of amino acids. This structure was, therefore, labeled HP5b, while HP5a was used as code for the sequence pyroGlu-Asp-Asp-Cys-Lys-OH. The latter peptide was published by PAUKOVITS et al. in 1980, but later turned out to be biologically inactive.

Based on studies with synthetic analogs of HP5b, the active compound seems to have a high structural specificity. Small chemical modifications usually lead to loss of biological effect. However, the peptide PyroGlu-Glu-Asp-Cys-Gly-Lys-OH (HP6a) is biologically active, but less than HP5b.

III. Synthesis

Liquid and solid-phase synthesis of HP5b using N-Boc/-OBzl/-SBzl/- -N-Cbz strategy was initially employed (PAUKOVITS and LAERUM 1982; PAUKOVITS et al. 1986). The synthesis produced complicated mixtures from which HP5b could be purified only with great difficulty. Stability problems (activity was reduced and even completely lost during the complicated purification procedure) were partly explained when it was found that unprotected HP5b was highly oxidation sensitive under physiological conditions.

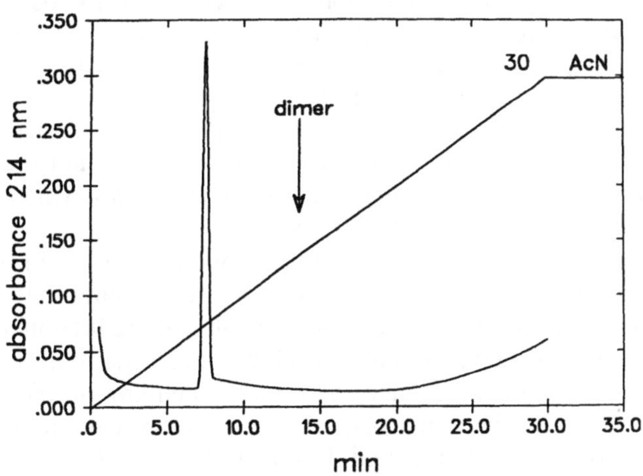

Fig. 3. Reverse-phase HPLC of the hemoregulatory peptide monomer (4.6 × 250 mm Beckman-Altex C18 column with acetonitrile gradient in 0.1% trifluoroacetic acid). The *arrow* indicates the elution position of the dimer. The inhibition-stimulation antagonism of the hemoregulatory peptide monomer and dimer (see text) requires the use of an absolutely dimer-free preparation in all bioassays of the monomer (and vice versa)

We have thus changed synthetic strategies, and most of the previous difficulties can be avoided (PAUKOVITS et al. 1987; ERIKSEN et al. 1987). Both new procedures are based on FMOC strategies avoiding use of strong mineral acids, and protection of the thiol group of cysteine during the entire procedure, including purification. ERIKSEN et al. (1987) used the disulfide-bonded cysteine and synthesized the HP5b-dimer directly, while PAUKOVITS et al. (1987) used tertiary butyl mercaptan as the second partner of the disulfide bridge. Both methods allow purification and storage of the synthetic product as oxidation-insensitive derivatives. The reduction step to obtain the HP5b monomer can then be performed either before a final purification step, or in situ, requiring only the evaporation of the highly volatile tertiary butyl mercaptan (PAUKOVITS et al. 1987, 1990a, b). These methods allow the preparation of essentially dimer-free HP5b monomer, as shown in Fig. 3.

To avoid dimerization, solution of the monomer in either 10^{-4} *M* 2-mercaptoethanol or 3 μ*M* Ca Na_2 ethylenediaminetetraacetic acid (EDTA) is recommended. The dimer, on the other hand, is more stable and can be dissolved in saline.

IV. Biological Activities on Normal Hematopoiesis

A common problem during the purification procedures was that the inhibitory effect was often not reproducible in parallel experiments and that it was easily lost. Some workers could not find any such inhibitor (LORD et al. 1977; HERMAN et al. 1978), while others found that the granulocyte extracts were partly inhibitory and partly stimulatory depending on how the fractions were treated (AARDAL et al. 1977). Even after the purification and synthesis of the hemoregulatory peptide the inhibitory effect was easily lost and replaced by a stimulatory effect (LAERUM and PAUKOVITS 1984a). This seemed to be due to an oxidation process and could be counteracted by the presence of SH compounds such as 2-mercaptoethanol.

In a recent publication this stimulatory substance was identified as the disulfide-bridged dimer of the hemoregulatory peptide (LAERUM et al. 1988a). This dimer has a strong stimulatory effect on myelopoiesis, which will be dealt with in more detail in later parts of this review.

Thus, it seems that the hemoregulatory peptide can occur in two different chemical versions, a monomer that is inhibitory and a dimer that is stimulatory. Given the present knowledge of oxidation and reduction processes in granulocytes, it is conceivable that granulocytes can maintain an adjustable equilibrium between the monomer and dimer depending on the actual needs for cell renewal in the bone marrow (see e.g., KLEBANOFF and CLARK 1978; S.J. WEISS et al. 1983; BANNAI and TATEISHI 1986; WATANABE and BANNAI 1987; LAERUM et al. 1988a).

1. Growth-Promoting Activity of HP5b Dimer

Pure synthetic dimer stimulates the formation of myelopoietic colonies in agar (CFU-GM) both in human and in mouse cells. This occurs as a dose-depen-

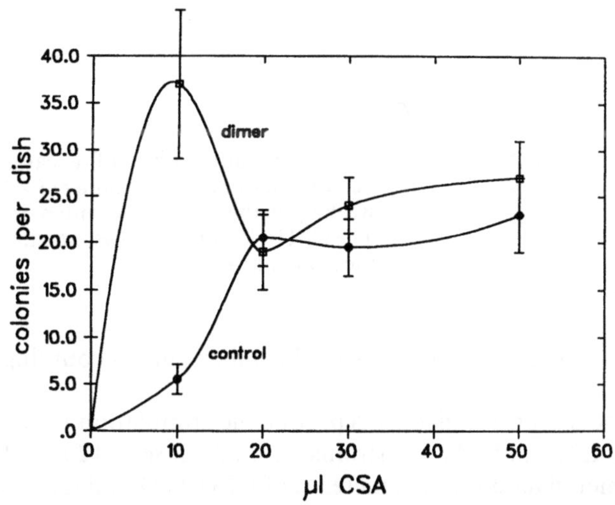

Fig. 4. Stimulatory effect of the hemoregulatory peptide dimer on murine CFU-GM showing its relation to the concentration of CSF in the cultures. The strongest dimer effect is observed at low CSF doses. Mouse endotoxin serum was used as CSA in these experiments, and dimer and CSA were added to the cultures simultaneously

dent enhancement of colony formation in the dose range 10^{-16}–10^{-5} *M*, with a saturation level reached at about 10^{-8} *M* (Laerum and Paukovits 1987; Laerum et al. 1988 a).

At low doses of colony-stimulating activity and in the linear dose range of CSF, up to tenfold increases of colony formation can be seen (Fig. 4). At higher CSF concentrations the effect is less pronounced. This enhancement of colony formation is dependent on the presence of a source of colony-stimulating factor, either as crude extract (mouse endotoxin-stimulated serum or mouse lung-conditioned medium; human placenta-conditioned medium) or as pure recombinant murine or human GM-CSF. The dimer stimulates all types of colonies, including granulocyte, macrophage, and mixed colonies. Still it is not a colony-stimulating factor itself, since the addition of dimer alone to agar cultures without CSF does not stimulate colony formation except for a few endogenous colonies (Laerum et al. 1988 a). When the dimer is reduced to monomer by use of 2-mercaptoethanol, the normal inhibitory monomer effect is restored.

The testing of hemoregulatory peptide dimer in vitro has in practice yielded variable results. The degree of stimulation is dependent on a relatively low background stimulation of colony formation by CSF and serum and is best at a relatively low plating efficiency (20–40 colonies/35-mm dish) (Fig. 4). Secondly, there are wide interexperimental variations; even under standardized conditions the effect can vary between no stimulation to ten times stimulation of colony formation. These problems seem to be associated with unknown variable culture conditions and will probably be solved by the use of serum-free cultures. As mentioned above, it cannot be excluded that bone

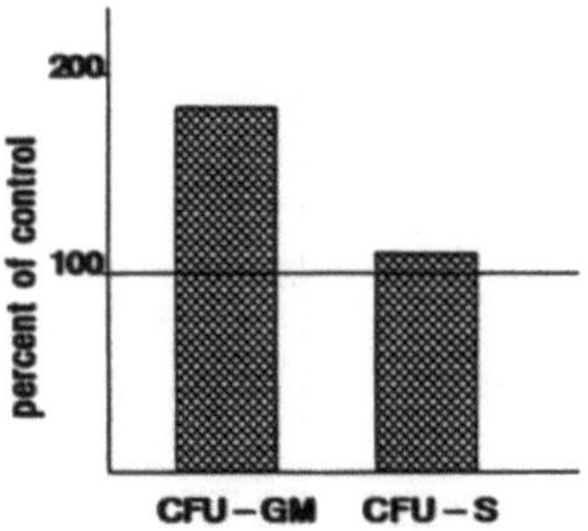

Fig. 5. Stimulatory effect of repeated injections of hemoregulatory peptide dimer (8 × 120 ng, twice weekly) on the CFU-GM content in murine spleen. The CFU-GM number was measured on day 28 by the agar culture method

marrow cells are capable of processing the dimer, thus accounting for the variable effects.

In vivo the dimer has given a reproducible stimulation of myelopoiesis in mice, in both C3H and C57/BL strains. A single dose of 120 ng, 1.2 µg, 12 µg, or 120 µg dimer gives a gradual increase of CFU-GM and CFU-S numbers in femur and spleen, reaching more than a doubling after 2–3 days. The increased levels usually persist for about 11 days. In contrast to in vivo actions of GM-CSF, the stimulation by dimer is stronger in femur than in spleen (LAERUM et al. 1989).

Continuous infusion with 12 ng/h or 1.2 µg/h for 6 days increases the femoral CFU-GM but not the CFU-S numbers after 13 days. Similarly, 2 weekly injections of 12 ng almost double the CFU-GM numbers and give a moderate increase of CFU-S after 28 days (Fig. 5). The effect of the dimer in vivo seems to be strongest at relatively low doses. O. SOLESVIK and N. PAPE (personal communication) have been able to stimulte CFU-GM in mice in vivo with doses down to 2 ng/mouse. Thus the dose-response relationship of the dimer in vivo seems to be nonlinear and the effect is seen in a restricted dose range.

In general, the in vivo effects of the dimer seem to be more stable than the in vitro effects. Both CFU-GM and CFU-S are stimulated, although the multipotent stem cells respond to a somewhat lesser extent. The consequences of this stem and progenitor cell stimulation have also been investigated (LAERUM et al. 1989). After a latency phase of some days, the bone marrow cellularity gradually increases, and after 1–2 weeks a granulocytosis is seen in peripheral blood. Also an increase of lymphocyte numbers has been observed.

2. Growth Inhibitory Activity of HP5b Monomer

In vitro the hemoregulatory peptide monomer gives a dose-dependent inhibition of mouse and human CFU-GM (LAERUM and PAUKOVITS 1984a; KREJA et al. 1986; LAERUM et al. 1987). Mouse CFU-GM are inhibited from 10^{-13} *M*, with an optimum around 10^{-11}–10^{-7} *M*. Human cells are inhibited from 10^{-11} *M*, with an optimum at 10^{-7} *M* (Fig. 6). Above this concentration, the inhibitory effect is gradually lost.

In single experiments (Fig. 6) up to 90% inhibition of colony formation can be seen at the optimal dose (KREJA et al. 1986). There is, however, a strong

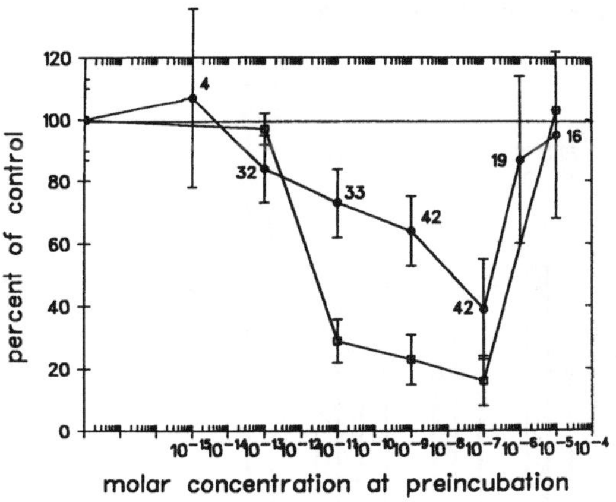

Fig. 6. Inhibitory effect of different concentrations of the hemoregulatory peptide monomer on human CFU-GM. The cells were exposed to the peptide for 1 h prior to addition of CSF and cultured for 12 days. The dose-response curve for human cells represents the mean of nine independent experiments with marrow from different donors (○) together with the result of an experiment with very strong response (□). (LAERUM et al. 1987)

interexperimental variation in vitro, where average inhibition in a large series of experiments (Fig. 6) was 30%–40% (LAERUM et al. 1987). The bell-shaped curve of the inhibitory effect has been most characteristic and is seen with crude granulocytic extracts, purified human material, and synthetic pentapeptide (KREJA et al. 1986).

In vivo there is a dose-dependent reduction of CFU-GM in mice with single injections, repeated injections, and continuous infusion by use of osmotic minipumps (LAERUM and PAUKOVITS 1984a, b) where up to 80% reduction of femoral progenitor numbers can be seen after 1 week of infusion. Also in vivo a broad, but restricted dose range for the inhibitory effect is seen (LAERUM et al. 1986); inhibitory action can be found down to nanogram doses given i.p. to mice. With 14 µg/h given as infusion over 19 days, a great increase of the bone marrow and peripheral granulocyte numbers was observed, possibly due to compensatory mechanisms (LAERUM and PAUKOVITS 1984b, 1985). Inhibitory effects are seen after i.v., i.p., and s.c. injections and, as mentioned, both after single injections and continuous infusion.

Multipotent stem cell (CFU-S) numbers are also reduced in this dose range although apparently to a lesser degree than the CFU-GM (LAERUM and PAUKOVITS 1984a). Therefore, CFU-GM seem to be the main target in vivo.

In the bone marrow myelopoietic cell proliferation was inhibited (LAERUM and PAUKOVITS 1984b, c). Both a reduction of DNA synthesis in myelopoietic cells and an accumulation of G-2 cells have been observed. The peptide, therefore, seems to affect all phases of the cell cycle, possibly by a general slowing down of cell cycle progression. Also a reduction of myelopoietic cell numbers is seen, followed by a reduction of granulocyte numbers in peripheral blood (Fig. 7). The granulocyte reduction can last for several weeks, due to inhibition at the progenitor cell level (LAERUM and PAUKOVITS 1984a, 1985).

So far in vivo tests on humans have not been performed. It is likely that human cells will also respond in vivo, based both on in vitro findings (LAERUM et al. 1987) and on the fact that the peptide was originally extracted from human granulocytes (PAUKOVITS and LAERUM 1982).

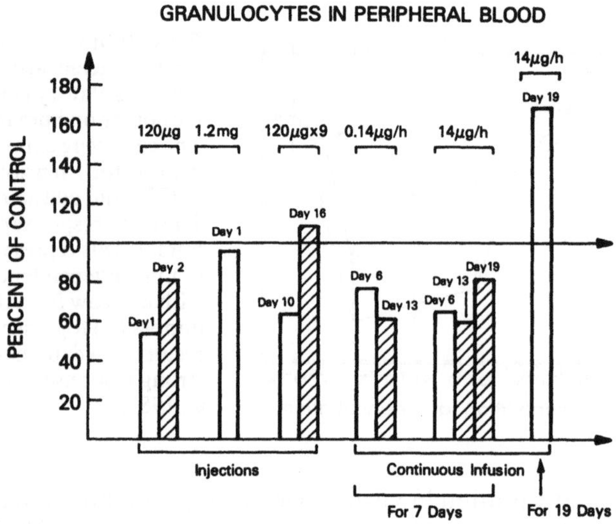

Fig. 7. Inhibitory effects of hemoregulatory peptide monomer on peripheral granulocyte numbers in mice. The peptide was administered in different ways at different doses, as indicated. Prolonged infusion (19 days) leads to granulocytosis, possibly through a compensatory mechanism. These data were obtained (1981) before optimization (see text) of synthetic procedures with a preparation of low specific activity (ca. 0.001). (LAERUM and PAUKOVITS 1985)

3. Effects on Leukemic Cell Lines

Dose-dependent inhibition by monomeric HP5b was observed with several leukemic cell lines. PAUKOVITS et al. (1986) have found inhibition with the myeloid cell line HL-60 when low passage number cultures were used for the experiments. The inhibitory dose range for clonogenic leukemic cells was higher by a factor of 100 than that for normal CFU-GM. Surprisingly the erythroleukemic cell line K-562 (V. OKULOV, personal communication) and the lymphoid RAJI cell line (PAUKOVITS et al., 1990 b) were also inhibited. The monocytoid cell line THP-1 was not inhibited (PAUKOVITS et al. 1986). Higher passage cultures of the HL-60 cell line were not sensitive to HP5b, which may also explain the negative results reported by FOA et al. (1983, 1987).

4. Specificity Tests and Activities not Related to Growth

The most extensive specificity tests have been performed with the monomer. The peptide had no effects on the early erythropoietic progenitor cells (BFU-E) in mice (KREJA et al. 1986). An inhibitory effect on the later erythropoietic progenitor cells (CFU-E) was found in adult mice, but not in fetal mice. Inhibitory doses were more than 1000 times higher than those which inhibited myelopoiesis in vitro (LAERUM et al. 1986).

Similarly, the monomer did not reduce in vivo erythropoiesis in doses which inhibited myelopoiesis (LAERUM and PAUKOVITS 1984b). Instead there was an increase of erythropoiesis, possibly due to channeling of cells through CFU-S to erythropoiesis due to inhibition at the CFU-GM level. Upon prolonged infusion or repeated injections a reduction of multipotent stem cell numbers occurs, and this is followed by a reduction of erythropoietic cell numbers (LAERUM and PAUKOVITS 1984a, b). Therefore, it seems that the monomer has a dose-dependent selective effect on myelopoiesis, but may at higher doses also influence the proliferation kinetics of other hematopoietic cell lineages. It is at present unclear if this effect occurs by direct action of the hemoregulatory peptide on the respective cells, or as a secondary consequence due to the interrelated network structure of the regulatory system of hematopoiesis. This also applies to the lymphocytes, whose numbers have been found to increase in the bone marrow and also in peripheral blood.

In recent unpublished investigations effects on other cell types have been studied. In vitro a slight inhibitory effect on [^{3}H]thymidine uptake of Ehrlich ascites and mouse mammary carcinoma cells at 10^{-4} *M* has been observed (P. LANGEN, personal communication). On the other hand, no direct effect of the monomer on colony formation of mouse B or T cells was found in capillary culture (H. R. MAURER, personal communication). Neither B- nor T-cell activation was found after monomer treatment in vitro (T. KALLAND, personal communication). Thymic cells in vitro show no influence of the monomer on [^{3}H]thymidine uptake (PAUKOVITS and HINTERBERGER 1978a).

Extensive tests with the monomer in hairless mice (h/h) in vivo have until now given no indication of inhibitory effects in other than hematopoietic organs. This includes the mitotic rate in epidermis, thymus, forestomach, glandular stomach, and the small and large intestine (O. NOME et al., in preparation).

The dimer also seems to have a dose-dependent selective effect on myelopoiesis both in vitro and in vivo. Preliminary data indicate that this dose range is narrower than for the monomer, since at high doses stimulating effects have been found on the epithelia in the skin, as well as in the forestomach and the gut (the authors, in preparation).

No species specificity has so far been found for the actions of monomer or dimer. Cells from mice, rats, and humans respond to the peptides. Similarly, the inhibitor has been extracted from human granulocytes (PAUKOVITS and LAERUM 1982), newborn calf spleen (KASTNER and MAURER 1984), rat bone marrow (PAUKOVITS and HINTERBERGER 1978a, b), and rat leukemia cells (RYTÖMAA 1969).

So far we have not been able to see any effects on cell differentiation. After 6 days incubation with the HL-60 cell line neither the monomer nor the dimer induced the formation of cells capable of reacting to phorbol ester treatment with an oxidative burst (NBT reduction). Similarly, incubation with Friend leukemia cells did not lead to the formation of benzidine-reactive cells (PAUKOVITS et al. 1990b). Granulocyte functions, such as adhesion, internalization, and degradation of both bacteria and fungi were not changed by the hemoregulatory peptide (LAERUM et al. 1987). These findings do not exclude that other, i.e., higher, doses may have effects on differentiated functions in various tissues. Our main conclusion is at present that the two types of the peptide have a dose-dependent selectivity and not a tissue-specific effect.

V. Effects on Perturbed Hematopoiesis

1. Inhibitory Effects of HP5b Monomer

a) Effects at the CFU-S Level

Under normal conditions the cycling of only a few CFU-S (LORD 1983) are sufficient to maintain the steady-state numbers of all the cells in the hematopoietic system. The loss of large numbers of hematopoietic cells due to irradiation or cytostatic drug application triggers many of the CFU-S into active proliferation. It has been observed that single injections of cytosine arabinoside (ara-C) (GUIGON and FRINDEL 1978) result in an extended proliferative wave, where approximately 50% of the CFU-S can be found in S-phase with suicide techniques. Multiple (4 ×) injections of high doses of ara-C (GUIGON 1987) induce sharp, highly synchronous waves of CFU-S proliferation with as many as 75% of the stem cells in S-phase at certain times.

Such proliferative activation of hematopoietic stem cells can cause severe long-term disturbances of hematopoiesis after tumor chemotherapy. It can be prevented by application of the HP5b monomer. This was shown by two techniques. Bone marrow from ara-C treated mice was incubated with HP5b monomer in vitro before it was subjected to suiciding and injected into irradiated hosts for CFU-S determination. HP5b treatment resulted in the prevention of CFU-S cycling. When HP5b was injected in vivo shortly before and after a single ara-C dose, the resulting wave of CFU-S proliferation was not observed (PAUKOVITS et al. 1990 a, b).

b) Protection Studies on Mice Treated with Cytostatic Drugs

The protocol of multiple ara-C injections (GUIGON 1987) was used which lead to multiple waves of CFU-S proliferation. Maximum stem cell damage was achieved by injecting the second, third, and fourth ara-C doses at times when the CFU-S which had survived the previous application(s) were in S-phase. After the fourth injection an insufficient number of CFU-S survive to allow hematological recovery, resulting in the death of 90% of the animals (relative survival = 1 in Fig. 8). A bone marrow transplant reconstituted the hematopoietic system, leading to reduced mortality (relative survival 2.7), and the residual mortality was interpreted as the nonhematological toxicity of ara-C. The application of HP5b monomer as a single 600ng dose before the fourth ara-C injection increased the relative survival to 2.1 (Fig. 8), not significantly different from the bone marrow grafted group. Thus HP5b was able to reduce the hematological component of ara-C toxicity (PAUKOVITS et al. 1990 a, b).

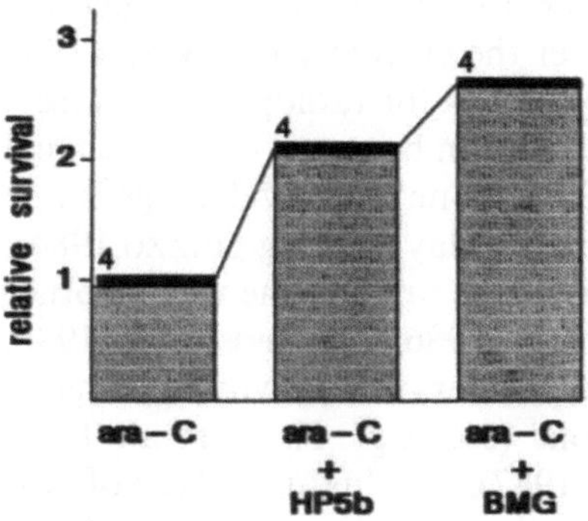

Fig. 8. Protective effect of HP5b-monomer on survival of mice treated with lethal doses of 4×900 mg/kg ara-C at 0, 7, 24, and 30 h. 600 ng HP5b was given at 26 h. Nonhematological ara-C toxicity was estimated by rescue with a bone marrow transplant given after the ara-C treatments. The peptide and transplant groups were not significantly different from each other, but from the ara-C alone group ($P < 0.001$, χ^2test)

Stem cell recruitment, leading to increased sensitivity of these cells to subsequent applications of cytostatic drugs, is considered to be the main source of long-term damage to the hamatopoietic system in cancer chemotherapy. The demonstration that HP5b, when given at or before the critical cytostatic drug application, can prevent stem cell damage may form the basis for clinical applications in which HP5b could be used (a) to reduce toxicity to the hematopoietic system, (b) to increase the dosage of cytostatic drugs while keeping myelotoxicity to tolerable levels, and (c) to make possible the use of potent drugs which so far did not enter clinical practice because of excessive myelotoxicity.

2. Possible Clinical Implications

In vivo actions of the HP5b monomer on myelogeneous leukemia have been tested by FOA et al. (1987). They found that a transplantable myelogenic leukemia in rats was temporarily suppressed, although the increase of survival was only of magnitude of some few days. Since colony formation by primary human myeloid leukemia cells in agar has been strongly inhibited by the monomer in some cases (M. GUIGON et al., personal communication), a suppressive effect on human leukemia is potentially achievable.

The main application of the monomer still seems to be for protection of the bone marrow against cytostatic side effects during cancer chemotherapy. Indeed, several laboratories have found that mice can be protected from lethal effects of various cytostatic, such as ara-C (PAUKOVITS et al. 1986, 1990a), cyclophosphamide, and whole body irradiation (O. SOLESVIK and N. PAPE, in preparation). The mechanism for this is probably that the cells are prevented from entering the cell cycle phases which are most susceptible to the therapeutic agents.

The combined use of the monomer before and the dimer after cytotoxic chemotherapy seems to be of even higher potential value for protecting and regenerating purposes in cytostatic- and irradiation-treated mice (O. SOLESVIK and N. PAPE, in preparation). Also a stimulation of the bone marrow in transplantation patients might be expected.

At present it is an open issue whether the hemoregulatory peptide monomer and dimer can give advantages over the currently used CSFs. The same applies to similar approaches using cytokines for radioprotection (see, e.g., NETA et al. 1986) or stimulation of regeneration by use of other chemical agents such as lithium (GALLICCHIO 1986). Since bone marrow damage is one of the main side effects in cancer chemotherapy today (see, e.g., PIZZO 1984), the peptides should be investigated for their effects in the same way as other biological response modifiers (see, e.g., MIHICH 1986; ZOUMBOS et al. 1986; WERNER et al. 1986; TALMADGE 1986). Since proliferative activity in hematopoiesis undergoes strong circadian variations (see, e.g., HAUS et al. 1983; LAERUM and AARDAL 1985; SMAALAND et al. 1987), this "natural" type of synchronization could also be used for protecting the marrow against side effects of cytostatics (HRUSHESKY 1984; LAERUM et al. 1988 b). It is possible that the protective effect of inhibitors or stimulators, such as the hemoregulatory peptides, can be substantially improved if coordinated with the naturally occurring waves of hematopoiesis.

A highly important and critical factor for clinical applications is to exclude toxic effects. Extensive toxicity studies have been performed with mouse and human bone marrow cells in vitro, both in suspension culture and in agar culture. So far, no toxic effects have been seen in vitro in the dose range 10^{-15}–10^{-4} *M*, with exposure times up to 7 days. This both applies to the monomer and the dimer (LAERUM and PAUKOVITS 1984a; LAERUM et al. 1988 a).

In vivo, more than 300 C3H mice have been injected with up to 9.1 mg monomer and 130 mice with up to 1 mg dimer, without any signs of toxicity as evidenced by systematic autopsy. This corresponds to 10^5–10^6 times the lower level for biological effect in vivo and in vitro. The extensive in vivo specificity tests mentioned above (CLAUSEN et al. 1986) also showed no signs of toxic effects. The LD_{50} dose remains to be established. Toxicity in man has not yet been determined.

Another important factor for the clinical applicability of the hemoregulatory peptide is that the protective effect should be specific for hematopoietic cells, i.e., that the peptide should not alter the sensitivity of transformed cells to cytostatic drugs. We have tested if HP5b could change the ara-C effect on peptide-sensitive (HL-60, RAJI) and -insensitive (Friend) transformed cell lines. The peptide did not change the ara-C sensitivity (IC_{50}) of these cell lines, even not in such cases (HL-60, RAJI) where the cells themselves were inhibited by HP5b (PAUKOVITS et al. 1990 a).

VI. Biochemical and Cellular Mechanisms of Action

According to the data presented above, both the hemoregulatory peptide monomer and dimer seem to exert their main effects on cell proliferation and with a selectivity for myelopoiesis. Since the available data indicate effects on all phases of cell cycle, including a slowing down of the S-phase (LAERUM and

MAURER 1973), the possibility for use of the peptides as synchronizing agents seems to be limited.

Other cellular functions do not seem to be involved at the optimal doses for inhibiting cell proliferation. This argues against the possibility that the peptides are general inhibitors or stimulators of myelopoietic functions. Evidence from granulocyte function studies directly support this (LAERUM et al. 1987).

PAUKOVITS and PAUKOVITS (1975a) have found that proteins on the surface of the target cells of the hemoregulatory peptide are necessary for its inhibitory action. These receptor sites could be destroyed by short trypsin treatment of the cells, but were rapidly regenerated in the absence of protein synthesis inhibitors (cycloheximide).

When bone marrow cells were incubated with biologically active trace amounts of hemoregulatory peptide the peptide disappeared from the medium, indicating a rapid uptake or binding by the cells (PAUKOVITS et al. 1987). By use of ^{3}H-labeled peptide, a linear uptake of the monomer has been seen into mouse bone marrow cells, but not into thymic or spleen cells (ERIKSEN et al. 1987). This indicates a rather selective binding of the substance to the target cells. Apart from this, we have no available data on the receptor or on internalization mechanisms.

The dimer seems to be an enhancer of colony formation, its activity being dependent on the presence of GM-CSF (LAERUM et al. 1988a). To what extent the dimer is dependent on other cofactors is at present not known. No information is at present available if the peptides compete with other molecules for certain receptors at the target cell surface.

B. Epidermal Inhibitory Pentapeptide

The epidermis consists of two functionally different layers; the basal cell layer, which contains cells that retain their capacity for dividing throughout life, and the differentiating cell layer, which consists of cells that are no longer able to divide and which gradually differentiate (keratinize) as they pass up to the surface. To maintain a constant thickness of the epidermis the rate of cell renewal in the basal cell layer has to be exactly adjusted to the rate of cell loss from the surface. Both theoretical considerations and experimental evidence (BULLOUGH and LAURENCE 1960; IVERSEN 1960) indicated that this balance was based on a negative feedback principle. Early experiments had confirmed that skin extracts contain a water-soluble mitotic inhibitor that seemed to be specific for the epidermis (BULLOUGH et al. 1967).

I. Purification Procedures

Lyophilized aqueous extracts of hairless mouse skin (HENNINGS et al. 1969) were the starting material in all purification experiments. Early attempts to isolate the active inhibitor by means of Sephadex chromatography at neutral pH were consistently unsuccessful, but in 1981 ISAKSSON-FORSÉN et al. showed that the inhibitory activity would pass a UM10 Amicon Diaflo Ultrafilter membrane when the skin extract was dissolved in 0.5 *M* acetic acid. This finding suggested that the inhibitor could be a fairly small molecule and the further purification procedures were based on this information. The following fractionation procedures were found to be best suited (REICHELT et al. 1987):

1. Freeze-dried aqueous extract of hairless mouse skin was treated with 1 *M* acetic acid for 2 h at 0–4°C.
2. Gel chromatography on Sephadex G-25 in 0.5 *M* acetic acid to separate high- from low-molecular-weight compounds.
3. Fractionation of active compounds on a Dowex 50 column in the H^+ form. The inhibitory activity was obtained by stepwise elution with water at pH 5, 0.2 *M* formic acid, 4 *M* formic acid, and 2 *M* NH_4OH. After freeze-drying, the inhibitory material was applied to an anion exchange Dowex-1 column. The inhibitory activity was not retained.
4. Passage through a coarse 50-μm C18 reverse-phase column (Bondapak C-18/Porasil B) in 20 m*M* trifluoroacetic acid at pH 2. Elution was performed with 60% (v/v) *n*-propanol in water.
5. The active fraction was further purified on Fractogel MG 2000 in 1 *M* acetic acid/40 m*M* HCl.
6. Final purification on a C18 reverse-phase column (Partisil M-9 10/25 ODS) with an *n*-propanol gradient in 20 m*M* TFA.

During the purification procedures all fractions were tested in hairless mice for their effect on epidermal G_2-M cell flux (the mitotic rate) as determined by the number of colcemid-arrested mitoses, giving an estimate of how many cells have passed from the G_2-phase into mitosis between colcemid injection and killing.

The amino acid composition was determined after total acid hydrolysis giving Glu^2, Asp^1 $Ser^{0.9}$, and Gly^1. The N-terminal amino acid was determined by acid opening of the pyroGlu ring, and by pyroglutamate amino peptidase. Further analyses indicated that the structure of the peptide was pyroGlu-Glu-Asp-Ser-GlyOH and the molecular weight 516.5 (REICHELT et al. 1987).

II. Biological Properties

The isolated, native pentapeptide inhibited epidermal mitoses at a dose of $25–75 \times 10^{-11}$ mol when given i.p. to hairless mice. Larger doses were stimulating in some experiments, and without any measurable effect in others (ELGJO and REICHELT 1984). The native, biological inhibitor was also tested in vitro for its effect on proliferation and differentiation of normal and transformed epidermal cells. Experiments with primary cultures of mouse epidermal cells showed that the pentapeptide transiently inhibited mitoses in this system at a concentration of 10^{-8} *M* (Fig. 9). The dose-response pattern was bell shaped, and lower and higher concentrations gave no significant effect.

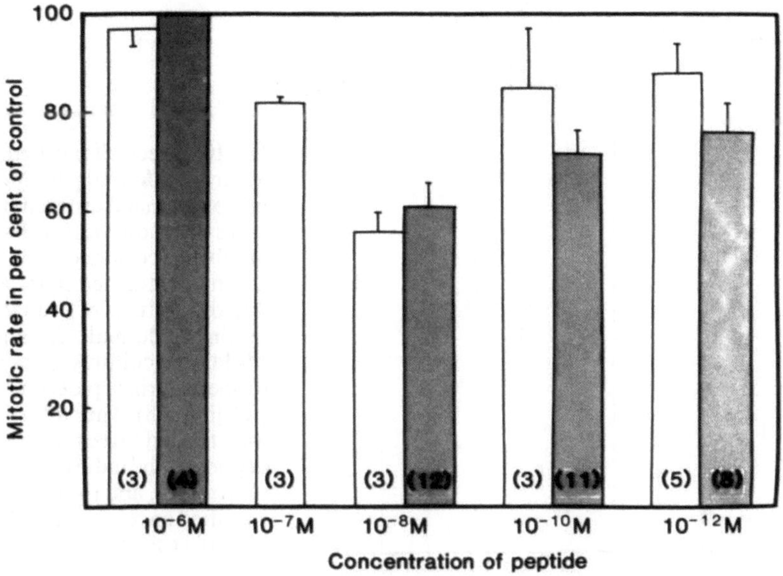

Fig. 9. Effects of purified natural epidermal pentapeptide on cultures of primary mouse epidermal cells grown for 3–4 days in calcium-containing medium (*open bars,* 0.2 m*M*; *shaded bars,* 1.2 m*M*). The mitotic rate was determined by means of vinblastine arrest. Vinblastine and epidermal pentapeptide were added at the same time. (ELGJO et al. 1986a)

When the pentapeptide was added to cultures of a transformed mouse epidermal cell line (line 308), a higher concentration (10^{-6} *M*) was needed to give inhibition (ELGJO et al. 1986a). Also, only cell cultures near confluence were inhibited at the tested doses.

The in vitro system was used to evaluate the effect of the pentapeptide on terminal differentiation, quantified by estimating the number of cornified envelopes among attached and unattached cells (ELGJO et al. 1986a). After repeated treatments over a 24-h period the number of cornified unattached cells was increased in both the primary cultures and in the 308 cell line (Fig. 10). Thus, the pentapeptide influenced both proliferation and differentiation when added to normal or transformed mouse epidermal cells.

The pentapeptide was synthesized (Peninsula Laboratories, CA, United States) together with several analogs. When tested in mice for their effect on cell turnover as indicated above, only the pentapeptide pGlu-Glu-Asp-Ser-GlyOH and a dipeptide that consisted of the two terminal amino acids pGlu-GlyOH were mitosis inhibitory over a fairly wide but low dose range (10^{-11}–10^{-14} mol/mouse) (ELGJO et al. 1986b). Here too, the dose-response pattern was bell shaped since either lower or higher doses had no effect on the mitotic rate (Fig. 11). The other analogs were either without effect or were inhibitory or stimulatory at one single concentration. The synthetic and the natural inhibitor were chromatographically identical in three different sys-

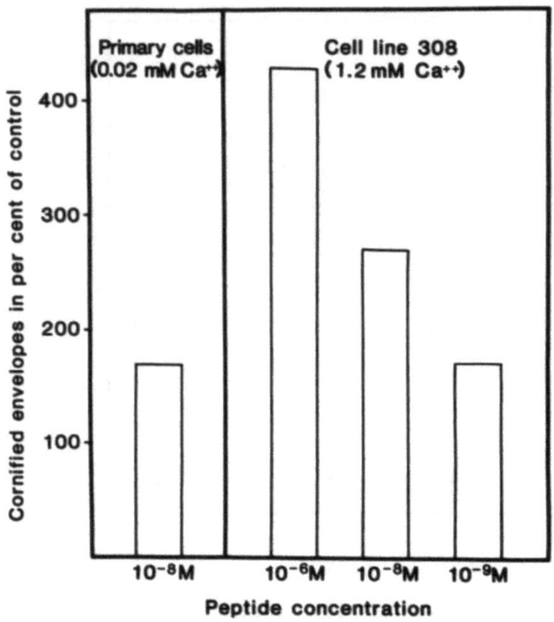

Fig. 10. Effects of purified natural epidermal pentapeptide on terminal differentiation as tested by estimating the number of cornified envelopes among unattached cells in 3-day-old cultures of primary murine epidermal cells, and in 14-day-old cultures of transformed murine epidermal cells (cell line 308). The cultures were treated three times with peptide at the indicated concentrations, and the number of cornified envelopes assessed at 24 h. (ELGJO et al. 1986 a)

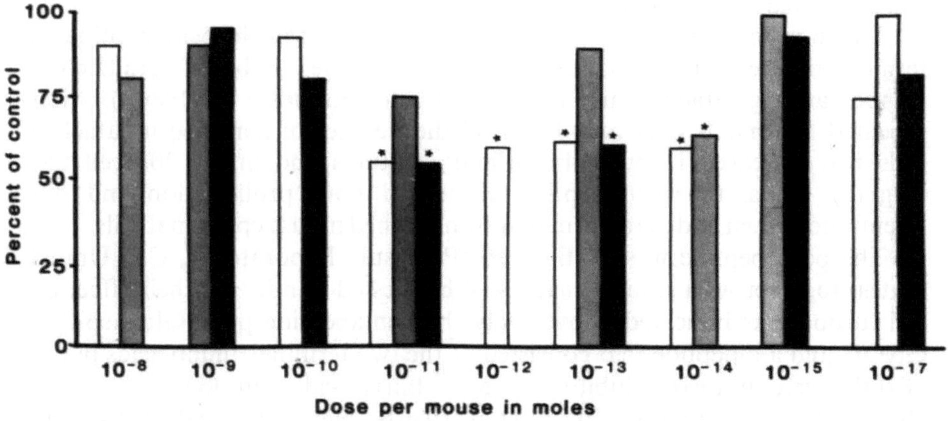

Fig. 11. Different doses of the synthetic epidermal pentapeptide were given i.p. to hairless mice together with colcemid. The animals were killed 3 h later and the number of colcemid-arrested mitoses was used as an estimate of the mitotic rate per 3 h (G_2-M flux) (ELGJO et al. 1986 b)

tems, and we assume they represent the same substance even though it has not been possible to perform mass spectrometry for technical reasons (REICHELT et al. 1987). All experiments described in the following were performed with the synthetic epidermal pentapeptide (EPP).

III. Long-Term Effects

The assay system that was used to test the various fractions during the purification procedures measured cell turnover at only one cell cycle transition, i.e., the G_2-M transition. Earlier experiments with crude or partially purified skin extracts had demonstrated that the effect of a single treatment was more complex and lasted for more than 3–4 h (ELGJO and CLAUSEN 1983). The long-term effect of a single treatment with EPP was therefore examined after i.p. injection of 10^{-13} mol into hairless mice. Both the mitotic rate and the labeling index revealed a biphasic pattern; an immediate inhibition of the epidermal cell proliferation was followed by a small overshoot (of the mitotic rate) or a return to normal values (of the labeling index), and then by a fairly long-lasting second period of reduced epidermal cell proliferation (Fig. 12). The second period of mitosis inhibition was followed by a short overshoot too, but by 30 h after the treatment the epidermal cell proliferation was back to normal values again (ELGJO and REICHELT 1989).

A single topical application with EPP in a water-miscible cream base was followed by a more complex alteration of the pattern of epidermal cell proliferation (ELGJO 1988). The mitotic rate oscillated with several troughs; the last one lasting for several days until the mitotic rate returned to normal values on day 4 (Fig. 13).

Since the reaction to a single treatment is of a fairly long duration, it is reasonable to assume that epidermal basal cells are susceptible to EPP at several cell cycle phase transitions. The minute dose of EPP that was given i.p. can hardly maintain a sufficient concentration in the epidermis for a long time. The pentapeptide is strongly charged and has a high affinity for other molecules like polyamines, probably by ionic binding to γ- and β-carboxyl groups (ELGJO et al. 1986b). Also, it will probably be broken down locally and systemically by peptidases. All this will tend to decrease the time that EPP is available at sufficient concentrations in the epidermis. The effect on epidermal cell proliferation that is seen after a delay of several hours is therefore likely to be a result of an effect on cells at the different cell cycle-phase transitions. Direct measurements with a double-labeling technique have shown that partially purified skin extracts inhibit cell turnover at several phase transitions (ELGJO and CLAUSEN 1983).

The effect of topically applied EPP is more difficult to interpret. We do not know how long locally applied EPP is available in the epidermis, what the local concentration is, or how it is metabolized in the epidermis. Also, we do not know yet how it interacts with other growth-modulating factors, or whether EPP modifies receptor affinity or availability for other factors that influence

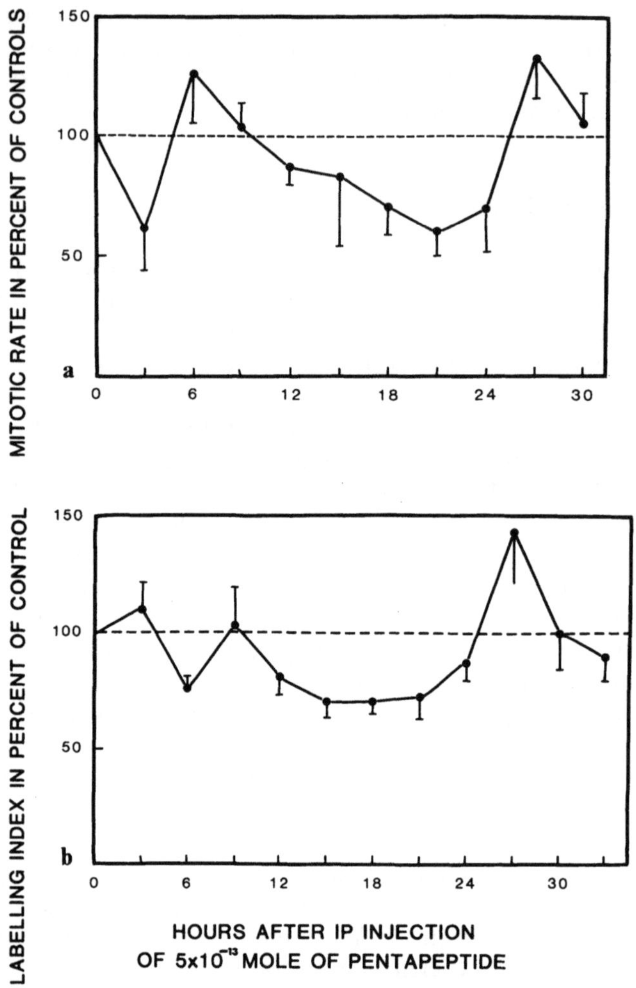

Fig. 12. The mitotic rate (**a**) and the labeling indices (**b**) were estimated at different times after a single i.p. injection of 10^{-13} mol synthetic epidermal pentapeptide. The mitotic rate was estimated at 3-h intervals with the colcemid method, and the labeling indices by counting labeled epidermal cells after injection of [^{3}H]thymidine. (ELGJO and REICHELT 1989)

epidermal cell proliferation. Radioactively labeled EPP, or antibodies against EPP, would be necessary to evaluate these factors.

IV. Repeated Treatments with the Epidermal Pentapeptide

Several attempts have been made to examine the effect of repeated treatment with EPP, but so far they have met with only small success. The reason is that one treatment with EPP is followed by a refractory period lasting for several

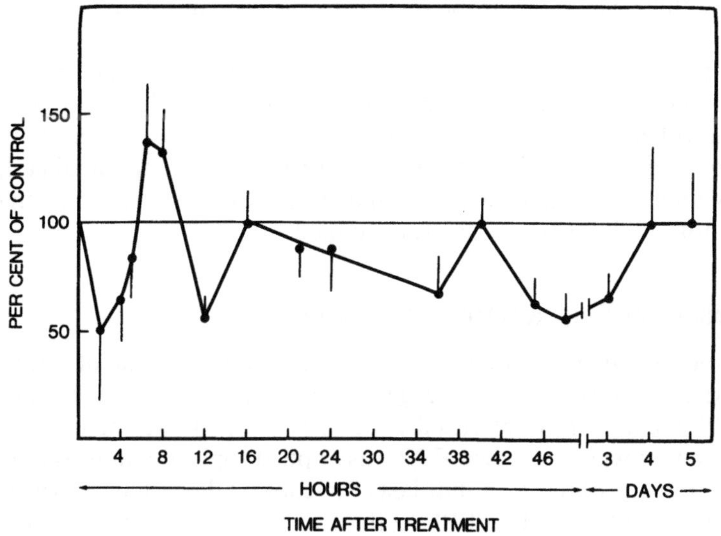

Fig. 13. The mitotic rate after a single topical application of a 0.02% pentapeptide cream. The controls were treated with the cream base. The mitotic rate was estimated by the colcemid method. (ELGJO and REICHELT 1989)

hours. Thus, one i.p. injection of a fairly large and ineffective dose of EPP prevents the inhibitory effect of a smaller and mitosis inhibitory dose given 1 h later (REICHELT et al. 1987). Similarly, repeated twice-weekly topical applications of an EPP cream are inhibitory only during the first 3 days. After that time, no significant effect occurs (ELGJO and REICHELT 1988). Further experiments to evaluate the effect of long-term treatment with EPP therefore have to wait until we know more about the duration of the refractory period induced by a single treatment. However, EPP is not unique in inducing a refractory period in its target cell population. Another, recently isolated mitosis-inhibiting peptide, pGlu-His-GlyOH, has the same effect on its target cells, the large bowel surface epithelium (SKRAASTAD et al. 1988; SKRAASTAD and REICHELT 1988, 1990).

V. Tissue Specificity

A high degree of tissue specificity or tissue preference would from a theoretical point of view seem to be a prerequisite for a locally acting growth regulator. This can easily be observed in the skin; after removal of the uppermost differentiated cells from the epidermis by means of adhesive tape, the rate of cell proliferation is transiently increased in the epidermal basal cell layer, but not in the adjacent connective tissue. At the optimal dose level EPP does not exert any effect on the rate of cell proliferation in the large or small intestinal epithelium, or in the thymus, but the squamous epithelium of the mouse forestomach reacts in the same manner as the epidermis (CLAUSEN et al. 1986).

Specificity therefore seems to be related to the type of differentiation, not to the organ as such.

Even though we have no data at the present time about the breakdown of EPP and similar peptides it seems probable that pyroglutamate aminopeptidase and other peptidases play an important role in this process. High doses of EPP, or of other peptides with pyroglutamate at the N-terminal end, would probably alter the breakdown of similar peptides by substrate competition. This has not yet been systematically examined.

VI. Epidermal Regeneration and Malignancy

Early experiments had shown that crude water extracts of mouse skin prevented or delayed epidermal regeneration after wounding (IVERSEN et al. 1974). EPP seems to have essentially the same effect. First, the water-miscible cream base that was used when EPP was applied topically induces a period of epidermal hyperproliferation that lasts for about 2 days. This hyperproliferation was inhibited by adding EPP to the cream base. It is therefore clear that EPP is active even when the rate of epidermal cell proliferation is increased, as in regeneration. Also, primary cultures of epidermal cells represent a population of hyperproliferative cells, when compared with the normal in situ situation. After addition of EPP, these cultures reacted in the same manner as epidermal cells in situ and at the same dose level.

Recently, the efficacy of EPP was tested in mice that had been treated with a minimal erythematous dose of UVB light (OLSEN and ELGJO 1989). Such a dose of UVB light induces a period of pronounced hyperproliferation in the epidermis after a delay of about 20 h. Repeated applications of this dose are followed by the development of skin cancer in mice. When EPP was applied in a water-miscible cream base immediately after irradiation with UVB, the resulting increase in the rate of epidermal cell proliferation was significantly inhibited.

The effect of EPP on squamous cell carcinomas has not been tested yet in vivo, but in vitro experiments referred to above (ELGJO et al. 1986a) showed that the biological inhibitor reduced the mitotic rate even in transformed mouse epidermal cells. Quite recently, we examined whether pretreatment with EPP would alter the tumor yield after a single application of a carcinogenic dose of MMNU. The results showed a significant increase in the tumor yield after such a pretreatment (IVERSEN et al. 1989). This effect was probably due to an accumulation of cells in the late G_1-phase and in the S-phase, so that more cells than normal were present in a susceptible phase when the carcinogen was applied.

VII. Species Specificity

Early experiments with crude skin extracts indicated that whatever had mitosis inhibitory activity in the extracts was not species specific (W. S. BULLOUGH et al. 1967). Experiments with the isolated, biological inhibitor (JENSEN and BOLUND 1986) showed that the material that had been isolated from mouse skin was mitosis inhibitory to human epidermal cells in culture. The synthetic pentapeptide is equally active in mouse epidermal cell cultures (see above) as in cultures of rat tongue epithelial cells (unpublished data). It is therefore reasonable to assume that the epidermal pentapeptide has no species specificity.

VIII. Toxicity

Like several other growth factors and peptide hormones, EPP has a very low optimal dose range (see Sect. B.II). Thus, it has no inhibitory effect when given in microgram doses. Even if this observation does not preclude a toxic effect on epidermal basal cells it makes it unlikely that the inhibition is due to a simple toxic effect on the proliferating cells. Autopsies of the treated mice have never revealed any pathological changes in other organs. However, the most important finding in this context is that the number of vital cells (line 308, mouse epidermal cells) is not decreased in vitro by EPP treatment, as estimated by means of staining with trypan blue. To test whether EPP has a more general toxic effect, it was added to cultures of fertilized mouse ova, and no alterations were recorded in the normal development (W. OLSEN, personal communication). Thus, none of the effects reported above can be ascribed to simple, toxic effects.

IX. Precursors

We have considered the possibility that EPP could represent an active fragment of a larger molecule. In order to identify such a larger compound, or some precursor molecule, we are at the present time attempting to identify the gene(s) that could code for the pentapeptide. Systematic search in available gene banks has not revealed any known sequence that could represent such a gene. Also, none of the identified growth factors or oncogene products contains fragments that have the same amino acid sequence as EPP (August 1988). We are also trying to make monoclonal antibodies against EPP for the purpose of identifying a putative precursor. This work is in progress, and at the present time we have no hard data that could either prove or disprove the assumption that EPP is a fragment of a larger molecule.

X. Possible Clinical Applications

No clinical trials have been performed with the epidermal pentapeptide, but from a theoretical point of view it could play a role in the future treatment of pathological hyperproliferation of squamous epithelia, like psoriasis. Before it is put to practical use, more research is necessary to obtain information about (a) the correct doses for systemic and topical use, (b) possible hypersensitizatioin by repeated applications, and (c) whether other factors are needed for EPP to be fully effective. So even though one may envisage several diseases that would seem suitable for treatment with EPP, much work is still needed before it can be tested in a clinical situation.

C. Conclusion

Crude tissue extracts often inhibit cell proliferation when added to cell cultures or when applied in vivo. Under selected conditions such inhibitions can display a high degree of tissue specificity (IVERSEN 1981). It was, however, difficult to relate observed effects of such multicomponent mixtures to the actions of postulated inhibitory factors, called chalones (BULLOUGH and LAURENCE 1960; IVERSEN 1960). Despite early attempts to purify the factors responsible for the inhibitory actions the problem persisted mostly because it was difficult to make clear distinctions between growth inhibition and cytotoxicity and because purification techniques and assay methods were not sufficiently powerful.

Nevertheless, the hypothesis of tissue-specific inhibitors regulating cell proliferation via negative feedback mechanisms was experimentally pursued by several research groups, finally leading to the isolation and characterization of two small chemically related peptides, the hemoregulatory and the epidermal inhibitory pentapeptide. Both peptides appear to be rather selective inhibitors of progenitor proliferation in their respective tissues, although their properties do not always coincide with theoretical requirements and definitions. It appears that both peptides are not cytotoxic and that their actions are reversible. The complexity and interdependence of regulatory networks makes some in vivo actions of these peptides look less specific than in vitro experiments would suggest. However, both factors have clearly preferred cellular targets, unlike many of the rather pleiotropic actions of other growth stimulators and inhibitors.

The results indicate that both peptides participate in the physiological regulation of proliferative activity in their respective tissue and that this may also be the case in situations deviating from the normal situation. Of special interest is the unique property of the hemoregulatory peptide to form a stimulatory dimer by oxidation of the inhibitory monomer. With respect to the functional repertoire of activated granulocytes (and other cells), the rapid and reversible redox switch from monomer to stimulatory dimer may constitute an almost instantaneously acting emergency mechanism for adjusting the production of myeloid cells to altered requirements.

It is our opinion that both the epidermal pentapeptide and the hemoregulatory peptide are members of a larger family of small peptides with inhibitory effects on cell proliferation in different target tissues. They may, therefore, constitute a regulatory system which is different from the other growth factors. Unfortunately, the research related to mechanisms of action, their precursors, and genes is still in its infancy. One reason may be that the peptides are rather labile and have been extremely difficult to purify and test. With the presently available synthetic peptides, it is expected that these problems will be overcome in the near future, providing both biologists and clinicians with potent nontoxic substances which can be used to alter proliferation profoundly in a cell-line-selective manner.

References

Aardal NP, Laerum OD, Paukovits WR, Maurer HR (1977) Inhibition of agar colony formation by partially purified granulocyte extracts (chalone). Virchows Arch [B] 24:27–39

Aardal NP, Laerum OD, Paukovits WR (1982) Biological properties of partially purified granulocyte extract (chalone) assayed in soft agar culture. Virchows Arch [B] 38:253–261

Balazs A, Sajgo M, Klupp T, Kemeny A (1980) Purification of an endopeptide to homogenity and the verification of its selective inhibitory action on myeloid cell proliferation. Cell Biol Int Rep 4:337–345

Bannai S, Tateishi N (1986) Role of membrane transport in metabolism and function of glutathione in mammals. J Membr Biol 98:1–8

Benestad HB, Hersleth IB (1984) Production of proliferation inhibitors by mature granulocytes. Blut 48:201–211

Benestad HB, Rytömaa T, Kiviniemi K (1973) The cell specific effect of the granulocyte chalone demonstrated with the diffusion chamber technique. Cell Tissue Kinet 6:147–154

Bjornson BH, Pincus SH, DiNapoli AM, Desforges JF (1984) Inhibition of CFU-GM and CFU-EOS by mature granulocytes. Blood 63:376–379

Bullough WS, Laurence EB (1960) The control of epidermal mitotic activity in the mouse. Proc Roy Soc London Ser B 151:517–536

Bullough WS, Laurence EB, Iversen OH, Elgjo K (1967) The vertebrate chalone. Nature 214:578–580

Chan SH (1971) Influence of serum inhibitors of colony development in vitro by bone marrow cells. Aust J Exp Biol Med Sci 49:553–564

Clausen OPF, Elgjo K, Laerum OD, Reichelt KL, Paukovits W, Iversen OH (1986) Tissue specificity of peptide growth factors. In: Serono symposium on biological regulation of cell proliferation, Milano (abstr no 29)

Elgjo K, Clausen OPF (1983) Proliferation-dependent effect of skin extracts on mouse epidermal cell flux at the G1-S, S-G2 and G2-M transitions. Virchows Arch [B] 42:143–151

Elgjo K, Reichelt KL (1984) Purification and characterization of a mitosis inhibiting epidermal peptide. Cell Biol Int Rep 8:379–382

Elgjo K, Reichelt KL (1989) Structure and function of growth inhibitory epidermal pentapeptide. Ann NY Acad Sci 548:197–203

Elgjo K, Reichelt KL, Hennings H, Michael D, Yuspa SH (1986a) Purified epidermal pentapeptide inhibits proliferation and enhances terminal differentiation in cultured mouse epidermal cells. J Invest Dermatol 87:555–558

Elgjo K, Reichelt KL, Edminson P, Moen E (1986b) Endogenous peptides in epidermal growth control. In: Baserga R, Foa P, Metcalf D, Polli EE (eds) Biological regulation of cell proliferation. Raven, New York, p 259

Eriksen JA, Schanche JS, Hestdal K, Jakobsen SE, Tveterås T, Johansen JH, Paukovits WR, Laerum OD (1987) Hemoregulatory peptide synthesis, purification of tritium labelled peptide and uptake of peptide in hematopoietic tissues in vitro. In: Najman A, Guigon M, Gorin NC, Mary JY (eds) The inhibitors of hematopoiesis. Libbey, London, pp 51–54 (Colloques INSERM, vol 162)

Fetsch J, Maurer HR (1987) A specific low molecular mass granulopoiesis inhibitor, isolated from calf spleen. In: Najman A, Guigon M, Gorin NC, Mary JY (eds) The inhibitors of hematopoiesis. Libbey, London, pp 55–58 (Colloques INSERM, vol 162)

Foa P, Lombardi L, Ciani A, Chillemi F, Maiolo AT, Cesan BM, Polli EE (1983) A synthetic pentapeptide inhibiting normal and leukaemic myelopoiesis in vitro. IRCS Med Sci: Libr Compend 11:272–273

Foa P, Chillemi F, Lombardi L, Lonati S, Maiolo A, Polli EE (1987) Inhibitory activity of a synthetic pentapeptide on leukaemic myelopoiesis both in vitro and in vivo in rats. Eur J Haematol 39:399–403

Gallicchio VS (1986) Lithium and CFU-mix after single-dose administration of cyclophosphamide. Exp Hematol 14:395–400

Guigon M (1987) Biological properties of low molecular weight pluripotent stem cell (CFU-S) inhibitors. In: Najman A, Guigon M, Gorin NC, Mary JY (eds) The inhibitors of hematopoiesis. Libbey, London, pp 241–251 (Colloque INSERM, vol 162)

Guigon M, Frindel E (1978) Inhibition of CFU-S entry into S-phase after irradiation and drug treatment. Biomedicine 29:176–178

Haus E, Lakatua DJ, Swoyer J, Sackett-Lundee L (1983) Chronobiology in hematology and immunology. Am J Anat 168:647–671

Havemann K, Gassel W-D, Laukel H (1979) Humoral factors modulating growth of granulocyte macrophage progenitor cells. Acta Haematol (Basel) 62:306–314

Hennings H, Iversen OH, Elgjo K (1969) Delayed inhibition of epidermal DNA synthesis after injection of an aqueous skin extract. Virchows Arch [B] 3:45–53

Herman SP, Golde DW, Cline MJ (1978) Neutrophil products that inhibit cell proliferation: relation to granulocytic "chalone". Blood 51:207–219

Hinterberger W, Frischauf H, Kletter K, Paukovits W, Bais P (1977) Humoral function of acute leukaemic blasts. Scand J Haematol 19:121–128

Hrushesky WJM (1984) Chemotherapy timing: an important variable in toxicity and response. In: Perry MC, Yrbro JW (eds) Toxicity of chemotherapy. Grune and Stratton, New York, pp 449–477

Isaksson-Forsén G, Elgjo K, Burton D, Iversen OH (1981) Partial purification of the epidermal G2 inhibitor based on an in vivo assay system. Cell Biol Int Rep 5:195–199

Iversen OH (1960) A homeostatic mechanism regulating the cell-number in epidermis. In: Masturzo A (ed) Proc 1st Congr Int Cybernetic Med. Gianno, Naples, p 420

Iversen OH (1976) The history of chalones. In: Houck JC (ed) Chalones. North-Holland. Amsterdam, pp 37–69

Iversen OH (1981) The chalones. In: Baserga R (ed) Tissue growth factors. Springer, Berlin Heidelberg New York, p 491 (Handbook of experimental pharmacology, vol 57)

Iversen OH, Bhangoo KS, Hansen K (1974) Control of epidermal cell renewal in the bat web. Virchows Arch [B] 16:157–197

Iversen OH, Elgjo K, Reichelt KL (1989) Enhancement of methylnitrosourea-induced skin tumorigenesis and carcinogenesis in hairless mice by pretreatment with a mitosis inhibiting epidermal pentapeptide. Carcinogenesis 10:241–244

Jensen PKA, Bolund L (1986) Changes in proliferating cell subpopulations and mitotic activity in human epidermal cultures treated with epithelial growth inhibitors. J Invest Dermatol 86:46–50

Kastner M, Maurer HR (1980) Pure bovine granulocytes as a source of granulopoiesis inhibitor (chalone). Hoppe-Seyler's Z Physiol Chem 361:197–200

Kastner M, Maurer HR (1984) Partial purification and characterization of an endogenous granulomonopoiesis inhibitor from calf spleen. Hoppe-Seyler's Z Physiol Chem 365:129–135

Klebanoff SJ, Clark RA (1978) The neutrophil: function and clinical disorders. Elsevier, Amsterdam

Kreja L, Haga P, Muller-Bérat N, Laerum OD, Sletvold O, Paukovits WR (1986) Effects of a hemoregulatory peptide (HP5b) on erythroid and myelopoietic colony formation in vitro. Scand J Haematol 37:79–86

Laerum OD, Aardal NP (1985) Rhythms in blood and bone marrow. In: Rietveld WJ (ed) Clinical aspects of chronobiology. Meduservice Hoechst, Leiden, pp 85–97
Laerum OD, Maurer R (1973) Proliferation kinetics of myelopoietic cells and macrophages in diffusion chambers after treatment with granulocyte extracts (chalone). Virchows Arch [B] 14:293–305
Laerum OD, Paukovits WR (1984a) Inhibitory effects of a synthetic pentapeptide on hemopoietic stem cells in vitro and in vivo. Exp Hematol 12:7–17
Laerum OD, Paukovits WR (1984b) Modulation of murine hemopoiesis in vivo by a synthetic hemoregulatory pentapeptide (HP5b). Differentiation 27:106–112
Laerum OD, Paukovits WR (1984 c) Modification of mouse hemopoietic cell proliferation in vivo by a hemoregulatory pentapeptide (HP5b). Virchows Arch [B] 46:333–348
Laerum OD, Paukovits WR (1985) Peripheral blood leukocyte alterations in mice induced by a hemoregulatory pentapeptide (HP5b). Leuk Res 9:1075–1084
Laerum OD, Paukovits WR (1987) Biological effects of myelopoiesis inhibitors. In: Najman A, Guigon M, Gorin NC, Mary JY (eds) The inhibitors of hematopoiesis. Libbey, London, pp 21–30 (Colloques INSERM, vol 162)
Laerum OD, Paukovits WR, Sletvold O (1986) Hemoregulatory peptide: biological aspects. In: Baserga R, Foa P, Metcalf D, Polli EE (eds) Biological regulation of cell proliferation. Raven, New York, pp 121–129
Laerum OD, Sletvold O, Bjerknes R, Paukovits WR (1987) A synthetic hemoregulatory peptide (HP5b) inhibits human myelopoietic colony formation (CFU-GM) but not leukocyte phagocytosis in vitro. Eur J Haematol 39:259–266
Laerum OD, Sletvold O, Bjerknes R, Eriksen JA, Johansen JH, Schanche JS, Tveteraas T, Paukovits WR (1988a) The dimer of hemoregulatory peptide (HP5b) stimulates mouse and human myelopoiesis in vitro. Exp Hematol 16:274–280
Laerum OD, Smaaland R, Sletvold O (1988b) Rhythms in blood and bone marrow: potential therapeutic implications. In: Lemmer B (ed) Cellular and biochemical aspects of chronopharmacology. Dekker, New York (in press)
Laerum OD, Sletvold O, Smaaland R, Paukovits WR (1989) Hemoregulatory peptide monomer and dimer: inhibitory and stimulatory effects on myelopoiesis in vitro and in vivo. (submitted for publication)
Lord BI (1983) Hemopoietic stem cells. In: Potten CS (ed) Stem cells: their identification and characterization. Churchill-Livingstone, Edinburgh, pp 118–154
Lord BI, Mori EG, Wright EH, Lajtha LG (1976) An inhibitor of haemopoietic cell proliferation in normal bone marrow. Br J Haematol 34:441–445
Lord BI, Mori KJ, Wright EG (1977) A stimulator of stem cell proliferation in regenerating bone marrow. Biomedicine 27:223–226
Maurer HR, Weiss G, Laerum OD (1976) Starting procedures for the isolation and purification of granulocyte chalone activities. Blut 33:161–170
Mihich E (1986) Future perspectives for biological response modifiers: a viewpoint. Semin Oncol 13:234–254
Neta R, Vogel SN, Oppenheim JJ, Douches SD (1986) Cytokines in radioprotection. Comparison of the radioprotective effects of IL-1 to IL-2, GM-CSF and IFN. Lymphokine Res 5:105–110
Olsen WM, Elgjo K (1989) UVB-induced epidermal hyperproliferation is partially inhibited by a single treatment with an endogenous epidermal pentapeptide. Carcinogenesis (to be published)
Paran M, Ichikawa Y, Sachs L (1969) Feedback inhibition of the development of macrophage and granulocyte colonies. II. Inhibition by granulocytes. Proc Natl Acad Sci USA 62:81–87
Paukovits WR (1971) Control of granulocyte production: separation and chemical identification of a specific inhibitor. Cell Tissue Kinet 4:539–547
Paukovits WR (1973) Granulopoiesis inhibiting factor (GIF): demonstration and preliminary chemical-biological characterization of a specific polypeptide (chalone). Natl Cancer Inst Monogr 38:147–155

Paukovits WR, Hinterberger W (1978a) Molecular weight and some chemical properties of the granulocyte chalone. Blut 37:7–18
Paukovits WR, Hinterberger W (1978b) Biochemical characterization of humoral factors regulating myelopoiesis. In: Rainer H (ed) Cell separation and cryobiology. Schattauer, Stuttgart, pp 75–78
Paukovits WR, Laerum OD (1982) Isolation and synthesis of a hemoregulatory peptide. Z Naturforsch [C] 37:1297–1300
Paukovits WR, Laerum OD (1984) Structural investigations on a peptide regulating hemopoiesis in vitro and in vivo. Hoppe-Seyler's Z Physiol Chem 365:303–311
Paukovits WR, Paukovits J (1975a) Mechanism of action of granulopoiesis inhibiting factor (chalone): I. Evidence for a receptor protein on bone marrow cells. Exp Pathol 10:343–352
Paukovits WR, Paukovits J (1975b) The solubility of granulopoiesis inhibiting factor. Experientia 31:1357–1358
Paukovits WR, Paukovits JB (1978) Peptide nature and proteolytic sensitivity of granulopoiesis inhibiting factor. IRCS Med Sci: Libr Compend 6:176
Paukovits WR, Paukovits JB, Laerum OD, Hinterberger W (1980) Granulopoiesis inhibiting factor (chalone): purification and chemical composition. IRCS Med Sci: Libr Compend 8:305–306
Paukovits WR, Laerum OD, Paukovits JB, Hinterberger W, Rogan AM (1983) Methods for the preparation of purified granulopoiesis-inhibiting factor (chalone). Hoppe-Seyler's Z Physiol Chem 364:383–396
Paukovits WR, Laerum OD, Guigon M (1986) Isolation, characterization and synthesis of a chalone-like hemoregulatory peptide. In: Baserga R, Foa P, Metcalf D, Polli EE (eds) Biological regulation of cell proliferation. Raven, New York, pp 111–119
Paukovits WR, Laerum OD, Paukovits JB, Guigon M, Schanche JS (1987) Regulatory peptides inhibiting granulopoiesis. In: Najman A, Guigon M, Gorin NC, Mary JY (eds) Libbey, London, pp 31–42 (Colloques INSERM, vol 162)
Paukovits WR, Guigon M, Binder KA, Hergl A, Laerum OD, Schulte-Hermann R (1990a) Prevention of hematotoxic side effects of cytostatic drugs by a synthetic hemoregulatory peptide. Cancer Res (in press)
Paukovits WR, Hergl A, Schulte-Hermann R (1990b) Hemoregulatory peptides: improved techniques for synthesis and characterization. Mol Pharm (to be published)
Pelus LM, Broxmeyer HE, Moore MAS (1981) Regulation of human myelopoiesis by prostaglandin E and lactoferrin. Cell Tissue Kinet 14:515–526
Perrins DJD, Whiernik G, Jones WA (1980) Granulocyte chalone assayed in vivo in the mouse. Acta Haematol 64:72–76
Pizzo PA (1984) Granulocytopenia and cancer therapy. Past problems, current solutions, future challenges. Cancer 54(11):2649–2661
Reichelt KL, Elgjo K, Edminson PD (1987) Isolation and structure of an epidermal mitosis inhibiting pentapeptide. Biochem Biophys Res Commun 146:1493–1501
Rytömaa T (1969) Granulocytic chalone and antichalone. In Vitro 4:47–57
Rytömaa T, Kiviniemi K (1968a) Control of granulocyte production. I. Chalone and antichalone, two specific humoral regulators. Cell Tissue Kinet 1:329–340
Rytömaa T, Kiviniemi K (1968b) Control of granulocyte production. II. Mode of action of chalone and antichalone. Cell Tissue Kinet 1:341–350
Rytömaa T, Kiviniemi K (1968c) Control of cell production in rat chloroleukaemia by means of the granulocytic chalone. Nature 220:136–137
Skraastad O, Reichelt KL (1988) An endogenous colon mitosis inhibitor and dietary calcium inhibit the increased colonic cell proliferation induced by cholic acid. Scand J Gastroenterol 23:801–807
Skraastad O, Reichelt KL (1990) Biological characteristics of a mitotic inhibitor in the mouse colonic epithelium. Carcinogenesis (in press)
Skraastad O, Fossli T, Edminson PD, Reichelt KL (1988) Purification and characterization of a mitosis inhibitory tripeptide from mouse intestinal extracts. Epithelia 1:107–119

Smaaland R, Sletvold O, Bjerknes R, Lote K, Laerum OD (1987) Circadian variations of cell cycle distribution in human bone marrow. In: International Society for Chronobiology Conference, Leiden (abstr no 17)
Talmadge JE (1986) Biological response modifiers: realizing their potential in cancer therapeutics. Trends Pharmacol Sci 7:277–281
Watanabe H, Bannai S (1987) Induction of cystine transport activity in mouse peritoneal macrophages. J Exp Med 165:628–640
Weiss P, Kavanau JL (1957) A model of growth and growth control in mathematical terms. J Gen Physiol 41:1–47
Weiss SJ, Lampert MB, Test ST (1983) Long-lived oxidants generated by human neutrophils. Characterization and bioactivity. Science 222:625–628
Werner GH, Floch F, Migliore-Samour D, Jolles P (1986) Immunomodulating peptides. Experientia 42:521–531
Zoumbos N, Raefsky E, Young N (1986) Lymphokines and hematopoiesis. Prog Hematol 14:201–227

Section C: Coordinate Actions of Growth Factors in Specific Tissues or Cells

CHAPTER 29

Coordinate Actions of Hematopoietic Growth Factors in Stimulation of Bone Marrow Function

M. A. S. Moore

A. Introduction

The proliferation and differentiation of bone marrow stem cell populations requires a balance between the need to meet the production of mature myeloid cells and the necessity of ensuring the retention of a population of self-renewing stem cells. This requires that the stem cell pool is protected from exhaustion by excessive differentiation pressure, providing a constant pool or reserve of stem cells throughout life. A number of models have been provided to explain this phenomenon. The stochastic model of Till and McCulloch (1980) envisages hematopoiesis engendered randomly (HER), whereas the deterministic model of Trentin (1970) envisaged the role of a hematopoietic inductive microenvironment (HIM). The current status of understanding of stem cell regulation requires a compromise between these two extreme views, in which stochastic events occur with a probability that can be modulated by external influences – a "loaded dice" concept. It is in this latter respect that hematopoietic growth factors may play a critical role.

B. Stem Cells, Growth Factors, and the Extracellular Matrix

In long-term bone marrow culture systems hematopoiesis occurs only if the stem cells and progenitor cells are in intimate contact with the bone marrow stromal cell populations, which are themselves phenotypically diverse. Evidence suggests that a nonrandom association occurs between different stromal cell types and the expression of different hematopoietic lineages. Marrow stromal cells do not secrete factors that are readily detected in suspension; however, purified stem cells (colony-forming units, CFUs), when allowed to attach to stroma, undergo proliferation and differentiation (Spooncer et al. 1985, 1986). Stem cells can also attach to metabolically dead stromal cells that have been fixed in glutaraldehyde and can undergo proliferation and differentiation (Roberts et al. 1987).

A critical component of the stromal cell environment is the extracellular matrix (ECM). This consists of fibronectin, laminin, collagen, and glycosaminoglycans (GAGs). The culture of bone marrow on an established ECM results in up to 30-fold greater adherent cellularity and enhanced total numbers of progenitor cells and differentiated progeny generated (Campbell et al.

1985). Inhibition of collagen synthesis results in decreased stem cell numbers. Stimulation of GAG synthesis with β-xylosides also promotes increased cellularity and stem cell numbers in long-term bone marrow culture (Spooncer et al. 1983). Campbell et al. (1985) postulated that the role of the extracellular matrix is to stimulate soluble growth factor production by stromal cells. A more recent concept is that the GAGs found in the ECM function to selectively retain hematopoietic growth factors within the stromal layers. Gordon et al. (1987a) showed that granulocyte-macrophage colony-stimulating factor (GM-CSF) can be eluted from cultured stromal layers and that exogenous GM-CSF binds to GAGs from bone marrow ECMs. The GAGs in question, heparan sulfate, hyaluronic acid, and chondroitin sulfate, may be the major binding protein. The diversity of GAGs may equip them for specific recognition events with respect to different species of hematopoietic growth factors. Furthermore, the expression of various extracellular matrix components together with complexes of stromal cells and growth factors may comprise the hematopoietic microenvironment (Dexter 1987). In this context, interleukin-3 (IL-3) but not GM-CSF or erythropoietin bind to intact cultured stromal layers, and GM-CSF binds to salt-extracted stroma unless hyaluronic acid is removed from the ECM. In contrast, fibroblast growth factor, IL-3, and erythropoietin bound to heparin-Sepharose beads, but GM-CSF did not (Gordon et al. 1988). Thus, the relative amounts of certain matrix glycoproteins in the marrow microenvironment may influence the relative concentrations of different growth factors.

Roberts et al. (1988) also showed that the major sulfated GAG in mouse bone marrow stroma, heparan sulfate, possessed the ability to bind IL-3 but also they demonstrated effective GM-CSF binding. The functional domains of the growth factors were not affected by binding to heparan sulfate since ECM preparations pre-incubated with GM-CSF or IL-3 supported the survival and proliferation of IL-3 or GM-CSF factor-dependent cell lines. The binding of the factors to heparan sulfate does not involve glycosylation sites since *Escherichia coli* recombinant factors bound as efficiently as glycosylated forms. The physiological relevance of this factor binding may involve protection of growth factors from enzymatic degradation in addition to localizing and concentrating specific factors at sites of production of specific target cells.

The extracellular matrix may also play a role in the release of mature cells from the bone marrow environment. Maturing erythroblasts express a receptor for fibronectin, a component of the ECM, and with maturation to the reticulocyte stage this receptor is lost, coinciding with the release of red cells into the circulation (Patel and Lodish 1986). Hemonectin is a matrix glycoprotein which is associated with specific adherence of maturing granulocytes with the adherence diminishing as cells mature from the metamyelocyte to the segmented stage (Campbell et al. 1987).

C. Hematopoietic Growth Factor Interactions with Early Stem Cells

The original characterization of CSFs involved stimulation of colony formation by early hematopoietic progenitors that were thought by some investigators to be true stem cells. If the definition of a true stem cell requires that the cell has extensive self-renewal capacity and pluripotentiality, then the conventional CFU-c populations directly stimulated by G, GM- or M-CSF acting alone are not stem cells. The search for assays for true stem cells began with characterization of the self-renewal and differentiation properties of CFU-s cells capable of forming hematopoietic colonies in the spleens of lethally irradiated mice. A subpopulation of CFU-s, forming colonies scored at day 12 after transplantation corresponded most closely with true stem cells capable of extensive, prolonged reconstitution of the immune and hematopoietic system of lethally irradiated mice. Recently, it has been shown that an even smaller subset of CFU-s constitutes true stem cells (BERTONCELLO et al. 1988). In vitro, two types of clonal assays have been developed that detect cells with stem cell properties – the blast colony assay of NAKAHATA and OGAWA (1982) and the high-proliferative-potential (HPP-CFU) assay of BRADLEY et al. (1982). In both assay systems a complex interplay of growth-stimulating, potentiating, and inhibiting factors has become evident.

I. Hematopoietic Growth Factor Interactions in the HPP-CFU Assay

Early studies involving mice treated with 5-fluorouracil (5-FU) suggested the necessity of a synergistic interaction in vitro between two factors required for hematopoietic colony formation, one being a direct hematopoietic colony stimulus (e.g., M-CSF, IL-3), and the other (which lacked colony-stimulating activity) acting as a synergistic activity (BRADLEY et al. 1982; STANLEY et al. 1986). Synergistic activity/Hemopoietin-1 (H-1) was identified in conditioned media or extracts of various human tissues and was also identified and purified to homogeneity from the human bladder cancer cell line, 5637 (JUBINSKY and STANLEY 1985). This latter cell line had also been used to identify, purify, and clone G- and GM-CSF (WELTE et al. 1985; PLATZER et al. 1985; GABRILOVE et al. 1986; SOUZA et al. 1986). The constitutive expression of H-1, which was shown to be a 17-kDa protein (JUBINSKY and STANLEY 1985), led us to attempt to purify, sequence, and clone this molecule, using similar strategies to those that we had employed to clone the G-CSF gene from 5637 cells (SOUZA et al. 1986). The identity of IL-1 and H-1/synergistic activity was established by (a) the demonstration that 5637 bladder cancer cells produced high levels of IL-1 active in the thymocyte comitogenesis assay, (b) synergistic activity and IL-1 copurified using various protein purification procedures, (c) abundant levels of IL-1 α and β mRNA can be detected in 5637 cells – comparable to levels in activated macrophages, (d) mRNA fractions from 5637 expressed in oocytes resulted in intracellular production of hematopoietic

Table 1. Morphology of HPP colonies developing in 5-FU marrow cultures in the presence of IL-1 and CSFs

Stimulus[a]	HPP-CFU/5 × 10⁴ 4 days post 5-FU marrow cells	Morphology (%)		
		G	GM	M
IL-1	4			
M-CSF	1			
M-CSF + IL-1	58	0	0	100
GM-CSF	8			
GM-CSF + IL-1	35	0	40	60
IL-3	2			
IL-3 + IL-1	16	0	50	50
G-CSF	0			
G-CSF + IL-1	18	0	70	30

[a] All factors were used at 100 U/ml with the exception of G-CSF, which was used at 1000 U/ml. HPP colonies >0.5 mm at day 8 and morphology assessed in situ on dried, fixed, and stained cultures. G, neutrophil; M, macrophage.

synergistic activity and thymocyte comitogenesis factor production, (e) synergistic activity and thymocyte comitogenesis activity were completely neutralized by monoclonal and polyclonal antibodies to IL-1, (f) rhIL-1 α and β at 0.1–10 U/ml synergized with G-, GM-, M-CSF, and IL-3 in stimulating high proliferative potential (HPP-CFU) in clonogenic assays of 5-FU-treated murine bone marrow (Moore et al. 1987a, b; Moore and Warren 1987; Warren and Moore 1988, Moore 1987c, 1988a, c).

In the HPP-CFU assay synergistic interactions between IL-1 and M-CSF led to colonies comprising exclusively macrophages (Table 1). Low numbers of HPP colonies developed with GM-CSF or IL-3 alone but synergism was evident following addition of IL-1, with colonies comprising either macrophage or neutrophil-macrophage in approximately equal proportions. G-CSF alone did not stimulate colony formation, but, with IL-1, substantial numbers of large colonies developed with a higher proportion containing neutrophils than seen with GM-CSF or IL-3 plus IL-1 (Moore et al. 1987a; Moore and Warren 1987). McNiece et al. (1988) reported two categories of HPP-CFU, one population, designated HPP-CFC-1, is stimulated by M-CSF in the presence of IL-1, resulting in the generation of a population of cells still capable of forming large colonies but no longer requiring IL-1. These HPP-CFC-2 colonies are stimulated by IL-3 directly and by the synergistic interaction of GM-CSF and M-CSF. HPP-CFC-1 colonies are relatively resistant to 5-FU and regenerate as early as 2–4 days postchemotherapy. HPP-CFC-2 colonies are depleted by 5-FU treatment and regeneration is not observed until 6–8 days post-5-FU.

IL-1 has a direct effect upon hematopoietic stem cells in addition to its ability to elicit production of various CSF species by accessory cell populations within hematopoietic tissues. In our original studies it was not possible to conclude that IL-1 was acting directly on early stem cells since accessory cell populations were not depleted from the target bone marrow cell population. In more recent studies we have established a linear dose response relationship between the numbers of HPP-CFU and the number of bone marrow cells plated, with a highly significant correlation ($r=0.97$) indicative of a single-hit phenomena.

II. Action of IL-1 in Short-Term Marrow Suspension Culture (Delta Assay)

The rationale behind the delta assay is to demonstrate the ability of hematopoietic growth factors to promote the survival, recruitment, or expansion of stem cells and/or progenitor cells in relatively short-term suspension culture systems. As originally developed, we utilized bone marrow from mice that had been treated with 5-FU for 24 h and then subjected to a 4- to 7-day suspension culture in the presence of IL-1 alone, CSFs alone, or combinations of IL-1 with various CSF species (Moore et al. 1987a; Moore and Warren 1987). At the end of the suspension culture phase, total cellularity and morphology was determined, and cells were cloned in semisolid culture, again in

Table 2. IL-1^{+} and CSF-induced amplification of CFU-GM in 7-day suspension cultures of 2.5×10^5 murine bone marrow cells obtained 24 h, post-5-fluorouracil treatment

Stimulus in suspension phase	Delta (CFU-GM output/input)			
	"Readout" stimulus in second clonogenic assay			
	G+IL-1	GM+IL-1	IL-3+IL-1	M-CSF-1+IL-1
Medium	0	0	0	0
rh-IL-1-α	190	53	39	61
rh-G-CSF	0	5	2	5
IL-1+G-CSF	255	91	41	41
mM-CSF	0	2	1	2
IL-1+M-CSF	175	47	38	80
mGM-CSF	40	17	15	34
IL-1+GM-CSF	280	58	49	54
mIL-3	110	17	18	33
IL-1+IL-3	510	178	120	115

Femoral bone marrow cells taken from B6D2F1 mice 24 h after a single i.v. injection of 5-FU (150 mg/kg) were incubated at 2.5×10^5 cells/ml in Iscove's modified Dulbecco's medium with 20% fetal calf serum in 24-well cluster plates containing 100 units rh-IL-1-α, 2000 U/ml rh-G-CSF, 1000 U/ml m-GM-CSF (purified from murine postendotoxin lung CM), 200 U m-IL-3 (purified from WEHI-3 cell line CM), or 1000 U M-CSF (purified from L-cell CM). After 7 days of incubation, cells were recovered and assayed for CFU-GM in agarose cultures stimulated by the various CSF species alone or in combination with IL-1. The delta value (CFU-GM output/input) was calculated on recovery from triplicate clonal assays from triplicate suspension cultures.

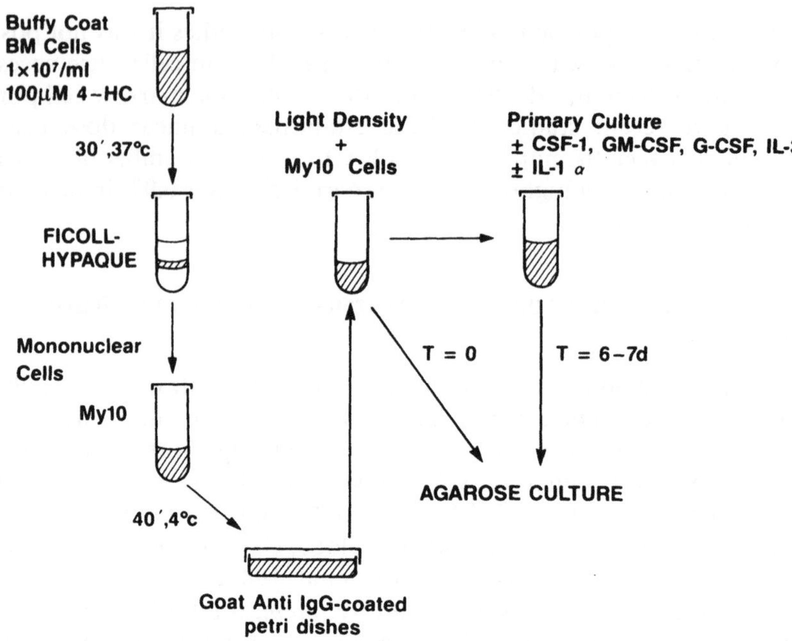

Fig. 1. Assay for CD34$^+$ 4-HC-resistant human stem cells. Normal human bone marrow is exposed to 4-HC and subjected to positive selection by "purging" with MY10 antibody. The selected cells are subject to colony assay in semisolid medium pre- and post- a 6-to 7-day incubation in suspension culture in the presence of CSFs plus or minus IL-1. The amplification of colony formation (delta) between input and output is determined in the presence of CSFs plus or minus IL-1 addition

the presence of IL-1 alone, CSFs alone, or combinations of IL-1 and CSFs. Table 2 shows that IL-1 caused an expansion of the numbers of CFU-GM recovered after 7 days of culture. Neither M-CSF (CSF-1) nor G-CSF alone supported survival or expansion of CFU-GM; however, the combination of IL-1 and CSFs demonstrated additive or synergistic effects on the expansion of these progenitors. GM-CSF or IL-3 alone caused some expansion of progenitor populations, but again the combination with IL-1 evidenced synergism.

Human systems based on in vitro purging with 4-hydroperoxycyclophosphamide (4HC) and positive selection by "panning" with MY10 monoclonal antibody for CD34$^+$ cells followed by 7-day suspension culture (Fig. 1) also demonstrated synergistic interactions between IL-1 and various CSF species. The most dramatic effect was observed with the combination of IL-1 and IL-3, where an up to 85-fold increase in progenitor cells was noted following 7-day suspension culture (Table 3). In contrast to the murine system, synergism between IL-1 and CSF-1 was not evident (Moore 1987c). IL-1 has been shown to be radioprotective in mice when administered both pre- and postirradiation (Neta et al. 1988). This radioprotective action may be mediated indirectly through induction of other radioprotective proteins, or directly by af-

Table 3. IL-1⁺ and CSF-induced amplification of CFU-GM in 7-day suspension cultures of 2.5×10^5 4-HC-treated, CD34⁺ human bone marrow cells

Stimulus in suspension phase	Delta (CFU-GM output/input)			
	"Readout" stimulus in second clonogenic assay			
	rh-G-CSF	rh-GM-CSF	rh-IL-3	rh-M-CSF
Medium	0	0	0	0
rh-IL-1-α	10	6	2	2
rh-G-CSF	8	9	2	2
IL-1 + G-CSF	8	9	2	1
rh-GM-CSF	8	9	2	2
IL-1 + GM-CSF	33	22	6	10
rh-IL-3	5	30	17	2
IL-1 + IL-3	15	85	15	1

Bone marrow cells, obtained with informed consent from normal volunteers, were separated over Ficoll-Paque, subjected to plastic adherence, and incubated for 30 min with 100 μ*M* 4-hydroperoxycyclophosphamide (4-HC). Cells were then treated with anti-Myl monoclonal antibody (anti-HPCA-1 Beckton Dickinson, Mountain View Ca.), on ice for 45 min, washed, and incubated for 1 h at 4° C on bacteriological-grade plastic petri dishes previously coated with goat anti-mouse IgG. Adherent CD34⁺ cells were harvested by vigorous pipetting. 2.5×10^5 harvested cells per well were incubated in Iscove's modified Dulbecco's medium (IMDM) plus 20% fetal calf serum in 24-well cluster plates containing test stimuli. In suspension phase, stimuli were 10 ng/ml rh-G-CSF, rh-GM-CSF, rh-IL-3 (Amgen), rh-M-CSF (Cetus), and rh-IL-1-α (Roche). 4-hc-purged, CD34-⁺ marrow cells were plated at 2×10^4 cells/ml in semisolid agarose culture in the presence of 10 ng/ml GM-CSF, G-CSF, CSF-1, or IL-3 alone or in combination with IL-1 both pre- and postsuspension culture. Colony formation was assessed after 12 days. The delta value (CFU-GM output/input) was calculated on recovery from triplicate clonal assays from duplicate suspension cultures.

fecting hematopoietic progenitors. We have shown a direct radioprotective effect of IL-1 in vitro in clonogenic and delta assays of human CD34⁺, 4HC-resistant stem cell populations (Laver et al. 1988 a). Such cells when irradiated in vitro with 100, 200, and 400 cGY and plated in the presence of G-CSF, GM-CSF, or IL-3 for direct colony formation showed Do's of CFU-GM survival of 103–116 (Table 4). The addition of IL-1 to the cultures in the presence of the CSFs almost doubled the Do, indicating a direct radioprotective effect on progenitors. To evaluate the effects of radiation on the pre-CFU-GM

Table 4. Do's of CFU-GM survival (cGY)

	G-CSF	GM-CSF	IL-3
Without IL-1	103	127	116
With IL-1	205	205	215

Do, the radiation dose that reduces survival to 37% of control on the straight portion of the survival curves.

Table 5. Effects of IL-1 in delta assay[a]

Radiation dose (cGY)	Without IL-1	With IL-1
0	3.3	6.04
100	0.3	3.3
200	0	2.8

Results expressed as fold increase compared with input.
[a] Mean on four experiments.

population we utilized a two-step assay in which $CD34^+$ cells were irradiated and cultured in a delta-suspension culture system in the presence of IL-3, or IL-3 plus IL-1. After 7 days, cells were assayed for colony-forming capacity in the presence of GM-CSF. The recovery value of colonies following 7 days suspension revealed up to tenfold relative increase in progenitor cell recovery in irradiated cultures exposed to IL-1 (Laver et al. 1988 a).

III. Hematopoietic Growth Factor Interactions in the Blast Cell Colony Assay

Hematopoietic blast cell colonies in both murine and human culture systems have high replating ability and self-renewal capacity generating various committed progeny (CFU-GEMM, BFU-E, CFU-GM) in a stochastic manner (Nakahata and Ogawa 1982; Koike et al. 1986, 1988). IL-3 provides a permissive milieu for the proliferation of early progenitors in culture but does not trigger stem cells in G_0 to begin proliferation. In vitro data suggest that blast cell colony formation requires less IL-3 than the multilineage colonies derived from recloning the blast colony cells (Koike et al. 1986). By selection in prolonged culture in the presence of low concentrations of IL-3 it has proved possible to select for early stem cell populations that are dormant in cell cycle and require only small amounts of IL-3 when activated into cell cycle. IL-3 itself was not considered to activate these cells into cell cycle, a phenomenon considered to be a stochastic event (Koike et al. 1986).

Interleukin 6 (IL-6) has been reported to have both a direct and indirect action in the stimulation of hematopoietic progenitors. Using human IL-6 in spleen or bone marrow cultures established from 5-FU-treated mice, Ikebuchi et al. (1987) showed that IL-3 and IL-6 acted synergistically to support multipotential progenitors in culture. Furthermore, the time course of total splenic blast cell colony formation was significantly shortened in cultures containing both factors relative to either factor alone. In 5-FU bone marrow cultures IL-6 alone failed to support colony formation and IL-3 stimulated only a few granulocyte-macrophage colonies. The combination of IL-3 and IL-6 acted synergistically to yield multilineage, and a variety of other types of colonies. In this system, IL-1-α also acted synergistically with IL-3 but the effect was smaller and no multiple-lineage colonies were seen. In secondary replating studies, IL-6 allowed recloning of blast colonies with a high efficiency; however, replating of the blast colony cells at limiting dilutions with IL-6 produced only a few neutrophil and macrophage colonies in contrast to many multilineage colonies observed with IL-3 (WONG et al. 1988). In a human blast colony assay using $CD34^+$ marrow cells, cultures were established for 2 weeks prior to addition of IL-3, IL-6, IL-1, or a combination of these factors and blast colony formation was assayed daily between 18 and 32 days of culture (LEARY et al. 1988). In the presence of IL-3 alone blast colonies continued to emerge between 21 and 27 days; however, with IL-3 plus IL-6, blast colonies were more frequent and developed earlier. IL-1 failed to augment IL-3-dependent blast colony formation in this system.

In serum-free cultures of murine blast cell colonies precursors, no colony growth was seen with IL-6, but the combination of IL-3 and IL-6 resulted in a significant increase in the number of colonies formed from multipotential progenitors in spleen and bone marrow of 5-FU-treated mice (KOIKE et al. 1988). This synergism was restricted to multipotential progenitors, since it was not observed on single lineage or oligolineage colonies formed by spleen cells of normal mice. These data suggest that IL-6 enhances the sensitivity of multipotential progenitors to IL-3, possibly by increasing the number of IL-3 receptors or upregulating their affinity on multipotential progenitors. IL-1 also failed to support colony formation in this serum-free system but enhanced IL-3-dependent colony growth and did so by a mechanism independent of IL-6.

In an innovative assay for human marrow blast colony-forming cells, GORDON et al. (1985) identified a population of early hematopoietic cells distinguished by their ability to adhere to preformed irradiated marrow stromal layers. The binding of the progenitor cells to the stromal cells was completed within 2 h and stromal layers of 9.6 cm^2 provided adhesion sites for at least 2000 blast colony-forming cells characterized by self-renewal potential, as well as a capacity to generate multipotential and lineage-committed colony-forming cells.

We have shown that 1–5 units of IL-1 induces GM-CSF production by adherent cell layers in murine bone marrow culture (LOVHAUG et al. 1986). IL-1 induced both transcriptional and translational activation of GM-CSF (and probably G-CSF and CSF-1) in endothelial cells (BAGBY et al. 1983; SIEFF et al. 1987) and fibroblasts, while mononuclear phagocytes are IL-1 inducible to produce CSFs (ZUCALI et al. 1988; FIBBE et al. 1986, 1988) which are probably G-CSF and CSF-1 species. In this context, the complexity of cytokine networks is further emphasized by the observations that CSF species can induce macrophage IL-1 production (MOORE et al. 1986) and both endothelial cells and fibroblasts (OPPENHEIM et al. 1986) can also produce, as well as respond to, IL-1. ISCOVE et al. (1988) reported a soluble activity from adherent marrow cells that cooperated with IL-3 in stimulating the growth of pluripotential hematopoietic precursors, and was shown to be distinct from IL-1 or M-CSF. This activity may be IL-6. mRNA for murine IL-6 is expressed in IL-1-treated mouse bone marrow stromal cells, activated T cells, and macrophage cell lines. It has a 65% nucleotide and 41% amino acid homology with human IL-6 and homology with G-CSF (CHIU et al. 1988). In this context, mIL-6 stimulated the survival and proliferation of the murine factor-dependent NFS-60 cell line that is stimulated by IL-3 and G-CSF.

IV. Inhibitory Influences on Hematopoietic Stem Cells and Progenitor Cells

Tumor necrosis factor (TNF), initially identified as an activity which induced hemorrhagic necrosis in tumor-bearing, BCG-primed mice challenged with endotoxin, has since been associated with a number of effects. Hematologically it activates polymorphonuclear leukocytes (Shalaby et al. 1985; Klebanoff et al. 1986) and can synergize with IL-1 in this respect (Wankowicz et al. 1988). It induces endogenous IL-1 production (Dinarello et al. 1986) and fibroblast or endothelial cell production of M-CSF, G-CSF, and GM-CSF (Koeffler et al. 1987; Oster et al. 1987; Zucali et al. 1988). TNF has been shown to inhibit proliferation and differentiation of human multipotential and erythroid cells in vitro (Moore et al. 1987a; Peetre et al. 1988). Recently it has been reported that administration of TNF in mice offers protection against the lethal effects of radiation (Neta et al. 1988). We have confirmed that TNF enhances recovery in sublethally irradiated mice (Slordal et al. 1989), and this effect is distinct from the effects of G or GM-CSF and IL-1, and is caused, in part, by cycle arrest of hematopoietic stem cells detected by the CFU-s assay.

Transforming growth factor-β1 (TGF-β1) is multifunctional in its influence on cellular proliferation but is clearly growth inhibitory in many different tissue systems including immunocompetent cell proliferation and differentiation. TGF-β is a potent inhibitor of hematopoietic cell growth (Ohta et al. 1987; Keller et al. 1988; Ruscetti et al. 1988). TGF-β appears selectively to inhibit early murine hematopoietic progenitor cell growth since it has little or no effect on differentiation induced by GM-CSF, G-CSF or M-CSF, or erythropoietin alone; it does inhibit IL-3-dependent growth of myeloid colonies and of IL-3 plus erythropoietin-dependent erythroid BFU-E (Keller et al. 1988). In human culture systems stimulated by either IL-3 or GM-CSF plus IL-1, TGF-β1 inhibited growth of BFU-E, CFU-E, and CFU-GEMM over a concentration range of 0.01–1.0 μg/ml (Ottmann and Pelus 1989). The inhibitory effect was progressively lost when addition was delayed for 40–120 h, suggesting a mode of action during early cell diversion. In contrast, growth of CFU-GM stimulated by plateau concentrations of rG-CSF, rGM-CSF, or rIL-3 was not inhibited; indeed slight potentiation was observed. In the HPP-CFU assay, TGF-β at picomolar concentrations was potently but reversibly suppressive to colony formation (Moore and Warren 1989, unpublished). A direct effect of TGF-β on early hematopoietic cells is supported by the observation that fluorescence activated cell sorting (FACS)-purified Thy-1$^+$ murine progenitor cells are growth inhibited by TGF-β. In a survey of a panel of hematopoietic cell lines, growth inhibition with TGF-β was seen only in lines that were IL-3 dependent for their growth, with the exception of the myelomonocytic leukemia WEHI-3, which is the only myeloid line known to produce IL-3 constitutively (KELLER et al. 1988). In long-term bone marrow culture the most primitive clonogenic hematopoietic cells residing in the adherent layer can be activated into cell cycle by addition of horse serum or known mesenchymal cell activators that influence growth factor production by stromal cells. TGF-β (5 µg/ml) can override these stimulatory events and delay by 2–3 days the entry of the most primitive stem cells into cycle (EAVES et al. 1987).

D. Synergistic Interactions Between IL-1, IL-3, and IL-5 in the Production and Activation of Eosinophils

The role of IL-5, IL-3, and GM-CSF in activation of the function of mature eosinophils has been documented in a number of studies. IL-5 maintains the viability of mature eosinophils in vitro (YAMAGUCHI et al. 1988). The activated eosinophils have a markedly increased capacity to adhere and to kill antibody-coated schistosomula of *Schistosoma mansoni* when tested in vitro (LOPEZ et al. 1988). GM-CSF also exerts a powerful stimulus to mature human eosinophils, enhancing antibody-dependent cell-mediated cytotoxicity (ADCC), phagocytosis of serum-opsonized yeast, and enhancing eosinophil survival by 9 h (LOPEZ et al. 1986). IL-3 also prolongs the viability of human eosinophils in vitro and increases leukotriene C4 in response to calcium ionophore. Furthermore, fresh eosinophils only killed 14% of antibody-coated *S. mansoni* larvae whereas following IL-3 exposure they killed 54% (ROTHENBERG et al. 1988).

The in vitro production of eosinophils from murine bone marrow precursors is influenced by IL-5 (SANDERSON et al. 1986) and to a lesser extent by IL-3 (IHLE et al. 1983), and in man and primates also by GM-CSF (DONAHUE et al. 1986; ANTMAN et al. 1988). In primary suspension cultures of bone marrow from normal or parasitized mice, IL-3 produced a modest stimulation of eosinophil production whereas IL-5 was a significantly better stimulus particularly in the parasitized model (WARREN and MOORE 1988). However, IL-3 was sevenfold more effective than IL-5 in generating eosinophil progenitors (CFU-eo) in this suspension culture system. Preincubation of bone marrow cells in suspension culture with IL-3, but not IL-5, increased the recovery of myeloid precursors responsive to G-CSF, GM-CSF, CSF-1, or IL-3 two- to fourfold while progenitor cells responsive to IL-5 were increased by 70-fold. Similar studies were undertaken using a primary culture phase in methylcellulose under limiting dilution, clonal conditions followed by secondary recloning with IL-3 or IL-5. Preincubation of bone marrow cells with IL-3 under clonal conditions resulted in a 50-fold increase in CFU-eo responding to IL-5 over input values (WARREN and MOORE 1988).

The bone marrow from mice pretreated with 5-FU is greatly depleted of progenitor cells capable of responding directly to IL-3 or IL-5. Suspension culture of such marrow with IL-5 failed to generate eosinophils or CFU-eo and cultures with IL-3 also failed to sustain input cellularity and few CFU-eo were generated. IL-1 also failed to stimulate the production of eosinophils or CFU-eo. A marked synergism was observed when IL-1 and IL-3 were combined in the suspension of preculture phase with a sixfold greater recovery of CFU-eo than induced by either factor alone. Furthermore, preculture of 5-FU-treated marrow cells with a combination of IL-1 and IL-3 resulted in a 260-fold increase of CFU-eo over input numbers (Warren and Moore 1988). Yamaguchi et al. (1988) also observed that spleen cells from 5-FU-treated mice did not produce eosinophil colonies in response to IL-5. In this model, primary blast colonies supported by IL-3 or G-CSF upon replating required IL-5 for eosinophil terminal differentiation, suggesting that IL-5 does not act on primitive hematopoietic cells, but upon blast cells induced by IL-3 or G-CSF. Taken together these data suggest that the concatenate action of IL-1, IL-3, and IL-5 is an absolute requirement for the in vitro generation of eosinophils from primitive hematopoietic stem cells and a model of the sequential interaction is presented in Fig. 2.

A selective eosinophilia and eosinophil activation is seen in idiopathic hypereosinophilic syndrome and the eosinophilia in patients with helminthic infections and allergic conditions appears to involve IL-5 production from IL-2-induced T cells (Enokihara et al. 1988; Lopez et al. 1988; Raghavachar et al. 1987). The synergism observed in vitro between IL-5, IL-3, and IL-1 may also play a role in the expansion of eosinophils where urgent mobilization is necessary at times of helminthic infection and allergic response. There is, however, another mechanism for selective production of eosinophils rather than neutrophils under conditions where, for example, IL-3 or GM-CSF levels may be elevated.

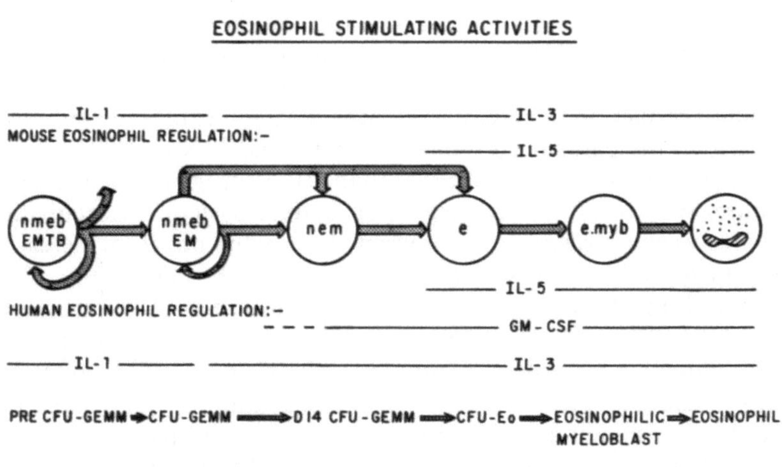

Fig. 2. Role of various interleukins and GM-CSF in the generation of eosinophils from primitive stem cells in mouse and man. Alphabetic designations on cell lineages refer to lineage potentiality (*n*, neutrophil; *m*, monocyte; *e*, eosinophil; *b*, basophil; *E*, erythroid; *M*egakaryocyte; *T*, *B*, lymphoid). *Arrow* indicates self-renewal and differentiation potential

In clonal assays of human bone marrow cells depleted of accessory cell populations and enriched for progenitor cells using immunoadherence "panning" and complement-mediated cytotoxicity, 50% inhibition of G-CSF-stimulated colonies was evident with as little as 5 units of TNF, a 2-log lower concentration than required to produce comparable inhibition of colony formation in GM-CSF-stimulated cultures (MOORE et al. 1987b). Further analysis of this TNF inhibition revealed that it was specific for neutrophils, and to a lesser extent monocytes, and that the eosinophil differentiation in-

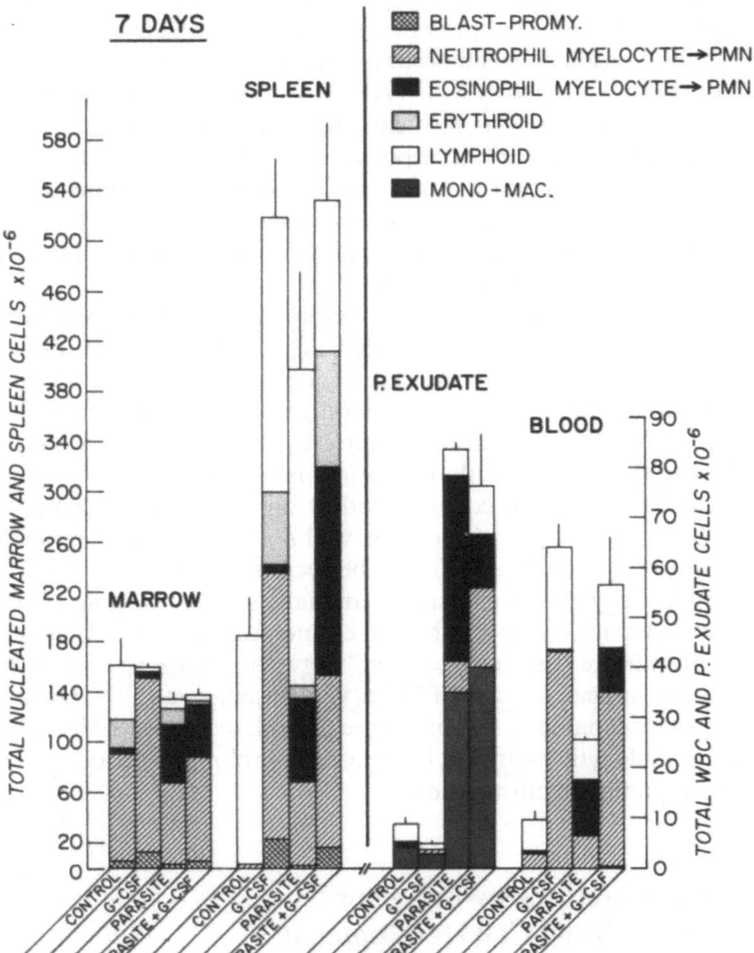

Fig. 3. Production of various hematopoietic cells in the spleen. Circulation and peritoneal exudate of BALB/c mice infected with *Schistosoma mansoni* which induced a marked eosinophilia. Control mice or eosinophilic mice were treated with 2 µg rhG-CSF i.p. daily for 7 days. Note the neutrophilia in G-CSF-treated mice in all tissue and the coexistence of neutrophilia and eosinophilia in G-CSF-treated, parasitized mice

duced by GM-CSF was completely refractory to even very high concentrations of TNF. These studies were confirmed in human long-term bone marrow cultures where G-CSF addition caused a substantial increase in neutrophil production over periods of 14 weeks and the addition of TNF with or without G-CSF administration caused rapid loss of neutrophil production in this system (MOORE et al. 1987b). The addition of GM-CSF protected against TNF inhibition in terms of generation of CFU-GM and total hematopoietic cells. Morphological analysis revealed that GM-CSF stimulated a population that was exclusively eosinophilic and TNF addition suppressed the capacity of GM-CSFs to promote neutrophil differentiation. This very striking difference in sensitivity of the neutrophil versus the eosinophil pathway to TNF inhibition provides yet another mechanism by which selective eosinophil production can occur in response to the coordinate interaction of IL-3 or GM-CSF together with TNF. This provides yet another example of the overlapping mechanisms that exist to induce the production of a single species of hematopoietic cell in response to a particular physiological requirement. An additional question can be raised concerning the potential competition that may exist under situations where a demand for neutrophil granulocyte production versus eosinophil granulocyte production exists, and where IL-5 and G-CSF may be induced.

The underlying stochastic nature of commitment and/or receptor display in hematopoietic differentiation would suggest that competition for differentiation is unlikely to exist: however, formal proof that this is valid in an in vivo situation has been lacking. We investigated one model in which marked chronic eosinophilia was induced in mice by infectioin with *S. mansoni* and animals were then treated with daily administrations of high doses of recombinant G-CSF. The subsequent production of mature eosinophils and neutrophils in the peripheral blood as well as the progenitor populations in the marrow were monitored. As can be seen in Fig. 3, G-CSF did induce a neutrophil leukocytosis in parasitized animals, yet there was no evidence of a significant reduction in the numbers of eosinophils generated in this situation. These data indicate that "competition" between lineage-specific growth factors does not appear to exist in an in vivo situation and the hematopoietic system appears capable of sustaining greatly enhanced levels of production of different granulocytic pathways for relatively prolonged periods (MOORE and WARREN, unpublished information).

E. Hematopoietic Growth Factors and Basophil/Mast Cell Development

YUNG et al. (1981), YUNG and MOORE (1982), and TERTIAN et al. (1981) were the first to report the development of a long-term bone marrow culture system supporting the continuous replication of Thy-1 cells with basophilic cytoplasmic vaculation (for review see MOORE 1988a–c). These cells were later identified as mast cells on the basis of their characteristic morphology at the light microscopic and ultrastructural level, reactivity with Alcian blue and toluidine blue, the monoamines and sulfated products contained in the cytoplasm, and the presence of IgE receptors. Normal mouse bone marrow or spleen, either fresh or following Dexter-type long-term culture, consistently produced indefinitely, exponentially growing lines of mast cells with a five- to tenfold increase in cell number with biweekly passage. As originally described, mitogen-stimulated spleen conditioned medium or conditioned medium from a murine myelomonocytic leukemia (WEHI-3) proved equally effective in sustaining mast cell proliferation. Spontaneous differentiation and proliferation of marrow-derived mast cells were also features of long-term cultures established from the Prosimian tree shrew *Tupia glis* (MOORE et al. 1979; MOORE and SHERIDAN 1979).

Mast cells are subdivided into mucosal, present in the gastrointestinal tract and characterized by chondroitin sulfate as the predominant proteoglycan, and serosal (connective tissue type) characteristically producing heparin proteoglycan. Basophils, in contrast to classic mast cells, possess lobulated nuclei, smaller-sized cytoplasmic granules, and a low histamine content. The detection of mast cell growth factor activity in lymphocyte condition medium indicated a role for interleukin-3; however, purified IL-3 preparation did not account fully for all mast-cell-stimulating activity in murine bone marrow culture (YUNG et al. 1981; YUNG and MOORE 1982). A second lymphokine was subsequently identified as synergizing with IL-3 in the promotion of mast cell proliferation and differentiation and this activity was named interleukin-4 (SMITH and RENNICK 1986). Mast cells induced by IL-3 predominantly express chondroitin sulfate and resemble the mucosal mast cell phenotype, whereas mast cells requiring IL-4 for in vitro growth express heparan sulfate and are of the connective tissue type (HAMAGUCHI et al. 1987).

In human systems, T-cell-derived factors have been implicated in basophil colony formation and a basophil-mast-cell-promoting activity was characterized as capable of supporting the proliferation of cells with lobulated nuclei and low histamine content, comparable to the circulating basophilic granulocyte population (ABBOUD and MOORE 1988). Biochemical characterization of this factor suggested an identity with IL-3 rather than IL-4. We have studied the effects of various recombinant human growth factors on the development of basophil/mast cells from normal marrow precursors (ABBOUD and MOORE 1988). Accessory cell-depleted bone marrow or CD34-positive bone marrow cells selected by monoclonal antibody panning were cultured in the presence of G-CSF, GM-CSF, IL-3, and IL-4 and cultures were passaged weekly with readdition of the specific growth factors. Basophil/mast cells characterized by Alcian blue-positive and safronin-negative features appeared after 1 week in culture in the presence of IL-3. They increased in percentage and absolute number over 2 weeks, reaching 40%–70% of all cells in culture. Small numbers of similar cells developed in the presence of GM-CSF. Neither G-CSF nor IL-4 alone stimulated basophil/mast cell development. The basophil/mast cells obtained in this system had the ultrastructural features, histamine content, and presence of surface IgE receptors characteristic of mast cells. Further maturation was promoted by the addition of IL-4, which on its own was not an effective mast cell differentiation-inducing agent in this culture system.

One of the major effects of IL-3 administration in normal mice was the appearance in the spleen of large numbers of incompletely granulated mast cells (METCALF 1989). Microinjection of IL-3 into 9- to 14-day-old mouse fetuses induced large numbers of mast cells in subdermal tissue, adrenals, periocular mesenchyme, and genital ridges within 48 h, attesting to the physiological role of IL-3 as a mast cell growth factor (DANIELS 1988). In primate in vivo studies, DONAHUE et al. (1988) showed that continuous infusion of IL-3 induced a modest, delayed leukocytosis of approximately 2-fold, but with an 18-fold increase in an unusual population of leukocytes containing toluidine blue-staining granules that differed from normal basophils in their hypogranulation and more diffuse chromatin.

The study of human basophil-mast cells has long been hampered by their biological heterogeneity, difficulty in isolation, and absence of a consistent and successful culture system. The availability of a bone marrow derived IL-3/IL-4-dependent basophil/mast cell culture system has permitted a more in-depth analysis of the role of mast cells in a variety of pathophysiological situations. In addition to the well-known role of mast cells in the allergic response and anaphylaxis, these cells have also been implicated in bone remodeling, stimulation of fibroblast proliferation, stimulation of collagenase and prostaglandin E production by rheumatoid synovial cells, activity in parasitic infections, and cytotoxicity against human tumor cells in vitro. Using in situ hybridization and the avidin-biotin-complex immunoperoxidase methods, STEFFEN et al. (1989) demonstrated the presence of TNF mRNA in the cytoplasm, and TNF protein in the granules, of individual human basophil/mast cells generated in the presence of IL-3. The production of high levels of TNF by these cells could explain many of their reported functions. Murine mast cell lines also produce IL-6 as determined by molecular analysis and immunoassay (HULTNER et al. 1988).

F. Preclinical In Vivo Experience with Hematopoietic Growth Factors

I. Murine Studies

The species crossreactivity of G-CSF has allowed us to study the action of rh-G-CSF purified from *E. coli* in mice. An advantage of murine systems resides in the availability of a spectrum of assays for committed progenitor cells and early pluripotential stem cells, whose total population size can readily be enumerated in different organs. In addition, it is possible precisely to quantitate total differentiated hematopoietic cell production in steady-state or myelosuppressed situations. This has allowed us to analyze the action of chronic and acute G-CSF therapy on different lineages and different levels of differentiation to determine if depletion of specific compartments occurs with a chronic neutrophil differentiation stimulus, or whether competition for restricted stem cell populations may compromise recovery of other lineages. In addition, the use of C3H/HeJ endotoxin-hyporesponsive mice in some of these studies provides a valuable control for possible effects of low doses of endotoxin.

A twofold increase in blood neutrophil levels was apparent 48 h after initiating G-CSF therapy in mice, with counts rising progressively to 26000/mm^3 by 5 days and fluctuating between 10000 and 54000/mm^3 for up to 5 weeks of continuous treatment. The average increase in neutrophils in C3H/HeJ mice was 8- to 10-fold and 50- to 60-fold in BALB/c mice. Metamyelocytes were present in the peripheral blood of treated mice but more immature stages were rare. The absolute numbers of monocytes fluctuated, but increased on average threefold from days 5–14, whereas absolute numbers of eosinophils remained unchanged. A persisting mild lymphocytosis was induced with a twofold increase from 4–14 days of treatment.

Following 7–14 days of G-CSF therapy, marrow cellularity remained unchanged but spleen cellularity increased progressively in all strains to a maximum average increase of three- to fivefold with maximum spleen weights of up to 800 mg. Morphologically, granulopoiesis predominated in marrow and spleen, being increased over sevenfold with respect to control by 14 days, at which time granulopoietic tissue comprised 5% of total body weight. This increase in granulopoiesis resulted in almost complete suppression of marrow lymphopoiesis and erythropoiesis. The latter was more than compensated for by an increase in splenic erythropoiesis, resulting in a more than twofold overall increase over control by 7 days. The total splenic lymphoid population and splenic B-lymphocyte clonogenic cells were unchanged and there was no evidence of stress-related thymic involution. Megakaryocyte numbers dramatically increased in the spleen but fell progressively in the marrow, and overall the total megakaryocyte populations in all strains increased threefold by day 7 and more than fourfold by day 14. Increase in CFU-GM in the spleen continued progressively, although marrow numbers remained in the control range. Total body CFU-GM responsive to IL-3 increased four- to sixfold between day 7 and day 14. CFU-GM responsive to G-CSF, M-CSF, and GM-CSF increased three- to fivefold by day 7. Eosinophil progenitors were unchanged by day 7, but like all other progenitors were increased more than twofold by day 14 relative to control. CFU-s increased two- to threefold relative to control at day 7 and 14. CFU-GEMM, BFU-E, and CFU-Meg greatly increased in the spleen but the decline in the marrow resulted in a transient decline in total numbers of these progenitors at day 7 followed by their rapid recovery to supranormal levels by day 14. Continued daily G-CSF treatment for up to 5 weeks was associated with a sustained increase in total myeloid mass and of CFU-GM in marrow and spleen CFU-s; CFU-GEMM, BFU-E, and CFU-Meg remained elevated in absolute numbers despite the chronic stimulation of granulopoiesis.

In order to investigate the potential of G-CSF in preventing episodes of neutropenia following high-dose chemotherapy with cyclophosphamide (CY), C3H/HeJ mice were subject to weekly injections of 200 mg/kg CY intraperitoneally followed by G-CSF therapy (1.75 μg × 2 daily) beginning 2 h after CY treatment and finishing 48 h before the second cycle of CY. This protocol prevented the subsequent nadirs of neutrophil counts between 4 and 5 days after CY treatment, and in all ten cycles of CY therapy G-CSF abrogated the neutrophil nadirs. The mice receiving CY alone had a substantial mortality evident by the eighth cycle of treatment. In eight cycles of CY treatment over the course of 70 days the G-CSF-treated animals were neutropenic (<1000 ANC/mm)^3 for only 3 days, whereas the untreated animals were neutropenic for 24 days. The time of recovery of absolute neutrophil counts (ANCs) to control levels was also substantially affected, with the non-G-CSF-treated groups requiring an additional 16 days to recovery. Neutropenia at levels <500 ANC/mm^3 indicated an even more significant difference, with such low levels observed on only 1 day out of 70 in the G-CSF + CY-treated mice, in contrast to 15 days in the CY-treated group.

This cyclophosphamide model allowed us to investigate the timing of initiation of G-CSF therapy following a single injection of a high dose of CY.

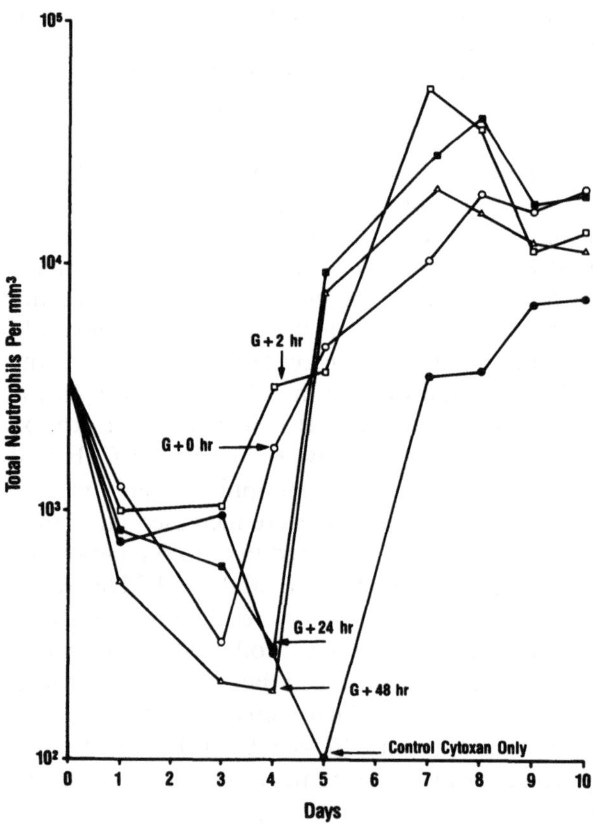

Fig. 4. Recovery of absolute neutrophil counts in C3H/HeJ mice following cyclophosphamide treatment (200 mg/kg i.p.) with rhG-CSF administration (1.75 μg twice daily for 10 days). G-CSF was administered concurrently with CY, or 2, 24, and 48 h after. Note the log scale of total neutrophils

Administration of G-CSF coincident with CY treatment resulted in an improvement in recovery of neutrophil counts, but this was substantially less than observed when G-CSF was administered 2 h after CY therapy (Fig. 4). Administration of G-CSF 24 and 48 h after CY was less effective than after 2 h, indicating that treatment early in the course of chemotherapy is likely to be more effective.

Myleran (MY) administered orally (100 mg/kg) was also associated with a significant reduction in neutrophil counts in bone marrow progenitor populations. G-CSF therapy initiated within 6 h of MY treatment led to an accelerated recovery of neutrophils so that by 1 week the neutrophil counts had returned to normal values, whereas the MY-treated controls remained profoundly neutropenic (ANC $<200/mm^3$). By 14 days MY + G-CSF-treated animals exhibited a neutrophil leukocytosis of between 10^4 and 2×10^4 mm^3 when the MY control animals had <100 neutrophils/mm^3.

1. In Vivo Interaction Between IL-1 and G-CSF in Mice Treated with 5-FU

In mice treated with 5-FU, G-CSF administration restores neutrophil counts to normal values some 5–6 days earlier than in mice not receiving the factor, but a period of profound neutropenia is still observed (MOORE and WARREN 1987). Administration of IL-1 alone, given postchemotherapy for 4–10 days twice daily at doses of 0.2 µg/mouse per day, reduced the severity of the neutrophil nadir, and accelerated the recovery of the neutrophil count to an extent greater than observed with G-CSF alone (Fig. 5). The combination of G-CSF and IL-1 administered after 5-FU therapy also resulted in accelerated hematopoietic reconstitution, although the results were additive rather than synergistic (MOORE and WARREN 1987). Analysis of total hematopoietic cell reconstitution in 5-FU-treated mice also showed the efficacy of IL-1 therapy or IL-1 plus G-CSF therapy in accelerating total recovery of erythroid as well as granulocytic elements in the marrow, spleen, and blood.

The potential of combination biotherapy as an effective means of accelerating hematopoietic cell differentiation raises the issue of whether premature exhaustion of the stem cells and progenitor cell population may occur. Measurement of colony-forming units (CFU-s, CFU-C, BFU-e, CFU-GEMM, CFU-Meg) on an incidence basis proved to be misleading because of the redistribution of hematopoietic precursors in regenerating murine tissues. By quantitation of total numbers of these cell populations in 5-FU-treated mice exposed to both IL-1 alone and IL-1 plus G-CSF, an absolute increase in pluripotential stem cells and progenitor populations was evident in bone marrow and spleen. The results indicate that the mechanism of action of these cytokines, both alone or in combination, is not simply mediated by accelerated differentiation, but involves an absolute expansion of the most primitive stem cell populations that can be measured, e.g., by the day 12 murine spleen colony (CFU-s) assay, or the high proliferative potential in vitro colony assay requiring IL-1 plus CSF as the read-out stimuli (MOORE et al. 1987a; MOORE and WARREN 1987). This is an important observation for

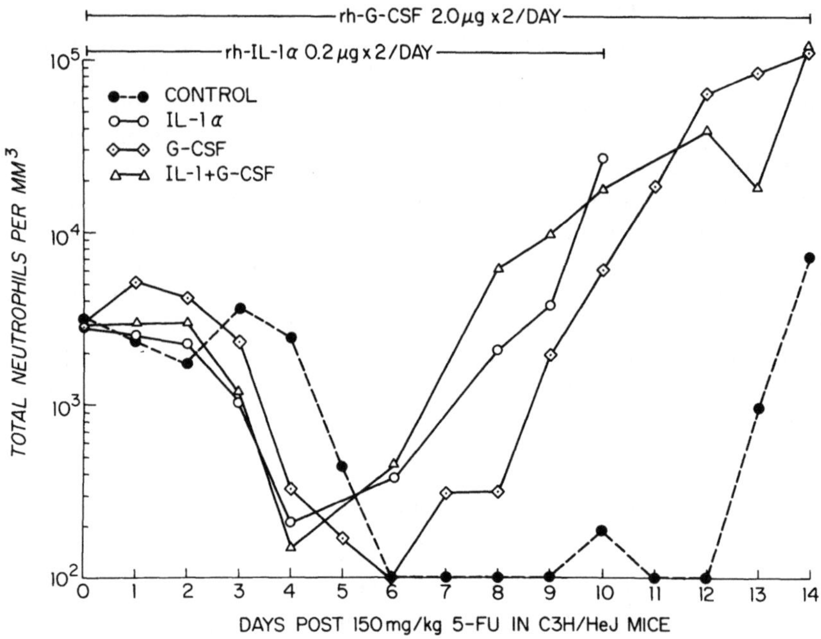

Fig. 5. Neutrophil counts in mice receiving 150 mg/kg 5-FU as a single dose i.v., and 0.2 μg rhIL-1-α i.p. × 2 daily for 10 days or combined with × 2 daily injections of 2.0 μg rhG-CSF given for the duration of the study or 2.0 μg rhG-CSF alone. All factor injections were begun 4 h after treatment with 5-FU. Control mice received 5-FU only. Results on groups of four C3H/HeJ mice. Note log scale

extrapolation of these studies to clinical situations because it lays to rest some of the concerns that combination biotherapy may accelerate exhaustion of the stem cell compartment in patients receiving myeloablative therapy.

The action of G-CSF alone in 5-FU-treated mice clearly does lead to accelerated recovery of neutrophils, hematopoietic cells, and all classes of progenitors and stem cells but recovery is much delayed compared with what is seen after high-dose cyclophosphamide produced with G-CSF alone, and synergistic effects were seen in the neutrophil recovery between 7 and 9 days postchemotherapy (Moore and Warren 1987). In view of the in vitro evidence that IL-1 alone does not support myeloid or erythroid differentiation, the most plausible explanation for the in vivo effects implicates a dual role for the cytokine induction of elevated production of G- and GM-CSF by an action on host accessory cells, and a direct stimulation of early stem cells with CSF receptor upregulation. The rate of regeneration of hematopoietic tissue following 5-FU is thus not limited by stem cell availability but rather by cytokine availability, with exogenous IL-1 overriding the deficiency in endogenous factor capable of directly activating stem cells.

The availability of CSF is also rate limiting but the potential to produce CSF upon IL-1 stimulation is certainly not totally ablated as evidenced by the response of the animal to IL-1 alone. However, the greater than additive ef-

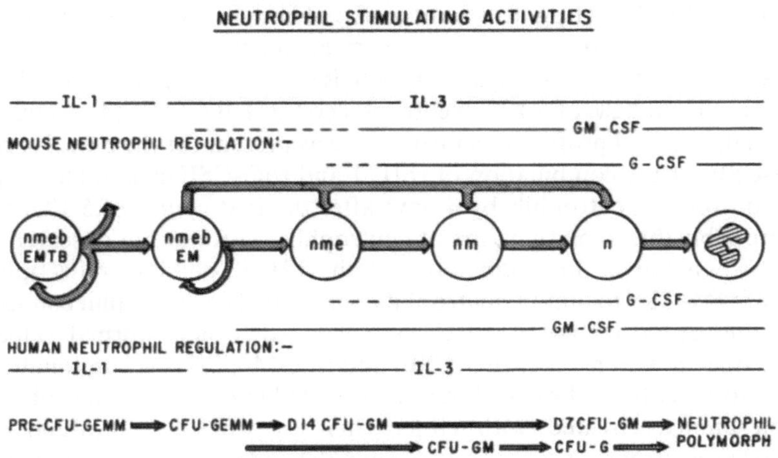

Fig. 6. Role of various interleukins and CSF species in the generation of neutrophils from primitive stem cells. Alphabetic designations of cell lineage as in Fig. 2

fects of the combination of IL-1 and G-CSF in vivo in granulopoiesis and total neutrophils generated strongly suggests that G-CSF production, even in response to IL-1, is less than optimal and the combination of both factors is the most effective way to promote maximum hematopoietic regeneration.

The observation that G-CSF can evoke a regenerative response involving early stem cells and progenitors and lineages other than neutrophil granulocyte was unexpected in view of the relatively limited spectrum of cells able to respond directly to G-CSF. This suggests that in vivo administration of G-CSF initiates regulatory cascades whereby the depletion of G-CSF-responsive populations by differentiation triggers endogenous mechanisms for recruitment and activation of earlier multilineage stem cell populations. It is attractive to postulate that IL-1 is involved in such a G-CSF-initiated cascade.

The in vivo observations indicate an important role for IL-1 in combination with specific hematopoietic growth factors in counteracting chemotherapy- or irradiation-induced myelosuppression. The generation of neutrophils from primitive stem cells can be seen as a sequential interaction of a variety of regulatory macromolecules with overlapping, additive, or synergistic actions (Fig. 6). Early pluripotential stem cells interact with IL-1 to acquire responsiveness to IL-3, GM-CSF, and G-CSF. Any of these CSFs can promote the generation of a functionally active, mature neutrophil, although concatenate interactions between G- and GM-CSF or IL-3 and G-CSF can specifically amplify the neutrophil response.

The potential of IL-1 in combination with CSF to allow intensification of chemotherapy has been tested in a murine tumor system. In CD8F1 female mice transplanted s.c. with spontaneous arising breast tumors, 5-FU at a dose of 130 mg/kg was administered weekly with or without IL-1 (0.1 μg at 8-h intervals) or GM-CSF (1 μg at 8-h intervals). These factors were given either alone or in combination. Tumor size was measured weekly and WBC and differentials were performed 4 days after each dose of 5-FU and again on the day of the subsequent dose of 5-FU. Neutrophil counts fell after the first course of 5-FU treatment and profound neutropenia was present for the 3 weeks. rhIL-1, rhG-CSF, or the combination of rhIL-1 and rhG-CSF caused an early accelerated drop in neutrophils by 4 days after the first course of 5-FU, treatment probably due to margination of neutrophils. This was not observed with mGM-CSF alone or in combination with rhIL-1, possibly reflecting the ability of GM-CSF to inhibit neutrophil migration. The neutrophil counts following the second course of 5-FU were elevated to near normal values in animals treated with rhIL-1, rhG-CSF, and rmGM-CSF or a combination of those factors. By the 3rd week of treatment, a striking difference was observed between animals treated with either rhG- or rmGM-CSF alone (they remained as neutropenic as the control animals) and the rhIL-1-treated group, which showed rapid recovery of neutrophil counts. A combination of rhIL-1 and rhG-CSF and particularly rhIL-1 plus mGM-CSF resulted in the most effective stimulation of neutrophil recovery in comparison to control groups. The survival of tumor-bearing mice was also affected by growth factor administratioin with 40% death in the 5-FU-treated animals, whereas rhIL-1-treated animals had an improved survival, showing a mortality rate of 95%. 5-FU administration to the animals resulted in a markedly reduced tumor mass.

II. Primate Studies

The effects of rh-G-CSF and rh-GM-CSF on cytopenias associated with repeated cycles of chemotherapy have been studied in primates treated with cyclophosphamide every 28 days for four cycles. Animals were pretreated with either G- or GM-CSF prior to cyclophosphamide, and daily thereafter through subsequent cycles. Other animals received G- or GM-CSF beginning 72 h following each cycle for 12-day intervals. The continuous rh-GM-CSF-treated primates had a neturophil nadir of $2–3 \times 10^3/mm^3$ on days 6–7 followed by a leukocytosis from $21–22 \times 10^3/mm^3$. Primates treated postchemotherapy with rh-GM-CSF had a more delayed recovery of neutrophils. These data suggested that continuous treatment was more effective with GM-CSF, but that postchemotherapy treatment with G-CSF was equally effective in ameliorating neutropenia (BONILLA et al. 1987).

In primate studies, DONAHUE et al. (1988) reported that continuous infusions of IL-3 at 20 μg/kg per day for 6–7 days in cynomologus monkeys elicited only a modest and delayed leukocytosis, predominantly involving eosinophils but also atypical basophils. The low magnitude and delay of its leukocyte response contrasts with that seen with G- or GM-CSF (DONAHUE et al. 1986; BONILLA et al. 1987), suggesting that IL-3 may expand the number of early progenitors but is not sufficient to generate mature cells that are released into the periphery. Administration of IL-3 and GM-CSF simultaneously did not promote any improved leukocytic response; indeed they may have suppressed the leukocytosis associated with GM-CSF treatment (O'REILLY, personal communication). However, the sequential administration of IL-3 for 7 days followed by infusion of low doses of GM-CSF (2 μg/kg per day for 4 days) elicited a dramatic leukocytosis far greater than seen at the respective dose of GM-CSF. GM-CSF administration before IL-3 was no more effective than IL-3 alone. More prolonged treatment with GM-CSF following IL-3 priming showed that the leukocytosis could be sustained at the enhanced level for up to 14 days.

In primates, administration of 150 mg/kg 5-FU resulted in profound neutropenia lasting for 30 days. Administration of rhIL-1 at 1 μg/kg for 2 days or for 7 days shortened the period of neutropenia to 17 days. Administration of rhIL-1 at the same dose for 14 days resulted in delayed recovery. In vitro studies showed that 2 and 7 days of IL-1 therapy resulted in a significantly increased frequency of marrow progenitors responsive to G-, GM-CSF, and IL-3 by day 14 after therapy. Animals treated for 14 days with IL-1 had a low frequency of progenitors on day 14 but recovered by day 21 (7 days after cessation of IL-1 therapy) (GASPARETTO et al. 1989).

G. Clinical Experience with G- and GM-CSF

The preclinical evaluation of G- and GM-CSF indicated a potential value of these hematopoietic growth factors in preventing iatrogenic myelosuppression in patients with malignant disease receiving high-dose chemotherapy and/or radiation therapy. Potential efficacy was based on the ability of these factors to stimulate bone marrow progenitor cells as well as activating the function of mature cell populations such as neutrophils, monocytes, and eosinophils.

I. CSFs in Chemotherapy-Induced Neutropenia

To date rhG-CSF has been studied in phase I and II trials in hundreds of cancer patients. The clinical study with rhG-CSF was completed at Memorial Sloan-Kettering in patients with transitional cell carcinoma of the urothelium who are at risk for chemotherapy-induced neutropenia (GABRILOVE et al. 1988). Two courses of rhG-CSF were administered. The first was given over a 6-day period prior to M-VAC (methotrexate, velban, adriamycin, and cisplatin) chemotherapy, and the second began on day 4 and continued through day 11 of the first chemotherapy cycle. Daily intravenous administration (over a 20- to 30-min period) of rhG-CSF was given during these two periods of therapy. Historical data had shown this chemotherapy regimen caused severe leukopenia and that 67% of the patients had their day 14 chemotherapy withheld because of the leukopenia. In the second cycle of M-VAC chemotherapy, no rhG-CSF was administered. In this first study, 22 patients (three–six per group) were treated at rhG-CSF dose levels of 1, 3, 10, 30, or 60 μg/kg per day prior to their chemotherapy. There was a mean increase in absolute neutrophil counts (ANCs) from pre-rhG-CSF levels. The response was dose dependent and ranged from an approximately 1.8- to 2-fold increase at the 1-μg/kg dose to 13-fold at the 60-μg/kg dose. The increase in WBC count was due predominantly to neutrophil granulocytes. The vast majority of the neutrophils were segmented with some bands and a few immature cells (metamyelocytes) observed at the higher doses. Differentials indicated that other hematopoietic cell types found in the peripheral blood remained stable during rhG-CSF administration. Blood chemistries were within normal limits with the exception of an elevation in leukocyte alkaline phosphatase and lactate dehydrogenase, which would be expected with the increase in neutrophil count. None of the side effects experienced with other biological response modifiers (fever, chills, nausea, headache, blood pressure changes, diarrhea, fatigue, local irritation, myalgia, hemorrhage, and confusion) were reported or observed in patients treated with rhG-CSF. The only side effect noted was that of mild medullary pain/discomfort in the iliac crest and/or a sense of fullness/discomfort in the sternum during the rhG-CSF administration. The medullary pain/discomfort was transient, not dose limiting, and did not occur in all patients. Patients were eligible for a second course of rhG-CSF following chemotherapy if the average absolute neutrophil count (ANC) for days 5 and 6 of their first course of treatment was greater than twofold the average of their pretreatment baseline ANC. The second course of rhG-CSF was administered beginning on day 4 following M-VAC chemotherapy (days 0 and 1) and continued through day 11. Day 14 counts were of importance because the patients were scheduled to receive more chemotherapy (methotrexate and vinblastine) on this day if they were not leukopenic.

All but the three patients treated at the lowest dose (1 μg/kg) qualified for a second course of rhG-CSF. Twelve patients were evaluated for their WBC profiles during the 8-day administration of rhG-CSF following chemotherapy, and for their day 14 WBC/ANC. In patients receiving both courses of rhG-CSF there were two profiles of response. Profile 1 was characterized by the absence of any significant leukopenia (ANC >1500 throughout days 1–14) and a normal WBC/ANC on day 14. The second profile was characterized by a brief nadir (1–4 days total; ANC <1500 but >500 for 0–2 days, ANC <500 for 3–4 days) between days 7–10 postchemotherapy. This was followed by a pronounced rebound by days 10 and/or 11 such that the WBC/ANCs were again normal by day 14. Four of the five patients in the second profile had prior pelvic radiation therapy for their disease, and one had no history of radiation or chemotherapy. There were no patients with a history of pelvic radiation in the first profile. All patients who received both courses of rhG-CSF had normal or elevated WBC/ANC on day 14 and qualified for their day 14 chemotherapy. Of five patients not receiving rhG-CSF postchemotherapy, all had ANCs ≦1200 on day 14 and were not eligible for the additional chemotherapy.

A phase I/II clinical study of G-CSF is currently ongoing at the Ludwig Institute for Cancer Research in Australia in patients with advanced malignancy (MORSTYN et al. 1988). G-CSF was given intravenously in doses of 1, 3, 10, 30, and 60 μg/kg on days 1–5 and days 10–18. Melphalan (25 mg/m^2) was administered on day 9. Three patients were treated at each dose level on each schedule. To date intravenous (i.v.) and subcutaneous (s.c.) administration of rhG-CSF elevated neutrophil counts, with a 3-μg/kg i.v. dose raising the neutrophil count to the same level as 1 μg/kg s.c. in these patients. In the variety of tumor types involved in this study of 15 patients there was a dramatic reduction in the post-melphalan duration of leukopenia. The rhG-CSF (which had a half-life in the second phase of 110 min) resulted in a tenfold increase in neutrophils with a transient acute (5-min) depression in neutrophil counts following intravenous administration. At the highest doses of G-CSF administered myelocytes appeared in the blood and there was also a rise in leukocytes and monocytes. The only toxicity observed was a slight bone pain.

BRONCHUD et al. (1987) studied continuous infusion of G-CSF in 12 patients with advanced small cell lung cancer (SCLC). This resulted in a rapid dose-dependent increase in peripheral neutrophil counts with functional normality of the cells as tested by mobility and bactericidal activity. In a phase II component of this analysis, chemotherapy was repeated every 3 weeks and G-CSF was given for 14 days on alternate cycles of chemotherapy. The CSF administration reduced the periods of absolute neutropenia considerably, with a return of neutrophil counts to normal or above normal within 2 weeks following the initiation of chemotherapy. There was a reduction in severe infected episodes during cycles of chemotherapy when G-CSF was administered. Again no febrile episodes, "flu-like symptoms," or changes in blood pressure were noted. In more detailed analysis (BRONCHUD et al. 1988) it was observed that G-CSF caused an early fall of neutrophils within the 1st h followed by a rapid influx of mature neutrophils into the circulating pool followed by stimulation of proliferation and differentiation of neutrophil precursors in the bone marrow.

The efficacy of rhGM-CSF in preventing myelosuppression following chemotherapy has also been studied in a number of clinical situations. HERRMANN et al. (1987) reported that in 17 patients with advanced cancer the comparison of i.v. bolus or continuous infusion of GM-CSF over a 5- to 9-day period resulted in a two- to fourfold increase in absolute neutrophil counts. In a more detailed study, ANTMAN et al. (1988) investigated continuous infusion of GM-CSF over 7 days at doses of 4–64 μg/kg in patients with inoperable or metastatic sarcomas receiving high-dose chemotherapy. In the first cycle of chemotherapy an increase in white cell count was observed relative to control populations. However, this was not obviously dose related and two patients died of sepsis during this cycle despite recovery of white cell counts. Out of 16 patients, 9 had an eosinophilia of greater than 10%. There was significant increase in platelets or reticulocytes but marrow cellularity was increased on biopsy. The second cycle of chemotherapy without the addition of GM-CSF showed a reduction in time for recovery of neutrophils to greater than 500/mm^3. The side effects of GM-CSF administered by continuous infusion at the higher doses included thrombi around the central venous catheters, fluid retention with pericardial and pleural effusions, fever, and arthralgia.

The mechanisms of thrombus formation may have resulted from enhanced MoI surface antigen expression, this being an adhesion molecule with a role in neutrophil-endothelium cell interactions. Possibly a local accumulation of activated neutrophils at the site of the GM-CSF administration led to binding to vascular endothelial and subsequent thrombus formation. The mechanism of fluid retention is less obvious, but it has been observed in clinical trials of tumor necrosis factor (TNF), and GM-CSF is known to induce macrophage production of TNF.

In nine patients with metastatic carcinoma treated with GM-CSF prior to chemotherapy, a rapid increase in expression of CD11b surface adhesion molecules on circulating granulocytes was seen within 30 min in some patients and levels remained elevated in all patients 12–24 h later (Socinski et al. 1988). At later stages, CD11b levels decreased, due in part to a rise in circulating immature granulocytes with lower CD11b density. CD11b is a member of a family of surface glycoproteins essential for adhesion-dependent granulocyte functions such as phagocytosis, aggregation, and chemotaxis. The transient leukopenia reported following GM-CSF therapy (Devereux et al. 1987) may be due to enhanced aggregation and margination of granulocytes secondary to rapid increases in CD11b surface expression induced by GM-CSF. On the one hand this surface adhesion molecule can promote enhanced granulocyte function but may prove detrimental in certain pathophysiological states such as sepsis. This could lead to activation of granulocytes marginated in the microvasculature, leading to endothelial damage and "capillary leak syndrome." This latter state has been observed in a number of patients receiving high-dose GM-CSF.

Notwithstanding these dose-limiting toxicities, GM-CSF administration after chemotherapy in the first cycle was associated with a significantly shorter period of neutropenia and a trend toward higher neutrophil nadirs than treatment with identical doses of chemotherapy in historical controls. Furthermore, the response rate among patients with sarcoma in the GM-CSF study was 79% vs. 52% in earlier studies using identical chemotherapy.

Aglietta et al. (1989) confirmed that GM-CSF is in vivo a powerful stimulator of myelopoiesis and added three new observations. First, GM-CSF delivers a striking proliferative signal to bone marrow committed progenitors CFU-GM and BFU-E while its effect on morphologically identifiable precursors (amplification compartment) is predominantly restricted to the granulopoietic lineage. Second, GM-CSF shortens the cell cycle time and the duration of S-phase of BM cycling cells and more than doubles their birth rate. Third, after discontinuation of GM-CSF, the BM proliferative activity drops rapidly to values lower than pretreatment levels. These data may provide the basis for a more rational and kinetically interrelated administration of cell-cycle-specific drugs and GM-CSF in cancer protocols.

Administration of GM-CSF in patients with myeloproliferative diseases or cancer causes a dose-dependent monocytosis; however, activation of monocyte-mediated tumorilytic properties was seen in only one of seven patients and GM-CSF did not stimulate IL-1 or TNF production by blood monocytes (KLEINERMAN et al. 1988). The monocytes induced in vivo could be activated in vitro by exposure to lipopolysaccharide, interferon-γ, or muramyl dipeptide. In an adjuvant therapy setting, rGM-CSF may need to be combined with other monocyte/macrophage-activating agents.

The clinical trials with M-CSF have not been initiated with recombinant material; however, extensive studies have been undertaken with highly purified M-CSF purified from human urine. MOTOYOSHI and TAKAKU (1988) have reported, in 44 patients receiving high-dose chemotherapy for urogenital malignancy, an accelerated recovery from the thrombocytopenia and neutropenia associated with this anticancer therapy. The mechanism of action of M-CSF in vivo remains obscure and probably involves indirect induction of other growth factors; for example, M-CSF can induce macrophage production of G-CSF.

Current thought concerning the high incidence of opportunistic infections and neoplasms in AIDS patients points to abnormalities of lymphocytes, monocytes, and neutrophils. The rationale for this phase I/II trial of rh-GM-CSF in 16 leukopenic AIDS patients was that by increasing the number and function of leukocytes one might reduce the morbidity and mortality associated with an immunocompromised host. Continuous i.v. administration of rhGM-CSF effected dose-dependent increases in the number of circulating leukocytes including neutrophils, eosinophils, and monocytes. Consistent with the findings of others, GM-CSF was well tolerated; side effects were mild and few, and included fever, chills, myalgia, and headache. These results indicate that leukopenic AIDS patients can benefit from treatment with rh-GM-CSF. However, long-term studies investigating the relationship between circulating leukocytes and morbidity/mortality from opportunistic infection are clearly indicated. The clinical response demonstrated by patients in this study has important therapeutic implications for patients other than those with AIDS; the successful regulation of myelopoiesis could maximize our ability to manage opportunistic infections in many immunocompromised patients.

II. CSFs in Autologous Bone Marrow Transplantation

GM-CSF has been investigated in a phase I trial in 19 patients with breast cancer and melanoma treated with high-dose combination chemotherapy with alkylating agents and autologous bone marrow support (BRANDT et al. 1988). GM-CSF accelerated myeloid recovery at tolerable doses, but with dose escalations there was a limitation due to side effects associated with myalgias and fluid retention. The severe drug-related toxic effects of GM-CSF were seen at dose levels of 32 mg/kg per day administered by continuous infusion. At more tolerable dose levels, compared with historical controls, there was a lower morbidity and mortality and bacteremia was observed in only 16% of GM-CSF-treated patients versus 35% of evaluable controls. The predominant cause of early morbidity and mortality after high-dose chemotherapy related to multiple organ failure syndrome with associated hepatotoxicity and nephrotoxicity. The administration of GM-CSF substantially reduced this form of organ damage. Probably the changed intestinal flora and alterations of the normal barrier functions of the gastrointestinal tract, which lead to bacteremia, play a major role in the multiple organ failure syndrome. This vicious cycle may be broken by GM-CSF stimulation of accelerated leukocyte recovery.

The functions of granulocytes generated in patients receiving autologous bone marrow transplants and GM-CSF were normal as assessed by margination capacity, phagocytosis, and hydrogen peroxide production (Peters et al. 1988). However, the migration of granulocytes to a sterile inflammatory site was markedly reduced during continuous GM-CSF infusion. This may predict increased difficulties with soft tissue infections in these patients. However, GM-CSF, while inhibitory to nondirected neutrophil migration in an agarose assay (Gasson et al. 1984), acts as a relatively potent chemoattractant for neutrophils and monocytes in Boyden chambers (Wang et al. 1987). G-CSF and M-CSF also induced directional migration of neutrophils and monocytes across polycarbonate or nitrocellulose filters (Wang et al. 1988).

Nemunaitis et al. (1988) also utilized autologous bone marrow rescue for patients with lymphoid leukemia and Hodgkin's or non-Hodgkin's lymphoma receiving high-dose chemotherapy and treatment with GM-CSF as 2-h infusions on a daily basis for 14 days. Absolute neutrophil counts recovered significantly earlier than in control patients. The myalgias and bone pain were dose-limiting toxicities and may be attributed to the production of prostaglandins by macrophages activated by GM-CSF.

In a phase II study of 12 patients with Hodgkin's disease after high-dose chemotherapy and autologous bone marrow transplantation, G-CSF was also found to be effective (at doses of up to 60 µg/kg per day beginning 24 h after bone marrow transplantation) in reducing the duration of significant leukopenia by approximately 25%. There was no change in the time to recovery of the platelet count (Taylor et al. 1988).

Use of urinary purified M-CSF as a post bone marrow transplant myelostimulating agent has been studied in 120 bone marrow transplant patients in a placebo-controlled trial (Masaoka et al. 1988). A slight but significant ($P = 0.05$) difference was observed in the time to recovery of absolute neutrophil counts > 500. As in the postchemotherapy study with this natural material the probability is that the effects observed are indirect, due to endogenous release of other hematopoietic growth factors.

III. CSFs in Myelodysplastic Syndromes

G- and GM-CSF have been explored in a variety of myelodysplastic situations in addition to the iatrogenic neutropenias. The use of GM-CSF was explored in a phase I/II study in a group of patients with aplastic anemia or myelodysplastic syndrome (MDS) (Antin et al. 1988). Doses of up to 10 µg/kg per day were given for periods of 7 days by 1- to 4-h infusions or up to 14 days with 12-h infusions. Temporary improvements were seen in the granulocyte, monocyte, and reticulocyte counts in five of seven patients with MDS, but two of seven patients had an increase in immature myeloid cells and myeloblasts. Again some transient toxicity was observed with bone pain, anorexia, myalgias, arthralgias, and low-grade fever. One patient with idiopathic agranulocytosis did not respond to the GM-CSF, and there were no changes in either neutrophil or monocyte counts throughout the duration of infusion. Interestingly, two individuals with MDS with a missing long arm of chromosome 5 showed the weakest response to exogenous administration of GM-CSF.

The concern that GM-CSF administration to MDS patients might accelerate progression to acute leukemia was raised in the instance of the two patients that showed increased numbers of circulating blast cells. Bone marrow biopsy showed not only increased cellularity but also maturation of all cell lines and the blast cell counts returned to pretreatment values following discontinuation of GM-CSF (ANTIN et al. 1988). In a further small group of patients with MDS, VADHAN-RAJ et al. (1987) reported that continuous i.v. infusion of GM-CSF at doses up to 10 µg/kg per day for 14 days produced a 5- to 70-fold increase in leukocytes and up to a 370-fold increase in granulocytes, with a minority of patients having an increase in platelets and erythropoiesis. Intriguingly, and in contrast to other studies, a decrease in blast cells in the bone marrow was observed in refractory anemia with excess of blasts. Following a 14-day rest period a second weekly cycle of GM-CSF was initiated and, while this did not result in eradication of the abnormal clone, there was no evidence of progression to acute leukemia. Analysis of premature chromosome condensation to determine the karyotypic nature of the mature cells developing in these patients suggested that these cells were indeed differentiated from the abnormal clone. BROXMEYER et al. (1988) in an in vitro analysis of the bone marrow of these patients were unable to demonstrate any change in the abnormal growth patterns of the CFU-GM following GM-CSF therapy. However, there was a marked enhancement of the cycle status of the leukemic CFU-GM as determined by thymidine suicide techniques in vitro. GANSER et al. (1988) reported that in 10 of 11 patients with MDS there was a dose-related increase in WBC, neutrophils, monocytes, and eosinophils following administration of GM-CSF. Four of these patients showed an increase in blast counts and of these subsequently receiving Ara-C two responded. MERTELSMANN and COZMAN (1988) reported that GM-CSF caused a worsening of leukemia in patients where the proportion of blast cells in the bone marrow exceeded 20%.

In a further study, rhGM-CSF was administered in escalating doses by 8-h i.v. administration to ten patients with MDSs (refractory anemia, refractory anemia with excess of blasts, and chronic myelomonocytic leukemia). All patients were severely cytopenic and the majority of patients had received no prior chemotherapy. A dose-dependent increase in leukocyte numbers was observed in nine of ten patients. No change in reticulocyte numbers was seen and only one patient experienced an increase in platelet counts (HOELZER et al.

1988). As an adverse side effect, an increase in leukemic blast cells occurred in four of ten patients but only in those cases with an initial blast count of 20% or more. Of these latter, two responded to subsequent low-dose Ara-C. The ultimate effect of GM-CSF on normal versus abnormal proliferation and differentiation in vitro depends on multiple factors and reflects the unique clonality of most malignant hemopathies.

The potential of CSFs, particularly IL-3 or GM-CSF acting on early multilineage precursors, to facilitate self-renewal rather than differentiation of leukemic or preleukemic cells, is an obvious limitation to the use of growth factors in leukemia or myelodysplasia. The morbidity and mortality due to infection and bleeding related to cytopenia is high in patients with MDS and, as has been pointed out by Vadhan-Raj et al. (1988), the potential risk of GM-CSF enhancing progression to frank leukemia has to be weighed against the beneficial effects of this therapy on hematopoiesis.

The clinical experience with G-CSF in MDS and acute myeloid leukemia is preliminary. In two of five cases of MDS, Kobayashi et al. (1988) observed responses with 6 days of intravenous infusion of up to 60 μg/kg per day of rhG-CSF. Four of five patients with MDS also responded to G-CSF in a small study reported by Negrin et al. (1988), with a striking rise in absolute neutrophil counts (6- to 35-fold) with a maximum response at 3 μg G-CSF/kg per day and with a decreased transfusion requirement. There was no evidence of elevation of blast counts nor of progression to acute myeloblastic leukemia (AML) in these cases. In one case report a patient with refractory AML showed a decrease in leukemic blast counts following G-CSF administration (Teshima et al. 1988).

Clearly, the decision to administer G- or GM-CSF to patients with myeloid hematopoietic malignancies, be they preleukemic or overtly leukemic, is a difficult one. The preclinical evidence would suggest that G-CSF is more likely to favor differentiation whereas GM-CSF is more likely to support leukemic blast cell self-renewal. However, the considerable heterogeneity of the MDS/AML spectrum renders interpretation of small studies difficult. The uniqueness of the clonal hemopathies suggests that in certain instances one may be anticipated. Furthermore, the autocrine ability of many acute may be anticipated; furthermore, the autocrine ability of many acute leukemias to produce growth factors such as GM-CSF or IL-1 adds an additional complexity to this area (Griffin 1988). At the present time an empirical approach may be most desirable, involving, where possible, preclinical in vitro evaluation of the response of the patient's bone marrow to short-term treatment with the various growth factors likely to be used in subsequent therapy. Clonogenic techniques have proved effective in determining proliferation and differentiation responses and short-term suspension cultures are also available, as are molecular techniques for detection of growth factor message expression within the leukemic blast cell population.

Aplastic anemia while not a neoplastic disorder nevertheless involves profound impairment of pluripotential stem cells and their progeny. This impairment may be intrinsic to the stem cell, may involve a suppressor mechanism which has been reported to involve activated T lymphocytes and interferon production (LAVER et al. 1988b), or may involve a defect in the hematopoietic microenvironment which in turn may be associated with defects in specific hematopoietic growth factor production. In this regard deficiencies in IL-1 production have been reported in most cases of aplastic anemia studied (GASCON et al. 1987). To date, CSF therapy in aplastic anemia has been restricted to GM-CSF. ANTIN et al. (1988) reported a temporary improvement in six of eight patients who had antithymocyte globulin (ATG) nonresponsive aplastic anemia using doses of 10 μg/kg GM-CSF for 7 or 14 days. The temporary response involved improvement in levels of granulocytes, monocytes, and reticulocytes. VADHAN-RAJ et al. (1988) reported a two- to eightfold increase in WBC, predominantly neutrophils, eosinophils, and monocytes, in ten patients with aplastic anemia receiving continuous infusions of GM-CSF over 14 days. There was no change in platelet or red cell numbers. GORDON-SMITH et al. (1988) observed only one response out of five patients with aplastic anemia receiving continuous infusions of GM-CSF, but this one patient showed an increase in neutrophils and a resolution of skin *Candida* lesions. These preliminary studies suggest that CSF may play a role in certain subsets of aplastic anemia but, in view of the inability of GM-CSF to stimulate the very earliest pluripotential stem cell population, conceivably the cell type involved in the majority of aplastic anemias, the future of single-agent growth factor therapy may be limited. Indeed, the administration of IL-1 or IL-6 together with factors such as IL-3, G-, and GM-CSF may eventually be the answer to ATG-nonresponsive aplastic anemias. In cynomolgus monkeys IL-3 did not increase the WBC count in the study of KRUMWIEH et al. (1988); however, a 10-day treatment with IL-3 followed by a suboptimal dose of GM-CSF resulted in a significant increase in WBC and platelets, again confirming the importance of priming or concatenate use of growth factors.

IV. In Vivo Studies of G-CSF in Congenital and Idiopathic Neutropenia

Cyclic neutropenia is an inherited disease of man and gray collie dogs, characterized by regular oscillations of the number of peripheral blood cells and of bone marrow progenitor cells. Cycling of serum or urine CSF levels has also been reported (MOORE et al. 1974; DUNN et al. 1982). However, induction or cure of cyclic neutropenia by bone marrow transplantation has suggested that this disorder is a disease of pluripotential hematopoietic stem cells. It should be noted that the hematopoietic stem cell itself can generate cells (macrophages, T cells) capable of hematopoietic growth factor production and thus it is not possible to exclude intrinsic growth factor production defects in the pathophysiology of cyclic neutropenia. In the dog model we have demonstrated that daily administration of rh-G-CSF (5 μg/kg × 2 daily for 30 days) causes an immediate (within 12 h) and persistent leukocytosis (<40000 WBC) in both cyclic and normal dogs due to a tenfold increase in the numbers of cir-

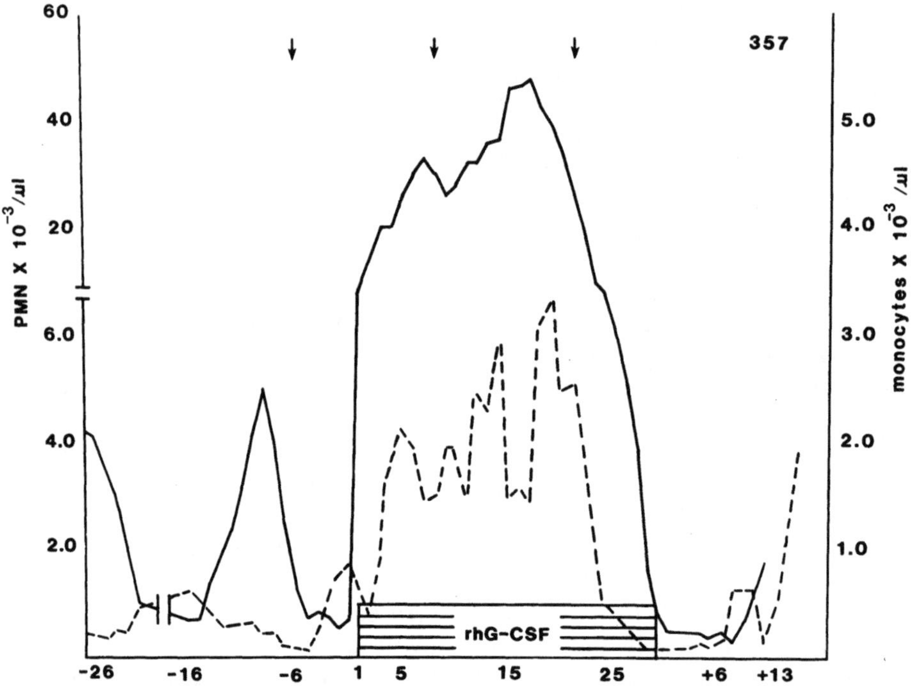

Fig. 7. Correction of cyclic hematopoiesis with rhG-CSF. Two CH and one normal dog were treated with rhG-CSF (5 µg/kg) twice daily by subcutaneous injection for 28 days. The days are numbered relative to the first day of rhG-CSF treatment, which was arbitrarily set as day 1. The *arrows* indicate the predicted day 1 (PMN < 1600) of the 14-day cycles based on the previous 17 cycles. The PMNs are indicated by the *solid line* and the monocytes by the *broken line*

culating neutrophils and monocytes (Lothrop et al. 1988). This therapy eliminated two predicted neutropenic episodes and suppressed the cycling of CFU-GM in the bone marrow (Figs. 7 and 8). Comparable therapy with rh-GM-CSF induced a monocytosis and neutrophilia in normal dogs but did not eliminate the recurrent neutropenia in cyclic dogs. Analysis of the serum levels of CSF indicated a striking periodicity with peak levels of G-CSF coinciding with the peaks of monocytes and the nadirs of neutrophils.

Hammond et al. (1988) have suggested that a lymphocyte subpopulation normally capable of generating GM-CSF in response to a monocyte-derived recruiting activity (possibly IL-1) is defective in patients with cyclic neutropenia.

A disorder of neutrophil regulation is also seen in chronic idiopathic neutropenia of man in which the peripheral neutrophil count is reduced to <2000 cells/mm^3 for prolonged periods, with maturation arrest of neutrophil granulocyte precursors in the bone marrow. Other hematopoietic cell counts are usually normal and antineutrophil antibodies are absent. Clinically, these patients experience recurrent episodes of life-threatening infections, ulcers of

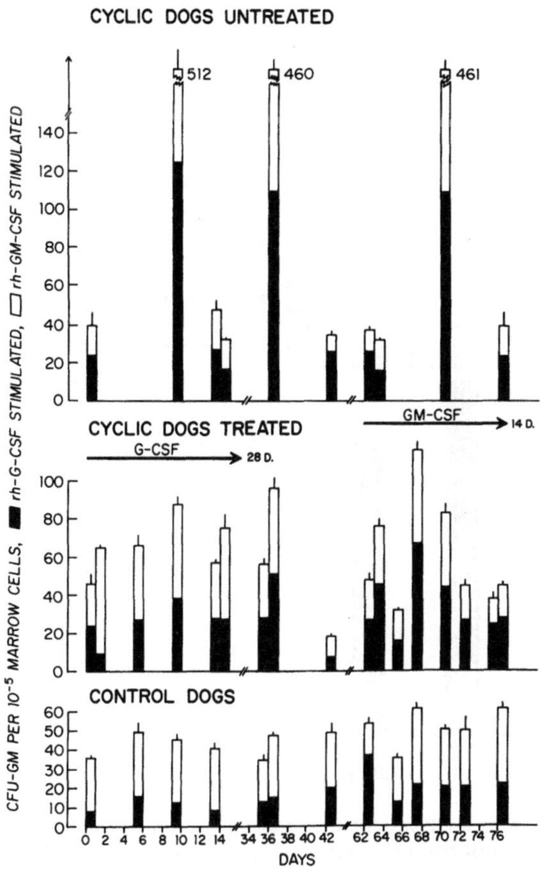

Fig. 8. Granulocyte/macrophage colony-forming units in cyclic hematopoietic dogs treated with rhG-CSF and rhGM-CSF. Bone marrow was prepared and CFU-GM assay performed using either 1000 units rhG-CSF or 1000 units rhGM-CSF as stimulus. Day 1 is the 1st day of rhG-CSF treatment and subsequent days are numbered consecutively

the mucous membrane, and periodontal disease. We have had the opportunity of studying a number of patients with this disorder utilizing primary clonogenic assay and in vitro suspension cultures of patient bone marrow cells (Jakubowski et al. 1989). The incidence of CFU-GM observed in primary bone marrow culture was always in the high range of normal with respect to each species of CSF (G-, GM-, M-, and IL-3), indicating that the neutropenia was not attributed to a shortage of appropriate myeloid progenitors. Furthermore, morphological studies showed normal neutrophil maturation within the colonies developing in the presence of G-, GM-CSF, and IL-3, with eosinophil maturation and macrophage development a significant feature with GM-CSF stimulation. In suspension cultures G-CSF was particularly effective in generating mature segmented neutrophils which were absent in input bone marrow, and did not develop spontaneously in over 3 weeks of culture. With G-CSF addition, myelocytes expanded in the 1st week of culture. By the 2nd and 3rd week high levels of production of segmented, functionally normal neutrophils were found (Fig. 9). Based upon these in vitro studies, patients were treated with subcutaneous G-CSF on a daily continuous basis. In the

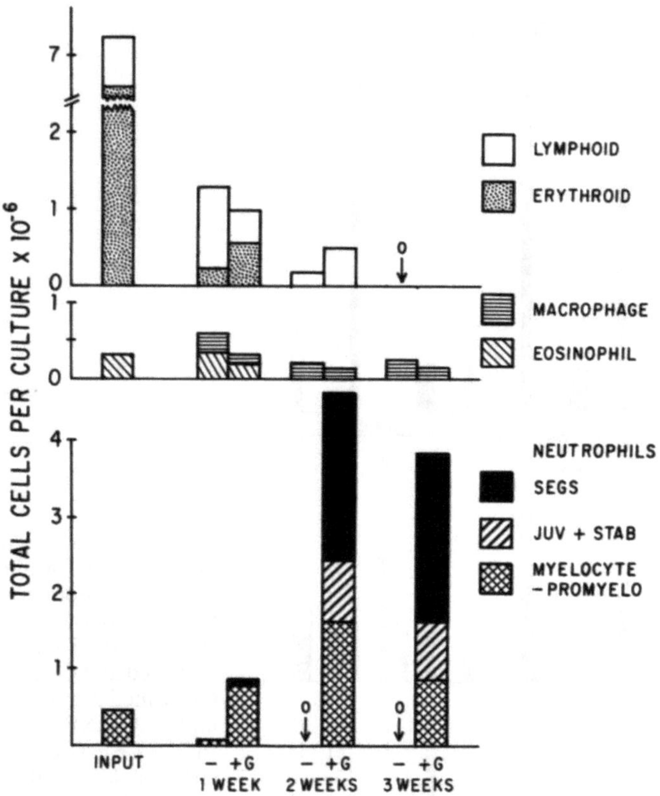

Fig. 9. In vitro culture of bone marrow from a patient with idiopathic neutropenia. The addition of rhG-CSF caused expansion of total hematopoietic cells and supported the differentiation of functionally normal segmented neutrophils

first patient studied, the neutrophil count rose rapidly, and by 20 days the patient's ANC was >1000 cells/mm^3 and plateaued in the range of 2000–3000/mm^3 with evidence of a 40-day cycle (observed even when the dose of G-CSF was increased). The neutrophils were functionally normal and the patient has remained on this continuous G-CSF treatment without any notable toxic side effects (Jakubowski et al. 1989).

A more severe form of neutropenia is found in patients with Kostmann syndrome (congenital agranulocytosis). In this disorder, marked by severe neutropenia and maturation arrest at the myelocyte level, five patients received G-CSF therapy, and the neutrophils increased from 0%–1% to 10%–72% with clinical resolution of preexisting infections (Bonilla et al. 1988). In four Kostmann syndrome patients treated with 3–30 μg/kg per day rh-GM-CSF for 42 days, a specific dose-dependent increase in granulocytes was seen, involving the eosinophil lineage in three cases and a neutrophil response in one (Welte et al. 1988).

H. Conclusions

Figure 10 summarizes some of the conclusions concerning regulation of early stem cells. By direct mechanisms, but via independent pathways, IL-1 and IL-6 can activate resting stem cell populations into cell cycle (MOORE et al. 1987a; DUKES et al. 1988) and induced responsiveness to other species of interleukins and to CSFs. Whether this involves upregulation of receptors, which then leads to cycle activation and differentiation, or vice versa, is not yet clear. IL-3 and IL-4 also appear to influence a subset of the stem cell pool, but not the most primitive population. Synergism has been demonstrated in this stem-cell-activating model between IL-1 and IL-3, IL-1 and IL-6, IL-6 and IL-3, and between both IL-1 and IL-6 with G-, GM-, and M-CSF. Other factors may also stimulate early stem cells. MOREAU et al. (1987) characterized a factor produced by human T-cell clones that triggered the proliferation of IL-3-sensitive murine cells lines, and also influenced the proliferation of early erythroid progenitors. This cytokine, termed HILDA (human interleukin stimulating DA1a cell line), was purified from human T-cell lines (GODARD et al. 1988) and from the 5637 bladder cancer cell line, and was distinguished from the IL-1, IL-6, G-CSF, and GM-CSF produced by these cells (GASCAN et al. 1989). The cDNA coding for HILDA has an identical sequence to the newly described leukemia inhibitory factor (LIF), which is reported to induce differentiation of the M1 murine myeloid leukemic cell line without apparent growth-promoting activity on normal human or murine CFU-GM (GOUGH et al. 1988).

Inhibitory influences are mediated by TNF, TGF-β, and possibly interferons (IFN) and synergism has been demonstrated between TNF and IFN. Hematopoietic growth factor cascades are also central to an understanding of the regulatory interaction both in vivo and in vitro. As shown in Fig. 11, macrophage-derived TNF and IL-1 induce CSF and interleukin production by fibroblasts, endothelial cells, and T lymphocytes. In vivo TNF stimulates, production of GM-CSF, M-CSF, and IL-1 (KAUSHANSKY et al. 1987) and G-CSF (KOEFFLER et al. 1987), and IL-1 induces G- and M-CSF (VOGEL et al. 1987).

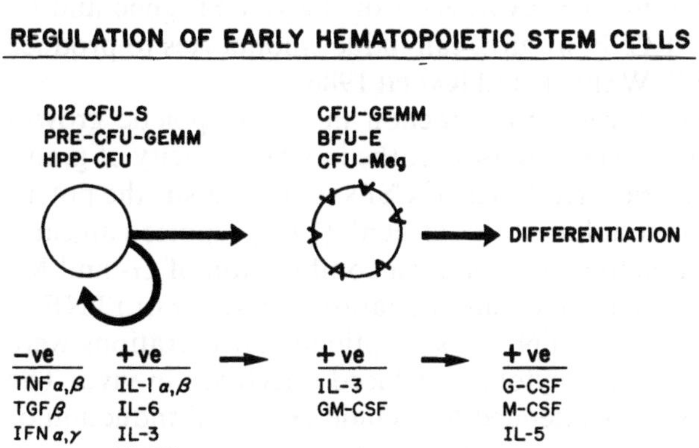

Fig. 10. Model of positive (^{+}ve) and negative (^{-}ve) growth factor influences on the self-renewal and differentiation of pluripotential stem cells as assessed in the in vivo D12 CFU-S assay, the high-proliferative potential CFU assay, and the pre-CFU-GEMM blast cell colony assay

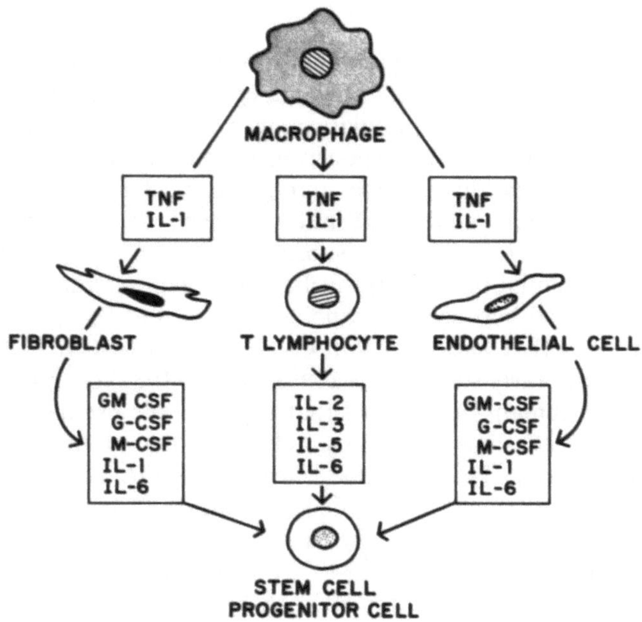

Fig. 11. Cytokine cascades involving macrophage production of TNF and IL-1, interactions with T cells and stromal cells, and the induction of the production of multiple species of interleukins and CSFs that influence bone marrow stem cell and progenitor cell proliferation and differentiation

Cascades of growth factor production, and the overlapping actions of the different CSFs and interleukins, require that the coordinate or independent regulation of CSF gene expression be determined in different tissues exposed to different stimuli. For example both M-CSF and G-CSF can be produced by stimulated monocytes but the genes are independently regulated (Vellenga et al. 1988). IL-3 and GM-CSF primarily induce monocyte expression of the TNF and M-CSF gene and secretion of M-CSF and TNF, whereas bacterial lipopolysaccharide primarily induces expression of the G-CSF gene and its bioactive protein product. M-CSF, in turn, can induce monocytes to produce interferon, TNF, and G-CSF (Warren and Ralph 1986).

Further insight into the cascade phenomenon in hematopoietic growth factor production is provided by comparison of the in vivo capacity of gram-negative endotoxin to induce elevated levels of CSFs in sera versus the potential of endotoxin-associated monokines. Vogel et al. (1987) reported an early (3-h peak) induction of CSF activity (probably a combination of G- and M-CSF) in mouse serum following in vivo administration of rIL-1-α or rTNF-α. Combined injections of IL-1-α and TNF at suboptimal concentrations were additive. Unlike endotoxin, neither cytokine induced interferon in vivo. The IL-1-regulatory cascade has been extended to include GM-CSF induction of IL-1 mRNA and bioactive protein in mature granulocytes (Lindemann et al.

1988). While controversial, this observation implies that neutrophil granulocytes may be involved in the amplification of their own precursors by establishing a positive feedback loop involving IL-1 stimulation of GM-CSF release from endothelial cells, fibroblasts, and T cells. The family of CSFs and interleukins provide a potent spectrum of biotherapeutic agents that can act individually, or in an additive or synergistic mode, to accelerate hematopoietic recovery following a variety of insults. Our ability to define the optimal conditions for administration of these factors in a multitude of clinical situations is the challenge of the future.

Acknowledgements. I thank my many colleagues who have collaborated in the studies cited, and Joreatha Jones for excellent secretarial assistance. This work was supported by the National Cancer Institute Grants CA20194, CA31780, CA22766, by American Cancer Society Grant CH-3K, and by the Gar Reichman Foundation.

References

Abboud M, Moore MAS (1988) Recombinant human IL-3 is a basophil/mast cell growth factor (Abstr 123). Exp Hematol 16:489

Aglietta M, Piacibello W, Sanavio F, Stacchini A, Apra F, Schena M, Mossetti C, Carnio F, Calgiras-Cappio F, Gavosto F (1989) Kinetics of human hemopoietic cells following in vivo administration of grnaulocyte-macrophage colony stimulating factor (GM-CSF). J Clin Invest (in press)

Antin JH, Smith BR, Holmes W, Rosenthal DS (1988) Phase I/II study of recombinant human granulocyte-macrophage colony-stimulating factor in aplastic anemia and myelodysplastic syndrome. Blood 72:705–713

Antman KS, Griffin JD, Elias A, Socinsky MA, Ryan L, Cannistra SA, Oette D, Whitley M, Frei E, Schnipper LE (1988) Effect of recombinant human granulocyte-macrophage colony-stimulating factor on chemotherapy-induced myelosuppression. N Engl J Med 319:593–598

Bagby GC, McCall E, Layman DL (1983) Regulation of colony stimulating activity production. Interactions of fibroblasts, mononuclear phagocytes and lactoferrin. J Clin Invest 71:340–349

Bagby GC, Dinarello CA, Wallace P (1986) Interleukin 1 stimulates granulocyte macrophage colony-stimulating activity release by vascular endothelial cells. J Clin Invest 78:1316–1325

Bertoncello I, Hodgson GS, Bradley TR (1988) Multiparameter analysis of transplantable hemopoietic stem cells. II. Stem cells of long-term bone marrow-reconstituted recipients. Exp Hematol 16:245–249

Beutler B, Cerami A (1986) Cachectin and TNF as two sides of the same biological coin. Nature 320:584

Blazer BR, Widmer MB, Soderling CCB, Gillis S, Vallera DA (1988) Enhanced survival but reduced engraftment in murine recipients of recombinant granulocyte/macrophage colony-stimulating factor following transplantation of T-cell-depleted histoincompatible bone marrow. Blood 72:1148–1154

Bonilla MA, Gillio AP, Potter GK, O'Reilly RJ, Souza LM, Welte K (1987) Effects of recombinant human G-CSF and GM-CSF on cytopenias associated with repeated cycles of chemotherapy in primates. Blood [Suppl 1] 70:130 a

Bonilla MA, Gillio AP, Ruggiero M, Kernan NA, Brochstein JA, Fumagalli L, Bordignon C, Vincent M, Welte K, Souza LM, O'Reilly RJ (1988) Correction of neutropenia in patients with congenital agranulocytosis with recombinant human granulocyte colony stimulating factor in vivo (Abstr 243). Exp Hematol 16:520

Bradley TR, Hodgson GS, Bertoncello I (1982) Characteristics of primitive macrophage progenitor cells with high proliferative potential: relationship to cells with marrow repopulating ability in 5-fluorouracil treated mouse bone marrow. In: Baum SJ, Ledney GD, van Bekkum DW (eds) Experimental hematology today. Springer, Berlin Heidelberg New York, pp 285–289

Brandt SJ, Peters WP, Atwater SK, Kurtzberg J, Borowitz MJ, Jones RB, Shpall EJ, Bast RC, Gilbert CJ, Oette DH (1988) Effect of recombinant human granulocyte-macrophage colony-stimulating factor on hematopoietic reconstitution after high-dose chemotherapy and autologous bone marrow transplantation. N Engl J Med 318:869–876

Bronchud MH, Scarffe JH, Thatcher N, Crowther D, Souza LM, Alton NK, Testa NG, Dexter TM (1987) Phase I/II study of recombinant human granulocyte colony-stimulating factor in patients receiving intensive chemotherapy for small cell lung cancer. Br J Cancer 56:809–813

Bronchud MH, Potter MR, Morgenstern G, Blasco MJ, Scarffe JH, Thatcher N, Crowther D, Souza LM, Alton NK, Testa NG, Dexter TM (1988) In vitro and in vivo analysis of the effects of recombinant human granulocyte colony-stimulating factor in patients. Br J Cancer 58:64–69

Broxmeyer HE, Cooper S, Williams DE, Hangoc G, Gutterman JU, Vadhan-Raj S (1988) Growth characteristics of marrow hematopoietic progenitor/precursor cells from patients on a phase I clinical trial with purified recombinant human granulocyte-macrophage colony-stimulating factor. Exp Hematol 16:594–602

Campbell AD, Wicha MS, Long M (1985) Extracellular matrix promotes the growth and differentiation of murine hematopoietic cells in vitro. J Clin Invest 75:2085–2090

Campbell AD, Long MW, Wicha MS (1987) Haemonectin, a bone marrow adhesion protein specific for cells of granulocyte lineage. Nature 329:744–746

Champlin RE, Nimer SD, Oette D, Golde DW (1988) Granulocyte-macrophage colony-stimulating factor (GM-CSF) treatment for aplastic anemia (AA) or agranulocytosis (Abstr 238). Exp Hematol 16:519

Chiu C-P, Moulds C, Coffman RL, Rennick D, Lee F (1988) Multiple biological activities are expressed by a mouse interleukin-6 cDNA clone isolated from bone marrow stromal cells. Proc Natl Acad Sci USA 85:7099–7103

Daniels E (1988) Preliminary observations of the direct effects of interleukin-3 (IL-3) on mouse fetal hemopoiesis in vivo (Abstr 207). Exp Hematol 16:511

Devereux S, Linch DC, Campos-Costa D, Spittle MF, Jelliffe AM (1987) Transient leukopenia induced by granulocyte-macrophage colony-stimulating factor. Lancet 2:1523

Dexter TM (1987) Commentary: growth factors involved in haemopoiesis. J Cell Sci 88:1–6

Dinarello CA, Cannon JG, Wolff SM, Bernheim HA, Beutler B, Cerami A, Figari IS, Palladino MA, O'Connor JV (1986) Tumor necrosis factor (cachectin) is an endogenous pyrogen and induces production of interleukin-1. J Exp Med 163:1433–1439

Donahue RE, Wang EA, Stone S (1986) Stimulation of haemopoiesis in primates by continuous infusion of recombinant human GM-CSF. Nature 321:872–874

Donahue RE, Sechra J, Metzger M, Lefebore D, Rock D, Carbone S, Nathan DG, Garnick M, Sehgal PK, Laston D, LaVillie E, McCoy J, Schendel PF, Norton C, Turner K, Yang Y-C, Clark SC (1988) Human IL-3 and GM-CSF act synergistically in stimulating hematopoiesis in primates. Science 241:1820–1823

Dukes PP, Zsebo KM, Ma A, Polk C (1988) IL-1 affects the cell cycle state of day-12 CFU-s in marrow cell cultures (Abstr 49). Exp Hematol 16:471

Dunn CDR, Jones JB, Lange RD, Wright EG, Moore MAS (1982) Production of presumptive humoral haematopoietic regulators in canine cyclic haematopoiesis. Cell Tissue Kinet 15:1–10

Eaves CJ, Cashman JD, Eaves AC (1987) Selective and reversible inhibition of very primitive hemopoietic cells by tumor growth factor beta (TGF-β). Blood [Suppl 1] 70:171 a

Enokihara H, Furusawa S, Kajitani H, Hamaguchi H, Saito K, Fukuda T, Shishido H (1988) Interleukin 2 stimulates T-cells from patients with eosinophilia to produce CFU-Eo growth stimulating factor. Br J Haematol 69:431–436

Fibbe WE, van Damme J, Billiau A (1986) Interleukin-1 (22-K factor) induces release of granulocyte-macrophage colony-stimulating activity from human mononuclear phagocytes. Blood 68:1316–1322

Fibbe WE, Damme J, Billiau A, Goselink HM, Voogt PJ, Eeden G, Ralph P, Altrock BW, Falkenburg JHF (1988) Interleukin 1 induces human marrow stromal cells in long-term culture to produce granulocyte colony-stimulating factor and macrophage colony-stimulating factor. Blood 71:430–435

Gabrilove JL, Welte K, Harris P, Platzer E, Lu L, Levi E, Mertelsmann R, Moore MAS (1986) Pluripoietin alpha: a second human hematopoietic colony-stimulating factor produced by the human bladder carcinoma cell line 5637. Proc Natl Acad Sci USA 83:2478–2482

Gabrilove JL, Jakubowski A, Scher H, Sternberg C, Wong G, Grous J, Yagoda A, Fain K, Moore MAS, Clarkson B, Oettgen HF, Alton K, Welte K, Souza L (1988) Effect of granulocyte colony-stimulting factor on neutropenia and associated morbidity due to chemotherapy for transitional-cell carcinoma of the urothelium. N Engl J Med 318:1414–1422

Ganser A, Ottmann OG, Volkers B, Walther F, Becher R, Bergmann L, Hoelzer D (1988) Treatment with recombinant human granulocyte-macrophage colony-stimulating factor (rhGM-CSF) in patients with myelodysplastic syndromes (MDS) (Abstr 241). Exp Hematol 16:520

Gascan H, Godard A, Praloran V, Naulet J, Moreau JF, Jacques Y, Soulillou JP (1989) Characterization of hilda/leukemia inhibitory factor secreted by human tumor cell lines. J Immunol (in press)

Gasparetto C, Laver J, Abboud M, Gillid A, Smith C, O'Reilly RJ, Moore MAS (1989) Effects of IL-1 on hematopoietic progenitors: evidence of stimulatory and inhibitory activities in a primate model. Blood 74:547–550

Gascon P, Scala G, Young N (1987) Interleukin 1 in hyper- and hypoproliferative bone marrow diseases: monocyte production in myelofibrosis and aplastic anemia. Blood [Suppl 1] 70:135 a

Gasson JC, Weisbart RH, Kaufman SE, Clark SC, Hewick RM, Wong GG, Golde DW (1984) Purified human granulocyte-macrophage colony-stimulating factor: direct action on neutrophils. Science 226:1339

Godard A, Gascan H, Naulet J, Peyrat M-A, Jacques Y, Soulillou J-P, Moreau J-F (1988) Biochemical characterization and purification of HILDA, a human lymphokine active on eosinophils and bone marrow cells. Blood 71:1618–1623

Gordon MY, Hibbin JA, Dowding C, Gordon-Smith EC, Goldman JM (1985) Separation of human blast progenitors from granulocytic, erythroid, megakaryocytic, and mixed colony-forming cells by "panning" on cultured marrow-derived stromal layers. Exp Hematol 13:937–940

Gordon MY, Riley GP, Watt SM, Greaves MF (1987 a) Compartmentalization of a hematopoietic growth factor (GM-CSF) by glycosaminoglycans in the bone marrow microenvironment. Nature 326:403–405

Gordon MY, Dowding CR, Riley GP, Greaves MF (1987 b) Characterisation of stroma-dependent blast colony-forming cells in human marrow. J Cell Physiol 130:150–156

Gordon MY, Bearpark AD, Clarke D, Healy LE (1988) Matrix glycoproteins may locally regulate the relative concentrations of different haemopoietic growth factors (Abstr 16). Exp Hematol 16:416

Gordon-Smith EC, Malkovska V, Myint AA, Gibson F, Milne A (1988) Growth factors and lymphokines in aplastic anaemia (Abstr 27). Exp Hematol 16:419

Gough NM, Gearing DP, King TA, Willson DJ, Hilton DJ, Nicola NA, Metcalf D (1988) Molecular cloning and expression of the human homologue of the murine gene encoding myeloid leukemia-inhibiting factor. Proc Natl Acad Sci USA 85:2623–2628

Griffin JD (1988) Role of colony stimulating factors in the biology of leukemia (Abstr 20). Exp Hematol 16:417

Groopman JE, Mitsuyasu RT, DeLeo MJ, Oette DH, Golde DW (1987) Effect of recombinant human granulocyte-macrophage colony-stimulating factor on myelopoiesis in the acquired immunodeficiency syndrome. N Engl J Med 317:593–598

Hamaguchi Y, Kanakura Y, Fujita J, Takeda S-I, Nakano T, Tauri S, Honjo T, Kitamura Y (1987) Interleukin 4 as an essential factor for in vitro clonal growth of murine connective tissue-type mast cells. J Exp Med 165:268–273

Hammond WP, Miller JE, McCall E (1988) Abnormal GM-CSF generation by lymphocytes in patients with cyclic hematopoiesis. Exp Hematol 16:389–393

Herrmann F, Schulz G, Lindemann A, Gerhards U, Oster W, Mertelsmann R (1987) A phase I clinical trial of recombinant human granulocyte-macrophage colony stimulating factor (rh GM-CSF) in cancer patients: initial results. Blood [Suppl 1] 70:230a

Herrmann F, Oster W, Meuer SC, Lindemann A, Mertelsmann RH (1988) Interleukin 1 stimulates T lymphocytes to produce granulocyte-monocyte colony-stimulating factor. J Clin Invest 81:1415–1418

Hoelzer D, Ganser A, Volkers B, Greher J, Walther F (1988) In vitro and in vivo action of recombinant human GM-CSF in patients with myelodysplastic syndrome. Blood Cells 14:551–559

Hultner L, Welle M, Szots H, Dormer P (1988) Expression of IL-6 mRNA and secretion of IL-6 in mouse bone marrow-derived IL-3-dependent mast cell lines and tumorigenic sublines (Abstr 187). Exp Hematol 16:506

Ihle JN, Keller J, Oroszan S, Henderson L, Copeland TD, Finch F, Prostowsky MD, Goldwasser E, Schrader JW, Palasynski E, Dy M, Label B (1983) Biological properties of homogeneous interleukin-3. I. Demonstration of WEHI-3 growth factor activity, mast cell growth factor activity, P cell-stimulating factor activity, colony stimulating factor activity and histamine producing cell-stimulating activity. J Immunol 131:282–287

Ikebuchi K, Wong GG, Clark SC, Ihle JN, Hirai Y, Ogawa M (1987) Interleukin 6 enhancement of interleukin 3-dependent proliferation of multipotential hemopoietic progenitors. Proc Natl Acad Sci USA 84:9035–9039

Iscove NN, Fagg B, Keller G (1988) A soluble activity from adherent marrow cells cooperates with IL-3 in stimulating growth of pluripotential hematopoietic precursors. Blood 71:953–957

Jakubowski AA, Souza L, Kelly F, Fain K, Budman D, Clarkson B, Bonilla MA, Moore MAS, Gabrilove JL (1989) Effects of human granulocyte colony stimulating factor on a patient with idiopathic neutropenia. N Engl J Med 320:38–42

Jubinsky PT, Stanley ER (1985) Purification of hemopoietin-1: a multilineage hemopoietic growth factor. Proc Natl Acad Sci USA 82:2764–2769

Kaushansky K, Broudy VC, Harlan JM, Asamson JW (1987) Tumor necrosis factor (TNF-α) and lymphotoxin (TNF-β) stimulate the production of GM-CSF, M-CSF, and IL-1 production in vivo (Abstr 552). Blood [Suppl 1] 70:174a

Keller JR, Mantel C, Sing GK, Ellingsworth LR, Ruscetti SK, Ruscetti FW (1988) Transforming growth factor β1 selectively regulates early murine hematopoietic progenitors and inhibits the growth of IL-3-dependent myeloid leukemia cell lines. J Exp Med 168:737–750

Klebanoff SJ, Vadas MA, Harlan JM, Sparks LJ, Gamble JR, Agosti JM, Waltersdorph AM (1986) Stimulation of neutrophils by tumor necrosis factor. J Immunol 36:4220

Kleinerman ES, Knowles RD, Lachman LB, Gutterman JU (1988) Effect of recombinant granulocyte/macrophage colony-stimulating factor on human monocyte activity in vitro and following intravenous administration. Cancer Res 48:2604–2609

Kobayashi Y, Okabe T, Urabe A, Takaku F (1988) Effect of recombinant human granulocyte colony stimulating factor (rHuG-CSF) on bone marrow failure cases (Abstr 223). Exp Hematol 16:515

Koeffler HP, Gasson J, Ranyard J, Souza L, Shepard M, Munker R (1987) Recombinant human TNF alpha stimulates production of granulocyte colony-stimulating factor. Blood 70:55

Koike K, Ihle JN, Ogawa M (1986) Declining sensitivity to interleukin 3 of murine multipotential hemopoietic progenitors during their development: application to a culture system that favors blast cell colony formation. J Clin Invest 77:894–899

Koike K, Nakahata T, Takagi M, Kobayashi T, Ishiguro A, Tsuji K, Naganuma K, Okano A, Akiyama Y, Akabane T (1988) Synergism of BSF-2/interleukin 6 and interleukin 3 on development of multipotential hemopoietic progenitors in serum-free culture. J Exp Med 168:879–890

Krumwieh D, Seiler FR, Welte K (1988) Changes of hematopoiesis in cynomolgus monkeys after application of human colony-stimulating factors (Abstr 364). Exp Hematol 16:551

Kurland JI, Pelus LM, Ralph P, Bockman RS, Moore MAS (1979) Induction of prostaglandin E synthesis in normal and neoplastic macrophages: role for colony-stimulating factor(s) distinct from effects on myeloid progenitor cell proliferation. Proc Natl Acad Sci USA 76:2326–2330

Laver J, Abboud M, Warren D, Moore MAS (1988a) Radioprotective effects of interleukin 1 on human hematopoietic progenitors (Abstr 96). Exp Hematol 16:482

Laver J, Castro-Malaspina H, Kernan NA, Levick J, Evans RL, O'Reilly RJ, Moore MAS (1988b) In vitro interferon-gamma production by cultured T-cells in severe aplastic anaemia: correlation with granulomonopoietic inhibition in patients who respond to anti-thymocyte globulin. Br J Haematol 69:545–550

Leary AG, Ikebuchi K, Hirai Y, Wong GG, Yang Y-C, Clark SC, Ogawa M (1988) Synergism between interleukin-6 and interleukin-3 in supporting proliferation of human hematopoietic stem cells: comparison with interleukin-1a. Blood 71:1759–1763

Lindemann A, Riedel D, Oster W, Meuer SC, Blohm D, Mertelsmann RH, Herrmann F (1988) Granulocyte/macrophage colony-stimulating factor induces interleukin 1 production by human polymorphonuclear neutrophils. J Immunol 140:837–839

Lopez AF, Williamson DJ, Gamble GR, Begley CG, Harlan JM, Klebanoff SJ, Waltersdorph A, Wong G, Clark SC, Vadas MA (1986) Recombinant human granulocyte-macrophage colony-stimulating factor stimulates in vitro mature human neutrophil and eosinophil function, surface receptor expression, and survival. J Clin Invest 78:1220–1228

Lopez AF, Sanderson CJ, Gamble JR, Campbell HD, Young L, Vadas MA (1988) Recombinant human interleukin 5 is a selective activator of human eosinophil function. J Exp Med 167:219–224

Lothrop CD, Warren DJ, Souza LM, Jones JB, Moore MAS (1988) Correction of canine cyclic hematopoiesis with recombinant human granulocyte colony stimulating factor. Blood 72:1324–1328

Lovhaug D, Pelus LM, Nordlie EM, Boyum A, Moore MAS (1986) Monocyte conditioned medium and interleukin 1 induce granulocyte-macrophage colony stimulating factor production in the adherent cell layer of murine bone marrow cultures. Exp Hematol 14:1037–1042

Masaoka T, Motoyoshi K, Takaku F, Moriyama Y, Ohira M, Kato S, Dohi H, Horiuchi A (1988) Human urinary macrophage-colony stimulating factor for bone marrow transplantation (Abstr 42). Exp Hematol 16:469

McNiece IK, Stewart FM, Deacon DM, Quesenberry PJ (1988) Synergistic interactions between hematopoietic growth factors as detected by in vitro mouse bone marrow colony formation. Exp Hematol 16:383–388

Mertelsmann R, Cozman D (1988) Colony stimulating factors in vivo and in vitro. Immunol Today 9:97–98
Metcalf D (1989) Haematopoietic growth factors I. Lancet 126:825–827
Moore MAS (1987a) G-CSF as a differentiation-inducing agent in normal and leukemic myelopoiesis. In: Aarbakke J, Chiang P, Koeffler HP (eds) The biology and pharmacology of tumor cell differentiation. Humana, Clifton, pp 29–49
Moore MAS (1987b) Growth and maturation factors in leukemia. In: Oldham RK (ed) Principles of cancer biotherapy. Raven, New York, pp 399–446
Moore MAS (1987c) Interactions between hematopoietic growth factors: the clinical role of combination biotherapy. In: Neth R, Gallo RC, Greaves MF, Kabisch H (eds) Modern trends in human leukemia 7. Springer, Berlin Heidelberg New York
Moore MAS (1988a) The use of hematopoietic growth and differentiation factors for bone marrow stimulation. In: Devita VT, Hellman S, Rosenberg SA (eds) Important advances in oncology 1988. Lippincott, Philadelphia, pp 31–54
Moore MAS (1988b) Interleukin 3: an overview. In: Schrader J (ed) Lymphokines, vol 15. Academic, New York, pp 219–280
Moore MAS (1988c) Combination biotherapy: synergistic, additive and concatenate interactions between CSFs and interleukins in hematopoiesis. In: Fortner JG, Rhoads JE (eds) Accomplishments in cancer research 1987. Lippincott, Philadelphia, pp 335–350
Moore MAS, Sheridan AP (1979) Pluripotential stem cell replication in continuous human, prosimian, and murine bone marrow culture. Blood Cells 5:297–311
Moore MAS, Warren DJ (1987) Interleukin-1 and G-CSF synergism: in vivo stimulation of stem cell recovery and hematopoietic regeneration following 5-fluorouracil treatment in mice. Proc Natl Acad Sci USA 84:7134–7138
Moore MAS, Warren DJ (1989) Combination biotherapy in vivo and in vitro with IL-1, IL-3, IL-5, G-CSF, and GM-CSF. In: Groopman J, Evans C, Golde D (eds) Mechanisms of action and therapeutic applications of biologicals in cancer and immune deficiency disorders. Liss, New York pp 77–88 (UCLA symposia on molecular and cellular biology, new series)
Moore MAS, Spitzer G, Metcalf D, Penington DG (1974) Monocyte production of colony stimulating factor in familial cyclic neutropenia. Br J Haematol 27:47–55
Moore MAS, Sheridan APC, Allen TD, Dexter TM (1979) Prolonged hematopoiesis in a primate bone marrow culture system: characteristics of stem cell production and the hematopoietic microenvironment. Blood 54:775–793
Moore MAS, Broxmeyer HE, Sheridan APC, Meyers PA, Jacobsen N, Winchester RJ (1980) Continuous human bone marrow culture: Ia antigen characterization of probable human pluripotential stem cells. Blood 55:682–690
Moore MAS, Warren DJ, Souza L (1987a) Synergistic interaction between interleukin-1 and CSFs in hematopoiesis. In: Gale RP, Golde DW (eds) Recent advances in leukemia and lymphoma. Liss, New York, pp 445–456 (UCLA symposium on leukemia)
Moore MAS, Welte K, Gabrilove J, Souza LM (1987b) Biological activities of recombinant human granulocyte-colony stimulating factor (rhG-CSF) and tumor necrosis factor: in vivo and in vitro analysis. In: Neth R, Gallo RC, Greaves MF, Kabisch H (eds) Modern trends in human leukemia 7. Springer, Berlin Heidelberg New York, pp 210–220
Moore RN, Oppenheim JJ, Farrar JJ (1986) Production of lymphocyte-activated with colony-stimulating factors. J Immunol 125:1302–1310
Moreau JF, Bonneville M, Godard A, Gascan H, Gruart V, Moore MA, Soulillou JP (1987) Characterization of a factor produced by human T cell clones exhibiting eosinophil-activating and burst-promoting activities. J Immunol 138:3844–3849
Morstyn G, Souza LM, Keech J, Sheridan W, Campbell L, Alton NK, Green M, Metcalf D, Fox R (1988) Effect of granulocyte colony stimulating factor on neutrophenia induced by cytotoxic chemotherapy. Lancet 1:667–672

Motoyoshi K, Takaku F (1988) Accelerated recovery from thrombocytopenia and neutropenia after anticancer chemotherapy by consecutive intravenous infusions of human monocytic colony-stimulating factor (Abstr 237). Exp Hematol 16:519
Munker R, Koeffler HP (1987) In vitro action of tumor necrosis factor on myeloid leukemia cells. Blood 69:1102
Munker R, DiPersio J, Koeffler HP (1987) Tumor necrosis factor: receptors on hematopoietic cells. Blood 70:1730
Nakahata T, Ogawa M (1982) Identification in culture of a class of hemopoietic colony-forming units with extensive capability to self-renew and generate multipotential hemopoietic colonies. Proc Natl Acad Sci USA 79:3843–3848
Naparstek E, Ohana M, Kedar E, Slavin S (1988) Continuous intravenous administration of GM-CSF enhances hemopoietic and immune recovery in marrow transplantation in mice (Abstr 156). Exp Hematol 16:498
Negrin RS, Haeuber DH, Nagler A, Souza LM, Greenberg PL (1988) Treatment of myelodysplastic syndromes with recombinant human granulocyte colony stimulating factor (Abstr 240). Exp Hematol 16:519
Nemunaitis J, Singer JW, Buckner CD, Hill R, Storb R, Thomas ED, Appelbaum FR (1988) Use of recombinant human granulocyte-macrophage colony-stimulating factor in autologous marrow transplantation for lymphoid malignancies. Blood 72:834–836
Neta R, Oppenheim JJ, Douches SD (1988) Interdependence of the radioprotective effects of human recombinant interleukin-1a, tumor necrosis factor a, granulocyte colony stimulating factor and murine recombinant granulocyte-macrophage colony stimulating factor. J Immunol 140:108
Nissen C, Stern A, Jones TH, Schaedelin J (1988) European double blind randomized trial of GM-CSF in patients after transplantation of T-cell depleted bone marrow (Abstr 235). Exp Hematol 16:518
Ohta M, Greenberger JS, Anklesaria P, Bassols A, Massague J (1987) Two forms of transforming growth factor-β distinguished by multipotential haematopoietic progenitor cells. Nature 329:539–541
Oppenheim JJ, Kovacs EJ, Matsushima K (1986) There is more than one interleukin 1. Immunol Today 7:45–55
Oster W, Lindemann A, Horn S, Mertelsmann R, Herrmann F (1987) Tumor necrosis factor (TNF)-alpha but not TNF-beta induces secretion of colony stimulating factor for macrophages (CSF-1) by human lymphocytes. Blood 70:1700
Ottmann OG, Pelus LM (1988) Differential proliferative effects of transforming growth factor-β on human hematopoietic progenitor cells. J Immunol 140:2661–2665
Ottmann OG, Welte K, Souza LM, Moore MAS (1987) Proliferative effects of a recombinant human granulocyte-colony stimulating factors (rG-CSF) on highly enriched hematopoietic progenitor cells. In: Neth R, Gallo RC, Greaves MF, Kabisch H (eds) Modern trends in human leukemia 7. Springer, Berlin Heidelberg New York, pp 244–247
Patel VP, Lodish HF (1986) Loss of adhesion of murine erythroleukemia cells to fibronectin during erythroid differentiation. Science 224:996–998
Peetre C, Gullberg U, Nilsson E, Olsson I (1988) Effects of recombinant tumor necrosis factor on proliferation and differentiation of leukemic and normal hematopoietic cells in vitro. J Clin Invest 78:1694
Peters WP, Stuart A, Affronti ML, Kim CS, Coleman RE (1988) Neutrophil migration is defective during recombinant human granulocyte-macrophage colony-stimulating factor infusion after autologous bone marrow transplantation in humans. Blood 72:1310–1315
Platzer E, Welte K, Gabrilove JL, Lu L, Harris P, Mertelsmann R, Moore MAS (1985) Biological activities of a human pluripotent hemopoietic colony stimulating factor on normal and leukemic cells. J Exp Med 162:1788–1801
Raghavachar A, Fleischer S, Frickhofen N, Heimpel H, Fleischer B (1987) T lymphocyte control of human eosinophilic granulopoiesis: clonal analysis in an idiopathic hypereosinophilic syndrome. J Immunol 139:3753–3758

Roberts RA, Spooncer E, Parkinson EK, Lord BI, Allen TD, Dexter TM (1987) Metabolically inactive 3T3 cells can substitute for marrow stromal cells to promote the proliferation and development of multipotent haemopoietic stem cells. J Cell Physiol 132:203–214

Roberts RA, Gallagher J, Spooncer E, Allen TD, Bloomfield F, Dexter TM (1988) Heparin sulfate bound growth factors: a mechanism for stromal cell mediated haemopoiesis. Nature 332:376–378

Rothenberg ME, Owen WF, Silberstein DS, Woods J, Soberman RJ, Austen FK, Stevens RL (1988) Human eosinophils have prolonged survival, enhanced functional properties and become hypodense when exposed to human interleukin 3. J Clin Invest 81:1986–1992

Ruscetti FW, Sing G, Ruscetti SK, Ellingsworth LE, Keller JR (1988) The role of transforming growth factor-β in the regulation of normal and leukemic hematopoiesis (Abstr 19). Exp Hematol 16:417

Sanderson CJ, O'Garra A, Warren DJ, Klaus GGD (1986) Eosinophil differentiation factor also has B-cell growth factor activity: proposed name interleukin 4. Proc Natl Acad Sci USA 83:437–441

Shalaby MR, Aggarwal BB, Rinderknecht E, Svedersky LP, Finkle BS, Palladino MA (1985) Activation of human polymorphonuclear neutrophil functions by interferon gamma and tumor necrosis factor. J Immunol 135:2069–2076

Sieff CA, Tsai S, Faller DV (1987) Interleukin 1 induces cultured human endothelial cell production of granulocyte-macrophage colony-stimulating factor. J Clin Invest 78:48–57

Slordal L, Warren DJ, Moore MAS (1989) Effect of recombinant murine tumor necrosis factor on hematopoietic reconstitution in sublethally irradiated mice. J Immunol 142:833–835

Smith CA, Rennick DM (1986) Characterization of a murine lymphokine distinct from interleukin 2 and interleukin 3 (IL-3) possessing a T-cell growth factor activity and a mast-cell growth factor activity that synergizes with IL-3. Proc Natl Acad Sci USA 83:1857–1861

Socinski MA, Cannistra SA, Sullivan R, Elias A, Antman K, Schnipper L, Griffin JD (1988) Granulocyte-macrophage colony-stimulating factor induces the expression of the CD11b surface adhesion molecule on human granulocytes in vivo. Blood 72:691–697

Souza LM, Boone TC, Gabrilove JL, Lai PH, Zsebo KM, Murdock DC, Chazin VR, Bruszewski J, Lu H, Chen KK, Barendt J, Platzer E, Moore MAS, Mertelsmann R, Welte K (1986) Recombinant human granulocyte colony-stimulating factor: effects on normal and leukemic myeloid cells. Science 232:61–65

Spooncer E, Gallagher JT, Krizsa F, Dexter TM (1983) The regulation of haemopoiesis in long term bone marrow cultures. IV. glycosaminoglycan synthesis and stimulation of haemopoiesis by beta-D-xylosides. J Cell Biol 96:510–514

Spooncer E, Lord BI, Dexter TM (1985) Defective ability to self-renew in vitro of highly purified primitive haematopoietic cells. Nature 316:62–64

Spooncer E, Heyworth CM, Dunn A, Dexter TM (1986) Self-renewal and differentiation of interleukin-3 dependent multipotent stem cells are modulated by stromal cells and serum factors. Differentiation 31:111–118

Stanley ER, Bartocci A, Patinkin D, et al. (1986) Regulation of very primitive multipotent hemopoietic cells by hemopoietin-1. Cell 45:667–671

Steffen M, Abboud M, Potter GK, Yung YP, Moore MAS (1989) Production of tumor necrosis factor by human mast cells. Immunology (in press)

Taylor K, Spitzer G, Jagannath S, Dicke K, Vincent M, Souza L (1988) rhG-CSF hastens granulocyte recovery in Hodgkin's disease after high-dose chemotherapy and autologous bone marrow transplant (Abstr 3). Exp Hematol 16:413

Tertian G, Yung YP, Guy-Grand D, Moore MAS (1981) Long term in vitro culture of murine mast cells I. Description of a growth factor dependent culture technique. J Immunol 127:788–794

Teshima H, Shibata H, Hiraoka A, Nakamura H, Masaoka T, Takaku F (1988) Clinical effects of recombinant human granulocyte colony-stimulating factor (Abstr 244). Exp Hematol 16:520
Till JE, McCulloch (1980) Haemopoietic stem cell differentiation. Biochim Biophys Acta 605:431–459
Trentin JJ (1970) Influence of hematopoietic organ stroma (hematopoietic inductive microenvironment) on stem cell differentiation. In: Gordon A (ed) Regulation of hematopoiesis. Appleton-Century-Crofts, New York, pp 161–186
Vadhan-Raj S, Keating M, LeMaistre A, Hittelman WN, McCredie K, Trujillo JM, Broxmeyer HE, Henney C, Gutterman JU (1987) Effects of recombinant human granulocyte-macrophage colony-stimulating factor in patients with myelodysplastic syndromes. N Engl J Med 317:1545–1552
Vadhan-Raj S, Buescher S, LeMaistre A, Ventura G, Lepe-Zuniga J, Broxmeyer HE, Gillis S, Gutterman JU (1988) Effects of recombinant human granulocyte-macrophage colony-stimulating factor in patients with aplastic anemia (Abstr 239). Exp Hematol 16:519
Vellenga E, Rambaldi A, Ernst TJ, Ostapovicz D, Griffin JD (1988) Independent regulation of M-CSF gene expression in human monocytes. Blood 71:1529–1532
Vogel SN, Douches SD, Kaufman EN, Neta R (1987) Induction of colony stimulating factor in vivo by recombinant interleukin 1a and recombinant tumor necrosis factor a. J Immunol 138:2143–2148
Wang JM, Chen ZG, Colella S, Bonilla MA, Welte K, Bordignon C, Mantovani A (1988) Chemotactic activity of recombinant human granulocyte colony-stimulating factor. Blood 72:1456–1460
Wankowicz Z, Megyeri P, Issekutz A (1988) Synergy between tumor necrosis factor alpha and interleukin-1 in the induction of polymorphonuclear leukocyte migration during inflammation. J Leukocyte Biol 43:349–356
Warren DJ, Moore MAS (1988) Synergism among interleukin 1, interleukin 3, and interleukin 5 in the production of eosinophils from primitive hemopoietic stem cells. J Immunol 140:94–99
Warren MK, Ralph P (1986) Macrophage growth factor CSF-1 stimulates human monocyte production of interferon, tumor necrosis factor, and colony stimulating activity. J Immunol 137:2281–2285
Welte K, Platzer E, Lu L, Gabrilove JL, Levi E, Mertelsmann R, Moore MAS (1985) Purification and biochemical characterization of human pluripotent hematopoietic colony-stimulating factor. Proc Natl Acad Sci USA 82:1526–1530
Welte K, Bonilla MA, Gillio AP, Boone TC, Potter GK, Gabrilove JL, Moore MAS, O'Reilly J, Souza LM (1987) Recombinant human granulocyte colony-stimulating factor: effects on hematopoiesis in normal and cyclophosphamide-treated primates. J Exp Med 1645:941–948
Welte K, Reiter A, Odenwald E, Muller W, Seidel J, Schulz G, Riehm H (1988) Initial results with granulocyte-macrophage colony stimulating factor in children with congenital agranulocytosis (Abstr 242). Exp Hematol 16:520
Wong GG, Witek-Giannotti JS, Temple PA, Kriz R, Ferenz C, Hewick RM, Clark SC, Ikebuchi K, Ogawa M (1988) Stimulation of murine hemopoietic colony formation by human IL-6. J Immunol 140:3040–3044
Yamaguchi Y, Suda T, Suda J, Eguchi M, Miura Y, Harada N, Tominaga A, Takatsu K (1988) Purified interleukin 5 supports the terminal differentiation and proliferation of murine eosinophilic precursors. J Exp Med 167:43–56
Yung YP, Moore MAS (1982) Long-term in vitro culture of murine mast cells III. Discrimination of mast cell growth factor and granulocyte CSF. J Immunol 129: 1256–1261
Yung YP, Eger R, Teritian G, Moore MAS (1981) Long term in vitro culture of murine mast cells II. Purification of a mast cell growth factor and its dissociation from TCGF. J Immunol 127:794–799

Zoumbos NC, Gascon P, Djeu JY, Young NS (1985) Interferon is a mediator of hematopoietic suppression in aplastic anemia in vitro and possibly in vivo. Proc Natl Acad Sci USA 82:188–192

Zucali JR, Broxmeyer H, Gross MA, Dinarello CA (1988) Recombinant human tumor necrosis factors alpha and beta stimulate fibroblasts to produce hemopoietic growth factors in vitro. J Immunol 140:840–850

CHAPTER 30

Peptide Growth Factors and the Nervous System

M. E. Gurney

Many of the growth factors acting on somatic or hematopoietic tissues also are expressed within the brain. Their roles in the embryogenesis of neural tissue and in the maintenance of its integrity in the adult are just beginning to be appreciated. The lineages giving rise to the cell types comprising neural tissues, to neurons in all their variety, and to astrocytes and oligodendrocytes have been described in outline. Many of the factors influencing proliferation of progenitor cells and the choices made as they traverse neural lineages have been identified. Surprisingly, the factors acting on neural linages are not unique to the nervous system and include platelet-derived growth factor (PDGF) and insulin-like growth factor type 1 (IGF-1). Progress at identifying growth factors acting on postmitotic neurons has been somewhat slower. Nerve growth factor (NGF) remains the only neuronal growth factor brought to a molecular level of analysis.

A. Embryogenesis of Neural Tissues

At gastrulation a sheet of ectodermal cells, the neural plate, becomes committed to the formation of neural tissue. As development continues, that sheet of cells folds into a tube and then sinks beneath the surface of the embryo. From it derives both central and peripheral nervous system structures (CNS and PNS, respectively). At closure, dorsal cells of the neural tube contribute to a transient structure termed the neural crest from which will derive PNS structures as well as some somatic tissues (Ledouarain 1982; Bronner-Fraser and Fraser 1988). The cephalic neural tube will form the brain and its continuation caudally will form the spinal cord. From neural crest cells will derive sympathetic, parasympathetic, and sensory nerve ganglia, the enteric nerve plexus, the glia cells of peripheral nerves which are known as Schwann cells, melanocytes in skin, adrenal medullary cells and other endocrine cells, as well as a variety of mesenchymal tissues including the skull, musculature in the tongue, and the aortic arch (Ledouarin 1982).

B. Progenitor Cells in the Neural Crest

Individual, premigratory neural crest cells are pluripotent and can contribute progeny to all of the tissues named above. Soon after neural tube closure, dye-marked progenitor cells can be seen to contribute to both the neural tube and the neural crest (BRONNER-FRASER and FRASER 1988). Descendants of single progenitors can include cells as diverse as Schwann cells, sensory neurons, and sympathetic neurons, or, in other instances, sensory neurons, adrenal medullary cells, and metanephric mesenchyme. Crest cells following a dorsolateral pathway of migration travel under the ectoderm and differentiate into melanocytes. Cells following a ventral pathway of migration populate the sensory ganglia, sympathetic chain, adrenals, and mesenchymal crest-derivatives. Different regions of the neural crest along the rostral-caudal axis of the embryo contribute to different structures (LEDOUARIN 1982). For example, neural crest cells caudal to somite 5 in chick embryos populate the sympathetic chain, while only crest cells at the level of somites 18–24 contribute progeny to the adrenal medulla. Enteric neurons of the gut largely derive from vagal neural crest rostral to somite 5 and vagal crest does not contribute to the sympathetic chain. Crest cells transplanted from one region to another along the length of the embryo adopt a migratory path and cell fates typical of the region into which they are transplanted (LEDOUARIN and TEILLET 1974; LEDOUARIN et al. 1975). Heterotopic transplantation shows that presumptive adrenomedullary crest from somites 18 to 24, when transplanted to a vagal position, populates the enteric ganglia and does not contribute to the adrenals (LEDOUARIN et al. 1975). Correspondingly, cervical crest taken from the mesencephalic or rhombencephalic levels and grafted into the presumptive adrenomedullary position between somites 18–24 appropriately populates the adrenals and sympathetic ganglia (LEDOUARIN and TEILLET 1974). Thus, the developmental fate of crest cells appears to be environment dependent and to emerge through interactions along their pathway of migration and at the site they settle in the developing embryo.

The preparation of neural crest cell-derived cultures takes advantage of the fact that the cells are a migratory population. Neural tubes are harvested from early embryos before the period of crest cell migration and are separated from other tissues by brief trypsinization. The crest cells remain adhered to the dorsal surface of the neural tube, and when the neural tube is allowed to adhere to a collagen substrate in vitro, within a few hours crest cells begin to migrate from the explant onto the culture substratum (COHEN and KONIGSBERG 1975). Their migration requires a fibronectin substratum (ROVASIO et al. 1983), which can be provided by deposition of fibronectin from serum in the culture medium. After a day or two, the neural tube is removed and the mesenchymal cell outgrowth remaining is for the most part crest derived. Such neural crest cultures recapitulate much of the timing of development and expression of cell phenotypes seen in vivo.

I. Melanocytes Are a Terminally Differentiated Cell Type

Among the cell types identifiable in crest cultures are pigmented melanocytes, unpigmented, small intensely fluorescent cells (SIF cells), and process-bearing neurons expressing a variety of neurotransmitter phenotypes (COHEN 1977;

COHEN and KONIGSBERG 1975; FAUQUET et al. 1981; MAXWELL et al. 1984). Manipulation of the culture conditions can influence the types of crest derivative obtained. In conditions that prevent crest cell dispersal from the tube explant (for example, a nonadhesive agar culture substrate), crest cells form clusters which remain attached to the surface of the neural tube and practically all differentiate as melanocytes (GLIMELIUS and WESTON 1981). Either contact with the neural tube or failure to migrate apparently programs the crest cells to enter the melanocyte lineage, as after 24 h of culture on the nonadhesive substrate the crest cells can be dispersed into secondary cultures and virtually all become pigmented. The pigmented phenotype is a stable one. Melanocytes can be cloned in vitro and then reintroduced into the somite of a young embryo, where their subsequent fate can be followed (BRONNER and COHEN 1979). Normally, presumptive melanocytes migrate dorsolaterally beneath the ectoderm. By depositing the melanocytes in the somite, they are directed into the ventral pathway of crest cell migration which is traversed by cells that will populate the sympathetic chain. Will cloned melanocytes transdifferentiate into neurons? The deposited melanocytes initiate migration, follow the ventral pathway of crest migraton, and populate appropriate crest-derived structures including sensory and sympathetic ganglia and the adrenal medulla; however, the injected melanocytes do not lose or alter their phenotype. Thus, the melanocyte represents a terminally committed crest derivative which is unresponsive to environmental cues along the pathway of migration which might influence lineage choice.

II. Heterogeneity of Cell Types Within the Neural Crest

Some restriction of cell fate probably occurs as crest cells begin their migration from the dorsal surface of the neural tube. Melanocyte progenitors segregate early, and hetereogeneity of the migrating population is further indicated by the heterogeneous expression of cell surface and cytoplasmic determinants in the migrating population. One marker for neural crest cells is a carbohydrate epitope recognized by the HNK-1 monoclonal antibody (ABO and BALCH 1981). The HNK-1 epitope is present on the neural cell adhesion molecule, N-CAM, the neuron-glial cell adhesion molecule, Ng-CAM, myelin-associated glycoprotein, MAG (KRUSE et al. 1984, 1985), and some glycolipids (CHOU et al. 1985). HNK-1 labels subpopulations of migrating neural crest cells, is present on peripheral neurons and Schwann cells, but is not expressed on melanocytes or on mesenchymal cells derived from crest. Approximately 25% of migrating crest cells carry a second marker reacting with the GIN1 antibody (BARBU et al. 1986). GIN1 antigen marks all satellite cells of peripheral ganglia, Schwann cells, and some sympathetic and sensory neurons. Additional subtypes of crest cells are distinguished by expression of GD3 ganglioside, which reacts with a monoclonal antibody designated R24 (PUKEL et al. 1982), and also by 9-O-acetylated GD3, which is recognized by monoclonal antibodies designated D1.1 and JONES (LEVINE et al. 1986; MENDEZ-OTERO et al. 1988). GD3 is present on migrating crest cells, sensory neurons, and perhaps enteric neurons (GIRDLESTONE and WESTON 1985; ROSNER et al. 1985). The antigen recognized by D1.1 and JONES marks migrating neural crest cells, sensory neurons, and adrenal chromaffin cells. Sympathetic and parasympathetic neurons are unlabeled.

III. SIF Cells Arise from HNK-1+ Progenitor Cells

Among the more interesting experiments with neural crest cultures is one by MAXWELL et al. (1988). They used flow cytometry to sort crest cells expressing either the HNK-1 antigen or GD3. Approximately 12% of the crest cells sorted express both markers, 32% are HNK-1$^+$, 7% are GD3$^+$, and 49% are double negative. HNK-1$^+$ cells differentiated into melanocytes, unpigmented cells, and catecholamine-positive cells when cultured, whereas HNK-1$^-$ cells gave rise to melanocytes and unpigmented cells, but few if any catecholamine-positive cells. GD3$^+$ and GD3$^-$ cells give rise to all three phenotypes. Thus, the experiment further indicates the segregation of developmental potential in the migrating crest cell population.

The catecholamine-containing cells obtained in neural crest cultures are SIF cells which lack expression of neuronal intermediate filaments (CHRISTIE et al. 1987). They apparently arise from a mitotically active precursor pool (KAHN and SIEBER-BLUM 1983). The mitogens for HNK-1$^+$ cells and SIF cells have not been identified. SIF cells are of particular interest as they appear to be a bipotential precursor cell within the sympathoadrenal lineage. Their differentiation from HNK-1$^+$ precursor cells is influenced by the culture substratum, particularly fibronectin and extracellular matrix gels, both of which suppress the melanocyte lineage and enhance development of SIF cells (LORING et al. 1982; SIEBER-BLUM et al. 1981; MAXWELL and FORBES 1987). Thus, contact of mitotically active HNK-1$^+$ cells with fibronectin during migration through the ventral pathway may induce their differentiation into SIF cells and determine in part commitment to the sympathoadrenal lineage. SIF cells can be differentiated in culture into either sympathetic neurons or adrenal chromaffin cells under the influence of exogenously added factors.

IV. SIF Cells Are Bipotential Progenitor Cells Within the Sympathoadrenal Lineage

The SIF cells seen in culture resemble a type of small, intensely fluorescent cell seen in vivo which populates sympathetic ganglia and extraadrenal chromaffin tissues. The ultrastructure of SIF cells is similar to that of adrenal medullary chromaffin cells. They secrete norepinephrine and dopamine, which, together with their resemblance to chromaffin cells, suggests that they might be part of a primitive adrenergic system distributed throughout the length of the sympathetic chain (COHEN 1977). Also contained within sympathetic ganglia are the principal adrenergic neurons which secrete norepinephrine and a small number of cholinergic neurons which innervate the exocrine sweat glands (PATTERSON 1987). The survival and growth of postmitotic sympathetic neurons is dependent on NGF both in vitro and in vivo (this volume). In the absence of NGF, sympathetic neurons die when placed in culture.

SIF cells are a minority cell population within sympathetic ganglia. They can be cultured from sympathetic ganglia with the addition of glucocorticoids to the medium (DOUPE et al. 1985a). The number of SIF cells obtained is low initially and then increases with time in culture, suggesting either that they are a slowly mitotic population or that SIF cells differentiate from a catecholamine-negative precursor under the influence of glucocorticoids. When glucocorticoids are withdrawn and replaced with NGF, neuronal differentiation is induced and the cells become dependent upon NGF for survival.

Adrenal chromaffin cells behave much like ganglionic SIF cells when placed in culture (DOUPE et al. 1985a). In the presence of micromolar glucocorticoid, adrenal chromaffin cells maintain a differentiated phenotype. Withdrawal of gluccorticoids and replacement with NGF has two effects. Initially, NGF is mitogenic and induces proliferation of the cultured chromaffin cells (LILLIEN and CLAUDE 1985). With long-term exposure to NGF, the chromaffin cells then withdraw from the cell cycle and undergo neuronal differentiation (UNSICKER et al. 1978; DOUPE et al. 1985a; LILLIEN and CLAUDE 1985). The adrenal cells differentiate neuritic processes and can form synapses (OGAWA et al. 1984). Basic fibroblast growth factor (bFGF) also induces neuronal differentiation by adrenal chromaffin cells (STEMPLE et al. 1988): The effect of bFGF is similarly blocked by glucocorticoids but, unlike NGF, bFGF does not support survival of chromaffin cell-derived adrenergic neurons. A cell line derived from an adrenal medullary tumor, the PC12 cell, has retained responsiveness to NGF and can also be induced to undergo neuronal differentiation in the presence of NGF (GREENE and TISCHLER 1976). Interestingly, neuronal differentiation induced by NGF is dominant to the genetic lesion which rendered the PC12 cell oncogenic. Thus, depending upon the cell type and its stage of differentiation, NGF can act as a mitogen for competent cells, or as a survival factor for cells that have withdrawn from the cell cycle. As the adrenal chromaffin cells differentiate into neurons under the influence of NGF, they pass through a stage that resembles the ganglionic SIF cell, which suggests to DOUPE et al. (1985b) that SIF cells represent a bipotential precursor cell within the sympathoadrenal lineage.

V. Does NGF Direct SIF Cells Toward Production of Sympathetic Neurons In Vivo?

Does NGF function in vivo to channel the sympathoadrenal lineage toward production of sympathetic neurons? The answer is both yes and no. Glucocorticoids may function in vivo to prevent NGF from inducing the neuronal differentiation of adrenal medullary cells. The adrenal gland contains two types of endocrine cells, an outer layer of cortical cells (which release glucocorticoids) surrounds the chromaffin cells within the adrenal medulla which release norepinephrine and epinephrine. Due to their proximity, adrenal chromaffin cells are probably bathed in vivo by relatively high concentrations of glucocorticoids. In culture, glucocorticoids block the induction of neuronal differentiation by NGF (DOUPE et al. 1985b). Administration of NGF to pregnant rats, however, can cause the replacement of adrenal chromaffin cells by sympathetic neurons (ALOE and LEVI-MONTALCINI 1979). So the answer is yes, NGF can induce adrenal chromaffin call transdifferentiation in vivo, but probably is prevented from doing so by adrenal glucocorticoid synthesis.

If the question is whether or not NGF is obligate for induction of SIF cell differentiation into adrenergic neurons within the sympathetic chain, the answer is probably no, but there may be room for doubt. First, more SIF cells are present in sympathetic ganglia early in development rather than later (LEMPINEN 1964) and administration of glucocorticoids at birth further increases their number (ERANKO et al. 1972). Administration of NGF to newborn rat pups increases the number of neurons in sympathetic ganglia at a time when substantial numbers of SIF cells are present, but is believed to do so by preventing naturally occurring neuronal cell death, rather than by influencing proliferation or differentiation of sympathetic neurons (HENDRY and CAMPBELL 1976). Second, crest-derived sensory neurons in the trigeminal ganglion do not express NGF receptors until they have withdrawn from the cell cycle, have differentiated a peripheral process, and have begun to innervate their synaptic targets (DAVIES et al. 1987). Thus it seems unlikely that NGF influences commitment of crest progenitors to production of sensory neurons. One might expect sympathetic neurons to follow the same developmental profile; however, sympathetic neurons differ from sensory neurons in the relationship between differentiation and withdrawal from the cell cycle. Sensory neurons are thought to differentiate after the terminal mitosis of precursor cells (ROHRER and THOENEN 1987), while sympathetic neurons express aspects of their differentiated phenotype at a time when they are mitotically active (ROTHMAN et al. 1980; ROHRER and THOENEN 1987). As SIF cells clearly express NGF receptors, are they an obligate intermediate in sympathetic neuron differentiation in vivo?

VI. Neurotransmitter Choice is Determined by Environmental Factors

Most sympathetic neurons secrete norepinephrine, although a minority which innervate the exocrine sweat glands are cholinergic. The sympathetic neurons that innervate the sweat glands pass through an adrenergic stage before they become cholinergic (LANDIS and KEEFE 1983). The switch from adrenergic to cholinergic can be manipulated in vitro under the influence of a purified, 45-kDa cholinergic-inducing factor (FUKADA 1985). The action of the cholinergic-inducing factor can be demonstrated by coculture of sympathetic neurons with heart cardiocytes. Single sympathetic neurons were shown to change their neurotransmitter phenotype from adrenergic to cholinergic

(FURSHPAN et al. 1976; REICHARDT and PATTERSON 1977), and eventually the inducing signal was traced to a soluble factor released into heart-cell-conditioned medium (PATTERSON and CHUN 1977). The factor has been purified 100000-fold to homogeneity (FUKADA 1985). The factor is highly glycosylated, is basic, and exhibits charge heterogeneity. An amino acid sequence is not yet available. SIF cells and adrenal chromaffin cells induced to undergo neuronal differentiation by NGF become adrenergic, but can also be induced to switch to a cholinergic phenotype under the influence of the cholinergic-inducing factor (DOUPE et al. 1985a, b). The factor may also enhance expression of a cholinergic phenotype by cultured spinal neurons (GIESS and WEBER 1984).

The choice of neurotransmitter type is a fundamental one which will affect the sign of the connection made by the neuron, i.e., excitatory or inhibitory, and also the transmitter released by the presynaptic neuron must match the neurotransmitter receptor expressed by the postsynaptic neuron if the connection is to be functional. The 45-kDa factor identified by Fukada and Patterson may be representative of a class of factors which influence phenotypic choices. A second, 20-kDa factor which elevates expression of choline acetyltransferase in spinal neuron cultures has been purified (MCMANAMAN et al. 1988) and other examples of substitution of one neurotransmitter phenotype for another are known (PATTERSON 1987).

C. CNS Progenitor Cells Give Rise to Both Neurons and Glial Cells

The control of progenitor cell proliferation and the genetic mechanisms which set aside CNS neurons as a postmitotic population for the life of the organism are little understood. All CNS neurons are generated by progenitor cells within specialized proliferative zones of the embryonic brain. These line the ventricular surfaces early in development, and, later, secondary proliferative zones form in the dentate gyrus of the hippocampus and the external granule layer of the cerebellum. The second two proliferative zones generate the granule neurons of the hippocampus and cerebellum, respectively.

The early neural tube and the ventricular proliferative zone of the cerebrum contain two types of cells. The majority of cells are mitotically active CNS progenitor cells which are organized into a pseudostratified epithelium. Existing side-by-side with the proliferative CNS progenitor cells is a class of differentiated glial cells, the radial glia (RAKIC 1972; LEVITT et al. 1981). These are astrocytic cells which express glial fibrillary acidic protein (GFAP) and comprise perhaps 1% of the cells in the neural tube or ventricular proliferative zones. Radial glial cells are generated early in development and remain mitotically active throughout the period of neurogenesis (LEVITT et al. 1981). Their contribution to gliogenesis is unclear. Gliogenesis occurs in two waves. In rats, the peak of astrocyte proliferation occurs during the perinatal period, while oligodendrocyte proliferation peaks at 1–2 weeks after birth (SKOFF 1980).

Astrocytes remain mitotically competent in the adult and proliferate in response to brain injury, perhaps under the influence of interleukin-1 (GIULIAN and LACHMAN 1985) or fibroblast growth factor (BERRY et al. 1983). Oligodendrocytes probably represent a terminally differentiated cell type that has withdrawn from the cell cycle (SKOFF 1980).

CNS progenitor cells in S-phase are attached by a basilar process to the lumen of the neural tube and by an apical process to the glial limitans, a lamina which forms the external surface of the developing spinal cord (SAUER 1953; HINDS and RUFFET 1971). The glial limitans is formed by the endfeet of radial glia. During S-phase, the nuclei of CNS progenitor cells migrate from the luminal surface to the glial limitans and then return to the lumen at the completion of DNA synthesis (SAUER 1953). Dividing cells retract their apical process, round up on the luminal surface, and divide. Daughter cells may then reenter S-phase or withdraw from the cell cycle. Those that become postmitotic lose their attachment to the ventricular surface and begin to migrate to the position they will occupy in the adult brain (HINDS and RUFFET 1971). The regrowth of apical processes by progenitor cells and the migration of postmitotic cells from the proliferative zone is guided by the radial glial fibers (RAKIC 1972). These span the thickness of the developing brain and provide a pathway of migration for postmitotic neuroblasts.

Perhaps surprisingly, glial and neuronal lineages are not segregated early in development. Generation of the various types of neurons forming the mammalian retina extends over several days of development (SIDMAN 1961). The retina is a layered structure comprising photoreceptors, bipolar cells, horizontal cells, amacrine cells, retinal ganglion cells, and Mueller cells (a specialized glial cell found only in the retina). Cell fate is correlated with time of withdrawal from the cell cycle (birthdate) and final position. Ganglion cells are born first and successive waves of mitoses add the different types of neurons to the retina layer-by-layer (SIDMAN 1961). Migration from the ventricular proliferative zone is radial, so daughter cells arising from a single progenitor cell form a retinal column (HOLT et al. 1988). Retinal progenitor cells have been marked by integration of retroviral genomes expressing a marker enzyme, *Escherichia coli* β-galactosidase (TURNER and CEPKO 1987) and by injection of membrane-impermeant dyes (HOLT et al. 1988). Both techniques indicate that retinal progenitor cells are pluripotent. The progeny of single progenitors can include all classes of retinal neurons, photoreceptors, and Mueller glial cells. The same observation is made in cortex using the retroviral marking technique (WALSH and CEPKO 1988). Progeny of single CNS progenitor cells can include both neurons and astrocytes, and, with longer survival times, even oligodendrocytes.

Several ways in which lineage might be linked to cell determination in the vertebrate brain can be envisaged. Cell type might be determined autonomously, within the progenitor cell or within the presumptive neuroblast, either through the counting of cell divisions or the segregation of cytoplasmic determinants. Alternatively, cell type might be determined by the extracellular microenvironment or through cellular interactions as is probably the case in the differentiation of neural crest-derivatives. The results obtained by marking single CNS progenitor cells in retina and cortex suggest that lineage is not a major determinant of cell type in the CNS. Several conclusions have been drawn by HOLT et al. (1988), TURNER and CEPKO (1987), and WALSH and CEPKO (1988). First, their data strongly argue that CNS progenitor cells are not committed to forming particular classes of progeny, nor are they even committed to neuronal as compared with glial lineages. Second, their data weakly argue that the cellular phenotype of progeny is probably not determined autonomously or assigned by counting cell divisions, but instead is likely to be due to position-dependent interactions with adjacent cells or environmental cues. In the glial lineage giving rise to a type of astrocyte and to oligodendrocytes it has been possible to identify factors which influence lineage choices.

D. Identification of CNS Progenitor Cells In Vitro

Surface expression of the disialoganglioside GD3 appears to mark CNS progenitor cells. GD3 is the major ganglioside present in early brain and its distribution has been studied by immunofluorescent staining (GOLDMAN et al. 1984; ROSNER et al. 1985). In both the chicken and the mouse embryos, $GD3^+$ cells are restricted to the proliferative zones of the developing CNS: the ventricular zone, the external granule cell layer of the cerebellum, and in mouse the dentate gyrus of the hippocampus. Early neuroblasts that have withdrawn from the cell cycle and mature neurons lack GD3 expression. Migrating neural crest cells are also $GD3^+$ as are cells in some endodermal and mesodermal epithelia. The distribution of GD3 in spinal cord is particularly striking (ROSNER et al. 1985). $GD3^+$ cells fill the proliferative zone adjacent to the central lumen of the spinal cord and are virtually absent from the ventral horn where postmitotic motor neurons have begun to settle. In the mature CNS, GD3 is expressed at high levels on retinal Mueller cells, on oligodendrocyte plasma membranes, but not on myelin, and is a marker for reactive astrocytes which have been induced by brain injury (SEYFRIED and YU 1985). Mitotic CNS progenitors do not express any markers characteristic of differentiated neurons. Ganglioside antigens, neurotransmitter biosynthetic enzymes, neuron-specific enolase, neuronal intermediate filaments, neuronal *src* protein, etc., are first expressed by neuroblasts on withdrawal from the cell cycle (EISENBARTH et al. 1979; ROTHMAN et al. 1980; TAPSCOTT et al. 1981; SORGE et al. 1984; MASUKO and SHIMADA 1983; YAMAMOTO et al. 1981).

When fetal brain is dissociated and placed in culture, astrocytes proliferate and form a confluent monolayer upon which grow neurons, glial progenitor cells, and oligodendrocytes. Markers which allow the identification of neurons in culture include expression of neuron-specific intermediate filaments, binding of tetanus toxin to the GM1 ganglioside, neuron-specific enolase, and other enzymes important for neurotransmitter metabolism (KENNEDY et al.; SCHMECHEL et al. 1980). Astrocytes are identified in culture by expression of GFAP, and can be further subdivided into two classes based on surface labeling with A2B5 monoclonal antibody, which reacts with GQ gangliosides (RAFF et al. 1983 a). A2B5-, $GFAP^+$ cells are referred to as type-1 astrocytes and $A2B5^+$, $GFAP^+$ cells are known as type-2 astrocytes. The type-1 and type-2 astrocyte phenotypes remain stable in culture with no indication that the two cell types are interconvertible (RAFF et al. 1983 a) Oligodendrocytes are identified in culture by surface expression of galactocerebroside (galC) and, as an indication of maturation, by expression of myelin basic protein, MBP (RAFF et al. 1978; BENEVISTE and MERRILL 1986). Formation of myelin by oligodendrocytes requires coculture with neurons. RAFF et al. (1983 b) identify a third cell-type, known as an O2A progenitor cell, that is $A2B5^+$, $vimentin^+$, $GFAP^-$, $galC^-$. The O2A cell is a bipotential glial progenitor that can generate either oligodendrocytes or type-2 astrocytes.

Type-1 and type-2 astrocytes are believed to correspond to the two types of astrocytes distinguished histologically in vivo, protoplasmic and fibrous astrocytes, respectively (MILLER and RAFF 1984). Protoplasmic astrocytes are found mainly in gray matter. Their processes are in close contact with endothelial cells lining brain capillaries and form the glial-limiting membranes surrounding the optic nerve and the external surface of the brain. Protoplasmic astrocytes have been shown to induce formation of tight junctions between the endothelial cells in an experimental model, and thus may induce formation of the blood-brain barrier (JANZER and RAFF 1987). Fibrous astrocytes are located mainly in white matter and optic nerve and are believed to contribute to the structure of nodes of Ranvier (FRENCH-CONSTANT and RAFF 1986). Type-1 astrocytes which are $GFAP^+$, $GD3^-$, $A2B5^-$ are the predominant astrocytic cell type obtained from fetal brain (MCCARTHY and DE VELLIS 1980; RAFF et al. 1983 a; GOLDMAN et al. 1984). They may derive from radial glial cells, although molecular markers with which to establish that derivation are unavailable. Epidermal growth factor (RAFF et al. 1983 a; GUENTERT-LAUBER and HONEGGER 1985), insulin-like growth factor type 1 (LENOIR and HONEGGER 1983; SHEMER et al. 1987), basic and acidic fibroblast growth factors (MORRISON and DE VELLIS 1983; KNISS and BURRY 1988; WALICKE and BAIRD 1988), and glial growth factor (BROCKES et al. 1980) all have mitogenic activity for cultured astrocytes. Which of these mitogens is important for type-1 astrocyte proliferation in vivo is unknown.

Young neural tubes or early cerebrum, harvested prior to the peak of neurogenesis, provide a source of pluripotent CNS progenitor cells for culture experiments. Although surface expression of GD3 should provide a marker for identifying CNS progenitor cells in vitro, these cells have been extremely difficult to study. Transient expression of GD3 is seen on both neurons and astrocytes immediately after plating fetal brain cultures (SUGAYA et al. 1987; GOLDMAN et al. 1986). This reinforces the view that GD3 expression marks progenitor cells capable of contributing to either glial or neuronal lineages; however, proliferation of CNS progenitors with production of new neurons is limited in culture (JUURLINK and FEDOROFF 1982; BUSE 1987; GENSBURGER et al. 1987). Knowledge of the mitogens to which pluripotent CNS progenitors respond or the signals for neuronal differentiation is lacking, but should become an area of considerable interest. Basic fibroblast growth factor (bFGF) and insulin-like growth factor type 1 (IGF 1) are candidate mitogens for these cells (GENSBURGER et al. 1987; LENOIR and HONEGGER 1983).

Progenitor cells of restricted potential which can give rise to either oligodendrocytes or type-2 astrocytes are present in fetal brain cultures (O2A progenitors), but are best studied in cultures of optic nerve (GOLDMAN et al. 1984; RAFF et al. 1983b). In fetal brain cultures, O2A progenitors form a population of small, A2B5$^+$, GD3$^+$ cells which rest on top of the type-1 astrocyte monolayer. They can be dislodged by shaking (MACCARTHY and DE VELLIS 1980) and subcultured as a relatively pure population. Their differentiation into either type-2 astrocytes or oligodendrocytes can be manipulated by the culture conditions (ABNEY et al. 1983; GOLDMAN et al. 1986; LEVI et al. 1986) as described for optic nerve cultures by RAFF et al. (1983b).

E. The O2A Glial Lineage

Glial cell lineages have been explored in detail by RAFF et al. (1983b). The work has focused on a relatively simple tissue, the optic nerve. As no neuronal cell bodies are present in the optic nerve, predominantly glial cell cultures can be prepared which contain a mixture of astrocytes, oligodendrocytes, and their progenitors.

Freshly dissociated optic nerves from 7-day-old rats contain small numbers of type-1 astrocytes and oligodendrocytes. Type-2 astrocytes are less than 0.1% of the dissociated cells (NOBLE and MURRAY 1984). Approximately one-third of the cells obtained are bipotential O2A progenitor cells that can be induced to differentiate into either type-2 astrocytes (A2B5$^+$, GFAP$^+$) or oligodendrocytes depending upon the culture conditions used. The O2A progenitors are A2B5$^+$, GFAP$^-$, galC$^-$ (RAFF et al. 1983b).

When grown in medium containing fetal calf serum, O2A progenitor cells differentiate into type-2 astrocytes as indicated by two lines of evidence (RAFF et al. 1983b). First, the number of A2B5$^+$, GFAP$^+$ cells increases slowly with time in culture. Yet, if A2B5 antibody and complement are used to eliminate A2B5$^+$ cells from optic nerve cultures, no type-2 astrocytes develop. Second, optic nerve cells can be labeled with A2B5 antibody before culture. After 2 days the majority of type-2 astrocytes that develop have residual A2B5 antibody on their surface, indicating that they arose from A2B5$^+$ cells in the original optic nerve cell suspension. Thus, the O2A lineage is channeled toward production of astrocytes in fetal calf serum. Few oligodendrocytes develop in these cultures.

O2A progenitors are induced to differentiate into oligodendrocytes if grown in a serum-free medium based on that of BOTTENSTEIN and SATO (1979). The medium consists of Dulbecco's modified Eagle's medium with added insulin, transferrin, albumin, progesterone, putrescine, selenium, thyroxine, triiodothyronine, and glucose. Two experiments show that oligodendrocytes develop from A2B5$^+$ progenitors (RAFF et al. 1983b). First, elimination of O2A progenitors with A2B5 antibody and complement prevents development of oligodendrocytes in the optic nerve cultures. Second, if freshly dissociated optic nerve cells are labeled with A2B5 antibody and then placed in culture, the galC$^+$ oligodendrocytes which subsequently develop retain residual A2B5 antibody on their surface. The same evidence was obtained by ABNEY et al. (1983), who showed that in fetal brain cultures oligodendrocytes also develop from an A2B5$^+$ precursor. Why serum-free medium favors production of oligodendrocytes by the O2A lineage is not clear from these experiments.

Thus, type-2 astrocytes and oligodendrocytes develop from A2B5$^+$ optic nerve cells, but the above experiments did not prove that the two lineages share a common precursor. Two additional lines of evidence indicate that a common glial progenitor can give rise to either cell type. First, in bulk cultures of optic nerve, cells of mixed GFAP$^+$, galC$^+$ phenotype can be created by initiating the cultures in fetal calf serum and then switching the cells to insulin-containing, serum-free medium (RAFF et al. 1984). If cultured 2 days in fetal calf serum and then switched to serum-free medium, approximately 50% of the process-bearing cells have a mixed GFAP$^+$, galC$^+$ phenotype. If cultured 3 days in fetal calf serum, galC$^+$ cells can no longer be induced in serum-free medium, implying that the type-2 astrocyte phenotype has been fixed. The opposite switch in phenotype from oligodendrocyte to type-2 astrocyte occurs somewhat less readily after transfer from serum-free to serum-containing medium. Secondly, single optic nerve cells have been cultured successfully in microcultures (TEMPLE and RAFF 1985). O2A progenitor cells are A2B5$^+$ process-bearing cells and can be induced to differentiate into either GFAP$^+$ or galC$^+$ cells depending upon the culture conditions. Interestingly, in bulk cultures grown in fetal calf serum approximately 90% of the A2B5$^+$ cells become GFAP$^+$, while in the microcultures only 16% become GFAP$^+$. In order to shift the majority to GFAP expression, it is necessary to increase the fetal calf serum to 20% or to use 10% fetal calf serum with conditioned medium from bulk cultures. This suggests that soluble factors released by cells in the bulk cultures influence the differentiation of O2A progenitors into type-2 astrocytes.

The relationship of type-1 astrocytes to the O2A lineage is unclear. Type-1 astrocytes are generated during development prior to the peaks of type-2 astrocyte and oligodendrocyte generation and probably represent a separate glial lineage (ABNEY et al. 1981; NOBLE and MURRAY 1984; RAFF et al. 1985, 1988). They may derive from radial glia which remain mitotically active throughout development (LEVITT et al. 1981), although the molecular markers with which to clarify the relationship between the radial glia and type-1 astrocytes are unavailable. The type-1 astrocytes play a crucial role in the development of the O2A lineage. They are felt both to time the developmental clock for glial generation (RAFF et al. 1985) and to be a source of mitogen, allowing expansion of the pool of O2A progenitors (NOBLE and MURRAY 1984; RICHARDSON et al. 1988). Both effects are believed due to secretion of PDGF by type-1 astrocytes.

I. PDGF is Mitogenic for O2A Progenitor Cells

When O2A progenitor cells from the optic nerve of embryonic day 17 rats (gestation is 21 days) are cultured in serum-free medium, little proliferation occurs and the majority differentiate into galC$^+$ oligodendrocytes within 1–2 days of culture (RAFF et al. 1983b). If the O2A cells are plated on a monolayer of type-1 astrocytes or in medium conditioned by type-1 astrocytes, however, the O2A progenitors proliferate and produce postmitotic oligodendrocytes for several weeks in culture (NOBLE and MURRAY 1984; RAFF et al. 1985). The first oligodendrocytes in the cultures appear after 4 days in vitro, which parallels the timing of their appearance in vivo soon after birth (RAFF et al. 1985).

PDGF secreted by the type-1 astrocytes is felt both to drive the clock and to stimulate O2A progenitor cell proliferation. Several lines of evidence support this contention. First, NOBLE et al. (1988) tested PDGF, EGF, TGF-α, TGF-β, IL-1, IL-2, and IL-3 for stimulation of O2A cell proliferation. Only PDGF was active in the assay. Both human and porcine PDGF were mitogenic for O2A progenitor cells and were active in the concentration range of 0.1–10 ng/ml. Second, RICHARDSON et al. (1988) demonstrated that the mitogen for O2A cells present in type-1 astrocyte-conditioned medium elutes on high-performance size exclusion chromatography with an apparent molecular weight equivalent to PDGF, and both RICHARDSON et al. (1988) and RAFF et al. (1988) show that antibody to PDGF inhibits the mitogenic activity of type-1 astrocyte-conditioned medium. Basic FGF is also a mitogen for O2A progenitor cells, but is not secreted in appreciable amounts by type-1 astrocytes (NOBLE et al. 1988). In addition to being a mitogen, PDGF also enhances motility of O2A precursors and may be a chemoattractant for O2A cells as well as for astrocytes (BRESSLER et al. 1985; NOBLE et al. 1988). By driving O2A cell division, PDGF apparently delays the differentiation of O2A progenitors into oligodendrocytes. Normally conversion of O2A progenitors to oligodendrocytes is complete within 72 h of culture (NOBLE and MURRAY 1984). PDGF delays, but does not prevent, differentiation of O2A progenitors into oligodendrocytes by 2–3 days in vitro depending upon the age of embryo from which the optic nerve cells were obtained (RAFF et al. 1988). The timing of oligodendrocyte appearance in PDGF-treated cultures is felt to match the timing of their appearance in vivo, and to be equivalent to the day of birth.

Is PDGF the O2A mitogen in vivo? PDGF consists of disulfide-linked dimers of homologous polypeptide chains designated A and B (ROSS et al. 1986). The dimer structure of PDGF is functionally important as reduction destroys biological activity. Cultured type-1 astrocytes express both A-chain and B-chain PDGF mRNAs, although the PDGF A chain predominates (RICHARDSON et al. 1988). In brain, the expression of A-chain mRNA is developmentally regulated. Little A-chain mRNA is seen prior to embryonic day 17 in the rat and then A-chain expression increases several fold over the next few days of development. This parallels the development of type-1 astrocytes and expansion of the O2A lineage in vivo (MILLER et al. 1985). Expression of B-chain mRNA is lower and shows little variation during development. PDGF initially was purified from platelets and subsequently was found in a variety of human glial cell lines (NISTER et al. 1984). Human glioma-derived PDGF consists of AA homodimers (NISTER et al. 1988), whereas human PDGF purified from platelets contains AB heterodimers, and purified porcine PDGF is a BB homodimer (ROSS et al. 1986). Both AB and BB dimers are equally potent mitogens for O2A progenitors (RICHARDSON et al. 1988; NOBLE et al. 1988).

The B-chain of PDGF was the first to be sequenced and was shown to be highly homologous to the v-*sis* oncogene of the simian sarcoma virus (SSV) (WATERFIELD et al. 1983). When injected into the brain, SSV induces formation of glioblastomas, presumably due to the autocrine effect of SSV-encoded PDGF (DEINHARDT 1980; WESTERMARK et al. 1985). Many glioblastomas of human origin also secrete PDGF and express PDGF receptors (NISTER et al. 1984); could they be derived from type-1 astrocytes or even adult O2A progenitors? PDGF apparently has little mitogenic activity for type-1 astrocytes, in contrast to its effect on O2A progenitors (BRESSLER et al. 1985; NOBLE et al. 1988). Many gliomas display a mixture of glial phenotypes; could they be derived from an undifferentiated O2A precursor (ZULCH and WECHSLER 1968)?

II. IGF-1 and CNTF Direct the O2A Lineage Toward Production of Oligodendrocytes or Type-2 Astrocytes

RAFF et al. (1983a) showed that O2A progenitor cells cultured in fetal calf serum differentiate into type-2 astrocytes, while in serum-free medium containing insulin, the lineage is channeled toward production of oligodendrocytes. MCMORRIS et al. (1986) subsequently showed that in mixed cultures of fetal brain development of oligodendrocytes is greatly stimulated by insulin and IGF-1, and that the effect of IGF-1 is antagonized by serum. In the presence of 10% fetal calf serum, IGF-1 increases the number of oligodendrocytes six fold, while in a serum-free medium supplemented with IGF-1, triiodothyronine, selenium and transferrin, oligodendrocyte numbers increase 60-fold. The increase in oligodendrocyte numbers could be due to stimulation of oligodendrocyte proliferation, proliferation of O2A progenitors, or induction of O2A progenitor cell differentiation into oligodendrocytes. MCMORRIS et al. (1986) favor the last interpretation. IGF-1 is more potent than insulin in the assay, suggesting that the stimulation of oligodendrocyte development is mediated by IGF-1 receptors. Both A2B5$^+$ cells (O2A progenitors?) and galC$^+$ cells, when isolated by fluorescence-activated cell sorting, express IGF-1 receptors, and thus are at least one target of IGF-1 action in the assay. Since insulin activates IGF-1 receptors at high concentration, its inclusion in the BOTTENSTEIN and SATO (1979) medium probably accounts for the induction of oligodendrocyte differentiation in serum-free optic nerve cultures (RAFF et al. 1983b). IGF-1 may also be a weak mitogen for immature oligodendrocytes (SANETO and DE VELLIS 1985). At relatively high concentrations (10–100 ng/ml) bFGF also supports slight increases in [H^3]thymidine incorporation and oligodendrocyte numbers (ECCLESTON and SILBERBERG 1985; SANETO and DE VELLIS 1985). Both IGF-1 and bFGF are present in brain (GOSPODAROWITZ et al. 1984) and cultured type-1 astrocytes express IGF-1 mRNA (BALLOTTI et al. 1987). Thus, type-1 astrocytes may be a source of two factors important for oligodendrocyte production by the O2A lineage: PDGF, which supports proliferation of O2A progenitors, and IGF-1, which promotes their maturation into oligodendrocytes.

Galactocerebroside (galC) expression is an early marker of oligodendrocyte maturation and precedes expression of the major protein of myelin, myelin-basic protein (MBP). BENVENISTE and MERRILL (1986) find that interleukin-2 (IL-2) enhances maturation of galC$^+$, MBP$^-$ oligodendrocytes to galC$^+$, MBP$^+$ cells. IL-2 is also weakly mitogenic. The effect of IL-2 on maturation and proliferation is in a range of concentration comparable to that required to support proliferation of T-lymphocytic cell lines.

Do IGF-1, bFGF, or IL-2 play a role in oligodendrocyte development and maturation in vivo? The stronger case can be made for IGF-1 (MCMORRIS et al. 1986). IGF-1 mediates the effects of growth hormone on tissues, and growth hormone is known to affect myelination in mice and rats. Administration to newborn rats of an antiserum to growth hormone results in hypomyelination, perhaps due to a reduction in the number of oligodendrocytes (PELTON et al. 1977). Second, a mutant mouse strain, the Snell dwarf mouse, which is deficient in growth hormone as well as several other pituitary hormones, myelinates poorly and the deficit is reversible by administration of growth hormone (NOGUCHI et al. 1982, 1985).

If culture in serum-free medium containing PDGF and insulin or IGF-1 expands the lineage and directs it toward production of immature, galC$^+$ oligodendrocytes, what directs the O2A lineage toward production of type-2 astrocytes? HUGHES and RAFF (1987) find that extracts of optic nerve from 3- to 4-week-old optic nerve stimulate generation of type-2 astrocytes by optic nerve cells grown in medium containing 0.5% fetal calf serum and insulin. Expression of GFAP by A2B5$^+$ cells is induced within 24 h after exposure to the optic nerve extract, but is transient. The GFAP-inducing activity was shown to have an apparent molecular weight of 20–25 kDa by high-performance liquid chromatography and by elution from a nonreducing, SDS-polyacrylamide gel. The GFAP-inducing factor is present at low levels in medium conditioned by type-1 astrocytes, while 50-fold more activity is released by lysing the cells (LILLIEN et al. 1988). Fetal calf serum inhibits IGF-1 stimulation of oligodendrocyte production, and also appears to stimulate production of the GFAP-inducing factor by type-1 astrocytes. Surprisingly, the GFAP-inducing factor is similar or identical to the ciliary neuronotrophic factor (CNTF), a previously characterized factor which influences the growth and survival of ciliary, sympathetic, and sensory ganglion neurons (BARBIN et al. 1984; HUGHES et al. 1988; LILLIEN et al. 1988). The dose responses for induction of GFAP expression by O2A progenitors and for stimulation of ciliary neuron survival are identical (HUGHES et al. 1988). CNTF has been purified from the embryonic chick eye (BARBIN et al. 1984) and rat sciatic nerve (MANTHORPE et al. 1986). It is an acidic protein and amino acid sequence information is not yet available.

F. Neural Growth Factors

The study of growth factors and their action on neurons has been largely the study of nerve growth factor (NGF). NGF was identified initially by LEVI-MONTALCINI and HAMBURGER in 1951. It was purified by COHEN in 1960 and its sequence was determined by ANGELETTI and BRADSHAW in 1971. NGF remains the only neuronal growth factor which has been brought to a molecular level of analysis. Why has progress on the identification of other neuronal growth factors been so slow?

With few exceptions, the growth factors that act on somatic tissues are mitogens. Many of the factors discussed above including EGF, bFGF, PDGF, IL-1, IL-2, and IGF-1 were followed by assaying their mitogenic activity through the course of the purification. Cellular proliferation in response to mitogenic factors is conveniently assayed in tissue culture by quantitating either DNA synthesis or an increase in cell number, and, often, the culture assays use cell lines chosen for their ability to respond to that mitogen. This allows standardization of the assay both within the laboratory and between laboratories. Day-to-day and week-to-week interassay variability is lessened by the use of a defined cell line and that cell line can be exchanged between laboratories.

Unlike the cells comprising somatic tissues, from which cell lines can be derived, mammalian neurons are postmitotic for the life of the organism. That has two consequences, First, culture assays must be performed with freshly dissociated primary neurons. Often only one cell type is under study, but it is obtained in a mixture of cells. Motor neurons, for example, are at most 1%–2% of the cells obtained in dissociations of embryonic spinal cord. Along with motor neurons, the mixture of cells includes CNS progenitor cells, perhaps several kinds of glia and glial progenitors, as well as other types of neurons including spinal cord interneurons, and sympathetic preganglionic neurons. Astrocytes and other nonneuronal cell types will influence neuronal growth and survival and may obviate the requirement for exogenously added factors. So not only is it difficult to study motor neurons among the background of other types of neurons, but the assay is also complicated by the nonneuronal cells. The studies of optic nerve and glial lineages by RAFF and coworkers progressed so rapidly in part because of the limited cellular heterogeneity of the tissue.

Assays of the effects of neuronal growth factors do not quantitate proliferation, but instead measure aspects of neuronal growth, and growth of the target neuron must be measured against the background of other cells and neurons present in the culture. Typically, measurement of neuronal growth relies on microscopic identification of neurons and then quantitation of their survival or of the number which have extended processes across the culture substratum. Thus, assays of neuronal growth factors have a low throughput in comparison to proliferation assays based on [^{3}H]thymidine incorporation, the assays are subjective, and they may not lend themselves to interlaboratory standardization.

The study of NGF has been so productive for two reasons. First, an apparent "accident-of-nature" has made NGF easy to obtain in large amounts in purified form, and, second, an essentially homogenous neuronal population has been identified which responds to NGF. The accident-of-nature is the presence of NGF in large amounts in the male mouse submaxillary salivary gland (COHEN 1960). LEVI-MONTALCINI and HAMBURGER (1951) initially identified the effect of NGF in an in vivo bioassay which measured the sprouting of processes by sympathetic and sensory neurons in chick embryos after implantation of a mouse sarcoma into the body wall. A culture assay subsequently was devised which quantitated the sprouting of processes from explanted fragments of sensory ganglia in response to extracts prepared from the sarcoma (LEVI-MONTALCINI et al. 1954). In chemically characterizing the active factor, COHEN and LEVI-MONTALCINI (1956) used crude snake venom phosphatase to digest nucleic acid in the sarcoma extract. To what must have been his surprise, the crude snake venom induced more sprouting from the sensory ganglia than did the sarcoma extract. Thus, the control became the experiment and led COHEN (1960) to ask if the mammalian homolog of the snake's venom gland, the salivary gland, might also contain the active factor. The male mouse salivary gland contains approximately 0.3% NGF by weight. Cohen discovered EGF in side fractions of the NGF purification (COHEN 1962).

Sympathetic neurons and a subtype of sensory neuron are exquisitely dependent upon NGF for survival and growth in culture and provide the cornerstone for routine assays of the factor. LEVI-MONTALCINI et al. (1954) assessed process extension from ganglia explants as a measure of neuronal growth. Alternatively, the survival of cultured sympathetic or sensory neurons can be assayed in dissociated cell cultures. Two major cell populations are obtained from dissociated sympathetic ganglia, sympathetic neurons and fibroblastic, ganglionic nonneuronal cells. The two cell types can be separated by size or by their differential adhesivity to tissue-culture plastic. Nonneuronal cells can also be eliminated by mitotic poisons or by selective media, which leaves an essentially pure population of neurons which are homogeneous in type.

Studies of other neuronal growth factors have been difficult because target neurons cannot be obtained so easily as a purified population, and the factors they respond to are present in minute quantities in tissue extracts. Frequently, the question tends to be is a neuron a neuron when placed in culture? Does it express appropriate ion channels, form synapses, release neurotransmitters, express neuronal cytoskeletal proteins or surface markers, etc.? Those questions can be asked of neurons cultured from CNS structures in the absence of knowledge of growth factors affecting those neurons simply by culturing the neurons at high density together with glial cells. Reaggregating cell cultures, or "minibrains," will recapitulate much of the anatomical organization and developmental milestones seen in vivo.

Happily, several new neurotrophic factors have been purified including CNTF (Barbin et al. 1984) and brain-derived growth factor (BDNF) (Barde et al. 1982), and their amino acid sequences should soon be available. A role for BDNF in vivo is suggested by recent experiments (Hofer and Barde 1988; Kalcheim et al. 1987). The impact of molecular cloning has been felt most strongly in the characterization of polypeptide growth factors, particularly those affecting hematopoiesis and immune function. In many cases, what initially appeared to be the effects of different factors have turned out to be the work of a single factor (Sporn and Roberts 1988). The examples given above reinforce the idea that growth factors are multifunctional proteins which are used again and again in different tissues to regulate different processes of growth, maturation, or repair.

Why then study neuronal growth factors other than NGF, or attempt to deal with complex mixtures of cell types in setting up a neurotrophic factor bioassay? Because neurons are postmitotic for the life of the organism, when lost to disease, they are not replaced. Thus, neurodegenerative diseases are permanently debilitating and presently have no therapy. In the adult, neuronal growth factors may regulate growth and repair, and as such have potential as therapeutics for treating neurodegenerative disease. In polio, for example, selective infection and the killing of spinal cord motor neurons leads to paralyis. Yet, if enough motor neurons remain after the acute phase of polio infection, these can sprout additional axonal processes and reinnervate muscle fibers that were denervated due to the death of infected motor neurons. Considerable recovery of function can ensue. Thus, neuronal growth factors, if they can enhance sprouting and repair processes, and if ways can be devised to deliver them to their site of action (behind the blood-brain barrier, for example), may be an important class of novel therapeutics. Particularly in the case of Alzheimer's dementia, which afflicts one to two million individuals in the aged population, new approaches to therapy are required.

The description of neural lineages and the factors directing them are the first step toward a new neurology. Potential mitogens might be identified which will allow the expansion of CNS progenitors in vitro, for subsequent transplantation in vivo as therapy for neurodegenerative disease. So little is known concerning the process of commitment to neuronal differentiation. Is differentiation the result of an intrinsic genetic program or is it modulated by environmental factors or positional cues? If programmed by the microenvironment, CNS progenitors when injected into a particular brain structure might differentiate appropriately. Could fetal human O2A progenitors be expanded in vitro with PDGF and then used to repair myelin damage in multiple sclerosis? The techniques required already have left the laboratory and entered the clinic. The use of adrenal or fetal neural transplants is actively being explored as a therapy for Parkinson's disease. Soluble factors which promote specific pathways of chemical differentiation potentially could augment expression of that phenotype by mature neurons. Could a cholinergic neuron be induced to overexpress biosynthetic enzymes required for synthesis of neurotransmitter to compensate for disease-induced deficits, i.e., loss of cholinergic neurons in Alzheimer's dementia or synaptic transmission failure in myasthenia gravis? Finally factors influencing the regression phase of neuronal development, i.e., those which promote neuronal survival, have obvious application in neurodegenerative disease or treatment of trauma. We are just seeing the end of the beginning.

References

Abney ER, Bartlett PP, Raff MC (1981) Astrocytes, ependymal cells, and oligodendrocytes develop on schedule in dissociated cell cultures of embryonic rat brain. Dev Biol 83:301–310

Abney ER, Williams BP, Raff MC (1983) Tracing the development of oligodendrocytes from precursor cells using monoclonal antibodies, fluorescence-activated cell sorting, and cell culture. Dev Biol 100:166–171

Abo T, Balch CM (1981) A differentiation antigen of human NK and K cells identified by a monoclonal antibody (HNK-1) J Immunol 127:1024–1029

Aloe L, Levi-Montalcini R (1979) Nerve growth factor-induced transformation of immature chromaffin cells in vivo into sympathetic neurons: effect of antiserum to nerve growth factor. Proc Natl Acad Sci USA 76:1246–1250

Angeletti RH, Bradshaw RA (1971) Nerve growth factor from mouse submaxillary gland: amino acid sequence. Proc Natl Acad Sci USA 68:2417–2420

Ballotti R, Nielsen FC, Pringle N, Kowalski A, Richardson WD, van Obberghen E, Gammeltoft S (1987) Insulin-like growth factor I in cultured rat astrocytes: expression of the gene and the receptor tyrosine kinase. EMBO J 6:3633–3639

Barbin G, Manthorpe M, Varon S (1984) Purification of the chick eye ciliary neuronotrophic factor. J Neurochem 43:1468–1478

Barbu M, Ziller C, Rong PM, LeDouarin NM (1986) Hetereogeneity in migrating neural crest cells revealed by a monoclonal antibody. J Neurosci 6:2215–2225

Barde Y-A, Edgar D, Thoenen H (1982) Purification of a new neurotrophic factor from mammalian brain. EMBO J 1:549–553

Benveniste EN, Merrill JE (1986) Stimulation of oligodendroglial proliferation and maturation by interleukin-2. Nature 321:610–613

Berry M, Maxwell WL, Logan A, Mathewson A, McConnell P, Ashhurst DE, Thomas GH (1983) Deposition of scar tissue in the central nervous system. Acta Neurochir [Suppl] (Wien) 32:31–53

Blum AS, Barnstable CJ (1987) O-Acetylation of a cell-surface carbohydrate creates discreate molecular patterns during neural development. Proc Natl Acad Sci USA 84:8716–8720

Bottenstein JE, Sato GH (1979) Growth of a rat neuroblastoma cell line in serum-free supplemented medium. Proc Natl Acad Sci USA 76:514–517

Bressler JP, Grotendorst GR, Levitov C, Hjelmeland LM (1985) Chemotaxis of rat brain astrocytes to platelet-derived growth factor. Brain Res 334:249–254

Brockes JP, Lemke GE, Balzer DR (1980) Purification and preliminary characterization of glial growth factor from the bovine pituitary. J Biol Chem 255:8374–8377

Bronner ME, Cohen AM (1979) Migratory patterns of cloned neural crest melanocytes injected into host chicken embryos. Proc Natl Acad Sci USA 76:1843–1847

Bronner-Fraser ME, Fraser SE (1988) Cell lineage analysis reveals multipotency of some avian neural crest cells. Nature 335:161–164

Buse E (1987) Ventricular cells from the mouse neural plate, stage Theiler 12, transform into different neuronal cell classes in vitro. Anat Embryol (Berl) 176:295–302
Chou KH, Ilyas AA, Evans JE, Quarles RH, Jungalwala FB (1985) Structure of a glycolipid reacting with monoclonal IgM in neuropathy and with HNK-1. Biochem Biophys Res Commun 128:383–388
Christie DS, Forbes ME, Maxwell GD (1987) Phenotypic properties of catecholamine-positive cells that differentiate in avian neural crest cultures. J Neurosci 7:3749–3763
Ciment G, Weston JA (1982) Early appearance in neural crest and crest-derived cells of an antigenic determinant present in avian neurons. Dev Biol 93:355–367
Cohen AM (1977) Independent expression of the adrenergic phenotype by neural crest cells in vitro. Proc Natl Acad Sci USA 74:2899–2903
Cohen AM, Konigsberg IR (1975) A clonal approach to the problem of neural crest determination. Dev Biol 46:262–280
Cohen S (1960) Purification of a nerve-growth promoting protein from the mouse salivary gland and its neuro-cytotoxic antiserum. Proc Natl Acad Sci USA 46:302–311
Cohen S (1962) Isolation of a mouse submaxillary gland protein accelerating incisor eruption and eyelid opening in the newborn animal. J Biol Chem 237:1555–1562
Cohen S, Levi-Montalcini R (1956) A nerve growth-stimulating factor isolated from snake venom. Proc Natl Acad Sci USA 9:571–574
Davies AM, Bandtlow C, Heumann R, Korsching S, Rohrer H, Thoenen H (1987) Timing and site of nerve growth factor synthesis in developing skin in relation to innervation and expression of the receptor. Nature 326:353–358
Deinhardt F (1980) Biology of primate retroviruses. In: Klein G (ed) Viral oncology. Raven, New York, pp 357–398
Doupe AJ, Landis SC, Patterson PH (1985 a) Environmental influences in the development of neural crest derivatives: glucocorticoids, growth factors, and chromaffin cell plasticity. J Neurosci 5:2119–2142
Doupe AJ, Patterson PH, Landis SC (1985 b) Small intensely fluorescent cells in culture: role of glucocorticoids and growth factors in their development and interconversions with other neural crest derivatives. J Neurosci 5:2143–2160
Eccleston PA, Silberberg DH (1985) Fibroblast growth factor is a mitogen for oligodendrocytes in vitro. Dev Brain Res 21:315–318
Eisenbarth GS, Walsh FS, Nirenberg M (1979) Monoclonal antibody to a plasma membrane antigen of neurons. Proc Natl Acad Sci USA 76:4913–4917
Eranko O, Eranko L, Hill CE, Burnstock G (1972) Hydrocortisone-induced increase in the number of small intensely fluorescent cells and their histochemically demonstrable catecholamine content in cultures of sympathetic ganglia of the newborn rat. Histochem J 4:49–58
Fauqeut M, Smith J, Ziller C, LeDouarin NM (1981) Differentiation of autonomic neuron precursors in vitro: cholinergic and adrenergic traits in cultured neural crest cells. J Neurosci 1:478–492
Ffrench-Constant C, Raff MC (1986) The oligodendrocyte-type-2 astrocyte cell lineage is specialized for myelination. Nature 323:335–338
Fukada K (1985) Purification and partial characterization of a cholinergic neuronal differentiation factor. Proc Natl Acad Sci USA 82:8795–8799
Furshpan EJ, MacLeish PR, O'Lague PH, Potter DD (1976) Chemical transmission between rat sympathetic neurons and cardiac myocytes developing in microcultures: evidence for cholinergic, adrenergic, and dual-function neurons. Proc Natl Acad Sci USA 73:4225–4229
Gensburger C, Labourdette G, Sensenbrenner M (1987) Brain basic fibroblast growth factor stimulates the proliferation of rat neuronal precursor cells in vitro. FEBS Lett 217:1–5
Giess MC, Weber MJ (1984) Acetylcholine metabolism in rat spinal cord cultures: regulation by a factor involved in the determination of the neurotransmitter phenotype of sympathetic neurons. J Neurosci 4:1442–1452

Girdlestone J, Weston JA (1985) Identification of early neuronal subpopulations in avian neural crest cultures. Dev Biol 109:274–287
Giulian D, Lachman LB (1985) Interleukin-1 stimulation of astroglial proliferation after brain injury. Science 228:497–499
Glimelius B, Weston JA (1981) Analysis of developmentally homogeneous neural crest cell populations in vitro. III. Role of culture environment in cluster formation and differentiation. Cell Diff 10:57–67
Goldman JE, Hirano M, Yu RK, Seyfried TN (1984) GD3 ganglioside is a glycolipid characteristic of immature neuroectodermal cells. J Neuroimmunol 7:179–192
Gospodarowicz D, Cheng J, Lui G-M, Baird A, Bohlen P (1984) Isolation of brain fibroblast growth factor by heparin-sepharose affinity chromatography: identity with pituitary fibroblast growth factor. Proc Natl Acad Sci USA 81:6963–6967
Greene LA, Tischler AS (1976) Establishment of a noradrenergic clonal line of rat adrenal pheochromocytoma cells which respond to nerve growth factor. Proc Natl Acad Sci USA 73:2424–2428
Guentert-Lauber B, Honegger P (1985) Responsiveness of astrocytes in serum-free aggregate cultures to epidermal growth factor: dependence on the cell cycle and the epidermal growth factor concentration. Dev Neurosci 7:286–295
Hendry A, Campbell J (1976) Morphometric analysis of rat superior cervical ganglion after axotomy and nerve growth factor treatment. J Neurocytol 5:351–360
Hinds JW, Ruffet TL (1971) Cell proliferation in the neural tube: an electron microscopic and Golgi analysis in the mouse cerebral vesicle. Z Zellforsch 115:226–264
Hofer MM, Barde Y-A (1988) Brain-derived neurotrophic factor prevents neuronal death in vivo. Nature 331:261–262
Holt CE, Bertsch TW, Ellis HM, Harris WA (1988) Cellular determination in the *Xenopus* retina is independent of lineage and birthdate. Neuron 1:15–26
Hughes SM, Raff MC (1987) An inducer protein may control the timing of fate switching in a bipotential glial progenitor cell in rat optic nerve. Devclopment 101:157–167
Hughes SM, Lillien LE, Raff MC, Rohrer H, Sendtner M (1988) Ciliary neurotrophic factor induces type-2 astrocyte differentiation in culture. Nature 335:70-73
Janzer RC, Raff MC (1987) Astrocytes induce blood-brain barrier properties in endothelial cells. Nature 325:253–257
Juurlink BHJ, Fedoroff S (1982) The development of mouse spinal cord in tissue culture. II. Development of neuronal precursor cells. In Vitro 18:179–182
Kahn CR, Sieber-Blum M (1983) Cultured quail neural crest cells attain competence for terminal differentiation into melanocytes before competence to terminal differentiation into adrenergic neurons. Dev Biol 95:232–238
Kalcheim C, Barde Y-A, Thoenen H, LeDouarin NM (1987) In vivo effect of brain-derived neutrotrophic factor on the survival of developing dorsal root ganglion cells. EMBO J 6:2871–2873
Kennedy PG, Lisak RP, Raff MC (1980) Cell type-specific markers for human glial and neuronal cells in culture. Lab Invest 43:342–351
Kniss DA, Burry RW (1988) Serum and fibroblast growth factor stimulate quiescent astrocytes to re-enter the cell cycle. Brain Res 439:281–288
Kruse J, Mailhammer R, Wernecke H, Faissner A, Sommer I, Goridis C, Schachner M (1984) Neural cell adhesion molecules and myelin-associated glycoprotein share a common carbohydrate moiety recognized by monoclonal antibodies L2 and HNK-1. Nature 311:153–155
Kruse J, Keilhauer G, Faissner A, Timpl R, Schachner M (1985) The J1 glycoprotein – a novel nervous system cell adhesion molecule of the L2/HNK-1 family. Nature 316:146–148
Landis SC, Keefe D (1983) Evidence for neurotransmitter plasticity in vivo: developmental changes in properties of cholinergic sympathetic neurons. Dev Biol 98: 349–372
LeDouarin NM (1982) The neural crest. Cambridge University Press, Cambridge

LeDouarin NM, Teillet M-A (1982) Experimental analysis of the migration and differentiation of neuroblasts of the autonomic nervous system and of neurectodermal mesenchymal derivatives, using a biological cell marking technique. Dev Biol 41:162–184
LeDouarin NM, Renaud D, Teillet M-A, LeDouarin GH (1975) Cholinergic differentiation of presumptive adrenergic neuroblasts in interspecific chimeras after heterotopic transplantations. Proc Natl Acad Sci USA 72:728–732
Lempinen M (1964) Extra-adrenal chromaffin tissue of the rat and the effect of cortical hormones on it. Acta Physiol Scand [Suppl 231] 62:1–91
Lenoir D, Honegger P (1983) Insulin-like growth factor I (IGF-I) stimulates DNA synthesis in fetal rat brain cell cultures. Dev Brain Res 7:205–231
Levi G, Gallo V, Ciotti MT (1986) Bipotential precursors of putative fibrous astrocytes and oligodendrocytes in rat cerebellar cultures express distinct surface features and "neuron-like" gamma-aminobutyric acid transport. Proc Natl Acad Sci USA 83:1504–1508
Levi-Montalcini R, Hamburger V (1951) Selective growth stimulating effects of mouse sarcoma on the sensory and sympathetic nervous system of the chick embryo. J Exp Zool 123:233–287
Levi-Montalcini R, Meyer H, Hamburger V (1954) In vitro experiments on the effects of mouse sarcoma 180 and 37 on the spinal and sympathetic ganglia of the chick embryo. Cancer Res 14:49–57
Levine JM, Beasley L, Stallcup WB (1986) Localization of a neurectoderm-associated cell surface antigen in the developing and adult rat. Dev Brain Res 27:211–222
Levitt P, Cooper ML, Rakic P (1981) Coexistence of neuronal and glial precursor cells in the cerebral ventricular zone of the fetal monkey: an ultrastructural immunoperoxidase analysis. J Neurosci 1:27–39
Lillien LE, Claude P (1985) Nerve growth factor is a mitogen for cultured chromaffin cells. Nature 317:632–634
Lillien LE, Sendtner M, Rohrer H, Hughes SM, Raff MC (1988) Type-2 astrocyte development in rat brain cultures is initiated by a CNTF-like protein produced by type-1 astrocytes. Neuron 1:485–494
Loring J, Glimelius B, Weston JA (1982) Extracellular matrix materials influence quail neural crest cell differentiation in vitro. Dev Biol 90:165–174
Manthorpe M, Skaper SD, Williams LR, Varon S (1986) Purification of adult rat sciatic nerve ciliary neuronotrophic factor. Brain Res 367:282–286
Masuko S, Shimada Y (1983) Neuronal cell-surface specific antigen(s) is expressed during the terminal mitosis of cells destined to become neuroblasts. Dev Biol 96:396–404
Maxwell GD, Forbes ME (1987) Exogenous basement-membrane-matrix stimulates adrenergic development in avian neural crest cultures. Development 101:767–776
Maxwell GD, Sietz PD, Jean S (1984) Somatostatin-like immunoreactivity is expressed in neural crest cultures. Dev Biol 101:357–366
Maxwell GD, Forbes ME, Christie DS (1988) Analysis of the development of cellular subsets present in the neural crest using cell sorting and cell culture. Neuron 1:557–568
McCarthy KD, DeVellis J (1980) Preparation of separate astroglial and oligodendroglial cell cultures from rat cerebral tissues. J Cell Biol 85:890–902
McManaman JL, Crawford FG, Stewart SS, Appel SH (1988) Purification of a skeletal muscle polypeptide which stimulates choline acetyltransferase activity in cultures spinal cord neurons. J Biol Chem 263:5890–5897
McMorris FA, Smith TM, DeSalvo S, Furlanetto RW (1986) Insulin-like growth factor I/somatomedin C: a potent inducer of oligodendrocyte development. Proc Natl Acad Sci USA 83:822–826
Mendez-Otero R, Schlosshauer B, Barnstable CJ, Constantine-Paton M (1988) A developmentally regulated antigen associated with neural cell and process migration. J Neurosci 8:564–579

Miller RH, Raff MC (1984) Fibrous and protoplasmic astrocytes are biochemically and developmentally distinct. J Neurosci 4:585–592

Miller RH, David S, Patel R, Abney ER, Raff MC (1985) A quantitative immunohistochemical study of macroglial cell development in the rat optic nerve: in vivo evidence for two distinct astrocyte lineages. Dev Biol 111:35–41

Morrison RS, de Vellis J (1981) Growth of purified astrocytes in a chemically defined medium. Proc Natl Acad Sci USA 78:7205–7209

Morrison RS, Sharma A, de Vellis J, Bradshaw RA (1986) Basic fibroblast growth factor supports the survival of cerebral cortical neurons in primary culture. Proc Natl Acad Sci USA 83:7537–7541

Morrison RS, Kornblum HI, Leslie FM, Bradshaw RA (1987) Trophic stimulation of cultured neurons from neonatal rat brain by epidermal growth factor. Science 238:72–75

Nister M, Heldin C-H, Wasteson A, Westermark B (1984) A glioma-derived analog to platelet-derived growth factor: demonstration of receptor competing activity and immunological cross-reactivity. Proc Natl Acad Sci USA 81:926–930

Nister M, Hammacher A, Mellstrom K, Siegbahn A, Ronnstrand L, Westermark B, Heldin C-H (1988) A glioma-derived PDGF A chain homodimer has different functional activities from a PDGF AB hetereodimer purified from human platelets. Cell 52:791–799

Noble M, Murray K (1984) Purified astrocytes promote the in vitro division of a bipotential glial progenitor cell. EMBO J 3:2243–2247

Noble M, Murray K, Stroobant P, Waterfield MD, Riddle P (1988) Platelet-derived growth factor promotes division and motility and inhibits premature differentiation of the oligodendrocyte-type-2 astrocyte progenitor cell. Nature 333:560–562

Noguchi T, Sugisaki T, Takamatsu K, Tsukada Y (1982b) Postnatal action of growth and thyroid hormones on the retarded cerebral myelinogenesis of snell dwarf mice. J Neurochem 39:257–263

Noguchi T, Sugisaki T, Tsukada Y (1985) Cerebral myelinogenesis in the Snell dwarf mouse: stimulatory effects of GH and T4 restricted to the first 20 days of postnatal life. Neurochem Res 10:767–778

Ogawa M, Ishikawa T, Irimajiri A (1984) Adrenal chromaffin cells form functional cholinergic synapses in culture. Nature 307:66–68

Patterson PH (1987) The molecular basis of phenotypic choices in the sympathoadrenal lineage. Ann NY Acad Sci 493:20–26

Patterson PH, Chun LLY (1974) The influence of non-neuronal cells on catecholamine and acetylcholine synthesis and accumulation in cultures of dissociated sympathetic neurons. Proc Natl Acad Sci USA 71:3607–3610

Patterson PH, Chun LLY (1977) The induction of acetylcholine synthesis in primary cultures of dissociated rat sympathetic neurons. I. Effects of conditioned medium. Dev Biol 56:263–280

Pelton EW, Grindeland RE, Young E, Bass NH (1977) Effects of immunologically induced growth hormone deficiency on myelinogenesis in developing rat cerebrum. Neurology (Minneap) 27:282–288

Pukel CS, Lloyd KO, Travassos LR, Dippold WR, Oettgen HF, Old LJ (1982) GD3 – a prominent ganglioside of human melanoma: detection and characterization by mouse monoclonal antibody. J Exp Med 155:1133–1147

Raff MC, Mirsky R, Fields KL, Lisak RP, Dorfman SH, Silberberg DH, Gregson NA, Leibowitz S, Kennedy MC (1978) Galactocerebroside is a specific cell-surface marker for oligodendrocytes in culture. Nature 274:813–816

Raff MC, Abney ER, Cohen J, Lindsay R, Nobel M (1983a) Two types of astrocytes in cultures of developing rat white matter: differences in morphology, surface gangliosides, and growth characteristics. J Neurosci 3:1289–1300

Raff MC, Miller RH, Noble M (1983b) A glial progenitor cell that develops in vitro into an astrocyte or an oligodendrocyte depending on culture medium. Nature 303:390–396

Raff MC, Williams BP, Miller RH (1984) The in vitro differentiation of a bipotential glial progenitor cell. EMBO J 3:1857–1864

Raff MC, Abney ER, Fok-Seang J (1985) Reconstitution of a developmental clock in vitro: a critical role for astrocytes in the timing of oligodendrocyte differentiation. Cell 42:61–69

Raff MC, Lillien LE, Richardson WD, Burne JF, Noble MD (1988) Platelet-derived growth factor from astrocytes drives the clock that times oligodendrocyte development in culture. Nature 333:562–565

Rakic P (1972) Mode of cell migration to the superficial layers of fetal monkey neocortex. J Comp Neurol 145:61–84

Reichardt LF, Patterson PH (1977) Neurotransmitter synthesis and uptake by isolated sympathetic neurons in microcultures. Nature 270:147–151

Richardson WD, Pringle N, Mosley MJ, Westermark B, Dubois-Dalcq M (1988) A role for platelet-derived growth factor in normal gliogenesis in the central nervous system. Cell 53:309–319

Rohrer H, Thoenen H (1987) Relationship between differentiation and terminal mitosis: chick sensory and ciliary neurons differentiate after terminal mitosis of precursor cells, whereas sympathetic neurons continue to divide after differentiation. J Neurosci 7:3739–3748

Rosner H, Al-Aqtum M, Henke-Fahle S (1985) Developmental expression of GD3 and polysialogangliosides in embryonic chicken nervous tissue reacting with monoclonal antiganglioside antibodies. Dev Brain Res 18:85–95

Ross R, Raines EW, Bowen-Pope DF (1986) The biology of platelet-derived growth factor. Cell 46:155–169

Rothman TP, Specht LA, Gershon MD, Joh TH, Teitelman G, Pickel VM, Reis DJ (1980) Catecholamine biosynthetic enzymes are expressed in replicating cells of the peripheral but not the central nervous system. Proc Natl Acad Sci USA 77:6221–6225

Rovasio R, Delouvee A, Timpl R, Yamada K, Thiery JP (1983) In vitro avian neural crest cells adhesion and migration: requirement for fibronectin in the extracellular matrix. J Cell Biol 96:462–473

Saneto RP, de Vellis J (1985) Characterization of cultured rat oligodendrocytes proliferating in a serum-free, chemically defined medium. Proc Natl Acad Sci USA 82:3509–3513

Sauer FC (1953) Mitosis in the neural tube. J Comp Neurol 62:377–405

Schmechel DE, Brightman MW, Marangos PJ (1980) Neurons switch from non-neuronal enolase to neuron-specific enolase during differentiation. Brain Res 190:195–214

Seyfried TN, Yu RK (1985) Ganglioside GD3: structure, cellular distribution, and possible function. Mol Cell Biochem 68:3–10

Shemer J, Raizada MK, Masters BA, Ota A, LeRoth D (1987) Insulin-like growth factor I receptors in neuronal and glial cells. J Biol Chem 262:7693–7699

Sidman RL (1961) Histogenesis of the mouse retina studied with thymidine-H^3. In: Smelser GK (ed) The structure of the eye. Academic, New York, pp 487–506

Sieber-Blum M, Sieber F, Yamada KM (1981) Cellular fibronectin promotes adrenergic differentiation of quail neural crest cells in vitro. Exp Cell Res 133:285–295

Skoff RP (1980) Neuroglia: a reevaluation of their origin and development. Pathol Res Pract 168:279–300

Sorge LK, Levy BT, Maness PF (1984) $pp60^{c\text{-}src}$ is developmentally regulated in the neural retina. Cell 36:249–257

Sporn MB, Roberts AB (1988) Peptide growth factors are multifunctional. Nature 332:217–219

Stemple DJ, Mahanthappa NK, Anderson DJ (1988) Basic FGF induces neuronal differentiation, cell division, and NGF dependence in chromaffin cells: a sequence of events in sympathetic development. Neuron 1:517–525

Sugaya E, Asou H, Itoh K, Ishige A, Sekiguchi K, Iizuka S, Sugimoto A, Abrurada M, Hosoya E, Takagi T, Kajiwara K, Komatsubara J, Hirano S (1987) Characteristics of primary cultured neurons from embryonic mutant El mouse cerebral cortex. Brain Res 406:270–274

Tapscott SJ, Bennet GS, Holtzer H (1981) Neuronal precursor cells in the chick neural tube express neurofilament proteins. Nature 292:836–838

Temple S, Raff MC (1985) Differentiation of a biopotential glial progenitor cell in a single cell microculture. Nature 313:17–23

Turner DL, Cepko CL (1987) A common progenitor for neurons and glia persists in rat retina late in development. Nature 238:131–136

Unsicker K, Krisch B, Otten U, Thoenen H (1978) Nerve growth factor-induced fiber outgrowth from isolated rat adrenal chromaffin cells: impairment by glucocorticoids. Proc Natl Acad Sci USA 75:3498–3502

Unsicker K, Reichert-Preibsch H, Schmidt R, Pettmann B, Labourdette G, Sensenbrenner M (1987) Astroglial and fibroblast growth factors have neurotrophic functions for cultured peripheral and central nervous system neurons. Proc Natl Acad Sci USA 84:5459–5463

Walicke PA, Baird A (1988) Neurotrophic effects of basic and acidic fibroblast growth factors are not mediated through glial cells. Brain Res 468:71–79

Walicke P, Cowan WM, Ueno N, Baird A, Guillemin R (1986) Fibroblast growth factor promotes survival of dissociated hippocampal neurons and enhances neurite extension. Proc Natl Sci USA 83:3012–3016

Walsh C, Cepko CL (1988) Clonally related cortical cells show several migration patterns. Science 241:1342–1345

Waterfield MD, Scrace GT, Whittle N, Stroobant P, Johnsson A, Wasteson A, Westermark B, Heldin C-H, Huang JS, Deuel TF (1983) Platelet-derived growth factor is structurally related to the putative transforming protein of $p28^{sis}$ of simian sarcoma virus. Nature 304:35–39

Yamamoto M, Steinbusch HWM, Jessell TM (1981) Differentiated properties of identified serotonin neurons in dissociated cultures of embryonic rat brain stem. J Cell Biol 91:142–152

Zulch KJ, Wechsler W (1968) Pathology and classification of gliomas. Prog Neurol Surg 2:1–84

CHAPTER 31

Role of Growth Factors in Cartilage and Bone Metabolism

J. PFEILSCHIFTER, L. BONEWALD, and G. R. MUNDY

Bone is subject to extensive modeling and remodeling processes throughout life. These processes start with the endochondral and membranous formation of bone tissue during embryonic life, dominate the period of postnatal bone growth, and persist as constant cycles of removal of old bone and replacement by new bone long after the bones have ceased to grow. Size, shape, and structure of bone is determined by these processes.

The cell which is essential for bone formation is the osteoblast. Active or inactive "resting" osteoblasts line the surfaces of preexisting bone or are generated during the process of bone induction. Osteoblasts produce the bone matrix, consisting largely of type I collagen and a variety of noncollagen proteins, which finally mineralize and become new bone.

There is increasing evidence that growth factors may have an essential role during all phases of bone development. Growth factors participate in the processes of chondrification and ossification during fetal life, in endochondral growth processes during postnatal growth, and in the constant remodeling of growing and adult bones throughout life. Growth factors may also provide the mechanism for mediating and modulating the effects of many systemic and local factors and forces on bone metabolism.

Bone induction during embryonic development and bone growth are intimately linked with the induction and the growth of cartilage. Modulation of chondrocyte proliferation and activity by growth factors is therefore likely to be equally important in growing cartilage. Although chondrocyte proliferation seems to be negligible during adult life, constant biochemical stimulation seems to be required to maintain the cells as differentiated chondrocytes and to continue the synthesis and deposition of the cartilage extracellular matrix. The continuous physiological turnover of the cartilage extracellular ground matrix may be largely determined by the same growth factors which are involved in the regulation of bone matrix turnover.

This article summarizes the potential role of those growth factors in bone and cartilage which are likely to have anabolic effects on growth and/or maintenance of bone and cartilage tissue. The precise role of many growth factors on these processes is still unknown. The main emphasis will be on the effects of the insulin-like growth factors (IGFs), platelet-derived growth factor (PDGF), and transforming growth factor-β (TGF-β), all of which have been shown to be present in skeletal tissue and are likely to have a role in the local regulation of bone and cartilage metabolism.

A. Origin of Growth Factors in Bone and Cartilage

When IGF was discovered it was first thought that it was produced outside bone and cartilage, and might exert its effects on bone and cartilage predominantly through systemic action (SALMON and DAUGHADAY 1957). However, it has become more and more obvious that many growth factors which have effects on bone and cartilage, including IGF, are produced in large amounts by cells in bone and cartilage and that local effects of these factors may be the predominant mode of action. Bovine bone matrix has been shown to contain a volume concentration of growth factors up to 20 times greater than serum (HAUSCHKA et al. 1986). IF G-I, IGF-II, PDGF, and TGF-β-like peptides have all been found in bone matrix in high concentrations (MOHAN et al. 1987; HAUSCHKA et al. 1986; SEYEDIN et al. 1986). Bovine bone matrix contains approximately 200 μg TGF-β/kg bone tissue, which makes bone a major storage site for this factor. A PDGF-like peptide has been detected in bone matrix in concentrations of 50 μg/kg bone tissue. IGF and TGF-β activity have been demonstrated in bone organ cultures in a variety of studies (STRACKE et al. 1984; CANALIS et al. 1988; CENTRELLA and CANALIS 1985; PFEILSCHIFTER and MUNDY 1987a,b; CENTRELLA et al. 1987a) and immunohistochemical studies have provided further evidence for the presence of the two factors in bone (ANDERSSON et al. 1986; ELLINGSWORTH et al. 1986: HEINE et al. 1987). Bone matrix not only contains the ubiquitous TGF-β 1, but also the 70% homologous TGF-β 2 (SEYEDIN et al. 1985, 1986 1987). This form seems to be more restricted in its tissue distribution than the TGF-β 1. Approximately 20% of the bone TGF-β can be ascribed to TGF-β 2 but its precise functional relationship to TGF-β 1 in bone is still unknown. Bone cultures also produce the homologous IGF-I and IGF-II (SLOOTWEG et al. 1987).

Cartilage cells and cartilage cultures have been shown to produce IGF-I (BURCH et al. 1986) and may also produce some related IGF-like peptides that have IGF-like activities but differ from the known IGFs in their molecular weight, such as cartilage-derived factor (CDF) (KATO et al. 1981a, b) and cartilage-derived growth factor (CDGF) (KLAGSBRUN et al. 1977; KLAGSBRUN and SMITH 1980; SULLIVAN and KLAGSBRUN 1985). Embryonic cartilage of cows and mice has been shown to stain intensively for TGF-β (ELLINGSWORTH et al. 1986; HEINE et al. 1987). Staining was particularly pronounced in cells of the articular cartilage which were in close association with the cartilage canals.

Apart from osteoblast-like cells and chondrocytes, other cells in bone and cartilage, including fibroblasts, blood-derived cells, and adjacent bone marrow cells are likely to produce a similar or even greater spectrum of growth factors that might act on osteoblasts and chondrocytes. Likewise, growth factor production by osteoblasts and chondrocytes might also affect growth and activity of many cells in the bone or cartilage environment. Many effects of growth factors on bone and cartilage metabolism may therefore be autocrine as well as paracrine.

While TGF-β and IGF-I and -II have been clearly demonstrated in cartilage and bone in close association with osteoblasts or chondrocytes, the source of PDGF in bone matrix is still unclear. PDGF activity is produced by some osteosarcoma cells (BETSHOLTZ et al. 1984; GRAVES et al. 1984; HELDIN et al. 1980, 1986) and may also be produced by normal osteoblastic cells (VALENTIN-OPRAN et al. 1987b), but PDGF-like activity has not yet been detected in bone cultures (CENTRELLA and CANALIS 1985). However, there is some evidence that PDGF-like peptides can remain mainly membrane associated (THIEL and HAFENRICHTER 1984; ROBBINS et al. 1985) and it may be that PDGF in bone is mainly associated with the cell membrane and is not secreted in detectable amounts into the culture medium. At least some of the PDGF-like activity in bone matrix may originate from blood-derived cells occurring in bone, such as macrophages (SHIMOKADO et al. 1985; MARTINET et al. 1985).

While there is ample evidence for the local production of growth factors in bone and cartilage, it is still unclear whether bone and cartilage might be also subject to the effects of circulating forms of these factors. Although platelets are a major storage site for PDGF (RAINES and ROSS 1982), PDGF activity in serum seems to be very low (ROSS et al. 1986). IGF-I and TGF-β occur in serum in high concentrations but they are both bound to one or several binding proteins which restrict their activity and may even restrict capillary passage (D'ERCOLE and WILKINS 1984; WILKINS and D'ERCOLE 1984; DROP et al. 1984; ROMANUS et al. 1986; CLEMMONS et al. 1986; ZAPF et al. 1979; KNAUER and SMITH 1980; ELGIN et al. 1987).

The original somatomedin hypothesis of SALMON and DAUGHADAY (1957) postulated a systemic effect of IGF-I on cartilage growth. IGFs were thought to be predominantly produced by the liver and to exert their effects via the circulation (ISAKSSON et al. 1985). This conclusion was mainly based on the fact that in many studies growth hormone had no detectable local effect on cartilage (HILL 1979; MANKIN et al. 1978; VETTER et al. 1986; HILL et al. 1983; ASH and FRANCIS 1975; ASHTON and FRANCIS 1977, 1978; MCQUILLAN et al. 1986) but was capable of inducing IGFs in vivo which subsequently stimulated cartilage growth in vitro (SALMON and DAUGHADAY 1957). This would speak against a major role for cartilage-derived IGF production by growth hormone. Meanwhile, however, a variety of studies in vitro and in vivo have demonstrated a local effect of growth hormone on cartilage growth (MADSEN et al. 1983; MEIER and SOLURSH 1972; EDEN et al. 1983; ISAKSSON et al. 1985; RUSSELL and SPENCER 1985; ISGAARD et al. 1986). SCHLECHTER et al. (1986a, b) could completely abolish the effects of locally infused growth hormone in the rat tibia by coinfusing an antiserum raised againt IGF, which suggests that growth hormone effects on cartilage are mediated through the local production of IGF. The discrepancy between these studies and the studies which failed to find any effect of growth hormone on cartilage cells in culture may be mainly due to the differences in culture conditions, since it would be difficult to induce sufficient amounts of active IGF-I under in vitro conditions to generate subsequent IGF effects.

While initial evaluation of the systemic effects of IGF-I on cartilage metabolism and bone growth were largely disappointing (UTHNE 1975; FRYK-

LUND et al. 1976; THORNGREN et al. 1977), later studies demonstrated small effects of systemic IGF-I application on bone growth (VAN BUUL-OFFERS and VAN DEN BRANDE 1979; HOLDER et al. 1981; SMEETS and VAN BUUL-OFFERS 1983; SCHOENLE et al. 1982). However, even infusions of up to 180 μg recombinant IGF-I/day resulted in only a slight growth-promoting activity, and it is not yet clear whether this effect on growth may have been due to direct effects on cartilage or indirect effects via the insulin-like actions of IGF which may have led to an increased food intake in these animals (SKOTTNER et al. 1987). The question whether IGF may act systemically is therefore not answered conclusively as yet, but it seems likely that locally produced IGF accounts for most of the effects of IGF on bone and cartilage metabolism.

While local production of growth factors may account for most of the effects of these factors on normal bone and cartilage metabolism, there may be some situations where extraskeletal sources could be important. This may be the case during fracture repair, where large amounts of both TGF-β and PDGF may be released from platelets, macrophages, and lymphocytes at the fracture site (SPORN et al. 1987). There is increasing evidence that both factors play a major role in wound healing (SPORN et al. 1983; ROSS et al. 1986; MUSTOE et al. 1987), and the effects of TGF-β and PDGF on matrix synthesis during fracture healing may be similar to their effects during wound healing.

For many growth factors release into the circulation may be more important for clearance than for delivery. Changes in plasma levels of growth factors might reflect changes in the metabolic activity of individual organs. Although all of the known growth factors in bone and cartilage are nonspecific in their target organs and effects, bone or cartilage may be the predominant source for certain growth factors at a certain stage of development, and it might be possible to evaluate metabolic activity in bone or cartilage by measuring systemic levels of locally produced factors. HALL and FILIPSSON (1975) indeed found a correlation between plasma IGF-I levels and growth velocity in a large population of normal children. However, ROSENFELD et al. (1981) found that IGF levels did not correlate with the growth rate in growth hormone-deficient children receiving replacement therapy. BENNETT et al. (1984a) failed to detect a correlation between the bone mineral density of the radius and IGF-I levels in women with postmenopausal osteoporosis, but observed a general decline of IGF levels with age which may be an indicator of decreased IGF production in old age. Whether growth factor production in bone and cartilage may have a significant effect on the plasma levels of these factors and whether plasma levels of these factors may be useful in evaluating metabolic activity in bone and cartilage cannot be conclusively answered yet.

B. Receptors for Growth Factors in Bone and Cartilage

The effects of IGF-I, PDGF, and TGF-β on bone and cartilage metabolism are mediated through specific receptors on osteoblasts and chondrocytes. IGF type I and type II receptors have been demonstrated in several studies on cartilage cells of different origin and on bone cells (POSTEL-VINAY et al. 1983; WATANABE et al. 1985; ZAPF et al. 1979; TRIPPEL et al. 1983; SESSIONS et al. 1987; SCHALCH et al. 1986; BENNETT et al. 1984b). Type I receptors bind preferentially IGF-I, and type II receptors IGF-II, but there is some crossreactivity for either of the two forms of IGF. Type I IGF receptors also share a considerable homology with the insulin receptor (CHERNAVSEK et al. 1981; KASUGA et al. 1981; CZECH et al. 1983; MASSAGUE and CZECH 1982) and insulin is known to bind to IGF-I receptors, although with approximately 100-fold lower affinity than IGF-I. The effects of insulin on cell proliferation and proteoglycan synthesis in cartilage cells which are observed at pharmacological concentrations of insulin (STUART et al. 1979; ASTON and FRANCIS 1978; ZINGG and FROESCH 1973; SALMON and DAUGHADAY 1957; SALMON et al. 1968) may indeed be due to the binding of insulin to the IGF Type I receptor and may also account for the stimulatory effect of insulin on cell proliferation and alkaline phosphatase activity in osteoblast-like cells (CANALIS 1980; HOCK et al. 1988a; SCHMID et al. 1983; ERNST and FROESCH 1987a). The effects of insulin on collagen synthesis in osteoblast-like cells, however, seem to be equimolar with the effects of IGF and mediated through high-affinity receptors for insulin on osteoblasts (CANALIS et al. 1977a, b; CANALIS 1980; SCHWARTZ et al. 1970; KREAM et al. 1985). In contrast to IGF type I receptors, type II receptors do not bind insulin even at excess concentrations (RECHLER and NISSLEY 1986).

PDGF receptors have been demonstrated on cartilage cells (BOWEN-POPE et al. 1985), and TGF-β receptors have been found in a variety of osteoblastic cells of different origin (ROBEY et al. 1987). Two of the three known classes of TGF-β receptors have a more than ten fold higher affinity for TGF-β 1, while the predominantly high molecular weight receptor type binds TGF-β 1 and 2 with a similar affinity (CHEIFETZ et al. 1987; SEYEDIN et al. 1987). No major differences between the effects of TGF-β 1 and 2 on bone cells have been reported yet (NODA and RODAN 1987; PFEILSCHIFTER et al. 1987a, b), which indicates that the effects of TGF-β on bone cells are probably mediated by the high molecular weight type TGF-β receptor.

C. Growth Factors in Bone Formation

IGF-I, TGF-β, and PDGF have all been shown to be potent stimulators of either proliferation or differentiated function of osteoblastic cells in vitro. Since the rate of bone formation is likely to be determined by the number of participating osteoblasts and the activity of the individual osteoblasts, factors which stimulate one or both parameters are potential contributors to new bone formation. IGF-I and PDGF are known stimulators of cell proliferation in many cell culture systems (ZAPF and FROESCH 1986a, b; ROSS et al. 1986) and it is not surprising therefore that they are also potent stimulators of cells of the osteoblastic lineage. IGF- and PDGF-induced increases in the proliferation of osteoblast-like cells have been observed in a variety of osteoblast-like cell cultures as well as in studies using organ cultures of bone (HOCK et al. 1988a; CANALIS 1981; CANALIS et al. 1981, 1988; KUROSE et al. 1987; ERNST and FROESCH 1987a; SCHMIDT et al. 1983; RODAN et al. 1987b; VALENTIN-OPRAN et al. 1987a, b). Histological studies in fetal rat calvariae, using autoradiography, have shown that IGF can preferentially increase DNA synthesis in the cell layer, which is assumed to contain most of the osteoprogenitor cells (HOCK et al. 1988).

TGF-β is known to have potent effects on the growth of most cells in vitro but its effects are more variable and include stimulation of cell proliferation as well as biphasic effects and inhibition of cell proliferation (SPORN et al. 1987). While TGF-β is a potent inhibitor of the proliferation of cells of epithelial origin and has not been shown yet to stimulate proliferation in epithelial cells, its effects on cell proliferation of mesenchymal cells in vitro are more complex and seem to depend on a variety of conditions, such as extracellular environment, presence or absence of other growth factors, and TGF-β concentration (SPORN et al. 1987; MOSES et al. 1980). TGF-β therefore seems to be rather a modulating than simply a stimulating factor for cell proliferation and might have a crucial role in coordinating the effects of other local growth factors such as IGF-I and PDGF. TGF-β is a stimulator of cell proliferation in bone organ cultures of neonatal rat calvariae (CENTRELLA et al. 1986; PFEILSCHIFTER and MUNDY 1987a, b) and also stimulates the proliferation of human osteoblast-like cells in vitro (D'SOUZA et al. 1987; HAKEDA et al. 1987) but has little (UMR 106) or potent inhibitory effects (ROS 17/2.8) on osteoblast-like osteosarcoma cells (PFEILSCHIFTER et al. 1987a, b; NODA and RODAN 1987; CENTRELLA et al. 1986). In cultures of fetal rat calvarial cells, TGF-β was shown to have biphasic effects on cell proliferation, being a potent stimulatory agent at high cell densities while being inhibitory at high concentrations in sparse cultures (CENTRELLA et al. 1986).

The predominant role of PDGF seems to be its stimulatory effect on osteoblastic cell proliferation. IGF-I and TGF-β, however, also have potent stimulatory effects on the formation of new bone matrix. Both stimulate the accumulation of collagen in bone cell and bone organ cultures (CANALIS 1980; CANALIS et al. 1977a, b 1988; KUROSE et al. 1987; CENTRELLA et al. 1986; PFEILSCHIFTER et al. 1987; NODA and RODAN 1987; D'SOUZA et al. 1987; ROBEY et al. 1987) and increase the rate of matrix apposition in cultures of fetal rat calvariae (HOCK et al. 1988, a, b). TGF-β has been shown to increase mRNA levels for type I procollagen in bone cells (CENTRELLA et al. 1986; NODA and RODAN 1987), and both TGF-β and IGF-I seem to increase collagen synthesis at least partially through transcriptional mechanisms. In fibroblast cultures, TGF-β has also been observed to stabilize the procollagen mRNA (RAGHOW et al. 1987) and may have a stimulatory effect on the processing of procollagen into mature forms (VARGA and JIMENEZ 1986). TGF-β and IGF seem to inhibit matrix degradation, and thereby contribute to the accumulation of bone matrix (BALLARD et al. 1980). In the case of TGF-β the inhibitory action on matrix degradation may be mainly due to a potent inhibition of plasminogen activator activity in bone which may decrease the generation of plasmin, a protease capable of degrading many matrix components and capable of activating latent collagenase (PFEILSCHIFTER, unpublished observations). Whether TGF-β may also directly inhibit collagenase production, as has been observed in some fibroblastic cells (EDWARDS et al. 1987), or whether it might even increase collagenase production, as has been found in a different study on fibroblastic cells (CHUA et al. 1985), has not been determined yet for bone cells.

Although TGF-β and IGF-I stimulate predominantly the production of type I collagen, they also increase the production of other matrix proteins. IGF stimulates the production of osteocalcin in human osteoblast-like cells in vitro. In the osteoblastic cell line ROS 17/2.8, TGF-β has been shown to have differential effects on the production of attachment proteins which are part of the extracellular bone matrix. While TGF-β stimulates osteopontin mRNA which encodes for a 44-kDa phosphoprotein which is abundant in bone matrix, it decreases mRNA levels for fibronectin (NODA and RODAN 1987). It is intriguing to speculate that despite their similar effects on collagen accumulation in bone cultures, TGF-β and IGF-I might have differential effects on the composition of other matrix proteins which could be of great importance for the remodeling of the matrix at a later stage. Such differential effects could be important for determining future sites of bone resorption. There is some evidence that osteoclasts are not capable of degrading unmineralized bone matrix (CHAMBERS et al. 1985). Since all bone surfaces are covered by a thin layer of unmineralized matrix, the removal of this layer may therefore be crucial for the initiation of osteoclastic bone resorption. Which part of the bone surface will be selected for resorption may in part depend on the composition of the matrix layer. Conversion of plasminogen into plasmin which seems to be the major physiological activator of latent collagenase is greatly facilitated by binding of plasminogen to certain matrix proteins (KNUDSEN et al. 1986). One of these matrix proteins seems to be thrombospondin, whose production has also been shown to be increased by TGF-β (PENTTINEN et al. 1988). Growth factors such as TGF-β may therefore have long-lasting effects on matrix turnover in bone.

Apart from stimulating the proliferation of cells in the osteoblastic lineage, TGF-β, IGF, or PDGF might also affect the differentiation of committed osteoblastic precursor cells into cells of a mature osteoblastic phenotype. At the present time this is difficult to evaluate since there are no specific markers for the different stages of osteoblastic development. However, alkaline phosphatase is known to be a marker for the mature osteoblastic phenotype and increases or decreases in the activity of this enzyme are thought to reflect the maturation state of osteoblastic cells (RODAN and RODAN 1983). Alkaline phosphatase activity has been shown to increase in cell cultures of osteoblastic cells upon treatment with IGF-I, which may support a role for IGF in the maturation of the osteoblast (KUROSE et al. 1987; VALENTIN-OPRAN et al. 1987b; SCHMID et al. 1983). The effects of TGF-β on alkaline phosphatase activity are inversely related to the effects of TGF-β on cell proliferation. Under those conditions where TGF-β stimulates cell proliferation alkaline phosphatase activity is decreased, while it is increased where TGF-β inhibits cell proliferation (CENTRELLA et al. 1986; D'SOUZA et al. 1987; PFEILSCHIFTER et al. 1987a, b; NODA and RODAN 1987; CHEN et al. 1987; SONE et al. 1987). Similar observations have been made with many osteotropic hormones which also have inverse effects on proliferation and alkaline phosphatase activity of osteoblast-like cells. This may indicate that the same cell may not be able to proliferate and express a differentiated phenotype at the same time. However, the increase in alkaline phosphatase activity in ROS 17/2.8 cells upon treatment with TGF-β seems to be independent from its inhibitory effects on cell growth (PFEILSCHIFTER et al. 1987; NODA and RODAN 1987), which suggests that proliferation and expression of a mature phenotype may be regulated independently rather than be interdependent. Whether proliferation might decrease or increase markers of cell maturation may also largely depend on the subpopulations within the cell or organ culture which are induced to proliferate. A predominant effect on early osteoblast precursor cells will increase the numbers of cells in the culture which express a less mature phenotype while the proliferation of more mature precursor cells will increase the percentage of cells which express a more mature phenotype and thus increase overall tissue expression of the more mature osteoblastic phenotype.

Apart from their direct effects on bone cell proliferation and matrix production, various additional effects have been reported for IGF-I, PDGF, and TGF-β on osteoblastic cells. IGF has been shown to stimulate glycogen synthesis in osteoblast-like cells in vitro (SCHMID et al. 1982, 1983). Both PDGF and TGF-β have been shown to induce the production of prostaglandin E_2 (PGE_2) in bone cultures (HARROD et al. 1986; CENTRELLA et al. 1986; TASHJIAN et al. 1982, 1985; SHUPNIK et al. 1982), which seems to be in part due to increased production by the osteoblast, although other mesenchymal cells are likely to contribute to the increase in PGE_2. Prostaglandins are local agents which are known to have potent effects on both bone formation and bone resorption. While they inhibit collagen synthesis at higher concentrations they have been shown to increase collagen production in bone cultures at lower concentrations (RAISZ and KOOLEMANS-BEYNEN 1974; CHYUN and RAISZ 1984), and they are also potent stimulators of osteoblastic cell proliferation (FEYEN et al. 1985). On the other hand, PGE_2 in particular is a potent bone resorbing agent in vitro (DIETRICH et al. 1975). An increase in bone resorption has been indeed observed with PDGF and TGF-β in cultures of neonatal mouse calvariae and seems to be due to an induction of prostaglandin synthesis (TASHJIAN et al. 1982, 1985). It is difficult to draw any definitive conclusions from the observed increase in bone resorption in mouse calvariae since the same effects have not been observed in cultures of fetal rat long bones, which is the major alternative system used to study bone resorption in vitro. Whether mouse calvariae have an increased sensitivity to the induction of prostaglandins or whether long bones have a decreased sensitivity is unclear. The fact that TGF-β and PDGF are capable of inducing the release of prostaglandins may, however, provide an additional mechanism for a more precise regulation of the effects of these factors on matrix synthesis and cell proliferation. There are several lines of evidence which suggest that TGF-β might be a potent long-term inhibitor of bone resorption by having direct and/or indirect effects on the generation of new osteoclasts. TGF-β has a delayed inhibitory effect on bone resorption in fetal rat long bone cultures which resembles the effects of hydroxyurea – a potent inhibitor of DNA synthesis capable of inhibiting the proliferation of osteoclast precursor cells (PFEILSCHIFTER et al. 1988). TGF-β may therefore inhibit the proliferation of osteoclast precursor cells which would result in a decreased number of osteoclasts and a decrease in bone resorption. Potent inhibitory effects of TGF-β on the formation of osteoclast-like cells have also been observed in long-term cultures of human bone marrow cells, where TGF-β was shown completely to inhibit the stimulating effects of many bone-resorbing agents on the generation of osteoclast-like cells (CHENU et al. 1987). There may be several ways by which TGF-β might decrease osteoclast formation. TGF-β may directly inhibit DNA synthesis in osteoclast precursor cells. Inhibitory effects of TGF-β on DNA synthesis have in fact been observed in a variety of marrow progenitor cell populations (ISHIBASHI et al. 1987). TGF-β may, however, additionally effect the size of the osteoclast precursor pool by preferentially switching the direction of precursor development from the monocyte lineage to the granulocyte lineage (CHENU et al. 1987). In addition, TGF-β may inhibit bone resorption through the above-mentioned inhibition of plasminogen activator activity. All bone-resorbing agents examined stimulate plasminogen activator activity in vitro and tissue plasminogen activator has been shown to stimulate bone resorption in vitro (HAMILTON et al. 1985; EVANS et al. 1987; HOEKMAN et al. 1987).

Most of the studies on the effect of TGF-β, IGF, and PDGF on bone metabolism are still preliminary and are restricted by the limitations of in vitro cultures and the lack of a model which would sufficiently mimic in vivo conditions. However, the in vitro studies strongly suggest that all three factors may have a major impact on the rate of local bone formation. While PDGF may mainly increase the overall rate of bone cell metabolism, IGF may be the major anabolic growth factor in bone, specifically increasing bone matrix apposition by having stimulatory effects on osteoblast cell proliferation, differentiation, and matrix synthesis. TGF-β may have similar effects on bone formation but its role in bone metabolism seems to be more complex. The many known biphasic effects of TGF-β and its capacity to have differential effects on the same cell population depending on environmental conditions, and to have strikingly different effects on different cell populations, make it an ideal coordinator of cellular processes at the tissue level. In bone, it may be involved in the coordination of the processes which are necessary for bone remodeling, and may provide a link between the phase of bone resorption and the subsequent phase of new bone formation. This is supported by the fact that TGF-β activity is increased in bone cultures during increases in bone resorption (PFEILSCHIFTER and MUNDY 1987a, b) and may be thus available in increased amounts at a time when the coordination of local growth factor activities is most needed for keeping a balance between degradation and subsequent formation of new bone. Apart from IGF, TGF-β, and PDGF, other local growth factors may contribute to bone formation. Bone matrix has been found to contain acidic and basic fibroblast growth factors (FGFs) (HAUSCHKA et al. 1987) which have stimulatory effects on the proliferation of osteoblast-like cells (CANALIS et al. 1987b; MCCARTHY et al. 1987; RODAN et al. 1987a) but inhibit collagen synthesis. Other growth stimulatory peptides that have been isolated from conditioned medium of bone cultures or from bone matrix include bone-derived growth factor (BDGF), which may be identical to β_2-microglobulin (CANALIS et al. 1987b), and skeletal growth factor (SGF), which appears to be identical to IGF-II (FARLEY and BAULINK 1982; HOWARD et al. 1981; MOHAN et al. 1988). Both β_2-microglobulin and SGF stimulate DNA and collagen synthesis in bone cells and organ cultures. Effects of β_2-microglobulin, however, require microgram per milliliter concentrations and this peptide therefore should not be regarded as a growth factor in the classical sense.

Many monokines, including interleukin-1 (IL-1), tumor necrosis factor-α and -β (TNF-α and -β), and interferon -γ (IFN-γ) have been shown to affect bone cell proliferation and collagen synthesis (GOWEN et al. 1985; CANALIS 1986; D.D. SMITH et al. 1987; BERTOLINI et al. 1986). IL-1 stimulates osteoblast-like cell proliferation and has been reported to have biphasic effects on collagen synthesis in bone organ cultures. Since osteoblasts are capable of producing IL-1-like material under certain conditions (JONCOURT et al. 1988), IL-1 may also have a role in the local regulation of bone turnover. While these factors may have a role in bone remodeling during inflammation or other pathological events, their role in physiological bone turnover remains conjectural. Epidermal growth factor (EGF) is known to stimulate bone resorption (TASHJIAN and LEVINE 1978; RAISZ et al. 1980) and has inhibitory effects on collagen synthesis in rodent calvarial cultures. It stimulates DNA-synthesis in fetal rat calvariae cultures and osteoblast-like cells (CANALIS and RAISZ 1979; NG et al. 1983). Like many other of the above-mentioned factors, it might have direct inhibitory effects on mature osteoblastic cells while increasing the pool of osteoblast progenitor cells. Such an increase in osteoblast progenitor cells could subsequently lead to an increase in bone formation. Whether growth factors which stimulate proliferation but inhibit function of mature osteoblasts may stimulate or inhibit bone formation may largely depend on the developmental stage of the cells exposed to these factors and the time for which they are exposed. At present, very little is known about the local regulation of growth factor ability in bone cells at the cellular level, and this will be one of the major challenges for future studies on the mechanisms of bone formation.

D. Growth Factors in Cartilage

While the evaluation of the role of growth factors on bone metabolism is still at an early stage, it is without doubt that IGF-I is the dominant growth-stimulating agent for cartilage during cartilage growth. IGF-I preparations are stimulators of chondrocyte proliferation and matrix synthesis in vitro and in vivo, and the lack of IGF production leads to a severe growth disorder, known as Laron dwarfism (LARON 1974; ZAPF et al. 1980). Stimulation of cartilage proteoglycan synthesis by IGF-I in vitro led to the discovery of the growth factor in its different variants and was for a long time the predominant bioassay for the evaluation of IGF activity (SALMON and DAUGHADAY 1957).

IGF-I has potent stimulatory effects on the proliferation of chondrocytes from various sources and species (DAUGHADAY and REEDER 1966; FROESCH et al. 1976; BOMBOY and SALMON 1980; JENNINGS et al. 1980; ASHTON and PHIZACKERLEY 1981; KATO et al. 1983; HILL et al. 1983; ASHTON and POCOCK 1983; HILL and MILNER 1984; ASHTON and OTREMSKI 1986; HIRAKI et al. 1986; SESSIONS et al. 1987) and was found to mediate the effects of growth hormone on the longitudinal growth of bones at the growth plate of these bones (SCHLECHTER et al. 1986a). It also stimulates proteoglycan synthesis in cartilage and chondrocyte cultures in all known types of cartilage and in every species examined (SALMON and DAUGHADAY 1957; PHILLIPPS et al. 1983; KEMP and HINTZ 1980; CORVOL et al. 1978; HALL 1970, 1972; FROESCH et al. 1976; ZINGG and FROESCH 1973; GUENTHER et al. 1982; VITTUR et al. 1983; SALMON and DUVALL 1970; SALMON et al. 1968; HIRAKI et al. 1985; STEVENS and HUSCALL 1981; MCQUILLAN et al. 1986; JENNINGS et al. 1980). Most of the studies on proteoglycan formation were performed using the incorporation of [^{35}S]-sulfate, which reflects mainly the synthesis of chondroitin sulfate proteoglycans. The biosynthesis of cartilage proteoglycans occurs through a process in which the formation of a core protein is followed by stepwise addition of monosaccharide units to growing chains which is catalyzed by six distinct glycosyltransferases. The final step is the addition of sulfate to the polysaccharide through the action of appropriate sulfotransferases. The observed increases in [^{35}S]sulfate uptake in cartilage cultures with IGF-I seem to reflect mainly an increase in the overall production of the core protein and therefore the number of glycosaminoglycan chains rather than an increase in the length or in the degree of sulfation of these chains. However, there may be also small increases in the size of the secreted proteoglycans which may increase under conditions where protein synthesis is not rate-limiting (KEMP and HINTZ 1980; STEVENS and HUSCALL 1981; VITTU et al. 1983). The effect of IGF on proteoglycan synthesis seems to be restricted to some classes of proteoglycans (HIRAKI et al. 1985) but little is known yet about the function of the different types of proteoglycans on cartilage structure and metabolism.

The stimulation of core protein synthesis is only part of the general anabolic action of IGF-I on cartilage metabolism. Various IGF-I preparations have been reported to stimulate protein synthesis in general (KEMP et al. 1984; SALMON and DUVALL 1970), amino acid transport (ADAMSON and ANAST 1966), RNA synthesis (SALMON and DUVALL 1970; EISENBARTH et al. 1973; SESSIONS et al. 1987), ornithine decarboxylase activity, which may be important for the synthesis of polyamines (BURCH and LEBOVITZ 1981), and collagen synthesis (DAUGHADAY and MARIZ 1962; GUENTHER et al. 1982; KATO et al. 1978; WILLIS and LIBERTI 1985).

The intracellular mechanism of the IGF-I action on cartilage still has to be elucidated. cAMP analogs and phosphodiesterase inhibitors have been reported to mimic the actions of IGF on cartilage (DREZNER et al. 1975; BOMBOY and SALMON 1980). However, cartilage membrane adenylate cyclase appears to be inhibited by IGF (TELL et al. 1973; STUART et al. 1982).

Most of the studies of the effects of IGF-I on cartlage metabolism have been performed with crude preparations of IGF from different sources and at a time where the relationship between the different preparations of IGF-like factors was still unknown. Meanwhile, since there are two major forms of IGF, termed IGF-I and IGF-II, the question is raised whether the two peptides may have different functions on cartilage metabolism. Indeed, there is increasing evidence that IGF-I is the predominant form of IGF during postnatal growth while IGF-II seems to be predominant during fetal life. In the rat the levels of basic somatomedin, the hormone thought to be closely related to human IGF-I, are low in the fetus and during the first weeks of extrauterine life (DAUGHADEY et al. 1982), whereas circulating levels of multiplication-stimulating activity (MSA), which is analogous to human IGF-II, are elevated in the fetus (MOSES et al. 1980). Similar observations have also been made for other species (GLUCKMAN and BUTLER 1983; HEY et al. 1988). Studies by VETTER et al. (1985, 1986) indicate that fetal rat chondrocytes may be more sensitive to IGF-II while the growth of adult chondrocytes may be more effectively stimulated by IGF-I. Since IGF-I is the form of IGF which is more closely regulated by growth hormone, this might also explain why fetal and early postnatal growth seems to be relatively unaffected by growth hormone (ISAKSSON et al. 1985).

Apart from the different sensitivity of cartilage cells to the different forms of IGF during fetal and postnatal growth and the fact that these forms may predominantly occur during a distinct period in life, both factors induce essentially the same qualitative responses in cartilage cells. However, the occurrence of different forms of growth factors during fetal and postnatal life may also be relevant with other growth factors, and data derived from studies on fetal cells or fetal organisms should always be treated with caution when making extrapolations to the adult organism.

In contrast to the vast number of studies on the effects of IGF on cartilage metabolism, there have been only few studies on the effects of PDGF and TGF-β on chondrocytes. A stimulation of chondrocyte proliferation has been observed in several studies with partially purified PDGF preparations, and may depend on the presence of other growth factors (HILL and MILNER 1984; PRINS et al. 1982a, b). PDGF has also been reported to stimulate sulfate incorporation into cartilage proteoglycans (HILL and MILNER 1984; PRINS et al. 1982a, b).

There are few studies on the effects of TGF-β on differentiated chrondrocytic cells. In chick epiphyseal chondrocytes. TGF-β was shown to stimulate proteoglycan synthesis while having no effect on collagen or DNA synthesis (O'KEEFE et al. 1987). In a different study, TGF-β, however, was found to inhibit proteoglycan and collagen production in chondrocytes (SKANTZE et al. 1985).

E. Growth Factors in Bone and Cartilage Induction

While the role of TGF-β in differentiated cartilage is still controversial, TGF-β is a potent promoter of the differentiation of mesenchymal cells into chondrocytes (SEYEDIN et al. 1985; POSER et al. 1986). In fact, two growth factors isolated from fetal bone and termed cartilage-inducing factors A and B (because of their potent effects on the induction of cartilage proteoglycans and type II collagen in cultures of embryonic rat muscle cells) have been subsequently found to be identical to TGF-β 1 and 2 (SEYEDIN et al. 1985, 1986, 1987). TGF-β may therefore have an important role in the induction of cartilage during fetal development. Immunohistochemical studies of TGF-β distribution in mouse embryos suggest that TGF-β may be involved in all stages of bone induction (HEINE et al. 1987). In the limbs of the embryo bone formation is preceded by a proliferation and condensation of mesenchymal cells while peripheral cells form a perichondrium and the future periosteum. Endochondral ossification starts by invasion of capillaries into the perichondrium and is accompanied by the appearance of cells that differentiate into osteoblasts. In these areas of active capillary growth and osteoblast generation intensive staining for TGF-β has been observed in mouse embryos. Strong TGF-β staining was also observed in areas of intramembraneous ossification of flat bones (HEINE et al. 1987).

Since the induction of bone involves a cascade of chemotactic, proliferative, and differentiative events, it is likely that a number of growth factors in addition to TGF-β have a role in bone induction. For many years, it has been known that demineralized bone matrix when implanted in soft tissue is capable of inducing new bone formation. One of the mediators extracted from demineralized bone matrix has been termed bone morphogenetic protein (BMP) (URIST et al. 1984). However, its relationship to TGF-β and whether it is a single factor or a mixture of different factors is unclear.

F. Regulation of Growth Factor Activity in Bone and Cartilage

Since TGF-β, IGF, and PDGF are probably all produced by bone and cartilage cells and bone and cartilage cells have high-affinity receptors for these factors, TGF-β, IGF, and PDGF might have autocrine regulatory effects on bone and cartilage cells. Indeed, there is some evidence that the inhibition of the autocrine growth factor activity in vitro by specific antibodies may affect the growth of cartilage and bone cells. BURCH et al. (1986) found that antibodies to IGF-I could prevent increases in cartilage weight of cultured embryonic chick cartilage, and ERNST et al. (1988) found a reduced proliferation rate in osteoblast-like cells after specific antibodies to TGF-β1 had been added.

Mechanisms for controlling autocrine effects of growth factors in bone and cartilage might include changes in the synthesis of the factor, changes in growth factor activation, and changes in receptor binding. These mechanisms may be used by local and systemic agents to modulate activity of bone and cartilage cells.

Changes in growth factor production seem to be of major importance in the regulation of the effects of IGF on cartilage growth. As already mentioned above, it is well known that growth hormone (GH) can increase IGF production in many tissues, including bone and cartilage (STRACKE et al. 1984; ERNST and FROESCH 1987a, b; SALMON and DAUGHADAY 1957; SCHLECHTER et al. 1986a). In fact, most if not all effects of GH on cartilage and bone may be mediated by the induction of IGF synthesis. GH-induced IGF production is essential for postnatal growth, and the importance of GH for normal longitudinal bone growth has been known for years (ASLING et al. 1948; HARRIS et al. 1972; HEANEY et al. 1972; THORNGREN et al. 1973). Preliminary studies indicate that many systemic and local osteotropic factors, such as PTH, insulin, and 1,25-dihydroxyvitamin D_3, might affect IGF synthesis in osteoblast-like cell cultures (VALENTIN-OPRAN et al. 1986; PFEILSCHIFTER, unpublished observations).

At the receptor level, insulin was found markedly to increase IGF binding to the type II receptor in rat chondrocytes (SESSIONS et al. 1987), which might be a mechanism for insulin to potentiate the action of IGF-II. Changes in IGF receptors also seem to occur in osteoblast-like cells in response to dexamethasone (BENNETT et al. 1984b).

A major control mechanism in the regulation of autocrine effects of TGF-β, IGF, and perhaps also PDGF in bone and cartilage metabolism, however, could be at the level of growth factor activation. Both TGF-β and IGF are known to exist largely as high molecular weight complexes bound to one or more binding proteins (PIRCHER et al. 1986; O'CONNOR-MCCOURT and WAKEFIELD 1987; NAKAMURA et al. 1986; D'ERCOLE and WILKINS 1984; DROP et al. 1984; ROMANUS et al. 1986; CLEMMONS et al. 1986; ZAPF et al. 1979; KNAUER and SMITH 1980), and binding proteins have also been described for PDGF (HUANG et al. 1984; RAINES et al. 1984). While there is some evidence that some IGF-binding proteins might restrict IGF effects (ZAPF et al. 1979; DROP et al. 1979; KNAUER and SMITH 1980), other binding proteins preparations have been shown to enhance the biological response to IGF (ELGIN et al. 1987). For some of these binding proteins, it is known that GH can stimulate their production (MARTIN and BAXTER 1986; SCOTT et al. 1985), which would provide an additional mechanism for GH to influence IGF activity in cartilage and bone. IGF-binding proteins are produced together with IGF in bone culture and might have complex effects on the regulation of IGF activity in bone (WONG et al. 1987).

As in many other tissues, TGF-β is released from bone cultures and bone cells in an inactive, latent form bound to a large molecular weight binding protein, which completey inhibits access of TGF-β to the TGF-β membrane receptors (PFEILSCHIFTER et al. 1987). This latency of TGF-β also explains why large amounts of TGF-β can accumulate in these cultures without having the effects observed with only a few percent of active TGF-β on the same cultures. With decreasing pH, TGF-β is exponentially activated (PFEILSCHIFTER et al. 1987a, b), which may be of some importance during the bone resorption process where the local pH in the vicinity of the osteoclasts is thought to be very low (BARON et al. 1985). Furthermore, osteoclasts may be able directly to activate TGF-β (OREFFO et al. 1988), probably through limited proteolytic action. However, it is likely that there are several other mechanisms for activation of TGF-β in bone and also in cartilage. Plasmin which may be generated in bone from plasminogen through conversion by tissue plasminogen activators may be capable of activating TGF-β (KESKI-OJA et al. 1987). Since tissue plasminogen activators are produced by osteoblasts and their activity is regulated by many osteotropic factors (HAMILTON et al. 1985), this might be an additional mechanism for local and systemic factors to influence TGF-β activation.

Apart from changes in the production rate, the activation, and the binding of growth factors, interactions between growth factors and the extracellular matrix may provide an important regulatory mechanism in the local control of bone formation. After being released from the cells, growth factors may bind to the extracellular matrix and be stored, thus becoming available to be released or activated at a later time. Calcified bone matrix is an ideal reservoir for the storage of vast amounts of growth factors (HAUSCHKA et al. 1986). Although there has been no direct proof yet, it is intriguing to speculate that the growth factors stored in the bone matrix may be released during the bone resorption process and modulate subsequent local events.

The fact that TGF-β activity is increased in the conditioned medium of bone cultures which have been stimulated to resorb and decreased in cultures where resorption was inhibited with calcitonin may be indeed due to the release of TGF-β from bone matrix during resorption (PFEILSCHIFTER and MUNDY 1987a, b). Bone matrix may in fact contain a large variety of growth factors, binding protein, activators, and inhibitors of activation which could provide a network of growth stimulatory and growth inhibitory elements which may be regulated by their attachment to different matrix proteins. Since cartilage matrix has been shown to contain growth factor activity, growth factors within the cartilage matrix may also have important functions for the maintenance of cartilage integrity.

The regulation of the effects of TGF-β, IGF, and PDGF on bone and cartilage may also include interactions among these factors themselves. TGF-β has been shown in several cell lines to stimulate the c-*sis* protooncogene, which encodes for the B-chain of PDGF, and to stimulate the production of PDGF-like material (LEOF et al. 1986). Whether this also might occur in bone or cartilage is not yet known. TGF-β might, however, also stimulate the production of IGF. In the osteoblast-like UMR-106 cells, TGF-β strongly induces the production of IGF-like material which crossreacts with a specific monoclonal antibody to IGF-I (PFEILSCHIFTER et al., unpublished results).

How difficult it is to determine if the effects of exogenous factors on bone and cartilage cells are direct or mediated by other local growth factors is exemplified by GH. It is still unclear whether GH has direct effects on bone and cartilage cells or whether all of its effects are mediated by the production of local IGF. The availability of specific antibodies against local growth factors and their receptors will be crucial to solve this problem in the future.

Although TGF-β, PDGF, and IGF may mediate some of the effects of local and systemic influences on bone and cartilage formation, it is likely that many effects of local and systemic agents on the proliferation and differentiated functions of osteoblasts and chondrocytes are direct effects. Local growth factors in bone and cartilage may provide a baseline level of activity in bone and cartilage cells, which is then directly or indirectly modulated by other agents. Modulation of local growth factor effects by systemic or other local factors may generate a complex pattern of antagonistic and synergistic effects, and combined effects of several local and systemic factors might lead to a potentiation as well as an inversion of the effects observed with the individual factors.

IGF has been shown to have synergistic effects with PDGF, EGF, and FGF on chondrocyte proliferation in vitro (HILL and MILNER 1984; KATO et al. 1983). IGF and triiodothyronine synergize in their effects on protein synthesis in chick chondrocytes (KEMP et al. 1984). TNF-α, basic FGF, and EGF have synergistic effects with TGF-β on the stimulation of cell proliferation in osteoblast-like cells from fetal rat calvariae at low stimulatory concentrations of TGF-β, while maximally stimulatory concentrations of TGF-β have the opposite effect and can decrease cell proliferation by these factors (CENTRELLA et al. 1987b).

G. Growth Factors and Cartilage Destruction

Although the pathophysiology of most cartilage diseases is unknown, there is strong evidence that locally produced factors, such as IL-1 and TNF are important inflammatory mediators and are likely to cause the sequence of events which leads to the persistence of synovitis and the destruction of cartilage in rheumatoid arthritis and other inflammatory diseases of joints (DUFF 1988; DECKER et al. 1984).

IL-1α and -β genes are activated in the synovial cells of patients with rheumatoid arthritis, and significant levels of IL-1 are detectable in joint effusions (DUFF 1988). Human synovial cells produce IL-1 after exposure to agents which may provoke certain types of arthritis (e.g., urate crystals) (DUFF 1988). IL-1 may primarily be produced by activated mononuclear cells within the synovial tissue, and by binding to receptors on synovial fibroblasts (BIRD and SAKLATVALA 1986) may be in part responsible for the drastic alteration in the composition of the synovial membrane which lines diarthrodial joints. Synovial fibroblasts proliferate in vitro in response to IL-1, and IL-1 may also contribute to the recruitment and activation of inflammatory cells. As a result, a pannus forms which is composed of proliferating fibroblasts, small blood vessels, collagen, and inflammatory cells.

IL-1 may also cause the subsequent invasion of the pannus into cartilage, subchondral bone, and tendons, which characterizes the final and most destructive stage of rheumatoid arthritis. A major part of this invasion is believed to result from the production of arachidonic acid metabolites and proteases by the pannus cells. IL-1 is known to induce PGE_2 release from synovial cells, which is probably due to IL-1-mediated stimulation of phospholipase A_2 activation (GILMAN et al. 1988). IL-1 also stimulates plasminogen activator production and collagenase production in synovial fibroblast-like cells, proteases which are directly or indirectly capable of degrading cartilage matrix proteins. In the case of plasminogen activator stimulation, the IL-1 activity can be inhibited by indomethacin (LEIZER et al. 1987).

As a potent bone resorbing agent, IL-1 is also likely to have a major role in the subchondral bone resorption and the characteristic erosions of rheumatoid arthritis.

Although most of the effects of IL-1 on synovial cells and cartilage destruction have been examined in vitro, similar effects have been observed after injection of IL-1 into rabbit knee joints. IL-1 injections induced a marked loss of articular cartilage glycosaminoglycans (GAGs), a great increase in synovial fluid GAG, and an acute synovitis together with lymphocytic foci and plasma cell infiltration (DINGLE et al. 1987).

Apart from IL-1, TNF may have similar effects on cartilage destruction in rheumatoid arthritis. Like IL-1, TNF stimulates the resorption and inhibits the synthesis of cartilage proteoglycans (SAKLATVALA 1986). TNF also stimulates synovial-cell production of PGE_2 and collagenase (DUFF 1988). Part of the effects of TNF on cartilage destruction may in fact be mediated by IL-1, since TNF is known to induce the release of IL-1 by monocytes (BEUTLER and CERAMI 1987). Besides TNF, IFN-γ and glucocorticoids are known modulators of cartilage metabolism in vitro.

While there is therefore ample evidence for a major role of IL-1 and other cytokines in pathological turnover of cartlage, it is less clear whether IL-1 may also be involved in the regulation of normal cartilage matrix turnover. However, it is known that chondrocytes express IL-1α as well as -β mRNA and IL-1 like activity has been detected in supernatants of bovine articular cartilage (DUFF 1988).

H. Growth Factors and Disorders of Bone and Cartilage

Our knowledge of the precise function of growth factors on bone growth and bone formation during remodeling is rapidly increasing due to the increased availability of purified or recombinant growth factors and continuous improvements of in vitro techniques to study their effects on bone and cartilage under defined conditions. Although it seems likely that growth factors such as TGF-β, IGF, and PDGF are essential for the formation and regulation of bone and cartilage throughout life, little is known yet about their potential role in skeletal diseases. Future studies will have to show whether the production, activation, or receptor binding of these factors may differ in diseases which are characterized by decreased bone formation such as osteoporosis.

I. Potential for Growth Factors as Therapeutic Agents in Diseases of Bone Loss

Some of the common diseases of bone loss are characterized by decreased bone formation. The two most striking examples are the bone loss associated with aging and the osteolytic lesions of the skeleton caused by tumor cells. Bone biologists still understand little of the defects in osteoblast biology which are responsible for this type of disordered bone formation. However, in these particular disorders the potential is present for the use of factors which stimulate proliferation of osteoblast precursors or differentiated function in committed osteoblast precursors as therapeutic agents. Such agents may have wide therapeutic usefulness and there is great interest in their identification. It seems likely that increased understanding of the endogenous factors which stimulate osteoblast activity may lead to new therapies in the treatment of these common diseases of bone loss.

Acknowledgements. We are grateful to Nancy Garrett for her help in compiling the references and preparation of the manuscript. Some of the work described here was supported by grants AR28149, CA40035, and RR01346 from the National Institutes of Health.

References

Adams SO, Nissley SP, Handwerger S, Rechler MM (1983) Developmental patterns of insulin-like growth factor-I and II synthesis and regulation in rat fibroblasts. Nature 302:150–153

Adamson LF, Anast CS (1966) Amino acid, potassium, and sulfate transport and incorporation by embryonic chick cartilage: the mechanism of the stimulatory effects of serum. Biochim Biophys Acta 121:10–20

Andersson I, Billig H, Fryklund L, Hansson HA, Isaksson O, Isgaard J, Nilsson A, Rozell B, Skottner A, Stemme S (1986) Localization of IGF-I in adult rats. Immunohistochemical studies. Acta Physiol Scand 126:311–312

Ash P, Francis MJO (1975) Response of isolated rabbit articular and epiphyseal chondrocytes to rat liver somatomedin. J Endocrinol 66:71–78

Ashton IK, Francis MJO (1977) An assay for plasma somatomedin: ^{3}H thymidine incorporation by isolated rabbit chondrocytes. J Endocrinol 74:205–212

Ashton IK, Francis MJO (1978) Response of chondrocytes isolated from human foetal cartilage to plasma somatomedin activity. J Endocrinol 76:473–477

Ashton IK, Otremski I (1986) In vitro effect of multiplication stimulating activity (MSA) on human fetal and postnatal cartilage. Early Human Dev 13:161–167

Ashton IK, Phizackerley S (1981) Human fetal cartilage response to plasma somatomedin activity in relation to gestatinal age. Calcif Tissue Int 33:205–209

Ashton IK, Pocock AE (1983) Action of multiplication-stimulating activity on ^{3}H thymidine incorporation in rabbit and human fetal chondrocytes in vitro. J Endocrinol 99:93–98

Asling CW, Becks H, Simpson ME, Li CH, Evans HM (1948) The effect of anterior hypophyseal growth hormones on epiphyseal closure in the third metacarpal of normal female rats. Anat Rec 101:23–31

Ballard FJ, Knowles SE, Wong SS, Bodner JB, Wood CM, Gunn JM (1980) Inhibition of protein breakdown in cultured cells is a consistent response to growth factors. FEBS Lett 114:209–212

Baron R, Neff L, Louvard D, Courtoy PJ (1985) Cell-mediated extracellular acidification and bone resorption: evidence for a low pH in resorbing lacunae and localization of a 100kD lysosomal membrane protein at the osteoclast ruffled border. J Cell Biol 101:2210–2222

Bennett AE, Wahner HW, Riggs BL, Hintz RL (1984a) Insulin-like growth factors I and II: aging and bone density in women. J Clin Endocrinol Metab 59:701–704

Bennett AE, Chen T, Feldman D, Hintz RL, Rosenfeld RG (1984b) Characterization of insulin-like growth factor I receptors on cultured rat bone cells: regulation of receptor concentration by glucocorticoids. Endocrinology 115:1577–1583

Bertolini DR, Nedwin GE, Bringman TS, Mundy GR (1986) Stimulation of bone resorption and inhibition of bone formation in vitro by human tumour necrosis factors. Nature 319:516–518

Betsholtz C, Westermark B, Ek B, Heldin CH (1984) Coexpression of a PDGF-like growth factor and PDGF receptor in a human osteosarcoma cells line: implications for autocrine receptor activation. Cell 39:447–457

Beutler B, Cerami A (1987) Cachectin: more than a tumor necrosis factor. N Engl J Med 316:379–385

Bird TA, Saklatvala J (1986) Identification of a common class of high affinity receptors for both types of porcine interleukin-1 on connective tissue cells. Nature 324:263–266

Bomboy JD, Salmon WD (1980) Effects of cyclic nucleotides on deoxyribonucleic acid synthesis in hypophysectomized rat cartilage: stimulation of thymidine incorporation and potentiation of the action of somatomedin by analogs of adenosine 3′,5′-monophosphate or a cyclic nucleotide phosphodiesterase inhibitor. Endocrinology 107:626–632

Bowen-Pope DF, Seifert RA, Ross R (1985) The platelet derived growth factor receptor. In: Boynton AL, Leffert HL (eds) Control of animal cell proliferation. Academic, New York, p 281

Burch WM, Lebovitz HE (1981) Hormonal activation of ornithine decarboxylase in embryonic chick pelvic cartilage. Am J Physiol 241:E454–E459

Burch WM, Weir S, van Wyk JJ (1986) Embryonic chick cartilage produces its own somatomedin-like peptide to stimulate cartilage growth in vitro. Endocrinology 119:1370–1376

Canalis E (1980) Effect of insulin-like growth factor I on DNA and protein synthesis in cultured rat calvaria. J Clin Invest 66:709–719

Canalis E (1981) Effect of platelet-derived growth factor on DNA and protein synthesis in cultured rat calvaria. Metabolism 30:970–975

Canalis E (1986) Interleukin-1 has independent effects on deoxyribonucleic releasing factor during early neonatal development in the rat. Endocrinology 118:74–81

Canalis E, Raisz LG (1979) Effect of epidermal growth factor on bone formation in vitro. Endocrinology 104:862–869

Canalis E, Dietrich JW, Maina DM, Raisz LG (1977a) Hormonal control of bone collagen synthesis in vitro, effects of insulin and glucagon. Endocrinology 100:668–674

Canalis E, Hintz RL, Dietrich JW, Maina DM, Raisz LG (1977b) Effect of somatomedin and growth hormone on bone collagen synthesis in vitro. Metab Clin Exp 26:1079–1087

Canalis E, McCarthy T, Centrella M (1987a) A bone-derived growth factor isolated from rat calvariae is beta 2 microglobulin. Endocrinology 121:1198–1200

Canalis E, Lorenzo J, Burgess WH, Maciag T (1987b) Effects of endothelial cell growth factor on bone remodelling in vitro. J Clin Invest 79:52–58

Canalis E, McCarthy T, Centrella M (1988) Isolation and characterization of insulin-like growth factor I (somatomedin-C) from culture of fetal rat calvariae. Endocrinology 122:22–27

Centrella M, Canalis E (1985) Transforming and nontransforming growth factors are present in medium conditioned by fetal rat calvariae. Proc Natl Acad Sci USA 82:7335–7339

Centrella M, Canalis E (1987) Isolation of EGF-dependent transforming growth factor (TGF beta-like) activity from culture medium conditioned by fetal rat calvariae. J Bone Min Res 2:29–36

Centrella M, Massague J, Canalis E (1986) Human platelet-derived trnsforming growth factor-beta stimulates parameters of bone growth in fetal rate calvariae. Endocrinology 119:2306–2312

Centrella M, McCarthy TL, Canalis E (1987a) Mitogenesis in fetal rat bone cells simultaneously exposed to type beta transforming growth factor and other growth regulators. FASEB J 1:312–317

Centrella M, McCarthy TL, Canalis E (1987b) Transforming growth factor beta is a bifunctional regulator of replication and collagen synthesis in osteoblast-enriched cell cultures from fetal rat bone. J Biol Chem 262:2869–2874

Chambers TJ, Darby JA, Fuller K (1985) Mammalian collagenase predisposes bone surfaces to osteoclastic resorption. Cell Tissue Res 241:671–675

Cheifetz S, Weatherbee JA, Tsang ML, Anderson JK, Mole JE, Lucas R, Massague J (1987) The transforming growth factor-beta system, a complex pattern of cross-reactive ligands and receptors. Cell 48:409–415

Chen TL, Mallory JB, Chang SL (1987) Effects of transforming growth factor (TGF beta) and its interaction with beta FGF and EGF in rat osteoblast-like (ROB) cells. J Bone Min Res 2:abstract 256

Chenu C, Pfeilschifter J, Mundy GR, Roodman GD (1987) Transforming growth factor beta inhibits formation of osteoclast-like cells in long-term human marrow cultures. J Bone Min Res 2:abstract 253

Chernausek SD, Jacobs S, van Wyk JJ (1981) Structural similarities between human receptors for somatomedin-C and insulin: analysis by affinity labeling. Biochemistry 20:7345–7350

Chua CC, Geiman DE, Keller GH, Ladda RL (1985) Induction of collagenase secretion in human fibroblast cultures by growth prompting factors. J Biol Chem 260:5213–5216

Chyun YS, Raisz LG (1984) Stimulation of bone formation by prostaglandin E2. Prostaglandins 27:97–103

Clemmons DR, Elgin RG, Han VK, Casella SJ, d'Ercole AJ, van Wyk JJ (1986) Cultured fibroblast monolayers secrete a protein that alters the cellular binding of somatomedin C/insulin-like growth factor I. J Clin Invest 77:1548–1558

Corvol MT, Dumontier MF, Rappaport R, Guyda H, Posner BI (1978) The effect of a slightly acidic somatomedin peptide (ILAs) on the sulphation of proteoglycans from articular and growth plate chondrocytes in culture. Acta Endocrinol (Copenh) 89:263–275

Czech MP, Oppenheimer CL, Massague J (1983) Interrelationships among receptor structures for insulin and peptide growth factors. Fed Proc 42:2598–2601

Daughaday WH, Mariz IK (1962) Conversion of proline-U-C^{14} to labeled hydroxyproline by rat cartilage in vitro: effects of hypophysectomy, growth hormone, and cortisol. J Lab Clin Med 59:741–752

Daughaday WH, Reeder C (1966) Synchronous activation of DNA synthesis in hypophysectomized rat cartilage by growth hormones. J Lab Clin Med 68:357–368

Daughaday WH, Parker KA, Borowsky S, Trivedi B, Kapadia M (1982) Measurement of somatomedin-related peptides in fetal, neonatal, and maternal rat serum by insulin-like growth factor I radioimmunoassay, IGF-II radioreceptor assay (RAA), and multiplication-stimulating activity RRA after acid-ethanol extraction. Endocrinology 110:575–581

Decker JL, Malone DG, Haraoui B, Wahl SM, Schrieber L, Klippel JH, Steinberg AD, Wilder RL (1984) Rheumatoid arthritis: evolving concepts of pathogenesis and treatment. Ann Intern Med 101:810–824

D'Ercole AJ, Wilkins JR (1984) Affinity labeled somatomedin-C binding proteins in rat sera. Endocrinology 114:1141–1144

Dietrich JW, Goodson JM, Raisz LG (1975) Stimulation of bone resorption by various prostaglandins in organ cultures. Prostaglandins 10:231–240

Dingle JT, Page-Thomas DP, King B, Bard DR (1987) In vivo studies of articular tissue damage mediated by catabolin/interleukin 1. Ann Rheum Dis 46:527–533

Drezner MK, Eisenbarth GS, Neelon FA, Lebovitz HE (1975) Stimulation of cartilage amino acid uptake by growth hormone deficient factors in serum. Mediation by adenosine 3′: 5′-monophosphate. Biochim Biophys Acta 381:384–396

Drop SLS, Valiquette G, Guyda HJ, Corvol MT, Poner BI (1979) Partial purification and characterization of a binding protein for insulin-like activity (IL-As) in human amniotic fluid: a possible inhibitor of insulin-like activity. Acta Endocrinol (Copenh) 90:505–518

Drop SLS, Kortleve DJ, Guyda HJ (1984) Isolation of a somatomedin-binding protein from preterm amniotic fluid. Development of a radioimmunoassay. J Clin Endocrinol Metab 59:889–907

D'Souza SM, Orcutt CM, Anglin AM, Ibbotson KJ (1987) Transforming growth factor beta: effects on human osteoblast-like cells. J Bone Min Res 2:abstract 133

Duff GW (1988) Arthritis and interleukins. Br J Rheumatol 27:2–5

Eden S, Isaksson OGP, Madsen K, Friberg U (1983) Specific binding of growth hormone to isolated chondrocytes from rabbit ear and epiphyseal plate. Endocrinology 112:1127–1129

Edwards DR, Murphy G, Reynolds JJ, Whitham SE, Docherty AJ, Angel P, Heath JK (1987) Transforming growth factor beta modulates the expression of collagenase and metalloproteinase inhibitor. EMBO J 6:1899–1904

Eisenbarth GS, Beuttel SC, Lebovitz HE (1973) Fatty acid inhibition of somatomedin (serum sulfation factor) stimulated protein and RNA synthesis in embryonic chicken cartilage. Biochim Biophys Acta 331:397–409

Elgin RG, Busby WH, Clemmons DR (1987) An insulin-like growth factor (IGF) binding protein enhances the biologic response to IGF-I. Proc Natl Acad Sci USA 84:3254–3258

Ellingsworth LR, Brennan JE, Fok K, Rosen DM, Bentz H, Piez KA, Seyedin SM (1986) Antibodies to the N-terminal portion of cartilage-inducing factor A and transforming growth factor beta. Immunohistochemical localization and association with differentiating cells. J Biol Chem 261:12362–12367

Ernst IM, Froesch ER (1987a) Osteoblast-like cells in a serum-free methylcellulose medium form colonies: effects of insulin and insulin-like growth factor I. Calcif Tissue Int 40:27–34

Ernst IM, Froesch ER (1987b) Growth hormone (GH) dependent stimulation of rat osteoblasts in culture is mediated partially by local synthesis of immuno-reactive insulin-like growth factor (IGF). J Bone Min Res 2:abstract 32

Ernst IM, Schmid C, Frankenfeldt C, Froesch ER (1988) Estradiol stimulation of osteoblast proliferation in vitro: mediator roles for TGF beta, PGE2, insulin-like growth factor (IGF) I? Calcif Tissue Int 42:abstract 117

Evans DB, Bunning RAD, van Damme J, Russell RGG (1987) Cytokines stimulate the production of plasminogen activator activity of human osteoblast-like cells. J Bone Min Res 2:abstract 144

Farley JR, Baylink DJ (1982) Purification of a skeletal growth factor from human bone. Biochemistry 21:3502–3507

Feyen JHM, DiBon A, van der Plas A, Lowik CW, Nijweide PJ (1985) Effects of exogenous prostanoids on the proliferation of osteoblast-like cells in vitro. Prostaglandins 30:827–840

Froesch ER, Zapf J, Audhya TK, Ben-Porath E, Segen BJ, Gibson KD (1976) Nonsuppressible insulin-like activity and thyroid hormones. Major pituitary dependent sulfation factors for chick embryo cartilage. Proc Natl Acad Sci USA 73:2904–2908

Fryklund L, Uthne K, Sievertsson H (1974) Identification of two somatomedin A active polypeptides and in vivo effects of a somatomedin A concentrate. Biochem Biophys Res Commun 61:957–962

Fryklund L, Skottner A, Sievertsson H, Hall K (1976) Somatomedins A and B. Isolation, chemistry and in vivo effects. In: Pecile A, Muller EE (eds) Growth hormone and related peptides. Excerpta Medica, Amsterdam, p 156
Gilman SC, Chang J, Zeigler PR; Uhl J, Mochan E (1988) Interleukin-1 activates phospholipase A2 in human synovial cells. Arthritis Rheum 31:126–130
Gluckman PD, Butler JH (1983) Parturition-related changes in insulin-like growth factors-I and II in the perinatal lamb. J Endocrinol 99:223–232
Gowen M, Wood DD, Russell RG (1985) Stimulation of the proliferation of human bone cells in vitro by human monocyte products with interleukin-1 activity. J Clin Invest 75:1223–1229
Graves DT, Owen AJ, Barth RK, Tempst P, Winoto A, Fors L, Hood LE, Antoniades HN (1984) Detection of c-*sis* transcripts and synthesis of PDGF-like proteins by human osteosarcoma cells. Science 226:972–974
Guenther HL, Guenther HE, Froesch ER, Fleisch H (1982) Effect of insulin-like growth factor on collagen and glycosaminoglycan synthesis by rabbit articular chondrocytes in culture. Experientia 38:979–981
Hakeda Y, Kurihara N, Hotta T, Yagyu H, Ikeda E, Miyakoshi S, Chavassieux PM, Meunier PJ, Kumegawa M (1987) Effects of transforming growth factor type beta (TGF beta) on human osteoblastic cell line. J Bone Min Res 2:abstract 131
Hall K (1970) Quantitative determination of the sulfation factor activity in human serum. Acta Endocrinol (Copenh) 63:338–350
Hall K (1972) Human somatomedin. Determination, occurrence, biological activity and purification. Acta Endocrinol [Suppl 162] (Copenh) 163:1–52
Hall K, Filipsson R (1975) Correlation between somatomedin A in serum and body height development in healthy children and children with certain growth disturbances. Acta Endocrinol (Copenh) 78:239–250
Hamilton JA, Lingelbach S, Partridge NC, Martin TJ (1985) Regulation of plasminogen activator production by bone resorbing hormones in normal and malignant osteoblasts. Endocrinology 116:2186–2191
Harris WH, Heaney RP, Jowsey J, Cockin J, Akins C, Graham J, Weinberg EH (1972) Growth hormone: the effect of skeletal renewal in the adult dog. I. Calcif Tissue Res 10:1–13
Harrod J, Hill DJ, Russell RGG (1986) Effect of transforming growth factor beta on osteoblast-like cells. Calcif Tiss Int 38:abstract 62
Hauschka PV, Mavrakos AE, Iafrati MD, Doleman SE, Klagsbrun M (1986) Growth factors in bone matrix. Isolation of multiple types by affinity chromatography on heparin Sepharose. J Biol Chem 261:12665–12674
Heaney RP, Harris WH, Cockin J, Weinberg EH (1972) Growth hormone: the effect of skeletal renewal in the adult dog. II. Calcif Tissue Res 10:14–22
Heine U, Munoz EF, Flanders KC, Ellingsworth LR, Lam HY, Thompson NL, Roberts AB, Sporn MB (1987) Role of transforming growth factor beta in the development of the mouse embryo. J Cell Biol 105:2861–2876
Heldin CH, Westermark B, Wasteson A (1980) Chemical and biological properties of a growth factor from human-cultured osteosarcoma cells: resemblance with platelet-derived growth factor. J Cell Physiol 105:235–246
Heldin CH, Johnsson A, Wennergren S, Wernstedt C, Betsholtz C, Westermark B (1986) A human osteosarcoma cell line secretes a growth factor structurally related to a homodimer of PDGF A-chains. Nature 319:511–514
Hey AW, Browne CA, Thornburn GD (1988) Purification and characterization of a fetal somatomedin from the sheep: similarity to insulin-like growth factor II. Endocrinology 122:12–21
Hill DJ (1979) Stimulation of cartilage zones of the calf costochondral growth plate in vitro by growth hormone dependent rat plasma somatomedin activity. J Endocrinol 83:219–227
Hill DJ, Milner RD (1984) Platelet-derived growth factor and multiplication stimulating activity II, but not multiplication stimulating activity III-2, stimulate 3H

thymidine and 35S sulphate incorporation by fetal rat costal cartilage in vitro. J Endocrinol 103:195–203
Hill DJ, Holder AT, Seid J, Preece MA, Tomlinson S, Milner RDG (1983) Increased thymidine incorporation into fetal rat cartilage in vitro in the presence of human somatomedin, epidermal growth factor and other growth factors. J Endocrinol 96:489–497
Hintz RL, Clemmons DR, Underwood LE, van Wyk JJ (1972) Competitive binding of somatomedin to the insulin receptors of adipocytes, chondrocytes, and liver membranes. Proc Natl Acad Sci USA 69:2351–2353
Hiraki Y, Yutani Y, Takigawa M, Kato Y, Suzuki F (1985) Differential effects of parathyroid hormone and somatomedin-like growth factors on the sizes of proteoglycan monomers and their synthesis in rabbit costal chondroytes in culture. Biochim Biophys Acta 845:445–453
Hiraki Y, Kato Y, Inoue H, Suzuki F (1986) Stimulation of DNA synthesis in quiescent rabbit chondrocytes in culture by limited exposure to somatomedin-like growth factors. Eur J Biochem 158:333–337
Hock JM, Centrella M, Canalis E (1988 a) Insulin-like growth factor I has independent effects on bone matrix formation and cell replication. Endocrinology 122:254–260
Hock JM, Centrella M, Canalis E (1988 b) Transforming growth factor beta (TGF-beta-1) stimulates bone matrix apposition and bone cell replication in cultured rat calvaria. Calcif Tissue Int 42:abstract 124
Hoekman K, Lowik C, van der Ruit M, Verheyen J, Bijvoet OLM (1987) Evidence for induction of bone resorption through osteoblast derived plasminogen activators. J Bone Min Res 2:abstract 17
Holder AT, Spencer EM, Preece MA (1981) Effect of bovine growth hormone and a partially pure preparation of somatomedin on various growth parameters in hypopituitary dwarf mice. J Endocrinol 89:275–282
Howard GA, Bottemiller BL, Turner RT, Rader JI, Baylink DJ (1981) Parathyroid hormone stimulates bone formation and resorption in organ culture: evidence for a coupling mechanism. Proc Natl Acad Sci USA 78:3204–3208
Huang JS, Huang SS, Deuel TF (1984) Specific covalent binding of platelet-derived growth factor to human plasma alpha 2-macroglobulin. Proc Natl Acad Sci USA 81:342–346
Isaksson OGP, Jansson JO, Gause IAM (1982) Growth hormone stimulates longitudinal bone growth directly. Science 216:1237–1239
Isaksson OGP, Eden S, Jansson JO (1985) Mode of action of pituitary growth hormone on target cells. Annu Rev Physiol 47:483–499
Isgaard J, Nilsson A, Lindahl A, Jansson JO, Isaksson OG (1986) Effects of local administration of GH and IGG-I on longitudinal bone growth in rats. Am J Physiol 250:E367–372
Ishibashi T, Miller SL, Burstein SA (1987) Type beta transforming growth factor is a potent inhibitor of murine megakaryocytopoiesis in vitro. Blood 69:1737–1741
Jennings J, Buchanan F, Freeman D, Garland JT (1980) Stimulation of chick embryo cartilage sulfate and thymidine uptake: comparison of human serum, purified somatomedins, and other growth factors. J Clin Endocrinol Metab 51:1166–1170
Joncourt F, Trechsel U, Guenther HL, Fleisch F (1988) Production of IL 1 activity by mouse and rat bone. Calcif Tissue Int 42:abstract 126
Kasuga M, Van Obberghen E, Nissley SP, Rechler MM (1981) Demonstration of two subtypes of insulin-like growth factor receptors by affinity cross-linking. J Biol Chem 256:5305–5308
Kato Y, Nasu N, Takase T, Suzuki F (1978) Demonstration of somatomedin activity of "multiplication-stimulating activity" in rabbit costal chondrocytes in culture. J Biochem 84:1001–1004
Kato Y, Nasu N, Takase T, Daikuhara Y, Suzuki F (1980 a) A serum-free medium supplemented with multiplication-stimulating activity (MSA) supports both proliferation and differentiation of chondrocytes in primary culture. Exp Cell Res 125:167–174

Kato Y, Nomura Y, Daikuhara Y, Nasu N, Tsuji M, Asada A, Suzuki F (1980b) Cartilage-derived factor (CDF). I. Stimulation of proteoglycan synthesis in rat and rabbit costal chondrocytes in culture. Exp Cell Res 130:73–81

Kato Y, Nomura Y, Tsuji M, Ohmae H, Kinoshita M, Hamamoto S, Suzuki F (1981a) Cartilage-derived factor (CDF). II. Somatomedin-like action on cultured chondrocytes. Exp Cell Res 132:339–347

Kato Y, Nomura Y, Tsuji M, Kinoshita M, Ohmae H, Suzuki F (1981b) Somatomedin-like peptide(s) isolated from fetal bovine cartilage (cartilage-derived factor): isolation and some properties. Proc Natl Acad Sci USA 78:6831–6835

Kato Y, Hiraki Y, Inoue H, Kinoshita M, Yutani Y, Suzuki F (1983) Differential and synergistic actions of somatomedin-like growth factors, fibroblast growth factor and epidermal growth factor in rabbit costal chondrocytes. Eur J Biochem 129:685–690

Kemp SF, Hintz RL (1980) The action of somatomedin on glycosaminoglycan synthesis in cultured chick chondrocytes. Endocrinology 106:744–749

Kemp SF, Mutchick M, Hintz RL (1984) Hormonal control of protein synthesis in chick chondrocytes: a comparison of effects of insulin, somatomedin C and triiodothyronine. Acta Endocrinol (Copenh) 107:179–184

Keski-Oja J, Lyons RM, Moses HL (1987) Inactive secreted form(s) of transforming growth factor beta (TGF beta). Activation by proteolysis. J Cell Biochem 11A: abstract 224

Klagsbrun M, Smith S (1980) Purfication of a cartilage-derived growth factor. J Biol Chem 255:10859–10866

Klagsbrun M, Langer R, Levenson R, Smith S, Lillehei C (1977) The stimulation of DNA synthesis and cell division in chondrocytes and 3T3 cells by a growth factor isolated from cartilage. Exp Cell Res 105:99–108

Knauer DJ, Smith GL (1980) Inhibition of biological activity of multiplication-stimulating activity by binding to its carrier protein. Proc Natl Acad Sci USA 77:7252–7256

Knudsen BS, Silverstein RL, Leung LL, Harpel PC, Nachman RL (1986) Binding of plasminogen to extracellular matrix. J Biol Chem 261:10765–10771

Kream BE, Smith MD, Canalis E, Raisz LG (1985) Characterization of the effect of insulin on collagen synthesis in fetal rat bone. Endocrinology 116:296–302

Kurose H, Seino Y, Yamaoka K, Tanaka H, Yabuuchi H (1987) Effects of somatomedin D and 1,25 dihydroxyvitamin D3 on clonal osteoblasts, MC3T3-E1. J Bone Min Res 2:abstract 22

Laron Z (1974) Syndrome of familial dwarfism and high plasma immunoreactive growth hormone. Isr J Med Sci 10:1247–1253

Leizer T, Clarris BJ, Ash PE, van Damme J, Saklatvala J, Hamilton JA (1987) Interleukin-1 beta and interleukin-1 alpha stimulate the plasminogen activator activity and prostaglandin E2 levels of human synovial cells. Arthritis Rheum 30:562–566

Leof EB, Proper JA, Goustin AS, Shipley DG, DiCorleto PE, Moses HL (1986) Induction of c-*sis* mRNA and activity similar to platelet-derived growth factor by transforming growth factor β: a proposed model for indirect mitogenesis involving autocrine activity. Proc Natl Acad Sci USA 83:2453–2457

Lund PK, Moats-Straats BM, Mynes MA, Simmons JG, Jansen M, d'Ercole AJ, van Wyk JJ (1986) Somatomedin-C/insulin-like growth factor-I and insulin-like growth factor-II mRNAs in rat fetal and adult tissues. J Biol Chem 261:14539–14544

Madsen K, Friberg U, Roos P, Eden S, Isaksson O (1983) Growth hormone stimulates the proliferation of cultured chondrocytes from rabbit ear and rat rib growth cartilage. Nature 304:545–547

Mankin HJ, Thrasher AZ, Weinberg EH, Harris WH (1978) Dissociation between the effect of bovine growth hormone in articular cartilage and in bone of the adult dog. J Bone Joint Surg [Am] 60a:1071–1075

Martin JL, Baxter RC (1986) Insulin-like growth factor binding protein from human plasma. J Biol Chem 261:8754–8760

Martinet Y, Bitterman PB, Mornex J-F, Grotendorst GR, Martin GR, Crystal RG (1985) Activated human monocytes express the c-*sis* proto-oncogene and release a mediator showing PDGF-like activity. Nature 319:158–160

Massague J, Czech MP (1982) The subunit structures of two distinct receptors for insulin-like growth factors I and II and their relationship to the insulin receptor. J Biol Chem 257:5038–5045

McCarthy T, Centrell M, Fox G, Arakawa T, Canalis E (1987) Endothelial cell growth factor (ECGF) and basic fibroblast growth factor (bFGF) independently regulate bone cell replication and type I collagen transcription and translation. J Bone Min Res 2:abstract 252

McQuillan DJ, Handley CJ, Campbell MA, Bolis S, Milway VE, Herington AC (1986) Stimulation of proteoglycan biosynthesis by serum and insulin-like growth factor-I in cultured bovine articular cartilage. Biochem J 240:423–430

Meier S, Solursh M (1972) The comparative effects of several mammalian growth hormones on sulfate incorporation into acid mucopolysaccharides by cultured chick embryo chondrocytes. Endocrinology 90:1447–1451

Mohan S, Linkhart TA, Jennings JC, Baylink DJ (1987) Identification and quantification of four distinct growth factors stored in human bone matrix. J Bone Min Res 2:abstract 44

Mohan S, Jennings JC, Linkhart TA, Baylink DJ (1988) Primary structure of human skeletal growth factor (SGF): homology with IGF-II. Biochim Biophys Acta 14:44–55 [Suppl 1]:abstract 598

Moses AC, Nissley SP, Short PA, Rechler MM, White RM, Knight AB, Higa OZ (1980) Increased levels of multiplication-stimulating activity, an insulin-like growth factor, in fetal rat serum. Proc Natl Acad Sci USA 77:3649–3653

Mustoe TA, Pierce GF, Thomason A, Gramates P, Sporn MB, Deuel TF (1987) Accelerated healing of incisional wounds in rats induced by transforming growth factor beta. Science 237:1333–1336

Nakamura T, Kitazawa T, Ichihara A (1986) Partial purification and characterization of masking protein for beta type transforming growth factor from rat platelets. Biochem Biophys Res Commun 141:176–184

Ng KW, Partridge NC, Niall M, Martin TJ (1983) Stimulation of DNA synthesis by epidermal growth factor in osteoblast-like cells. Calcif Tissue Int 35:624–628

Noda M, Rodan GA (1987) Type beta transforming growth factor (TGF beta) regulation of alkaline phosphatase expression and other phenotype-related mRNAs in osteoblastic rat osteosarcoma cells. J Cell Physiol 133:426–437

O'Connor-McCourt MD, Wakefield LM (1987) Latent transforming growth factor-beta in serum. A specific complex with alpha 2-macroglobulin. J Biol Chem 262:14090–14099

O'Keefe RJ, Rosfer RN, Brand JS (1987) Effects of transforming growth factor beta on chick epiphyseal chondrocytes. J Bone Min Res 2:abstract 28

Oreffo ROC, Mundy GR, Bonewald LF (1988) Osteoclasts activate latent transforming growth factor beta and vitamin A treatment increases TGF beta activation. Calcif Tiss Int 42:abstract 56

Padgett RW, St Johnston RD, Gelbart NM (1987) A transcript from a *Drosophila* pattern gene predicts a protein homologous to the transforming growth factor beta family. Nature 325:81–84

Penttinen RP, Kobayashi S, Bornstein P (1988) Transforming growth factor beta increases mRNA for matrix proteins both in the presence and in the absence of changes in mRNA stability. Proc Natl Acad Sci USA 85:1105–1108

Pfeilschifter J, Mundy GR (1987a) Modulation of type beta transforming growth factor activity in bone cultures by osteotropic hormones. Proc Natl Acad Sci USA 82:2024–2028

Pfeilschifter J, Mundy GR (1987b) Transforming growth factor beta stimulates osteoblast activity and is released during the bone resorption process. In: Cohn DV, Martin TJ, Meunier PJ (eds) Calcium regulation and bone metabolism, vol 9. Excerpta Medica, Amsterdam, p 450

Pfeilschifter J, d'Souza SM, Mundy GR (1987a) Effects of transforming growth factor beta on osteoblastic osteosarcoma cells. Endocrinology 121:212–218

Pfeilschifter J, Bonewald L, Mundy GR (1987b) TGF beta is released from bone with one or more binding proteins which regulate its activity. J Bone Min Res 2:abstract 249

Pfeilschifter J, Seyedin S, Mundy GR (1988) Transforming growth factor β inhibits bone resorption in fetal rat long bone culture. J Clin Invest 82:680–685

Phillips LS, Weiss LJ, Matheson K (1983) Circulating growth factor studies in growth plate versus resting cartilage in vitro. II. Recognition of growth factors in rat serum. Endocrinology 113:1494–1502

Pircher R, Jullien P, Lawrence DA (1986) Beta transforming growth factor is stored in human blood platelets as a latent high molecular weight complex. Biochem Biophys Res Commun 136:30–37

Poser JW, Coppinger WJ, Lucas DS (1986) Transforming growth factor beta induces neonatal rat muscle cells to become chondrocytes. Calcif Tissue Int 38:abstract 79A

Postal-Vinay MC, Corvol MT, Lang F, Fraud F, Guyda H, Posner B (1983) Receptors for insulin-like growth factors in rabbit articular and growth plate chondrocytes in culture. Exp Cell Res 148:105–116

Prins APA, Lipman JM, McDevitt CA, Sokoloff L (1982a) Effect of purified growth factors on rabbit articular chondrocytes in monolayer culture. II. Sulfated proteoglycan synthesis. Arthritis Rheum 25:1228–1238

Prins APA, Lipman JM, McDevitt CA, Sokoloff L (1982b) Effect of purified growth factors on rabbit articular chondrocytes in monolayer culture. I. DNA synthesis. Arthritis Rheum 25:1217–1227

Raghow R, Postlethwaite AE, Keski-Oja J, Moses HL, Kang AH (1987) Transforming growth factor beta increases steady state levels of type 1 procollagen and fibronectin messenger RNAs posttranscriptionally in cultured human dermal fibroblasts. J Clin Invest 79:1285–1288

Raines EW, Ross R (1982) Platelet-derived growth factor. I. High yield purification and evidence for multiple forms. J Biol Chem 257:5154–5160

Raines EW, Bowen-Pope DP, Ross R (1984) Plasma binding proteins for platelet-derived growth factor that inhibit its binding to cell-surface receptors. Proc Natl Acad Sci USA 81:3424–3428

Raisz LG, Koolemans-Beynen AR (1974) Inhibition of bone collagen synthesis by prostaglandin E2 in organ culture. Prostaglandins 8:377–385

Raisz LG, Simmons HA, Sandberg AL, Canalis E (1980) Direct stimulation of bone resorption by epidermal growth factor. Endocrinology 107:270–273

Rechler MM, Nissley SP (1986) Insulin-like growth factor (IGF)/somatomedin receptor subtypes: structure, function, and relationships to insulin receptors and IGF carrier proteins. Horm Res 24:152–159

Rinderknecht E, Humbel RE (1978a) The amino acid sequence of human insulin-like growth factor I and its structural homology with proinsulin. J Biol Chem 253:2769–2776

Rinderknecht E, Humbel RE (1978b) Primary structure of human insulin-like growth factor II. FEBS Lett 89:283–286

Robbins KC, Leal F, Pierce JH, Aaronson SA (1985) The v-*sis*/PDGF-2 transforming gene product localizes to cell membranes but is not a secretory protein. EMBO J 4:1783–1792

Robey PG, Young MF, Flanders KC, Roche NS, Kondaiah P, Reddi AH, Termine JD, Sporn MB, Roberts AB (1987) Osteoblasts synthesize and respond to transforming growth factor-type beta (TGF beta) in vitro. J Cell Biol 105:457–463

Rodan GA, Rodan SB (1983) Expression of the osteoblastic phenotype. In: Peck WA (ed) Bone and mineral research 2. Elsevier, Amsterdam, p 244

Rodan SB, Wesolowski G, Thomas K, Rodan GA (1987a) Growth stimulation of rat calvaria osteoblastic cells by acidic fibroblast growth factor. Endocrinology 121:1917–1923

Rodan SB, Wesolowski G, Thomas K, Rodan GA (1987b) Effect of growth factors on calvaria and ROS 17/2.8 cells. Calcif Tissue Int 41:abstract OP10
Romanus JA, Terrell JE, Yang YW, Nissley SP, Rechler MM (1986) Insulin-like growth factor carrier proteins in neonatal and adult rat serum are immunologically different: demonstration using a new radioimmunoassay for the carrier protein from BRL-3A rat liver cells. Endocrinology 118:1743–1758
Rosenfeld R, Kempf SF, Hintz RL (1981) Constancy of somatomedin response to growth hormone treatment of hypopituitary dwarfism, and lack of correlation with growth rate. J Clin Endocrinol Metab 53:611–617
Ross R, Raines EW, Bowen-Pope DF (1986) The biology of platelet-derived growth factor. Cell 46:155–169
Russell SM, Spencer EM (1985) Local injection of human or rat growth hormone or of purified human somatomedin-C stimulate unilateral tibial epiphyseal growth in hypophysectomized rats. Endocrinology 116:2563–2567
Saklatvala J (1986) Tumour necrosis factor alpha stimulates resorption and inhibits synthesis of proteoglycan in cartilage. Nature 322:547–549
Salmon WD, Daughaday WH (1957) A hormonally controlled serum factor which stimulates sulfate incorporation by cartilage in vitro. J Lab Clin Med 49:825–836
Salmon WD, DuVall MR (1970) A serum fraction with "sulfation factor activity" stimulates in vitro incorporation of leucine and sulfate into protein-polysaccharide complexes, uridine into RNA, and thymidine into DNA of costal cartilage from hypophysectomized rats. Endorcinology 86:721–727
Salmon WD, von Hagen MJ, Thompson EY (1967) Effects of puromycin and actinomycin in vitro on sulfate incorporation by cartilage of the rat and its stimulation by serum sulfation factor and insulin. Endocrinology 80:999–1005
Salmon WD, DuVall MR, Thompson EY (1968) Stimulation of insulin in vitro of incorporation of (^{35}S) sulfate and (^{14}C) leucine into protein-polysaccharide complexes, (^{3}H) uridine into RNA and (^{3}H) thymidine into DNA of costal cartilage from hypophysectomized rats. Endocrinology 82:493–499
Schalch DS, Sessions CM, Farley AC, Masakawa A, Emler CA, Cills DG (1986) Interaction of insulin-like growth factor I/somatomedin-C with cultured rat chondrocytes: receptor binding an internalization. Endocrinology 118:1590–1597
Schlechter NL, Russell SM, Spencer EM, Nicoll CS (1986a) Evidence suggesting that the direct growth-promoting effect of growth hormone on cartilage in vivo is mediated by local production of somatomedin. Proc Natl Acad Sci USA 83:7932–7934
Schlechter NL, Russell SM, Greenberg S, Spencer EM, Nicoll CS (1986b) A direct growth effect of growth hormone in rat hindlimb shown by arterial infusion. Am J Physiol 250:E231–235
Schmid C, Steiner T, Froesch ER (1982) Parathormone promotes glycogen formation from ^{14}C glucose in cultured osteoblast-like cells. FEBS Lett 148:31–34
Schmid C, Steiner T, Froesch ER (1983) Insulin-like growth factors stimulate synthesis of nucleic acids and glycogen in cultured calvaria cells. Calcif Tissue Int 35:578–585
Schoenle E, Zapf J, Humbel RE, Froesch ER (1982) Insulin-like growth factor I stimulates growth in hypophysectomized rats. Nature 296:252–253
Schwartz PL, Wettenhall REH, Troeddel MA, Bornstein J (1970) A long term effect of insulin on collagen synthesis by newborn rat bone in vitro. Diabetes 19:465–466
Scott CD, Martin JL, Baxter RC (1985) Rat hepatocyte insulin-like growth factor I and binding protein: effect of growth hormone in vitro and in vivo. Endocrinology 116:1102–1107
Sessions SB, Emler CA, Schalch DS (1987) Interaction of insulin-like growth factor II with rat chondrocytes: receptor binding, internalization, and degradation. Endocrinology 120:2108–2116
Seyedin SM, Thomas TC, Thompson AY, Rosen DM, Piez KA (1985) Purification and characterzation of two cartilage-inducng factors from bovine demineralized bone. Proc Natl Acad Sci USA 82:2267–2271

Seyedin SM, Thompson AY, Bentz H, Rosen DM, McPherson JM, Conti A, Siegel NR, Galluppi GR, Piez KA (1986) Cartilage-inducing factor-A. Apparent identity to transforming growth factor beta. J Biol Chem 261:5693–5695

Seyedin SM, Segarini PR, Rosen DM, Thompson AY, Bentz H, Graycar J (1987) Cartilage-inducing factor-B is a unique protein structurally and functionally related to transforming growth factor-beta. J Biol Chem 262:1946–1949

Shimokado K, Raines EW, Madtes DK, Barrett TB, Benditt EP, Ross R (1985) A significant part of macrophage-derived growth factor consists of at least two forms of PDGF. Cell 43:277–286

Shupnik MA, Antoniades HN, Tashjian AH (1982) Platelet-derived growth factor increases prostaglandin production and decreases epidermal growth factor receptors in human osteosarcoma cells. Life Sci 30:347–353

Skantze KA, Brinckerhoff CE, Collier JP (1985) Use of agarose culture to measure the effect of transforming growth factor beta and epidermal growth factor on rabbit articular chondrocytes. Cancer Res 45:4416–4421

Skottner A, Clark RG, Robinson IC, Fryklund L (1987) Recombinant human insulin-like growth factor: testing the somatomedin hypothesis in hypophysectomized rats. J Endocrinol 112:123–132

Slootweg MC, Duursma SA (1987) Growth hormone promotes differentiation of rat calvaria cells in primary culture. J Bone Min Res 2:abstract 54

Slootweg MC, Herrmann-Erlee MPM, van der Meer SC, van Buul-Offers SC, de Poorter TL, Duursma SA (1987) Fetal mouse osteoblasts respond to physiological concentrations of growth hormone and produce insulin-like growth factors. Calcif Tissue Int 41:abstract OP11

Smeets T, van Buul-Offers S (1983) The influence of growth hormone, somatomedins, prolactin and thyroxine on the morphology of the proximal tibial epiphysis and growth plate of Snell dwarf mice. Growth 47:160–173

Smith DD, Gowen M, Mundy GR (1987) Effects of interferon-gamma and other cytokines on collagen synthesis in fetal rat bone cultures. Endocrinology 120:2494–2499

Smith EP, Sadler TW, d'Ercole AJ (1987) Somatomedins/insulin-like growth factors, their receptors and binding proteins are present during mouse embryogenesis. Development 101:73–82

Solursh M, Meier S (1973) A conditioned medium (CM) factor produced by chondrocytes that promotes their own differentiation. Dev Biol 30:279–289

Sone T, Yamamoto I, Kitamura N, Aoki J, Torizuka K (1987) Effects of cytokines on osteoblastic MC3T3-E1 cells. J Bone Min Res 2:abstract 128

Sporn MB, Roberts AB (1986) Peptide growth factors and inflammation, tissue repair, and cancer. J Clin Invest 78:329–332

Sporn MB, Roberts AB, Shull JH, Smith JM, Ward JM, Sodek J (1983) Polypeptide transforming growth factors isolated from bovine sources and used for wound healing in vivo. Science 219:1329–1331

Sporn MB, Roberts AB, Wakefield LM, de Crombrugghe B (1987) Some recent advances in the chemistry and biology of transforming growth factor beta. J Cell Biol 105:1039–1045

Stevens RL, Hascall VC (1981) Characterization of proteoglycans synthesized by rat chondrosarcoma chondrocytes treated with multiplication-stimulating activity and insulin. J Biol Chem 256:2053–2058

Stevens RL, Nissley SP, Kimura JH, Rechler MM, Caplan AI, Hascall VC (1981) Effects of insulin and multiplication-stimulating activity on proteoglycan biosynthesis in chondrocytes from the Swarm rat chondrosarcoma. J Biol Chem 256:2045–2052

Stevens RL, Austen KF, Nissley SP (1982) Insulin-induced increase in insulin-binding to cultured chondrosarcoma chondrocytes. J Biol Chem 258:2940–2944

Stracke H, Schulz A, Moeller D, Rossol S, Schatz H (1984) Effect of growth hormone on osteoblasts and demonstration of somatomedin-C/IGF I in bone organ culture. Acta Endocrinol (Copenh) 107:16–24

Stuart CA, Furlanetto RW, Lebovitz HE (1979) The insulin receptor of embryonic chicken cartilage. Endocrinology 105:1293–1302

Stuart CA, Vesely DL, Provom SA, Furlanetto RW (1982) Cyclic nucleotides and somatomedin action in cartilage. Endocrinology 111:553–558

Sullivan R, Klagsbrun M (1985) Purification of cartilage-derived growth factor by heparin affinity chromatography. J Biol Chem 260:2399–2403

Tashjian AH, Levine L (1978) Epidermal growth factor stimulates prostaglandin production and bone resorption in cultured mouse calvaria. Biochem Biophys Res Commun 85:966–975

Tashjian AH, Hohmann EL, Antoniades HN, Levine L (1982) Platelet-derived growth factor stimulates bone resorption via a prostaglandin-mediated mechanism. Endocrinology 111:118–124

Tashjian AH, Voelkel EF, Lazzaro M, Singer FR, Roberts AB, Derynck R, Winkler ME, Levine L (1985) Alpha and beta human transforming growth factors stimulate prostaglandin production and bone resorption in cultured mouse calvaria. Proc Natl Acad Sci USA 82:4535–4538

Tell GPE, Cuatrecasas P, van Wyk JJ, Hintz RL (1973) Somatomedin: inhibition of adenylate cyclase activity in subcellular membranes of various tissues. Science 180:312–314

Thiel HJ, Hafenrichter R (1984) Simian sarcoma virus transformation-specific glycopeptide: immunological relationship to human platelet-derived growth factor. Virology 136:414–424

Thorngren KG, Hamsson LI, Menandes-Sellman K, Stenstrom A (1973) Effect of dose and administration period of growth hormone on longitudinal bone growth in the hypophysectomized rat. Acta Endocrinol (Copenh) 74:1–23

Thorngren KG, Hansson LI, Fryklund L, Sievertsson H (1977) Human somatomedin A and longitudinal bone growth in the hypophysectomized rat. Mol Cell Endocrinol 6:217–221

Trippel SB, Van Wyk JJ, Foster MB, Svoboda ME (1983) Characterization of a specific somatomedin-C receptor on isolated bovine growth plate chondrocytes. Endocrinology 112:2128–2136

Tsuji M, Kato Y, Nomura Y,Kinoshita M, Kumahara Y, Suzuki F (1983) Effect of multiplication-stimulating activity (MSA) on the cyclic AMP level and proteoglycan synthesis in cultured chondrocytes. Acta Endocrinol (Copenh) 104:117–122

Urist MR, Huo YK, Brownell AG, Hohl WM, Buyske J, Lietze A, Tempst P, Hunkapiller M, DeLange RJ (1984) Purification of bovine bone morphogenetic protein by hydroxyapatite chromatography. Proc Natl Acad Sci USA 81:371–375

Uthne K (1975) Preliminary studies of somatomedin in vitro and in vivo in rats. Adv Metab Dis 8:115–126

Valentin-Opran A, Sainte-Marie LG, Vicard E, Chenu C, Delmas PD, Saez S, Meunier PJ (1986) Androgens increase osteoblast-stimulating activity of human breast cancer cells in vitro. J Bone Min Res 1:abstract 77

Valentin-Opran A, Bonewald L, Delmas PP, Graves D, Meunier PJ, Saez S, Mundy GR (1987a) Transforming growth factor beta (TGF-beta), platelet-derived growth factor (PDGF) and insulin-like growth factor (IGF-I) are involved in metastatic breast cancer bone formation. Calcif Tissue Int 41:abstract OP12

Valentin-Opran A, Delgado R, Valente T, Mundy GR, Graves DT (1987b) Autocrine production of platelet-derived growth factor (PDGF)-like peptides by cultured normal human bone cells. J Bone Min Res 2:abstract 254

Van Buul S, van den Brande JL (1979) The Snell-dwarf mouse. II. Sulphate and thymidine incorporation in costal cartilage and somatomedin levels before and during growth hormone and thyroxine therapy. Acta Endocrinol (Copenh) 89:646–658

Van den Brande JL, van Buul Offers S (1979) Effect of growth hormone and peptide fractions containing somatomedin activity on growth and cartilage metabolism of Snell dwarf mice. Acta Endocrinol (Copenh) 92:242–257

Varga J, Jimenez SA (1986) Stimulation of normal human fibroblast collagen production and processing by transforming growth factor beta. Biochem Biophys Res Commun 138:974–980
Vetter U, Helbring G, Heit W, Pirsig W, Sterzig K, Heinze E (1985) Clonal proliferation and cell density of chondrocytes isolated from human fetal epiphyseal, human adult articular and nasal septal cartilage. Influence of hormones and growth factors. Growth 49:229–245
Vetter U, Zapf J, Heit W, Helbing G, Heinze E, Froesch R, Teller WM (1986) Human fetal and adult chondrocytes. J Clin Invest 77:1903–1908
Vittur F, Dumontier MF, Stagni N, Corvol M (1983) In vitro biosynthesis by articular chondrocytes of a specific low molecular size proteoglycan pool. FEBS Lett 153:187–193
Watanabe N, Rosenfeld RG, Hintz RL, Dollar LA, Smith RL (1985) Characterization of a specific insulin-like growth factor-I/somatomedin-C receptor on high density, primary monolayer cultures of bovine articular chondrocytes: regulation of receptor concentration by somatomedin, insulin and growth hormone. J Endocrinol 107:275–283
Wettenhall REH, Schwartz PL, Bornstein J (1969) Actions of insulin and growth hormone on collagen and chondroitin sulfate synthesis in bone organ cultures. Diabetes 18:280–284
Wilkins JR, d'Ercole AJ (1985) Affinity-labeled plasma somatomedin-C/insulin-like growth factor I binding proteins. J Clin Invest 75:1350–1358
Willis DH, Liberti JP (1985) Post-receptor actions of somatomedin on chondrocyte collagen biosynthesis. Biochim Biophys Acta 844:72–80
Wong GL, Kotliar D, Schlaeger D, VanderPol C (1987) Somatomedin C-like and somatomedin C binding activity are present in and secreted by mouse osteoblasts. J Bone Min Res 2:abstract 4
Zapf J, Froesch ER (1986a) Pathophysiological and clinical aspects of the insulin-like growth factors. Horm Res 24:160–165
Zapf J, Froesch ER (1986b) Insulin-like growth factors/somatomedins: structure, secretion, biological actions and physiological role. Horm Res 24:121–130
Zapf J, Schoenle E, Jagars G, Sand I, Grunwald J, Froesch ER (1979) Inhibition of the action of nonsuppressible insulin-like activity on isolated rat fat cells by binding to its carrier protein. J Clin Invest 63:1077–1084
Zapf J, Morell B, Walter H, Laron Z, Froesch ER (1980) Serum levels of insulin-like growth factor (IGF) and its carrier protein in various metabolic disorders. Acta Endocrinol (Copenh) 95:505–517
Zapf J, Schoenle E, Froesch ER (1987) Insulin-like growth factors I and II. Some biological actions and receptor binding characteristics of two purified constituents of nonsuppressible insulin-like activity of human serum. Eur J Biochem 87:285–296
Zingg AE, Froesch ER (1973) Effects on partially purified preparations with nonsuppressible insulin-like activity (NSILA-S) on sulfate incorporation into rat and chicken cartilage. Diabetologia 9:472–476

CHAPTER 32

Role of Lymphokines in the Immune System

E. S. VITETTA and W. E. PAUL

A. Introduction

The immune system has evolued to recognize and respond to foreign substances that are often associated with pathogenic agents. This response, if successful, leads to the elimination of the foreign substance and thus removes the associated pathogen from the body. The immune response consists of the expansion and differentiation of cells bearing clonally expressed surface receptors capable of binding the foreign substance. This leads to the secretion of specific antibodies and the production of a set of potent polypeptides that regulate the activation, growth, and differentiation of the cells of the immune system. These polypeptides also mediate the inflammatory responses that are elicited as a result of interactions among immunocompetent cells.

I. Growth Regulation in the Immune System

It can be argued that nowhere in an adult animal is cell growth more important than in the immune system. Thus, lymphocytes are clonal in nature, i.e., their receptors can bind a single antigenic epitope. Since in the unimmunized individual the frequency of lymphocytes bearing cell surface receptors specific for any given epitope is very low, if the immune response to a replicating pathogen (which contains many epitopes) is to be effective, it must be prompt and of sufficient magnitude to eliminate that pathogen. This can only be achieved if specific cells are stimulated to undergo rapid expansion and differentiation. However, such growth must be carefully controlled; stimulated lymphocyte clones must expand rapidly enough so that their progeny and their secreted products can eliminate the pathogen. Yet, this expansion must be kept within bounds so that the response to one pathogen does not impair the response to another, the production of antibody ceases once the pathogen is eliminated, and the generation of antibody which is crossreactive with normal tissues does not occur.

The central role of growth and subsequent differentiation in the immune system makes the substances that regulate these processes of special importance. This chapter is devoted to a consideration of the general role played by secreted regulatory polypeptides in the immune system. To appreciate their function, it will be necessary to review the major cellular components of the immune system and the functions mediated by cells and their products.

II. Organization of the Immune System

The immune system consists of lymphocytes, macrophages, and related "antigen-presenting cells" (APCs) such as splenic dendritic cells, specialized endothelial cells, and epidermal Langerhans' cells. These cells are found in the fixed lymphoid tissues such as the spleen, lymph nodes, and the gut-associated lymphoid tissues (Peyer's patches of the intestine), in the blood, and in a recirculating pool in the lymphatics. Lymphocytes are generated in the thymus and bone marrow.

The lymphocytes are responsible for the specific responses of the immune system by virtue of the fact that they express antigen-specific cell surface receptors. These cells exist in two broad classes; T lymphocytes and B lymphocytes.

III. T Lymphocytes

T lymphocytes mediate both regulatory and effector functions in the immune system. They recognize foreign substances on the surface of a specialized set of antigen-presenting cells (APCs) (UNANUE 1984). In general, the receptors on T cells recognize peptides bound to class I or class II major histocompatibility complex (MHC) molecules (SCHWARTZ 1985). The processing of proteins to their constituent peptides occurs within the APC and the expression of peptide-MHC complexes is limited to the surface of such cells. Thus, the specificity of T-cell receptors ensures that T cells recognize cell-bound molecules.

T cells that are stimulated as a result of recognition of peptide-MHC complexes can either lyse or help the cells with which they interact. In general, "cytotoxic T cells" (T_cs) are members of the $CD8^+$ subset of T cells and generally recognize antigenic peptides associated with class I MHC molecules. In contrast, $CD4^+$ T cells regulate immune functions by helping B cells secrete antibody in response to antigenic challenge, by activating macrophages to increased cytotoxic activity, by enhancing the growth and differentiation of T_cs, and by mediating a variety of other functions such as delayed-type hypersensitivity. Many of these actions of $CD4^+$ T cells are mediated through the production of the soluble regulatory proteins that are the subject of this chapter. It should be noted that production of these factors is not limited to $CD4^+$ T cells; some lymphokines are also produced by $CD8^+$ T cells.

1. T_{H1} and T_{H2} Cells

$CD4^+$ T cells are often referred to as helper T (T_H) cells based on the ability of some of these cells to enhance the growth of antigen-stimulated B cells (SINGER and HODES 1983). It has recently been shown that long-term lines or clones of mouse $CD4^+$ T cells can be subdivided into at least two major groups based on the pattern of lymphokines that they secrete upon activation and the functions they mediate (MOSMANN et al. 1986b). These cells have been designated T_{H1} and T_{H2}, respectively. A third group of T-cell clones (T_{H0}) has also been described based on the lymphokines they secrete (FITCH et al., personal communication). It appears that this subdivision, which will be described in greater detail below, is also exhibited by normal $CD4^+$ T cells (ARTHUR and MASON 1986; RUDD et al. 1987; BOTTOMLY 1988; BIRKELAND et al. 1988). It remains to be determined whether T_{H1}, T_{H2}, and T_{H0} cells arise from a single precursor and whether or not they represent different stages during T-cell differentiation.

IV. B Cells

The other major set of lymphocytes are the B cells. These cells are derived from precursor cells in hematopoietic tissues. They express specific membrane receptors for antigens. These receptors are immunoglobulin (Ig) molecules which differ from secreted Igs (antibodies) in that they contain transmembrane and intracytoplasmic regions which anchor them in the plasma membrane. The membrane and secreted forms of Ig are derived from the same Ig gene through an alternative pattern of RNA splicing (WALL and KUEHL 1983). Resting B cells generally express both IgM and IgD on their membrane; these Igs have the same antigen-combining site.

B cells are the precursors of antibody-secreting cells. Upon activation, B cells may develop either into antibody-secreting cells or into "memory cells." Memory cells are responsible for the rapid production of antibody upon secondary challenge with antigen. Hence, when memory cells bind antigen, antibody is made more promptly and in larger amounts than in the primary response. In addition, antibody produced in such secondary responses generally has a higher affinity for antigen than does antibody made in the course of primary responses. Immunological memory is an important hallmark of an effective immune system, although it is still not clear how memory cells are generated.

V. Receptor-Mediated Signaling

Both B cells and T cells are activated as a result of the binding of antigen (or, in the case of T cells, of peptide-MHC complexes) to their receptors. This activation requires the crosslinking of receptors on the cell surface. This implies that stimulatory antigens must express more than one copy of the same antigenic determinant. In T cells, the receptor is associated with a complex of membrane proteins designated the CD3 (or T3) complex which consists of γ, δ, ε, ζ, and η chains (CLEVERS et al. 1988). Although not established, it seems likely that the CD3 complex plays an important role in the transmission of signals generated by crosslinking of the T-cell receptor. A comparable complex may exist for B cells, but this has not yet been definitively demonstrated.

Receptor crosslinking in both cell types causes activation of the inositol phospholipid metabolic pathway with elevations in the concentrations of cytosolic-free calcium and diacylglycerol (WEISS et al. 1986; CAMBIER and

RANSOM 1987). The latter activates protein kinase C. There is strong evidence that elevation of cytosolic calcium and activation of protein kinase C play important roles in the activation of both B cells and T cells. Nonetheless, receptor crosslinking by itself is generally not sufficient to cause the proliferation and differentiation of either B or T cells. For both cell types, lymphokines are required for growth and differentiation.

VI. Cognate T-Cell–B-Cell Interactions

Soluble proteins that regulate lymphocyte growth and differentiation are generally secreted by T cells. These factors are needed not only to stimulate the growth of B cells whose receptors have been crosslinked by a multivalent antigen, but play an important role in the stimulation of growth and differentiation of B cells whose receptors have bound an antigen that has no more than one copy of any individual antigenic epitope (HOWARD and PAUL 1983). Such B cells are not directly stimulated by the binding of antigen to their receptors. However, they endocytose and process antigen that had been bound to their receptors. This results in the expression of peptides derived from that antigen in a complex with an MHC class II molecule. T cells with receptors specific for that peptide-MHC complex recognize the antigen on the B-cell surface (CHESNUT and GREY 1986) and the interaction that ensues leads to the activation of the B cells (VITETTA et al. 1989). Such activation requires a number of lymphokines produced by the interacting T cells. Activation of this type is referred to as cognate interaction and it involves close cell-cell contact and directed transfer of lymphokines (POO et al. 1988). Similar intimate interactions occur between T cells and macrophages and lead to the secretion of factors that regulate the function of macrophages. It is very likely that other cell types that express class II MHC molecules and that can process and present antigen also participate in similar cognate interactions with T cells. In these interactions, the T cell and the partner cell (the APC) are found in apposition to one another as a result of specific recognition of peptide-MHC conjugates on the surface of the APC.

VII. Secreted T-Cell Regulatory Proteins (Lymphokines)

Many of the functions of T cells described in this overview of the immune system are mediated by the action of T-cell-derived lymphokines. Those molecules and the mechanisms through which they are postulated to work are described in detail in several of the chapters of this volume. In this chapter, we will briefly review those factors that have been fully purified and cloned and will describe only their most salient characteristics. The factors that will be considered are interleukin-2 (IL-2), interleukin-4 (IL-4), interleukin-5 (IL-5), interleukin-6 (IL-6), lymphotoxin (LT or TNF-β), and interferon-γ (IFN-γ). In addition, interleukin-1 (IL-1) plays an important role in the regulation of many immune phenomena although it is not generally a T-cell product.

It should be pointed out that one characteristic of these factors is that they are very pleiotropic. Thus, each factor can stimulate cells of distinct types and can cause a variety of responses, even in the same cell type. Furthermore, the same response in the same cell type can be elicited by different lymphokines. Thus, the lymphokine system displays both pleiotropism and redundancy; the physiological significance of this characteristic of the immune system remains to be established.

1. Functions of Selected Lymphokines

a) Interleukin-2

IL-2 is a T-cell-derived lymphokine of 18 kDa that plays a variety of roles in the immune system (SMITH 1984; TANIGUCHI 1988). It was first recognized as a potent T-cell growth factor. IL-2 regulates the growth of T cells that secrete it, providing a striking natural example of regulated autocrine-stimulated growth. In addition to its action as a stimulator of T-cell growth, IL-2 can also regulate growth and differentiation of activated B cells. IL-2 acts through a cell surface receptor that consists of at least two distinct polypeptide chains, p55 and p70. Both p55 and p70 independently bind IL-2, although with relatively low affinity. Cells that express p70 only are activated to divide as a result of the binding of IL-2. In contrast, cells that express p55 only are not stimulated by IL-2. Cells that express both p70 and p55 display a high-affinity receptor and are often exquisitely sensitive to IL-2-mediated growth stimulation.

b) Interleukin-4

IL-4 is a T-cell-derived lymphokine that is 19 kDa in size (PAUL and OHARA 1987). It was first recognized because of its action as a costimulator of proliferation of mouse B cells that had been stimulated with anti-Ig (HOWARD and PAUL 1983). IL-4 has several other actions on B cells (reviewed in VITETTA et al. 1989) including the ability to increase the level of class II MHC and CD23 molecules. It has a powerful effect in regulating Ig class switching to IgG_1, and IgE in vitro and to IgE in vivo (ISAKSON et al. 1982; COFFMAN and CARTY 1986; FINKELMAN et al. 1986). IL-4 functions as a T-cell growth factor in a manner similar to that of IL-2. IL-4 acts to costimulate the growth of in vitro mast cell lines and of precursors of erythroid, myeloid, and megakaryocytic cells. IL-4 also stimulates macrophage activation. The IL-4 receptor exists in low copy number on most cells of the hematopoietic lineage and on several other cell types.

c) Interleukin-5

IL-5 is a T-cell-derived lymphokine of 55 kDa that was first recognized for its ability to replace T-cell help in anti-sheep erythrocyte antibody responses and was originally designated T-cell-replacing factor (TRF) (HAMAOKA and ONO 1986; SANDERSON et al. 1988). It has striking effects in regulating differentiation of eosinophils. In mice, IL-5 also has potent action in regulating growth of B-cell blasts and of B-lymphoma cell lines (SWAIN et al. 1983); it also promotes Ig secretion by normal B cells and by B-cell lines. IL-5 has been reported to have little activity on human B cells but is a potent differentiation factor for human eosinophils (SANDERSON et al. 1988).

d) Interleukin-6

IL-6 is a highly pleiotropic factor made by a variety of cell types. It also has a broad range of activities (KISHIMOTO and HIRANO 1988). It regulates the growth of plasmacytomas and hybridomas and is a potent stimulant of Ig secretion. IL-6 also acts as a hepatocyte-stimulating factor and has been designated IFN-β2. It also has potent effects on the stimulation of very immature hematopoietic stem cells.

e) Lymphotoxin (TNF-β)

Lymphotoxin (TNF-β) is a glycoprotein with an M_r of 25000 principally, but not exclusively, made by T_{H1}, cells (PAUL and RUDDLE 1988). It has homology to tumor necrosis factor (TNF-α) (NEDOSPASOV et al. 1986). The genes encoding lymphotoxin and TNF-α are closely linked and the two factors utilize a common receptor on their target cells (AGGARWAL et al. 1985). Lymphotoxin was first recognized by virtue of its ability to inhibit the growth of rat embryo fibroblasts. It is cytotoxic for cell lines of a variety of types and has been reported to kill LPS-activated but not resting B lymphocytes (POWELL et al. 1985). It also induces growth in some cell lines and can induce cellular differentiation of certain cell lines whose growth rate is not affected by lymphotoxin.

f) Interferon-γ

Interferon-γ IFN-γ is secreted by activated T_{H1} cells as well as by $CD8^+$ T_c cells and by natural killer (NK) cells. It mediates a variety of important functions in the immune system. It upregulates the expression of both class I and class II MHC molecules on a variety of cell types, with the striking exception of B-lymphocyte expression, where IFN-γ not only fails to induce expression of class II MHC molecules, but inhibits the inducing effects of IL-4. IFN-γ can activate macrophages by enhancing their capacity to lyse a variety of tumor cell targets and to destroy cells that have ingested parasites (PACE et al. 1983; TRINCHIERI and PERUSIA 1985). IFN-γ has been reported to synergize with IL-1 and IL-2 in the generation of anti-sheep erythrocyte antibody responses and promotes Ig secretion by resting or activated mouse and human B cells (LIEBSON et al. 1982; LEIBSON et al. 1984; SIDMAN et al. 1984). IFN-γ also promotes the production of murine I_g of the IgG_{2a} class in LPS (SNAPPER and PAUL 1987) or T-cell-stimulated (STEVENS et al. 1988) B cells. On the other hand, it inhibits many immunological functions. In addition to its antiproliferative effects, IFN-γ opposes many of the actions of IL-4 on B cells, including induction of class II MHC molecules (OLIVER et al. 1987; MOND et al. 1986b), costimulation of B-cell proliferation (MOND et al. 1986a), and switching to the secretion of IgG_1 and IgE (COFFMAN and CARTY 1986; REYNOLDS et al. 1987; SNAPPER and PAUL 1987).

VIII. Lymphokines Produced by T_{H1} and T_{H2} Cells; Implications for Immune Functions

As noted above, most long-term mouse T-cell lines can be divided into T_{H1}, T_{H2}, and T_{H0} types based on the patterns of lymphokines they produce. Both T_{H1} and T_{H2} cells produce IL-3, CM-CSF, and TNF. T_{H1} cells secrete IL-2 and IFN-γ, but not IL-4 and IL-5, while T_{H2} cells show the opposite pattern (CHERWINSKI et al. 1987). T_{H3} cells secrete IFN-γ, IL-4, and often IL-2 (FITCH et al., personal communication). As will be discussed in greater detail below, this pattern of lymphokine production separates the $CD4^+$ T cells into at least two populations based on their functional capacities. T_{H2} cells secrete a set of lymphokines that may play an important role in mediating allergic inflammatory responses and in antiparasite immunity. Among these, IL-3 is the major mast cell growth factor, thus regulating the number of the cells that release vasoactive amines. IL-4 is the principal regulator of Ig class switching to IgE and thus controls the amount of the Ig that sensitizes mast cells and basophils for antigen-stimulated triggering. IL-5 is the major regulator of eosinophil differentiation; eosinophils are prominent in allergic fluids and play an important role in the elimination of parasites. A similar argument can be made for T_{H1} cells in the mediation of cellular immune and cytotoxic T-cell responses. IFN-γ regulates the level of expression of MHC molecules on many cell types, thus enhancing the capacity of T cells to recognize and respond to virus-infected and other altered cells. IFN-γ also exhibits potent action as a macrophage activation factor, implying that the cells which produce it play a major role in immune response to intracellular microorganisms. As mentioned earlier, a third type of $CD4^+$ clone (T_{H3}) which secretes IFN-γ and IL-4 has been described. These cells can provide help for activated B cells in certain experimental systems in mice and humans (FERNANDEZ-BOTRAN et al. unpublished observations). This concept that immunological functions are regulated through the pattern of lymhokines produced under a given set of circumstances will be developed in greater detail in Sect. D of this chapter.

B. Role of Lymphokines in the Immune Response

I. T-Cell Subsets

Activated T cells play a fundamental role in the regulation of humoral and cellular immune responses through direct cell/cell contact and/or release of lymphokines.

Three types of functionally distinct T cells have been described; T helper (T_H) cells, T suppressor (T_s) cells and T cytotoxic (T_c) cells. T_H cells mediate a variety of immunological functions and generally regulate the immune response. T_H cells mediate delayed-type hypersensitivity, help for B-cell-mediated antibody responses, and are involved in the activation of T_c cells. T_s cells generally downregulate the immune response and T_c cells are responsible for killing virus-infected cells, tumor cells, and other cells expressing foreign antigens on their surface. In general, T cells mediate their responses both by cell/cell contact and by secreted lymphokines and the regulation of T-cell activation and proliferation is controlled in part via lymphokine feedback circuits.

Functional heterogeneity exists among T cells and different subsets can be distinguished based on their expression of cell surface markers. T cells that help antibody responses express Ly1 and L3T4 in mice and CD4 in humans, while T cells that mediate cytotoxicity or suppression express Ly2,3 in mice and CD8 in humans.

In the past few years, with the development of techniques to clone and propagate functionally active, antigen-specific T_H cells in vitro (KIMOTO and FATHMAN 1980), their heterogeneity has become even more apparent. It has recently been reported that the pattern of lymphokines secreted by clones of established T_H cells can delineate at least two types of functionally distinct cells. MOSMANN et al. (1986b), studied a large panel of clones of L3T4$^+$ *MHC*-restricted T_H cells raised against a variety of antigens in a number of mouse strains. Virtually all clones fell into one of two groups depending on the pattern of lymphokines secreted after mitogenic or antigenic stimulation, and the type of "help" they provided to B cells. Cells from one type of clone, designated as T_{H1}, secreted IL-2, IFN-γ, and IL-3, but not IL-4, and some T_{H1} clones, but not others, could help B cells. The other type of clone T_{H2}, secreted IL-4, IL-5, and IL-3, but not IL-2 or IFN-γ, and provided excellent help to B cells to induce the secretion of IgG_1 and IgE. Thus, functional subsets of T_H cells can be distinguished based on their pattern of secreted lymphokines. Furthermore, differential secretion of lymphokines is at least partially responsible for the functional differences between the two types of T_H clones. Subsequently, the list of lymphokines produced by each type of T_H cell has been extended (CHERWINSKI et al. 1987). An alternative classification, proposed by Bottomly and colleagues (KIM et al. 1985; KILLAR et al. 1987; RASMUSSEN et al. 1988), divides T_H cells into "helper" and "inflammatory." These two subtypes are, for the most part, equivalent to the T_{H2} and T_{H1} subtypes, respectively, described by Mosmann et al. (MOSMANN et al. 1986b; BOTTOMLY 1988). Finally, some clones (T_{H0}) have been described which secrete lymphokines usually produced by only one or the other type of clone (NOELLE et al., submitted for publication; FERNANDEZ-BOTRAN, unpublished observations; FITCH et al., personal communication). Thus, it remains possible that either the delineation of all clones into T_{H1}, T_{H2}, or T_{H0} types is generally, but not always, the rule and/or that these clones represent T_H cells "arrested" at different stages of differentiation, i.e., that normal T_H cells of one type can give rise to T_H cells of the other type. Hence, these rare clones may represent "transitional" cells or precursor cells.

II. Functional Differences Between T_{H1} and T_{H2} Cells

Evidence has accumulated that clones representing the T_{H1} and T_{H2} types of T_H cells defined by MOSMANN et al. (1986b) are functionally different. T_{H1} but not T_{H2} cells can mediate antigen-specific delayed-type hypersensitivity reactions (CHER and MOSMANN 1987). Both T_{H1} and T_{H2} cells provide help to B cells for antibody responses, although T_{H2} clones are generally better helper cells than T_{H1} clones. In fact, some T_{H1} clones cannot provide help at all, while others can induce B cells to secrete IgM, IgG_3, and IgG_{2a} antibody. There is therefore a discrepancy with regard to the role of T_{H1} cells in providing help to B cells that will subsequently secrete antibodies. Three possible reasons for this discrepancy can be postulated: (a) IFN-γ (secreted by activated T_{H1} cells) often provides negative signals to B cells (MOND et al. 1986a, b; RABIN et al. 1986; COFFMAN and CARTY 1986; OLIVER et al. 1987; REYNOLDS et al. 1987); it is unclear why IFN-γ is sometimes, but not always, inhibitory. It is possible that different clones of T_{H1} cells secrete different amounts of IFN-γ and that the high secretors prevent B-cell help; (b) B cells which respond to T_{H1} clones may represent distinct subsets (e.g., memory versus virgin B cells) or cells at different stages of activation (KILLAR et al. 1987; STEVENS et al. 1988; MOSMANN et al. 1986b). A number of studies have now suggested that the ability of T_H cells to help B cells is, in fact, critically dependent on the activation state of the B cells. (c) Some T_{H1} clones, but not others, may secrete low levels of IL-5, a critical growth factor for B cells (BOTTOMLY 1988). The reason for the conflicting evidence concerning the helper activity of T_{H1} clones remains to be clarified.

III. Surface Markers of the Different T_H-Cell Subtypes

Although the characteristic pattern of lymphokine secretion can be used as a functional "marker" to distinguish murine T_{H1} and T_{H2} cells (MOSMANN et al. 1986b; CHERWINSKI et al. 1987), no cell surface marker has been described that is present *exclusively* on one type C, T_{H1} or T_{H2}) of murine T_H cell. However, antibodies have been reported to dinstinguish between functionally different T_H cells in several species. ARTHUR and MASON (1986) reported that the antibody MRC-OX22 could separate rat T_H cells into those responsible for the majority of IL-2 production ($OX22^+$) and those responsible for B-cell help ($OX22^-$). MRC-OX22 antibodies recognize an epitope on the common leukocyte antigen or T-200 molecule (CD45R) (SPICKETT et al. 1983). Based on recent reports, antibodies recognizing epitopes on human (RUDD et al. 1987) and mouse (BOTTOMLY 1988; BIRKELAND et al. 1988) CD45R molecules also appear to distinguish between functionally different T_H cells. Although the ontogenetic relationship between $CD45R^+$ and $CD45R^-$ T_H cells is not known, antigen or other factors may cause the conversion of one T_H-cell type into another. If this were the case, there would be a predictable sequence of lymphokines secreted during an immune response and these lymphokines would regulate the order of Ig isotypes secreted and the types of effector cells activated.

IV. Proliferative Response of Clones of T_{H1} and T_{H2} Cells

T_H-cell clones do not constitutively secrete lymphokines but are dependent upon antigenic or mitogenic stimulation to do so (MOSMANN et al. 1986b; FERNANDEZ-BOTRAN et al. 1986a; CHERWINSKI et al. 1987). Another important consequence of T_H-cell activation after antigen presentation is the acquisition of a state of "competence" which is characterized by the responsiveness of the T_H cells to the proliferative effects of some of the secreted lymphokines, particularly IL-2 (CANTRELL and SMITH 1983; ROBB et al. 1981; FERNANDEZ-BOTRAN et al. 1986a) and IL-4 (FERNANDENZ-BOTRAN et al. 1986a, 1988). T_H cells are stimulated by the binding of their receptors to either anti-TcR antibodies or antigen/class II (on an APC) (reviewed in ALCOVER et al. 1987; UNANUE 1984). This results in the induction of lymphokine secretion and the expression of cell surface receptors (R) for IL-2 (IL-2R) and IL-4 (IL-4R) (reviewed in VITETTA et al. 1987; MOSMANN and COFFMAN 1987). The secreted lymphokines act via autocrine and paracrine mechanisms to induce the proliferation and expansion of the activated T_H cells as well as proliferation and differentiation of the B cells which they help (VITETTA et al. 1984; FERNANDEZ-BOTRAN et al. 1986a, 1988; LICHTMAN et al. 1987; KUPPER et al. 1987; GREENBAUM et al. 1988).

V. Regulation of the Activation of T_{H1} and T_{H2} Cells

The different types of T_H cells differ in their sensitvity to the stimulatory or inhibitory effects of a variety of lymphokines (IL-4, IL-1, IFN-γ) (KURT-JONES et al. 1987; GREENBAUM et al. 1988; FERNANDEZ-BOTRAN et al. 1988), raising the possibility that lymphokine-mediated circuits regulate the activation, expansion, and eventual downregulation of each particular type of T_H cell. In this regard, IL-2 was originally defined as a lymphokine which mediated the proliferation of activated T cells (reviewed in SMITH 1980), and IL-4 was defined as its counterpart for activated B cells (HOWARD et al. 1982). However, IL-4 can also mediate the proliferation of some T-cell lines (MOSMANN et al. 1986a; LEE et al. 1986; GRABSTEIN et al. 1986a; FERNANDEZ-BOTRAN et al. 1986b; SEVERINSON et al. 1987) or resting T cells [in combination with phorbol myristate acetate (PMA)] (HU-LI et al. 1987). The delineation of the different T_H-cell subsets by MOSMANN et al. (MOSMANN et al. 1986b; CHERWINSKI et al. 1987) and the characterization of their lymphokines prompted investigations of how antigen-induced proliferation was mediated in T_{H2} cells, which were reported to lack the ability to secrete IL-2. A number of laboratories (FERNANDEZ-BOTRAN et al. 1986a; LICHTMAN et al. 1987; KUPPER et al. 1987) reported that when stimulated in vitro by antigen on APCs, T_{H2} cells could proliferate in response to IL-4 in an autocrine fashion. In T_{H1} cells, IL-2-mediated proliferation was also an autocrine event (LICHTMAN et al. 1987). FERNANDEZ-BOTRAN et al. (FERNANDEZ-BOTRAN et al. 1986a, 1988) reported that activated T_{H2} cells could also utilize IL-2 as a paracrine growth factor and that their responsiveness to both IL-2 and IL-4 increased after antigenic stimulation. These results suggest that the proliferation of one T_H-cell type can be modulated by lymphokines secreted by the other type of T_H cell. When *both* IL-2 and IL-4 are added to cultures of T_{H1} or T_{H2} cells, the proliferative response of both cell types is synergistic, suggesting that the presence of both lymphokines might be required for an optimal response (FERNANDEZ-BOTRAN et al. 1988).

T_{H1}-derived IFN-γ also has a regulatory effect on the proliferative response of T_{H2} cells. IFN-γ is generally inhibitory for IL-4-mediated activities on B cells (MOND et al. 1986a, b; RABIN et al. 1986; COFFMAN and CARTY 1986; OLIVER et al. 1987; FEYNOLDS et al. 1987). IFN-γ also inhibits the IL-2- and IL-4-mediated proliferation of T_{H2}, but not T_{H1} cells (GAJEWSKI and FITCH 1988; FERNANDEZ-BOTRAN et al. 1988).

In summary, there are a number of mechanisms whereby the initial stimulation of one or both types of T_H cells could be amplified and subsequently downregulated. T_H cells would become stimulated after antigen presentation by APCs, leading to lymphokine secretion and the expression of receptors for these lymphokines. The secreted lymphokines would then act via autocrine or paracrine mechanisms to induce the expansion of clones of activated T cells and proliferation and differentiation of activated B cells. Eventually, when antigen concentrations decrease, T-cell activation would become self-limiting by downregulation of surface IL-2Rs and IL-4Rs and by the cessation of lymphokine secretion. In addition, it has become clear that lymphokines produced by one T_H-cell type (i.e., IFN-γ) can downregulate the growth of the other T_H-cell type (GAJEWSKI and FITCH 1988; FERNANDEZ-BOTRAN et al. 1988).

VI. T_{H1} and T_{H2} Cells In Vivo

The characterization of the two types of T_H cells has been carried out in vitro using established clones of cells (MOSMANN et al. 1986b; CHERWINSKI et al. 1987; CHER and MOSMANN 1987; KILLAR et al. 1987). However, studies correlating the phenotype of normal T cells (Cd45R-low and CD45R-high) with their pattern of lymphokine secretion and other functional characteristics provide strong evidence that analogs of T_{H1} and T_{H2} cells exist in vivo (BOTTOMLY 1988; BIRKELAND et al. 1988). However, as mentioned previously, it is still not clear whether T_{H1} and T_{H2} cells represent two different lineages which differentiate from a common precursor, or whether they represent different stages of the T_H cell in the same differentiation pathway, although there is evidence to support the latter possibility (CLEMENT et al. 1988). Studies by SWAIN et al. (1988) and POWERS et al. (1988) suggest that precursors of T_{H2} cells require a maturation event before they can be stimulated to secrete IL-4 or IL-5. The delayed development of IL-4-producing cells (T_{H2}) after repeated antigenic exposure suggests a potential role for T_{H2} cells in the generation of immunological memory and/or in the immune response against antigens that are persistent or frequently encountered (POWERS et al. 1988).

C. Action of Lymphokines on Macrophages

Macrophages obtained from exudates induced by sterile irritants have a limited capacity to lyse tumor cells and microbial agents. In contrast, macrophages recovered in the course of in vivo immune responses to infectious agents such as *Listeria monocytogenes* or even to soluble protein antigens have the capacity to lyse a wide variety of bacteria, parasites, and neoplastic target cells (MACKANESS 1962). Such immunologically mediated macrophage activation is induced by activated T cells through the local production of macrophage-activating factors. In general, macrophage activation occurs in the context of T-cell recognition of antigen-MHC products on macrophages. It is presumed that in such cognate macrophage-T-cell interactions a mutual activation of the T cell and the macrophage occurs. Whether activation of bystander macrophages (i.e., macrophages which are themselves not involved in direct interactions with activated T lymphocytes) also occurs is uncertain, but seems likely in view of the large number of macrophages that become activated in physiological immune responses.

The identification and characterization of the macrophage activation factors that T cells produce has been a long-standing goal. The initial effort toward accomplishing this goal was the demonstration that T cells secreted a product that inhibited the migration of macrophages in vitro (DAVID 1966; BLOOM and BENNETT 1966). This material was designated migration inhibition factor (MIF). It was clear that production of MIF was an in vitro correlate of delayed-type hypersensitivity but the molecular characterization of MIF and the determination of its role in macrophage activation proved to be a very difficult matter. It has now been demonstrated that IFN-γ is a macrophage-activating factor (ROBERTS and VASIL 1982). Furthermore, monoclonal antibodies to IFN-γ remove the bulk of macrophage activation activity from supernatants (SNs) of in vitro-stimulated normal T cells. Administration of anti-IFN-γ antibody leads to the death of mice that receive a sublethal inoculation of *Listeria* organisms (BUCHMEIER and SCHREIBER 1985), strongly indicating that IFN-γ has a critical in vivo role in protection against *Listeria* infection, presumably because of its capacity to activate macrophages.

Nonetheless, although IFN-γ is a T-cell-derived macrophage activation factor, it is not the only T-cell product with such function. Both granulocyte-macrophage colony-stimulating factor (GM-CSF) (GRABSTEIN et al. 1986) and IL-4 (CRAWFORD et al. 1987) have considerable macrophage-activating potential, particularly for the lysis of tumor cells. In contrast, the induction of resistance to infection with *Leishmania major* requires the action of both IFN-γ and a second T-cell product. IL-4, GM-CSF, and IL-2 each have the capacity to complement IFN-γ in the induction of antimicrobicidal activity to *Leishmania* (BELOSEVIC et al. 1988). These results indicate that the T-cell-derived macrophage activation factors are not identical in their function. A detailed understanding of the distinctive properties of these activation factors and their relative physiological roles in antimicrobial and antitumor effects is clearly needed.

D. Actions of Lymphokines in B-Cell Responses

The responses of B lymphocytes to antigenic substances can be subdivided into stages of activation from the resting state, proliferation, differentiation into antibody-secreting cells, and Ig class switching. This complex set of responses can be initiated in B cells either through cognate interactions with specific T cells or as the result of crosslinking of their cell surface receptors (see Sect. A.V, A.VI). In both cases, the responses require the participation of soluble T-cell-derived lymphokines.

I. Activation

Soluble anti-Ig antibody is capable of binding to membrane Ig on essentially all B cells and thus provides a useful model for studying B-cell stimulation. Resting B cells cultured at low density will enter the G1-phase of the cell cycle in response to stimulation with anti-Ig but will generally fail to divide. Entry of such cells into the S-phase of the cell cycle requires a costimulant, such as IL-4 (HOWARD and PAUL 1983). For IL-4 to be effective in the B-cell response to anti-Ig, it is required at the outset of the stimulation. This indicates that one of its important actions occur on the resting or recently activated cell. Indeed, IL-4 acts directly on resting B cells (OLIVER et al. 1987) to regulate class II MHC molecules (NOELLE et al. 1984), upregulate CD23 (FcεRII) expression (KIKUTANI et al. 1986; HUDAK et al. 1987), and increase the numbers of IL-4 receptors on resting murine B cells (OHARA and PAUL 1988). Thus, IL-4 can be classified as a "cocompetence factor" in that it acts, together with anti-Ig, on resting B cells and allows those cells to become competent to develop into dividing cells. This function of IL-4 on B cells, and many of its other actions on such cells, is inhibitable by IFN-γ (RABIN et al. 1986). Human B cells can also be stimulated to divide with anti-IgM and recombinant human IL-4 (YOKOTA et al. 1986), but other lymphokines are also costimulatory (with anti-Ig) on resting B cells. These include the IFNs (MORIKAWA et al. 1987) and B-cell growth factors (BCGFs) (MURAGUCHI et al. 1983).

In cognate T-cell/B-cell interactions as in anti-Ig-mediated stimulation, it is often assumed that the the same general rules for growth regulation pertain except that the T cells contribute an "activation" stimulus that is in some way equivalent to the stimulus delivered by receptor crosslinking (by anti-Ig) or by LPS. The nature of the T-cell-mediated stimulus has not yet been elucidated. Indeed, it has not been convincingly determined whether this stimulus is due to the local production of an "activating lymphokine," to signals generated by membrane contact, or by the interaction of cellular ligands and receptors. However, it might be imagined that costimulants, such as IL-4, would be required together with such T-cell-derived activating stimulants to cause the B cell to divide. Although this is quite plausible, IL-4 does not appear to be an exclusive costimulant in cognate interactions since T-cell clones that do not produce IL-4 (i.e., T_{H1} lines) can, in some cases, stimulate resting B cells (COFFMAN et al. 1988; STEVENS et al. 1988). This may indicate that several T-cell products can mediate this function or that no soluble costimulant is required at this activation step.

II. Growth Stimulation

In many other cellular systems, entry of activated cells into S-phase depends upon the action of a specific growth factor. For T cells, IL-2 and, in some cases, IL-4, express this function. For B cells, no single potent growth factor that acts in the first round of cell division of previously resting B cells has been described. Thus, B-cell blasts produced by stimulating resting B cells with anti-Ig and IL-4 and harvested at 24 h show only modest proliferative responses to IL-4 alone (RABIN et al. 1986). Both IL-2 and IL-5 have been reported to be active as mouse B-cell growth factors (ZUBLER et al. 1984; PRAKASH et al. 1985; LOWENTHAL et al. 1985; SWAIN et al. 1983), but their stimulatory activity is rather modest on early B-cell blasts in the mouse. IL-2, but not IL-5, has been reported to be active on human B cells (SANDERSON et al. 1988). Although IL-6 is a potent growth factor for plasmacytomas and hybridomas, it appears to have no activity as a growth stimulant for normal human B cells (KISHIMOTO and HIRANO 1988). On the other hand, B-cell blasts divide vigorously in response to stimulation with additional anti-Ig or with LPS (RABIN et al. 1986). This may reflect the ability of receptor crosslinking or LPS to act as direct growth stimulants or it may be due to the production of an autocrine growth factor by appropriately stimulated activated B cells. If activation of blasts by receptor crosslinking is a result of autocrine growth factor production, this factor has not been identified. Alternatively, the stimulation of growth may reflect the synergistic actions of several growth factors.

Mouse B-cell blasts that have been stimulated over several days (ZUBLER et al. 1984) and activated human B cells (WALDMANN et al. 1984) express high-affinity receptors for IL-2. IL-2 often delivers potent growth-promoting stimuli to such cells, suggesting that it may act as a growth factor for activated B cells. Indeed, as has been pointed out earlier, mouse antibody responses to sheep erythrocytes are stimulated in vitro, in the absence of T cells, by antigen together with a mixture of IL-1, IL-2, and IFN-γ (MARRACK et al. 1982). This may reflect the fact that responses to sheep erythrocytes are a property of memory B cells and that the growth stimulatory responses of such cells may differ from those of "virgin" B cells.

III. Differentiation of B Cells into Antibody-Producing Cells

Ig secretion by B cells is tightly regulated. Among the important molecular changes involved in the development of B cells into the antibody secretory state is the change in the pattern of μ-chain mRNA expression from that predominantly encoding the membrane form of the chain to that predominantly specifying the secretory form (WALL and KUEHL 1983). This is accompanied by an increase in the total amount of mRNA for the secreted form of the μ-chain and by the appearance of J-chain mRNA and protein (KOSHLAND

1983). In normal mouse B cells that have been activated with anti-IgM and IL-4, this transition can be induced by IL-5 and IL-2 (NAKANISHI et al. 1984). In this model, IL-5 is required early in the culture period while the addition of IL-2 can be delayed. Furthermore, BLACKMAN et al. (1986) have analyzed a B lymphoma that already expresses a high ratio of the secretory to the membrane form of μ mRNA but fails to secrete IgM or to express J-chain mRNA. These cells are induced to express J-chain mRNA and to secrete IgM by the addition of IL-2 alone. This indicates that IL-2 is a late-acting factor responsible for the final steps of differentiation to Ig secretion in some B cells. Another factor that has been shown to be a potent stimulant of Ig secretion is IL-6 (KISHIMOTO and HIRANO 1988). Since IL-6 has been shown to act on plasmacytomas and hybridomas, it is an excellent candidate to be a late-acting differentiation factor in Ig secretion.

IV. Lymphokine Regulation of Ig Class Switching

The biological functions of Igs are determined by the structure of their constant regions. Thus, distinct classes have specialized biological functions and regulation of class expression is a central feature of the regulation of immune responses. It has been known for some time that T cells play an important role in determining the class of Ig produced in the course of an immune response. Recent studies have established that Ig class expression is largely regulated in vitro by the action of lymphokines. LPS-stimulated B cells normally secrete large amounts of IgM and substantial amounts of IgG_3 and IgG_{2b}. Addition of IL-4 strikingly enhances production of IgG_1 (VITETTA et al. 1985) and of IgE (COFFMAN et al. 1986) and suppresses production of the other isotypes. In contrast, addition of IFN-γ enhances production of IgG_{2a} and suppresses production of IgG_3 and IgG_{2b} (SNAPPER and PAUL 1987). The effects of IL-4 and of IFN-γ appear to occur prior to switching since the lymphokine can be removed before cells have expressed Ig of the "new" isotype on their membranes as surface receptors (SNAPPER and PAUL 1987) and before detectable functional mRNA for this isotype appears (JONES et al. 1983). Furthermore, clonal assays indicate that the frequency of precursors that can give rise to IgG_1- and IgE-producing cells is markedly increased by IL-4, reaching levels as high as 70% of the precursors (LAYTON et al. 1984; LEBMAN and COFFMAN 1988; BERGSTEDT-LINDQVIST et al. 1988). The findings that the lymphokines act prior to the switch and that precursor frequencies as high as 70% can be achieved supports the conclusion that the lymphokines determine the specific class of Ig to which a precursor cell will switch.

Recent studies have shown that IL-4 acts to regulate IgE production in vivo as well as in vitro (FINKELMAN et al. 1986). Administration of monoclonal anti-IL-4 antibody to mice strikingly inhibits IgE resposes to infection with *Nippostrongylus brasiliensis* or to injection with anti-IgD antibodies. On the other hand, such treatment fails to suppress IgG_1 responses, indicating that IgG_1 production may be controlled in vivo by factors other than or in addition to IL-4. Indeed, while the capacity of T_{H2} clones to induce IgE production is fully inhibited by anti-IL-4, the stimulation of IgG_1 production by these clones is partially resistant to the action of anti-IL-4 (HAUSER et al. 1989). Efforts are now in progress to clarify the regulation of IgA production by T-cell-derived lymphokines, although it has recently been reported that in mouse B cells both TGF-β (COFFMAN, personal communicaton) and IL-5 (COFFMAN et al. 1988) are involved.

V. B-Cell Growth and Development Control by Action of T-Cell-Derived Lymphokines

The results outlined in the sections above on the activation, proliferation, differentiation, and Ig class-switching events in B cells indicate that soluble T-cell-derived polypeptides play a critical role in these events. It is still not clear that the lymphokines thus far identified are the major physiologically acting factors; indeed, the possibility that combinations of lymphokines are physiologically of critical importance is only beginning to be explored. The pleiotropism of lymphokine function and the corollary that several lymphokines may express the same function suggest that the concerted action of lymphokines will prove to be key to the T-cell regulation of B-cell responses, as well as to the control of other cellular responses in the immune and hematopoietic systems.

E. Conclusions

In conclusion, we will attempt to combine well-known facts about the immune system and how it responds to different pathogenic organisms with a contemporary view of cellular interactions, lymphokine secretion, and effector mechanisms mediated by different cells and lymphokines. For the sake of brevity, we will use only a few specific examples in our discussion and will not attempt to reference the many well-known and important papers that provided their factual basis. Although this synthesis is incomplete and speculative, it brings to light the issues which we must understand and address over the next decade.

For the individual to mount an effective immune response against a particular pathogen, a number of cellular and molecular events must be tightly regulated. These include the isotypes of antibody secreted by the progeny of the activated B cells, the generation of memory cells, and the activation of effector cells. An attractive working hypothesis to explain how cellular interactions are initiated and regulated is that the type of antigen and its route of entry will determine the site and mechanism by which it is metabolized by the host. This, in turn, will influence the types, numbers, and perhaps the sequence in which T_{H1} and T_{H2} cells are activated. Once activated, these T_H cells can then regulate, via cell contact and/or release of lymphokines, the isotype of antibody secreted by specific B cells as well as the additional types of effector mechanisms generated, i.e., mast cell activation, recruitment of eosinophils, macrophage activation, delayed-type hypersensitivity (DTH), and the generation of T_s and T_c cells.

I. Lymphoid Organs

After local invasion of the body by pathogens or other antigens, cells in the draining lymph nodes interact to initiate an immune response. Within the lymph nodes, the B cells are sequestered in follicles, while the T cells (which enter the node through postcapillary venules) take up residence in paracortical areas. Soluble circulating antigen, antigen from the tissues, and antigen engulfed by macrophages enter the lymph node through the afferent lymphatics or through the capillaries. In the node, antigen is ingested by macrophages at the periphery of the cortex and in the germinal centers of the follicle, afferent sinuses, diffuse cortex, and medulla. In contrast to macrophages, dendritic cells may not process antigen efficiently, but may bind antigen fragments released by macrophages. Antigen not ingested by APCs collects in the primary follicle and is retained on the follicular dendritic cells (FDCs) in close proximity to the follicular B cells. There are several major mechanisms by which antigen from the blood and lymph can initiate cellular interactions in the lymph node. (a) When antigen concentrations are high, the macrophages within the nodes are very effective at ingesting organisms, degrading them in lysosomes, and either presenting antigenic fragments to T_H cells or releasing antigen fragments which then bind to dendritic cells or B cells. Because of their efficiency in taking up foreign material, macrophages may play a predominant role both as APCs and as a source of released antigen fragments. (b) The dendritic cells in the follicles very effectively capture immune complexes and retain this material on their plasma membrane for significant periods. Small antigen-containing portions of the membrane of the FDCs pinch off and presumably can be taken up by B cells in the follicle. FDCs serve as reservoirs for native antigen and hence may be the major source of antigen for the follicular B cells. It is also possible that FDCs present paucivalent epitopes to B cells in a multivalent form since high antigen concentrations on the surface of FDCs would be operationally multivalent. This would result in more effective crosslinking of sIg molecules on B cells. (c) In contrast to macrophages, antigen-specific B cells have an advantage in capturing antigen because of both their clonally expressed sIg receptors and their proximity to FDCs. Hence, when antigen concentrations are low, the B cells may be the more efficient APCs. Furthermore, the B cells may play a role in clonally expanding T cells activated by dendritic cells.

Taken together, mechanisms a–c above suggest that, in the simplest model, B cells would bind native antigen, while T cells would bind processed antigen, on the surface of clustered macrophages and dendritic cells at the periphery of the germinal centers. Here, clonal expansion of both the B and T_H cells should occur after antigen-specific, *MHC*-restricted interactions. The mechanism by which the antigen-specific T cells travel across the paracortical areas to the germinal centers is not known, but this could be mediated by chemotactic factors secreted by APCs. After interactions between T_H cells, B cells, and APCs, the activated T_H cells secrete lymphokines that are responsible for the growth of both T cells and B cells as well as for the differentiation of the B cells into antibody-producing cells. Fully differentiated plasma cells migrate to the medulla of the lymph node and secrete antibody into the efferent lymphatics. Antibody secreted from all the regional lymph nodes is collected in the lymphatic system and enters the bloodstream via the thoracic duct. In addition, memory cells in the node are probably generated in the germinal centers by the dividing B cells. Perhaps when the supply of antigen is exhausted, cycling B cells return to their G_0 state and either remain in the follicles or leave the nodes in the efferent lymph and travel to other areas of the body (e.g., bone marrow). Factors influencing the migration of memory B cells to other locations may include the induction of homing receptors and/or the numbers of memory cells in the nodes, i.e., for a node to return to its resting state, memory cells would be released into the lymph. The isotype of antibody produced by plasma cells (e.g., sIgG versus sIgA, etc.) is probably controlled by the types of memory cells and T_H cells in the different lymphoid tissues which are activated. The T-cell-derived lymphokines from T_{H1} or T_{H2} cells could mediate switching or induce B cells already expressing other isotypes to differentiate into plasma cells. Lymphokines would play a major role in regulating the secretion of different isotypes of antibody. Hence, IL-4 and IL-5 (secreted by T_{H2} cells) would be involved in IgG_1, IgE, and IgA antibody production while IFN-γ (secreted by T_{H1} cells) would be involved in IgG_{2a} antibody production. The growth and terminal differentiation of plasma cells may also require IL-6. In order for the T_{H1} and T_{H2} cells to interact in regulating the B cells, lymphokines from one type of T_H cell could upregulate receptors for lymphokines secreted by the other type of T_H cell. Since evidence is mounting that T_{H1} and T_{H2} cells might represent different stages of the same lineage, it is possile that low levels of antigen (on APCs) activate T_{H1} cells and that T_{H2} cells require repeated antigenic stimulation to be activated. Alternatively, T_{H2} cells might reside near the follicles while T_{H1} cells would cluster around macrophages in paracortical areas. Furthermore, antigen could induce T_{H1} cells to become T_{H2} cells. This would imply that the type of T cell activated would depend on the dose and/or rate of clearance (versus persistence) of antigen as well as the location of the different types of T cells in the different areas of the node. For example, in the case of minor viral infections (where antigen levels in the circulation would initially be low), T_{H1} cells, T_s cells, and T_c cells might be preferentially activated whereas, in the case of parasites (where antigen persists), T_{H2} cells may be activated or T_{H1} cells would be induced to differentiate into T_{H2} cells.

II. Immune Responses Against Bacterial Antigens

Bacteria are highly diverse and can enter the host by many different routes. Hence, the type of immune response to bacteria may be dependent on the portal of entry into the body. For example, in the gut and mouth, bacteria must penetrate mucosal or epithelial surfaces to enter the bloodstream. Any type of immune response which prevents attachment of bacteria to mucosa or epithelial cells will effectively prevent systemic infection. On the other hand, it is desirable *not* to damage the protective lining of the gut, mouth, etc. IgA that does not fix complement can fulfill these criteria by preventing attachment of bacteria in the gut without causing tissue damage. Hence, the generation of IgA-secreting plasma cells would be highly desirable in Peyer's patches. Bacteria that do gain entry into the bloodstream are cleared by macrophages or Kupffer cells or lysed by circulating opsonizing antibodies (e.g., IgM, IgG) and complement. As a consequence of ingestion of antigen, macrophages can secrete a wide variety of mediators which influence both immunocytes and other cells. Hence, in the case of bacterial infections, three types of immunity are probably operative: (a) phagocytosis of bacterial microorganisms by macrophages or neutrophils occurring both prior to the generation of an antibody response or, more efficiently, after the coating of these organisms with antibody and/or complement components. The opsonization and digestion of bacteria by macrophages results in destruction of the organism and also allows the macrophages to present fragments of bacterial antigens to T cells; (b) B cells might also be rapidly activated in a polyclonal manner by endotoxin or polysaccharides from the bacterial cell wall. (c) T_H cells would be activated by B-cell clones presenting TD epitopes from bacterial fragments or bacterial products, e.g., toxins. Since, in the mouse, antibody-mediated phagocytosis of bacteria, bacteriolysis, and the prevention of attachment of bacteria to mucosal and epithelial surfaces can be carried out by particular antibody isotypes, the secretion of IgG_1 and IgA antibodies may be dependent upon IL-4 and IL-5 secreted by T_{H2} cells. Thus, activation of T_{H2} cells is probably of major importance in the generation of antibodies which protect the host against systemic bacterial infections. In contrast, IgG_3 responses (e.g., to bacterial polysaccharides) may be relatively thymus independent.

III. Immune Response to Viral Antigens

Viruses, unlike most bacteria, are obligate intracellular pathogens; their extracellular concentrations may, in the early stages of infection, be low. Furthermore, during cellular interactions in general, virus might be transferred from one cell to another without reaching high concentrations in the blood. Viruses can be neutralized or eliminated by a wide variety of immune mechanisms including antibody-mediated neutralization (IgM, IgG, or IgA), complement-mediated neutralization, opsonization by macrophages, and cell-mediated cytotoxicity by T_c cells or NK cells. However, until viral antigens are synthesized and expressed on infected target cells or until virions are present in sufficient quantity in the circulation to be ingested by macrophages, degraded, and presented to T cells, humoral immunity, as well as the generation of T_c cells, would not occur. Assuming that sufficient antigenic exposure did occur, class I-restricted T_c cells or NK cells activated by IL-2-secreting T_{H1} cells may play a major role in killing cells either expressing viral antigens and/or actively producing virions. Since T_{H1} cells also secrete IFN-γ, this should facilitate IgG_{2a} secretion by B cells and, in the mouse, IgG_{2a} appears to be protective in the case of some viral infections. Furthermore, IFN-γ (also secreted by T_{H1} cells) is an effective antiviral agent and is involved in the activation of macrophages. In contrast, in other viral infections where IgA or IgG_1 play major roles, T_{H2} cells would be activated by antigen and APC.

IV. Immune Response to Parasites

Parasites differ enormously in their organization, complexity, life cycles, and pathogenic mechanisms. From many studies in exerimental animals, it has been established that different effector mechanisms are active against different parasites and that it is often difficult to predict which mechanism is the predominant one in a particular parasitic infection. Studies on immunity against parasites have suggested that in many cases the humoral antibody response plays a protective role. In the mouse, the two Ig isotypes most often involved in mediating antiparasitic responses are IgG_1 and IgE. IgE also plays a major role in antiparasite immunity in humans. Among the effector cells involved in immunity to parasites, eosinophils play a fundamental role in mediating parasite killing through interaction with parasites coated with specific antibody of the IgE and IgG_1 isotypes. This interaction results in the release of granules from the eosinophils which contain substances toxic to parasites. Activated macrophages and neutrophils are also capable of killing parasites and, in some instances, mast cells are also involved. Based on these effector mechanisms which involve IgG_1 and IgE antibodies, one would predict that the T_{H2} cells (which secrete IL-4 and IL-5) play a central role in the regulation of humoral immunity against many parasites. In particular, the persistence of parasites might be sufficient to induce the secretion of the large amounts of IL-4 necessary for the induction of IgE secretion (IgE secretion, in contrast to IgG_1 secretion, requires very high levels of IL-4 in mitogen-driven systems). IL-5 could act by recruiting eosinophils which would be further activated by factors derived from IL-4-activated mast cells. Mast cells would mediate parasite killing through their interaction with IgE- and IgG_1-coated organisms. IL-5 might also induce IgA secretion in local environments where parasites reside (e.g., gut).

In addition to protective antibodies, the host can defend itself against some intracellular parasites by the development of DTH responses and the activation of macrophages, both of which correlate with protective immunity and healing of parasite-induced lesions. Thus, T_{H1} cells may be involved in these responses since T_{H1} cells mediate DTH and secrete IFN-γ, which is a major macrophage-activating factor.

V. Concluding Remarks

Based on the considerations described above, information from three areas of research will be important to increase our understanding of which lymphokines and cellular interactions are involved in protective immune responses. These areas include: (a) the effect of dose, route, and type of antigen on the activation and regulation of subsets of T_{H1} and T_{H2} cells; (b) the mechanism by which anatomical barriers influence cellular interactions in vivo; and (c) the cellular and molecular mechanisms by which B cells interact with and are regulated by T_{H1} and T_{H2} cells and their lymphokines to generate specific antibodies of the appropriate isotypes. The disciplines of cellular and molecular immunology represent powerful tools for further elucidating mechanisms involved in cellular interactions. However, the major insight necessary to further our understanding of the immune response in both physiological and pathological states in the intact animal will involve more information about the pathogens themselves, the nature of protective immunity to different types of pathogens, and the activation and regulation of the immune network to elicit these protective responses.

Acknowledgements. We thank Ms. Gerry-Ann Cheek, Ms. Nelletta Stephens, and Ms. Shirley Starves for expert secretarial assistance.

References

Aggarwal B, Eessalu TE, Hass P (1985) Characterization of receptors for human tumor necrosis factor and their regulation by gamma-interferon. Nature 318:665–667

Alcover A, Ramarli D, Richardson NE, Chang H-C, Reinherz EL (1987) Functional and molecular aspects of human T lymphocyte activation via T3-Ti and T11 pathways. Immunol Rev 95:5

Arthur RP, Mason D (1986) T cells that help B cell responses to soluble antigen are distinguishable from those producing interleukin-2 on mitogenic or allogenic stimulation. J Exp Med 163:774–786

Belosevic M, Davis CE, Meltzer MS, Nacy CA (1988) Regulation of activated macrophage anti-microbial activities. Identification of lymphokines that cooperate with IFN-gamma for induction of resistance of infection. J Immunol 141:890–896

Bergstedt-Lindqvist S, Moon H-B, Persson U, Moller G, Heusser C, Severinson E (1988) Interleukin-4 instructs uncommitted B lymphocytes to switch to IgG_1 and IgE. Eur J Immunol 18:1073–1077

Birkeland ML, Metlay J, Sanders VM, Fernandez-Botran R, Vitetta ES, Steinman RM, Pure E (1988) Epitopes on CD45R (T200) molecules define differentiation antigens on murine B and T lymphocytes. J Mol Cell Immunol 4:71–85

Blackman MA, Tigger MA, Minnie ME, Koshland ME (1986) A model system for peptide hormone action in differentiation: interleukin-2 induces a B lymphoma to transcribe the J chain gene. Cell 47:609–617

Bloom BR, Bennett B (1966) Mechanism of a reaction in vitro associated with delayed-type hypersensitivity. Science 153:80–82

Bottomly K (1988) A functional dichotomy in CD4+ T lymphocytes. Immunol Today 9:268

Buchmeier NA, Schreiber RD (1985) Requirement of endogenous interferon-gamma production for resolution of *Listeria monocytogenes* infection. Proc Natl Acad Sci USA 82:7404–7413

Cambier JC, Ransom JT (1987) Molecular mechanisms of transmembrane signaling in B lymphocytes. Annu Rev Immunol 5:175–199

Cantrell DA, Smith KA (1983) Transient expression of interleukin-2 receptors: consequences for T cell growth. J Exp Med 158:1895

Cher DJ, Mosmann TR (1987) Two types of murine helper T cell clone. II. Delayed type hypersensitivity is mediated by T_{H1} clones. J Immunol 138:3688

Cherwinski HM, Schumacher JH, Brown KD, Mosmann TR (1987) Two types of mouse helper T cell clone. III. Further differences in lymphokine synthesis between T_{H1} and T_{H2} clones revealed by RNA hybridization, functionally monospecific bioassays and monoclonal antibodies. J Exp Med 166:1229–1233

Chesnut RW, Grey HM (1986) Antigen presentation by B cells and its significance in T-B interactions. Adv Immunol 39:51–94

Clement LT, Yamashita N, Martin AM (1988) The functionally distinct subpopulations of human $CD4^+$ helper/inducer T lymphocytes defined by anti-CD45R antibodies derive sequentially from a differentiation pathway that is regulated by activation-dependent post-thymic differentiation. J Immunol 141:1464–1470

Clevers H, Alarcon B, Wileman T, Terhorst C (1988) The T cell receptor/CD3 complex: a dynamic protein ensemble. Annu Rev Immunol 6:629–662

Coffman RL, Carty J (1986) A T cell activity that enhances polyclonal IgE production and its inhibition by interferon-gamma. J Immunol 136:949–954

Coffman RL, Ohara J, Bond MW, Carty J, Zlotnik A, Paul WE (1986) B cell stimulatory factor-1 enhances the IgE response of lipopolysaccharide-activated B cells. J Immunol 136:4538–4541

Coffman RL, Seymour BW, Lebman DA, Hiraki DD, Christiansen JA, Shrader B, Cherwinski HM, et al. (1988) The role of helper T cell products in mouse B cell differentiation and isotype regulation. Immunol Rev 102:5–28

Crawford RM, Finbloom DS, Ohara J, Paul WE, Meltzer MS (1987) B cell stimulatory factor-1 (interleukin 4) activates macrophages for increased tumoricidal activity and expression of Ia antigens. J Immunol 139:135–141

David JR (1966) Delayed hypersensitivity in vitro: its mediation by cell-free substances formed by lymphoid cell-antigen interaction. Proc Natl Acad Sci USA 56:72–77

Fernandez-Botran R, Sanders VM, Oliver KG, Chen YW, Krammer PH, Uhr JW, Vitetta ES (1986a) Interleukin 4 mediates autocrine growth of helper T cells after antigenic stimulation. Proc Natl Acad Sci USA 83:9689–9893

Fernandez-Botran R, Krammer PH, Diamantstein T, Uhr JW, Vitetta ES (1986b) B cell-stimulatory factor 1 (BSF-1) promotes growth of helper T cell lines. J Exp Med 164:580–593

Fernandez-Botran R, Sanders VM, Mosmann T, Vitetta ES (1988) Lymphokine-mediated regulation of the proliferative response of clones of T_{H1} and T_{H2} cells. J Exp Med 168:543–558

Finkelman FD, Katona IM, Urban JF Jr, Snapper CM, Ohara J, Paul WE (1986) Suppression of in vivo polyclonal IgE responses by monoclonal antibody to the lymphokine B-cell stimulatory factor 1. Proc Natl Acad Sci USA 83:9675–9678

Gajewski TF, Fitch FW (1988) Anti-proliferative effect of IFN-gamma in immune regulation. I. IFN-gamma inhibits the proliferation of T_{H2} but not T_{H1} murine helper T lymphocyte clones. J Immunol 140:4245

Grabstein KH, Eisenman J, Mochizuki D, Shanebeck K, Coulon P, Hopp T, March C, Gillis S (1986a) Purification to homogeneity of B cell-stimulating factor. A molecule that stimulates proliferation of multiple lymphokine-dependent cell lines. J Exp Med 163:1405

Grabstein KH, Urdal DL, Tushinshi RJ, Mochizuki DY, Price VL, Cantrell MA, Gillis S, Conlon PJ (1986b) Induction of macrophage tumorcidal activity by granulocyte-macrophage colony-stimulating factor. Science 232:506–608

Greenbaum LA, Horowitz JB, Woods A, Pasqualini T, Reich EP, Bottomly K (1988) Autocrine growth of CD4+ T cells. Differential effects of IL-1 on helper and inflammatory T cells. J Immunol 140:1555–1560

Hamaoka T, Ono S (1986) Regulation of B cell differentiation. Annu Rev Immunol 4:167–204

Hauser C, Snapper CM, Ohara J, Paul WE, Katz SI (1989) T helper cells grown with hapten-modified cultured Langerhan's cells produce interleukin-4 and stimulate IgE production by B cells. Eur J Immunol 19:245–251

Howard M, Paul WE (1983) Regulation of B-cell growth and differentiation by soluble factors. Annu Rev Immunol 1:307–333
Howard M, Farrar J, Hilfiker M, Johnson B, Takatsu K, Hamaoka T, Paul WE (1982) Identification of a T cell-derived B cell growth factor distinct from interleukin 2. J Exp Med 155:914–923
Hu-Li J, Shevach EM, Mizuguchi J, Ohara J, Mosmann T, Paul WE (1987) B cell-stimulatory factor-1 (interleukin-4) is a potent costimulant for normal resting T lymphocytes. J Exp Med 165:157
Hudak SA, Gollnick SO, Conrad DH, Kehry MR (1987) Murine B cell stimulatory factor-1 (interleukin-4) increases expression of the Fc receptor for IgE on mouse B cells. Proc Natl Acad Sci USA 84:4606–4610
Isakson PC, Pure E, Vitetta ES, Krammer PH (1982) T cell-derived B cell differentiation factor(s). Effect on the isotype switch of murine B cells. J Exp Med 155:734–748
Jones S, Chen YW, Isakson P, Layton J, Pure E, Word C, Krammer PH, Tucker P, Vitetta ES (1983) Effect of T cell-derived lymphokines containing B cell differentiation factor(s) for IgG (BCDF gamma) on gamma-specific mRNA in murine B cells. J Immunol 131:3049–3051
Kikutani H, Inui S, Sato R, Barsumian EL, Owaki H, Yamasaki K, Kaisho T et al. (1986) Molecular structure of human lymphocyte receptor for immunoglobulin E. Cell 47:657–665
Killar L, MacDonald G, West J, Woods A, Bottomly K (1987) Cloned, Ia-restricted T cells that do not produce interleukin 4(IL-4)/B cell stimulatory factor 1 (BSF-1) fail to help antigen-specific B cells. J Immunol 138:1674–1679
Kim J, Woods A, Becker-Dunn E, Bottomly K (1985) Distinct functional phenotypes of cloned Ia-restricted helper T cells. J Exp Med 162:188–201
Kimoto M, Fathman CG (1980) Antigen-reactive T cell clones. I. Transcomplementing hybrid I-A-region gene products function effectively in antigen presentation. J Exp Med 152:759
Kishimoto T, Hirano T (1988) Molecular regulation of B lymphocyte response. Annu Rev Immunol 6:485–512
Koshland ME (1983) The coming of age of the immunoglobulin J chain. Annu Rev Immunol 3:425–453
Kupper T, Horowitz M, Leo F, Robb R, Flood PM (1987) Autocrine growth of T cells independent of interleukin-2: identification of interleukin-4 (IL-4, BSF-1) as an autocrine growth factor for a cloned antigen-specific helper T cell. J Immunol 138:4280
Kurt-Jones EA, Hamberg S, Ohara J, Paul WE, Abbas AK (1987) Heterogeneity of helper/inducer T lymphocytes. I. Lymphokine production and lymphokine responsiveness. J Exp Med 166:1774
Layton JE, Vitetta ES, Uhr JW, Krammer PH (1984) Clonal analysis of B cells induced to secrete IgG by T cell-derived lymphokine(s). J Exp Med 160:1850–1863
Lebman DA, Coffman RL (1988) Interleukin-4 causes isotype switching to IgE in T cell-stimulated clonal B cell cultures. J Exp Med 168:853–862
Lee F, Yokota T, Otsuka T, Meyerson P, Villaret D, Coffman R, Mosmann T et al. (1986) Isolation and characterization of a mouse interleukin cDNA clone that expresses B-cell stimulatory factor 1 activities and T-cell- and mast-cell-stimulating activities. Proc Natl Acad Sci USA 83:2061–2065
Leibson HJ, Gefter M, Zlotnik A, Marrack P, Kappler JW (1984) Role of gamma-interferon in antibody-producing responses. Nature 309:799–801
Lichtman AH, Kurt-Jones EA, Abbas AK (1987) B-cell stimulatory factor 1 and not interleukin 2 is the autocrine growth factor for some helper T lymphocytes. Proc Natl Acad Sci USA 84:824–827
Liebson HJ, Marrack P, Kappler J (1982) B cell helper factors. II. Synergy among three helper factors in the response of T cell- and macrophage-depleted B cells. J Immunol 129:1398–1402

Lowenthal JW, Zubler RH, Nabholz M, MacDonald RH (1985) Similarities between interleukin-2 receptor number and affinity on activated B and T lymphocytes. Nature 315:669

Mackaness GB (1962) Cellular resistance to infection. J Exp Med 116:381–406

Marrack P, Graham SD Jr, Kushnir E, Leibson HJ, Roehm N, Kappler JW (1982) Nonspecific factors in B cell responses. Immunol Rev 63:33–49

Mond JJ, Finkelman FD, Sarma C, Ohara J, Serrate S (1986a) Recombinant interferon-gamma inhibits the B cell proliferative response stimulated by soluble but not Sepharose-bound anti-immunoglobulin antibody. J Immunol 135:2513

Mond JJ, Carman J, Sarma C, Ohara J, Finkelman FD (1986b) Interferon-gamma suppresses B cell stimulation factor (BSF-1) induction of class II MHC determinants on B cells. J Immunol 137:3534–3537

Morikawa K, Kubagawa H, Suzuki T, Cooper MD (1987) Recombinant interferon-alpha, -beta, and -gamma enhance the proliferative response of human B cells. J Immunol 139:761–766

Mosmann TR, Coffman RL (1987) Two types of mouse helper T cell clone: implications for immune regulation. Immunol Today 8:233

Mosmann TR, Bond MW, Coffman RL, Ohara J, Paul WE (1986a) T-cell and mast cell lines respond to B-cell stimulatory factor 1. Proc Natl Acad Sci USA 83:5654–5658

Mosmann TR, Cherwinski H, Bond MW, Giedlin MA, Coffman RL (1986b) Two types of murine helper T cell clone. I. Definition according to profiles of lymphokine activities and secreted proteins. J Immunol 136:2348–2357

Muraguchi A, Butler JL, Kehrl JH, Fauci AS (1983) Differential sensitivity of human B cell subsets to activation signals delivered by a monoclonal B cell growth factor. J Exp Med 157:530

Nakanishi K, Cohen DI, Blackman M, Nielsen E, Ohara J, Hamaoka T, Koshland ME, Paul WE (1984) Ig RNA expression in normal B cells stimulated with anti-IgM antibody and T cell-derived growth and differentiation factors. J Exp Med 160:1736–1751

Nedospasov SA, Hirt B, Shakov AN, Dobrynin VN, Kawashima E, Accolla RS, Jongenell CV (1986) The genes for tumor necrosis factor (TNF-alpha) and lymphotoxin (TNF-beta) are tandemly arranged on chromosome 17 of the mouse. Nucleic Acids Res 14:7713–7725

Noelle R, Krammer PH, Ohara J, Uhr JW, Vitetta ES (1984) Increased expression of Ia antigens on resting B cells: an additional role for B-cell growth factor. Proc Natl Acad Sci USA 81:6149–6153

Ohara J, Paul WE (1988) Upregulation of interleukin 4/B cell stimulatory factor 1 receptor expression. Proc Natl Acad Sci USA 85:8221–8225

Oliver K, Krammer PH, Tucker PW, Vitetta ES (1987) The effects of cytokines and adherent cells on the interleukin 4-mediated induction of Ia antigens on resting B cells. Cell Immunol 106:428–436

Pace J, Russell S, Torres B, Johnson H, Gray P (1983) Recombinant mouse interferon-gamma induces the primary step in macrophage activation for tumor cell killing. J Immunol 130:2011

Paul NL, Ruddle NH (1988) Lymphotoxin. Annu Rev Immunol 6:407–438

Paul WE, Ohara J (1987) B-cell stimulatory factor-1/interleukin 4. Annu Rev Immunol 5:429–459

Poo WJ, Conrad L, Janeway CA Jr (1988) Receptor-directed focusing of lymphokine release by helper T cells. Nature 332:378

Powell MB, Conta BS, Ruddle NH (1985) The differential inhibitory effect of lymphotoxin and immune interferon on normal and malignant lymphoid cells. Lymphokine Res 4:13–26

Powers GD, Abbas AK, Miller RA (1988) Frequencies of IL-2- and IL-4-secreting T cells in naive and antigen-stimulated lymphocyte populations. J Immunol 140: 3353–3357

Prakash S, Robb RJ, Stout RD, Parker DC (1985) Induction of high affinity IL 2 receptors on B cells responding to anti-Ig and T cell-derived helper factors. J Immunol 135:117–122
Rabin EM, Mond JJ, Ohara J, Paul WE (1986) Interferon-gamma inhibits the action of B cell stimulatory factor (BSF)-1 on resting B cells. J Immunol 137:1573–1576
Rasmussen R, Takatsu K, Harada N, Takahashi T, Bottomly K (1988) T cell-dependent hapten-specific and polyclonal B cell responses require release of interleukin 5. J Immunol 140:705–712
Reynolds DS, Boom WH, Abbas AK (1987) Inhibition of B lymphocyte activation by interferon-gamma. J Immunol 139–767–773
Robb RJ, Munck A, Smith KA (1981) T cell growth factor receptors. Quantitation, specificity and biological relevance. J Exp Med 154:1455
Roberts WK, Vasil A (1982) Evidence for the identity of murine gamma interferon and macrophage-activating factor. J Interferon Res 2:519–532
Rudd CE, Morimoto C, Wong LL, Schlossman SF (1987) The subdivision of the T4 (CD4) subset on the basis of the differential expression of L-C/T200 antigens. J Exp Med 166:1758
Sanderson CJ, Campbell HD, Young IG (1988) Molecular and cellular biology of eosinophil differentiation factor (interleukin-5) and its effects on human and mouse B cells. Immunol Rev 102:29–50
Schwartz RH (1985) T lymphocyte recognition of antigen in association with gene products of the major histocompatibility complex. Ann Rev Immunol 3:237–261
Severinson E, Naito T, Tokumoto H, Fukushima D, Hirano A, Hanna K, Honjo T (1987) Interleukin-4 (IgG_1 induction factor): a multifunctional lymphokine acting also on T cells. Eur J Immunol 17:67–72
Sidman CL, Marshall JD, Schultz LD, Gray PW, Johnson HM (1984) Interferon-gamma is one of several direct B cell-maturing lymphokines. Nature 309:801
Singer A, Hodes RJ (1983) Mechanisms of T cell-B cell interaction. Annu Rev Immunol 1:211–241
Smith KA (1980) T cell growth factor. Immunol Rev 51:337
Smith KA (1984) Interleukin-2. Annu Rev Immunol 2:319–333
Snapper CM, Paul WE (1987) Interferon-gamma and B cell stimulatory factor-1 reciprocally regulate Ig isotype production. Science 236:944–947
Spickett GP, Brandon MR, Mason DW, Williams AF, Woollett GR (1983) MRC OX-22, a monoclonal antibody that labels a new subset of T lymphocytes and reacts with the high molecular form of the leukocyte-common antigen. J Exp Med 158:795
Stevens TL, Bossie A, Sanders VM, Fernandez-Botran R, Coffman RL, Mosmann TR, Vitetta ES (1988) Regulation of antibody isotype secretion by subsets of antigen-specific helper T cells. Nature 334:255–258
Swain SL, Howard M, Kappler J, Marrack P, Watson J, Booth R, Wetzel GD, Dutton RW (1983) Evidence for two distinct classes of murine B cell growth factors with activities in different functional assays. J Exp Med 158:822–835
Swain SL, McKenzie DT, Dutton RW, Tonkonogy SL, English M (1988) The role of IL4 and IL5: characterization of a distinct helper T cell subset that makes IL4 and IL5 (Th2) and requires priming before induction of lymphokine secretion. Immunol Rev 102:77–105
Taniguchi T (1988) Regulation of cytokine gene expression. Annu Rev Immunol 6:439–464
Trinchieri G, Perusia B (1985) Immune interferon: a pleiotropic lymphokine with multiple effects. Immunol Today 6:131
Unanue ER (1984) Antigen-presenting function of the macrophage. Annu Rev Immunol 2:395–428
Vitetta ES, Brooks K, Chen YW, Isakson P, Jones S, Layton J, Mishra GC, Pure E, Weiss E, Word C, Yuan D, Tucker P, Uhr JW, Krammer PH (1984) T-cell-derived lymphokines that induce IgM and IgG secretion in activated murine B cells. Immunol Rev 78:137–157

Vitetta ES, Ohara J, Myers CD, Layton JE, Krammer PH, Paul WE (1985) Serological, biochemical, and functional identity of B cell-stimulatory factor 1 and B cell differentiation factor for IgG1. J Exp Med 162:1726–1731

Vitetta ES, Bossie A, Fernandez Botran R, Myers CD, Oliver KG, Sanders VM, Stevens TL (1987) Interaction and activation of antigen-specific T and B cells. Immunol Rev 99:193–239

Vitetta ES, Fernandez-Botran R, Myers CD, Sanders VM (1989) Cellular interactions in the humoral immune response. Adv Immunol (in press)

Waldmann TA, Goldman CL, Robb RJ, Depper JM, Leonard WJ, Sharrow SO, Bongiovanni KF, Korsmeyer SJ, Greene WC (1984) Expression of interleukin-2 receptors on activated human B cells. J Exp Med 160:1450

Wall R, Kuehl M (1983) Biosynthesis and regulation of immunoglobulins. Annu Rev Immunol 1:393–422

Weiss A, Imboden J, Hardy K, Manger B, Terhorst C, Stobo J (1986) The role of the T3/antigen receptor complex in T cell activation. Annu Rev Immunol 4:593–619

Yokota R, Otsuka T, Mosmann T, Banchereau J, DeFrance T, Blanchard D, DeVries J, Lee F, Arai L (1986) Isolation and characterization of a human interleukin cDNA clone homologous to mouse BSF-1 which expresses B cell- and T cell-stimulating activities. Proc Natl Acad Sci USA 83:5894

Zubler RH, Lowenthal JW, Erard F, Hashimoto N, Devos R, MacDonald HR (1984) Activated B cells express receptors for, and proliferate in response to, pure interleukin-2. J Exp Med 160:1170

CHAPTER 33

Coordinate Actions of Growth Factors in Monocytes/Macrophages *

C. F. NATHAN

A. Introduction

A proposal to review the known actions of peptide growth factors on mononuclear phagocytes runs immediately into the obstacle of specifying which cytokines ought to be considered peptide growth factors. At this juncture, it is difficult even to define "cytokine." Cytokine research is in transition. A flood of new facts has swept away old concepts; new ideas have not yet arisen with the power to reveal the logical relations among recent discoveries or predict the next ones. Thus, it seems necessary to explain the philosophical perspective from which this review has been written.

I will define "cytokine" restrictively as a protein or glycoprotein, nonimmunoglobulin in nature, released by living cells of the host, which reaches other cells of the host chiefly by local diffusion, sometimes by cell-to-cell contact, and sometimes via the circulation, whereupon it acts in solution, in extremely low (typically picomolar) concentrations, to regulate their function. This definition is designed to exclude nonprotein autocoids, endocrine hormones, products of dead cells, products of invading organisms, extracellular matrix components, and effector molecules acting on invading organisms, all of which might otherwise qualify. Some use "monokine" or "lymphokine" in contexts in which the predominant or sole source of a cytokine is presumed to be mononuclear phagocytes or lymphocytes, respectively. However, no cytokine is known that is produced exclusively by mononuclear phagocytes, and the source of cytokines in vivo is not often known with certainty. Thus there no longer seems to be any reason to use the term "monokine," and the term "lymphokine" may sometimes be misleading.

Historically, four types of cytokines were studied by as many sets of investigators. Virologists were interested in factors that interfered with viral replication (interferons). Cell biologists focused on factors that affected the growth and differentiation of cells from solid tissues (peptide growth factors). The appetite of immunologists for complexity was satisfied by confining their attention to factors thought to arise chiefly from and affect lymphocytes and macrophages (interleukins). Hematologists pursued factors that drive marrow cells to proliferate and differentiate into colonies (colony-stimulating factors). It was assumed that cytokines functioned largely within the category in which

* Preparation of this article and some of the research reviewed herein was supported by NIH grants CA43610 and CA45218.

they first came to light. As a corollary, the name of a cytokine was taken to reflect its principal action. Each cytokine was thought to arise chiefly from one type of cell, and its actions, if multiple, were supposed to be related to each other in some way that seemed logical.

These notions are defunct. Most cytokines arise from more than one type of cell. Most have bioactivities so diverse as to seem unrelated. Almost all bioactivities manifest by a given cytokine can be exerted by several. Almost all cytokines can induce differentiation, and can either promote or inhibit the growth of various cells. For example, interleukin-1 (IL-1) is a colony-stimulating factor (CSF); interferon-γ (IFN-γ) and IL-2 are peptide growth factors; granulocyte-macrophage-CSF (CSF-GM), IFN-γ, and transforming growth factor-β (TGF-β) are interleukins; the interleukins tumor necrosis factor-α (TNF-α) and IL-6 are interferons and peptide growth factors; nerve growth factor (NGF) has CSF-G activity, and so on. Groups of cytokines share receptors, and we are baffled why this situation has evolved: IL-1α and IL-1β; TNF-α, variant TNF-α, and TNF-β; the IL-6 family and IFN-β1; the IFN-α family and IFN-β1; TGF-β1, TGF-β2, and TGF-β1.2; epidermal growth factor (EGF) and TGF-α; and platelet-derived growth factor (PDGF)-AA, PDGF-BB, and PDGF-AB. Most telling, *the structure of a cytokine, and its known actions, do not permit us to predict its additional actions.* Since most if not all cytokines can act as peptide growth factors, the scope of this chapter must be correspondingly broad.

Previous concepts failed to prepare us for another striking set of recent findings: the production of cytokines and their receptors in cells of the fetus and placenta. We have no idea why CSF-M is produced in prodigious quantities by uterine epithelium and bound by placenta; what an IFN-α and TNF-α are doing in amniotic fluid and placenta; why a PDGF gene is transcribed in *Xenopus* embryos; or exactly what part TGF-β is playing at specific times and places in the developing embryo.

Thus, we are compelled to ask in earnest a question which until recently would have seemed purely rhetorical: What are cytokines for? The speculation offered below represents the philosophical basis of this chapter.

Cytokines comprise a third set of hormones, distinct from, interacting with, and occupying a position intermediate between two others: endocrine secretions and neurotransmitters. The influence of endocrine secretions is systemic. Neurotransmitters are normally confined to anatomically defined synaptic spaces. Between these two extremes, the niche of cytokines is normally a zone of tissue, variable in size and location. At times, however, cytokines act in either of the other two modes – spilling into the circulation, or signaling only those cells with which the producer cell is in contact.

In this view, the function of cytokines is to coordinate the (re)modeling of tissues. Tissues are (re)modeled de novo in ontogeny; constitutively in lymphohematopoiesis and the turnover of epithelia, bone and other tissues; and inducibly in response to wounding and infection. This chapter deals only with the last form of remodeling.

The remodeling that accompanies wounding and infection begins with the constitution of a new but temporary tissue within an existing one. The new tissue comprises different cells at different times; the cellular constituents may include platelets, polymorphonuclear leukocytes, monocytes/macrophages, dendritic cells, B cells/plasma cells, T cells, NK cells, and mast cells. This "immune response tissue" is put together by a variable combination of resident cells, immigrant cells, and cells arising locally through proliferation. As it forms, the immune response tissue may invade, occupy, destroy, and indeed liquefy any other. In an evolutionary sense, it is worth sacrificing a small volume of tissue to save the whole. These processes go on not only in infected tissue, but also in wounded tissue that must be presumed by the immune system to be infected, in normal tissue mistaken by the immune system for infected, and in the drainage perimeter, i.e., lymph nodes and spleen. Recruitment, proliferation, and differentiation of the cellular constituents of the immune response tissue lead to variable anatomic constructs: for example, abscesses, granulomas, granulation tissue, germinal centers, fatty streaks, and pannus. Their construction must proceed alongside the destruction of existing tissue, and is normally followed in due course by a brake on proliferation, the disbandment and/or autolysis of the mobilized elements, angiogenesis, restoration of the preexisting architecture, and/or formation of a scar.

Coordination of these processes is the province of cytokines. From this perspective, it begins to make sense that many of them function both in fetus and adult, in de novo, constitutive, and inducible (re)modeling.

In the complex milieux envisioned above, what determines the specificity of action of cytokines? Of course, a major factor must be the cellular distribution of cytokine receptors. However, additional determinants are increasingly recognized. These include the solid-phase microenvironment of the responding cell, notably its contact with extracellular matrix proteins and neighboring cells. For example, the ability of neutrophils to respond to the cytokines TNF-α, TNF-β, CSF-G, and CSF-GM with a large respiratory burst is contingent on ligation of their integrins, especially the CD11/CD18 adhesion receptors by suitable extracellular matrix proteins (NATHAN 1987b; NATHAN et al. 1989). Likewise, responses to cytokines are affected by the fluid phase microenvironment, including nonprotein autocoids, plasma proteins, metabolites, and microbial products. An example is the synergistic dependence of TNF-α on both bacterial products (ROTHSTEIN and SCHREIBER 1988) and complement component C5a (ROTHSTEIN et al. 1988) to induce hemorrhagic necrosis of skin in the mouse. Finally, the actions of one cytokine are contingent on the mix of other cytokines present, and the sequence of their appearance.

These notions of cytokine function draw a stark contrast between the effects of cytokines in isolated, reductionist assay systems in vitro, and their integrated effects in situ. How, then, can we determine which cytokine actions are physiologic? For in vitro observations, we can evaluate the magnitude of a given effect, its independence of specialized conditions, the potency with which it is induced, and its relevance to the known functions of the cell. In vivo, we need to know cytokine levels in cells and body fluids in health and disease; the distribution of cytokine receptors; and the effects of administration of the cytokine, of anticytokine antibodies, and of anticytokine receptor antibodies. Finally, it is extremely helpful to identify cytokine deficiency states, and to study the effects of replacement therapy in these conditions (e.g., NATHAN et al. 1986).

The following sections track the macrophage as it responds to cytokines that coordinate the construction of the "immune response tissue" and the destruction and repair of surrounding tissues. It will become apparent that our knowledge of these important processes is rudimentary.

B. Migration

In vitro, the substances chemotactic for mononuclear phagocytes (usually tested with monocytes) include several cytokines: TGF-β (WAHL et al. 1987), CSF-GM (WANG et al. 1987), and CSF-M (WANG et al. 1988). The potency of TGF-β as a monocyte chemotactin is extraordinary, with activity in the femtomolar range. At higher concentrations, TGF-β, like most chemotactins, inhibits migration. Since macrophages can be induced to release TGF-β (ASSOIAN et al. 1987), the question must be considered whether TGF-β of macrophage origin may mediate some of the promigratory effects of other substances, such as CSF-GM and CSF-M, or the antimigratory effects of others, such as migration inhibition factor (DAVID 1966; BLOOM and BENNETT 1966), IFN-γ (THURMAN et al. 1985), or complement components and coagulation factors (BIANCO et al. 1976). Likewise, it must be considered whether some of the incompletely defined macrophage chemotactic factors from antigen- or mitogen-stimulated lymphocytes (e.g., WARD et al. 1969) or other cells may actually by TGF-β in its active or latent forms.

In vivo, cytokines whose injection elicits accumulation of mononuclear phagocytes include IFN-γ (NATHAN et al. 1986), TGF-β (MUSTOE et al. 1987), TNF-β (AVERBOOK et al. 1987), and CSF-M (DONAHUE, personal communication). In these situations it is not possible to infer whether immigration of monocytes was induced by the agent injected or by another agent released in response to the injection. It would be of interest to test the effects of a panel of of anticytokine antibodies on the accumulation of mononuclear phagocytes in different types of inflammatory states, including infections, wounds, and delayed-type hypersensitivity reactions. Only a few such studies have been reported. A monoclonal antibody to IFN-γ, for example, diminished mouse footpad swelling in response to intradermal injection of bacterial lipopolysaccharide, which otherwise elicits mononuclear phagocyte accumulation (HEREMANS et al. 1987). In experimental murine listeriosis, the accumulation of mononuclear phagocytes in the liver was diminished by anti-TNF-α antibody (HAVEL, personal communication), and similar results were seen in mycobacterial infection (KINDLER et al. 1989). In experimental murine cerebral malaria, macrophage accumulation in the spleen was blocked by administration of antibodies to both IL-3 and CSF-GM, but not by antibodies to either alone (GRAU et al. 1988). Antibodies to other cytokines, such as TGF-β, were not tested in these studies. It is not clear whether the mAbs neutralized a chemotactic cytokine or a cytokine necessary for the generation of a chemotactin.

The mechanisms by which cytokines recruit monocytes into an inflammatory site are unknown. Some of the steps may include enhanced expression of adhesion receptors and/or their ligands on monocytes and subjacent endothelial cells (ROSEN and GORDON 1987); vasodilatation mediated by cytokine-induced, macrophage-derived smooth muscle relaxing factors, such as nitric oxide (STUEHR et al. 1989a), hypothetically loosening the interendothelial junctions; tipping of the local protease-antiprotease balance in favor of proteases by enhancing the release of reactive oxygen intermediates (SWAIM and PIZZO 1988), thus facilitating migration of monocytes between endothelial cells and across basement membranes; stimulation of migration up a chemical gradient, a process involving chemosensory mechanisms and their control over actin and its regulatory proteins as well as Ca^{2+} and its regulatory proteins; and immobilization of cells at the top of the gradient. Cytokines influence several of these processes in vitro.

Thus, there is no doubt that cytokines have the capacity to recruit monocytes to an inflammatory site. However, there is no proof that they normally do so; nor, if they do, how; nor any evidence to suggest what proportion of the signal to recruit monocytes into an exudate is contributed by cytokines and what proportion by eicosanoids, complement and coagulation components, matrix fragments, neutrophil granule constituents, and other substances.

C. Extramedullary Proliferation [1]

The study of proliferation of macrophages in extramedullary sites has a controversial history (STEWART 1980; NATHAN and COHN 1985). For years it was debated whether the markedly increased number of mononuclear phagocytes in inflammatory sites arose chiefly by immigration of monocytes from the blood or by local replication. It is now widely held that both processes contribute substantially. The controversy has moved on to whether the cells that proliferate in extravascular sites are mostly immature cells that have themselves recently arrived from the blood, or mature macrophages that have reentered cell cycle after residing for some time in the tissues. The latter view clashes with the widely held belief that mature macrophages are terminally differentiated.

Macrophage progenitors were first quantified in uninflamed mouse tissues by STEWART et al. (1975), using colony growth in liquid media in the presence of L-cell-conditioned medium, a source of CSF-M. Less than 0.3% of the cells

[1] MOORE has reviewed the growth of macrophage colonies from bone marrow cells in Chap. 29; that topic is not considered here.

in resident macrophage populations from the peritoneal and pleural cavities gave rise to colonies. With inflammation, the proportion increased to 10%–20%, most likely reflecting the recruitment of precursor cells from marrow to blood to tissue. In contrast, as many as 30% of resident mouse pulmonary alveolar macrophages proliferated in response to CSF-GM (CHEN et al. 1988 b) or IL-3 (CHEN et al. 1988 c). However, IL-3 did not promote proliferation of resident peritoneal macrophages or blood monocytes (CHEN et al. 1988 c).

The studies of CHEN et al. (1988 b, c) are among the few to provide compelling evidence in support of the proposition that a substantial proportion of mature macrophages from uninflamed tissues can undergo extensive proliferation in response to cytokines. Such evidence consists in the demonstration of an increase over time in the number of cells above that initially cultured. Other kinds of evidence must be evaluated with some skepticism. Visual impressions of cell replication, for example, can be misleading. Macrophages derived by prolonged cultivation of blood monocytes (NAKAGAWARA et al. 1981) or CSF-treated macrophages from the tissues (AMPEL et al. 1986) increase rapidly and massively in diameter and protein content per cell. As monolayers become confluent, some cells are displaced and pile up; others clump or fuse, giving the impression of replicative clusters. The uptake of radiolabeled thymidine into DNA may increase markedly, along with the labeling index (GENDELMAN et al. 1988). Yet, this is not necessarily accompanied by a net increase in the cell population. Instead, it may in part reflect dramatic alterations in the secretion of endogenous thymidine, altering the specific activity of the exogenous tracer (YOSHIDA and NATHAN, in preparation); balanced cell birth and death; unscheduled DNA repair; and/or extensive proliferation of mitochondria. Finally, macrophage cultures devoid of CSF-M die off (TUSHINSKI et al. 1982). Exogenous CSF-M, CSF-GM, TNF-α, or IFN-γ tend to counterbalance this trend and, in addition, increase the percentage of viable macrophages which remain adherent (REED et al. 1987). Thus, if cytokine-treated populations are compared with untreated controls at any given time, and especially if adherent cells alone are enumerated (CHEN et al. 1988 a), decreased cell loss in the treated group may be interpreted as cell growth.

In conclusion, a subpopulation of mononuclear phagocytes from the blood or from inflamed tissues has considerable replicative potential in response to cytokines. However, extensive replicative potential of the mature tissue macrophage from uninflamed tissues has apparently been demonstrated only for the pulmonary alveolar population in the mouse. In vivo, there is as yet no evidence that would led us judge the relative importance of cytokines in the expansion of the mononuclear phagocyte population through extramedullary replication.

D. Changes in Shape

The culture of human blood monocytes as adherent cells on glass is accompanied by a dramatic evolution of their morphology. The cells rapidly increase in diameter and protein content per cell; the latter value doubles between day 0 and day 5, triples again by day 13, and doubles yet again by day 29 (NAKAGAWARA et al. 1981) as the cells adopt an increasingly epithelioid appearance. After about 7–10 days, increasing proportions of multinucleated giant cells arise by cell fusion (SCHLESINGER et al. 1984). These changes are of interest because they may mirror the differentiation of monocytes into various types of tissue macrophages, such as those in the peritoneum, liver (Kupffer cells), pulmonary alveoli, or the granulomata characteristic of sarcoidosis and tuberculosis. Superimposed on this consistent picture are other shape changes that are more variable. These include conversion from the irregular, ameboid, fan shape of the monocyte and early macrophage, either toward circumferentially spread cells that resemble fried eggs or toward spindle-shaped cells that resemble fibroblasts and are often arrayed with their long axes in parallel.

Cultures of mononuclear phagocytes may be contaminated by T cells, NK cells, fibroblasts, or endothelial cells; all can produce cytokines that affect macrophages. In addition, mononuclear phagocytes can produce many of the cytokines to which they themselves respond, and may be stimulated to do so by culture conditions, such as adherence to fibronectin, exposure to traces of bacterial lipopolysaccharide (THORENS et al. 1987), endocytosis of modified serum proteins, or binding of complement components (GOODMAN et al. 1982). Finally, the serum usually used in the culture of mononuclear phagocytes may contain CSF-M or TGF-β. Thus, it is possible that some of the apparently spontaneous shape changes described above are mediated by cytokines. The impact of anticytokine antibodies on the morphologic features of differentiation of mononuclear phagocytes has not been systematically evaluated; there is no evidence that endogenous cytokines orchestrate these changes.

On the other hand, there is no doubt that exogenous cytokines have a profound impact on the morphology of mononuclear phagocytes in vitro. IFN-γ induces circumferential spreading (NATHAN et al. 1983); IFN-α and IFN-β retard spreading (NATHAN et al. 1984); CSF-GM and CSF-M induce the fibroblastic arrays described above (NATHAN et al. 1984). With mouse peritoneal macrophages, IFN-γ, TNF-α, CSF-M, and CSF-GM increase spreading (AMPEL et al. 1986; REED et al. 1987), while TGF-β retards it (TSUNAWAKI et al. 1988). Finally, IFN-γ hastens the seemingly spontaneous formation of multinucleated giant cells from human blood monocytes (WEINBERG et al. 1985).

The significance of these in vitro observations is unknown. As yet, there is no evidence to support the notion that cytokines control the differentiation of monocytes in vivo into the morphologically varied populations observed in different tissues.

E. Endocytosis, Cell Surface Receptors, and Antigens

I. Endocytic Receptors

Using latex beads, starch granules, carbon particles, or formalinized bacteria – particles whose uptake is not known to be mediated by specific receptors – the maximum phagocytic rate and capacity of macrophages are each so high that they are technically difficult to measure by quantitative assays in which particle dose is not limiting (NATHAN and TERRY 1977). The few such measurements reported have not revealed much impact of cytokines on these parameters (e.g., TSUNAWAKI et al. 1988). This seems to support the view that the scavenger function of macrophages is constitutive, in contrast to the cytotoxic functions (see below).

On the other hand, certain particles or macromolecular complexes are recognized by specific macrophage receptors whose capacity to mediate the binding or ingestion of their ligands is qualitatively controlled by cytokines. A striking example concerns receptors for the Fc piece of IgG (FcR). IFN-γ markedly enhances the expression of FcR (PERUSSIA et al. 1983; GUYRE et al. 1983; WEINSHANK et al. 1988) and hence the ability of macrophages to bind and engulf IgG-coated particles. The phenomenon occurs in vivo in man. Administration of rIFN-γ to cancer patients dramatically induces FcR on their monocytes and neutrophils (GUYRE, personal communication). An expected corollary would be for IFN-γ to enhance the ability of macrophages to secrete eicosanoids and reactive oxygen intermediates upon binding of immune complexes, and to be better able to injure antibody-coated microbes and tumor cells. Indeed, IFN-γ appears to be an essential factor within the conditioned medium of mitogen-stimulated lymphocytes for enhancement of macrophage-mediated antibody-dependent cytotoxicity (RALPH et al. 1988). Surprisingly, similar experiments have apparently not been reported with antibody-coated microorganisms.

An equally striking example, but whose physiologic import is unclear, concerns the receptor for complement component C3bi (complement receptor type 3, or CR3, consisting of the CD11b/CD18 heterodimer). Human macrophage CR3 binds C3bi-coated erythrocytes and mediates their ingestion. If the macrophages are treated with IFN-γ, binding is preserved, but ingestion is prevented; the inhibition is overcome by fibronectin (WRIGHT et al. 1986). TNF-α promotes the mobilization of CR3 from internal membranes to the plasma membrane of human monocytes (MILLER et al. 1987).

IFN-γ markedly decreases the expression of another endocytic receptor on mononuclear phagocytes, that for mannosylated or fucosylated glycoproteins (MOKOENA and GORDON 1985). The physiologic significance of this response is unknown.

An unidentified cytokine can downregulate the so-called scavenger receptor on macrophages, which mediates the uptake of modified low-density lipoproteins (FOGELMAN et al. 1983).

II. Other Surface Antigens

At least three cytokines – IFN-γ (STEEG et al. 1982), IL-4 (CRAWFORD et al. 1987), and CSF-GM (ALVARO-GRACIA et al. 1989) – increase the expression of immune response associated (Ia) antigens (class II antigens of the major histocompatibility complex) on mononuclear phagocytes. TGF-β antagonizes the increase seen with IFN-γ (CZARNIECKI et al. 1988). The IFN-γ-induced increase in Ia antigen expression on murine macrophages in response to IFN-γ may be mediated by another cytokine of macrophage origin (WALKER et al. 1984). In order to undergo clonal expansion in response to antigen, T lymphocytes must bind via their antigen-receptor to a complex formed between Ia molecules on another cell and an immunogenic fragment of the antigen itself. Thus, increased Ia antigen expression should enhance the ability of macrophages to serve as antigen-presenting cells. However, IFN-γ induces many other changes in macrophages besides increased Ia antigen expression, some of which may be inimical to the clonal expansion of T lymphocytes, such as release of eicosanoids (HAMILTON et al. 1985b; BORASCHI et al. 1985), reactive oxygen intermediates (NATHAN et al. 1983), and reactive oxides of nitrogen (STUEHR and MARLETTA 1987; STUEHR et al. 1989a). In addition, it must be emphasized that the macrophage is not unique as an antigen-presenting cell. T lymphocytes that have not yet undergone clonal expansion require stimulation by a distinct Ia-positive cell, the dendritic cell, in order to divide; mononuclear phagocytes are poor substitutes for dendritic cells in this role (INABA and STEINMAN 1984). On the other hand, T cells that have already been activated can be induced to proliferate further by contact with a wide variety of Ia-positive cells, including but not limited to the macrophage. Finally, the ability of IFN-γ to induce Ia antigens on macrophages can be compromised when the cells have first been parasitized by an infectious agent (REINER et al. 1988), which mirrors a physiologic situation.

Intact antigens are generally not recognized by T cells; rather T cells see immunogenic fragments in the form of peptides bound to major histocompatibility complex molecules on antigen-presenting cells. The generation of immunogenic peptides from microbes or proteins is called antigen processing. The macrophage is orders of magnitude more efficient at the combined processes of antigen processing and presentation than the fibroblast (UNANUE and ALLEN 1987). It has not been determined what proportion of this superiority is attributable to more efficient antigen processing and what proportion to the higher expression of Ia antigens. There is no experimental evidence that establishes the superiority of macrophages in antigen processing per se. Indeed, assays that quantify antigen processing per se are not yet available. The impact of cytokines on antigen processing by macrophages is a central question in immunology that has not yet been settled.

IFN-γ, TNF-α, and TNF-β also increase class I major histocompatibility complex antigens on a variety of cell types, including macrophages. These antigens must be bound by most cytolytic T cells in order for them to kill infected target cells. It is widely held that one of the central physiologic roles of

IFN-γ secretion during the T-cell response to infectious agents is to increase class I expression on such cells so that, if they become infected, they may be more susceptible to lysis by cytolytic T cells. Because mononuclear phagocytes ingest or are invaded by a wide range of pathogens, increased expression of histocompatibility complex antigens on mononuclear phagocytes may be important in the host inflammatory and immune response to infection, not only by viruses but also by intracellular bacterial and protozoal pathogens (KAUFMANN et al. 1987). However, there is little information on the ability of cytokines to increase class I antigen expression on mononuclear phagocytes after they have become infected. In addition, interferons can make host cells resistant to the cytolytic action of lymphocytes (TRINCHIERI and SANTOLI 1978). This possibility has apparently not been tested with mononuclear phagocytes.

Thus, the ability of cytokines, especially IFN-γ, to induce class II and class I major histocompatibility antigens on mononuclear phagocytes is often considered one of the most important physiologic actions of this cytokine. However, the impact of this induction on the overall efficiency of the immune response in situ is difficult to judge from the information available from in vitro assays.

IFN-γ enhances macrophage expression of the integrin LFA-1 (STRASSMAN et al. 1985). This may promote the migration of mononuclear phagocytes into inflamed tissues and their binding to mammalian target cells, such as tumors. IFN-γ also enhances the expression of IL-2 receptors on mononuclear phagocytes (HERRMANN et al. 1985), whereupon IL-2 can act, apparently independently of IFN-γ, in enhancing the cytotoxic function of macrophages (e.g., RALPH et al. 1988).

IFN-γ augments the ability of mouse pulmonary macrophages both to bind fluoresceinated derivatives of bacterial lipopolysaccharide (LPS) and to respond to LPS with enhanced antitumor activity (AKAGAWA and TOKUNAGA 1985). Whether this reflects increased expression of specific LPS receptors (HAMPTON et al. 1988) remains to be tested.

Tissue factor-like activity is induced on human monocytes by exposure to rTNF-α (CONKLING et al. 1988).

F. Secretion

Mononuclear phagocytes secrete over 100 different biologically active molecules (NATHAN 1987a). In vitro, the secretion of most of these can be affected by cytokines: cytokines induce the secretion of some, capacitate the secretion of others in response to further signals, and suppress the secretion of still others. The literature on this subject is so voluminous that it would serve little purpose to attempt to catalog these effects. Selected examples are presented below from several of the major molecular classes of macrophage secretory products.

I. Cytokines

Macrophages can secrete nearly two dozen cytokines, many of them in a manner affected by cytokines. Only a few examples of cytokine effects on cytokine secretion by macrophages will be mentioned. One of the most striking is the induction of IL-6 in monocytes by IL-1 (BAUER et al. 1988). This is of interest because the ratio of IL-6 synthesis in IL-1-treated versus control monocytes is extremely high; IL-1-induced IL-6 amounts to as much as 0.7% of total secreted proteins; TNF-α, the effects of which usually parallel those of IL-1 (NATHAN 1987a), has no effect in this system, nor does IFN-γ or IL-2; and the mature macrophage, in contrast to the monocyte, is refractory to this action of IL-1. This phenomenon may reflect one of the mechanisms by which IL-1 augments the acute-phase and immune responses.

IFN-γ promotes LPS-stimulated TNF-α expression by both transcriptional and posttranscriptional means (BEUTLER et al. 1986; COLLART et al. 1986). This is of interest in view of the synergistic antiproliferative, tumoricidal, and antischistosomal actions of IFN-γ plus TNF-α (ESPARZA et al. 1987). Likewise, CSF-GM (HORIGUCHI et al. 1987), IL-3 (VELLENGA et al. 1988), and TNF-α (LU et al. 1988) stimulate human monocyte secretion of CSF-M; CSF-GM and IL-3 synergize with CSF-M in promoting the proliferation of mouse pulmonary alveolar macrophages (CHEN et al. 1988c). CSF-GM promotes IL-1 release from human monocytes (SISSON and DINARELLO 1988); IL-1 interacts synergistically with CSF-GM in stimulation of marrow progenitors. Finally, as an example of a suppressive effect, IFN-γ blocks the ability of macrophages to release CSF-GM (THORENS et al. 1987).

II. Complement Components and Other Proteases

IFN-γ markedly induces C2 and factor B gene transcription in human monocytes (STRUNK et al. 1985). The concentration of C2 in extracellular fluid may be rate-limiting for the classical path of complement activation. Thus, its production by activated macrophages in inflammatory sites may be important, despite the production of vastly greater quantities by the liver. IFN-γ also increases macrophage production of plasminogen activator (COLLART et al. 1986), a key enzyme in the migration of macrophages into inflammatory sites and in the degradation of inflammatory mediators and extracellular matrix proteins, directly and in conjunction with macrophage elastase.

III. Sterols

IFN-γ induces 1α-vitamin D_3 hydroxylase in alveolar macrophages from normal subjects (KOEFFLER et al. 1985). Both the lymphocytes and the alveolar macrophages in subjects with sarcoidosis appear spontaneously to secrete IFN-γ (ROBINSON et al. 1985). IFN-γ appears to activate the macrophages in situ, in that sarcoidosis macrophages display the same enhanced capacity to release reactive oxygen intermediates as is manifest by normal donors' alveolar macrophages after treatment with IFN-γ in vitro (FELS et al. 1987). Finally, alveolar macrophages from sarcoidosis subjects spontaneously 1-α-hydroxylate 25-hydroxyvitamin D_3 (ADAMS and GACOD 1985); this is probably a major source of the 1,25-dihydroxyvitamin D_3 that contributes to the hypercalcemia of sarcoidosis. Thus, this is an example of a cytokine action on macrophages which is likely to be of pathophysiologic significance.

IV. Reactive Intermediates of Oxygen and of Nitrogen

The dramatic impact of cytokines on the production of the two classes of inorganic secretory products of macrophages is discussed below in the context of macrophage activation.

G. Activation

Most research concerned with the effects of cytokines on mononuclear phagocytes deals with the phenomenon of "activation." Unfortunately, the term has no single meaning for specialists in the field. "Activation" is used to refer either to an induced increase in the capacity of the macrophage to carry out one of its chief functions, or, alternatively, to an increase in any given measurable property. The latter usage was dominant in the 1970s but receded as it became clear that various features of the macrophage (proliferation, migration, endocytosis, receptors, antigens, secretion, cytotoxicity) can change independently of each other in response to different stimuli; a cell deemed activated according to one parameter is not necessarily activated according to another. In the 1980s more workers began to reserve the term "activation" for acquisition of enhanced capacity to perform one of the chief physiologic functions of the cell. However, there was no agreement on which of the cell's chief functions would be the referent.

The chief physiologic functions of mononuclear phagocytes can be grouped into five categories: killing of microbial pathogens; killing of host cells or host-like cells, such as neoplasms and transplants; contribution to the generation of inflammatory and immune responses; promotion of wound healing; and scavenging of senescent cells and other debris not associated with wounds. The impact of cytokines on each of these functions will be considered in turn, with most emphasis on the first.

I. Killing of Microbial Pathogens

Many investigators use "macrophage activation" to refer specifically to enhancement of the antimicrobial capacity of these cells. This is how the term was introduced historically, through the work of Metchnikoff, Lurie, Middlebrook, and Mackaness. Moreover, this is the only sense of the term in which the capacity of macrophages to undergo activation is known to be essential to life (NATHAN 1986). This class of macrophage functions shares with the other cytotoxic functions the property of being profoundly regulated by cytokines. Thus, while the ability of macrophages to kill the agents of nonopportunistic infections is constitutive, their ability to kill many opportunistic microorganisms is inducible. Speculations on the implications of this situation are offered below in the discussion of macrophage deactivation (Sect. VIII).

Cytokines reported to enhance macrophage antimicrobial activity include IFN-γ, CSF-GM, CSF-M, and TNF-α (reviewed in NATHAN and YOSHIDA 1988; MURRAY 1988; NATHAN et al. 1988). The effects of IFN-γ in this regard will be discussed in most detail, because this is one of the few instances in which a chain of evidence is nearly complete in support of the proposition that a described effect of a cytokine on macrophages is physiologic.

Mackaness showed that cell-mediated immunity to intracellular microbial pathogens could be broken down into an immunologically specific component and a nonspecific component. Lymphcytes reacted to the specific microbial antigen by conferring on macrophages an enhanced antimicrobial activity that was manifest not only toward microbes bearing the specific antigen but toward antigenically unrelated microbes as well (MACKANESS 1969). The evidence that a soluble glycoprotein from lymphocytes was responsible for activating macrophages in this regard (NATHAN et al. 1973; FOWLES et al. 1973) led to a prolonged search for the responsible cytokines, a search that still continues.

One of the cytokines released upon activation of T lymphocytes by antigen is IFN-γ. The capacity of an individual's lymphocytes to secrete IFN-γ in response to antigen mounts in parallel with other manifestations of cell-mediated immunity or delayed-type hypersensitivity; media conditioned by such lymphocytes can enhance the antimicrobial activity of macrophages as well as a functionally related property, their capacity to secrete reactive oxygen intermediates (NATHAN et al. 1983).

In many reports, media conditioned by antigen- or mitogen-stimulated polyclonal lymphocyte populations from man or mouse have lost their ability to activate macrophages by the above criteria after incubation with monoclonal antibodies that neutralize IFN-γ (e.g., NATHAN et al. 1983; MURRAY et al. 1983, 1985; ROTHERMEL et al. 1983; WILSON and WESTALL 1985; SCHREIBER et al. 1985), although in some experimental systems additional macrophage-activating activity remains after such procedures (NACY et al. 1985). Copurification of macrophage-activating factor in highly enriched preparations of native IFN-γ makes it extremely unlikely that the monoclonal antibodies discussed above fortuitously crossreacted with a cytokine physicochemically distinct from IFN-γ (NATHAN et al. 1983).

The next link in the chain of evidence was the association of macrophage-activating factor with pure, recombinant IFN-γ. Since the initial reports (NATHAN et al. 1983; MURRAY et al. 1983), this observation has been extended to macrophage killing of at least 22 species of microbial pathogens (MURRAY 1988). The human macrophage populations which have been activated in vitro with IFN-γ have been derived from the blood (NATHAN et al. 1983; MURRAY et al. 1983), pulmonary alveoli (FELS et al. 1987; MURRAY et al.

1987a), and peritoneal cavity (LAMPERI and CAROZZI 1988). The potency of rIFN-γ as macrophage-activating factor is noteworthy, with half-maximal effective concentrations often in the picomolar or subpicomolar range (NATHAN et al. 1983), on the order of several hundred molecules per macrophage.

Thus the secretion of IFN-γ, and its action on macrophage antimicrobial activity, are consistent with Mackaness' analysis, that macrophage activation can be antigen specific in its elicitation, but is antigen nonspecific in its expression.

The next step was the demonstration that IFN-γ activates macrophages in experimental animals. rIFN-γ was injected into mice intravenously, intramuscularly, or intraperitoneally. Peritoneal macrophages were subsequently explanted and tested for antitoxoplasma and antileishmania activity and the capacity to secrete reactive oxygen intermediates (MURRAY et al. 1985). rIFN-γ proved highly effective in activating macrophages after injection by all three routes. It appeared that resident tissue macrophages rather than newly immigrant monocytes underwent activation, because there was no evidence for recruitment of new cells into the peritoneal cavity. Extremely low doses of rIFN-γ were effective. Similar results were recorded by REED (1988), who tested the ability of the explanted macrophages to kill a third type of protozoan pathogen, *Trypanosoma cruzi*, and by BRUMMER et al. (1988), who tested the ability of explanted mouse pulmonary macrophages to kill the fungi *Paracoccidiodes brasiliensis* and *Blastomyces dermatitidis*.

Other studies have addressed the fate of rIFN-γ-treated mice during infection with organisms that parasitize macrophages: *Listeria monocytogenes* (KIDERLEN et al. 1984), *Toxoplasma gondii* (McCABE et al. 1984), *Mycobacterium intracellulare* (EDWARDS et al. 1986), and *Leishmania donovani* (MURRAY et al. 1987b). In each case, rIFN-γ restrained the advancement of the infection. Most impressively, injection of rIFN-γ as sole therapy cured mice otherwise lethally infected by *Trypanosoma cruzi*, a protozoan which parasitizes macrophages as well as several other cell types (REED 1988).

The challenge remained whether macrophage activation by IFN-γ was demonstrable not only in mice, but also in man. The first opportunity to address this came in conjunction with phase I trials of rIFN-γ in cancer patients. This was not an optimal setting, because it was not feasible to take serial samples of tissue macrophages from such patients. Only blood monocytes were available for study. Monocytes already have features seen in tissue macrophages after activation, such as a capacity to release large amounts of reactive oxygen intermediates and to kill efficiently a variety of microorganisms. Only modest increases in these properties could be expected. Nonetheless, 11 of 13 patients responded to rIFN-γ with clear-cut enhancement of H_2O_2-releasing capacity in their monocytes (NATHAN et al. 1985). This observation was subsequently confirmed (MURRAY et al. 1987a; MALUISH et al. 1988) and extended to a demonstration of enhanced antimicrobial activity in monocytes of rIFN-γ-treated patients suffering from acquired immunodeficiency syndrome (MURRAY et al. 1987a).

The first clear-cut evidence for a physiologic role of endogenous IFN-γ in macrophage activation came from experiments in which injection of monoclonal antibody to IFN-γ led to marked increases in the number of *Listeria monocytogenes* in the organs of infected mice. Many of the antibody-treated mice died, while control animals eliminated the infection and survived (Buchmeier and Schreiber 1985). Anti-IFN-γ monoclonal antibody had a similar effect on murine infection with *Toxoplasma gondi* (Suzuki et al. 1988).

The issue remains whether low doses of rIFN-γ would reverse features of disease associated with deficiency of endogenous IFN-γ and with failure of tissue macrophages to be activated. Such demonstrations are now underway in two of the most prevalent of the major, nonviral infectious diseases of the world, in both of which macrophages are the predominant host cell for the infectious agent: lepromatous leprosy and visceral leishmaniasis.

In lepromatous leprosy, exuberant growth of *Mycobacterium leprae* within macrophages marks their activation as inadequate, in striking contrast to the paucibacillary macrophages of tuberculoid leprosy. There may be no intrinsic defect in the ability of macrophages of lepromatous individuals to respond to macrophage-activating factor, because in vitro such cells are indistinguishable from normals' in their oxidative response to rIFN-γ. Instead, the lymphocytes of lepromatous subjects may generate inadequate amounts of macrophage-activating factor at sites of infection with *M. leprae*. Thus, in contrast to tuberculoid patients, lepromatous subjects are anergic upon skin testing with *M. leprae* and mobilize few lymphocytes in their lesions, especially cells bearing helper markers. The keratinocytes overlying lepromatous lesions are negative when stained for HLA-DR antigens, in contrast to those above tuberculoid lesions. Since rIFN-γ regulates keratinocyte HLA-DR expression in vitro, this observation is consistent with a deficiency of endogenous IFN-γ in lepromatous lesions. Finally, exposure of lepromatous patients' blood lymphocytes to *M. leprae* in vitro fails to elicit release of macrophage-activating factor or IFN-γ. The foregoing evidence was reviewed by Nathan et al. (1986).

We therefore asked whether rIFN-γ could convert some features of lepromatous lesions toward the tuberculoid pole, which in the natural setting is associated with a more favorable and often self-limited clinical course, and which has been a goal of earlier trials of immunotherapy. Because there was no prior experience with a recombinant product of the immune system in a nonviral infectious disease in man, and because leprosy patients are prone to immunologically mediated exacerbations, we chose a conservative approach: intralesional injection of rIFN-γ in extremely low doses. Intradermal injection of 1 or 10 μg rIFN-γ daily for 3 days induced local induration, infiltration by monocytes and T cells, proliferation of keratinocytes, a decrease in epidermal Langerhans cells, and intense display of HLA-DR antigens on keratinocytes as well as on cells of the dermal infiltrate. Except perhaps for the change in Langerhans cells, these are all features of delayed-type hypersensitivity reactions. In addition, in most subjects there was an apparent reduction in the number of bacteria within macrophages in the injected sites, suggesting that either the resident macrophages or the recruited monocytes had been activated (NATHAN et al. 1986). These observations have been confirmed (SAMUEL et al. 1987). Experiments are now underway to see if administration of rIFN-γ can bring about a systemic reduction in intracellular bacteria.

Visceral leishmaniasis is also predominantly an infection of macrophages. In this disease, as in lepromatous leprosy, there is a deficiency in IFN-γ production by T lymphocytes in response to the specific antigen (CARVALHO et al. 1985). Addition of low doses of rIFN-γ to standard chemotherapy in children with advanced disease refractory to chemotherapy alone appears to have cured the majority (BADARO and JOHNSON, personal communication). This appears to represent the clinical correction of an endogenous deficiency state in the production of a macrophage-activating factor, and as such constitutes a powerful argument for macrophage activation being a physiologic action of a cytokine.

None of the foregoing militates against the existence of additional macrophage-activating factors besides IFN-γ. Other macrophage-activating factors may require different conditions for production than used in most of the in vitro experiments summarized above; may have been present in too low a concentration to be detected by those bioassays; may require IFN-γ as a cofactor; may be inhibited in their action by other substances in the same media, such as TGF-β (TSUNAWAKI et al. 1988); or may activate macrophages for enhanced ability to kill different microbes than those tested. In fact, all these explanations seem to be applicable. Other macrophage-activating factors (TNF-α, CSF-GM, CSF-M, and factors not yet completely characterized) have been discovered in the medium of cloned lymphoid cell lines, tumors, or hybridomas, as products of nonlymphoid cells, while screening recombinant cytokines cloned on the basis of their other bioactivities, or while screening presumptive cytokines cloned on the basis of their differential expression in activated lymphocytes. Finally, several cytokines have apparently not yet been screened for their ability to activate macrophages, including IL-5, IL-6, IL-7, MIP-1α, MIP-1β, MIP-2, and NAF. There may well be a macrophage-activating factor among these.

How do other cytokines compare with IFN-γ as a macrophage activator? Definitive comparison is premature. IFN-γ was the first product of T lymphocytes to be cloned; some of the apparent differences between rIFN-γ and other cytokines may mean nothing more than that rIFN-γ has been available for experimentation for a longer time.

The preliminary information available at this juncture suggests that, compared with IFN-γ, other macrophage-activating factors generally confer on the macrophage a more limited antimicrobial spectrum. For example, rTNF-α and rTNF-β enhance the ability of macrophages to kill *T. cruzi* but not *T. gondi* (DE TITTO et al. 1986; SHPARBER et al., unpublished observations); rCSF-GM activates macrophages toward *T. cruzi* (REED et al. 1987), *L. donovani* (WEISER et al. 1987), and *M. avium* (BERMUDEZ and YOUNG 1987), but not *T. gondi* (NATHAN et al. 1984; COLEMAN et al. 1988) or *L. pneumophila*

(JENSEN et al. 1988); both rCSF-GM and rCSF-M enhance mouse peritoneal macrophage H_2O_2-releasing capacity, but to a lesser extent than rIFN-γ or rTNF-α (REED et al. 1987). In some instances, the non-IFN-γ macrophage-activating factors incude a lesser degree of inhibition of a given pathogen (REED et al. 1987); require higher concentrations; or require the joint participation of IFN-γ, as with the contribution of TNF-α toward induction of macrophage antischistosome activity (ESPARZA et al. 1987) and with unidentified cytokines in the induction of resistance of macrophages to infection by leishmania (BELSOEVIC et al. 1988). It is particularly difficult to judge if a factor like CSF-M is a macrophage activator when it exerts a profound trophic influence on the survival of macrophages in vitro under the conditions of the experiment. Finally, the sort of evidence marshalled above for the ability of IFN-γ to activate macrophages in vivo, and especially in man, is not in hand for other macrophage-activating factors. In fact, treatment of experimental murine leishmaniasis with recombinant CSF-GM has exacerbated the disease (GREIL et al. 1988), in contrast to results with recombinant IFN-γ (MURRAY et al. 1987b).

What might be the physiologic significance of the fact that multiple cytokines can enhance macrophage antimicrobial activity? Several speculations can be considered.

First, production of IFN-γ appears to be limited to T lymphocytes, NK cells, and possibly alveolar macrophages (ROBINSON et al. 1985). The capacity to produce CSF-GM and TNF-α is distributed much more widely. For example, CSF-GM is a product not only of antigen-stimulated T lymphocytes, but also of endothelial cells, fibroblasts, keratinocytes, and macrophages themselves. While the production of appreciable levels of IFN-γ in an infected tissue site presumably requires the immigration and clonal expansion of lymphocytes, CSF-GM might be released by cells already present. In addition, in vitro, rCSF-GM induces a maximal elevation in H_2O_2-releasing capacity in mouse peritoneal macrophages (REED et al. 1987) and increases antileishmanial activity in human macrophages (WEISER et al. 1987) at least a day earlier than does rIFN-γ. For both reasons, macrophage activation in situ might ensue more quickly in response to CSF-GM than to IFN-γ. Similar arguments can be made for TNF-α.

Second, if macrophage activation is important in the ability of the host to reject certain invading organisms, it can be anticipated that some pathogens will have evolved mechanisms for suppression of the release and/or action of macrophage-activating factors (e.g., DING et al. 1987; SIBLEY et al. 1988). A repertoire of activators arising from various cells in response to different signals and acting through distinct receptors might constitute a countermeasure, also selected by evolution.

Third, different cytokines might enhance the various antimicrobial mechanisms of the macrophage to different degrees, resulting in different spectra of antimicrobial activity, according to the susceptibility of individual pathogens. The production of reactive oxygen intermediates is one important antimicrobial mechanism of the activated macrophage. However, it is clear that the macrophage makes use of major, respiratory burst-independent antimicrobial mechanisms as well (NATHAN 1983; CATTERALL et al. 1986). Unfortunately, aside from a cytostatic action presumably due to reactive nitrogen intermediates against *Cryptococcus neoformans* (GRANGER et al. 1988), little evidence for the role of specific respiratory burst-independent mechanisms is currently in hand. Thus, our limited understanding of how the macrophage kills microorganisms makes the above speculation difficult to test.

II. Killing of Host-Type Cells

The ability of mononuclear phagocytes to lyse neoplastic cells or inhibit their growth is profoundly influenced by cytokines, and forms the basis for the second of the two most widely used definitions of macrophage activation. Similar mechanisms presumably underlie the expression of cytotoxicity toward tumor cells on the one hand, and, on the other hand, toward allografts, xenografts, and normal host cells caught up in autoimmune reactions or implicated as innocent bystanders in other inflammatory processes. The following discussion is confined to antitumor activity, which has been most intensively studied.

The experimental settings in which antitumor cell activity is measured are diverse, and so are the known biochemical mechanisms that contribute to cytotoxicity. Thus it is not surprising that many cytokines can activate macrophages by this criterion. In assays conducted without antitarget cell antibody or secretagogues, monocytes or macrophages isolated from normal tissues usually exert little or no cytotoxicity. Substantial cytotoxic activity is manifest following prior exposure of the effector cells to IFN-γ (SCHREIBER et al. 1983), IFN-α, IFN-β (VARESIO et al. 1984), IL-1α, TNF-α (PHILIP and EPSTEIN 1986), CSF-GM (GRABSTEIN et al. 1986), CSF-M (RALPH and NAKOINZ 1987), or IL-4 (CRAWFORD et al. 1987). In many of these systems, a stimulus in addition to the cytokine must also be present. This second signal is usually bacterial lipopolysaccharide, but in some experiments the second signal has been bacteria that lack lipopolysaccharide (SCHREIBER et al. 1983), or other cytokines (ESPARZA et al. 1987). In assays in which antitarget cell antibody is included, the list of macrophage-activating factors includes IFN-γ, IFN-α, IFN-β, and IL-2 (whose effect was not blocked by antibody to IFN-γ), but not IL-1, IL-4, or CSF-M (RALPH et al. 1988).

The mechanisms for the cytokine effects are incompletely understood, but are likely to involve alterations both in the surface receptors through which macrophages bind tumor cells and in the secretion of the molecules that inflict injury. In the case of antibody-dependent, cell-mediated cytotoxicity, increased expression of Fc receptors is likely to be of major importance. In the absence of antibody, increased expression of adhesion receptors, including LFA-1 (STRASSMAN et al. 1986), may play a role in the ability of cytokine-treated macrophages to bind to tumor cells.

Well-defined molecules released by activated macrophages and known to kill tumor cells or inhibit their growth are presently limited to the following:

reactive oxygen intermediates (NATHAN 1982), reactive nitrogen intermediates (HIBBS et al. 1987a), TNF-α (ROTHSTEIN and SCHREIBER 1988), IL-1α/β (PHILIP and EPSTEIN 1986; ICHINOSE et al. 1988), and arginase (CURRIE et al. 1979). These will be discussed in turn.

Reactive oxygen intermediates are inorganic compounds that are intermediate in the reduction of molecular oxygen to water. The best characterized are superoxide, hydrogen peroxide, and hydroxyl radical; in phagocytes that contain myeloperoxidase, such as human monocytes, the term may embrace the peroxidase-catalyzed oxidation products of halides as well, that is, hypohalous acids (KLEBANOFF 1988). The molecular targets of these redox-active agents include DNA (SCHRAUFSTATTER et al. 1988), key enzymes (HYSLOP et al. 1988), lipids (ZOELLER et al. 1988), nonprotein thiols (NATHAN et al. 1981), and α-keto acids (O'DONNELL-TORMEY et al. 1987). The cytokines known to enhance the capacity of mononuclear phagocytes to release reactive oxygen intermediates are IFN-γ, TNF-α, TNF-β, CSF-GM, CSF-M, and possibly migration inhibition factor (MIF) (NATHAN et al. 1983, 1984; MURRAY et al. 1983; REED et al. 1987; COLEMAN et al. 1988). Reactive oxygen intermediates play a major role in antibody-dependent, cell-mediated cytotoxicity, in spontaneous cytotoxicity toward tumor cells coated with the cytophilic eosinophil peroxidase, or in cytotoxicity triggered by the secretagogic phorbol diesters (NATHAN 1982). Moreover, hydrogen peroxide can interact synergistically with unidentified protein(s) released from activated macrophages to kill tumor cells (ADAMS et al. 1981). However, when neither antibody, peroxidase, nor phorbol diesters are added, reactiven oxygen intermediates do not appear to play a major role in cytotoxicity toward tumor cells in vitro.

Reactive nitrogen intermediates are inorganic oxides of nitrogen derived from the guanido nitrogens of L-arginine; they can be considered intermediates in the sequential oxidation of ammonia to nitrate. Products of this pathway in activated macrophages include nitric oxide (MARLETTA et al. 1988; HIBBS et al. 1988; STUEHR et al. 1989a) and nitrite (STUEHR and MARLETTA 1987; HIBBS et al. 1987a). Reactive nitrogen intermediates are responsible for the cytostatic effects of macrophages on certain tumor cell lines in vitro (HIBBS et al. 1987a; STUEHR and NATHAN 1989). The basis of cytostasis includes inactivation of several Fe-S cluster enzymes in the respiratory pathway (*cis*-aconitase, and compounds I and II in the mitochondrial electron transport chain). There appears to be an additional target enzyme(s) in the pathway of DNA synthesis. The cytostatic lesion is converted to a cytolytic one if glucose is withheld, interfering with the glycolytic pathway (HIBBS et al. 1978b). Under the conditions commonly used for assaying macrophage-mediated antitumor activity in vitro, glucose is often depleted prior to target cell death, largely through the prodigious glucose consumption of activated macrophages, especially under the influence of bacterial lipopolysaccharide (NATHAN, unpublished observations). Thus, it is likely that reactive nitrogen intermediates contribute substantially to antitumor effects of activated macrophages in many in vitro studies. When 12 cytokines were tested as single agents, the only one that could enhance the ability of mouse macrophages to release reactive nitrogen intermediates was IFN-γ. However, when combinations of agents were tested, TNF-α and TNF-β could synergistically enhance the effect of IFN-γ. In addition, bacterial lipopolysaccharide alone was effective, and it could interact synergistically with IFN-γ, or with IFN-α/β (DING et al. 1988). The mechanism of action of IFN-γ in augmenting cytotoxicity by reactive nitrogen intermediates is through induction of the L-arginine-dependent enzyme that synthesizes them (STUEHR et al. 1989b).

The cytokines that enhance the capacity of mononuclear phagocytes to release the cytotoxic factors TNF-α and IL-α/β include IFN-γ, TNF-α, and IL-1α/β (COLLART et al. 1986; PHILIP and EPSTEIN 1986). Release of TNF-α is both necessary and sufficient for macrophage killing of some tumor cells (ROTHSTEIN and SCHREIBER 1988), but these particular targets are exceptionally sensitive to TNF-α. In general, reagent rTNF-α kills or inhibits the growth of about one-third of otherwise unselected human tumor cell lines that grow in vitro, has no discernible effect on one-third, and promotes the growth of the remainder (SUGARMAN et al. 1985). The cytotoxic effects if IL-1α/β on tumor cells have generally not been marked (ONOZAKI et al. 1985). It is not understood how either TNF-α or IL-1α/β exerts cytotoxic actions. In summary, it is not clear that cytokine-enhanced secretion of cytokines is a major component of macrophage activation for cytotoxicity toward most tumors.

In contrast to arginine-dependent cytotoxicity mediated by reactive nitrogen intermediates, it is also possible for macrophages to kill some tumor cells by depleting arginine, an essential amino acid, from the medium. This occurs through secretion of arginase by the macrophages. Arginine depletion has been detected in lymph draining some tissues rich in activated macrophages, and thus this mechanism of cytotoxicity may be physiologic (CURRIE et al. 1979). However, the number of tumors susceptible to inhibition in this manner appears to be quite limited; many workers have been unable to confirm a role for arginine in macrophage-mediated cytotoxicity toward the tumor cells used in their laboratories. It has apparently not been demonstrated that cytokines can regulate the ability of macrophages to secrete arginase.

III. Promotion of Wound Healing

It was noted above that cytokines can affect the ability of macrophages to secrete cytokines that might be involved in wound healing, such as PDGF, TGF-β, TGF-α, TNF-α, IL-1α, IL-1β, and fibroblast growth factor (FGF). Elegant studies by RAPPOLEE et al. (1988) have documented the production of several of these cytokines by wound-associated macrophages. However, there is almost no experimental evidence that bears directly on the impact of cytokines on macrophages as it affects their role in wound healing.

IV. Generation of Inflammatory and Immune Responses

Previous sections have discussed the effect of cytokines on the ability of macrophages to process and present antigen; to secrete the T-cell growth factors IL-1α, IL-1β, TNF-α, and IL-6; and to release pro-inflammatory enzymes, cytokines, and other autocoids, including plasminogen activator, leukotriene C, IL-1α, IL-1β, TNF-α, and IL-6. However, cytokines can also lead to decreased release of inflammatory mediators (e.g., exposure to TGF-β decreases macrophage release of TNF-α; ESPEVIK et al. 1987), and, as noted above, increased release of agents that inhibit inflammatory processes or lymphocyte responses. Thus it is difficult to integrate the information from in vitro experiments to predict the physiologic consequences of cytokine exposure on the mediation of inflammatory and immune responses by mononuclear phagocytes in vivo.

V. Scavenging of Senescent Cells

Our limited knowledge regarding the impact of cytokines on endocytosis through receptors for Fc, C3bi, and advanced glycosylation end products, discussed above, bears on this topic. Senescent cells may be targetted to macrophages by autoantibodies or by accumulation of advanced glycosylation end products (VLASSARA et al. 1987). However, there are few studies dealing directly with the effect of cytokines on the scavenger function of macrophages. The scavenger function is probably essential for life. It must be executed continually for the disposal of massive numbers of erythrocytes and polymorphonuclear leukocytes, as well as other cells. Thus, it is to be expected that the capacity to perform this function is constitutive. An inducible increase in capacity to scavenge ("activation" of the scavenger function) is probably not a prominent feature of the response of the macrophage to external signals.

H. Deactivation

The capacity to activate macrophages is essential to the life of the host. Why is such a crucial faculty not constitutive – why must it be induced? The answer is probably that the biochemistry of microbial killing poses a threat to normal tissues. Thus, there may be selective pressure on both the host and its parasites to be able to block or reverse the activation of macrophages. Indeed, there are soluble factors of both host cell and microbial origin that can deactivate macrophages. The following discussion deals with cytokines that have this property.

The first host cell-derived macrophage-deactivating factor to be studied in detail was detected as an activity in medium conditioned by all mouse tumor cells tested, as well as by some nontransformed murine cells (SZURO-SUDOL and NATHAN 1982). The purification of this polypeptide from medium conditioned by P815 mastocytoma cells is underway in the author's laboratory. Complete purification, the production of specific antibodies, and the isolation of cDNA encoding this factor will permit tests of the hypothesis that the same polypeptide comprises the principal deactivating factor in medium conditioned by other mouse tumors, some nonneoplastic mouse cells, and human tumor lines. This macrophage deactivation factor exerts a gradual, profound, relatively selective, nontoxic, and slowly reversible suppressive effect on the ability of macrophages to undergo a respiratory burst and to kill the protozoal pathogens *Toxoplasma gondi* and *Leishmania donovani* (SZURO-SUDOL and NATHAN 1982; SZURO-SUDOL et al. 1983). The deactivating factor antagonizes the ability of rIFN-γ to induce Ia antigen expression and enhanced respiratory burst capacity in macrophages (TSUNAWAKI and NATHAN 1986; NATHAN and TSUNAWAKI 1986). The action of this macrophage deactivation factor on human mononuclear phagocytes, its presence in tumor-bearing hosts, its full range of bioactivities, its relation to other cytokines, and its physiologic function all remain to be demonstrated.

Recent studies have led to the identification of a second class of potent macrophage-deactivating factors. It was considered that the scavenger activities of macrophages, as in wound healing, could be detrimental, if ingestion of debris should trigger a respiratory burst that could injure growing endothelial cells, fibroblasts, and other tissue elements. This speculation led to a screening of cytokines believed to be involved in wound healing, for possible action as macrophage-deactivating factors. Among 12 cytokines tested, potent respiratory burst-suppressing activity was detected only in one highly homologous pair, TGF-β1 and TGF-β2. Their half-maximal inhibitory concentrations were 0.6 and 4.8 pM, respectively, suggestive of a possible physiologic role for the described action (TSUNAWAKI et al. 1988). The characterization of TGF-β as a mediator of wound healing, fibrosis, and angiogenesis, as well as a potent suppressor of T-cell responses, is summarized elsewhere in this volume. These observations suggest a way for several cell types to restrain the activated macrophage. Antigen-stimulated T cells, degranulating platelets, some tumor cells, and activated macrophages themselves are all sources of TGF-β. The previously described macrophage-deactivating factor from tumor cells, discussed above, appears to be distinct from TGF-β, based on physicochemical and immunochemical evidence, as well as the contrasting effects of the two cytokines on the respiratory burst enzyme, to be described below.

Up to this point, our speculations about macrophage deactivation have been confined to the site of a tumor or wound. Let us consider next the problem of how the host may restrict macrophage activation to an inflammatory site. The question arises because of the systemic circulation of macrophage-activating factors leaking out of such sites into the circulation, as evidenced by activation of blood monocytes following injection of traces of rIFN-γ in the skin of patients with lepromatous leprosy (NATHAN et al. 1986). Thus, while it seems critical that macrophages in an inflammatory site should be responsive to IFN-γ, it may be equally important that the response of macrophages residing in other tissues should be damped.

Experiments on the change in IFN-binding sites during macrophage maturation in vitro suggest a possible mechanism for anatomic compartmentalization of macrophage activation in response to IFN-γ (YOSHIDA et al. 1988). Fresh monocytes bound iodinated rIFN-γ with kinetics suggesting a single class of receptors for IFN-γ, in agreement with previous work by other investigators. However, after the cells differentiated into macrophages, they expressed both a high-affinity and a low-affinity class of IFN-γ receptors. There were also maturation-dependent changes in the response of the cells to IFN-α and -β. While IFN-α and -β had no effect in our assays with monocytes, or were toxic at high doses, they were weak activators of the mature macrophage, by the dual criteria of ability to inhibit the growth of *T. gondii* and to release increased amounts of H_2O_2. Surprisingly, at extremely low doses, IFN-α and -β blocked macrophage activation by low doses of IFN-γ. This effect appears to operate at the level of competition for binding to the high-affinity class of IFN-γ receptors. Thus, IFN-α and -β were able quantitatively to displace IFN-γ from the high-affinity sites, without affecting the binding of IFN-γ to the low-affinity sites. Indeed, IFN-α and -β bound to the high-affinity site with higher affinity than IFN-γ. Trace levels of IFN-α are produced constitutively by many stromal and parenchymal cells (TOVEY et al. 1987). Thus, activation of resident tissue macrophages by traces of IFN-γ circulating or diffusing from inflammatory sites might well be blocked. In the inflammatory sites themselves, however, macrophage activation should not be blocked, for the following reasons: (a) High concentrations of IFN-γ activate mature macrophages through the low-affinity receptor. (b) High concentrations of IFN-α or -β activate mature macrophages through their own low-affinity receptors. (c) Younger monocytes and macrophages enter inflammatory lesions; such cells lack the high-affinity regulatory site and are not subject to this putative damping mechanism. Hence, damping should be restricted to uninflamed tissues – those containing mature macrophages and low concentrations of all three types of IFN.

What are the biochemical mechanisms of activation and deactivation? We know only that there are probably at least four regulatory loci. When the respiratory burst is triggered, and macrophages are permeabilized with selected detergents, the respiratory burst ceases, and can be reconstituted with NADPH, as detected by dual wavelength spectroscopy of the superoxide-dependent reduction of acetylated cytochrome c. The rate of superoxide release is measured as a function of NADPH concentration, permitting estimations of the V_{max} and K_m of the oxidase. Using this technique, it has been shown that, compared with granulocytes, activated macrophages have about half the V_{max}, but virtually the same K_m for NADPH (about 0.05 m*M*), well below the concentration of NADPH in macrophage cell water (0.6 m*M*). The surprising finding is that nonactivated macrophages have the same V_{max} as activated cells. Their lesser respiratory burst is due chiefly to an order of magnitude higher K_m for NADPH. When activated macrophages are treated with MDF, they show modest changes in V_{max}, but, again, an order of magnitude increase in K_m, so that they come to resemble the nonactivated macrophage. In short, in this situation, activation and deactivation appear to exert nearly mirror image effects, not on the quantity of oxidase, but on its affinity for reducing cosubstrate after triggering (TSUNAWAKI and NATHAN 1986). In contrast, deactivation induced by TGF-β leaves no imprint in this assay. That is, the oxidase functions in lysates from TGF-β-treated cells with the same kinetics as in activated macrophages (TSUNAWAKI et al. 1989). The gross defect seen in intact TGF-β-treated cells must be more proximal, or is reversed by permeabilization. It is likely that resident, activated, tumor factor-deactivated, and TGF-β-deactivated macrophages each differ in their ability to transduce signals to assemble an active oxidase complex from its dormant components. Whatever the defects in deactivated macrophages, they are relatively specific. Many other biochemical changes have been excluded, which could plausibly regulate the secretion of reactive oxygen intermediates, and the secretion of many other macrophage products is unaffected.

I. Mechanisms of Action of Cytokines on Macrophages

Perhaps the most active frontier in the study of mononuclear phagocytes over the next decade will be the detailed analysis of the mechanism of action of cytokines. The availability of pure, recombinant cytokines, known to induce specific gene products (e.g., γIP-10, LUSTER and RAVETCH 1987) or secretory products (e.g., nitric oxide, STUEHR et al. 1989) of biomedical relevance in readily cultured, nearly pure mammalian cell populations, constitutes an attractive system for exploring basic issues in cell and molecular biology that are not amenable to study in simpler systems. No doubt it will soon be possible to describe most if not all of the key molecules involved in connecting an externally applied cytokine to the eventual alteration in one of the chief physiologic functions of the mononuclear phagocyte. At present, several such chains are under construction, but none is linked from end to end.

The characterization of cytokine receptors, their purification, and the cloning of their cDNAs have been described elsewhere in this volume. In several instances, mononuclear phagocytes were the chief cell type used for receptor isolation. Similar progress can be anticipated in the near future with regard to receptors for other cytokines with prominent actions on mononuclear phagocytes.

Little is yet known about signal transduction mechanisms in macrophages in response to cytokines. Most of the available evidence concerns IFN-γ. Changes in Na^+/H^+ antiport and the sustained generation of diacylglycerol from a source other than polyphosphoinositides are early effects of IFN-γ on mononuclear phagocytes and are under intensive study (ADAMS and HAMILTON 1987). A transient increase in activity of protein kinase C peaks about 4 h after IFN-γ treatment (HAMILTON et al. 1987a) in one experimental system, while the most prominent effect of IFN-γ on macrophage protein kinase C in another system was its translocation to a membrane fraction (FAN et al. 1988). The differences between these observations have not been reconciled, nor has any firm evidence been presented for the involvement of protein kinase C in mediation of an action of IFN-γ in macrophages. Some (CELADA and SCHREIBER 1986) but not all (RADZIOCH and VARESIO 1988) effects of IFN-γ have been inhibited by nonspecific pharmacologic agents whose actions include inhibition of protein kinase C.

Cytokine-response control elements in DNA are currently being identified in many cells; no doubt these studies will soon include the macrophage. Many cytokine-induced genes have been identified in mononuclear phagocytes, including protooncogenes, genes for cytokines or presumptive cytokines, and genes encoding some of the well-known secretory products of the macrophage. As yet no specific functions have been definitively ascribed to the cytokine-induced protooncogene products in macrophages. No novel principles have yet emerged that set apart gene regulation in macrophages from that in other cell types.

J. Autocrine Effects

The fact that mononuclear phagocytes can produce many of the cytokines to which they respond (e.g., IL-1, IL-6, TNF-α, CSF-GM, CSF-M, IFN-α, TGF-β) has repeatedly prompted the speculation that autocrine effects may be prominent in the physiologic responses of these cells. However, this has rarely been demonstrated in a definitive manner. Two considerations should be kept in mind when evaluating the evidence for autocrine actions of cytokines on macrophages.

First, the distinction must be made between results at the population level and at the single cell level. Given the heterogeneity of macrophage populations (FORSTER and LANDY 1981), the autocrine hypothesis might require that the same cell produce a cytokine and respond to it. Otherwise, it becomes semantic whether the effects should be classed as autocrine or paracrine. Evidence for the autocrine hypothesis at the single cell level is not yet available.

Second, at the population level, an autocrine loop is not established merely by demonstrating the ability of mononuclear phagocytes to produce a cytokine and to respond to additions of the same cytokine. Instead, it would be necessary to show that the cells respond to the cytokine when they produce it. This need not be the case. Some signals for cytokine production may themselves block the ability to macrophages to respond to the same cytokine, forestalling an autocrine effect. The two best-characterized examples pertain to CSF-M and TNF-α. In both cases, bacterial LPS triggers the release of the cytokines from mononuclear phagocytes. However, prior to cytokine release, LPS triggers internalization of receptors for the same cytokines, precluding a biologic response (GUILBERT and STANLEY 1984; DING et al. 1989).

K. Polymorphonuclear Leukocytes

Not just mononuclear phagocytes, but all phagocytic leukocytes, are responsive to cytokines. A brief comment on this point will put the discussion of mononuclear phagocytes into better perspective, although a full treatment of cytokine actions on polymorphonuclear leukocytes (PMNs) is beyond the scope of this chapter.

Until recently, PMNs seemed to stand outside the networks by which other cells of the immune system regulate each other. To the contrary, and quite apart from the ability of CSFs to augment the number of circulating PMNs, recent studies have revealed that cytokines of mononuclear cell origin can modulate many functions of mature PMNs, including their survival; migration; expression of cell adhesion molecules and receptors for complement, IgG, IgA, and formylated peptides; phagocytosis; antitumor activity; secretion of eicosanoids and lysomal contents; and respiratory burst capacity. Moreover, certain cytokines can themselves trigger a massive respiratory burst in neutrophils that are adherent to proteins of extracellular matrix. Thus, the situation with cytokine effects on PMNs parallels that with mononuclear phagocytes.

These insights are already leading to clinical applications, such as administration of rIFN-γ for the enhancement of antimicrobial activity and respiratory burst capacity in neutrophils as well as monocytes of patients with chronic granulomatous disease (EZEKOWITZ et al. 1988; SECHLER et al. 1988). As little as two closely spaced injections of rIFN-γ can lead to an increase in respiratory burst capacity in circulating PMNs which is sustained for up to 5 weeks before it wanes. This response time is roughly two orders of magnitude longer than the half-life of the cytokine, and an order of magnitude longer than the half-life of the cells in which the response is expressed. This suggests that rIFN-γ can reprogram gene expression in a population of precursor cells, an effect which is manifest at the gene product level, not in the precursor itself, but in its progeny, and which lasts through several replicative cycles in the precursor before fading. These observations not only carry clinical promise; they suggest that novel aspects of gene regulation may be waiting to be revealed through further study of cytokine action on phagocytes.

L. Conclusions

Cytokines comprise the protein subset of autocoids. They can exert their influence over anatomic distances as great as those of endocrine hormones or as small as those of neurotransmitters, but usually intermediate between the two. Cytokines individually and collectively induce such an extraordinarily broad set of biochemical and functional responses in such a wide array of cells that it has become increasingly difficult to discern either a commonality in their functions or a basis for their classification. In particular, compelling reasons are no longer evident for restricting the term "peptide growth factors" to the subset of cytokines which used to bear that designation.

The philosophical stance adopted in this chapter is that the cardinal function shared by most if not all cytokines is to regulate the (re)modeling of tissues, including the "immune response tissue." The ability to form the "immune response tissue" is essential for survival of the host. The "immune response tissue" can assemble transiently in and destroy any other tissue in the course of forestalling or combatting infection. The mononuclear phagocyte is a key regulatory and effector cell in the "immune response tissue."

Accordingly, the goal of this chapter has been to survey actions of all known cytokines on mature mononuclear phagocytes, excluding only the effects on their precursors. Published evidence is so voluminous that this goal, while perhaps within sight, is not within reach.

In any case, very little is yet known about the challenging topic defined by the editors, the coordinate actions of cytokines.

In fact, as yet, there are few actions of cytokines on macrophages for which the evidence is sufficient to establish the effects as physiologic. At present, physiologic control by cytokines seems to be best documented for certain macrophage secretory products and for the activation of macrophages for enhanced antimicrobial activity. Clinical trials aimed at inducing human macrophage activation with cytokines both contribute to and depend on this evidence.

Increasingly powerful research tools will soon extend our knowledge of physiologic cytokine actions to many other aspects of mononuclear phagocyte function, and at the same time reveal the mechanisms of action of cytokines on this migratory and multipotential class of cells.

References

Adams DO, Hamilton TA (1987) Molecular transduction mechanisms by which interferon gamma and other signals regulate macrophage development. Immunol Rev 97:5–28

Adams DO, Johnson WJ, Fiorito E, Nathan CF (1981) Hydrogen peroxide and cytolytic factor can interact in effecting cytolysis of neoplastic targets. J Immunol 127:1973–1977

Adams JS, Gacod MA (1985) Characterization of 1-α-hydroxylase of vitamin D3 sterols by cultured alveolar macrophages from patients with sarcoidosis. J Exp Med 161:755–765

Akagawa KS, Tokunaga T (1985) Lack of binding of bacterial lipopolysaccharide to mouse lung macrophages and restoration of binding by gamma interferon. J Exp Med 162:1444–1459

Alvaro-Gracia JM, Zvaifler NJ, Firestein GS (1989) Cytokines in chronic inflammatory arthritis. IV. GM-CSF mediated induction of class II MHC antigen on human monocytes: a possible role in rheumatoid arthritis. J Exp Med (170: 865–875)

Ampel NM, Wing EJ, Waheed A, Shadduck RK (1986) Stimulatory effects of purified macrophage colony-stimulating factor on murine resident peritoneal macrophages. Cell Immunol 97:344–356

Assoian RK, Fleurdelys BE, Stevenson HC, Miller PJ, Madtes DK, Raines EW, Ross R, Sporn MB (1987) Expression and secretion of type β transforming growth factor by activated human macrophages. Proc Natl Acad Sci USA 84:6020–6024

Averbook BJ, Yamamoto RS, Ulich TR, Jeffes EW, Masunaka I, Granger GA (1987) Purified native and recombinant human alpha lymphotoxin [tumor necrosis factor (TNF)-beta] induces inflammatory reactions in normal skin. J Clin Immunol 7:333–340

Bauer J, Ganter U, Geiger T, Jacobshagen U, Hirano T, Matsuda T, Kishimoto T, Andus T, Acs G, Gerok W, Ciliberto G (1988) Regulation of interleukin-6 expression in cultured human blood monocytes and monocyte-derived macrophages. Blood 72:1134–1140

Belsoevic M, Davis CE, Meltzer MS, Nacy CA (1988) Regulation of activated macrophage antimicrobial activities. Identification of lymphokines that cooperate with IFN-gamma for induction of resistance to infection. J Immunol 141:890

Bermudez LE, Young LS (1987) Granulocyte-macrophage colony stimulating factor activates human macrophages to kill *Mycobacterium avium* complex (Abstr). Clin Res 35:612 A

Beutler B, Tkacenko V, Milsark I, Krochin N, Cerami A (1986) Effect of gamma interferon on cachectin expression by mononuclear phagocytes. Reversal of the lps_d (endotoxin resistance) phenotype. J Exp Med 164:1791–1796

Bianco C, Eden A, Cohn ZA (1976) The induction of macrophage spreading: role of coagulation factors and the complement system. J Exp Med 144:1531

Bloom BR, Bennett B (1966) Mechanism of a reaction in vitro associated with delayed type hypersensitivity. Science 153:80–82

Boraschi D, Censini S, Bartalini M, Tagliabue A (1985) Regulation of arachidonic acid metabolism in macrophages by immune and nonimmune interferons. J Immunol 135:502–505

Brummer E, Hanson LH, Restrepo A, Stevens DA (1988) In vivo and in vitro activation of pulmonary macrophages by IFN-γ for enhanced killing of *Paracoccidiodes brasiliensis* or *Blastomyces dermatitidis*. J Immunol 140:2786–2789

Buchmeier NA, Schreiber RD (1985) Requirement of endogenous interferon-gamma production for resolution of *Listeria monocytogenes* infection. Proc Natl Acad Sci USA 82:7404–7408

Carvalho EM, Badaro R, Reed SG, Jones TC, Johnson WD Jr (1985) Absence of gamma interferon and interleukin 2 production during active visceral leishmaniasis. J Clin Invest 76:2066–2069

Catterall JR, Sharma SD, Remington JS (1986) Oxygen-independent killing by alveolar macrophages. J Exp Med 163:1113–1131

Celada A, Schreiber RD (1986) Role of protein kinase C and intracellular calcium mobilization in the induction of macrophage tumoricidal activity by interferon gamma. J Immunol 137:2375–2379

Chen BD-M, Clark CR, Chou T-H (1988a) Granulocyte/macrophage colony-stimulating factor stimulates monocyte and tissue macrophage proliferation and enhances their responsiveness to macrophage colony-stimulated factor. Blood 71:997–1002

Chen BD-M, Mueller M, Chou T-H (1988b) Role of granulocyte/macrophage colony-stimulating factor in the regulation of murine alveolar macrophage proliferation and differentiation. J Immunol 141:139–144

Chen BD-M, Mueller M, Olencki T (1988c) Interleukin-3 (IL-3) stimulates the clonal growth of pulmonary alveolar macrophage of the mouse: role of IL-3 in the regulation of macrophage production outside the bone marrow. Blood 72:685–690

Coleman DL, Chodakewitz JA, Bartiss AH, Mellors JW (1988) Granulocyte-macrophage colony-stimulating factor enhances selective effector functions of tissue-derived macrophages. Blood 72:573–578

Collart MA, Belin D, Vassalli J-D, de Kossodo S, Vassalli P (1986) γ-Interferon enhances macrophage transcription of the tumor necrosis factor/cachectin, interleukin-1, and urokinase genes, which are controlled by short lived repressors. J Exp Med 164:2113–2118

Conkling PR, Greenburg CS, Weinberg JB (1988) Tumor necrosis factor induces tissue factor-like activity in human leukemia cell line U937 and peripheral blood monocytes. Blood 72:128–133

Crawford RM, Finbloom DS, Ohara J, Paul WE, Meltzer MS (1987) B cell stimulatory factor-1 (interleukin 4) activates macrophages for increased tumoricidal activity and expression of Ia antigens. J Immunol 139:135

Currie GA, Basham C (1978) Differential arginine dependence and the selective cytotoxic effects of activated macrophages for malignant cells in vitro. Br J Cancer 38:653–659

Currie GA, Gyure L, Cifuentes L (1979) Microenvironmental arginine depletion by macrophages in vivo. Br J Cancer 39:613–620

Czarniecki CW, Chiu HH, Wong GH, McCabe SM, Palladino MA (1988) Transforming growth factor-beta 1 modulates the expression of class II histocompatibility antigens on human cells. J Immunol 140:4217–4223

David JR (1966) Delayed hypersensitivity in vitro: its mediation by cell-free substances formed by lymphoid cell-antigen interaction. Proc Natl Acad Sci USA 56:72–77

De Titto E, Catteral JR, Remington JS (1986) Activity of recombinant tumor necrosis factor on *Toxoplasma gondii* and *Trypanosoma cruzi*. J Immunol 137:1342

Ding A, Nathan CF (1987) Trace levels of bacterial lipopolysaccharide prevent interferon-γ or tumor necrosis factor-α from enhancing mouse peritoneal macrophage respiratory burst capacity. J Immunol 139:1971–1977

Ding A, Nathan CF, Stuehr DJ (1988) Release of reactive nitrogen intermediates and reactive oxygen intermediates from mouse peritoneal macrophages: comparison of activating cytokines and evidence for independent production. J Immunol 141:2407–2412

Ding A, Sanchez E, Srimal S, Nathan CF (1989) Macrophages rapidly internalize their tumor necrosis factor receptors in response to bacterial lipopolysaccharide. J Biol Chem 264:3924–3929

Edwards CK III, Hedegaard HB, Zlotnik A, Gahgadharam PR, Johnston RB Jr, Pabst MJ (1986) Chronic infection due to *Mycobacterium intracellulare* in mice: association with macrophage release of prostaglandin E2 and reversal by injection of indomethacin, muramyl dipeptide, and interferon-γ. J Immunol 136:1820–1827

Esparza I, Mannel D, Ruppel A, Falk W, Krammer PH (1987) Interferon-γ and lymphotoxin or tumor necrosis factor act synergistically to induce macrophage killing of tumor cells and schistosomula of *Schistosoma mansoni*. J Exp Med 166:589–594

Espevik T, Figari IS, Shalaby MR, Lackides GA, Lewis GD, Shepard HM, Palladino MA Jr (1987) Inhibition of cytokine production by cyclosporin A and transforming growth factor-β. J Exp Med 166:571–576

Ezekowitz RAB, Dinauer MC, Jaffe HS, Orkin SH, Newburger PE (1988) Partial correction of the phagocyte defect in patients with X-linked chronic granulomatous disease by subcutaneous interferon gamma. N Engl J Med 319:146

Fan XD, Goldberg M, Bloom BR (1988) Interferon-gamma-induced transcriptional activation is mediated by protein kinase C. Proc Natl Acad Sci USA 85:5122–5125

Fels AOS, Nathan CF, Cohn ZA (1987) H_2O_2 release by alveolar macrophages from sarcoid patients and by alveolar macrophages from normals after exposure to recombinant interferons αA, β, γ and 1,25-dihydroxyvitamin D3. J Clin Invest 80:381–386

Fogelman AM, Seager J, Groopman JE, Berliner JA, Haberland ME, Edwards PA, Golde DW (1983) Lymphokines secreted by an established lymphocyte line modulate receptor-mediated endocytosis in macrophages derived from human monocytes. J Immunol 131:2368–2373

Forster O, Landy M (eds) (1981) Heterogeneity of mononuclear phagocytes. Academic, London

Fowles RE, Fajardo IM, Leibowitch JL, David JR (1973) The enhancement of macrophage bacteriostasis by products of activated lymphocytes. J Exp Med 138:952–964

Gendelman H, Orenstein JM, Martin MA, Ferrua C, Mitra R, Phipps T, Wahl LA, Lane HC, Fauci HS, Burke DS, Skillman D, Meltzer MS (1988) Efficient isolation and propagation of human immunodeficiency virus on recombinant colony-stimulating factor 1-treated monocytes. J Exp Med 167:1428

Goodman MG, Chenoweth DE, Weigle WO (1982) Induction of interleukin 1 secretion and enhancement of humoral immunity by binding of human C5a to macrophage surface C5a receptors. J Exp Med 156:912

Grabstein KH, Urdal DL, Tushinski RJ, Mochizuki DY, Price BL, Cantrell MA, Gillis S, Conlon PJ (1986) Induction of macrophage tumoricidal activity by granulocyte-macrophage colony-stimulating factor. Science 232:506

Granger DL, Hibbs JB Jr, Perfect JR, Durack DT (1988) Specific amino acid (L-arginine) requirement for the microbiostatic activity of murine macrophages. J Clin Invest 81:1129

Grau GE, Kindler V, Piguet P-F, Lambert P-H, Vassalli P (1988) Prevention of experimental cerebral malaria by anticytokine antibodies. Interleukin 3 and granulocyte macrophage colony-stimulating factor are intermediates in increased tumor necrosis factor production and macrophage accumulation. J Exp Med 168:1499–1504

Greil J, Bodendorfer B, Rollinghoff M, Solbach W (1988) Application of recombinant granulocyte-macrophage colony-stimulating factor has a detrimental effect in experimental murine leishmaniasis. Eur J Immunol 18:1527–1533

Guilbert LJ, Stanley ER (1984) Modulation of receptors for the colony-stimulating factor, CSF-1, by bacterial lipopolysaccharide and CSF-1. J Immunol Methods 73:17–28

Guyre PM, Morganelli PM, Miller R (1983) Recombinant interferon gamma increases immunoglobulin Fc receptors on cultured human mononuclear phagocytes. J Clin Invest 72:393–397

Hamilton TA, Bacton DL, Somers SD, Gray PW, Adams DO (1985a) Interferon-gamma modulates protein kinase C activity in murine peritoneal macrophages. J Biol Chem 260:1378–1381

Hamilton TA, Rigsbee JE, Scott WA, Adams DO (1985b) Gamma interferon enhances the secretion of arachidonic acid metabolites from murine peritoneal macrophages stimulated with phorbol diesters. J Immunol 134:2631–2636

Hampton RY, Golenbock DT, Raetz CRH (1988) Lipid A binding sites in membranes of macrophage tumor cells. J Biol Chem 263:14802–14807

Heremans H, Dijkmans R, Sobis H, Vandekerckhove F, Billiau A (1987) Regulation by interferons of the local inflammatory response to bacterial lipopolysaccharide. J Immunol 138:4175–4179

Herrmann R, Cannistra SA, Levine H, Griffin JD (1985) Expression of interleukin 2 receptors and binding of interleukin 2 by gamma interferon-induced human leukemic and normal monocytic cells. J Exp Med 162:1111–1116

Hibbs JB Jr, Taintor RR, Vavrin Z (1987a) Macrophage cytotoxicity: role for L-arginine deiminase and imino nitrogen oxidation to nitrite. Science 235:473

Hibbs JB Jr, Vavrin Z, Taintor RR (1987b) L-Arginine is required for expression of the activated macrophage effector mechanism causing selective metabolic inhibition in target cells. J Immunol 138:550–565

Hibbs JB Jr, Taintor RR, Vavrin Z, Rachlin EM (1988) Nitric oxide: a cytotoxic activated macrophage effector molecule. Biochem Biophys Res Commun 157:87–94

Horiguchi J, Warren MK, Kufe D (1987) Expression of the macrophage-specific colony-stimulating factor in human monocytes treated with granulocyte-macrophage colony-stimulating factor. Blood 69:1259–1261

Hyslop PA, Hinshaw DB, Halsey WA Jr, Schraufstatter IU, Sauerheber RD, Spragg RG, Jackson JH, Cochrane CH (1988) Mechanisms of oxidant-mediated cell injury. The glycolytic and mitochondrial pathways of ADP phosphorylation are major intracellular targets inactivated by hydrogen peroxide. J Biol Chem 263:1665–1675

Ichinose Y, Bakouche O, Tsao JY, Fidler IJ (1988) Tumor necrosis factor and IL-1 associated with plasma membranes of activated human monocytes lyse monokine-sensitive but not monokine-resistant tumor cells whereas viable activated monocytes lyse both. J Immunol 141:512

Inaba K, Steinman RM (1984) Resting and sensitized T lymphocytes exhibit distinct stimulatory (antigen presenting cell) requirements for growth and lymphokine release. J Exp Med 160:1717–1735

Jensen WA, Rose RM, Burke RA Jr, Anton K, Remold HG (1988) Cytokine activation of anti-bacterial activity in human mononuclear phagocytes: comparison of recombinant interferon-γ and granulocyte macrophage stimulating factor. Cell Immunol 117:369–377

Kaufmann SH, Hug E, Vath U, DeLibero G (1987) Specific lysis of *Listeria monocytogenes*-infected macrophages by class II-restricted L3T4+ T cells. Eur J Immunol 17:237–246

Kiderlen AF, Kaufmann SHE, Lohmann-Matthes M-L (1984) Protection of mice against the intracellular bacterium *Listeria monocytogenes* by recombinant immune interferon. Eur J Immunol 14:964–967

Kindler V, Sappino AP, Grau GE, Piguet PF, Vassalli P (1989) The inducing role of tumor necrosis factor in the development of bactericidal granulomas during BCG infection. Cell 56:731–740

Klebanoff SJ (1988) Phagocytic cells: products of oxygen metabolism. In: Gallin JI, Goldstein IM, Snyderman R (eds) Inflammation: basic principles and clinical correlates. Raven, New York, pp 391–444

Koeffler HP, Reichel H, Bishop JE, Norman AW (1985) Gamma-interferon stimulates production of 1,25-dihydroxyvitamin D3 by normal human macrophages. Biochem Biophys Res Commun 127:596–603

Lamperi S, Carozzi S (1988) Interferon-γ (IFN-γ) as in vitro enhancing factor of peritoneal macrophage defective bactericidal activity during continuous ambulatory peritoneal dialysis (CAPD). Am J Kidney Dis 11:225–230

Lu L, Walker D, Graham CD, Waheed A, Shadduck RK, Broxmeyer HE (1988) Enhancement of release from MHC class II antigen-positive monocytes of hematopoietic colony stimulating factors CSF-1 and G-CSF by recombinant human tumor necrosis factor-alpha: synergism with recombinant human interferon-gamma. Blood 72:34–41

Luster AD, Ravetch JV (1987) Biochemical characterization of a γ interferon-inducible cytokine. J Exp Med 166:1084–1097

Mackaness GB (1969) The influence of immunologically committed lymphoid cells on macrophage activity in vivo. J Exp Med 129:973–992

Maluish AE, Urba WJ, Longo DL, Overton WR, Coggin D, Crisp ER, Williams R, Sherwin SA, Gordon K, Steis RG (1988) The determination of an immunologically active dose of interferon-gamma in patients with melanoma. J Clin Oncol 6:434–445

Marletta MA, Yoon PS, Iyengar R, Leaf CD, Wishnok JS (1988) Macrophage oxidation of L-arginine to nitrite and nitrate: nitric oxide is an intermediate. Biochemistry 27:8706–8711

McCabe RE, Luft BJ, Remington JS (1984) Effect of murine interferon gamma on murine toxoplasmosis. J Infect Dis 150:961–962

Miller LJ, Bainton DF, Borregaard N, Springer TA (1987) Stimulated mobilization of monocyte Mac-1 and p150,95 adhesion proteins from an intracellular vesicular compartment to the cell surface. J Clin Invest 80:535–544

Mokoena T, Gordon S (1985) Human macrophage activation. Modulation of mannosyl, fucosyl receptor activity in vitro by lymphokines, gamma and alpha interferons, and dexamethasone. J Clin Invest 75:624–631

Murray HW (1988) Interferon-gamma, the activated macrophage, and host defense against microbial challenge. Ann Intern Med 108:595–608

Murray HW, Rubin BY, Rothermel CD (1983) Killing of intracellular *Leishmania donovani* by lymphokine-stimulated human mononuclear phagocytes. Evidence that interferon-γ is the stimulating lymphokine. J Clin Invest 72:1506–1510

Murray HW, Spitalny GL, Nathan CF (1985) Activation of mouse peritoneal macrophages in vitro and in vivo by interferon-gamma. J Immunol 134:1619–1622

Murray HW, Scavuzzo D, Jacobs JL, et al. (1987a) In vitro and in vivo activation of human mononuclear phagocytes by gamma interferon: studies with normal and AIDS monocytes. J Immunol 138:2457–2462

Murray HW, Stern JJ, Welte K, Rubin BY, Carriero SM, Nathan CF (1987b) Experimental visceral leishmaniasis: production of interleukin 2 and interferon-γ, tissue immune reaction, and response to treatment with interleukin 2 and interferon-γ. J Immunol 138:2290–2297

Mustoe TA, Pierce GF, Thomason A, Gramates P, Sporn MB, Deuel TF (1987) Accelerated healing of incisional wounds in rats induced by transforming growth factor-beta. Science 237:1333–1336

Nacy CA, Fortier AH, Meltzer MS, Buchmeier NA, Schreiber RD (1985) Macrophage activation to kill *Leishmania major*: activation of macrophages for intracellular destruction of amastigotes can be induced by both recombinant interferon-gamma and non-interferon lymphokines. J Immunol 135:3505–3511

Nakagawara A, Nathan CF, Cohn ZA (1981) Hydrogen peroxide metabolism in human monocytes during differentiation in vitro. J Clin Invest 68:1243–1252

Nathan CF (1982) Secretion of oxygen intermediates: role in effector functions of activated macrophages. Fed Proc 41:2206–2211

Nathan CF (1983) Macrophage microbicidal mechanisms. Trans R Soc Trop Med Hyg 77:620–630

Nathan CF (1986) Macrophage activation: some questions. Ann Inst Pasteur 137C:345–351

Nathan CF (1987a) Secretory products of macrophages. J Clin Invest 79:319–326

Nathan CF (1987b) Neutrophil activation on biological surfaces: massive release of hydrogen peroxide in response to products of macrophages and lymphocytes. J Clin Invest 80:1550–1560

Nathan CF (1989) Respiratory burst of adherent human neutrophils: triggering by colony stimulating factors CSF-GM and CSF-G. Blood 73:301–306

Nathan CF, Cohn ZA (1985) Cellular components of inflammation: monocytes and macrophages. In: Kelley WN, Harris ED Jr, Ruddy S, Sledge CB (eds) Textbook of rheumatology. Saunders, Philadelphia, pp 144–168

Nathan CF, Terry WD (1977) Decreased phagocytosis by macrophages from BCG-treated mice: induction of the phagocytic defect in normal macrophages given BCG in vitro. Cell Immunol 29:295–311

Nathan CF, Tsunawaki S (1986) Secretion of toxic oxygen products by macrophages: regulatory cytokines and their effects on the oxidase. Ciba Found Symp 118:211–230

Nathan CF, Yoshida R (1988) Cytokines: interferon-γ. In: Gallin J, Goldstein I, Snyderman R (eds) Inflammation: basic principles and clinical correlates. Raven, New York, pp 229–251

Nathan CF, Remold HG, David JR (1973) Characterization of a lymphocyte factor which alters macrophage functions. J Exp Med 137:275–290

Nathan CF, Arrick BA, Murray HW, DeSantis NM, Cohn ZA (1981) Tumor cell antioxidant defenses: inhibition of the glutathione redox cycle enhances macrophage-mediated cytolysis. J Exp Med 153:766–782

Nathan CF, Murray HW, Weibe ME, Rubin BY (1983) Identification of interferon-γ as the lymphokine that activates human macrophage oxidative metabolism and antimicrobial activity. J Exp Med 158:670–689

Nathan CF, Prendergast TJ, Weiber JE, Stanley ER, Platzer E, Remold HG, Welte K, Rubin BY, Murray HW (1984) Activation of human macrophages: comparison of other cytokines with interferon-γ. J Exp Med 160:600–605

Nathan CF, Horowitz CR, de la Harpe J, Vadhan-Raj S, Sherwin SA, Oettgen HF, Krown SE (1985) Administration of recombinant interferon-γ to cancer patients enhances monocyte secretion of hydrogen peroxide. Proc Natl Acad Sci USA 82:8686–8690

Nathan CF, Kaplan G, Levis WR, Nusrat A, Witmer MD, Sherwin SA, Job CK, Horowitz CR, Steinman RM, Cohn ZA (1986) Local and systemic effects of low doses of recombinant interferon-γ after intradermal injection in patients with lepromatous leprosy. N Engl J Med 315:6–14

Nathan C, Campanelli D, Ding A, de la Harpe J, Gabay J, Srimal S, Stuehr D, Tsunawaki S, Yoshida R (1988) Regulatory and effector molecules for cytotoxicity by phagocytes. In: Schwarz M (ed) Proceedings of the Centennary of the Institut Pasteur. Elsevier, Amsterdam, pp 267–282

O'Donnell-Tormey J, Nathan CF, Lanks K, Deboer C, de la Harpe J (1987) Secretion of pyruvate: an antioxidant defense of mammalian cells. J Exp Med 165:500–514
Onozaki K, Matsushima K, Aggarwal BB, Oppenheim JJ (1985) Human interleukin 1 is a cytocidal factor for several tumor cell lines. J Immunol 135:3962
Perussia B, Dayton ET, Lazarus R, Fanning V, Trinchieri G (1983) Immune interferon induces the receptor for monomeric IgG_1 on human monocytic and myeloid cells. J Exp Med 158:1092
Philip R, Epstein LB (1986) Tumour necrosis factor as immunomodulator and mediator of monocyte cytotoxicity induced by itself, γ-interferon and interleukin-1. Nature 323:86–89
Radzioch D, Varesio L (1988) Protein kinase C inhibitors block the activation of macrophages by IFN-β but not by IFN-γ. J Immunol 140:1259–1263
Ralph P, Nakoinz I (1987) Stimulation of macrophage tumoricidal activity by the growth and differentiation factor CSF-1. Cell Immunol 105:270–279
Ralph P, Nakoinz I, Rennick D (1988) Role of interleukin 2, interleukin 4, and α, β, and γ interferon in stimulating macrophage antibody-dependent tumoricidal activity. J Exp Med 167:712–717
Rappolee DA, Mark D, Banda MJ, Werb Z (1988) Wound macrophages express transforming growth factor-α and other growth factors in vivo: analysis by mRNA phenotyping. Science 241:708–712
Reed SG (1988) In vivo administration of recombinant IFN-γ induces macrophage activation, and prevents acute disease, immune suppression, and death in experimental *Trypanosoma cruzi* infections. J Immunol 140:4342–4347
Reed SG, Nathan CF, Pihl DL, Rodricks P, Shanebeck K, Conlon PJ, Grabstein KH (1987) Recombinant granulocyte-macrophage colony stimulating factor activates macrophages to inhibit *Trypanosoma cruzi* and release hydrogen peroxide: comparison to interferon-γ. J Exp Med 166:1734–1746
Reiner NE, Ng W, Ma T, McMaster WR (1988) Kinetics of γ interferon binding and induction of histocompatibility complex class II mRNA in *Leishmania*-infected macrophages. Proc Natl Acad Sci USA 85:4330–4334
Robinson BW, McLemore TL, Crystal RG (1985) Gamma interferon is spontaneously released by alveolar macrophages and lung T lymphocytes in patients with pulmonary sarcoidosis. J Clin Invest 75:1488–1495
Rosen H, Gordon S (1987) Monoclonal antibody to the murine type 3 complement receptor inhibits adhesion of myelomonocytic cells in vitro and inflammatory cell recruitment in vivo. J Exp Med 166:1685–1701
Rothermel CD, Rubin BY, Murray HW (1983) γ-Interferon is the factor in lymphokine that activates human macrophages to inhibit intracellular *Chlamydia psittaci* replication. J Immunol 131:2542–2544
Rothstein JL, Schreiber H (1988) Synergy between tumor necrosis factor and bacterial products causes hemorrhagic necrosis and lethal shock in normal mice. Proc Natl Acad Sci USA 85:607–611
Rothstein JL, Flint TF, Schreiber H (1988) Tumor necrosis factor/cachectin. Induction of hemorrhagic necrosis in normal tissue requires the fifth component of complement (C5). J Exp Med 168:2007–2021
Samuel NM, Grange JM, Samuel S, Lucas S, Owilli OM, Adalla S, Leigh IM, Navarrette C (1987) A study of the effects of intradermal administration of recombinant gamma interferon in lepromatous leprosy patients. Lepr Rev 58:389–400
Schlesinger L, Musson RA, Johnston RB Jr (1984) Functional and biochemical studies of multinucleated giant cells derived from the culture of human monocytes. J Exp Med 159:1289–1294
Schraufstatter I, Hyslop PA, Jackson JH, Cochrane CG (1988) Oxidant-induced DNA damage of target cells. J Clin Invest 82:1040–1050
Schreiber RD, Pace JL, Russell SW, Altman A, Katz DH (1983) Macrophage-activating factor produced by a T cell hybridoma: physicochemical and biosynthetic resemblance to γ-interferon. J Immunol 131:826

Schreiber RD, Hicks LJ, Celada A, Buchmeier NA, Gray PW (1985) Monoclonal antibodies to murine γ-interferon which differentially modulate macrophage activation and antiviral activity. J Immunol 134:1609–1618

Sechler JMG, Malech HL, White CJ, Gallin JI (1988) Recombinant human interferon-γ reconstitutes defective phagocyte function in patients with chronic granulomatous disease of childhood. Proc Natl Acad Sci USA 85:4874–4878

Sibley LD, Hunter SW, Brennan PJ, Krahenbuhl JL (1988) Mycobacterial lipoarabinomannan inhibits gamma interferon-mediated activation of macrophages. Infect Immun 56:1232–1236

Sisson SD, Dinarello CD (1988) Production of interleukin-1α, interleukin-1β, and tumor necrosis factor by human mononuclear cells stimulated with granulocyte-macrophage colony-stimulating factor. Blood 72:1368–1374

Steeg PS, Moore RN, Johnson M, Oppenheim JJ (1982) Regulation of murine macrophage Ia antigen expression by a lymphokine with immune interferon activity. J Exp Med 156:1780–1793

Stewart CC (1980) Formation of colonies by mononuclear phagocytes outside the bone marrow. In Mononuclear Phagocytes: Functional Aspects. R. van Furth, ed. Martinus Nijhoff, Boston 377–416

Stewart CC, Lin H, Adles C (1975) Proliferation and colony-forming ability of peritoneal exudate cells in liquid culture. J Exp Med 141:1114–1132

Strassman G, Springer TA, Adams DO (1985) Studies on antigens associated with the activation of murine mononuclear phagocytes: kinetics of and requirements for induction of lymphocyte function-associated (LFA-1) antigen in vitro. J Immunol 135:147–151

Strassman G, Springer TA, Somers SD, Adams DO (1986) Mechanisms of tumor cell capture by activated macrophages: evidence for involvement of lymphocyte function-associated (LFA)-1 antigen. J Immunol 136:4328–4333

Strunk RC, Cole FS, Perlmutter DH, Colten HR (1985) γ-Interferon increases expression of class III complement genes C2 and factor B in human monocytes and in murine fibroblasts transfected with human C2 and factor B genes. J Biol Chem 260:15280–15285

Stuehr DJ, Marletta MA (1987) Induction of nitrite/nitrate synthesis in murine macrophages by BCG infection, lymphokines, or interferon-γ. J Immunol 139:518

Stuehr DJ, Nathan CF (1989) Nitric oxide: a macrophage product responsible for cytostasis and respiratory inhibition in tumor target cells. J Exp Med 169:1543–1555

Stuehr DJ, Gross SS, Sakuma I, Levi R, Nathan CF (1989a) Activated murine macrophages secrete a metabolite of arginine with the bioactivity of endothelium-derived relaxing factor and the chemical reactivity of nitric oxide. J Exp Med 169:1011–1020

Stuehr DJ, Kwon NS, Gross SS, Thiel BA, Levi R, Nathan CF (1989b) Synthesis of nitrogen oxides from L-arginine by macrophage cytosol: requirement for inducible and constitutive components. Biochem Biophys Res Commun 161:420–426

Sugarman BJ, Aggarwal BB, Hass PE, Figari IS, Palladino MA Jr, Shepard HM (1985) Recombinant human tumor necrosis factor-alpha: effects on proliferation of normal and transformed cells in vivo. Science 230:943–945

Suzuki Y, Orellana MA, Schreiber RD, Remington JS (1988) Interferon-γ: the major mediator of resistance against *Toxoplasma gondii*. Science 240:516–518

Swaim MW, Pizzo SV (1988) Methionine sulfoxide and the oxidative regulation of plasma proteinase inhibitors. J Leuk Biol 43:365–380

Szuro-Sudol A, Nathan CF (1982) Suppression of macrophage oxidative metabolism by products of malignant and nonmalignant cells. J Exp Med 156:945–961

Szuro-Sudol A, Murray HW, Nathan CF (1983) Suppression of macrophage microbicidal activity by a tumor cell product. J Immunol 131:384–387

Thorens B, Mermod J-J, Vassalli P (1987) Phagocytosis and inflammatory stimuli induce GM-CSF mRNA in macrophages through posttranscriptional regulation. Cell 48:671–679

Thurman GB, Braude IA, Gray PW, Oldham RK, Stevenson HC (1985) MIF-like activity of natural and recombinant human interferon-gamma and their neutralization by monoclonal antibody. J Immunol 134:305–309
Tovey MG, Streuli M, Gresser I, Gugenheim J, Blanchard B, Guymarho J, Vignaux F, Gigou M (1987) Interferon messenger RNA is produced constitutively in the organs of normal individuals. Proc Natl Acad Sci USA 84:5038
Trinchieri G, Santoli D (1978) Anti-viral activity induced by culturing lymphocytes with tumor-derived or virus-transformed cells. Enhancement of human natural killer cell activity by interferon and antagonistic inhibition of susceptibility of target cells to lysis. J Exp Med 147:1314–1334
Tsunawaki S, Nathan CF (1986) Macrophage deactivation: altered kinetic properties of the superoxide-producing enzyme after exposure to tumor cell-conditioned medium. J Exp Med 164:1319–1331
Tsunawaki S, Sporn M, Ding A, Nathan CF (1988) Macrophage deactivation by transforming growth factor-β. Nature 334:260–262
Tsunawaki S, Sporn M, Nathan CF (1989) Comparison of transforming growth factor β and a macrophage-deactivating polypeptide from tumor cells. J Immunol 142:3462–3468
Tushinski RJ, Oliver IT, Guilbert LJ, Tynan PW, Warner JR, Stanley ER (1982) Survival of mononuclear phagocytes depends on a lineage-specific growth factor that the differentiated cells selectively destroy. Cell 28:71–81
Unanue ER, Allen PM (1987) The basis for the immunoregulatory role of macrophages and other accessory cells. Science 236:551–557
Varesio L, Blasi E, Thurman GB, Talmadge JE, Wiltrout RH, Herberman RB (1984) Potent activation of mouse macrophages by recombinant interferon-gamma. Cancer Res 44:4465–4469
Vellenga E, Rambaldi A, Ernst TJ, Ostapovicz D, Griffin JD (1988) Independent regulation of M-CSF and G-CSF gene expression in human monocytes. Blood 71:1529–1532
Vlassara H, Valinsky J, Brownlee M, Cerami C, Nishimoto S, Cerami A (1987) Advanced glycosylation endproducts on erythrocyte cell surface induce receptor-mediated phagocytosis by macrophages: a model for turnover of aging cells. J Exp Med 166:539–549
Wahl SM, Hunt DA, Wakefield LM, McCartney-Francis N, Wahl LM, Roberts AB, Sporn MB (1987) Transforming growth factor type β induces monocyte chemotaxis and growth factor production. Proc Natl Acad Sci USA 84:5788–5792
Walker EB, Maino V, Sanchez-Lanier M, Warner N, Stewart C (1984) Murine gamma interferon activates the release of a macrophage-derived Ia-inducing factor that transfers Ia inductive capacity. J Exp Med 159:1532–1547
Wang JM, Collela S, Allavena P, Mantovani A (1987) Chemotactic activity of human recombinant granulocyte-macrophage colony-stimulating factor. Immunology 60:439–444
Wang JM, Griffin JD, Rambaldi A, Chen ZG, Mantovani A (1988) Induction of monocyte migration by recombinant macrophage colony-stimulating factor. J Immunol 141:575–579
Ward PA, Remold HG, David JR (1969) Leukotactic factor produced by sensitized leukocytes. Science 163:1079–1082
Weinberg JB, Hobbs MM, Misukonis MA (1985) Phenotypic characterization of gamma interferon-induced human monocyte polykaryons. Blood 66:1241–1246
Weinshank RL, Luster AD, Ravetch JV (1988) Function and regulation of a murine macrophage-specific IgG Fc receptor. J Exp Med 167:1909
Weiser WY, van Niel A, Clark SC, David JR, Remold HG (1987) Recombinant human granulocyte/macrophage colony-stimulating factor activates intracellular killing of *Leishmania donovani* by human monocyte-derived macrophages. J Exp Med 166:1436–1446
Wilson CB, Westall J (1985) Activation of neonatal and adult human macrophages by alpha, beta, and gamma interferons. Infect Immun 49:351–356

Wright SD, Detmers PA, Jong MT, Meyer BC (1986) Interferon-gamma depresses binding of ligand by Cdb and C3bi receptors on cultured human monocytes, an effect reversed by fibronectin. J Exp Med 163:1245–1259
Yoshida R, Murray HW, Nathan CF (1988) Two classes of interferon-γ receptors on human macrophages: blockade of the high-affinity sites by interferon-α or -β. J Exp Med 167:1171–1185
Zoeller RA, Morand OH, Raetz CR (1988) A possible role for plasmalogens in protecting animal cells against photosensitized killing. J Biol Chem 263:11 590–11 596

CHAPTER 34

Extracellular Matrices, Cells, and Growth Factors

G. R. MARTIN and A. C. SANK

A. Introduction

The importance of cell-matrix interactions in development has been long recognized. The major functions of the matrix are to provide physical support to the tissues and to maintain cellular viability. Stability and homeostasis are achieved in this way. Extracellular matrices are structurally diverse but they vary in a tissue-specific fashion. Thus cartilage, tendon, bone, basement membranes, etc., contain different matrix proteins optimized for their individual functions and for their particular cells. Current concepts indicate that cells form a continuum with their matrix (HAY 1981), having specific receptors that bind to the various proteins that comprise the matrix. These contacts are instructive for the cells and influence their migration, morphology, growth, and differentiation. Such receptors also permit the cells to monitor the composition of the matrix and allow an autocrine regulation of matrix production.

Several studies suggest that matrix molecules can alter the growth of cells with some components increasing and others decreasing proliferation. Some matrix proteins are directly mitogenic, either alone or in combination with other growth factors. These matrix molecules might have a role as immobilized growth stimulators during development and regeneration. Such molecules may also be important in controlling the migration of cells during development and regeneration.

Growth factors are produced and are active in conditions, such as development, wound healing, cancer, and inflammation, where tissue and cell-matrix interactions are destroyed (SPORN and ROBERTS 1986). Surprisingly, the genes for matrix-degrading enzymes are among the early genes activated by a variety of growth factors (see below). These enzymes may be required in formed tissues to free cells from matrix contacts to allow the cells to migrate and proliferate. Thus growth factors, cells, and matrix show complex interactions reflecting the Yin and Yang of proliferation and homeostasis.

This review will present a summary of the information on molecular mechanisms involing cell-matrix interactions including the structure of matrix proteins, their cell attachment domains, cellular receptors for matrix proteins, and the role the matrix proteins may play in controlling cellular proliferation.

B. Nature of Extracellular Matrices

I. Collagens

Extracellular matrices are composed of one or more collagens, glycoproteins, and proteoglycans. Collagens are the critical structural element of the matrix, providing a scaffolding to which other matrix proteins are attached. Some 13 different collagens have been identified which share many features but differ genetically, structurally, and functionally (MAYNE and BURGESON 1987). Some collagens (types I, II, III, V) form fibrous structures while the others form sheets (type IV), microfilaments (type VI), or straps (type VII) (MARTIN et al. 1985). Some, such as collagen I, the major protein in skin, bone and tendon, and collagen II, the major cartilage collagen, are abundant proteins. However, other collagens, including types III, VI, IX, and X, are minor components of the matrix and appear to serve specialized functions. These include type IX, which links collagen II fibers together in cartilage, and type VI, which creates microfilaments which join various tissue structures. Collagen IV, while not an abundant protein, is the major protein structural component in basement membranes, the matrix for epithelial and endothelial cells, muscle, and peripheral nerves.

While it might appear redundant for the genome to encode and the body to produce many collagens whose function was long thought to be only structural, each collagen has the ability to interact with different matrix proteins and different cellular receptors (KLEINMAN et al. 1981). Furthermore, the removal of these collagens may be achieved by different collagenases, as shown for the interstitial collagens (types I–III) and basement membrane collagen (type IV) (LIOTTA et al. 1979), allowing selective remodeling of matrix structures. Thus, the collagen family serves diverse functions. As discussed below, these functions may include terminating the proliferation of cells by blocking their ability to respond to growth factors.

II. Glycoproteins

Many glycoproteins have been identified as matrix components. The two most studied are fibronectin (YAMADA 1983; RUOSLAHTI 1988a) and laminin (MARTIN and TIMPL 1987). Fibronectin is found in fibrous connective tissues and in serum, while laminin is only found in basement membranes. A detailed model for fibronectin (M_r = 440000) has been proposed (Fig. 1a), showing it as a dimeric molecule whose chains are linked at one end by disulfide bonds (YAMADA 1983). Each chain has a modular structure with distinct functional domains distributed along the chains.

Laminin (Fig. 1B) (MARTIN and TIMPL 1987) is a large (M_r = 900000) cross-shaped protein composed of three chains. These chains have been cloned (SASAKI et al. 1988) and sequenced and this information, as well as studies on proteolytic fragments of the proteins, predict a multidomain structure with a globule-rod motif for the short arms of the cross and a coiled-coil structure forming the long arm. Both fibronectin and laminin bind to a number of other matrix proteins and membrane receptors and have multiple biological activities.

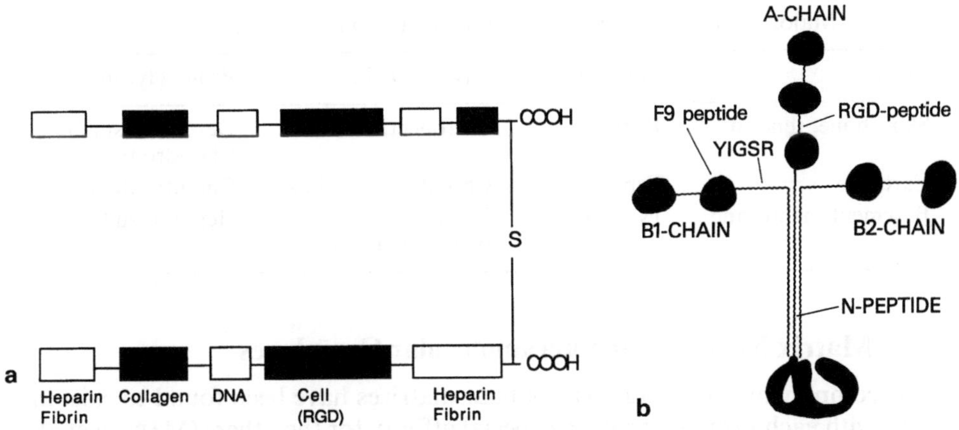

Fig. 1. a Model of fibronectin. Fibronectin has two chains joined by disulfide bonds. Each chain has a series of domains with different activities. The cell attachment domain contains a tripeptide, RGD, which is part of the receptor recognition site. **b** Model of laminin – a major mediator of cell interactions with basement membrane. Laminin is formed of three chains (B1, B2, and A) which are associated in a cross-like structure. Laminin is a multidomain molecule with a variety of biological activities. Four sites are identified on laminin which appear to be biologically active. The F9 site (CHARONIS et al. 1988) prepared as a synthetic peptide shows both heparin- and cell binding activity. Similarly, the YIGSR site serves as a cell attachment site for many epithelial cells (GRAF et al. 1987). The RGD site was examined because of the sequence arginine-glycine-aspartic acid (R-G-D), which is known to be part of the cell attachment site in fibronectin and some other proteins. Synthetic peptides from this region have the ability to support the attachment of endothelial as well as a variety of other cells (TASHIRO et al., unpublished). Finally, the long arm of the molecule contains a site(s) which induces neurite formation by cultured neural cells (EDGAR et al. 1984) and cell attachment (AUMAILLEY et al. 1987; GOODMAN et al. 1987). Presumably cells possess different receptors for each of these sites

III. Proteoglycans

Proteoglycans, highly charged macromolecules, containing sulfated glycosaminoglycan chains, are found in all matrices. These form a diverse family, arising as proteins to which the carbohydrate chains are attached as posttranslational modifications (HASSELL et al. 1986). In some cases, as with the cartilage proteoglycan, the final product has much more carbohydrate than protein. Proteoglycans bind to other matrix molecules, as well as to cells, through both their protein and carbohydrate chains (RUOSLAHTI 1988b). In terms of specific functions (EVERED and WHELAN 1986), the major proteoglycan of cartilage, by virtue of its large negative charge, binds water and maintains the cartilage resilient and incompressible. In contrast, the large heparan sulfate proteoglycan of basement membranes is fixed to the matrix by strong nonionic forces. Its heparan sulfate chains form a charged barrier on the surface of blood vessels and glomeruli and this prevents the passage of proteins across the basement membranes that line these sites. The functions of other proteoglycans are less well defined.

Table 1. Major matrix components found interacting in tissues

Tissue	Collagen	Glycoprotein	Proteoglycan
Skin, bone, tendon	Types I, III	Fibronectin	Dermatan sulfate Chondroitin sulfate
Cartilage	Type II	Chondronectin, link	Chondroitin sulfate
Basement membrane	Type IV	Laminin Nidogen/entactin	Heparan sulfate

IV. Matrix Molecules in Supramolecular Complexes

The components of several extracellular matrices have been found to exist in sets, with each protein showing a special affinity for the others (MARTIN et al. 1984; KLEINMAN et al. 1983; Table 1). Thus, laminin binds to nidogen, heparan sulfate proteoglycan, and collagen IV, the components of basement membranes, but not to other matrix proteins (KLEINMAN et al. 1983; TIMPL and DZIADEK 1986). Also the components of cartilage including cartilage proteoglycans, link protein, and hyaluronic acid bind to one another to form large aggregate structures (HASSELL et al. 1986). Thus, there may be considerable specificity in the interaction of matrix proteins that explain in part their tissue colocalization. In addition, the cells which are resident in a specific matrix have receptors for the components of the matrix. This combination of contacts elicits different cellular responses than those obtained from the individual components, as shown, for example, in comparing laminin with a combination of basement membrane proteins (KLEINMAN et al. 1986).

C. Cell-Matrix Interactions

I. Fibronectin

Matrix molecules are well known for their ability to support the attachment and growth of cells in culture. Initial studies were concentrated on fibronectin and showed that this protein was able to mediate the attachment of fibroblasts and other mesenchymal cells to plastic and to collagen-coated surfaces, and to affect their growth, migration, and differentiation (see KLEINMAN et al. 1981 for a review of this work). Binding to collagen occurs at a specific sequence in the collagen molecule near the collagenase cleavage site (KLEINMAN et al. 1981) in a region which lacks proline and hydroxyproline and thus may have a less stable helix.

A different domain in fibronectin, some distance from the collagen-binding domain, contains the cell-binding site (RUOSLAHTI 1988a). This region includes a sequence of four amino acids: arginine-glycine-aspartic acid-serine (RGDS), which is an important part of the receptor-binding site (Table 1). Synthetic RGDS peptides are able to support cell attachment and are also able to detach cells from substrates of fibronectin by competing for the fibronectin receptor. Synthetic RGDS peptides show a rather low affinity for the receptor and a second part of the cell-binding domain confers additional affinity and specificity in receptor binding (OBARA et al. 1988). Other cell-binding sites may also exist on fibronectin.

II. Laminin

Laminin is unique to and a major component of basement membranes, where it plays a role analogous to fibronectin's but for epithelial and endothelial cells, nerves, and muscle (MARTIN and TIMPL 1987). Laminin is a multidomain molecule, containing three chains arranged in a cross-shaped structure (Fig. 1 B). The structure is stabilized by disulfide bonds and by a coiled-coil structure formed by the three chains as they extend down the long arm of the molecule. Laminin supports the attachment of cells, stimulates their migration, and induces their differentiation (KLEINMAN et al. 1985). It has been shown to have a key role in axon formation by nerve cells (VAN EVERCOOREN et al. 1982). It can increase the invasive activity of tumor cells, probably by increasing the production of collagenase IV (TURPEENNIEMI-HUJANEN et al. 1986).

Not unexpectedly, given its multiple activities, four distinct cell-binding sites have been identified in the molecule, with synthetic peptides corresponding to these sequences showing biological activity (Fig. 1 B). The F9-peptide occurs in a globular domain on the B1-chain, is basic, and exhibits heparin- and cell-binding activity (CHARONIS et al. 1988). The adjacent domain with homology to epidermal growth factor (EGF) has a rod-like structure with many disulfide bonds and contains the sequence tyrosine-isoleucine-glycine-serine-arginine (YIGSR), which when present in synthetic peptides shows cell- and receptor-binding activity (GRAF et al. 1987). Cyclic forms of this peptide show higher activity than linear peptides as expected from the many disulfide bonds that occur in this domain of the native molecule. However, even cyclic YIGSR peptides are much less potent than laminin on a molar basis. It is possible that the two adjacent binding sites act in concert in the native molecule.

A third cell attachment site in laminin is located in the short arm of the A-chain and contains an RGD sequence (SASAKI et al. 1988). Preliminary studies show that synthetic peptides to amino acid sequences in this region are active in endothelial cell attachment (GRANT et al., unpublished). Finally, there is a fourth region of laminin, part of the coiled-coil domain near the end of the long chain, that contains the site in laminin for neurite formation and also forms a strong attachment site for many cells (AUMAILLEY et al. 1987; GOODMAN et al. 1987). A unique integrin receptor for laminin has been identified and may be specific for this region (GEHLSEN et al. 1988; TOMASELLI et al. 1988).

As noted earlier, laminin has diverse effects on cells and studies with proteolytic fragments established that separate portions of the molecule are able to elicit different cellular responses (reviewed in MARTIN and TIMPL

1987). The ability of a cell to interact with more than one receptor may allow for multiple responses or a certain cell-binding site might be available in the basement membranes in one tissue but not another. Also binding to one cell attachment site might cause the receptor for a second site to appear and when the two receptors are occupied could stabilize a particular phenotype, such as the capillary-like structures formed by endothelial cells (KUBOTA et al. 1988). Thus, laminin has the potential to transmit complex signals to a variety of cells.

III. Collagen

Many researchers have found that cells in culture grow better and maintain a differentiated phenotype longer on a collagen substrate. The binding of collagen to cells via high-affinity receptors was demonstrated some time ago (GOLDBERG 1979; RUBIN et al. 1981) and cells can attach and spread directly on collagen. Collagen IV has been demonstrated to have more than one attachment site occurring in different parts of the molecule and including both helical and nonhelical domains (AUMAILLEY and TIMPL 1986; HERBST et al. 1988). A possible function was suggested by the observation that a collagen receptor induced on U-937 cells by 1,25-dihydroxyvitamin D (POLLA et al. 1987) was associated with the acquisition of an adherent phenotype by the cells. The affinity of endothelial cells for collagen IV is greater than for laminin (HERBST et al. 1988).

IV. Matrix Receptors

The receptor for fibronectin is part of a large family of receptors termed the integrins (HYNES 1987; RUOSLAHTI 1988a). Integrin receptors have also been identified for collagen (DEDHAR et al. 1987) and for laminin (TOMASELLI et al. 1988; GEHLSEN et al. 1988), while some integrin receptors recognize more than one ligand (reviewed in RUOSLAHTI 1988a; TAKADA et al. 1988). The integrins are integral membrane proteins, dimeric molecules composed of an α- and a β-subunit which are held together by noncovalent interactions. Each subunit has a transmembrane domain adjacent to its C-terminus. To date, three subfamilies of integrins have been recognized, which are distinguished by distinct but homologous β-subunits. These β-subunits combine with different α-subunits to generate distinct receptors. The fibronectin receptor is concentrated in focal adhesions and it serves to organize the cytoskeleton around these sites. The loss of focal adhesions with a reduction in fibronectin binding is characteristic of transformed and proliferating cells (YAMADA 1983). A 67-kDa laminin-binding receptor which is membrane associated has been described as a laminin receptor. Cloning and sequencing revealed still another laminin receptor (M_r = 32000) (Sequi-Real, YAMADA, unpublished) which is active in the attachment of these cells to laminin and is increased in malignant cells.

D. Role of Matrix Molecules in Cell Growth

I. Storage Sites for Growth Factors

The matrix deposited by cells on culture dishes has a striking ability to enhance the growth of other cells seeded upon it (GOSPODAROWICZ et al. 1978). Different phenomena may contribute to these effects. Extracellular matrix increases the attachment and survival of cells and this increases the yield of cells in culture. Also cells often express a more differentiated phenotype on matrix than on plastic (KLEINMAN et al. 1981), which could lead to a different expression of growth factor receptors on their surface. In addition, matrix molecules are able to bind and store growth factors and thus act as a depot. This has been most clearly shown by the heparin-binding class of growth factors, including the fibroblast growth factors (VLODAVSKY et al. 1987), which bind to the heparan sulfate present in basement membranes and in other structures. The growth factors bound in this way may be used directly by the cells or released by enzymes such as heparinase produced by these cells.

Certain matrix molecules appear to alter the requirement of proliferating cells for growth factors. For example, mammary epithelial cells grown on collagen IV are able to proliferate in defined media lacking EGF (SALOMON et al. 1981). Also corneal endothelial cells show a greater response to EGF when plated on a collagen substrate and respond better to fibroblast growth factor (FGF) when plated on plastic, suggesting differential expression of growth factor receptors (GOSPODAROWICZ et al. 1978).

II. Mitogenic Activities of Fibronectin and Laminin

Both fibronectin and laminin have mitogenic activity. Fibronectin is able to induce growth-arrested human fibroblasts to synthesize DNA and to divide in the presence of plasma (BITTERMAN et al. 1983). A discrimination between mitogenic and attachment effects was achieved by adding the proteins or fragments derived from them in solution, by the kinetics of the response, and by the molecular changes elicited by these proteins. Fibronectin acts early in the G1-phase of the cell cycle and it can act synergistically to induce DNA synthesis with insulin, but not with platelet-derived growth-factor (PDGF) or FGF. A subsequent study showed that fibronectin induces the transcription of certain protooncogenes, including those for *myc* and *fos* and increases the level of histone mRNA (DIKE and FARMER 1988). Such responses are characteristic of cells treated with growth factors such as PDGF and FGF, the so called "competence" factors, as are the early site of action in the cell cycle and the synergism with "progression" factors.

In a separate study, it was found that a chymotrypsin fragment of fibronectin, but not the native molecule, stimulates the proliferation of hamster fibroblasts (HUMPHRIES and AYAD 1983). This fragment is derived from the gelatin-binding domain of fibronectin and is located some distance from the cell-binding site. The fact that the fragment but not the intact molecule exhibited activity suggested that the mitogenic activity was cryptic in normal tissues and released by proteolysis. The degradation of fibronectin with the release of a growth factor domain could occur in a wound and serve to promote healing. A similar proteolytic fragment of fibronectin was also observed to cause morphological transformation of normal cells in culture (DEPETRO et al. 1981) and in these studies could have been acting as a mitogen.

Laminin also promotes cell growth (TERRANOVA et al. 1986; PANAYOTOU et al. 1989). Laminin, as well as a proteolytic fragment (P1) containing the region of the molecule where the three short arms intersect, stimulates the proliferation of epithelial cells. These regions of the short arms contain rod-like segments with homology to EGF and this portion of the B1-chain contains the YIGSR cell attachment site. Since a line of cells devoid of EGF receptors was not induced to proliferate by laminin, the EGF receptor was implicated in the mitogenic response to laminin (PANAYOTOU et al. 1989). However, no direct competition in receptor binding was observed between laminin and EGF. Thus is is not clear how the mitogenic signal from laminin is transmitted. The results suggest that laminin, through at least one portion of the molecule, is able to exert a mitogenic signal, probably of the "competence" type.

Laminin and fibronectin as well as fragments derived from these proteins, are able to induce the directed movement of cells when presented to the cells either in solution or as substrate (MCCARTHY and FURCHT 1984; HERBST et al. 1988; KLEINMAN et al. 1985). While the molecular events in the cell that elicit directed cell movement are not well defined, they may include some of the responses elicited by "competence" growth factors, since many growth factors show chemotactic activity (GROTENDORST and MARTIN 1986). Thus, it may be that a major role of laminin and fibronectin during development is to stimulate and guide cellular migrations.

III. Termination of Proliferation by Collagen

Though widely separated, fibroblasts in formed tissues such as skin are quiescent. Even when removed from the skin, fibrolasts only slowly acquire the ability to proliferate when isolated and placed in culture (TERRACIO et al. 1988). The failure of these cells to proliferate even in the presence of high concentrations of serum has been shown to be associated with a low expression of the PDGF receptor (TERRACIO et al. 1988). Reexpression of the PDGF receptor was observed when the cells began to proliferate in culture.

Recent studies suggest that the binding of cells to collagen causes the cells to stop proliferating, possibly by decreasing the cells' ability to respond to growth factors. When collagen is used as a coating on the surface of tissue culture dishes, it enhances the attachment and growth of cells. In contrast, when the cells are placed either in or on a gel of native collagen fibers, they respond differently. Here the cells attach to the collagen fibers, become elongated, and contract the gel to a fraction of its original size (BELL et al. 1979; SARBER et al. 1981; EHRLICH 1988). Also even though the cells are widely separated from one another in the contracted gel and are in the presence of high concentrations of serum or defined growth factors, they stop proliferating and stop col-

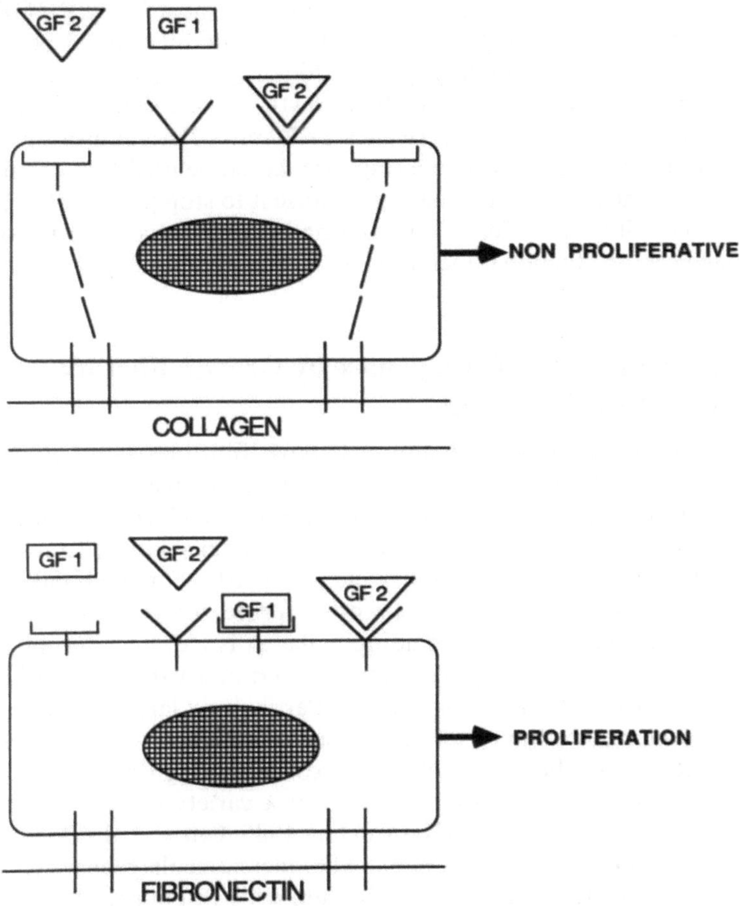

Fig. 2. Model for the regulation of cell proliferation by collagen through its receptor. A fibroblast-like cell is shown expressing receptors for both competence and progression types of growth factors. Some studies on cells in collagen gels are consistent with collagen binding to its cellular receptors and causing the downregulation of competence (PDGF) receptors. Under these conditions the cell ceases DNA synthesis and is not able to respond to PDGF. As discussed elsewhere in this paper, the production of collagenase is an early response to growth factors and would be expected to degrade collagen bound to the cell. Freeing the collagen receptors would allow the reexpression of PDGF receptors and restore the responsiveness of the cell to this growth factor

lagen synthesis (Sarber et al. 1981; Nusgens et al. 1984; Yoshizato et al. 1986; Weinberg and Bell 1985; Rhudy and McPherson 1988). It should be noted that these changes occur only after the contracture of the gel and not, for example, with mutant cells which are enclosed in the gel but do not contract it (Mauck et al. 1988).

A model to explain these results is shown in Fig. 2. It proposes that there is a linkage between the cell surface collagen receptors and the receptors for one of the competence class of growth factors, most likely PDGF. This linkage could be direct, or possibly by the transmodulation of receptors via their phosphorylation. Binding of a cell to collagen converts it to a quiescent state by causing the withdrawl of the growth factor receptor. These contacts also reduce the production of collagen probably by mechanisms not related to the antimitogenic effects. These responses could be used by a cell following either development or wound healing to cause it to stop proliferating and as a signal to the cell that sufficient collagen had been produced. It is notable that collagenase production is increased under these conditions (NUSGENS et al. 1984).

E. Induction of Collagenase by Growth Factors – Role in Proliferation

According to the scheme outlined above, fibroblasts and other cells would be held in a quiescent state in tissues with their collagen receptors attached to collagen. Presumably, a limited degradation of collagen would free these receptors and would be necessary to restore the cells' ability to respond to growth factors. Collagenase is the key enzyme needed for the breakdown of matrix, since collagen is resistant to other proteases (GROSS 1981). Indeed, the degradation of matrix including collagen is elevated during growth, during wound healing, during regeneration, and in a variety of pathological conditions where cells are proliferating. Particularly large amounts of collagenase are produced by tumor cells and in sites of inflammation (KRANE et al. 1988).

Recent studies have helped to explain the high production of collagenase by proliferating cells by showing that a variety of growth factors including PDGF, FGF, and EGF (BAUER et al. 1985; CHUA et al. 1985) as well as phorbol esters (BRINCKERHOFF et al. 1986) increase collagenase syntheses. Phorbol esters activate some of the same mitogenic responses as growth factors and have been used to elucidate the regulation of the collagenase gene (Fig. 3). Genes such as collagenase and stomelysin contain a characteristic sequence (ATGAGTCAGA in the case of the human collagenase gene) which enhances their transcription in cells treated with phorbol esters (ANGEL et al. 1987a, b). Collagenase gene activation is rapid and rather transient and dependent on the production of nuclear-transacting proteins. Current concept on the response of cells to growth factors suggest that the binding of certain growth factors to their receptors activates phospholipase C, which cleaves phosphatidyl inositol generating diacyl glycerol, the natural homolog of phorbol esters. Diacyl glycerol, as well as phorbol esters, activates protein kinase C, which amplifies the mitogenic signal by phosphorylating various intracellular proteins (NISHIZUKA 1986). These events induce the transcription of various genes necessary for proliferation including *fos* (VERMA 1986).

The *fos* protein has a special role in activating the transcription of the collagenase gene (SCHONTHAL et al. 1988). *Fos* forms a complex with *jun*, a protooncogene protein with a high affinity for the enhancer present in collagenase and other genes responsive to phorbol esters. Binding of the *fos-jun*

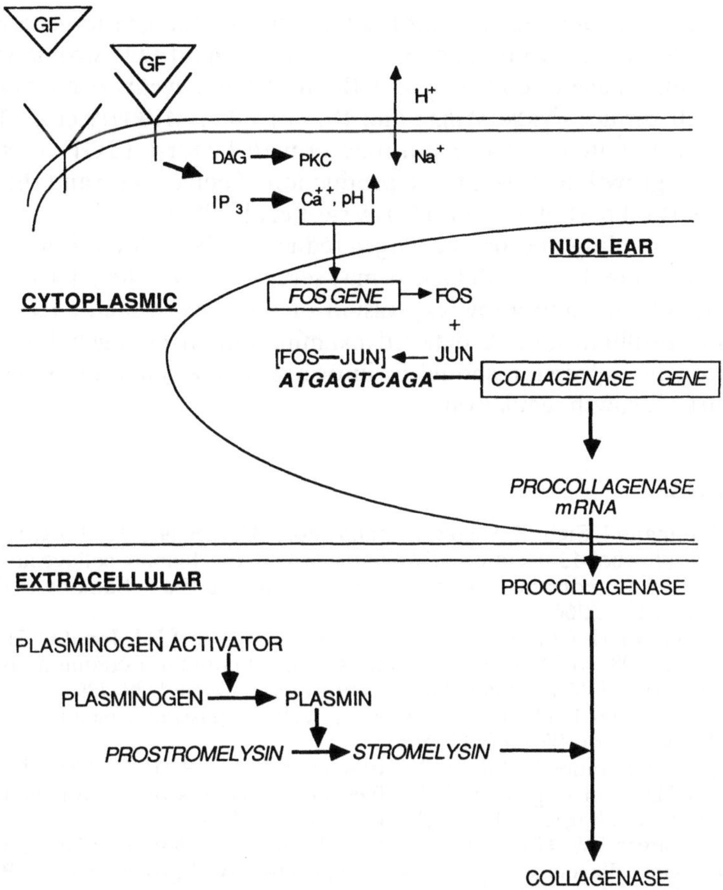

Fig. 3. Induction of collagenase and stromelysin transcription by growth factors through a *fos-jun* complex and the coordinated activation of these proteases. The diagram shows some intracellular events following the reaction of a growth factor with its receptor. These included the activation of phospholipase and the cleavage of phosphatidyl inositol to generate diacylglycerol (*DAG*) and inositol triphosphate (IP_3). DAG, as well as phorbol esters, activate protein kinase C, which phosphorylates various proteins, altering their activity. Under these conditions, there is an activation of certain protooncogenes including *fos*. *Fos* combines with *jun*, another protooncogene, and the *fos-jun* complex binds to an octanucleotide sequence, the phorbol-responsive element on the collagenase and stromelysin genes causing their transcription. Both enzymes are produced in precursor form and their activation involves a concerted process involving plasminogen activation and plasminogen. The activation process probably occurs on the external surface of cells

complex to this site activates the transcription of these genes in the appropriate cells (Chiu et al. 1988). Thus factors that activate protein kinase C induce *fos* and the transcription of the gene for collagenase and certain other proteins.

Enzymes like stromelysin and collagenase may act in a concerted fashion since collagenase and stromelysin are produced as proenzymes. Their activation involves plasminogen activator acting on plasminogen to generate plasmin which activates stromelysin. Some studies suggest that stromelysin converts procollagenase to collagenase with all of these events occurring on the external cell surface (Moscatelli and Rifkin 1988; Murphy et al. 1987). It should also be noted that transforming growth factor-β (TGF-β) is able to suppress the growth factor-induced production of collagenase and this may be a very important part of its action (Edwards et al. 1987).

Since the binding of cells to collagen reduces their proliferation, the induction of collagenase by growth factors may serve to release the cell from its contacts with collagen and allow expression of the growth factor receptors required for proliferation. A detailed examination of collagen binding and protooncogene expression should be informative in elucidating other aspects of this part of growth regulation.

References

Angel P, Baumann I, Stein B, Delius H, Rahmsdorf HJ, Herrlich P (1987a) 12-O-Tetra decanoyl-phorbol-13-acetate (TPA) induction of the human collagenase gene is mediated by an inducible enhancer element located in the 5′ flanking region. Mol Cell Biol 7:2256–2266

Angel P, Imagawa M, Chiu R, Stein B, Inbra RJ, Rahmsdorf HJ, Jonat C, Herrlich P, Karin MB (1987b) Phorbol ester-inducible genes contain a common *cis* element recognized by a TPA modulated *trans*-acting factor. Cell 49:729–739

Aumailley M, Timpl R (1986) Attachment of cells to basement membrane collagen type IV. J Cell Biol 103:1569–1575

Aumailley M, Nurcombe V, Edgar D, Paulsson M, Timpl R (1987) The cellular interactions of laminin fragments. Cell adhesion correlates with two fragment-specific high affinity binding sites. J Biol Chem 262:11532–11538

Bauer EA, Cooper TW, Huang JS, Altman J, Deuel TF (1985) Stimulation of in vitro human skin collagenase expression by platelet-derived growth factor. Proc Natl Acad Sci USA 82:4132–4136

Bell E, Ivarsson B, Merrill C (1979) Production of a tissue-like structure by contracture of collagen lattices by human fibroblasts of different proliferative potential in vitro. Proc Natl Acad Sci USA 76:1274–1278

Bitterman PB, Rennard SI, Adelberg S, Crystal RG (1983) Role of fibronectin as a growth factor for fibroblasts. J Cell Biol 97:1925–1932

Brinckerhoff CE, Plucrnska M, Sheldon ZA, O'Connor GT (1986) Half-life of synovial cell collagenase mRNA is modulated by phorbol myristic acetate but not by all-*trans*-retinoic acid or dexamethasone. Biochemistry 25:6378–6384

Charonis AS, Skubitz APN, Koliakos GG, Reger L, Dege J, Vogel AM, Wohlueter R, Furcht LT (1988) A novel synthetic peptide of the B1 chain of laminin with heparin-binding and cell-adhesion promoting activities. J Cell Biol 107:1253–1260

Chiu R, Boyle WJ, Meek J, Smeal T, Hunter T, Karin M (1988) The C-*fos* protein interacts with c-*jun*/AP-1 to stimulate transcription of AP-1 responsive genes. Cell 54:541–552

Chua CC, Geiman DE, Keller GH, Ladda RL (1985) Induction of collagenase secretion in human fibroblast cultures by growth promoting factors. J Biol Chem 260:5213–5216

Dedhar S, Ruoslahti E, Pierschbacher MD (1987) A cell surface receptor complex for collagen type I recognizes the Arg-Gly-Asp sequence. J Cell Biol 104:585–592

De Petro G, Barlali S, Vartio T, Vaheri A (1981) Transformation-enhancing activity of gelatin-binding fragments of fibronectin. Proc Natl Acad Sci USA 78:4965–4969

Dike LE, Farmer SR (1988) Cell adhesion induces expression of growth-associated genes in suspension-arrested fibroblasts. Proc Natl Acad Sci USA 85:6792–6796

Edgar D, Timpl R, Thoenen H (1984) The heparin-binding domain of laminin is responsible for its effects on neurite out growth and neuronal survival. EMBO J 3:1463–1468

Edwards DR, Murphy G, Reynolds JJ, Whitham SE, Docherty AJB, Angel P, Heath JK (1987) Transforming growth factor beta modulates the expression of collagenase and metalloproteinase inhibitor. EMBO J 6:1899–1904

Ehrlich HP (1988) The role of connective tissue matrix in wound and fracture healing. In: Barbul A, Pines E, Caldwell M, Hunt TK (eds) Growth factors and other aspects of wound healing. Liss, New York, pp 243–258

Evered D, Whelan J (eds) (1986) Function of the proteoglycans. Ciba Found Symp 124

Gehlsen KR, Dillmer L, Engvall E, Ruoslahti E (1988) The human laminin receptor. Science 241:1228–1229

Goldberg B (1979) Binding of soluble type I collagen molecules to the fibroblast plasma membrane. Cell 16:265–275

Goldberg GI, Wilhelm SM, Kronberger A, Bauer EA, Grant GA, Eisen AZ (1986) Human fibroblast collagenase. Complete primary structure and homology to an oncogeny transformation-induced rat protein. J Biol Chem 261:6600–6605

Goodman SL, Deutzmann R, von der Mark K (1987) Two distinct cell binding domains in laminin can independently promote non-neuronal cell adhesion and spreading. J Cell Biol 105:589–598

Gospodarowicz D, Greenburg G, Birdwell CR (1978) Determination of cell shape by the extracellular matrix and its correlation with the control of cellular growth. Cancer Res 38:4155–4171

Graf J, Iwamoto Y, Sasaki M, Martin GR, Kleinman HK, Robey FA, Yamada Y (1987) Identification of an amino acid sequence in laminin mediating cell attachment, chemotaxis and receptor binding. Cell 48:989–996

Gross J (1981) An essay on the biological degradation of collagen. In: Hay ED (ed) Cell biology of extracellular matrix. Plenum, New York, pp 217–258

Grotendorst GR, Martin GR (1986) Chemotaxis in wound healing and fibrosis. Rheumatol Annu 10:385–403

Hassell JR, Kimura JH, Hascall VC (1986) Proteoglycan core protein families. Annu Rev Biochem 55:539–568

Hay ED (1981) Collagen and embryonic development. In: Hay ED (ed) Cell biology of extracellular matrix. Plenum, New York, pp 379–409

Herbst TJ, McCarthy JB, Tsilibary EC, Furcht LT (1988) Differential effects of laminin, intact type IV collagen and specific domains of type IV collagen on endothelial cell adhesion and migration. J Cell Biol 106:1365–1373

Humphries JJ, Akujama SK, Komoriya A, Olden K, Yamada KM (1986) Identification of an alternatively spliced site in human plasma fibronectin that mediates cell type specific adhesion. J Cell Biol 103:2637–2647

Humphries MJ, Ayad SR (1983) Stimulation of DNA synthesis by cathepsin D digests of fibronectin. Nature 305:811–813

Hynes RO (1987) Integrins: a family of cell surface receptors. Cell 48:549–554

Kleinman HK, Klebe RJ, Martin GR (1981) Role of collagenous matrices in the adhesion and growth of cells. J Cell Biol 88:473–485

Kleinman HK, McGarvey ML, Hassell JR, Martin GR (1983) Formation of a supramolecular complex is involved in the reconstitution of basement membrane components. Biochemistry 22:4969–4974

Kleinman HK, Cannon FB, Laurie G, Hassell JR, Aumailley M, Terranova VP, Martin GR, Dubois-Dalcq M (1985) Biological activities of laminin. J Cell Biochem 27:317–325

Kleinman HK, McGarvey ML, Hassell JR, Star VL, Cannon FB, Laune GW, Martin GR (1986) Basement membrane complexes with biological activity. Biochemistry 25:312–318

Krane SM, Amento EP, Goldberg MB, Goldring SR, Stephenson ML (1988) Modulation of matrix synthesis and degradation in joint inflammation. In: Glauert EL (ed) The control of tissue damage. Elsevier, Amsterdam, pp 179–195
Kubota Y, Kleinman HK, Martin GR, Lawley TJ (1988) Role of laminin and basement membrane in the morphological differentiation of human endothelial cells into capillary-like structures. J Cell Biol 107:1589–1598
Liotta LA, Abe S, Gehron-Robey P, Martin GR (1979) Preferential digestion of a basement membrane collagen by an enzyme derived from a metastatic murine tumor. Proc Natl Acad Sci USA 76:2268–2272
Martin GR, Timpl R (1987) Laminin and other basement membrane components. Annu Rev Cell Biol 3:57–87
Martin GR, Kleinman HK, Terranova VP, Ledbetter S, Hassell JR (1984) The regulation of basement membrane formation and cell-matrix interactions by defined supramolecular complexes. Ciba Found Symp 108:197–212
Martin GR, Timpl R, Muller PK, Kuhn K (1985) The genetically distinct collagens. Trends Biochem Sci 10:285–287
Mauck C, van der Mark K, Helle O, Mollenhauer J, Paffle M, Krieg T (1988) A defective cell surface collagen-binding protein in dermatosparatic sheet fibroblasts. J Cell Biol 106:205–211
Mayne R, Burgeson RE (1987) Structure and function of collagen types. Academic, New York
McCarthy JB, Furcht LT (1984) Laminin and fibronectin promote the haeptotactic migration of B16 mouse melanoma cells in vitro. J Cell Biol 98:1474–1480
Moscatelli D, Rifkin DB (1988) Membrane and matrix localization of proteinases: a common theme in tumor cell invasion and angiogenesis. Biochim Biophys Acta 948:67–85
Murphy G, Cockett MI, Stephens PE, Smith BJ, Docherty AJP (1987) Stromelysin is an activator of procollagenase. A study with natural and recombinant enzymes. Biochem J 248:265–268
Nishizuka Y (1986) Protein kinases in signal transduction. In: Bradshaw RA, Printes S (eds) Oncogenes and growth factors. Elsevier, Amsterdam, pp 248–254
Nusgens B, Merrill C, Lapiere C, Bell E (1984) Collagen biosynthesis by cells in a tissue equivalent matrix in vitro. Coll Relat Res 4:351–364
Obara M, Kang MS, Yamada KM (1988) Site directed mutagenesis of the cell binding domain of human fibronectin: separable, synergistic sites mediate adhesive function. Cell 53:649–657
Panayotou G, End P, Aumailley M, Timpl R, Engel J (1989) Domains of laminin with growth-factor activity. Cell 56:93–101
Polla BS, Healy AM, Byrne M, Krane SM (1987) 1,25-Dihydroxyvitamin D_3 induces collagen binding to the human monocyte line A937. J Clin Invest 80:962–969
Rhudy RW, McPherson JM (1988) Influence of the extracellular matrix on the proliferative response of human skin fibroblasts to serum and purified platelet-derived growth factor. J Cell Physiol 137:185–191
Rubin K, Hook M, Obrink B, Timpl R (1981) Substrate adhesion of rat hepatocytes: mechanism of attachment to collagen substrates. Cell 24:463–470
Ruoslahti E (1988 a) Fibronectin and its receptors. Annu Rev Biochem 57:375–413
Ruoslahti E (1988 b) Structure and biology of proteoglycans. Annu Rev Cell Biol 4:229–255
Salomon DS, Liotta LA, Kidwell WR (1981) Differential response to growth factor by rat mammary epithelium plated on different collagen substrata in serum-free medium. Proc Natl Acad Sci USA 78:382–386
Sarber R, Hull B, Merrill C, Soranno T, Bell E (1981) Regulation of proliferation of fibroblasts of low and high population doubling levels grown in collagen lattices. Mech Ageing Dev 17:107–117
Sasaki M, Kleinman HK, Huber H, Deutzmann R, Yamada Y (1988) Laminin, a multidomain protein: the A chain has a unique globular domain and homology with the

basement membrane proteoglycan and the laminin B chains. J Biol Chem 263:16536–16544
Schonthal A, Herrlich P, Rahmsdorf HJ, Ponta H (1988) Requirement for *fos* gene expression in the transcriptional activation of collagenase by other oncogenes and phorbolesters. Cell 54:325–334
Sporn MB, Roberts AB (1986) Peptide growth factors and inflammation, tissue repair and cancer. J Clin Invest 78:329–332
Takada Y, Wayner EA, Carter WG, Hemmler ME (1988) Extracellular matrix receptors, ECMRII and ECMRI, for collagen and fibronectin correspond to VLA-2 and VLA-3 with VLA family of heterodimers. J Cell Biochem 37:385–393
Terracio L, Ronnstrand L, Tingstrom A, Rubin K, Claesson-Welsh L, Funa K, Heldin C (1988) Induction of platelet-derived growth factor receptor expression in smooth muscle cells and fibroblasts upon tissue culturing. J Cell Biol 107:1947–1957
Terranova VP, Aumailley M, Sultan LH, Martin GR, Kleinman HK (1986) Regulation of cell attachment and growth by fibronectin and laminin. J Cell Physiol 27:473–479
Timpl R, Dziadek M (1986) Structures, development and molecular pathology of basement membranes. Int Rev Exp Pathol 29:1–112
Tomaselli JJ, Damsky CH, Reichardt LF (1988) Purification and characterization of mammalian integrins expressed by a rat neuronal cell line (PC12): evidence that they function as α/β heterodimeric receptors for laminin and type IV collagen. J Cell Biol 107:1241–1251
Turpeenniemi-Hujanen T, Thorgeirsson UP, Rao CN, Liotta LA (1986) Laminin increase the release of type IV collagenase. J Biol Chem 261:1883–1889
Van Evercooren A, Kleinman HK, Ohno S, Marangos P, Schwartz JP, Dubois-Dalcq ME (1982) Nerve growth factor, laminin and fibronectin promote neurite growth in human fetal sensory ganglia cultures. J Neurosci Res 8:179–194
Verma IM (1986) Proto-oncogene *fos*: a multifaceted gene. In: Bradshaw RA, Prentes S (eds) Oncogenes and growth factors. Elsevier, Amsterdam, pp 67–73
Vlodavsky I, Folkman J, Sullivan R, Fridman R, Michaeli I, Sasse J, Klagsbrun M (1987) Endothelial cell derived basic fibroblast growth factor: synthesis and deposition into subendothelial extracellular matrix. Proc Natl Acad Sci USA 84:2292–2296
Weinberg CB, Bell E (1985) Regulation of proliferation of bovine endothelial cells, smooth muscle cells and adventitial fibroblasts in collagen lattices. J Cell Physiol 122:410–414
Yamada KM (1983) Cell surface interactions with extracellular materials. Annu Rev Biochem 52:761–799
Yoshizato K, Taira T, Shioya N (1986) Collagen dependent growth suppression and changes in the shape of human dermal fibroblasts. Ann Plast Surg 13:9–14

Section D: Processes Regulated by Growth Factors

CHAPTER 35

Induction of Proteases and Protease Inhibitors by Growth Factors*

D. E. MULLINS and D. B. RIFKIN

A. Introduction

Most polypeptide growth factors have been identified and assayed on the basis of their ability to stimulate DNA synthesis and cell division. The binding of growth factors to specific cell surface receptors initiates a cascade of biochemical reactions which alter many aspects of cellular physiology, change the pattern of gene expression, and, ultimately, trigger DNA synthesis and cell division. In addition to changes in those processes associated with the initiation of cell division, other effects of growth factors on cells include alterations in the cytoskeleton associated with modifications in cell shape, alterations in ion transport, phosphorylation of specific cellular proteins, and activation of certain genes, including, in some cases, those coding for proteases and protease inhibitors. At this time it is not clear whether all of these growth factor-induced changes are necessary for cell division or whether they indicate that growth factors affect cellular physiology in ways which are independent of their growth-promoting effects.

A survey of the recent literature indicates that the stimulation of the production of proteases and protease inhibitors is a common effect of growth factors. Seven well-characterized growth factors have been reported to induce such activities in a variety of target cells. In this review we discuss these experimental data with respect to each of these growth factors. Within the discussion of each growth factor there is a certain unavoidable overlap, because some studies utilized more than one growth factor. In the conclusion, we discuss some of the common findings and attempt to integrate these data into a framework which suggests some testable hypotheses on growth factor induction of proteases and inhibitors.

B. Fibroblast Growth Factor

Two distinct forms of fibroblast growth factor (FGF) have been purified, characterized, and cloned (THOMAS et al. 1984; ESCH et al. 1985; JAYE et al. 1986; ABRAHAM et al. 1986; SOMMER et al. 1987). These two proteins are most easily distinguished on the basis of differences in their isoelectric points and their affinities for heparin. Acidic FGF (aFGF) has a pI of 4.5, while basic

* Supported in part by grants from the American Cancer Society and the National Institutes of Health to D. B. R.

FGF (bFGF) has a pI of greater than 9. Both proteins bind tightly to heparin, but aFGF is eluted with 1 *M* NaCl, while bFGF is eluted with 1.5 *M* NaCl (KLAGSBRUN and SHING 1985). This allows the acidic and basic forms to be separated easily by affinity chromatography on heparin-Sepharose.

The acidic and basic FGFs are products of different genes located on different chromosomes (MERGIA et al. 1986) but share many biological properties, including the ability to increase protease production in certain target cells. We will review several studies which demonstrate that aFGF and bFGF can function as potent stimulators of plasminogen activator (PA) and collagenase activities.

RIFKIN and coworkers have studied the mechanism of endothelial cell invasion and capillary growth during angiogenesis. They hypothesized that the endothelial cells forming a new capillary sprout must produce degradative enzymes in order to penetrate through the tissue barriers surrounding the growing blood vessel and, additionally, that these enzymes would be induced by angiogenic factors. Many other types of invasive cells elaborate PA and collagenase which act in concert to degrade the connective tissue matrices invading cells must penetrate (MOSCATELLI and RIFKIN 1988). PA converts the zymogen plasminogen to plasmin, a protease with broad substrate specificity. Plasmin, in turn, can activate procollagenase (WERB et al. 1977). Therefore, the ability of various angiogenic preparations to stimulate endothelial cell PA and procollagenase production was measured.

Several angiogenic preparations were found to increase PA and procollagenase production in cultured bovine capillary endothelial (BCE) cells (GROSS et al. 1983b). A similar protease-inducing activity was identified by Presta and coworkers in extracts of human placenta (PRESTA et al. 1985) and was subsequently purified by MOSCATELLI et al. (1986b) and identified as bFGF. Basic FGF stimulated the production of both forms of PA, the urokinase-type PA (u-PA), and the tissue-type PA (t-PA), as well as procollagenase in BCE cells at concentrations ranging from 0.1 to 10 ng/ml and was mitogenic for BCE cells in the same concentration range.

MOSCATELLI et al. (1986a) surveyed a wide variety of cultured normal and tumor cell lines from several different species for the production of PA-inducing (bFGF) activity. Extracts from normal human, bovine, and chicken cells, as well as four human tumor cell lines, stimulated the production of PA by BCE cells. The PA-inducing activity of the extracts was abolished by incubation with antibodies against bFGF.

Likewise, acidic FGF has also been studied for its ability to stimulate PA and procollagenase activities in cultured BCE cells. Other workers have shown that the biological activity of aFGF is greatly augmented by heparin (SCHREIBER et al. 1985). BCE cells treated with aFGF in the absence of heparin produced increased levels of PA relative to control cells. However, this required concentrations of aFGF 30-fold higher than those required for the induction of PA by bFGF. In the presence of heparin aFGF was as effective as bFGF in inducing PA activity. Acidic FGF also stimulated the production of BCE cell collagenase activity, with the induction of this enzyme showing a less-pronounced dependence on heparin than the PA induction (PRESTA and MULLINS, unpublished observations).

In other studies MIRA-Y-LOPEZ et al. (1983) discovered a PA-inducing activity in extracts of bovine pituitary. They found that rat mammary gland tumors, induced by mammary tumor virus (MTV), dimethylbenzanthracene (DMBA), or nitrosomethylurea (NMU), produced higher levels of both cell-associated and secreted PA activities than normal mammary gland in organ culture. The addition of prolactin (2.2×10^{-7} *M*) to the organ cultures resulted in 1.5- to 3.6-fold increases in PA secretion in 22 of 27 cultures. In most MTV- and DMBA-induced tumors a crude pituitary gland extract produced an additional two- to fourfold increase in PA secretion over the prolactin-induced levels.

Pituitary gland is a rich source of FGF and, in a subsequent study, MIRA-Y-LOPEZ et al. (1986) identified bFGF as the pituitary factor responsible for PA induction in several breast tumor cell lines. They observed that treatment of cultured T47D, MTW9/PL, and SC115 breast tumor cells with pituitary extract induced a four- to fivefold increase in secreted PA activity. In contrast to their earlier study, neither prolactin nor growth hormone had any effect on PA levels. The addition of anti-bFGF IgG to the pituitary extract abolished its PA-inducing activity, indicating that the active factor was, indeed, bFGF. Three additional tumor cell lines, a human mammary carcinoma (MDA-MB-231), a mouse squamous carcinoma (COLO 16), and a human ovarian carcinoma (COLO 110), as well as human embryo fibroblast line also responded to bFGF with increases in PA activity.

The regulation of PA activity in BCE cells was characterized by SAKSELA et al. (1987), who found that bFGF and transforming growth factor-β1 (TGF-β1) have opposite effects on PA production. As was seen by GROSS et al. (1983b) and MOSCATELLI et al. (1986b), untreated BCE cells produced a low but measurable amount of PA activity which increased in a dose-dependent manner within 2–4 h after the addition of bFGF. TGF-β1 reduced both the basal level as well as the bFGF-induced level of PA activity in BCE cells in a dose- and time-dependent manner, with half-maximal inhibition of PA induction requiring only 50 pg/ml TGF-β1.

Interestingly, these authors found that the response of the cells to bFGF or TGF-β1 was dependent on the order and time of addition of each factor. If bFGF was added 3–6 h before TGF-β1, the TGF-β1-mediated inhibition of PA activity was abrogated. However, when TGF-β1 was added to the cells 3–6 h prior to bFGF, PA levels were greatly reduced.

These results suggested that the TGF-β1 inhibition might be affected through the induction of a PA inhibitor (PAI). Therefore, the production of PA and type 1 inhibitor of PA (PAI-1), the major inhibitor of both u-PA and t-PA, by the cells was examined by direct fibrin zymography (to detect PA) (GRANELLI-PIPERNO and REICH 1978) and by reverse fibrin zymography (to detect PA inhibitors) (LOSKUTOFF et al. 1983). These techniques take advantage of the fact that u-PA, t-PA, and PAI-1 are all stable to electrophoresis in SDS gels. They allow PA activity or PA inhibitory activity in an SDS polyacrylamide gel to be visualized in an opaque fibrin indicator gel. In direct zymography, PA diffuses from the SDS polyacrylamide gel into an opaque fibrin indicator gel containing plasminogen. Plasmin is formed beneath the original PA band and cleaves the fibrin, resulting in a clear lysis zone. In reverse zymography, the fibrin indicator gel contains PA and plasminogen. Diffusion of PAI-1 into the indicator gel prevents plasmin-mediated lysis of the indicator gel, yielding an opaque band corresponding to PAI-1 in an otherwise clear gel. BCE cells were found to produce both u-PA and t-PA. When cells were treated with bFGF, the amount of u-PA in the cell extracts increased, and when cells were treated with TGF-β1 the PA levels decreased concomitantly with the appearance of increased amounts of PAI-1. Reverse fibrin autography revealed that even untreated BCE cells produce a PA inhibitor of M_r 47000 whose level is increased somewhat by bFGF and markedly by TGF-β1.

In addition to inducing PAI-1, TGF-β1 also reduced the production of u-PA. Quantitative immunoprecipation of metabolically radiolabeled medium from BCE cells indicated that exposure to bFGF resulted in a 2.1-fold increase in secreted u-PA, while exposure to TGF-β1 caused an 8.3-fold decrease in u-PA, relative to medium from untreated cells. Immunoprecipitation with antibodies against PAI-1 confirmed the results obtained by reverse zymography. Treatment of the cells with bFGF resulted in a 2.2-fold increase in secreted PAI-1, while treatment with TGF-β1 brought about a 5.8-fold increase. Therefore, the inhibitory effect of TGF-β1 on BCE cell PA production is the result of both induction of PAI-1 and a decrease in BCE cell PA production.

Another member of the FGF family, K-FGF, which is structurally and functionally related to bFGF, also induces increases in PA production in BCE cells (DELLI-BOVI et al. 1988). It is likely that other members of the FGF family will be found to increase protease production when tested.

C. Transforming Growth Factor-β

TGF-β is an M_r 25000 homodimeric protein which is produced by several cell types and which has multiple effects on target cells. TGF-β has a pronounced effect on cell proliferation. In general, TGF-β inhibits proliferation in epithelial cells but may stimulate cell division in mesenchymal cells, depending upon culture conditions. TGF-β is also a potent stimulator of the synthesis of proteins which make up the extracellular matrix (ECM). The ECM is composed primarily of collagen, fibronectin, laminin, elastin, glycosaminoglycans, and other glycoproteins. These macromolecules are degraded by cell-derived proteases, including collagenase, plasmin (generated through the action of PA), and stromelysin. Cells attach to the ECM by interactions with specific matrix proteins, and this interaction is believed to be critical in regulating histogenesis, cell proliferation, and differentiation. Proteolytic degradation of the ECM, such as that associated with tissue remodeling and malignancy, may modify or abolish ECM/cell interactions. In this section we will discuss several studies which indicate that TGF-β stimulates the production of inhibitors of these proteases, as well as certain proteases. The induction of protease inhibitors may be one mechanism by which TGF-β protects the ECM.

THALACKER and NILSEN-HAMILTON (1987) conducted an extensive series of experiments aimed at elucidating the mechanism whereby TGF-β regulates cellular proliferation. They examined the effect of TGF-β on the incorporation of [^{35}S]methionine into secreted proteins from five cell lines, each derived from a different tissue and different species. They found that TGF-β increased the synthesis of an M_r 48000 protein identified as PAI-1 by immunoprecipitation with antibodies against authentic bovine PAI-1. In lung epithelial cells the levels of PAI-1 increased after only 2 h of treatment with TGF-β. Epidermal growth factor (EGF) also increased the level of PAI-1 synthesis, and TGF-β and EGF had synergistic effects on the production of this inhibitor.

Because TGF-β inhibited DNA synthesis and cell proliferation in lung epithelial cells, the authors attempted to determine whether the TGF-β induction of PAI-1 and inhibition of DNA synthesis were causally linked. A comparison of the time course of TGF-β induction of PAI-1 with the time course of TGF-β inhibition of [^{3}H]thymidine incorporation in lung epithelial cells suggested that PAI-1 production could be responsible for the subsequent inhibition of DNA synthesis. PAI-1 was rapidly induced with the maximal rate of synthesis occurring 5 h after exposure to TGF-β and then declined. Inhibition of [^{3}H]thymidine incorporation did not reach a maximum until 24 h after the addition of TGF-β to cells. However, a direct role for PAI-1 in inhibition of DNA synthesis was ruled out by the finding that the tumor promoter PMA stimulated both PAI-1 synthesis and DNA synthesis in lung epithelial cells.

LAIHO et al. (1986b) studied TGF-β stimulation of PA levels in six strains of human skin fibroblasts. No detectable PA activity was observed in the untreated cultures, but two of the cultures produced PA when treated with TGF-β. One culture was selected for further study and showed a dose- and time-dependent increase in PA activity upon treatment with TGF-β at concentrations ranging from 0.5 to 20 ng/ml. EGF, platelet-derived growth factor (PDGF), and insulin were ineffective in inducing PA synthesis in these cells. The addition of the plasmin inhibitor aprotinin to the culture medium abolished the PA activity, whereas the addition of plasmin enhanced PA activity. This finding suggested that PA was being secreted in a proteolytically inactive proenzyme form which required activation by plasmin (WUN et al. 1982).

The use of reverse fibrin zymography indicated that the cells also produced a PA inhibitor. Untreated cultures produced a protease inhibitory activity of M_r 60000 whose synthesis was enhanced by TGF-β. This M_r 60000 band may represent a u-PA/PAI-1 complex (see below).

LAIHO et al. (1986a) extended their research on TGF-β modulation of fibrinolytic activity by studying two cultured human embryonic lung fibroblast lines, HEL-299 and WI-38. These cell lines secrete both u-PA and t-PA. In contrast to their earlier findings with human skin fibroblasts, they found that upon treatment with TGF-β HEL-299 and WI-38 cells decreased their production of PA. This inhibitory effect was abolished if the cells were treated with cycloheximide, suggesting that the decrease in PA activity might be due to the secretion of a TGF-β-induced PA inhibitor. Reverse fibrin zymography of conditioned medium from TGF-β-treated cells, in fact, revealed the presence of PAI-1. However, complexes of PA/PAI-1 were not detected by reverse fibrin zymography, suggesting that PAI-1 inactivation of PA was not the principle reason for the lower PA activity of TGF-β-treated cultures.

Metabolic radiolabeling of untreated or TGF-β-treated cultures indicated that TGF-β stimulated the synthesis of an M_r-47000 protein. Antibodies against PAI-1 recognized this M_r-47000 protein and abolished the inhibitory activity of TGF-β-treated culture supernatants in reverse fibrin zymography. Analysis of cell extracts and ECM from radiolabeled cultures indicated that PAI-1 did not accumulate intracellularly but was deposited into the substratum. This deposition of PAI-1 in the ECM may serve to protect the matrix from PA-mediated proteolysis.

TGF-β and EGF have antagonistic effects in WI-38 cells. EGF stimulates the production of both u-PA and t-PA in a dose-dependent fashion but does not stimulate production of PAI-1 (LAIHO et al. 1986a). In TGF-β-treated cultures, increasing amounts of EGF could overcome the initial inhibition of PA production, yielding a dose-dependent increase in PA activity. Likewise, in EGF-treated cultures, increasing amounts of TGF-β abolished PA activity and produced a net PA inhibitory activity in the culture medium.

In another study LUND et al. (1987) investigated the molecular biology of the TGF-β induction of PAI-1 in WI-38 cells. Northern blot analysis of TGF-β-treated WI-38 cells revealed the presence of two mRNAs of 3.4 and 2.4 kb which hybridized to a full-length PAI-1 cDNA. Five nanograms TGF-β per milliliter (the dose which maximally stimulated PAI-1 production) gave a measurable increase in PAI-1 mRNA levels within 1 h of treatment and a maximal increase of approximately 50-fold at 8 h. It was not determined whether this increase was due to an enhancement of transcription or a stabilization of PAI-1 mRNA. Interestingly, treatment of the cells with 10 µg/ml of cycloheximide was as effective as TGF-β in inducing PAI-1 mRNA transcription, but treatment with TGF-β and cycloheximide together did not yield additive or synergistic increases in mRNA levels.

The TGF-β-induced increase in PAI-1 mRNA levels was reflected in a 50-fold increase in the synthesis of PAI-1 protein. The increase was detectable within 1 h and reached a maximum in cell extracts after 10 h, closely following the kinetics of mRNA production. PAI-1 continued to accumulate in the conditioned medium for up to 48 h. Immunocytochemical staining of TGF-β-treated WI-38 cells localized intracellular PAI-1 to the pericellular region and several unidentified "patch-like" areas throughout the cytoplasm. Untreated cells showed only faint straining in the perinuclear region.

LAIHO et al. (1987) also studied the effect of TGF-β on the accumulation of PAI-1 in the ECM of HT1080 human fibrosarcoma cells. TGF-β stimulated total protein synthesis, especially production of PAI-1, while decreasing the production of u-PA by approximately 20%. The total synthesis of ECM proteins was increased 2.5-fold by exposure to TGF-β, which PAI-1 incorporation into the ECM increasing 4-fold. EGF, TGF-α, PDGF, and insulin were ineffective in stimulating the accumulation of ECM proteins and had no effect on the production of PA or PAI-1 in these cells.

The kinetics of PAI-1 production and accumulative were measured by following the synthesis, localization, and turnover of [^{35}S]methionine-labeled PAI-1 Treatment of HT1080 cells with 1 ng/ml TGF-β resulted in the appearance of [^{35}S]PAI-1 in the ECM within 4 h after treatment and in the conditioned medium within 8 h. PAI-1 continued to accumulate in the ECM for 24 h and in the conditioned medium for 48 h. When ^{35}S-labeled TGF-β-treated cultures were incubated in serum-free medium without TGF-β, 60% of the PAI-1 was removed from the ECM by 24 h. The loss of PAI-1 from the ECM occurred concomitantly with an increase of PAI-1 in the medium.

Laiho et al. (1987) demonstrated that u-PA at concentrations of 30–150 IU/ml could remove PAI-1 from the culture medium and from the ECM of intact TGF-β-treated HT1080 cells growing in monolayer culture. The addition of 150 IU/ml u-PA to culture medium of TGF-β-treated cells reduced the amount of M_r-47000 PAI-1 by 90%. Virtually all of the M_r 47000 PAI-1 was removed from the ECM of untreated cells. The loss of the M_r 47000 protein was paralleled by an increase in a protein of M_r 62000 which immunologically crossreacted with PAI-1 and u-PA, indicating that this band represented a PAI-1/u-PA complex. Therefore, the PAI-1 in the ECM of HT1080 cells is capable of interacting with its substrate, u-PA.

Edwards et al. (1987) examined the ability of several growth factors, including TGF-β, to stimulate the production of two ECM-degrading proteases and an inhibitor of these proteases. Using MRC-5 human fetal lung fibroblasts, they focussed on the regulation of the collagenase, stromelysin, and tissue inhibitor of metalloproteinase (TIMP) genes by several growth factors, including TGF-β. Collagenase and stromelysin were selected for study because they are known to degrade ECM components. Collagenase specifically cleaves the collagen triple helix, which is then susceptible to degradation by more general proteases, including stromelysin. TIMP, which has been isolated from several types of tissue, is an effective inhibitor of both collagenase and stromelysin.

MRC-5 fibroblasts were treated for 12 h with serum, EGF, bFGF, embryonal carcinoma-derived growth factor (ECDGF), TGF-β, or TGF-β in combination with EGF, bFGF, or ECDGF. mRNA was extracted from these cells and analyzed by northern blot analysis. All of the growth factors tested increased TIMP mRNA levels except TGF-β alone, while treatment with TGF-β in combination with EGF, bFGF, or ECDGF had a synergistic effect on TIMP mRNA levels. Collagenase mRNA levels were also increased by EGF, bFGF, and ECDGF but not by TGF-β. However, in contrast to its effect on TIMP mRNA expression, TGF-β in combination with EGF, bFGF, or ECDGF inhibited growth factor-stimulated collagenase mRNA induction.

Stromelysin mRNA levels are slightly elevated by EGF, bFGF, and ECDGF but were not elevated by TGF-β. TGF-β in combination with EGF and bFGF had no effect on the induction of stromelysin mRNA, but TGF-β did reduce the ECDGF-induced increase. Nuclear runoff transcription assays demonstrated that growth factor regulation of mRNA levels occurred at the transcriptional level. An increase in mRNA transport or stability might also contribute to the observed increases in mRNA levels, but these possibilities were not investigated.

The growth factor-induced modulation of collagenase and TIMP mRNA levels was reflected in both the level of activity and the amount of protein secreted by the cells. Treatment of the cells with TGF-β alone had no effect on the level of secreted collagenase activity, while treatment with TGF-β in combination with either EGF or bFGF blocked the growth factor induction of collagenase synthesis. Both EGF and bFGF by themselves increased collagenase levels in serum-free conditioned medium. Both TIMP protein concentration and TIMP inhibitory activity against collagenase were increased after exposure of cells to EGF and bFGF, and the inclusion of TGF-β augmented their effects. Stromelysin activity was not detectable in any samples. Interestingly, neither bFGF nor bFGF plus TGF-β affected the mRNA levels of three principle ECM proteins, fibronectin, procollagen α1(1), or procollagen α2(1), suggesting that growth factor control of ECM deposition is exerted in these cells through regulation of degradative activities rather than through synthesis of matrix components. These studies, then, may provide an explanation for the ability of TGF-β to enhance ECM deposition.

D. Platelet-Derived Growth Factor

PDGF is a potent mitogen and chemotactic agent for cells of mesenchymal origin and probably plays an important role in organogenesis, wound healing, and certain pathological conditions, including atherosclerosis and cancer. In wound healing, PDGF is released at sites of injury from platelets and macrophages and functions to attract fibroblasts into the wound and stimulate their proliferation.

Wound healing is also characterized by extensive deposition and remodeling of connective tissue matrices within newly formed tissue. Because collagenase had been shown to be involved in tissue remodeling in wound healing, BAUER et al. (1985) investigated whether PDGF would stimulate the production of this activity in cultured human skin fibroblasts (HSFs). They found that PDGF elicited a dose-responsive increase in collagenase production, measured as immunoreactive collagenase in an ELISA. Half-maximal stimulation was obtained with 20–40 ng/ml PDGF, while maximal stimulation was achieved with 90–120 ng/ml.

The release of collagenase by untreated cultures and by PDGF-treated cultures was equivalent for the first 8–10 h after treatment. However, after this time the rate of accumulation of collagenase in the medium nearly doubled in the treated cultures. This effect of PDGF was reversible. Collagenase accumulation decreased to control levels 4-6 h after removal of PDGF from the cultures. The PDGF-stimulated release of collagenase occurred under conditions in which cell division did not occur and was accompanied by an increase in collagenase mRNA levels, suggesting that the growth factor acts at a pretranslational level.

Major excreted protein (MEP) is an M_r-39000 protein produced by virally transformed mouse 3T3 cells and by BALB/c 3T3 cells in response to treatment with growth factors, concanavalin A, or tumor promoters (GOTTESMAN 1978; NILSEN-HAMILTON et al. 1981; GOTTESMAN and SOBEL 1980). MEP contains mannose-6-phosphate, binds to the mannose-6-phosphate receptor, and displays acid protease activity (GAL and GOTTESMAN 1986). It has recently been shown to be identical to the lysosomal protease, cathepsin L (MASON et al. 1988).

Expression of MEP mRNA in BALB/c 3T3 cells was found by FRICK et al. (1985) to be regulated by PDGF. Exposure of cells to PDGF resulted in a dose- and time-dependent increase in MEP mRNA levels. MEP mRNA levels were increased relative to untreated cells 3–4 h after treatment with PDGF and continued to increase for at least 12 h. The addition of cycloheximide to the culture medium completely inhibited the PDGF-stimulated increase in MEP mRNA, suggesting that MEP synthesis may depend upon the production of a positive regulatory protein.

Similar results were obtained by RABIN et al. (1986), who found that in mouse 3T3 A-31/714 cells MEP transcription was increased eightfold after addition of PDGF. They also observed that treatment of the cells with cycloheximide blocked the increase in MEP transcription.

Density-arrested ST2-3T3 cells, a spontaneously transformed cell line derived from BALB/c 3T3 cells, were found to have MEP mRNA levels four times higher that untreated BALB/c 3T3 cells. However, treatment of ST2-3T3 cells with PDGF had no effect on MEP mRNA levels (FRICK et al. 1985). The authors speculated that these cells may be producing a PDGF-like molecule which stimulates MEP mRNA expression through an autocrine mechanism.

E. Epidermal Growth Factor

EGF has been found to induce several proteases and protease inhibitors in cultured mammalian cell lines. The studies summarized below indicate that EGF may play a regulatory role in several processes involving cell-mediated proteolysis.

LEE and WEINSTEIN (1978) provided one of the earliest demonstrations that growth factors can increase the levels of proteolytic activity. Based on their observation that treatment of mammalian cell lines with the tumor promoter phorbol myristate acetate (PMA) caused several phenotypic changes reminiscent of those observed in transformed cells (WEINSTEIN and WIGLER 1977), including increased production of PA (WIGLER and WEINSTEIN 1976), they hypothesized that tumor promoters work through the signal transduction pathways which regulate cell proliferation, and would, therefore, be expected to have similar biological activities as growth factors. In support of this hypothesis, they presented data demonstrating that both EGF and PMA stimulated PA levels in HeLa cells. EGF elicited a dose-dependent and time-dependent increase in PA activity with maximal induction occurring with 10 ng/ml EGF. PA activity was detected both in the culture medium and in the cell layer. Treatment of the cells with either actinomycin D or cycloheximide abolished the effect. When cells were treated with submaximal doses of PMA and EGF together, the level of PA production was equivalent to the sum of the activities measured separately. However, addition of PMA to cells treated with the optimal dose of EGF caused no additional increase in PA synthesis. Many laboratories have subsequently demonstrated that PMA and growth factors have many similar effects on cultured cells. However, the reasons for these similarities are not clear.

The coordinate synthesis of both a protease and a protease inhibitor may be a mechanism by which cellular proteolytic activity is finely regulated. Such a system was studied by EATON and BAKER (1983), who observed a coordinate increase in the secretion of both PA and a PA inhibitor, protease nexin, by cultured human foreskin fibroblasts treated with EGF, thrombin, or PMA. These three agents, each of which is a fibroblast mitogen, elicited a dose-dependent increase in the secretion of the inactive proenzyme form of PA. These concentrations of EGF, thrombin, or PMA also increased production of protease nexin, which inhibits both PA and plasmin. The addition of plasmin to conditioned medium containing the proenzyme form of PA activated the PA only if the protease nexin in the conditioned medium was inactivated prior to addition of the plasmin. Active PA could not be generated in the presence of active protease nexin. Therefore, the coordinate induction of both PA and protease nexin may serve to regulate and limit proteolysis by inhibiting PA and plasmin directly and by inhibiting plasmin activation of the proenyzme form of PA.

GROSS et al. (1983a) have studied the mechanism responsible for inactivation of EGF receptors during EGF-induced receptor downregulation. Previous work suggested that PA was responsible for degradation of the acetylcholine receptor in chicken myoblasts (HATZFIELD et al. 1982). GROSS et al. (1983a) investigated whether a similar mechanism was responsible for inactivation of the EGF receptor. Treatment of A431 cells with 50 n*M* EGF resulted in a time-dependent increase in PA production and a concomitant decrease in cell surface EGF-binding activity. Both PA production and the loss of EGF binding were dose responsive, with half-maximal stimulation of PA activity requiring 15–20 n*M* EGF and half-maximal inhibition of EGF binding occurring at 10–15 n*M* EGF. Removal of EGF from A431 cultures resulted in a decrease in PA activity and a restoration of approximately 80% of the original EGF-binding activity in 20 h. Inhibition of PA synthesis by cycloheximide and dexamethasone blocked both the EGF-stimulated increase in PA activity and the decrease in receptor binding. Treatment of A431 cells with the protease inhibitor leupeptin inhibited EGF-stimulated PA activity and prevented the decrease in EGF binding. The lysosomotropic agent chloroquine, on the other hand, had no effect on EGF-stimulated PA activity or EGF binding.

In order to determine whether receptor inactivation was due to PA itself or to plasmin, A431 cells were treated with EGF in plasminogen-free medium. Under these conditions EGF still elicited a dose-dependent increase in PA production and in EGF receptor inactivation. Subsequent experiments showed that plasmin activity had little effect on EGF binding. These results indicate that EGF receptor inactivation may occur through the direct action of PA.

STOPPELLI et al. (1986) investigated the regulation of PA production by EGF in two human tumor cell lines, A1251 cells, derived from a kidney carcinoma, and A431 cells, derived from an epidermoid carcinoma of the vulva. They compared the abilities of PMA and EGF to stimulate PA production in these cell lines. PMA (10 n*M*) stimulated PA production in both cell lines. However, EGF (50 n*M*) induced PA synthesis only in the A431 cells and not in A1251 cells. Maximal induction of PA production in A431 cells was obtained with 20 n*M* EGF and was detectable in cell lysates 1–2 h after treatment and in the medium 4–6 h after treatment. Although PMA and EGF are both mitogens for A1251 cells, neither agent stimulated DNA synthesis under the conditions used for PA induction.

The ability of these cells to respond to EGF correlated with the number of EGF receptors on the cell surface. A1251 cells were found to have three- to sixfold fewer EGF receptors than A431 cells, and A431 clone 18, which has tenfold fewer receptors than the parental cell line, does not produce PA in response to EGF treatment. It should be noted that A431 cells possess an unusually high number of EGF receptors, while A1251 and A431 clone 18 cells have receptor numbers more typical of normal cells.

The PA produced by the A431 cells was u-PA, determined by the inhibition of enzymatic activity with anti-u-PA IgG. Northern blot analysis showed that EGF increased u-PA mRNA levels two- to threefold in A431 cells but not in A431 clone 18 cells. Cycloheximide caused a superinduction of u-PA mRNA in A431 cells, suggesting that u-PA mRNA production may be controlled by a short-lived negative regulator. PDGF (10 ng/ml) treatment also increased PA mRNA levels in A1251 cells two- to threefold but had no effect on A431 cells. This result was unexpected because epithelial cells normally do not bind PDGF. The authors speculate that tumorigenesis may result in the production of proteins, including, perhaps, the PDGF receptor, not produced by the normal parent cell.

Human urothelial cell PA production is also regulated by EGF. DUBEAU et al. (1988) studied PA production in vitro by normal human urothelial cells and by transformed human urothelial cells isolated from transitional cell tumors. They found that normal urothelial cells in the log phase of growth produced both u-PA and t-PA, with t-PA being the predominant form of the enzyme. The tumor cells produced only u-PA.

When the normal cells reached confluence, t-PA activity fell to undectable levels, while u-PA activity was unchanged. However, exposure to EGF resulted in a rapid increase in t-PA levels in both the conditioned medium and in the cell extracts. The EGF-stimulated increase was apparent within 6 h of exposure and continued for at least 48 h. Both t-PA and u-PA mRNA levels increased concomitantly with t-PA enzymatic activity, suggesting that the increased activity was due to increased enzyme synthesis rather than to a decrease in the production of a PA inhibitor.

In contrast to the normal cells, the transformed urothelial cells did not display a density-dependent change in PA expression. Exposure of confluent transformed cells to EGF had no effect on t-PA mRNA levels and caused a slight decrease in u-PA mRNA levels.

In histological sections of normal urothelium, t-PA was confined to the mature urothelial cells close to the lumen of the bladder. The authors speculated that the behavior of these cells might be similar to that of confluent normal urothelial cells in vitro and that their production of t-PA may be stimulated by the high levels of EGF in urine. It was also noted that t-PA production may be a marker for mature urothelial cells, because neither immature urothelial cells located near the basement membrane nor tumor cells, which do not display mature differentiated characteristics, produce this form of the enzyme.

CHIANG and NILSEN-HAMILTON (1986) observed opposite effects of EGF and TGF-β on the expression of two proteins secreted by mouse 3T3 cells. EGF (10 ng/ml) stimulated incorporation of [^{35}S]methionine into the lysosomal thiol proteinase MEP, and into mitogen-related protein (MRP), a glycoprotein related to prolactin. TGF-β, on the other hand, reduced [^{35}S]methionine incorporation into both MRP and MEP in EGF-treated cells, as well as in untreated cells. Neither EGF nor TGF-β affected the overall level of [^{35}S]methionine incorporation in this cell line. EGF (10 ng/ml) stimulated DNA synthesis in 3T3 cells nearly threefold. TGF-β, alone or in combination with EGF, had no effect on DNA synthesis.

Growth factor induction of procollagenase in human neonatal skin fibroblast cells was studied by CHUA et al. (1985). Treatment of these cells with EGF (100 ng/ml), TGF-β (2.5 ng/ml), or PDGF (10 µg/ml) resulted in an increased level of [^{35}S]methionine incorporation into two secreted proteins of M_r 60000 and 55000. Immunoprecipitation with antibodies against human fibroblast collagenase removed two proteins of M_r 55000 and 43000 from the medium of untreated cells and four proteins of M_r 60000, 55000, 50000, and 43000 from the medium of EGF-treated fibroblasts. The M_r-60000 and -55000 proteins have been shown previously to be procollagenase forms. The M_r-50000 and -45000 proteins have been identified as collagenases. The M_r-43000 protein observed by CHUA et al. (1985) is probably also active collagenase. Immunoprecipitates from the EGF-treated cultures contained 3.5-times as much radioactivity as did immunoprecipitates from control cells. Likewise, collagenase activity in the conditioned medium from EGF-treated cells was 3.2 times higher than that from untreated cells. Type I collagen was cleaved by this activity to yield the characteristic TC^A and TC^B degradation products. The authors speculated that the collagenase released upon exposure to growth factors may play a role in mitogenesis, possibly by degrading the extracellular matrix, thus releasing cells from their substratum and from surrounding cells.

Two laboratories have examined the role of EGF in bone metabolism. Because bone matrix is composed largely of collagen, collagen-degrading enzymes have been presumed to initiate bone resorption. Mouse osteoblasts synthesize procollagenase when treated with parathyroid hormone, a known bone resorbing hormone. Because plasmin, generated from plasminogen by PA, can activate procollagenase, HAMILTON et al. (1984) characterized the production of PA, which could subsequently generate collagenolytic activity, by UMR 106-01 rat osteogenic sarcoma cells and osteoblast-rich calvarial cells treated with several hormones. Production of PA by UMR 106-01 cells was stimulated by EGF (1.7×10^{-9} *M*), parathyroid hormone (3×10^{-8} *M*), and PGE_2 (10^{-7} *M*), all of which stimulate bone resorption in organ culture. PGE_2 gave the largest stimulation, with PA activity increasing nearly four fold over control. PTH was nearly as effective as PGE2, while EGF increased PA levels approximately 50% over control. In addition, several vitamin D metabolites also stimulated PA production. Hormones which do not stimulate bone resorption, including glucagon, adrenocorticotropin, and insulin, did not stimulate PA production by UMR 106-01 cells. Treatment of osteoblast-rich calvarial cells with EGF at concentrations of 1.7×10^{-11} *M* to 1.7×10^{-9} *M* as well as PTH, PGE_2, and 1,25-dihydroxyvitamin D_3 also stimulated PA production.

The authors noted that each of the agents which induce PA activity in the UMR 106-01 cells works through a different mechanism. EGF activates the intrinsic tyrosine kinase activity of its cell surface receptor, PGE_2 and PTH stimulate adenylate cyclase activity, and dihydroxyvitamin D_3 acts as a steroid hormone. They speculated that all three signal transduction pathways share a common element which regulates PA expression. Conversely, it is possible that the PA promoter has sequences which respond to each of these agents.

CHIKUMA et al. (1984) also studied the effect of EGF on collagen-degrading enzymes in bone cells. They pointed out that dipeptidyl-aminopeptidase (DAP) and collagenase-like peptidase (CL-peptidase) activities are high beneath resorbing deciduous teeth and in developing granuloma tissue in rats, suggesting a role for these enzymes in bone metabolism. They examined the effect of EGF on expression of these enzymes in MC3T3-E1 cells, derived from newborn mouse calvaria. After 4 days of treatment, EGF elicited an increase in both DAP and CL-peptidase activity at concentrations of EGF ranging from 0.4 to 50 ng/ml for DAP and 2.0 to 50 ng/ml for CL-peptidase. These concentrations of EGF also caused a decrease in the hydroxyproline content of the cultures, while the total amount of protein per culture increased slightly in EGF-treated cells.

Taken together these two studies indicate that EGF may induce several different proteolytic enzymes which influence the collagenolysis associated with bone metabolism.

F. Interleukin-1

Interleukin-1 (IL-1) is a cytokine produced primarily by monocytes and macrophages which has multiple effects on various target cells. Originally, IL-1 was identified as a lymphocyte-activating factor and, accordingly, has been assayed by measuring its ability to stimulate thymocyte proliferation. IL-1 has also been found to stimulate protease production by several cell types, and this property probably plays a role in the pathophysiology of certain diseases. In this section we will discuss studies which indicate that IL-1 is involved in hemostasis and in several diseases in which tissue degradation is a feature, including basal cell carcinoma, bronchial carcinoma, glomerulonephritis, and arthritis.

I. Hemostasis

Disseminated intravascular coagulation is often associated with allergic or immunological reactions. Because IL-1 is produced by monocytes and macrophages which are involved in these reactions, BEVILACQUA and et al. (1984) hypothesized that IL-1 might stimulate procoagulant activity. They treated cultured human umbilical vein endothelial cells (HUVECs) with IL-1 (10 U/ml) and found that there was a tenfold increase in the levels of procoagulant activity as measured by decreased clotting times. This activity was found in cell extracts and on the surface of endothelial cells but was not detected in conditioned medium. IL-1 failed to induce procoagulant activity in either human skin fibroblasts or a human lymphoblastoid B-cell line, indicating that the induction may be specific for endothelial cells. By using sera from patients deficient in various clotting factors and using an antibody against apoprotein III, it was shown that the IL-1-induced procoagulant activity acts via the extrinsic coagulation pathway and was probably tissue factor.

A second mechanism by which IL-1 promotes coagulation is through the induction of the type 1 inhibitor of plasminogen activator. Several laboratories have reported that exposure to IL-1 results in an increase in PAI-1 synthesis by endothelial cells. EMEIS and KOOISTRA (1986) reported that exposure to IL-1 resulted in a time- and dose-dependent increase in PAI-1 in HUVEC. Injection of rats with lipopolysaccharide (LPS), as well as treatment of HUVEC with LPS in vitro, also caused an increase in the production of a PA inhibitor. Neither IL-1 nor LPS decreased t-PA levels, as judged by immunoassay. LPS and IL-1 were not effective in inducing PAI-1 in HepG2 cells or primary hepatocytes, suggesting a target specificity for endothelium.

NACHMAN et al. (1986) also reported that recombinant IL-1β caused a tenfold increase in secretion of PAI-1 by HUVEC. They detected no PA activity in either treated or untreated HUVEC, indicating that even basal levels of PA activity are masked by PAI-1. When t-PA antigen levels were measured by ELISA, no differences were detected in either IL-1-treated or untreated HUVEC.

The IL-1 induction of endothelial cell PAI-1 was also observed by BEVILACQUA et al. (1986). They pointed out that the induction of procoagulant activity and the induction of PAI-1 activity have different time courses. Procoagulant activity is maximally induced 6 h after IL-1 treatment and returns to basal levels within 24 h. PAI-1 activity, on the other hand, was only slightly higher than control levels at 6 h and continued to increase for at least 72 h after treatment. Unlike EMEIS and KOOISTRA (1986) and NACHMAN et al. (1986), they found that IL-1 treatment caused a dose-dependent decrease in t-PA secretion.

These reports indicate that IL-1 may contribute to a procoagulant state by inducing a substance resembling tissue factor, by inducing production of PAI-1 by endothelial cells, and, possibly, by decreasing t-PA levels.

II. Cancer

GOSLEN et al. (1985) hypothesized a role for IL-1 in tumor cell invasion based on their studies with basal cell carcinoma (BCC). They had observed previously that explant cultures of BCC produce collagenolytic activity and that the enzyme could be localized by immunocytochemistry to connective tissue elements within the tumor stroma rather than to tumor cells themselves (BAUER et al. 1977). In this study extracts of BCC were found to elicit a doubling in the production of procollagenase activity and immunoreactive collagenase by normal human skin fibroblasts in culture. Neither extracts of normal epidermal tissue nor extracts of stromal tissue were effective in inducing collagenase. Furthermore, histological examination of the tumors revealed less than 5% contamination with stromal or inflammatory cells. Because IL-1 had been reported to induce collagenase in other cells, these authors assayed the BCC extract for IL-1 activity in the thymocyte proliferation assay. Pooled BCC extracts from five patients caused a twofold increase in thymocyte activation and induced a seven- to tenfold increase in HSF collagenase activity. Partial characterization of the collagenase-inducing activity showed that it had an apparent molecular weight of 19000, similar to that of IL-1. These data, although not definitive, suggest the activity may be due to IL-1. Another possible candidate, however, is bFGF.

Similar results were obtained by DAYER et al. (1985b) studying mechanisms of lung tissue destruction and remodeling in human bronchial carcinoma. They isolated alveolar macrophages from the lungs of patients with either squamous carcinoma or adenocarcinoma and maintained them in culture for 48 h. Conditioned medium from these cultures stimulated the production of collagenase activity and PGE_2 by normal human skin fibroblasts. Because blood macrophages were known to produce IL-1, which has the same biological activity as the conditioned medium, the authors hypothesized that the alveolar macrophages also produced IL-1. Molecular sieve chromatography indicated that most of the PGE_2- and collagenase-stimulating activity eluted with an apparent molecular weight of 18000, similar to that of IL-1.

III. Glomerulonephritis

MARTIN et al. (1986) postulated a role for IL-1 in glomerulonephritis based on studies of mesangial cell neutral protease production. Macrophage-like cells are often seen in the glomeruli in experimental models of nephritis, where they are believed to contribute to increased cellular proliferation and destruction of the basement membrane. Because mesangial cells were known to secrete a neutral protease (NP) capable of degrading ECM components, MARTIN et al. (1986) investigated whether macrophage-derived products would stimulate the production of this NP activity which might cause destruction of the glomerular basement membrane.

They found that cultured mesangial cells secreted an NP with gelatinase activity. Coculture of rat mesangial cells with macrophages, which do not secrete appreciable amounts of NP, resulted in a three- to fivefold increase in NP activity, and the level of activity released correlated with the number of macrophages present. Macrophage-conditioned medium, which also stimulated mesangial cell NP production, was fractionated by molecular sieve chromatography. The active fraction was found to have an apparent molecular weight of 10000–20000 and stimulated [^{3}H]thymidine uptake in the thymocyte proliferation assay, suggesting that the protease-inducing factor might be IL-1. Pure rat and macrophage-derived IL-1 did, in fact, stimulate mesangial cell protease production in a dose-dependent manner.

IV. Arthritis

Rheumatoid arthritis is characterized by inflammation of the synovial membrane lining the joint and the subsequent formation of a pannus, a granulation tissue which proliferates to involve large areas of the synovium. The pannus is composed, in part, of synovial fibroblasts and blood-derived cells which secrete degradative enzymes. This results in the destruction of articular cartilage and disruption of normal joint anatomy and function.

Articular cartilage is composed primarily of collagen and proteoglycans, and collagenase and proteoglycanases are among the enzymes released by the pannus cells. The regulation of the production of proteases in arthritis is not completely understood, but there is evidence that IL-1 may induce the production of these enzymes in both synovial cells and in chondrocytes, which are constituents of the cartilage itself. Much of the work which will be discussed in this section used macrophage-conditioned medium as the source of the protease-inducing activity. However, homogeneous purified IL-1 and recombinant IL-1α and IL-1β are now available and induce the same biological activities as the macrophage-conditioned medium (DAYER et al. 1986).

Many laboratories have reported that rheumatoid synovium produces elevated levels of collagenase. DAYER et al. (1977, 1979, 1980) demonstrated that a factor produced by lymphocytes and macrophages stimulated the production of collagenase by rheumatoid synovial cells. This factor was partially purified from the conditioned medium of a murine macrophage cell line and found to be identical with IL-1 (MIZEL et al. 1981). MCCROSKERY et al. (1985) observed that treatment of human rheumatoid synovial fibroblasts with partially purified macrophage factor increased the incorporation of [^{35}S]methionine into several proteins, two of which could be specifically immunoprecipitated with antibodies against human skin fibroblast collagenase. The two proteins represented the glycosylated and non-glycosylated forms of the enzyme.

Several different proteases have been identified in conditioned medium from IL-1-treated chondrocytes or cartilage organ cultures, including collagenase, proteoglycanases, and PA. There is evidence that IL-1 is synthesized by both synovial cells and chondrocytes, which may result in autocrine stimulation of protease production.

TOWLE et al. (1987) have purified IL-1 from the conditioned medium of synovial cell cultures. This material, as well as human and murine recombinant IL-1, stimulated the production of a protease by cartilage organ cultures. The protease, which has been purified and migrates with an apparent M_r- 55000 of -57000 on SDS-PAGE, degrades proteoglycans and also activates procollagenase. This group has also shown that cartilage produces a factor with an activity identical to that of IL-1. In addition, RNA isolated from human chondrocytes was shown to contain both IL-1α and IL-1β mRNA using the S1 nuclease protection assay (OLLIVIERRE et al. 1986). As with synovial cells, treatment of cartilage cultures with synovial cell or cartilage-conditioned medium, as well as recombinant mouse IL-1, induced the collagenase-activating protein.

PASTERNAK et al. (1986, 1987) also observed a stimulation of the production of collagenolytic and proteoglycan-degrading enzymes in rabbit chondrocytes treated with either human IL-1 or P388D1 murine macrophage-conditioned medium. These two enzymatic activities had different apparent molecular weights by gel filtration chromatography, but both were inhibited by ethylenediaminetetraacetic acid (EDTA), phenanthroline, and α_2-macroglobulin but not by PMSF, tosyllysylchloromethylketone (TLCK), pepstatin, or α_1-antitrypsin. GOWEN et al. (1984) obtained virtually identical results using monocyte-conditioned medium and human articular chondrocytes. In addition, they showed that human osteoblasts do not produce collagenase in response to IL-1, suggesting that they do not contribute to IL-1-mediated cartilage destruction in arthritis.

STEPHENSON et al. (1987) also demonstrated that human recombinant IL-1β stimulated collagenase activity in human articular chondrocytes and increased the levels of procollagenase mRNA, suggesting the IL-1 regulation of collagenase activity occurred at a pretranslational step.

PHADKE (1987) noted that the effect of IL-1 could be potentiated by aFGF and bFGF. He observed that purified human IL-1 was only 30%–40% as effective as rabbit macrophage-conditioned medium in stimulating collagenase and neutral protease activities in rabbit articular chondrocytes. Therefore, he hypothesized that the conditioned medium contained a factor which augmented the activity of IL-1, and he tested the ability of several well-characterized growth factors to potentiate the protease-inducing activity of IL-1. PDGF and multiplication-stimulating activity were ineffective. However, both aFGF and bFGF augmented the IL-1 activity. Basic FGF potentiated the activity of IL-1 in a dose-dependent manner at concentrations ranging from 10 to 500 ng. Basic FGF alone was ineffective in stimulating chondrocyte protease activity.

Several reports have indicated that the collagenase found in the synovial fluid of arthritic joints is secreted in an inactive form. This implies that PA may be important in arthritis, because it generates plasmin which, in turn, can activate procollagenase. Several laboratories have reported the presence of PA in synovial fluid, and MOCHAN et al. (1986b) provided evidence that PA production is regulated by IL-1. They reported that partially purified human monocyte IL-1 caused a dose-dependent increase in the production of PA by cultured synovial cells from patients suffering from either rheumatoid arthritis or osteoarthritis. The induction of PA was rapid, with activity being detected within 3 h of treatment. Antibodies to IL-1 abolished this effect, and treatment of the cells with either actinomycin D or cycloheximide prevented IL-1 induction of PA. They also obtained evidence that the induction was mediated by PGE_2 and cAMP (MOCHAN et al. 1986a). Indomethacin (an inhibitor of prostaglandin synthesis) caused a dose-dependent inhibition of IL-1-induced PA activity, while theophylline (a phosphodiesterase inhibitor), forskolin (a stimulator of adenylase cyclase), and dibutyryl cAMP (a cAMP analog) potentiated the effect of IL-1. LEIZER et al. (1987) confirmed these data using purified IL-1β and recombinant IL-1α.

BUNNING et al. (1987) have provided evidence that PA is produced by chondrocytes. They found that human articular chondrocytes produced PA molecules with apparent molecular weights of 50000, 65000, and 100000. The M_r-50000 species was identified as u-PA, the M_r-65000 species as t-PA, and the M_r-100000 species was presumed to be an enzyme/inhibitor complex. Treatment of chondrocytes with human IL-1 enhanced the production of t-PA and the M_r-100000 band in both cell extracts and conditioned medium. These results are in contrast to those obtained by SCHNYDER et al. (1987), who found that human recombinant IL-1α and IL-1β caused a dramatic reduction in PA activity in rabbit chondrocytes. Both groups used purified IL-1 and measured PA by enzymatic assays. The finding by BUNNING et al. (1987) that IL-1 enhanced the production of a species possibly representing a PA/inhibitor complex suggests that the decrease in PA activity measured by SCHNYDER et al. (1987) may be due to induction of a PA inhibitor, perhaps PAI-1.

Regulation of stromelysin expression by growth factors, including IL-1, and other agents was studied by FRISCH and RULEY (1987). Stromelysin is a broad-spectrum metalloproteinase believed to be involved in the degradation of ECM. Because stromelysin activity is associated with conditions such as rheumatoid arthritis, wound healing, tumor invasion, and inflammation where growth factors and cytokines are involved, FRISCH and RULEY (1987) reasoned that these agents would be positive regulators of stromelysin gene expression, while antiinflammatory agents would be negative regulators. Using cultured rabbit synovial fibroblasts, they showed that IL-1, PMA, and EGF induced expression of stromelysin mRNA as measured by northern blot analysis. PMA (50 ng/ml) elicited a 120-fold increase in mRNA levels, while IL-1 (4 ng/ml) and EGF (50 ng/ml) resulted in 63-fold and 5-fold increases, respectively. Tumor necrosis factor (TNF), another inflammatory mediator, had no effect. Dexamethasone (2 μM) reduced both the basal and stimulated levels of stromelysin mRNA expression in these cells. Interestingly, isobutylmethylxanthine and cholera toxin, agents which increase intracellular levels of cAMP, also stimulated expression of stromelysin, suggesting that a cAMP-stimulated protein kinase may be involved in signal transduction.

To elucidate the mechanism by which stromelysin is induced, Frisch and Ruley studied the regulatory regions of the gene. A stromelysin genomic clone was isolated from a rabbit genomic library and the transcription start site was mapped. They then isolated DNA fragments of 610 base pairs and 700 base pairs, containing the transcription start site and upstream regions. These were fused upstream from a chloramphenicol acetyltransferase (CAT) coding sequence and transfected into rabbit synovial fibroblasts. The expression of CAT mRNA was low in untreated cells but was induced 20-fold by PMA, 8-fold by IL-1, and by EGF and cholera toxin. Furthermore, expression of this construct mimicked expression of the natural gene in that dexamethasone reduced the induction of CAT mRNA by each of these agents. Although the degree of induction of CAT mRNA was lower than that observed for stromelysin mRNA, these experiments localized the regulatory elements for this gene, making it possible to study growth factor induction of stromelysin in greater detail.

G. Tumor Necrosis Factor

For many years investigators have studied the ability of conditioned medium from activated macrophages specifically to kill tumor cells. The active factor was purified from macrophage-conditioned medium and called tumor necrosis factor (Aggarwal et al. 1985). Independently, other investigators studying weight loss in chronically infected or tumor-bearing mice isolated a factor from the serum of these animals which inhibited uptake and storage of triglyceride by adipocytes (Beutler et al. 1985b). This factor was termed cachectin and was subsequently found to be identical to TNF (Beutler et al. 1985a).

Two laboratories have reported that TNF stimulates the production of collagenase. Dyer et al. (1985a) investigated whether TNF might induce collagenase and PGE_2 production in fibroblasts. They reasoned that the association of TNF with conditions such as chronic infection and septic shock and its production by activated macrophages made it a candidate for an inflammatory mediator. Because another cytokine associated with inflammation, IL-1, induces collagenase and PGE_2 in synovial cells and skin fibroblasts, a similar role was postulated for TNF.

Indeed, treatment of human synovial fibroblasts with TNF (0.3–3 n*M*) resulted in an increase in the production of both collagenase and PGE_2. The amounts of collagenase and PGE_2 produced are comparable to those induced by IL-1. Interferon-α was also tested and found to be ineffective in stimulating collagenase activity and only slightly effective in stimulating PGE_2 production. Similar results were obtained when human skin fibroblasts were treated with TNF. TNF elicited an increase in PGE_2 production at concentrations ranging from 0.3 to 3 n*M* but elicited collagenase activity only at 3 n*M*. TNF proved to be somewhat less effective than IL-1 in skin fibroblasts. The ability of TNF to induce protease synthesis suggests that this cytokine may have more diverse functions than previously recognized.

NAKAGAWA et al. (1987) investigated the ability of TNF to stimulate collagenase synthesis in inflammatory conditions. They had previously shown that collagen has a high turnover rate in granulation tissue and, therefore, they postulated that a cytokine associated with inflammatory conditions might induce collagenase synthesis (NAKAGAWA et al. 1981). Using rat carrageenin-induced granulation tissue, they looked for TNF-induced collagenase and gelatinase activities in organ culture. The organ culture medium contained both collagenase and gelatinase activities on days 3 and 4 of culture. If the cultures were supplemented with TNF at concentrations of 0.1–100 units/ml, the levels of these activities were increased. However, the authors could not prove that TNF directly stimulated production of these proteases and suggested that alternatively TNF may be stimulating collagenase and gelatinase synthesis indirectly, perhaps through a pathway involving IL-1.

H. Colony-Stimulating Factor 1

Colony-stimulating factor 1 (CSF-1), also called murine macrophage colony-stimulating factor, promotes the growth and differentiation of macrophages from bone marrow precursor cells. LIN and GORDON (1979) have shown that CSF-1 stimulates the production of PA by mouse bone marrow-derived macrophages (BMDMs). BMDMs were cultured in L-cell-conditioned medium, which contains CSF-1. Cells exposed continuously to L-cell-conditioned medium produced twice as much PA activity as did cells maintained for 3 days in medium lacking L-cell conditioning. Interestingly, if the conditioned medium was withdrawn for 2 days, then added back for 1 day, the cells produced four times as much PA activity as did cells cultured without L-cell-conditioned medium for 3 days.

To determine whether CSF-1 was responsible for stimulating PA activity, BMDMs were exposed to L-cell-conditioned medium or purified CSF-1 for 2.5 or 4 h. Treatment with both preparations resulted in a dose-dependent increase in secreted PA activity within 2.5 h. Thioglycollate-elicited macrophages, but not resident peritoneal macrophages, also produced increased amounts of PA activity when exposed to L-cell-conditioned medium.

HUME and GORDON (1984) studied the correlation between CSF-1-stimulated DNA synthesis and PA production in mouse BMDMs. Conditioned medium from mouse L cells promoted [^{3}H]thymidine uptake in BMDMs and stimulated the production of PA activity. The induced PA activity was determined to be u-PA and was found to be present in increased amounts in both the conditioned medium and the cell layer. Dexamethasone and PGE_1 inhibited the effect by causing a dose-dependent decrease in CSF-1 stimulation of DNA synthesis and PA production.

Having demonstrated a correlation between CSF-1-induced DNA synthesis and PA production, this group attempted to determine whether PA activity was necessary for cell division. At the time this study was performed, specific inhibitors of PA were not widely available. Therefore, they examined whether the product of PA activity, plasmin, was needed for DNA synthesis. BMDMs were cultured in medium containing plasminogen-depleted serum, which will not support the generation of plasmin. CSF-1-treated cells incorporated [^{3}H]thymidine at levels equivalent to cells cultured in plasminogen-containing serum, indicating that the generation of plasmin from plasminogen was not required. Furthermore, in order to rule out the possibility that small amounts of cell-associated plasminogen might be converted to plasmin, the plasmin inhibitors trasylol and soybean trypsin inhibitor were included in the culture medium but had no effect on DNA synthesis. However, the possibility remains that PA may be acting directly upon a substrate other than plasminogen. What relationship, if any, exists between PA production and DNA synthesis remains to be elucidated.

I. Discussion

The studies described in this chapter clearly indicate that the induction of proteases by growth factors is widespread with virtually every known growth factor capable of enhancing the synthesis of specific proteases in specific cell types. The physiological significance of this relationship is not clear and may, in fact, differ depending upon which cell type and growth factor are considered.

SPORN and ROBERTS (1988) have recently pointed out that growth factors may have functions independent of their mitogenic activities. Some of these may involve cellular activities which require require protease synthesis. For example, the production of PA by endothelial cells may relate to the antithrombogenic nature of these cells. Therefore, the stimulation of PA synthesis in endothelial cells by endogenous bFGF (SATO and RIFKIN 1988) may be one of the mechanisms regulating fibrinolysis. Likewise, other migratory cells or cells which are engaged in tissue remodeling may have their proteolytic balance determined by specific growth factors or inflammatory mediators.

Another explanation for the observed induction of proteases by growth factors may be a requirement for proteolytic activity for a specific step in cell division. In addition to the correlation of protease induction following exposure to growth factors, several other observations indicate a causal relationship between growth and protease production. For example, several genes whose rapid transcription is induced after growth factor exposure have been shown to code for proteases or protease inhibitors. Also, the restriction of protease production to specific phases of the cell cycle may demonstrate a specific requirement of proteolytic activity during cell division.

The major remaining question is what role do proteases fulfill in cell division. There is no conclusive evidence that proteases directly stimulate cell division as had been proposed earlier. It is more likely that proteases may be necessary for the completion of cell division. An attractive proposal is that the stimulation of protease production by growth factors reflects a requirement for limited degradation of the ECM and/or cell-ECM contacts for division to take place. This implies that the shape change which occurs when cells go through mitosis requires the degradation of specific adhesion points. This proteolysis must be localized since general proteolysis would be destructive to tissue cytoarchitecture. The binding and activation of proteases on the plasma membrane achieves this as does the large excess of proteolytic inhibitors which are usually produced by stimulated cells.

In addition to producing increased amounts of growth factors, tumor cells and transformed cells are often characterized by increased levels of protease production. It is appealing to relate the heightened levels of protease synthesis in these cells to the constitutive production of growth factors. Cell division in normal cells may be controlled by the temporal modulation of growth factor levels.

The demonstration of a causal role for proteases in cell division and a requirement for their regulation by growth factors will undoubtedly require the construction of specifically genetically engineered cells or animals in which the production of proteases, their inhibitors, and growth factors can be experimentally manipulated. The analysis of the phenotypes of these cells or organisms as a function of specific gene expression and action will define these potential interrelationships.

References

Abraham JA, Mergia A, Whang JL, Tumolo A, Friedman J, Hjerrild KA, Gospodarovicz D, Fiddes JC (1986) Nucleotide sequence of a bovine clone encoding the angiogenic protein, basic fibroblast growth factor. Science 233:545–548

Aggarwal BB, Kohr WJ, Hass PE, Moffat B, Spencer SA, Henzel WJ, Bringman TS, Nedwin GE, Goeddel DV, Harkins RN (1985) Human tumor necrosis factor. Production, purification, and characterization. J Biol Chem 260:2345–2354

Bauer EA, Gordon JM, Reddick ME, Eisen AZ (1977) Quantitation and immunocytochemical localization of human skin collagenase in basal cell carcinoma. J Invest Dermatol 69:363–367

Bauer EA, Cooper TW, Huang JS, Altman J, Deuel TF (1985) Stimulation of in vitro human skin collagenase expression by platelet-derived growth factor. Proc Natl Acad Sci USA 82:4132–4136

Beutler B, Greenwald D, Hulmes JD, Chang M, Pan Y-CE, Mathison J, Ulevitch R, Cerami A (1985a) Identity of tumour necrosis factor and the macrophage-secreted factor cachectin. Nature 316:552–554

Beutler B, Mahoney J, Le Trang N, Pekala P, Cerami A (1985b) Purification of cachectin, a lipoprotein lipase-suppressing hormone secreted by endotoxin-induced RAW 264.7 cells. J Exp Med 161:984–995

Bevilacqua MP, Pober JS, Majeau GR, Cotran RS, Gimbrone MA Jr (1984) Interleukin 1 (IL-1) induces biosynthesis and cell surface expression of procoagulant activity in human vascular endothelial cells. J Exp Med 160:618–623

Bevilaqua MP, Schleef RR, Gimbrone MA Jr, Loskutoff DJ (1986) Regulation of the fibrinolytic system of cultured human vascular endothelium by interleukin 1. J Clin Invest 78:587–591

Bunning RAD, Crawford A, Richardson HJ, Opdenakker G, Van Damme J, Russell RGG (1987) Interleukin 1 preferentially stimulates the production of tissue-type plasminogen activator by human articular chondrocytes. Biochim Biophys Acta 924:473–482

Chiang C-P, Nilsen-Hamilton M (1986) Opposite and selective effects of epidermal growth factor and human platelet transforming growth factor-β on the production of secreted proteins by murine 3T3 cells and human fibroblasts. J Biol Chem 261:10478–10481

Chikuma T, Kato T, Hiramatsu M, Kanayama S, Kumegawa M (1984) Effect of epidermal growth factor on dipeptidyl-aminopeptidase and collagenase-like peptidase activities in cloned osteoblastic cells. J Biochem 95:283–286

Chua CC, Geiman DE, Keller GH, Ladda RL (1985) Induction of collagenase secretion in human fibroblast cultures by growth promoting factors. J Biol Chem 260:5213–5216

Dayer J-M, Russell RGG, Krane SM (1977) Collagenase production by rheumatoid synovial cells: stimulation by a human lymphocyte factor. Science 195:181–183

Dayer J-M, Breard J, Chess L, Krane SM (1979) Participation of monocyte-macrophages and lymphocytes in the production of a factor that stimulates collagenase and prostaglandin release by rheumatoid synovial cells. J Clin Invest 64:1386–1392

Dayer J-M, Passwell JH, Schneeberger EE, Krane SM (1980) Interactions among rheumatoid synovial cells and monocyte-macrophages: production of collagenase-stimulating factor by human monocytes exposed to concanavalin a or immunoglobulin Fc fragments. J Immunol 124:1712–1720

Dayer J-M, Beutler B, Cerami A (1985a) Cachectin/tumor necrosis factor stimulates collagenase and prostaglandin E_2 production by human synovial cells and dermal fibroblasts. J Exp Med 162:2163–2168

Dayer J-M, Sundstrom L, Polla BS, Junod AF (1985b) Cultured human alveolar macrophages from smokers with lung cancer: resolution of factors that stimulate fibroblast proliferation, production of collagenase, prostaglandin E_2. J Leuk Biol 37:641–649

Dayer J-M, de Rochemonteix B, Burrus B, Demczuk S, Dinarello CA (1986) Human recombinant interleukin 1 stimulates collagenase and prostaglandin E2 production by human synovial cells. J Clin Invest 77:645–648

Delli-Bovi P, Curatola AM, Newman KM, Sato Y, Moscatelli D, Hewick RM, Rifkin DB, Basilico C (1988) Processing, secretion, and biological properties of a novel growth factor of the fibroblast growth factor family with oncogenic potential. Mol Cell Biol 8:2933–2941

Dubeau L, Jones PA, Rideout WM, Laug WE (1988) Differential regulation of plasminogen activators by epidermal growth factor in normal and neoplastic human urothelium. Cancer Res 48:5552–5556

Eaton DL, Baker JB (1983) Phorbol ester and mitogens stimulate human fibroblast secretions of plasmin-activatable plasminogen activator and protease nexin, an antiactivator/antiplasmin. J Cell Biol 97:323–328

Edwards DR, Murphy G, Reynolds JJ, Whitham SE, Docherty JP, Angel P, Heath JK (1987) Transforming growth factor β modulates the expression of collagenase and metalloproteinase inhibitor. EMBO J 6:1899–1904

Emeis JJ, Kooistra T (1986) Interleukin 1 and lipopolysaccharide induce an inhibitor of tissue-type plasminogen activator in vivo and in cultured endothelial cells. J Exp Med 163:1260–1266

Esch F, Baird A, Ling N, Ueno N, Hill F, Denoroy L, Klepper R, Gospodarowicz D, Bohlen P, Guillemin R (1985) Primary structure of bovine pituitary basic fibroblast growth factor (FGF) and comparison with the amino-terminal sequence of bovine brain acidic FGF. Proc Natl Acad Sci USA 82:6507–6511
Frick KK, Doherty PJ, Gottesman MM, Scher CD (1985) Regulation of the transcript for a lysosomal protein: evidence for a gene program modified by platelet-derived growth factor. Mol Cell Biol 5:2582–2589
Frisch SM, Ruley E (1987) Transcription from the stromelysin promoter is induced by interleukin-1 and repressed by dexamethasone. J Biol Chem 262:16 300–16 304
Gal S, Gottesman MM (1986) The major excreted protein of transformed fibroblasts is an activatable acid-protease. J Biol Chem 261:1760–1765
Goslen JB, Eisen AZ, Bauer EA (1985) Stimulation of skin fibroblast collagenase production by a cytokine derived from basal cell carcinomas. J Clin Invest 85:161–164
Gottesman MM (1978) Transformation-dependent secretion of a low molecular weight protein by murine fibroblasts. Proc Natl Acad Sci USA 75:2767–2771
Gottesman MM, Sobel ME (1980) Tumor promoters and Kirsten sarcoma virus increase synthesis of a secreted glycoprotein by regulating levels of translatable mRNA. Cell 19:449–455
Gowen M, Wood DD, Ihrie EJ, Meats JE, Russell GG (1984) Stimulation by human interleukin 1 of cartilage breakdown and production of collagenase and proteoglycanase by human chondrocytes but not by human osteoblasts in vitro. Biochim Biophys Acta 797:186–193
Granelli-Piperno A, Reich E (1978) A study of proteases and protease inhibitor complexes in biological fluids. J Exp Med 148:223–234
Gross JL, Krupp MN, Rifkin DB, Lane MD (1983a) Down-regulation of epidermal growth factor receptor correlates with plasminogen activator activity in human A431 epidermoid carcinoma cells. Proc Natl Acad Sci USA 80:2276–2280
Gross JL, Moscatelli D, Rifkin DB (1983b) Increased capillary endothelial cell protease activity in response to angiogenic stimuli in vitro. Proc Natl Acad Sci USA 80:2623–2627
Hamilton JA, Lingelbach SR, Partridge NC, Martin TJ (1984) Stimulation of plasminogen activator in osteoblast-like cells by bone-resorbing hormones. Biochem Biophys Res Commun 122:230–236
Hatzfield J, Miskin R, Reich E (1982) Acetylcholine receptor: effects of proteolysis on receptor metabolism. J Cell Biol 92:176–182
Hume DA, Gordon S (1984) The correlation between plasminogen activator activity and thymidine incorporation in mouse bone marrow-derived macrophages. Exp Cell Res 150:347–355
Jaye M, Howk R, Burgess W, Ricca GA, Chiu I-M, Ravera MW, O'Brien S, Modi WS, Maciag T, Drohan WN (1986) Human endothelial cell growth factor: cloning, nucleotide sequence, and chromosome localization. Science 233:541–544
Klagsbrun M, Shing Y (1985) Heparin affinity of anionic and cationic capillary endothelial cell growth factors: analysis of hypothalamus-derived growth factors and fibroblast growth factors. Proc Natl Acad Sci USA 82:805–809
Laiho M, Saksela O, Andreasen PA, Keski-Oja J (1986a) Enhanced production and extracellular deposition of the endothelial-type plasminogen activator inhibitor in cultured human lung fibroblasts by transforming growth factor-β. J Cell Biol 103:2403–2410
Laiho M, Salsela O, Keski-Oja J (1986b) Transforming growth factor β alters plasminogen activator activity in human skin fibroblasts. Exp Cell Res 164:399–407
Laiho M, Saksela O, Keski-Oja J (1987) Transforming growth factor β induction of type-1 plasminogen activator inhibitor. J Biol Chem 262:17 467–17 474
Lee L-S, Weinstein IB (1978) Epidermal growth factor, like phorbol esters, induces plasminogen activator in HeLa cells. Nature 274:696–697

Leizer T, Clarris BJ, Ash PE, van Damme J, Saklatvala J, Hamilton JA (1987) Interleukin-1β and interleukin-1α stimulate the plasminogen activator activity and prostaglandin E_2 levels of human synovial cells. Arthritis Rheum 30:562–566

Lin H-S, Gordon S (1979) Secretion of plasminogen activator by bone marrow-derived mononuclear phagocytes and its enhancement by colony-stimulating factor. J Exp Med 150:231–245

Loskutoff DJ, van Mourik JA, Erickson LA, Lawrence D (1983) Detection of an unusually stable fibrinolytic inhibitor produced by bovine endothelial cells. Proc Natl Acad Sci USA 80:2956–2960

Lund LR, Riccio A, Andreasen PA, Nielsen LS, Kristensen P, Laiho M, Saksela O, Blasi F, Dano K (1987) Transforming growth factor-β is a strong and fast acting positive regulator of the level of type-1 plasminogen activator inhibitor mRNA in WI-38 human lung fibroblasts. EMBO J 6:1281–1286

Martin J, Lovett DH, Diethard G, Sterzel RB, Davies M (1986) Enhancement of glomerular mesangial cell neutral proteinase secretion by macrophages: role of interleukin 1. J Immunol 137:525–529

Mason RW, Gal S, Gottesman MM (1987) The identification of the major excreted protein (MEP) from a transformed mouse fibroblast line as a catalytically active precursor form of cathepsin L. Biochem J 248:449–454

McCroskery PA, Arai S, Amento EP, Krane SM (1985) Stimulation of procollagenase synthesis in human rheumatoid synovial fibroblasts by mononuclear cell factor/interleukin 1. FEBS Lett 191:7–12

Mergia A, Eddy R, Abraham JA, Fiddes JC, Shows TB (1986) The genes for basic and acidic fibroblast growth factors are on different human chromosomes. Biochem Biophys Res Commun 138:644–651

Mira-y-Lopez R, Reich E, Ossowski L (1983) Modulation of plasminogen activator in rodent mammary tumors by hormones and other effectors. Cancer Res 43:5467–5477

Mira-y-Lopez R, Joseph-Silverstein J, Rifkin DB, Ossowski L (1986) Identification of a pituitary factor responsible for enhancement of plasminogen activator activity in breast tumor cells. Proc Natl Acad Sci USA 83:7780–7784

Mizel SB, Dayer J-M, Krane SM, Mergenhagen SE (1981) Stimulation of rheumatoid synovial cell collagenase and prostaglandin production by partially purified lymphocyte-activating factor (interleukin 1). Proc Natl Acad Sci USA 78:2474–2477

Mochan E, Uhl J, Newton R (1986a) Evidence that interleukin-1 induction of synovial cell plasminogen activator is mediated via prostaglandin E_2 and cAMP. Arthritis Rheum 29:1078–1084

Mochan E, Uhl J, Newton R (1986b) Interleukin 1 stimulation of synovial cell plasminogen activator production. J Rheumatol 13:15–19

Moscatelli D, Rifkin DB (1988) Membrane and matrix localization of proteinases: a common theme in tumor cell invasion and angiogensis. Biochim Biophys Acta 948:67–86

Moscatelli D, Presta M, Joseph-Silverstein J, Rifkin DB (1986a) Both normal and tumor cells produce basic fibroblast growth factor. J Cell Physiol 129:273–276

Moscatelli D, Presta M, Rifkin DB (1986b) Purification of a factor from human placenta that stimulates capillary endothelial cell protease production, DNA synthesis, and migration. Proc Natl Acad Sci USA 83:2091–2095

Murphy G, Reynolds JJ, Werb Z (1985) Biosynthesis of tissue inhibitor of metalloproteinases by human fibroblasts in culture. J Biol Chem 260:3079–3083

Nachman RL, Hajjar KA, Silverstein RL, Dinarello CA (1986) Interleukin 1 induces endothelial cell synthesis of plasminogen activator inhibitor. J Exp Med 163: 1595–1600

Nakagawa H, Isaji M, Hayashi M, Tsurufuji S (1981) Selective inhibition of collagen breakdown by protease inhibitors in granulation tissue in rats. J Biochem 89: 1081–1090

Nakagawa H, Kitagawa H, Aikawa Y (1987) Tumor necrosis factor stimulates gelatinase and collagenase production by granulation tissue in culture. Biochim Biophys Acta 142:791–797

Nilsen-Hamilton M, Hamilton RT, Allen R, Massoglia SL (1981) Stimulation of the release of two glycoproteins from mouse 3T3 cells by growth factors and by agents that increase intralysosomal pH. Biochem Biophys Res Commun 101:411–417

Ollivierre F, Gubler U, Towle CA, Laurencin C, Treadwell BV (1986) Expression of IL-1 genes in human and bovine chondrocytes: a mechanism for autocrine control of cartilage matrix degradation. Biochem Biophys Res Commun 141:904–911

Pasternak RD, Hubbs SJ, Caccese RG, Marks RL, Conaty JM, DiPasquale G (1986) Interleukin-1 stimulates the secretion of proteoglycan- and collagen-degrading proteases by rabbit articular chondrocytes. Clin Immunol Immunopathol 41:351–367

Pasternak RD, Hubbs SJ, Caccese RG, Marks RL, Conaty JM, DiPasquale G (1987) Interleukin-1 induces chondrocyte protease production: the development of collagenase inhibitors. Agents Actions 21:328–330

Phadke K (1987) Fibroblast growth factor enhances the interleukin-1-mediated chondrocytic protease release. Biochem Biophys Res Commun 142:448–453

Presta M, Mignatti P, Mullins DE, Moscatelli DA (1985) Human placental tissue stimulates bovine capillary endothelial cell growth, migration and protease production. Biosci Rep 5:783–790

Rabin MS, Doherty PJ, Gottesman MM (1986) The tumor promoter phorbol 12-myristate 13-acetate induces a program of altered gene expression similar to that induced by platelet-derived growth factor and transforming oncogenes. Proc Natl Acad Sci USA 83:357–360

Saksela O, Moscatelli D, Rifkin DB (1987) The opposing effects of basic fibroblast growth factor and transforming growth factor beta on the regulation of plasminogen activator activity in capillary endothelial cells. J Cell Biol 105:957–963

Sato Y, Rifkin DB (1988) Autocrine activities of basic fibroblast growth factor: regulation of endothelial cell movement, plasminogen activator synthesis, and DNA synthesis. J Cell Biol 107:1199–1206

Schnyder J, Payne T, Dinarello CA (1987) Human monocyte or recombinant interleukin 1s are specific for the secretion of a metalloproteinase from chondrocytes. J Immunol 138:496–503

Schreiber AB, Kenney J, Kowalski WJ, Friesel R, Mehlman T, Maciag T (1985) Interaction of endothelial cell growth factor with heparin: characterization by receptor and antibody recognition. Proc Natl Acad Sci USA 82:6138–6142

Sommer A, Brewer MT, Thompson RC, Moscatelli D, Presta M, Rifkin DB (1987) A form of human basic fibroblast growth factor with an extended amino terminus. Biochem Biophys Res Commun 144:543–550

Sporn MB, Roberts AB (1988) Peptide growth factors are multifunctional. Nature 332:217–219

Stephenson ML, Goldring MB, Birkhead JR, Krane SM, Rahmsdorf HJ, Angel P (1987) Stimulation of procollagenase synthesis parallels increases in cellular procollagenase mRNA in human articular chondrocytes exposed to recombinant interleukin 1β or phorbol ester. Biochem Biophys Res Commun 144:583–590

Stopelli MP, Verde P, Grimaldi G, Locatelli E, Blasi F (1986) Increase in urokinase plasminogen activator mRNA synthesis in human carcinoma cells is a primary effect of the potent tumor promoter, phorbol myristate acetate. J Cell Biol 102: 1235–1241

Thalacker FW, Nilsen-Hamilton M (1987) Specific induction of secreted proteins by transforming growth factor-β and 12-*O*-tetradecanoylphorbol-13-acetate. J Biol Chem 262:2283–2290

Thomas KA, Rios-Candelore M, Fitzpatrick S (1984) Purification and characterization of acidic fibroblast growth factor from bovine brain. Proc Acad Natl Sci USA 81:357–361

Towle CA, Trice ME, Ollivierre F, Awbrey BJ, Treadwell BV (1987) Regulation of cartilage remodeling by IL-1: evidence for autocrine synthesis of IL-1 by chondrocytes. J Rheumatol 14:11–13

Weinstein IB, Wigler M (1977) Cell culture studies provide new information on tumour promoters. Nature 270:659–660

Werb Z, Mainardi CL, Vater CA, Harris ED (1977) Endogenous activation of latent collagenase by rheumatoid synovial cells. N Engl J Med 296:1017–1023

Wigler M, Weinstein IB (1976) Tumour promoter induces plasminogen activator. Nature 259:232–233

Wun T-C, Ossowski L, Reich E (1982) A proenzyme form of human urokinase. J Biol Chem 257:7262–7268

CHAPTER 36

Inflammation and Repair

H. L. WONG and S. M. WAHL

A. Introduction

Injuries to tissue, whether due to mechanical, chemical, immunological, or thermal insults, initiate an orderly but complex series of cellular and biochemical interactions that lead to the formation of new tissue and the eventual repair of the wound. The cellular components of this pathway include hematopoietic cells such as platelets, polymorphonuclear leukocytes, lymphocytes, and monocytes, and mesenchymal cells such as fibroblasts and endothelial cells. These cells migrate to the site of tissue damage in a sequence determined by soluble factors that are released at the site of injury through a wide variety of mechanisms. Among these are tissue breakdown, blood coagulation, and cellular release. Each cell type has a specific function to perform such as degradation and resorption of damaged tissue, protection against infection, or deposition of new extracellular matrix.

Only the presence of the macrophage appears to be required throughout the tissue repair process. This requirement has been demonstrated by a number of studies in which depletion of other inflammatory cells had no appreciable effect on the primary tissue repair process. On the other hand, impairment of macrophage recruitment and function through a combination of ablative methods caused marked reduction in wound debridement and a delay in the onset of fibrosis during tissue repair (LEIBOVICH and ROSS 1975). Furthermore, attenuation of the inflammatory response, a macrophage-dependent process which precedes tissue repair, also retards the healing process (SANDBERG 1964). This chapter will deal with the roles that cellular components play in the inflammatory response and tissue repair. As we proceed, it will become apparent that these cellular components release a variety of soluble mediators that perform a multitude of functions leading to tissue and matrix biosynthesis (see Table 1).

The evolution and progression of tissue repair can be divided into three phases: inflammatory, proliferative, and remodeling. It is important to note that the events associated with each phase are not always separable from one another and, indeed, each is involved in the regulation of events that both precede and follow.

Table 1. Mediators of inflammation and tissue repair

Platelet-derived factors

- Polypeptide growth factors
  - Platelet-derived growth factor
  - Transforming growth factor-β
  - Basic fibroblast growth factor
  - Platelet factor 4
- Coagulation factors
  - Plasminogen
  - von Willebrand factor
  - Fibrinogen
- Vasoactive factors
  - Serotonin
  - Adenosine di- and triphosphates
- Neutral proteases and other enzymes
- Arachidonic acid metabolites
  - 12-HPETE
  - 12-HETE
  - 12,20-diHETE
  - TXA_2
- Adhesive and matrix proteins
  - Fibronectin

Neutrophil-derived factors

- Reactive oxygen intermediates
  - Superoxide
  - Hydrogen peroxide
  - Hydroxyl radical
- Adhesive and matrix proteins
- Proteoglycans
- Neutral proteases and other enzymes
- Bioactive lipids
  - Platelet-activating factor
  - Arachidonic acid metabolites

Fibroblast-derived factors

- Polypeptide growth factors
  - Interleukin-1
  - Interleukin-6
  - Platelet-derived growth factor
  - Granulocyte/macrophage colony-stimulating factor
- Adhesive and matrix proteins
  - Collagens
  - Fibronectin
  - Elastin
- Proteoglycans
  - Hyaluronic acid
  - Chondroitin sulfate
  - Dermatan sulfate
  - Heparin and heparan sulfate
  - Keratan sulfate
- Neutral proteases
- Bioactive lipids

Macrophage-derived factors

- Polypeptide growth factors
  - Platelet-derived growth factor
  - Transforming growth factor-α
  - Transforming growth factor-β
  - Interleukin-1
  - Interleukin-6
  - Interleukin-8
  - Tumor necrosis factor-α
  - Interferon-α
  - Interferon-β
  - Granulocyte/macrophage colony-stimulating factor
  - Basic fibroblast growth factor
  - Fibroblast-activating factor(s)
  - Factor-inducing monocytopoiesis
- Bioactive lipids
  - Arachidonic acid metabolites
  - Platelet-activating factor
- Neutral proteases and other enzymes
- Adhesive and matrix proteins
- Proteoglycans
- Reactive oxygen intermediates
  - Superoxide
  - Hydrogen peroxide
  - Hydroxyl radical

T-lymphocyte-derived factors

- Polypeptide growth factors
  - Interleukin-2
  - Interleukin-3
  - Interleukin-6
  - Granulocyte/macrophage colony-stimulating factor
  - Fibroblast-activating factor(s)
  - Interferon-γ
  - Lymphocyte-derived chemotactic factor for fibroblasts

Endothelial cell-derived factors

- Polypeptide growth factors
  - Interleukin-1
  - Interleukin-6
  - Tumor necrosis factor-α
  - Platelet-derived growth factor
  - Granulocyte/macrophage colony-stimulating factor
  - Basic fibroblast growth factor
- Adhesive and matrix proteins
- Proteoglycans
- Coagulation factors
  - von Willebrand factor
  - Factors V, VII, IX, and X
  - Tissue factor

B. Inflammatory Phase: Inflammatory Cell Recruitment and Function

A classical inflammatory response begins immediately after injury has occurred. Generally, inflammation is a localized reaction which is protective in nature. Whether initiated by damage to the vascular endothelium or deposition of immunologically active substances, its function is to eliminate or isolate injured tissue or pathogenic agents such as bacteria and other foreign bodies.

Inflammatory cells provide the physiological groundwork on which fibroblasts and endothelial cells restore damaged tissue. Their appearance at the injury site follows a sequence determined by a host of soluble mediators. Platelets appear initially and are followed by neutrophils and finally by macrophages and lymphocytes. Each cell type that enters the wound site is capable of interactions with cell types that precede them and also secretes soluble factors which contribute to the recruitment of each successive cell type.

I. Platelets

Platelets begin to adhere to the vascular endothelium in response to a defect on the endothelial surface or through the activation of the alternate pathway of the complement cascade. In the case of vascular injury, adhesion occurs through two different mechanisms, one involving direct interactions with polymerized collagen fibrils and the other involving glycoprotein-Ib receptors for von Willebrand factor (DE GROOT et al. 1988). During the adhesion process, activated platelets begin to form aggregates. The blood coagulation process is initiated through the interaction of the platelet surface membrane with various components of the plasma coagulation pathway (e.g., fibrinogen, factors V a, VII, IX, and X). The activation of these factors causes a 1000-fold increase in the rate of thrombin formation and, thus, the formation of blood clots. If bacterial antigens are present, studies suggest that vascular endothelial cells provide an alternate source of thrombin activation (STERN et al. 1985; NAWROTH and STERN 1986). Bacterial products or the inflammatory mediators induced by these products cause changes on the endothelial cell surface similar to those seen on activated platelets which lead to thrombin formation. Coagulation is initiated when platelets are activated indirectly through the interaction with endothelial cell-derived thrombin. Several components of the coagulation process such as kallikrein (KAPLAN et al. 1972; SCHAPIRA et al. 1982), fragments of both fibrin and Hagemann factor (GRAHAM et al.1965), and plasminogen activator (KAPLAN et al. 1973) have been reported to be chemotactic for leukocytes in vitro. Whether they exert the same physiological effects in vivo has yet to be determined.

Platelets contribute a host of known chemotactic factors which include vasoactive substances, lipids, growth factors, or enzymes that cleave complement components into active polypeptide fragments. Serum derived from plasma, which had been clotted in the presence of platelets, was strongly chemotactic for neutrophils and macrophages (HIRSCH 1960). Sources of the chemotactic activity included the complement polypeptide C5a, a product cleaved from complement component C5 through the action of neutral proteases secreted by activated platelets (for review, see WEKSLER 1988). Subsequently, C5a is cleaved to C5a-des-arg, which is a more potent chemoattractant for monocytes than for neutrophils (MARDER et al. 1985). Normally, C5a is inactivated by serum components at a very rapid rate; however, C5a generated within platelet aggregates and fibrin clots is partially protected from degradation and thereby may cause more prolonged effects.

Arachidonic acid liberated during platelet aggregation and activation can be converted to 12-hydroperoxyeicosatetraenoic acid (12-HPETE) by platelet-derived lipoxygenase. 12-HPETE is inherently unstable and is quickly reduced to 12-HETE, a relatively weak chemotactic factor. However, 12-HPETE can also be converted to 12,20-diHETE, which is also an extremely potent chemoattractant for neutrophils (MARCUS et al. 1984). Since platelet lipoxygenase cannot be inactivated by its own enzymatic products, as is the case with other lipoxygenases, it can thus maintain a localized and concentrated chemotactic stimulus at the wound site. Another arachidonic acid metabolite, thromboxane A_2 (TXA_2), can cause further platelet aggregation and also leads to the release of platelet granules.

Platelets contain two types of storage granules that can release their contents into the surrounding environment. In addition to those described above, these granules contain a variety of other compounds that perform a wide range of functions. Stored are vasoactive agents such as serotonin, a vasoconstrictor; adenosine di- and triphosphates and calcium, promoters of aggregation and adhesion; and numerous procoagulants such as von Wilebrand factor, plasminogen, and fibrinogen. Platelet granules also contain large pools of adhesive proteins that are involved in cell-cell interactions. Peptide growth factors such as platelet-derived growth factor (PDGF) (DEUEL et al. 1982), platelet factor 4 (PF4) (DEUEL et al. 1981), and transforming growth factor-β (TGF-β) (ASSOIAN and SPORN 1986) are also stored. PDGF was initially described as a factor that induced the proliferation of vascular and smooth muscle cells at relatively high molar concentrations (ROSS et al. 1974). However, at picomolar ranges where growth is not induced, potent chemotactic activity of monocytes and neutrophils may occur (DEUEL et al. 1982). PF4 is a highly cationic protein that is rapidly taken up by the endothelial wall (GOLDBERG et al. 1980). It has antiheparin activity and has been reported to stimulate histamine release from basophils (BRINDLEY et al. 1983). Platelets are the richest source of TGF-β, which was originally observed to be a product of virally transformed cells that induced phenotypic transformation of nonneoplastic cells in culture (for review, see SPORN et al. 1987). It is capable, though, of inducing leukocyte chemotaxis at femtomolar concentrations (WAHL et al. 1987a) and is now known to be produced by numerous cell types including lymphocytes (KERHL et al. 1986) and monocytes (ASSOIAN et al. 1987).

II. Neutrophils

Following (and likely dependent upon) platelet activation, migration of leukocytes and macrophages from the surrounding tissue commences. The early cell infiltrate is composed predominantly of polymorphonuclear cells (PMNs), mainly neutrophils. These cells play a key role in protecting tissue against infections through phagocytosis, the antibacterial effects of oxygen radicals, and the activation of complement.

Neutrophils are recruited to the site of the injured tissue by numerous chemotactic factors released during platelet activation and aggregation and blood coagulation. The coagulation- and platelet-derived factors thus far described not only exhibit chemotactic activity, but also cause activtion and adhesion of leukocytes. Adhesion to endothelium may be facilitated by increased levels of CR3 receptors on neutrophil surfaces (BUCHANAN et al. 1982; ANDERSEN et al. 1984). A dynamic interaction, which promotes adherence of these cells to one another, appears to occur between neutrophils and endothelial cells of the postcapillary venules. This enables diapedesis of the former cell into damaged tissues. In response to a concentration gradient of chemotactic signals, neutrophils traverse the endothelium. This process of directed migration is dependent upon interaction between chemotactic ligand and receptor which, in turn, causes a calcium-dependent transduction signal for hydrolysis of phosphoinositol and leads to subsequent motility responses (for review, see SNYDERMAN and UHLING 1988). As neutrophils reach the source of chemoattractant, the gradient is no longer sensed and directed motion terminates.

Activation of recruited neutrophils causes induction of or heightened phagocytic activity, oxygen metabolism and respiratory burst activity, and release of additional soluble mediators of cell recruitment, activation, and coagulation. Neutrophils begin to degranulate and secrete toxic reactive oxygen intermediates. For example, the release of O_2^-, H_2O_2, and other oxygen species augments their capacity to eliminate infectious organisms. Whereas the formation of these highly reactive species is necessary for microbicidal activity, free radical release can, on the other hand, also cause further damage to surrounding tissues (HALLIWELL 1987).

Neutrophils secrete several important mediators which aid in the subsequent recruitment of macrophages and connective tissue cells. Among these are platelet-activating factor (PAF) (LOTNER et al. 1980), arachidonic acid (WELSH et al. 1981), thromboxane B_2 (TXB_2) (WELSH et al. 1981), leukotriene B_4 (LTB_4) (WILLIAMS et al. 1985), 5-HETE (GOETZL et al. 1980) and the proteoglycans chondroitin sulfate and heparin sulfate (PARMLEY et al. 1983). In addition, neutrophils also secrete enzymes such as collagenase and elastase, which cause degradation of matrix structures, leading to the generation of chemotactic fragments of collagen and elastin, respectively (for review, see POSTLETHWAITE and KANG 1988).

Although active in the inflammatory process, neutrophils are not required for normal wound healing. In experimental animals depleted of neutrophils by antineutrophil serum, wound healing processes including recruitment of macrophages, fibroblasts, and endothelial cells appeared to be unimpaired (SIMPSON and ROSS 1972). Moreover, wounds healed at the same rate as in normal animals with no observable difference in rates of matrix formation or connective tissue proliferation. Therefore, unless there is overt infection, neutrophils apparently are not an absolute requirement for successful wound healing.

III. Monocytes/Macrophages

With or without the initial accumulation of neutrophils, a marked increase in the number of infiltrating mononuclear cells occurs. Influenced by chemotactic signals, blood monocytes traverse the endothelial barrier and migrate into the inflamed or damaged tissue site. Mediating this interaction are surface adhesion proteins that form part of the monocyte CDw18 complex (WALLIS et al. 1985). Studies indicate that this complex may be identical to or within the same family of leukocyte differentiation antigens which share a common β-subunit but have distinct α-subunits. Glycoprotein complexes within this family include LFA-1 and C3bi (SANCHEZ-MADRID et al. 1983). These molecules have all been implicated in leukocyte functions that are dependent upon adhesive interactions. Indeed, it has been reported that patients whose leukocytes lack a 150-kDa component of the CDw18 complex suffer from recurrent bacterial infections in the absence of pus formation (BOWEN et al. 1982). Furthermore, these cells were incapable of migrating into extravascular tissues.

Monocytes behave in many ways similar to neutrophils by responding to chemoattractants, phagocytosing particulate stimuli, and releasing various enzymes. However, monocytes have the added capacity to mature into macrophages with concomitant acquisition of unique morphological and functional features. Among these are development of secondary lysosomes, increased vacuolation, enhanced Golgi apparatus, extensive rough endoplasmic reticulum, and surface expression of HLA-DR antigens.

Localized within the inflammatory site, these activated macrophages release oxygen intermediates and arachidonic acid products and undergo increased phagocytic and lysosomal activity. The early release of products derived from cyclooxygenase (prostaglandins) and lipoxygenase (leukotrienes) metabolism augments the inflammatory response further. Enhanced macrophage phagocytosis occurs via increased expression of Fc receptors (GUYRE et al. 1983). During Fc receptor-mediated phagocytosis, macrophages release large amounts of oxygen and arachidonic acid metabolities whereas C3 receptor-mediated phagocytosis does not result in the release of these inflammatory mediators (WRIGHT and SILVERSTEIN 1983). Macrophages also possess a family of surface proteins which can directly bind to organisms such as *Escherichia coli* in the absence of immunoglobulin or complement (WRIGHT and JONG 1986; BULLOCK and WRIGHT 1987). This suggests a mechanism by which macrophages can recognize pathogens prior to the onset of adaptive immunity and, hence, a rapid route by which infections can be controlled.

Induction of monocytopenia, through a combination of systemic exposure to hydrocortisone to eliminate circulating macrophage precursors and local exposure to an antimacrophage serum to eliminate resident macrophages, caused reductions in certain elements of the wound-healing process (LEIBOVICH and ROSS 1975). Under these conditions, fibroblast and endothelial cell recruitment and proliferation, capillary regeneration, and matrix biosynthesis were all affected. Furthermore, debridement of damaged tissue was markedly reduced with the accumulation of large amounts of fibrin. These observations suggested that macrophages have two critical functions that are absolutely required during tissue repair: (a) macrophags are the principle phagocytic cell responsible for clearance of tissue debris such as dead and damaged cells, fibrin, and matrix components; and (b) macrophages are required for the accumulation and proliferation of connective tissue cells.

As a major secretory cell (for review, see NATHAN 1987), the activated macrophage produces a large number of mediators such as neutral proteases, hydrolytic enzymes, complement components, coagulation factors, and polypeptides. Neutral proteases play important roles in matrix degradation and the promotion of coagulation. This leads to the further activation of platelets and neutrophils which, in turn, upregulates the entire inflammatory and wound repair process. For example, plasminogen activator (UNKELESS et al. 1974) converts plasminogen to plasmin, an enzyme that degrades fibrin and activates complement and Hagemann factor. Other neutral proteases include collagenase (WAHL et al. 1975) and elastase (WERB and GORDON 1975), both of which contribute to the breakdown of matrix components. Furthermore, it is clear that macrophages also produce various inhibitors to these enzymes and, therefore, can regulate the inflammatory and tissue repair responses (VASSALI et al. 1984; TAKEMURA et al. 1986).

Activated macrophages also secrete bioactive lipids such as TXA_2 and other arachidonic acid metabolities. Besides having the usual effect of promoting inflammation and cell recruitment, one of these molecules, prostaglandin E_2 (PGE_2), can act as a feedback inhibitor of many inflammatory processes (GOLDYNE and STOBO 1981). Arachidonic metabolites can also act as chemotactic agents for fibroblast populations (MENSING and CZARNETOZKI 1984). Also chemotactic for fibroblasts is fibronectin (SEPPA et al. 1980), a product of macrophages but also of numerous other cell types.

Additional polypeptide factors that promote recruitment of other inflammatory cells (including additional monocytes) and the activation and function of fibroblasts and endothelial cells make up a large part of the macrophage's secretory capability. For instance, a novel chemotactic macrophage peptide has recently been described (LARSEN et al. 1989). Neutrophil activating peptide-1 (NAP-1) or interleukin-8 (IL-8) is a small ($M_r = 6500$) protein that recruits neutrophil and lymphocyte populations. Furthermore, a molecule termed factor-inducing monocytopoiesis (FIM) is responsible for the two-to threefold increase in the number of blood monocytes in the general circulation following the initiation of an inflammatory response (VAN WAARDE et al. 1977; SLUITER et al. 1983). This apparently represents increased differentiation of precursors in transit from the bone marrow to the site of inflammation and may be a major influence in the transition from a neutrophilic to a monocytic infiltrate. A greater number of monocytes in the circulation would, of course, mean a greater infiltration of monocytes at the wound site. FIM is produced by macrophages at the site of inflammation and demonstrates that these cells can regulate the pool of monocytes by increasing maturation of bone marrow progenitors (SLUITER et al. 1983, 1987). This positive regulatory loop apparently is terminated during the waning stages of the inflammatory response by another soluble factor called monocyte production inhibitor (MPI) (VAN WAARDE et al. 1978).

Of particular interest is the relationship between FIM and the colony-stimulating factors (CSFs). CSFs are a family hematopoietic differentiation factors which were originally distinguished from each other by the type of colonies that were produced from bone marrow progenitors in soft agar. For example, granulocyte/macrophage-CSF (GM-CSF), a peptide which induces the differentiation of single cell progenitors into granulocyte and macrophage colonies, is produced by a variety of cells, including macrophages (SULLIVAN et al. 1983). Other factors in this family include macrophage-CSF (M-CSF), granulocyte-CSF (G-CSF), and interleukin-3 (IL-3). Whether FIM and any of the macrophage inducing-CSFs are related molecules is presently unclear since only the cloning of FIM would answer this question. They are all similar in that they induce the differentiation of monocytes from bone marrow progenitors and have similar molecular weights (approximately 20000). On the other hand, FIM, appears to be distinct from M-CSF (SLUITER et al. 1983) and is not produced by lymphocytes, though these cells are potent producers of GM-CSF (SLUITER et al. 1987). FIM differs mainly from GM-CSF in that FIM appears to have no additional biological activities that affect inflammation. For example, GM-CSF inhibits neutrophil migration after their stimulation by known chemoattractants. This probably leads to the localization of reactive cells at the site of tissue injury (WEISBART et al. 1979). Inhibition may be caused by the increased expression of adhesion-promoting glycoproteins on leukocyte cell surfaces (ARNAOUT et al. 1986). On the other hand, GM-CSF alone does not induce neutrophils to release oxygen metabolities; however, it synergizes with C5a and LTB_4 to produce augmented superoxide release (WEISBART et al. 1985). Lastly, increased phagocytic activity in neutrophils has also been reported after their exposure to GM-CSF (FLEISCHMANN et al. 1986).

The activities listed above serve to augment the inflammatory response caused by either tissue injury or deposition of immunologically active components. In addition, these processes can also cause the release of GM-CSF, as well as other types of CSF, by a number of cell types. For example, GM-CSF can be released by activated T cells and by macrophages, fibroblasts, and endothelial cells in response to polypeptide factors such as IL-1 and TNF-α (BROUDY et al. 1987; MUNKER et al. 1986; ZUCALI et al. 1986; METCALF 1985). GM-CSF can cause the release of reactive oxygen intermediates and increased microbicidal activity in monocytes (SMITH et al. 1989). Furthermore, GM-CSF can also stimulate macrophage gene expression of IL-1 and other growth factors with concomitant release of translated protein (WONG et al., manuscript in preparation). Finally, as is the case with FIM, GM-CSF may also augment the inflammatory response by simply increasing the pool of circulating blood monocytes through increased hematopoiesis of bone marrow progenitors.

Since inflammation is an important component of the wound-healing response, factors that amplify the recruitment and function of inflammatory cells would also logically amplify tissue repair. Besides FIM and GM-CSF, macrophages synthesize and release a host of other peptides that promote inflammation. Originally described as an activator of lymphocyte proliferation and an endogenous pyrogen, interleukin-1 (IL-1) is one of the best-characterized macrophage-derived factors (for review, see DINARELLO 1988). Both forms of IL-1 (IL-1α and IL-1β) have numerous effects on neutrophils which include the promotion of degranulation (KLEMPNER et al. 1978; SMITH et al. 1985), secretion of reactive oxygen metabolities (KLEMPNER et al. 1979), and release of TXA_2 (CONTI et al. 1985). All lead to further activation of coagulation and inflammation with the subsequent recruitment of additional monocytes. IL-1 also regulates certain inflammatory and connective tissue cell functions and characteristics. For example, IL-1 causes increased of leukocytes to the endothelial surface (BEVILAQUA et al. 1985; CAVENDER et al. 1987), which may facilitate cell-cell interactions and stimulate endothelial cells to produce more platelet-activating factor and arachidonic acid metabolites (DEJANA et al. 1987). Furthermore, expression of class I major histocompatibility antigens on endothelial cells are also upregulated (COLLINS et al. 1986) as is release of soluble mediators including GM-CSF (BAGBY et al. 1986). Arachidonic acid metabolites are also released by fibroblasts following exposure to IL-1 (ESTES et al. 1984), although no reports to date suggest that IL-1 induces fibroblast chemotaxis.

Tumor necrosis factor-α (TNF-α) or cachectin is an interesting 157-amino-acid molecule ($M_r = 17\,000$) because it shares many biological properties with IL-1, especially those relating to the acute-phase and wound-healing responses. Both peptides induce fever and bone resorption and exhibit antitumor activity, albeit to different targets (for review, see DINARELLO 1986). TNF-α has no more than a 3% amino acid sequence homology with the two forms of IL-1. Moreover, TNF-α and IL-1 act through independent receptors and there is no competition for binding (DINARELLO 1986). Lastly, it is apparent that IL-1 can induce the release of TNF-α from mononuclear cells and vice versa (PHILIP and EPSTEIN 1986). TNF-α increases production of arachidonic acid metabolites, platelet-activating factor (BEUTLER and CERAMI 1986), and IL-1 (NAWROTH et al. 1986) by endothelial cells. It also augments leukocyte adhesion, procoagulant activity, and release of plasminogen activator inhibitor. TNF-α is not chemotactic for either fibroblasts or endothelial cells. It is, however, a potent, though reversible, inhibitor of vascular endothelial cell growth in vitro (SATO et al. 1986).

Finally, another macrophage-derived polypeptide with IL-1-like activities is interferon-β2 or interleukin-6 (IL-6). Though originally described to be a promoter of mouse B-cell hybridoma and plasmacytoma cell line growth

(AARDEN et al. 1985; LANSDORP et al. 1986), IL-6 is now known to induce T-cell activation in an IL-1-type proliferation assay (HOUSSIAU et al. 1988a), release of acute-phase proteins from hepatocytes (BAUMANN and MULLER-EBERHARD 1987), and fever (HELLE et al. 1988). In addition, it increases expression of class I HLA antigens on fibroblasts (MAY et al. 1986). Recently, IL-1 and IL-6 have been shown to synergize with one another in the induction of T-cell proliferation and acute-phase proteins (HELLE et al. 1988). Since IL-6 has been found to be a contaminant in nonrecombinant preparations of IL-1 (HELLE et al. 1988) and, together with the observations that IL-1 induces IL-6 production in fibroblasts (ZILBERSTEIN et al. 1986), thymocytes, and endothelial cells (HELLE et al. 1988), it is possible that many of the biological activities associated with IL-1 and, for that matter, TNF-α, are due to IL-6 or a combination of IL-1 and IL-6.

Macrophage-derived polypeptide growth factors not only amplify the proliferation of certain cell types, they are also chemotactic for inflammatory and matrix-producing cells. For example, platelet-derived growth factor (PDGF), though originally described as a 27- to 31-kDa glycoprotein product of activated platelets, is also a product of activated macrophages (SHIMOKADO et al. 1985). PDGF exists as a dimer of two distinct polypeptide chains, A and B. Natural PDGF, purified from platelets, is composed predominantly of the AB isoform but AA and BB dimers have been isolated from other sources (STROOBANT and WATERFIELD 1984; HELDIN et al. 1986). The receptor has been described as a membrane glycoprotein (GLENN et al. 1982; HELDIN et al. 1988) which may exist in two classes (GROUWALD et al. 1988). Recently, however, the receptor has been shown to be a dimer of two subunits, either α or β (ROSS 1988 personal communication). The α-subunit binds only the A-chain of the PDGF dimer whereas the β-subunit binds the B-chain. Therefore, three classes of receptor would exist. $\alpha\alpha$, $\alpha\beta$, and $\beta\beta$. Which PDGF isoforms each receptor would bind is, of course, dependent upon the composition of the individual receptor.

At low concentrations, PDGF (especially the BB isoform) displays chemotactic properties for neutrophils, fibroblasts, and macrophages (DEUEL et al. 1982, DEUEL and HUANG 1984). This observation is controversial since recent evidence demonstrates minimal chemotactic activity for monocytes occurs in the presence of lymphocytes whereas no activity is observed in purified monocyte populations. Some form of cell-cell interaction appears to be required (ROSS 1988, personal communication).

Another group of growth factors with chemotactic properties are the type β-transforming growth factors (TGF-β). They are a ubiquitous family of molecules that are produced by a wide variety of neoplastic and normal cells such as T lymphocytes, platelets, and macrophages. The two best-characterized homodimeric forms, TGF-β1 and TGF-β2, show approximately 70% amino acid sequence homology and appear in numerous mammalian tissues. Expression of the β1- and β2-chains in the same cell can give rise to a heterodimer, TGF-β1.2, which has only been identified in porcine platelets (CHEIFETZ et al. 1987). Three types of surface receptors have been identified to date (for review, see MASSAGUÉ et al. 1987): type I (65 kDa), type II (85–95 kDa), and type III (250–350 kDa). Types I and II have higher affinity for TGF-β1 than β2 and type III receptors bind both β1 and β2 with equally high affinity. Recently, evidence suggests that the β1-chain in the β1.2 heterodimer causes increased binding with type I and II receptors and competitive inhibition of β2 (CHEIFETZ et al. 1988).

Monocytes constitutively express TGF-β mRNA, but do not actively secrete translated protein unless activated. Protein release is not accompanied by an increase in the steady-state level of mRNA (ASSOIAN et al. 1987). TGF-β, however, appears to upregulate its own gene expression as well as genes that code for other polypeptide growth factors that are critical for wound healing (WAHL et al. 1989; MCCARTNEY-FRANCIS et al., manuscript in preparation). Picomolar concentrations of TGF-β will induce macrophage gene expression of TNF-α, IL-1, basic fibroblast growth factor, and PDGF. Furthermore, TGF-β causes chemotaxis of both monocyte (WAHL et al. 1987a) and fibroblast populations (POSTLETHWAITE et al. 1987). TGF-β, therefore, appears to have a particularly important role in the inflammatory and tissue repair process.

The ability of TGF-β to influence monocyte function points to an important immunoregulatory aspect of cellular function at the localized inflammatory site. Activated macrophages secrete TGF-β, which then stimulates the same or neighboring macrophages to generate polypeptide growth factors, including TGF-β. A key step in this activation sequence is that TGF-β is released by macrophages and other cells in latent form. It is postulated that macrophages, once activated, enzymatically cleave the latent form to release functional 25-kDa homodimeric molecules. TGF-β can also downregulate certain macrophage functions such as release of reactive oxygen intermediates while it can stimulate the release of growth factors such as IL-1 and PDGF (TSUNAWAKI et al. 1988). Phagocytosis, on the other hand, appears to be unaffected.

TGF-β also has the ability to inhibit the proliferation of T lymphocytes at femtomolar concentrations. The exact mechanism by which this occurs is not clear since production of the lymphocyte growth factor, interleukin-2, or its receptor is not affected (WAHL et al. 1988a). Furthermore, macrophage-derived TGF-β effectively suppresses lymphocyte proliferation in chronic inflammatory lesions in vivo (WAHL et al. 1988b). Coupled with its ability to suppress T-cell proliferation and function, TGF-β may inhibit the ability of reactive oxygen metablites and various T-cell products (e.g., IFN-γ) to perpetuate the inflammatory response. Clearance of damaged tissue through phagocytic action and subsequent repair would proceed unimpeded through the action of macrophage-derived products.

Though TGF-α shares a common nomenclature with TGF-β, these two macrophage-derived peptides display no functional or structural similarities. TGF-α does, however, have substantial amino acid sequence homology with epidermal growth factor (EGF). This homology is reflected in the observations that both TGF-α and EGF have similar biological functions

(MARQUARDT et al. 1984). Moreover, TGF-α apparently does not have a separate receptor and mediates its effect through binding with the EGF receptor (TODARO et al. 1980). EGF was originally described as a product of submaxillary glands (COHEN 1962) or a component of human urine (COHEN and CARPENTER 1975) and was shown to stimulate epidermal growth and keratinization directly. EGF stimulates a variety of biological phenomena including angiogenesis in rabbit cornea models (GOSPODAROWICZ et al. 1979) and proliferation and differentiation of various mesenchymal cell types (for review, see CARPENTER and COHEN 1979). Whether TGF-α and EGF mediate the same biological effects in vivo is at the present time unclear but some reports show they both increase the rate of angiogenesis and wound repair (SCHREIBER et al. 1986; SCHULTZ et al. 1987), although TGF-α appears to be more potent. The presence of EGF in saliva may explain the significance for the tendency of rodents and other mammals to lick their open wounds, thereby aiding the healing process (NIALL et al. 1982).

The macrophage-derived factors discussed thus far have been, for the most part, both mitogenic and chemotactic for a variety of cell types. Macrophage-derived interferon-α (IFN-α), on the other hand, appears to inhibit the chemotaxis of contact-inhibited populations of fibroblasts which have been exposed to known chemoattractants such as PDGF (ADELMANN-GRILL et al. 1987). Interestingly, IFN-α does not inhibit chemotaxis of actively growing fibroblast populations (i.e., low-density conditions). This activity may have a significant consequence because nongrowing cells may be more capable of matrix synthesis since cellular resources need not be devoted to the production of "housekeeping" materials required for proliferation. Matrix-producing cells would then be localized around the wound area. Therefore, this may also allow for the preferential selection of fibroblast populations with unique phenotypic characteristics that are compatible with the wound-healing process.

The secretory repertoire of the activated macrophage is an incredibly complex array of molecules and activities. That specific macrophage products may have numerous activities and that a specific activity can be shared by numerous products (some produced by other cells) compounds this complexity. As discussed above, nearly every biological activity possessed by IL-1 is shared with the structurally unrelated TNF-α molecule. Why there are multiple molecules for specific functions is not clearly known, but perhaps a redundancy factor has evolved over time for these important biological processes. Moreover, recent studies have identified a novel heparin-binding peptide secreted by endotoxin-stimulated macrophages with inflammatory and neutrophil-recruiting activities (WOLPE et al. 1988). Interestingly, this peptide shares N-terminal sequence homology with certain recently described low molecular weight T-cell-specific products such as RANTES (SCHALL et al. 1988) and TCA3.0 (BURD et al. 1987). This suggests that these proteins may make up a previously undescribed family of small immune factors.

Another level of complexity is that macrophage-derived products can affect the production and release of these same products by other cell types. For example, both IL-1 and TNF-α stimulate fibroblasts to release GM-CSF, arachidonc acid metabolites, and collagenase (BEUTLER and CERAMI 1985). Furthermore, these factors act synergistically or they may regulate the function of each other. TGF-β inhibits the mitogenic effect of bFGF on vascular endothelial cells whereas it potentiates bFGF-induced growth of osteoblasts (GOSPODAROWICZ et al. 1987a, b). Additionally, PDGF, in order to accelerate wound healing, has been shown to require additional factors found in partially purified preparations of PDGF (LYNCH et al. 1987). Purified preparations of PDGF, which are devoid of contaminating proteins and other undefined factors, fail to promote wound healing.

IV. Lymphocyte Function and Regulation

Along with macrophages, T lymphocytes are also present in the inflammatory cell infiltrate that occurs in the wound-healing process. They also play an important role in promoting healing through the release of several soluble mediators that either directly stimulate fibroblast recruitment, proliferation, and biosynthesis of collagen and other matrix components or indirectly by augmenting monocyte secretion of recruitment or growth factors. However, in the absence of a defined antigenic stimulus, the role of the lymphocyte has not been clearly established.

Physiological activation of resting T cells occurs through the interaction of two separate signals (for review, see ISAKOV et al. 1986; WEISS et al. 1986). One is recognition by the T-cell receptor of specific antigen in conjunction with major histocompatibility complex gene products present on the surface of antigen-presenting accessory cells. This stimulus causes the cleavage of membrane-associated phosphotidylinositolbiphosphate into two products, 1,4,5-inositol-triphosphate and diacylglycerol, which are responsible for increasing the intracellular calcium ion concentration and activating protein kinase C, respectively. Cytoplasmic free calcium and activation of protein kinase C are the critical parameters in T-cell activation. Indeed, antigen interaction with T-cell receptor can be completely bypassed through calcium ionophores, which allow free passage of Ca ions through the cell membrane, and phorbol esters, which can directly activate protein kinase C. The second signal is mediated by the monokine IL-1, which is released by accessory cells. Precisely how IL-1 contributes to the biochemical pathway leading to T-cell activation is at the moment unclear. Interestingly, in addition to activating protein kinase C, phorbol esters can mediate the second signal, substituting for IL-1.

After activation, two lymphocyte genes necessary for the continued growth and proliferation of T cells are expressed. One gene codes for the cytokine IL-2 which imparts the proliferative signal and the other codes for the receptor for IL-2. Unlike most other growth factor receptors, functional IL-2 receptors are not expressed on the surfaces of resting cells. Only upon activation will T cells begin transcription of the IL-2 receptor gene. With the availability of its receptor for binding and subsequent signal transduction, IL-

2 is now capable of acting in an autocrine fashion in order to expand the clone(s) of activated T cells. IL-2 receptors fall into high- and low-affinity classes. The low-affinity receptor consists of a 55-kDa glycoprotein (p55) and does not appear to be essential for biological activation. The biologically relevant high-affinity receptor has been postulated to be a complex of p55 noncovalently associated with a separate 70- to 75-kDa glycoprotein (p70).

IL-2 also appears to affect macrophages since they express IL-2 receptors after activation (HERMANN et al. 1985; WAHL et al. 1987b). The p55–p70 complex has recently been detected on activated peripheral blood monocytes (SHARON et al. 1988). Microbicidal activity such as cytotoxicity and superoxide radical generation are increased following exposure to IL-2 (WAHL et al. 1987b). In addition, increased production of IL-1 can also be observed (MCCARTNEY-FRANCIS and WAHL 1988, personal communication). Thus, the ability of IL-2 to modulate the function of activated macrophages contributes to the complex network of soluble regulatory signals involved in inflammation and wound repair. Although increased macrophage function as the result of exposure to IL-2 would suggest an increased capability to repair wounds, there is to date no direct experimental evidence to support this notion. Following i.p. injection of IL-2, however, fibrosis occurs in the peritoneal cavity, although this may be attributed to indirect monocyte stimulation through direct T-cell activation (ORTALDO and LONGO 1988).

Once activated, T cells produce a wide array of factors that affect inflammation and wound healing. Some, such as GM-CSF and TNF-α, are also produced by activated macrophages and have previously been discussed. On the other hand, other factors appear to be unique to T cells, such as IL-3. IL-3 was originally described as a 20- to 26-kDa peptide with the ability to promote early steps in T-cell maturation (for review, see IHLE et al. 1982). Since then, many other properties, similar to those associated with IL-2, have been identified with this molecule. These include stimulation of differentiation and growth of various lymphocyte cell lines. IL-3 can promote inflammation by increasing the pool of T lymphocytes available for activation. In addition, IL-3 also has the ability to induce the differentiation of bone marrow progenitors into various hematopoietic cells (e.g., erythrocytes, monocytes, and granulocytes), either alone or in combination with other factors such as GM-CSF (WILLIAMS et al. 1987). Hence, IL-3 is also known as a multi-CSF (for review, see SIEFF 1987). Finally, IL-3 can also affect mature cells like monocytes by activating various genes coding for a variety of growth and inflammatory factors such as IL-1 and TNF-α (WONG et al., manuscript in preparation).

Interferon-γ (IFN-γ) is also synthesized and released by activated T cells. Though macrophages have been reported only to secrete α- or β-type IFNs, recent evidence suggesting IFN-γ release by pulmonary alveolar macrophages from patients with sarcoidosis will require further investigation at the single-cell and DNA levels (ROBINSON et al. 1985). IFN-γ has a wide range of effects on both the inflammatory and tissue repair responses. Neutrophils exhibit marginally increased abilities to phagocytose unopsonized particles

(SHALLABY et al. 1985) and to display a respiratory burst (BERTON et al. 1986). The effects of IFN-γ on macrophages are more numerous and wide-ranging (for review, see NATHAN and YOSHIDA 1988). Macrophages display increased phagocytosis, respiratory burst activity, antimicrobial and antitumor activity, surface expression of class II HLA antigens and IL-2 receptors, and the ability to secrete a number of mediators including IL-1, TNF-α, GM-CSF, proteases, and arachidonic acid metabolites. A number of other genes are also expressed, one of which is referred to as IP-10 (LUSTER et al. 1985). The predicted amino acid sequence of the translation product demonstrates significant homology with a new family of proteins associated with inflammation and chemotaxis that includes PF4. IP-10 protein has also been shown to be released from keratinocytes, endothelial cells, and fibroblasts after exposure to IFN-γ (LUSTER and RAVETCH 1987).

There are at least two established effects of IFN-γ on the differentiation and function of T-cell populations. First, endogenously produced IFN may be essential for the maturation of precursors into cytolytic T cells (LANDOLFO et al. 1985). Though these cells may not play a direct role in inflammation or tissue repair, their action may promote the unresolved inflammation that is characteristic of certain pathological and autoimmune states. For example, destruction of antigenically altered tissue (whether virally or through some autoimmune mechanism) can cause the onset or contribute to unresolved inflammation and tissue repair. The second effect of IFN-γ is to enhance the production of TNF-β (lymphotoxin) following exposure to IL-2 (NEDWIN et al. 1985). With similar biological actions to TNF-α, TNF-β plays a direct role in promoting inflammation and repair.

Finally, IFN-γ may promote inflammation through the expression of class II histocompatibility antigens on endothelial cell surfaces (POBER et al. 1983). This action may endow endothelial cells at the site of inflammation the ability to present antigen to or bind primed T cells. In addition, IFN-γ may also induce the expression of an undefined antigen on endothelial cells that facilitates the emigration of circulating mononuclear cells to the inflammatory site (DUIJVESTIJN et al. 1986).

T cells also produce unique factors which regulate fibroblast chemotaxis and function. Antigen- or mitogen-stimulated T cells produce a factor that induces a chemotactic response in fibroblasts (POSTLETHWAITE et al. 1976). Lymphocyte-derived chemotactic factor for fibroblasts (LDCF-F) is a 22-kDa protein that is derived from a high molecular weight, biologically inactive precursor molecule (POSTLETHWAITE and KANG 1983). Latent LDCF-F appears to be converted to the active form by either trypsin or extracts from sonically ruptured macrophages. In some ways, this is similar to TGF-β in that the latter is also released in an inactive form and requires enzymatic cleavage in order to unmask biological activity. Activated T cells and T-cell lines also release a 43-kDa acidic protein (pI, 5.0–5.5) which stimulates proliferation of fibroblasts (WAHL and GATELY 1983).

Finally, T cells release a factor which inhibits fibroblast migration (ROLA-PLESZCYNSKI et al. 1982). Termed fibroblast inhibition factor (FIF), it may be responsible for the localization and accumulation of fibroblasts at the site of inflammation and tissue damage. Whether this factor is related to a trypsin-sensitive inhibitor found in normal human serum is not known. The serum inhibitor has a high molecular weight (approximately 210000), is not cytotoxic, and will not inhibit proliferation or protein synthesis (OCHS et al. 1987).

Inflammation is the critical process that sets in motion the complex interactions that occur among various cell types during the middle and late stages of wound healing. Platelets, PMNs, macrophages, and T cells all contribute a host of chemotactic and proliferative mediators that are not only vital to the inflammatory process but also to tissue repair and regeneration. The next two stages of the healing response emphasize connective tissue recruitment and function with the deposition of newly synthesized matrix components and the generation of new blood vessels.

C. Proliferative Phase

The regulated migration and proliferation of connective tissue cells characterizes the proliferative phase of wound healing. Fibroblastic cells migrate from uninjured tissue to areas of damage and inflammation and endothelial cells from the nearby vasculature into the wound area where the process of blood vessel formation or angiogenesis begins. These cellular traffic patterns are controlled by chemotactic factors that have been produced during the inflammatory phase of the response. Following connective tissue recruitment, the formation of granulation tissue with its characteristic increase in the biosynthesis of extracellular matrix components ensues.

I. Regulation of Fibroblast Proliferation

Although mesenchymal cell accumulation becomes visibly apparent only in the latter stages of the inflammatory response, the basis for their migration and subsequent growth was established in the early stages. Repair commences as the host continues to sequester and degrade the inflammatory stimulus, damaged tissue, and extracellular matrix components. At the same time, fibroblasts synthesize new matrix components and endothelial cells initiate neovascularization.

Expansion of the fibroblast population is the result of both recruitment and proliferation due to locally generated chemotactic and growth factors, respectively. Proliferation is influenced by many factors among which the polypeptide growth hormones are the most critical. Macrophages secrete a whole family of molecules called fibroblast-activating factors (FAFs) and fibroblast growth factors (FGFs). These peptides induce the proliferation of a variety of cell types including fibroblasts and endothelial cells and, as a result, augment the wound-healing process. Mitogen- or endotoxin-stimulated macrophages produce a series of FAFs ranging in molecular weight from 0.5 to 40 kDa. FAFs are specific in their action since some cause proliferation of certain smooth muscle cell lines whereas others activate different cell lines (for review, see TURK et al. 1987). This is in contrast to other peptide growth factors such as IL-1 and TNF, which exert their functions on a broad range of tissues and cell lines. The FGFs are a family of structurally related peptides ($M_r = 16000$) that have similar bioactivities, including the ability to act as potent mitogens and differentiation factors for a wide range of mesodermal- and ectoneurodermal-derived cell types (for review, see GOSPODAROWICZ et al. 1987a, b). Due to their affinity for heparin, they have also been termed heparin-binding growh factors (HBGFs). Basic FGF (bFGF, pI 9.6) was first identified by its ability to stimulate proliferation and phenotypic transformation of BALB/c-derived 3T3 fibroblasts. Acidic FGF (aFGF, pI 5.6) was described to be a myoblast mitogen. There is approximately 55% amino acid sequence homology between the two types. bFGF has been detected in macrophages (BAIRD et al. 1985) and numerous other normal and transformed cell types. On the other hand, aFGF has only been identified in brain, retina, bone matrix, and osteosarcoma cells. aFGF appears to be 30–100 times less potent than bFGF and apparently contributes only 8% and 0.15% of the total mitogenic activity isolated from crude brain or retinal extract, respectively. FGF receptors from a wide variety of cell types have been characterized and all share approximate molecular weights (130–165 kDa) which may represent different degrees of glycosylation. In all cases, both FGFs bind to the same receptor although the basic form may bind with higher affinity. This may explain the difference in observed potencies.

Historically, mitogenic factors for fibroblasts have been divided into two distinct groups based on their actions: competence and progression (PLEDGER et al. 1977; SCHER et al. 1979). These purported factors were complementary since neither class of factor caused proliferation by itself. Competence factors did not stimulate DNA synthesis but activated quiescent cells in the G_0- or G_1-phase of the mitotic cycle, enabling them to become responsive to the action of progression factors. Progression factors caused cells to progress from G_0- or G_1- to the S-phase and initiated DNA synthesis (SCHER et al. 1979). Factors such as PDGF, FAF, and bFGF were defined as competence signals (SCHER et al. 1979; GOSPODAROWICZ et al. 1987b), whereas progression signals were delivered by a number of other peptides. For example, endotoxin-activated or alveolar macrophages produce unique molecules which stimulate DNA synthesis in the presence of competence factors (BITTERMAN et al. 1982). Another progression factor is EGF (MULLER et al. 1984). Since activated macrophages are able to release both competence and progression factors, fibroblasts would be stimulated to proliferate in the absence of any other leukocyte- or plasma-derived component. This would likely explain why macrophages can sustain a wound repair reaction by themselves whereas other cell types could not produce both competence and progression signals.

Some of the known factors that stimulate fibroblast proliferation, such as TNF-α, IL-1, and TGF-β, have yet to be classified as either competence or progression signals since their exact modes of action have yet to be characterized. Although fibroblasts proliferate following exposure to IL-1 (SCHMIDT et al. 1982), it is uncertain whether IL-1 acts to increase expression of cell surface receptors for other growth factors such as EGF and PDGF. TNF-α, as well as TGF-β, promotes increased appearance of EGF receptors on fibroblasts (Sugarman et al. 1985; Palombella et al. 1987; Assoian et al. 1984). It has recently been shown, however, that both IL-1 and TNF-α appear to stimulate growth through the induction of PDGF production by fibroblasts (Raines et al. 1989). Treatment with PDGF-specific antibody completely inhibits proliferation following exposure to IL-1. Growth of fibroblasts occurs after a lag-phase since this time period was necessary for PDGF to be synthesized and released, and to exert its effects in an autocrine manner. TGF-β also promotes fibroblast proliferation in this manner (Moses et al. 1985; Keski-Oja et al. 1986). Whether IL-1, TNF-α, and TGF-β act as progression factors after PDGF confers competence is unclear.

Lastly, the action of IFN-γ is still controversial since conflicting reports have shown inhibition (Duncan and Berman 1985), stimulation (Brinckerhoff and Guyre 1985), or no effect (de la Maza et al. 1985) on the proliferation of human fibroblasts. Precisely why these divergent results occur is not known.

Other growth factors derived from both lymphoid and nonlymphoid cells have also been found to regulate the tissue repair process. For example, nerve growth factor (NGF), purified from mouse submandibular glands, caused accelerated wound contraction when topically applied to surgically induced wounds in mice (Li et al. 1980). Later, it was determined that NGF was chemotactic for polymorphonuclear leukocytes both in vitro (Gee et al. 1983) and in vivo (Boyle et al. 1985). The effect was absolutely dependent on a serine esterase activity located on the γ-subunit of the NGF molecule and not due to the generation of the chemotactic peptide C5a.

Somatomedins have also been reported to play a role in wound healing. These are a family of polypeptide hormones with mitogenic and insulin-like activities (hence they are also known as insulin-like growth factors or IGFs) (Rinderknecht and Humbel 1978 a, b). Somatomedin C, or IGF-I, is a basic polypeptide whereas the neutral somatomedin is termed IGF-II. PDGF has been shown to stimulate IGF-I production in cultured fibroblasts and aortic smooth muscle cells (Ross et al. 1986). This may provide an autocrine pathway for modulation of fibroblast function.

There appear to be two types of receptors for the IGFs (Kasuga et al. 1981). The type 1 receptor is structurally homologous to the insulin receptor and has the highest affinity for IGF-I whereas the type 2 receptor has higher affinity for IGF-II. Since fibroblasts, endothelial cells, and macrophages express both types of receptors on their cell surfaces, the IGFs are thought to be important in regulating the functions of these cell types. IGF-I and -II are classified as progression factors for fibroblast growth (Stiles et al. 1979). It was reported that exogenous IGF-I hastened the healing of wounds in diabetic rats (Scheiwiller et al. 1986). Furthermore, fluid obtained from healing human, rabbit, and rat wounds contained significant amounts of IGF-I (Spencer et al. 1988). In addition, IGF-I stimulates collagen synthesis by fibroblasts (Spencer et al. 1988).

II. Extracellular Matrix Synthesis

With the recruitment, accumulation, and proliferation of fibroblasts, extracellular matrix components are synthesized at the site of the healing wound. The extracellular matrix consists of fibrillar proteins (e.g., collagen) in addition to an amorphous ground substance that fills intercellular and interfibrillar spaces. Assessment of wound healing is usually based on measurements of the collagen content and the tensile strength of these fibers. Studies have shown that, as healing progresses, the crosslinking of collagen fibers correlates with increased tensile strength (PEACOCK and MADDEN 1966). The amorphous substance is composed mainly of glycoproteins and proteoglycans (e.g., fibronectin, laminin, and chondronectin). Each component of the matrix has a specific function which is critical to the integrity of the structure and the restoration of the wound site. For example, fibrillar proteins such as collagen, elastin, and reticulin act as structural components whereas the major glycoproteins serve as adhesion factors which link cells to the collagenous framework. In addition, proteoglycans have numerous functions which range from hydration of cartilage tissue to the provision of filtration barriers in basement membranes.

1. Collagen

The most abundant of matrix components is collagen. To date, 11 distinct types of collagen in connective tissue have been identified and characterized (types I–XI) (for review see MILLER 1985). They are composed of three constituent polypeptide α-chains which come together to form a single collagen monomer unit. Triple helical and globular regions are common features of all collagen types. Collagen can be generally classified into fibrillar and nonfibrillar types. Fibrillar collagen displays the characteristic cross-striations of connective tissue and the required tensile strength needed to hold tissues and cells together. Nonfibrillar collagens form filtration barriers and the structural framework needed for tissue binding. The major fibrillar collagens are types I–III, whereas types IV–XI represent the nonfibrillar collagens. Type I is found in most connective tissues (skin, bone, ligament, tendon) and is the body's most abundant form. Smaller amounts of type III are found in all tissues where type I is found. Type II is present only in hyaline cartilage. Fibroblasts synthesize only types I and III whereas chondrocytes can produce all three types of fibrillar collagen. The biosynthesis of collagen is accompanied by an unusually large number of cotranslational and posttranscriptional modifications that occur intra- and extracellularly. Intracellular modification of translated α-chains results in the formation of a triple-helical procollagen molecule. Fibrillar collagens (types I, II, and III) result after further extracellular processing.

Activated macrophages secrete soluble mediators that promote the synthesis of collagen by fibroblasts. Both TNF-α and IL-1 stimulate the production of collagen (BEUTLER and CERAMI 1986; KRANE et al. 1985) and collagenase (BEUTLER and CERAMI 1986; POSTLETHWAITE et al. 1983). Upregulation of collagen synthesis appears to be controlled at the level of mRNA (KRANE et al. 1985; AGELLI et al. 1987). Whether this is due to increased transcription of the collagen gene or increased stability of the mRNA molecule itself is not clear. On the other hand, recent studies have shown that under certain conditions IL-1 can cause an inhibition in the amount of types I and III collagen mRNA produced by chondrocytes and also the subsequent translation into protein and release (GOLDRING and KRANE 1987). PGE_2, which is also induced by IL-1, causes downregulation in the synthesis of collagen types I and III through the action of cAMP as a second messenger.

TGF-β has also been identified as a positive regulator of collagen biosynthesis. When injected subcutaneously into newborn mice, rapid fibrosis consistent with rapid increases in the production of collagen and granulation tissue occurs (ROBERTS et al. 1986). Studies have shown that, like IL-1, TGF-β causes augmentation in the amount of collagen protein and mRNA produced in fibroblasts (IGNOTZ and MASSAGUE 1986; IGNOTZ et al. 1987). In the murine system, transcriptional control is apparently exerted through some unknown mechanism involving a nuclear factor I DNA-binding site located on the collagen α-chain promoter sequence (ROSSI et al. 1988). Deletion of a 3-basepair (bp) sequence within this promoter site results in the loss of collagen mRNA upregulation after exposure to TGF-β.

TGF-β also causes increased incorporation of newly synthesized collagen into the extracellular matrix (IGNOTZ and MASSAGUE 1986; ROBERTS et al. 1986). Contraction of the collagen matrix in a healing wound functions to reduce the size of the original wound (ALLEN and SCHOR 1983). This phenomenon is apparently caused by the tension exerted by fibroblasts on collagen fibrils. TGF-β has been shown to stimulate the contraction of collagen gels in vitro, suggesting that this growth factor is a regulator of wound contraction in vivo (MONTESANO and ORCI 1988). TGF-β may improve the efficiency by which fibroblasts transmit mechanical forces to the matrix or through the increased production of matrix components and matrix receptors.

PDGF has also been described to stimulate fibroblast collagen biosynthesis and deposition (NARAYANAN and PAGE 1983), collagenase production (BAUER et al. 1985), and release of IGFs (ROSS et al. 1986). The fibroblast-stimulating and chemotactic activities appear to reside on separate structural sites on the PDGF peptide backbone (WILLIAMS et al. 1983).

Mechanisms must exist to downregulate the production of matrix proteins in order to prevent uncontrolled deposition of extracellular matrix components. Generally, IFN-γ suppresses overall protein synthesis in most responsive cells. However, its action on human fibroblasts is unique in that there is selective suppression of collagen synthesis with little or no suppression of overall protein synthesis (DUNCAN and BERMAN 1985; JIMINEZ et al. 1984). Suppression was pretranslational since procollagen mRNA levels were apparently decreased (STEPHENSON et al. 1985).

2. Proteoglycans

Proteoglycans largely make up the amorphous ground substance present in the intercellular and interfibrillar spaces of the extracellular matrix. These are large macromolecules in which glycosaminoglycan (GAG) chains are covalently linked to a protein core. Examples of GAG chains are hyaluronic acid, chondroitin sulfate, dermatan sulfate, heparin sulfate, and keratan sulfate. Studies of wound healing have shown that hyaluronic acid is produced during the early phases of repair and serves to promote recruitment of inflammatory cells and fibroblasts. Factors that stimulate collagen biosynthesis, such as IL-1 and TGF-β, also stimulate production of hyaluronic acid and other GAG chains (IGNOTZ and MASSAGUE 1986; WHITESIDE et al. 1986; POSTLETHWAITE and KANG 1988; BASSOLS and MASSAGUE 1988).

3. Fibronectin

The fibronectins are high molecular weight glycoproteins that are found in the pericellular and intercellular matrices, basement membranes, and various body fluids. Fibronectin is closely associated with fibroblasts, endothelial cells, platelets, and macrophages and appears to be involved in cellular interactions such as the promotion of cell movement, attachment, spreading, and coagulation. For example, the fibronectin matrix forms a scaffold on which collagen fibrils are organized (McDONALD et al. 1982). Both plasma- and cell-derived fibronectin are potent chemoattractants for fibroblasts (SEPPA et al. 1980). In addition, they promote platelet spreading on exposed collagen fibrils and fibrin clots (GRINNELL et al. 1979). Fibronectin biosynthesis is upregulated by both IL-1 and TGF-β (IGNOTZ and MASSAGUE 1986; IGNOTZ et al. 1987; KRANE et al. 1985).

Extracellular matrix components that are synthesized and deposited within the healing wound site will eventually lead to the repair of the damaged area. Specifc tissues and organs, of course, have unique compositions and characteristics which are produced by the proportional synthesis and deposition of these matrix components. Exactly how the body stores and recalls the proper "blueprints" or templates by which to direct this proportional synthesis is unknown. However, there are certain clues as to how regulation occurs. For example, tissues are composed of different collagen types and mechanisms must exist that are capable of altering the expression and production of the various collagens. Types I and III are both produced early in the fibrotic response. As matrix formation progresses, production of type III declines while type I becomes the dominant collagen produced. The same fibroblast populations are responsible for the production of both types of collagen but the exact mechanism responsible for switching is unknown. Recent evidence suggests that PGE_2 and IL-1 promote this switching mechanism (AGELLI et al. 1987; STEINMAN et al. 1982). Cortisol, antiinflammatory steroids, and parathyroid hormone can cause a specific decrease in the intracellular concentration of type I procollagen mRNA (OIKASINEN and RYHANEN 1981; KREAM et al. 1980). Whether this is due to decreased rates of transcription or increased degradation is not clear. Collagen synthesis can also be regulated at the translational level (TOLSTOSHEV et al. 1981). During the development of fetal sheep skin, synthesis of collagen falls off by tenfold whereas mRNA levels remain constant.

III. Endothelial Cell Function and Angiogenesis

Soon after the arrival of the first fibroblasts, new blood vessels begin to develop and invade the site of healing. Normally, endothelial cells line the interiors of all blood vessels and are attached via interaction with the underlying basement membrane. Endothelial cells are capable of secreting a whole array of matrix components such as collagen, fibronectin, and proteoglycan GAG chains. Moreover, endothelial cells display potent pro- and anticoagulant activities depending upon the extracellular conditions. For example, endothelial cells are intrinsically noncoagulant since platelets do not normally adhere unless an injury is present. In addition, factors exist on endothelial membranes that block thrombosis. These include heparan sulfate and heparin (BUONASSISI 1973; MARCUM and ROSENBERG 1984). On the other hand, endothelial cells constitutively secrete molecules that promote platelet adhesion (e.g., von Willebrand factor) (JAFFE et al. 1974) and additional procoagulant factors (e.g., factor V) (CERVENY et al. 1984) after stimulation. Furthermore, one of these procoagulants, factor X, is capable of stimulating endothelial cells to release PDGF (ROSS 1988, personal communication).

Studies have shown that the eventual phenotype of endothelial cells is determined by the nature of the matrix on which they attach (MACIAG et al. 1982; MADRI and STENN 1982). For example, endothelial cells plated on type I collagen proliferate and form continuous sheets. On the other hand, those plated on type IV collagen cease to proliferate and, instead, form tube-like structures. Such phenomena may be important in initiating the spread of capillaries throughout the wound and inflammatory site and their integration into tissue elements. Endothelial cells can also contribute to matrix formation and inflammation because of their ability to secrete both collagen and collagenase. Various types of collagen are synthesized depending upon the type of endothelial cell. For example, human umbilical vein endothelium secretes types IV and V into the cellular matrix (SAGE and BORNSTEIN 1982) whereas bovine aortic endothelium secretes types III, IV, V, and VIII (SAGE et al. 1981). Activated endothelial cells can also remodel basement membranes since they release collagenases capable of degrading various collagen types (MOSCATELLI et al. 1980; KALEBIC et al. 1983). They can also produce collagenase inhibitors (HERRON et al. 1986) and thus serve as possible regulatory points during the waning stages of inflammation and wound repair. IFN-γ has also been reported to stimulate the rate at which endothelial cells secrete an organized matrix of proteoglycan GAG chains (MONTESANO et al. 1984).

Several macrophage-derived polypeptide growth factors that regulate fibroblasts are also effective in modulating endothelial cell growth and function. Such factors include IL-1, TNF-α, TGF-β, and bFGF. Macrophages also secrete proteases, prostaglandins, and a unique molecule termed macrophage angiogenesis factor (MAF) which are chemotactic for endothelial cells (SHAHABUDDIN et al. 1985).

bFGF is probably the best-characterized macrophage-derived angiogenic polypeptide. It induces proliferation, chemotaxis, and mediator release (reviewed in GOSPODAROWICZ et al. 1987a, b). Capillary endothelial cells are induced to invade three-dimensional collagen matrices with subsequent organization into prevascular tubules. Successful migration appears to involve the initial disruption and breakdown of the extracellular matrix surrounding preexisting blood vessels. The elaboration of proteases, such as collagenase and plasminogen activator, by bFGF-stimulated endothelial cells may help contribute to matrix degradation (GROSS et al. 1983).

Cultured endothelial cells synthesize sizeable amounts of bFGF but apparently do not secrete translated protein into the culture medium (VLODAVSKY et al. 1987). Instead, the protein remains internalized or membrane bound. Based on cDNA data, translated bFGF may not contain a hydrophobic signal sequence, an element critical for transmembrane transport. Release may occur during tissue damage caused by trauma or diapedesis of inflammatory cells. Therefore, blood vessel-derived bFGF can stimulate blood vessel growth in an autocrine manner.

An endothelial cell growth factor, distinct from bFGF, has also been isolated from human platelets and has been demonstrated to have a molecular weight of approximately 45 kDa (MIYAZONO et al. 1987). It appears to be an acidic protein consisting of a single polypeptide chain. This factor did not bind heparin, nor did heparin potentiate its growth activity. The exact relationship between this peptide and other heparin-binding mitogenic factors has yet to be determined. The relevance of this factor in actual wound-healing responses, however, is also subject to question since its release from activated platelets has yet to be demonstrated. Purification of this peptide was accomplished through the use of platelet lysates rather than the use of supernatants from activated platelet populations.

IL-1 and TNF-α appear to inhibit endothelial cell growth in vitro (OOI et al. 1983), whereas they are potent angiogenic agents in vivo (FRATER-SCHRODER et al. 1987; LEIBOVICH et al. 1987). A critical difference between IL-1 and both bFGF and TNF-α is that the latter two molecules induce capillary formation in the near absence of an inflammatory response. Since inflammation in itself is a potent stimulator of angiogenesis, it is unclear whether TNF-α or IL-1 causes angiogenesis through the stimulated infiltration of macrophages which in turn produce bFGF and other angiogenic factors.

TGF-β also stimulates angiogenesis in the presence of an inflammatory response (ROBERTS et al. 1986). On the other hand, reports show TGF-β to be a transitory inhibitor of both endothelial cell migration and proliferation (HEIMARK et al. 1986). TGF-β is capable of stimulating both angiogenesis and fibrosis, but fails to stimulate endothelial cell growth, which suggests that this molecule may initially downregulate tissue repair. Apparently, complex interactions occur which regulate the way these factors affect endothelial cell function and the entire tissue repair response in general. Since activated endothelial cells produce proteases that may contribute to clot dissolution, one can postulate that the early release of TGF-β into the wound may initially promote clot formation because of its inhibitory effects on bFGF. Once the action of TGF-β wanes, bFGF-induced angiogenesis would then proceed followed by clot dissolution and healing.

D. Remodeling Phase: Matrix Turnover and Fibrotic Disorders

Normal wound healing is characterized by the formation of granulation or scar tissue. As inflammatory cells and fibroblasts disappear from the scar, it becomes increasingly acellular in nature. Normal regulatory pathways cause the controlled shutdown of matrix and blood vessel formation. This basically differs from normal homeostatic remodeling in which senescent, rather than damaged matrix components are coordinately removed and replaced. Both IL-1 and TNF-α promote necrosis and cell growth either directly or through the stimulated release of other soluble mediators (e.g., collagenase) and the increased expression of cell surface receptors. These paradoxical functions suggest that IL-1 and TNF-α play critical roles in tissue remodeling. A novel membrane-associated receptor on monocytes/macrophages has been identified which recognizes proteins modified by advanced glycosylation end products (VLASSARA et al. 1985). These modifications represent nonenzymatic additions of glucose to amino groups present in the polypeptide backbone of tissue proteins. They subsequently undergo dehydration and accumulate over time and may serve as biological markers of protein age. Recently, it has been demonstrated that both IL-1 and TNF-α are released after this receptor-ligand interaction (VLASSARA et al. 1988). It is possible that binding with receptors on monocytes would cause preferential removal of these modified proteins and concomitant synthesis of new matrix components through the action of IL-1 and TNF-α. Precisely how this system would function in a wound-healing response is not clear but regulatory imbalances would indeed result in pathological consequences.

Several human diseases such as pulmonary fibrosis, scleroderma, rheumatoid arthritis, and cirrhosis are characterized by the inappropriate deposition of fibrotic tissue or an unresolved inflammatory response. Continued activation of the mononuclear cell infiltrate by factors produced during inflammation causes the chronic secretion of inflammatory cytokines which, in turn, amplify and perpetuate the inflammatory lesion. In this setting, infiltrating macrophages continue to produce the same factors that are generated during the acute stages of the response. The continued effect of these molecules on various cells, however, causes alterations in the outcome of the host response. Macrophage products such as IL-1 contribute to the accumulation and activation of lymphocytes which, in turn, causes lymphokine secretion (e.g., IFN-γ, IL-2, and GM-CSF), resulting in further macrophage activation and inflammatory mediator release. As this cycle continues, fibroblasts and endothelial cells also accumulate and, due to constant exposure to growth and angiogenic factors, produce excessive amounts of collagen and other extracellular matrix components. Enzymes such as collagenase, a product of both macrophages and mesenchymal cells, are released and break down these newly formed matrix components into peptide fragments. These fragments lead to additional recruitment and activation of both macrophages and mesenchymal cells which lead to the synthesis and deposition of more collagen.

Animal models have been developed that mimic the unregulated tissue repair process. For example, a system has been developed where hepatic granulomas and chronic polyarthritis are induced by bacterial cell wall antigens (for review, see WAHL 1988). Specifically, deposition of group A streptococcal cell walls in the livers of genetically susceptible LEW/N rats results in a prolonged attempt by the host to repair the injury. This eventually leads to the development of chronic inflammation and fibrotic lesions. Analysis of the cellular makeup of liver granulomas reveals the lesions are composed primarily of T cells and macrophages. T cells isolated from such tissue elaborate a variety of lymphokines including IL-2, CSFs, LDCF, and TNF. These mediators cause enhanced recruitment and activation of additional inflammatory cells which produce factors such as IL-1, TNF-α, and PGE_2. Together, T cells and macrophages produce chemotactic signals for both fibroblasts and endothelial cells. The dependence upon T cells for the evolution of fibrotic lesions was demonstrated through the use of athymic nude rats (ALLEN et al. 1985). Injection of streptococcal cell walls (SCW) into athymic rats that lacked functional T-cell populations resulted in the absence of granuloma formation whereas normal fibrotic development was seen in their euthymic littermates. Further evidence of the T-cell-dependent nature of granuloma formation was obtained through the use of the T-cell-specific immunosuppressant cyclosporin A (CsA). Rats injected with both CsA and SCW showed no development of hepatic granulomas (YOCUM et al. 1986; WAHL et al. 1986a, b). Furthermore, lymphocytes treated in vivo or in vitro with CsA showed markedly reduced abilities to secrete lymphokines, including fibroblast growth factors. Therefore, in the absence of functional T-cell populations, SCW antigens cannot induce the development of fibrotic lesions.

After the initial deposition of cell wall antigen, inflammation and subsequent tissue repair proceeds normally. However, since bacterial cell walls are poorly degraded, there is a continued source of activation for macrophages and lymphocytes; therefore, the inflammatory response is prolonged. The cell wall antigens are compartmentalized and sequestered by macrophages and other leukocytes until a granuloma develops. During the ensuing weeks, there is increased activation and proliferation of fibroblasts and endothelial cells leading to matrix deposition and the formation of fibrous nodules in association with the granulation tissue. This process leads to impaired organ function since the unresolved inflammation results in a fibrous replacement of the hepatic parenchyma.

The ability of rats to display this fibrotic pathology is apparently controlled by non-major histocompatibility genes that are inherited as a dominant or codominant trait. Treatment of F344/N strain rats, which share major histocompatibility genes with LEW/N rats, with bacterial cell walls does not result in chronic arthritic or fibrotic responses (WILDER et al. 1982; 1983). The mechanism(s) by which these genetic differences account for susceptibility to fibrosis is unclear but attention has been focused on the macrophage and its ability to release inflammatory mediators. For example, susceptible rat strains release greater amounts of PGE_2 and LTB_4 compared with rats from resistant strains (YOSHINO et al. 1985). Furthermore, when macrophages were isolated from susceptible and resistant strains, it was observed that LEW/N macrophages generated higher levels of arachidonic acid metabolites and IL-1 compared with F344/N macrophages (FELDMAN et al. 1987). Whether the lower levels of inflammatory mediators and growth factors can prevent a chronic response in F344/N rats is not clear, but that macrophages do appear to play a role in the development of susceptibility is evident.

Macrophages appear to be vital in the pathology of many fibrotic diseases. Fluids isolated from experimental animals or human patients with chronic inflammatory or fibrotic diseases show abnormally high levels of fibroblast-activating factors. For example, extensive fibrosis in schistosomiasis has been shown to be the result of increased fibroblast activation due to the fibroblast-stimulating activity produced by egg granuloma-derived macrophages (WYLER et al. 1984). In addition, synovial fluids from arthritic patients have been shown to spontaneously produce high levels of fibroblast-activating factors (WAHL et al. 1985), including IL-1 (WOOD et al. 1983; NOURI et al. 1984), TNF-α (MAHMOOD et al. 1987), IL-6 (HOUSSIAU et al. 1988b), and PDGF (ROSS 1988, personal communication). Not to be left out, however, is the role of T cells in fibrotic disorders since IFN-γ has also been demonstrated to be present in synovial fluids of arthritic patients (DEGRE et al. 1983).

E. Concluding Remarks

Tissue repair is characterized by a complex cascade of cellular and biochemical events that is under the precise regulatory control of soluble mediators (see Fig. 1). Repair begins after an acute inflammatory response is initiated by a variety of mechanisms such as activation of platelets, complement, or coagulation. Subsequently, chemotactic factors are elaborated and inflammatory leukocytes are recruited to the site of tissue injury. Eventually, more chemotactic and activation signals are elaborated which induce the migration of mesenchymal cells such as fibroblasts and endothelial cells. Their activation leads to the synthesis and deposition of extracellular matrix components and the formation of new blood vessels. During this spurt of matrix synthesis, there is constant degradation of existing or damaged matrix components.

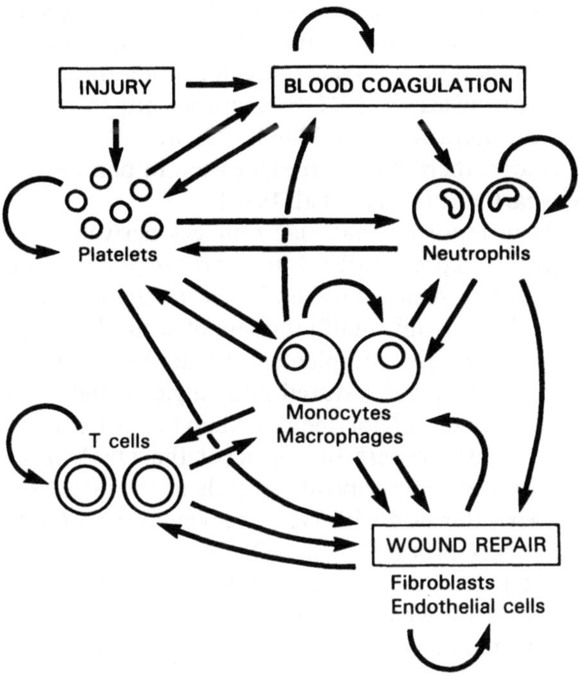

Fig. 1. Wound repair involves a complex series of interactions among numerous cell types which are depicted in this schematic as a *series of arrows*. These interactions are made possible through the release or recognition of soluble factors that can act in both a paracrine and autocrine fashion. The central figure in this sequence is the monocyte/macrophage since it is capable of producing all the critical factors involved in tissue repair. See Table 1 for a listing of factors that are involved in this process

Tissue fibrosis is characterized by the accumulation of excessive fibroblasts followed by the unregulated deposition of collagen and other matrix materials. Understanding the events associated with these processes and how they are regulated may lead to comprehension of the pathogenesis of a number of disease states associated with unresolved inflammation and fibrosis. Therapeutic approaches and methodologies that can modify or control the repair process may be uncovered. Since cell proliferation is a critical element in inflammation and repair, the use of growth factors (e.g., bFGF, TGF-β, and PDGF) offers a promising approach to the manipulation of these processes. A major question is whether these factors are actually expressed in wounds in vivo and whether they play a physiologically relevant role in repair. In this regard, numerous wound models have been developed over the years to examine the therapeutic value of exogenous growth factors, whether applied topically or embedded in slow-release capsules or collagen sponges. For the most part, growth factors have been shown to increase the rate of wound healing, whether measured as a function of wound size, collagen content, histological staining, wound strength, vascularization, or formation of granulation tissue (e.g., SCHREIBER et al. 1986; SCHULTZ et al. 1987; JENNISCHE et al. 1987; LYNCH et al. 1987; MUSTOE et al. 1987; THOMPSON et al. 1988). In addition, studies have shown the presence of mitogenic and angiogenic activities in fluids derived from wounds (BANDA et al. 1982; SPENCER et al. 1988) and from synovial joints of arthritic patients (see above). Lastly, expression of receptors for these growth factors has also been shown to increase on the surfaces of connective tissue present in inflamed synovium, therefore suggesting a role for growth factors in the stimulation of connective cell proliferation that accompanies chronic inflammatory diseases. PDGF receptors, for instance, appear in the tunica media of proliferating blood vessels as well as on cells present in the stroma. Furthermore, there was high expression in pannus tissue close to infiltrated bone and cartilage (Rubin et al. 1988).

Recently, however, it has been reported that macrophages derived from the wound-healing environment do express mRNA for various polypeptide growth factors (Rappolee et al. 1988). A novel method of RNA analysis was used in which total RNA was isolated from small numbers of wound macrophages and reverse transcribed to obtain complementary double-stranded DNA (cDNA) fragments. The cDNA was denatured into single strands and then amplified through a series of sequential polymerase chain reactions using growth factor-specific oligonucleotide primers that bracket the DNA fragment of interest. These primers hybridize to opposite strands of the target sequence and are oriented so primer extension with DNA polymerase proceeds across the region between the two primers. This effectively doubles the amount of target DNA and, after denaturation, the extension products are capable of binding additional primers, which allows for repeated extension cycles. Thus, the target DNA sequence is expanded exponentially and can then be visualized using standard electrophoretic techniques. Through this procedure, macrophages isolated from healing wounds were shown to express mRNA for TGF-α, TGF-β, PDGF, EGF, and IGF-I.

Identifying the proper sequence and combination in which these growth factors act may enable the use of recombinant forms of these molecules for therapeutic and clinical purposes. In addition, antagonists, either in the form of antibodies or structural analogs, could also be developed for the benefit of patients with certain diseases. However, clinical application of recombinant molecules to modify inflammatory and tissue repair responses should be tempered with the experience gained during recent anticancer trials involving IL-2. Systemic exposure of patients to recombinant forms of IL-2 resulted in numerous side effects, ranging from mild to life threatening (Rosenberg et al. 1985; Rosenberg and Lotze 1986). Unfortunately, IL-2 not only affected designated target cells but also caused unwanted stimulation of "bystander" cells and tissues, thereby leading to the observed deleterious side effects. Use of this approach must be replaced with more sophisticated, subtler, and, therefore, more specific methods. Unlike the disseminated nature of metastic cancers, the cellular and biochemical interactions that occur during inflammation and tissue repair in specific sites such as an arthritic joint or wound are, for the most part, localized phenomena. Restricted exposure could easily be accomplished through the use of localized injections or slow release delivery vehicles. However, not all unresolved inflammatory and wound-healing responses are limited to defined regions. Therefore, strategies must be devised that would limit the effect of therapeutic agents to relevant target cells or tissue.

Acknowledgments. We gratefully appreciate the efforts of Drs. Edward Janoff and Carl Manthey for critically reviewing this manuscript.

References

Aarden LA, Lansdorp PM, DeGroot ER (1985) A growth factor for B cell hybridomas produced by human monocytes. Lymphokines 10:175–186

Adelmann-Grill BC, Heinz R, Wach F, Krieg T (1987) Inhibition of fibroblast chemotaxis by recombinant human interferon-γ and interferon-α. J Cell Physiol 130:270–275

Agelli M, Sobel ME, Wahl SM (1987) Cytokine modulation of fibroblast collagen production. Fed Proc 46:924

Allen JB, Malone DG, Wahl SM, Calandra GB, Wilder RL (1985) The role of the thymus in streptococcal cell wall-induced arthritis and hepatic granuloma formation: comparative studies of pathology and cell wall distribution in athymic and euthymic rats. J Clin Invest 76:1042–1056

Allen TD, Schor SL (1983) The contraction of collagen matrices by dermal fibroblasts. J Ultrastruct Res 83:205–219

Andersen DC, Schmalsteig FC, Kohl MA, Tosi MF, Dana N, Buffone GJ, Hughes BJ, Brinkley BR, Dickey WD, Abramson JS, Springer T, Boxer LA, Hollers JM, Smith CW (1984) Abnormalities of polymorphonuclear leukocyte function associated with heritable deficiency of a high molecular weight surface glycoprotein (gp 138): common relationship to diminished cell adherence. J Clin Invest 74:536–551

Arnaout MA, Wang EA, Clark SC, Sieff CA (1986) Human recombinant granulocyte-macrophage colony-stimulating factor increases cell-cell adhesion and surface expression of adhesion-promoting surface glycoproteins on mature granulocytes. J Clin Invest 78:597–601

Assoian RK, Sporn MB (1986) Type beta transforming growth factor in human platelets: release during platelet degranulation and action on vascular smooth muscle cells. J Cell Biol 102:1217–1223

Assoian RK, Frolik CK, Roberts AB, Miller DM, Sporn MB (1984) Transforming growth factor-beta controls receptor levels for epidermal growth factor in NRK fibroblasts. Cell 36:35–41

Assoian RK, Fleurdelys BE, Stevenson HC, Miller PJ, Madtes DK, Raines EW, Ross R, Sporn MB (1987) Expression and secretion of type β transforming growth factor by activated human macrophages. Proc Natl Acad Sci USA 84:6020–6024

Bagby GC Jr, Dinarello CA, Wallace P, Wagner C, Kefeneider S, McCall E (1986) Interleukin 1 stimulates granulocyte-macrophage colony-stimulating activity release by vascular endothelial cells. J Clin Invest 78:1316–1323

Baird A, Mormede P, Bohlen P (1985) Immunoreactive fibroblast growth factor in cells of peritoneal exudate suggests its identity with macrophage-derived growth factor. Biochem Biophys Res Commun 126:358–364

Banda MJ, Knighton DR, Hunt TK, Werb Z (1982) Isolation of a nonmitogenic angiogenesis factor from wound fluid. Proc Natl Acad Sci USA 79:7773–7777

Bassols A, Massague J (1988) Transforming growth factor β regulates the expression and structure of extracellular matrix chondroitin/dermatan sulfate proteoglycans. J Biol Chem 263:3039–3045

Bauer EA, Cooper TW, Huang JS, Altman J, Deuel TF (1985) Stimulation of in vitro human skin collagenase expression by platelet-derived growth factor. Proc Natl Acad Sci USA 82:4132–4136

Baumann H, Muller-Eberhard U (1987) Synthesis of hemopexin and cysteine protease inhibitor is coordinately regulated by HSF-II and interferon-β_2 in rat hepatoma cells. Biochem Biophys Res Commun 146:1218–1228

Berton G, Zeni L, Cassatella MA, Rossi F (1986) Gamma interferon is able to enhance the oxidative metabolism of human neutrophils. Biochem Biophys Res Commun 138:1276–1282

Beutler BA, Cerami A (1985) Recombinant interleukin 1 suppresses lipoprotein lipase activity in 3T3-L1 cells. J Immunol 135:3969–3971
Beutler BA, Cerami A (1986) Cachectin and tumor necrosis factor as two sides of the same biological coin. Nature 320:584–588
Bevilacqua MP, Pober JS, Wheeler ME, Cotran RS, Gimbrone MA (1985) Interleukin 1 acts on cultured human vascular endothelium to increase the adhesion of polymorphonuclear leukocytes, monocytes and related leukocyte cell lines. J Clin Invest 76:2003–2011
Bitterman PB, Rennard SI, Hunninghake GW, Crystal RG (1982) Human alveolar macrophage growth factor for fibroblasts: regulation and partial characterization. J Clin Invest 70:806–822
Bowen TJ, Ochs HD, Altman LC, Price TH, van Epps DE, Brautigan DL, Rosen RE, Perkins WD, Babior BM, Klebanoff SJ, Wedgewood RJ (1982) Severe recurrent bacterial infections associated with defective adherence and chemotaxis in two patients with neutrophils deficient in a cell-associated glycoprotein. J Pediatr 101:932
Boyle MDP, Lawman MJP, Gee AP, Young MC (1985) Nerve growth factor: a chemotactic factor for polymorphonuclear leukocytes in vivo. J Immunol 134:564
Brinckerhoff CE, Guyre PM (1985) Increased proliferation of human synovial fibroblasts treated with immune interferon. J Immunol 134:3142–3146
Broudy VC, Kaushansky K, Segal GM, Harlan JM, Adamson JW (1987) Tumor necrosis factor stimulates human endothelial cells to produce granulocyte/macrophage colony-stimulating factor. Proc Natl Acad Sci USA 83:7467–7471
Buchanan MR, Crowley CA, Rosin RE, Gimbrone MA Jr, Babior BM (1982) Studies on the interaction between gp180–deficient neutrophils and vascular endothelium. Blood 60:160–165
Bullock WE, Wright SD (1987) The role of adherence promoting receptors CR3, LFA-1 and p150,95 in binding of *Histoplasma capsulatum* by human macrophages. J Exp Med 165:195–210
Buonassisi V (1973) Sulfated mucopolysaccharide synthesis and secretion in endothelial cell cultures. Exp Cell Res 76:363–368
Burd PR, Freeman GJ, Wilson SD, Berman M, DeKruyff R, Billings PR, Dorf ME (1987) Cloning and characterization of a novel T cell activation gene. J Immunol 139:3126
Carpenter G, Cohen S (1979) Epidermal growth factor. Annu Rev Biochem 48:193–216
Cavender D, Haskard D, Foster N, Ziff M (1987) Superinduction of T lymphocyte-endothelial cell (EC) binding by treatment of EC with interleukin 1 and protein synthesis inhibitors. J Immunol 138:2149–2154
Cerveny TJ, Fass DN, Mann KG (1984) Synthesis of coagulation factor V by cultured aortic endothelium. Blood 63:1467–1474
Cheifetz S, Weatherbee JA, Tsang ML-S, Anderson JK, Mole JE, Lucas R, Massague J (1987) The transforming growth factor-β system, a complex pattern of cross-reactive ligands and receptors. Cell 48:409–415
Cheifetz S, Bassols A, Stanley K, Ohta M, Greenberger J, Massague J (1988) Heterodimeric transforming growth factor β: biological properties and interaction with three types of cell surface receptor. J Biol Chem 263:10783–10789
Cohen S (1962) Purification of a nerve-growth promoting protein from the mouse salivary gland and its neuro-cytotoxic antiserum. Proc Natl Acad Sci USA 46:302–311
Cohen S, Carpenter G (1975) Human epidermal growth factor: isolation and chemical and biological properties. Proc Natl Acad Sci USA 72:1317–1321
Collins T, Lapierre LA, Fiers W, Strominger JL, Pober JS (1986) Recombinant human tumor necrosis factor increases mRNA levels and surface expression of HLA-A,B antigens in vascular endothelial cells and dermal fibroblasts in vitro. Proc Natl Acad Sci USA 83:466–450

Conti P, Cifone MG, Alesse E, Fieschi G, Angeletti PU (1985) Augmentation of thromboxane production in vitro by polymorphonuclears and macrophages exposed to IL1/LP. Agents Action 17:390–391

De Groot PG, Ottenhof-Rovers M, Mornik JA, Sixma JJ (1988) Evidence that the primary binding site of von Willebrand factor that mediates platelet adhesion on subendothelium is not collagen. J Clin Invest 82:65–73

Degre M, Mellbye OJ, Clarke-Jenssen O (1983) Immune interferon in serum and synovial fluid in rheumatoid arthritis and related disorders. Ann Rheum Dis 42:672–676

Dejana E, Brevario F, Erroi A, Bussolino F, Mussoni L, Gramse M, Pintucci G, Casali B, Dinarello CA, van Damme J, Montovani A (1987) Modulation of endothelial cell function by different molecular species of IL-1. Blood 69:695–699

De la Maza LM, Peterson EM, Fennie CW, Czarniecki CW (1985) The antichlamydial and anti-proliferation activities of recombinant murine interferon-gamma are not dependent upon tryptophan concentrations. J Immunol 135:198–200

Deuel TF, Huang JS (1984) Platelet-derived growth factor: structure, function and roles in normal and transformed cells. J Clin Invest 74:669–676

Deuel TF, Senior KM, Chang D, Griffin GL, Heinrickson RL, Kaiser ET (1981) Platelet factor 4 is chemotactic for neutrophils and monocytes. Proc Natl Acad Sci USA 78:4584–4587

Deuel TF, Senior KM, Huang JS, et al. (1982) Chemotaxis of monocytes and neutrophils to platelet-derived growth factor. J Clin Invest 69:1046–1049

Dinarello CA (1986) IL-1: amino acid sequences, multiple biological activities and comparison with tumor necrosis factor (cachectin). Year Immunol 2:68–89

Dinarello CA (1988) Cytokines: interleukin-1 and tumor necrosis factor (cachectin) In: Gallin JI, Goldstein IM, Snyderman R (eds) Inflammation: basic principles and clinical correlates. Raven, New York, pp 195–208

Duijvestijn AM, Schreiber AB, Butcher EC (1986) Interferon-γ regulates an antigen specific for endothelial cells involved in lymphocyte traffic. Proc Natl Acad Sci USA 83:9114–9118

Duncan MR, Berman B (1985) Gamma interferon is the lymphokine and beta interferon is the monokine responsible for inhibition of fibroblast collagen production and late but not early fibroblast proliferation. J Exp Med 162:516–527

Estes JE, Pledger WJ, Gillespie GY (1984) Macrophage-derived growth factor for fibroblasts and interleukin-1 are distinct entities. J Leuk Biol 35:115–129

Feldman G, Allen J, Swisher J, Pluznick D, Wahl L, Wahl S (1987) Susceptibility to streptococcal cell wall (SCW)-induced polyarthritis is associated with differential macrophage activation (Abstr) J Leuk Biol 42:344

Fleischmann J, Golde DW, Weisbart RH, Gasson JC (1986) Granulocyte-macrophage colony-stimulating factor enhances phagocytosis of bacteria by human neutrophils. Blood 68:708–711

Frater-Schroeder M, Risau W, Hallmann R, Gautschi P, Bohlen P (1987)Tumor necrosis factor type α, a potent inhibitor of endothelial cell growth in vitro, is angiogenic in vivo. Proc Natl Acad Sci USA 84:5277–5281

Gee AP, Boyle MDP, Munger KL, Lawman MJP, Young M (1983) Nerve growth factor: stimulation of polymorphonuclear chemotaxis in vitro. Proc Natl Acad Sci USA 80:721–728

Glenn K, Bowen-Pope DF, Ross R (1982) Platelet-derived growth factor. III. Identification of a platelet-derived growth factor receptor by affinity labelling. J Biol Chem 257:5172–5176

Goetzl EJ, Brash AP, Tauber AI, Oates JA, Hubbard WC (1980) Modulation of human neutrophil function by monohydroxyeicosatetraenoic acids. Immunology 39:491–501

Goldberg ID, Stemerman MB, Handin RI (1980) Vascular permeation of platelet factor 4 alters endothelial injury. Science 209:611–612

Goldring MB, Krane SM (1987) Modulation by recombinant interleukin-1 of synthesis of types I and III collagens and associated procollagen mRNA levels in cultured human cells. J Biol Chem 262:16724–16729

Goldyne ME, Stobo JD (1981) Immunoregulatory role of prostaglandins and related lipids. Crit Rev Immunol 2:189–223

Gospodarowicz D, Bialecki H, Tharkal TK (1979) The angiogenic activity of the fibroblast and epidermal growth factor. Exp Eye Res 28:501–514

Gospodarowicz D, Ferrara N, Schweigerer L, Neufeld G (1987a) Structural characterization and biological functions of fibroblast growth factor. Endocr Rev 8:95–114

Gospodarowicz D, Neufeld G, Schweigerer L (1987b) Fibroblast growth factor: structural and biological properties. J Cell Physiol [Suppl] 5:15–26

Graham R, Ebert RH, Ratnoff OD, Moses JM (1965) Pathogenesis of inflammation. II. In vivo observations of the inflammatory effects of activated Hageman factor and brandykinin. J Exp Med 121:807

Grinnell F, Feld M, Snell W (1979) The influence of cold insoluble globulin on platelet morphological response to substrata. Cell Biol Int Rep 3:585

Gross JL, Moscatelli D, Rifkin DB (1983) Increased capillary endothelial cell protease activity in response to angiogenic stimuli in vitro. Proc Natl Acad Sci USA 80:2623–2627

Gronwald RGK, Grant FJ, Haldeman BA, Hart CE, O'Hara PJ, Hagen FS, Ross R, Bowen-Pope DF, Murray MJ (1988) Cloning and expression of a cDNA coding for the human platelet-derived growth factor receptor: evidence for more than one receptor class. Proc Natl Acad Sci USA 85:3435–3439

Guyre PM, Morganelli PM, Miller R (1983) Recombinant immune interferon increases immunoglobulin G Fc receptors on cultured human mononuclear phagocytes. J Clin Invest 72:393

Halliwell B (1987) Oxidants and human disease: some new concepts. FASEB J 1:358–364

Heimark RL, Twardzik DR, Schwartz SM (1986) Inhibition of endothelial regeneration by type-beta transforming growth factor from platelets. Science 233:1078–1080

Heldin C-H, Johnsson A, Wennergren S, Wernstadt C, Betsholtz C, Westermark B (1986) A human osteosarcoma cell line secretes a growth factor structurally related to a homodimer of PDGF A-chains. Nature 319:511–514

Heldin C-H, Backstrom G, Ostman A, Hammacher A, Rounstrand L, Rubin K, Nister M, Westermark B (1988) Binding of different dimeric forms of PDGF to human fibroblasts: evidence for two separate receptor types. EMBO J 7:1387–1393

Helle M, Brakenhoff JPJ, DeGroot ER, Aarden LA (1988) Interleukin 6 is involved in interleukin 1-induced activities. Eur J Immunol 18:957–959

Hermann F, Cannistra SA, Levine H, Griffin JD (1985) Expression of interleukin 2 receptors on human peripheral blood monocytes. J Exp Med 162:1111

Herron GS, Banda MJ, Clark EJ, Gavrilovic J, Werb Z (1986) Secretion of metalloproteinases by stimulated capillary endothelial cells. II. Expression of collagenase and stromelysin activities is regulated by endogenous inhibitors. J Biol Chem 261:2814–2818

Hirsch JG (1960) Comparative bactericidal activities of blood serum and plasma serum. J Exp Med 112:15–22

Houssiau FA, Coulie PG, Olive D, van Snick J (1988a) Synergistic activation of human T cells by interleukin 1 and interleukin 6. Eur J Immunol 18:653–656

Houssiau FA, DeVogelaer JP, van Damme J, Nagant de Deuxchaisnes C, van Snick J (1988b) Interleukin 6 in synovial fluid and serum of patients with rheumatoid arthritis and other inflammatory arthritides. Arthritis Rheum 31:784–788

Ignotz RA, Massague J (1986) Transforming growth factor-β stimulates the expression of fibronectin and collagen and their incorporation into the extracellular matrix. J Biol Chem 261:4337–4345

Ignotz RA, Endo T, Massague J (1987) Regulation of fibronectin and type I collagen mRNA levels by transforming growth factor-β. J Biol Chem 262:6443–6446

Ihle JN, Rebal L, Keller J, Lee JC, Happel AJ (1982) Interleukin 3: possible roles in the regulation of lymphocyte differentiation and growth. Immunol Rev 63:5–32

Isakov N, Wolfgang S, Altman A (1986) Signal transduction and intracellular events in T-lymphocyte activation. Immunol Today 7:271–275

Jaffe EA, Hoyer LW, Nachman RL (1974) Synthesis of von Willebrand factor by cultured human endothelial cells. Proc Natl Acad Sci USA 71:1906–1909

Jennische E, Skottner A, Hansson H-A (1987) Dynamic changes in insulin-like growth factor I immunoreactivity correlate to repair events in rat ear after freeze thaw injury. Exp Mol Pathol 47:193–201

Jiminez SA, Freundlich B, Rosenbloom J (1984) Selective inhibition of human diploid fibroblast collagen synthesis by interferons. J Clin Invest 74:1112–1116

Kalebic T, Garbisa S, Glaser B, Liotta LA (1983) Basement membrane collagen: degradation by migrating endothelial cells. Science 221:281–283

Kaplan AP, Kay AB, Austen KF (1972) A prealbumin activator of prekallikrein. III. Appearance of chemotactic activity for neutrophils. J Exp Med 135:81–97

Kaplan AP, Goetzl EJ, Austen KF (1973) The fibrinolytic pathway of human plasma. II. The generation of chemotactic activity by activation of plasminogen proactivator. J Clin Invest 52:2591–2595

Kasuga M, van Obberghen E, Nissley SP, Rechler MM (1981) Demonstration of two subtypes of insulin-like growth factor receptors by affinity crosslinking. J Biol Chem 256:5305–5308

Kehrl JH, Wakefield LM, Roberts AB, Jakowlew S, Alvarez-Mon M, Derynck R, Sporn MB, Fauci AS (1986) Production of transforming growth factor-beta by human T lymphocytes and its potential role in the regulation of T cell growth. J Exp Med 163:1037–1050

Keski-Oja J, Leof EB, Lyons RM, Coffey RJ, Moses HL (1986) Transforming growth factors and control of neoplastic growth. J Cell Biochem 33:95–107

Klempner MS, Dinarello CA, Gallin JI (1978) Human leukocyte pyrogen induces release of specific granule contents from human neutrophils. J Clin Invest 61:1330–1336

Klempner MS, Dinarello CA, Henderson WR, Gallin JI (1979) Stimulation of neutrophil oxygen-dependent metabolism by human leukocytic pyrogen. J Clin Invest 64:996–1002

Krane SM, Dayer JM, Simon LS, Byrne LS (1985) Mononuclear cell-conditioned medium containing mononuclear cell factor (MCF), homologous with interleukin-1, stimulates collagen and fibronectin synthesis by adherent rheumatoid synovial cells: effects of prostaglandin E_2 and indomethacin. Coll Relat Res 5:99–117

Kream BE, Rowe DW, Gworek SC, Raiz LG (1980) Parathyroid hormone alters collagen synthesis and procollagen mRNA levels in fetal rat calvaria. Proc Natl Acad Sci USA 77:5654–5658

Landolfo S, Cofano F, Giovarelli M, Prat M, Cavallo G, Forni G (1985) Inhibition of interferon may suppress allograft reactivity by T lymphocytes in vitro and in vivo. Science 229:176–179

Lansdorp PM, Aarden LA, Calafat J, Zilberstein WP (1986) A growth factor-dependent B cell hybridoma. Curr Top Microbiol Immunol 132:105–113

Larsen CG, Anderson AO, Apella E, Oppenheim JJ, Matsushima K (1989) The neutrophil-activating protein (NAP-1) is also chemotactic for T lymphocytes. Science 243:1464–1466

Leibovich SJ, Ross R (1975) The role of macrophages in repair: a study with hydrocortisone and anti-macrophage serum. Am J Pathol 78:71–91

Leibovich SJ, Polverini PJ, Shepard HM, Wiseman DM, Shively V, Nuseir N (1987) Macrophage-induced angiogenesis is mediated by tumor necrosis factor-alpha. Nature 328:630–632

Li AKC, Koroly MJ, Schattlenkech ME, Malt RA, Young M (1980) Nerve growth factor: acceleration of the rate of wound healing. Proc Natl Acad Sci USA 77:4379–4381

Lotner GZ, Lynch JM, Betz SJ, Henson PM (1980) Human neutrophil-derived platelet activating factor. J Immunol 124:676–684
Luster AD, Ravetch JV (1987) Biochemical characterization of a γ-interferon-inducible cytokine (IP-10) J Exp Med 166:1084–1097
Luster AD, Unkeless JC, Ravetch JV (1985) Gamma interferon transcriptionally regulates an early-response gene containing homology to platelet proteins. Nature 315:673–676
Lynch SE, Nixon JC, Colvin RB, Antoniades HN (1987) Role of platelet-derived growth factor in wound healing: synergistic effects with other growth factors. Proc Natl Acad Sci USA 84:7696–7700
Maciag T, Kadish J, Wilkins L, Sternerman MB, Weinstein R (1982) Organizational behavior of human umbilical vein endothelial cells. J Cell Biol 94:511–520
Madri JA, Stenn KS (1982) Aortic endothelial cell migration. I. Matrix requirements and composition. Am J Pathol 106:180–186
Mahmood T, Busch HM, Racis SP, Krey PR (1987) Lymphokines in inflammatory arthritis (Abstr) Arthritis Rheum [Suppl 4] 30:S83
Marcum JA, Rosenberg RD (1984) Anticoagulantly active heparin-like molecules from vascular tissue. Biochemistry 23:1730–1737
Marcus AJ, Safier LB, Ullman HL, Broekman MJ, Islam N, Oglesby T, Gorman RR (1984) 12s,20-Dihydroxyeicosatetraeonic acid: a new eicosanoid synthesized by neutrophils from 12s-hydroxyeicosatetraeonic acid produced by thrombin- or collagen-stimulated platelets. Proc Natl Acad Sci USA 81:903–907
Marder SR, Chenoweth DE, Goldstein IM, Perez HD (1985) Chemotactic responses of human peripheral blood monocytes to the complement derived proteins C5a and C5a des arg. J Immunol 134:3325–3331
Marquardt H, Hunkapiller MW, Hood LE, Todaro GJ (1984) Rat transforming growth factor type I: structure and relation to epidermal growth factor. Science 223:1079–1082
Martin BM, Gimbrone MA Jr, Unanue ER, Cotran RS (1981) Stimulation of non-lymphoid mesenchymal cell proliferation by a macrophage-derived growth factor. J Imunol 126:1510–1515
Massague J, Cheifetz S, Ignotz R, Boyd FT (1987) Multiple type-beta transforming growth factors and their receptors. J Cell Physiol [Suppl] 5:43–48
May LT, Helfgott DC, Seghal PB (1986) Anti-β-interferon antibodies inhibit the increased expression of HLA-B7 mRNA in tumor necrosis factor-treated human fibroblasts: structural studies of the $\beta 2$ interferon involved. Proc Natl Acad Sci USA 83:8957–8961
McDonald JA, Kelly DG, Broekelmann J (1982) Role of fibronectin in collagen deposition: Fab, to the gelatin-binding domain of fibronectin inhibits both fibronectin and collagen organization in fibroblast extracellular matrix. J Cell Biol 92:485
Mensing H, Czarnetozki BM (1984) Leukotriene B_4 induces in vitro fibroblast chemotaxis. J Invest Dermatol 82:9–12
Metcalf D (1985) The granulocyte-macrophage colony-stimulating factors. Science 229:16–22
Miller JM (1985) The structure of fibril-forming collagens. Ann NY Acad Sci 460:1–13
Miyazono K, Okabe T, Urabe A, Takaku F, Heldin CH (1987) Purification and properties of an endothelial cell growth factor from human platelets. J Biol Chem 262:4098–4103
Montesano R, Orci L (1988) Transforming growth factor β stimulates collagen-matrix contraction by fibroblasts: implications for wound healing. Proc Natl Acad Sci USA 85:4894–4897
Montesano R, Mossez A, Ryser J-E, Orci L, Vassali P (1984) Leukocyte interleukins induce cultured endothelial cells to produce a highly organized glycosaminoglycan-rich pericellular matrix. J Cell Biol 99:1706–1715
Moscatelli D, Jaffe E, Rifkin DB (1980) Tetradecanoyl phorbol acetate stimulates latent collagenase production by cultured human endothelial cells. Cell 20:343–351

Moses HL, Tucker RF, Leof EB, Coffey RJ Jr, Halper J, Shipley GD (1985) Type beta transforming growth factor is a growth stimulator and a growth inhibitor. Cancer Cells 3:65–71

Muggli R, Baumgartner HR (1973) Collagen induced platelet aggregation: requirement for tropocollagen multimers. Thromb Res 3:715

Muller R, Bravo R, Burckhardt J, Curran T (1984) Induction of c-*fos* gene and protein by growth factors precedes activation of c-*myc*. Nature 312:716–720

Munker R, Gasson J, Ogawa M, Koeffler HP (1986) Recombinant human necrosis factor induces production of granulocyte-macrophage colony stimulating factor mRNA and protein from lung fibroblasts and vascular endothelial cells in vitro. Nature 323:79–82

Mustoe TA, Pierce GF, Thomason A, Gramates P, Sporn MB, Deuel TF (1987) Accelerated wound healing of incisional wounds in rats by transforming growth factor-β. Science 237:1333–1338

Narayanan AS, Page RC (1983) Biosynthesis and regulation of type V collagen in diploid human fibroblasts. J Biol Chem 258:11 694–11 699

Nathan CF (1987) Secretory products of macrophages. J Clin Invest 79:319–326

Nathan CF, Yoshida R (1988) Cytokines: interferon-γ. In: Gallin JI, Goldstein IM, Snyderman R (eds) Inflammation: basic principles and clinical correlates. Raven, New York, pp 229–251

Nawroth P, Stern D (1986) Modulation of endothelial cell hemostatic properties by tumor necrosis factor. J Exp Med 164:740–749

Nawroth PP, Bark I, Handley D, Cassimeris J, Ches L, Stern D (1986) Tumor necrosis factor/cachectin interacts with endothelial cell receptors to induce release of interleukin 1. J Exp Med 163:1363–1375

Nedwin GE, Svedersky LP, Bringman TS, Palladino MA Jr, Goeddel DV (1985) Effect of interleukin 2, interferon-gamma and mitogens on the production of tumor necrosis factors alpha and beta. J Immunol 135:2492–2497

Niall M, Ryan GB, O'Brien BJ (1982) The effect of epidermal growth factor on wound healing in mice. J Surg Res 33:164–169

Nouri AME, Parrayi GS, Goodman SM (1984) Cytokines and the chronic inflammation of rheumatic disease. I. The presence of interleukin 1 in synovial fluids. Clin Exp Immunol 55:295

Ochs ME, Postlethwaite AE, Kang AH (1987) Identification of a protein in sera of normal individuals that inhibits fibroblast chemotactic and random migration in vitro. J Invest Dermatol 88:183

Oikasinen J, Ryhanen L (1981) Cortisol decreases the concentration of translatable type-I procollagen mRNA species in the developing chick-embryo calvaria. Biochem J 198:519–524

Ooi BS, MacCarthy EP, Hsu A, Ooi YM (1983) Human mononuclear cell modulation of endothelial cell proliferation. J Lab Clin Med 102:428–433

Ortaldo JR, Longo DL (1988) Human natural lymphocyte effector cells: definition, analysis of activity, and clinical effectiveness. JNCI 80:999–1010

Palombella VJ, Yamashino DJ, Maxfield FR, Decker SJ, Vilcek J (1987) Tumor necrosis factor increases the number of epidermal growth factor receptors on human fibroblasts. J Biol Chem 262:1950–1954

Parmley RT, Hurst RE, Takagi M, Spicer SS, Austin RL (1983) Glycosaminoglycans in human neutrophils and leukemic myoblasts: ultrastructural, cytochemical, immunological and biochemical characterization. Blood 61:257–266

Peacock EE, Madden JW (1966) Some studies on the effect of β-aminopropionitrate on collagen in healing wounds. Surgery 60:7–12

Philip R, Epstein LB (1986) Tumor necrosis factor as immunomodulator and mediator of monocyte cytotoxicity induced by itself, γ-interferon and interleukin-1. Nature 323:86–89

Pledger WJ, Stiles CD, Antoniades HN, Scher CD (1977) Induction of DNA synthesis in Balb/c 3T3 cells by serum components: reevaluation of the commitment process. Proc Natl Acad Sci USA 74:4481–4485

Pober JS, Gimbrone MA Jr, Cotran RS, Reiss CS, Burakoff SJ, Fiers W, Ault KA (1983) Ia expression by vascular endothelium is inducible by activated T cells and human γ interferon. J Exp Med 157:1339–1353
Pohlman TH, Stanness KA, Beatty PG, Ochs HD, Harlan JM (1986) An endothelial cell surface factor(s) induced in vitro by lipopolysaccharide, interleukin 1, and tumor necrosis factor-alpha increases neutrophil adherence by a CDw18-dependent mechanism. J Immunol 136:4548–4553
Postlethwaite AE, Kang AH (1983) Latent lymphokines: isolation of human latent lymphocyte-derived chemotactic factor for fibroblasts. In: Oppenheim JJ, Cohen S (eds) Interleukins, lymphokines and cytokines. Proceedings of the third international lymphokine workshop. Academic, New York, pp 535–541
Postlethwaite AE, Kang AH (1988) Fibroblasts. In: Gallin JI, Goldstein IM, Snyderman R (eds) Inflammation: basic principles and clinical correlates. Raven, New York, pp 577–597
Postlethwaite AE, Snyderman R, Kang AH (1976) The chemotactic attraction of human fibroblasts to a lymphocyte-derived factor. J Exp Med 144:1188–1203
Postlethwaite AE, Lachman LB, Mainardi CL, Kang AH (1983) Stimulation of fibroblast collagenase production by human interleukin-1. J Exp Med 157:801–806
Postlethwaite AE, Keski-Oja J, Moses HL, Kang AH (1987) Stimulation of the chemotactic migration of human fibroblasts by transforming growth factor β. J Exp Med 165:251–256
Raines EW, Dower SK, Ross R (1989) IL-1 mitogenic activity for fibroblasts and smooth muscle cells is due to PDGF-AA. Science (in press)
Rappolee DA, Mark D, Banda MJ, Werb Z (1988) Wound macrophages express TGF-α and other growth factors in vivo: analysis by mRNA phenotyping. Science 241:708–712
Rinderknecht E, Humbel RE (1978a) Primary structure of human insulin-like growth factor II. FEBS Lett 89:283–286
Rinderknecht E, Humbel RE (1978b) The amino acid sequence of insulin-like growth factor I and its structural homology with proinsulin. J Biol Chem 253:2769–2776
Roberts AB, Sporn MB, Assoian RK, Smith JM, Roche NS, Wakefield LM, Heine UI, Liotta LA, Falanga V, Kehrl JH, Fauci AS (1986) Transforming growth factor type β: rapid induction of fibrosis and angiogenesis in vivo and stimulation of collagen formation in vitro. Proc Natl Acad Sci USA 83:4167–4171
Robinson BW, McLemore TL, Crystal RG (1985) Gamma interferon is spontaneously released by alveolar macrophages and lung T lymphocytes in patients with pulmonary sarcoidosis. J Clin Invest 75:1488–1495
Rola-Pleszczynski M, Lieu H, Hamel J, Lemaire I (1982) Stimulated human lymphocytes produce a soluble factor which inhibits fibroblast migration. Cell Immunol 74:104–110
Rosenberg SA, Lotze MT (1986) Cancer immunotherapy using interleukin-2 and interleukin-2-activated lymphocytes. Annu Rev Immunol 4:681–709
Rosenberg SA, Lotze MT, Muul LM, Leitman S, Chang AE, Ettinghausen SE, Matory YL et al. (1985) Observations on the systemic administration of autologous lymphokine-activated killer cells and recombinant interleukin-2 to patients with metastatic cancer. N Engl J Med 313:1485–1492
Ross R, Glomset B, Kariya B, Harker L (1974) A platelet-dependent serum factor that stimulates the proliferation of arterial smooth muscle cells in vitro. Proc Natl Acad Sci USA 71:1207–1210
Ross R, Raines EW, Bowen-Pope DF (1986) The biology of platelet-derived growth factor. Cell 46:155–169
Rossi P, Karsenty G, Roberts MB, Roche NS, Sporn MB, deCrombrugghe B (1988) A nuclear factor 1 binding site mediates the transcriptional activation of a type I collagen promoter by transforming growth factor-beta. Cell 52:405–414
Rubin K, Terracio L, Ronnstrand L, Heldin CH, Klareskog L (1988) Expression of platelet-derived growth factor receptors is induced on connective tissue cells during chronic synovial inflammation. Scand J Immunol 27:285–294

Sage H, Bornstein PL (1982) Endothelial cells from umbilical vein and hemangioendothelioma secrete basement membrane largely to the exclusion of interstitial collagens. Arteriosclerosis 2:27–36

Sage H, Pritzel P, Bornstein P (1981) Characterization of cell matrix-associated collagens synthesized by aortic endothelial cells in culture. Biochem 20:436–442

Sanchez-Madrid F, Nagey JA, Robbins E, Simon P, Springer TA (1983) A human leukocyte antigen family with distinct α-subunits and a common β-subunit: the lymphocyte function associated antigen (LFA-1), the C3bi complement receptor (OKM1/Mac-1) and the p150,95 molecule. J Exp Med 158:586

Sandberg N (1964) Time relationship between administration of cortisone and wound healing in rats. Acta Chir Scand 127:446–455

Sato N, Goto T, Haranaka K, Satoni N, Nariuchi H, Mano-Hirano Y, Sawasaki Y (1986) Actions of tumor necrosis factor on cultured vascular endothelial cells: morphologic modulation, growth inhibition and cytotoxicity. JNCI 76:113

Schall TJ, Jongstra J, Dyer BJ, Jorgensen J, Clayberger C, Davis MM, Krensky AM (1988) A human T cell specific molecule is a member of a new gene family. J Immunol 141:1018–1025

Schapira M, Despland E, Scott CF, Boxer LA, Colman RW (1982) Purified human plasma kallikrein aggregates human blood neutrophils. J Clin Invest 69:1199–1201

Scheiwiller E, Guler HP, Merryweather J, Scandella C, Maerki W, Zapf J, Froesch ER (1986) Growth restoration of insulin-deficient diabetic rats by recombinant human insulin-like growth factor I. Nature 323:169–171

Scher CD, Shephard RC, Antoniades HN, Stiles CD (1979) Platelet-derived growth factor and the regulation of the mammalian fibroblast cell cycle. Biochem Biophys Acta 560:212–241

Schmidt JA, Mizel SB, Cohen D, Green I (1982) Interleukin 1, a potential regulator of fibroblast proliferation. J Immunol 128:2177–2182

Schreiber AB, Winkler ME, Derynck R (1986) Transforming growth factor-α: a more potent angiogenic mediator than epidermal growth factor. Science 232:1250–1253

Schultz GS, White M, Mitchell R, Brown G, Lynch J, Twardzik DR, Todaro GJ (1987) Epithelial wound healing by transforming growth factor-α and vaccinia growth factor. Science 235:350–352

Seppa H, Seppa S, Yamada KM (1980) The cell binding fragment of fibronectin and platelet-derived growth factor are chemoattractants for fibroblasts. J Cell Biol 87:323 a

Shahabuddin S, Kumar S, West D, Arnold F (1985) A study of angiogenesis factors from five different sources using a radioimmunoassay. Int J Cancer 35:87–91

Shallaby MR, Aggarwal BB, Rinderknecht E, Sverdersky LP, Finkle BS, Palladino MA Jr (1985) Activation of human polymorphonuclear neutrophil function by interferon gamma and tumor necrosis factors. J Immunol 135:2069–2073

Sharon M, Siegel JP, Tosato G, Yodoi J, Gerrard TL, Leonard WJ (1988) The human interleukin 2 receptor β chain (p70). Direct identification, partial purification and patterns of expression on peripheral blood mononuclear cells. J Exp Med 167: 1265–1270

Shimokado K, Raines EW, Madtes DK, Barrett TB, Benditt EP, Ross R (1985) A significant part of macrophage-derived growth factor consists of at least two forms of PDGF. Cell 43:277–286

Sieff CA (1987) Hematopoietic growth factors. J Clin Invest 79:1549–1557

Simpson DW, Ross R (1972) The neutrophilic leukocyte in wound repair: a study with anti-neutrophil serum. J Clin Invest 51:2009–2023

Sluiter W, Elzenga-Claasen I, Hulsing-Hesselink E, van Furth R (1983) Presence of the factor increasing monocytopoiesis (FIM) in rabbit peripheral blood during acute inflammation. J Reticuloendothel Soc 34:235–252

Sluiter W, Hulsing-Hesselink E, Elzenga-Classen I, van Hemsberger-Oomens LWM, van der Voort van der Kleij-van Andel A, van Furth R (1987) Macrophages as origin of factor increasing monocytopoiesis. J Exp Med 166:909–922

Smith P, Lamerson C, Wong HL, Wahl SM (1989) Granulocyte-macrophage colony-stimulting factor augmentation of leukocyte effector function. J Immunol (in press)

Smith RJ, Speziale SC, Bowman BJ (1985) Properties of interleukin-1 as a complete secretogogue for human neutrophils. Biochem Res Commun 130:1233–1240

Snyderman R, Uhling RJ (1988) Stimulus-response coupling mechanisms. In: Gallin JI, Goldstein IM, Snyderman R (eds) Inflammation: basic principles and clinical correlate. Raven, New York, pp 309–323

Spencer EM, Skover G, Hunt TK (1988) Somatomedins: do they play a pivotal role in wound healing? In: Barbul A, Pines E, Caldwell M, Hunt TK (eds) Growth factors and other aspects of wound healing: biological and clinical implications. Liss, New York, pp 103–116

Sporn MB, Roberts AB, Wakefield LM, de Crombrugghe B (1987) Some recent advances in the chemistry and biology of transforming growth factor-beta. J Cell Biol 105:1039–1045

Steinman BU, Abe S, Martin GR (1982) Modulation of type I and type II collagen production in normal and mutant skin fibroblasts by cell density, prostaglandin E_2 and epidermal growth factor. Coll Relat Res 2:185–195

Stephenson ML, Krane SM, Amento EP, McCroskery PA, Byrne M (1985) Immune interferon inhibits collagen synthesis by rheumatoid synovial cells associated with decreased levels of procollagen mRNAs. FEBS Lett 180:43–50

Stern D, Bank I, Nawroth P, Cassimeris J, Kisiel W, Fenton JW, Dinarello CA, Chess L, Jaffe EA (1985) Self regulation of procoagulant events on the endothelial cell surface. J Exp Med 162:1223–1228

Stiles CD, Capone GT, Scher CD, Antoniades HN, Wyk JJ, Pledger WJ (1979) Dual control of cell growth by somatomedins and platelet-derived growth factor. Proc Natl Acad Sci USA 76:1279–1283

Stroobant P, Waterfield MD (1984) The c-*sis* gene encodes a precursor of the B chain of platelet-derived growth factor. EMBO J 3:921–928

Sugarman BJ, Aggarwal BB, Figari IS, Palladino MA, Shepard HM (1985) Recombinant human tumor necrosis factor-α: effects on proliferation of normal and transformed cells. Science 230:943–945

Sullivan R, Gans PJ, McCarrol LA (1983) The synthesis and secretion of granulocyte-monocyte colony stimulating activity (CSA) by isolated human monocytes: kinetics of the response to bacterial endotoxin. J Immunol 130:800–807

Takemura S, Rossing TH, Perlmutter DH (1986) A lymphokine regulates expression of alpha-1-protease inhibitor in human monocytes and macrophages. J Clin Invest 77:1207–1213

Thompson JA, Anderson KD, DiPietro JM, Zwiebel JA, Zametta M, Anderson WF, Maciag T (1988) Site-directed neovessel formation in vivo. Science 241:1349–1352

Todaro GJ, Fryling C, DeLarco JE (1980) Transforming growth factors produced by certain human tumor cells: polypeptides that interact with epidermal growth factor receptors. Proc Natl Acad Sci USA 77:5258–5262

Tolstoshev P, Haber R, Trapne BC II, Crystal RB (1981) Procollagen mRNA levels and activity and collagen synthesis during fetal development of sheep lung and skin. J Biol Chem 256:9672–9679

Tsunawaki S, Sporn M, Ding A, Nathan C (1988) Deactivation of macrophages by transforming growth factor-β. Nature 334:260–262

Turk CW, Dohlman TG, Goetzl EJ (1987) Immunological mediators of wound healing and fibrosis. J Cell Physiol [Suppl] 5:89–93

Unkeless JC, Gordon SC, Reich E (1974) Secretion of plasminogen activator by stimulated macrophages. J Exp Med 139:834–850

Van Furth R, Diesselhoff-den Dulk MMC, Mattie H (1973) Quantitative study on the production and kinetics of mononuclear phagocytes during an acute inflammatory reaction. J Exp Med 138:1314–1330

Van Waarde D, Hulsing-Hesselink E, Sandkuyl LA, van Furth R (1977) Humoral regulation of monocytopoiesis during the early phase of an inflammatory reaction caused by particulate substances. Blood 50:141–153

Van Waarde D, Hulsing-Hesselink E, van Furth R (1978) Humoral control of monocytopoiesis by an activator and an inhibitor. Agents Actions 8:423–437

Vassali JD, Dayer JM, Wohlwend A, Belin D (1984) Concomitant secretion of prourokinase and of a plasminogen activator inhibitor by cultured human monocytes-macrophages. J Exp Med 159:1653–1668

Vlassara H, Brownlee M, Cerami A (1985) High affinity receptor mediated uptake and degradation of glucose modified proteins: a potential mechanism for the removal of senescent macromolecules. Proc Natl Acad Sci USA 82:5588–5592

Vlassara H, Brownlee M, Manogue KR, Dinarello CA, Pasagian A (1988) Cachectin/TNF and IL-1 induced by glucose-modified proteins: role in normal tissue remodeling. Science 240:1546–1548

Vlodavsky I, Fridman R, Sullivan R, Sasse J, Klagsburn M (1987) Aortic endothelial cells synthesize basic fibroblast growth factor which remains cell associated and platelet-derived growth factor-like protein which is secreted. J Cell Physiol 131:402–408

Wahl LM, Wahl SM, Mergenhagen SE, Martin GR (1975) Collagenase production by lymphokine activated macrophages. Science 187:261–263

Wahl SM (1988) Mononuclear cells and fibrosis. In: Gallin JI, Goldstein IM, Snyderman R (eds) Inflammation: basic principles and clinical correlates. Raven, New York, pp 841–860

Wahl SM, Gately CL (1983) Modulation of fibroblast growth by a lymphokine of human T-cell and continuous T-cell line origin. J Immunol 130:1226–1230

Wahl SM, Malone DG, Wilder RL (1985) Spontaneous production of fibroblast activating factor(s) by synovial inflammatory cells. A potential mechanism for enhanced tissue destruction. J Exp Med 161:210–222

Wahl SM, Allen JB, Dougherty S, Evequoz E, Pluznik DH, Wilder RL, Hand AR, Wahl LM (1986a) T lymphocyte-dependent evolution of bacterial cell wall-induced hepatic granulomas. J Immunol 137:2199–2209

Wahl SM, Hunt DA, Allen JB, Wilder RL, Paglia L, Hand AR (1986b) Bacterial cell wall induced hepatic granulomas. An in vivo model of T cell dependent fibrosis. J Exp Med 163:884–902

Wahl SM, Hunt D, Wakefield L, McCartney-Francis N, Wahl L, Roberts A, Sporn M (1987a) Transforming growth factor beta (TGF-β) induces monocyte chemotaxis and growth factor production. Proc Natl Acad Sci USA 84:5788–5792

Wahl SM, McCartney-Francis N, Hunt DA, Smith PD, Wahl LM, Katona IM (1987b) Monocyte interleukin 2 receptor gene expression and interleukin 2 augmentation of microbicidal activity. J Immunol 139:1342–1347

Wahl SM, Hunt DA, Wong HL, Dougherty S, McCartney-Francis N, Wahl LM, Ellingsworth L, Schmidt JA, Hall G, Roberts AB, Sporn MB (1988a) Transforming growth factor-β is a potent immunosuppressive agent that inhibits IL-1-dependent lymphocyte proliferation. J Immunol 140:3026–3032

Wahl SM, Hunt DA, Bansal G, McCartney-Francis N, Ellingsworth L, Allen JB (1988b) Bacterial cell wall-induced immunosuppression: role of transforming growth factor beta. J Exp Med 168:1403–1417

Wahl SM, Wong HL, McCartney-Francis N (1989) Role of growth factors in inflammation and repair. J Cell Biochem (40:193–199)

Wallis WJ, Beatty PG, Ochs HD, Harlan JM (1985) Human monocyte adherence to cultured vascular endothelium: monoclonal antibody-defined mechanisms. J Immunol 35:2323–2330

Weisbart RH, Golde DW, Spolter L, Eggena P, Rinkerknecht H (1979) Neutrophil migration inhibition factor from T lymphocytes (NIF-T): a new lymphokine. Clin Immunol Immunopathol 14:441–448

Weisbart RH, Golde DW, Clark SC, Wong GG, Gasson JC (1985) Human granulocyte-macrophage colony-stimulating factor is a neutrophil activator. Nature 314:361–363

Weiss A, Imboden J, Hardy K, Manger B, Terhorst C, Stobo J (1986) The role of T3/antigen receptor complex in T-cell activation. Annu Rev Immunol 4:593–619

Weksler BB (1988) Platelets. In: Gallin JI, Goldstein IM, Snyderman R (eds) Inflammation: basic principles and clinical correlates. Raven, New York, pp 543–558
Welsh CE, Waite BM, Thomas MJ, DeChatelet LR (1981) Release and metabolism of arachidonic acid in human neutrophils. J Biol Chem 256:7228–7234
Werb Z, Gordon S (1975) Elastase secretion by stimulated macrophages: characterization and regulation. J Exp Med 142:361–377
Whiteside TL, Wonall JG, Prince RK, Buckingham RB, Rodnan GP (1986) Soluble mediators from mononuclear cells increase the synthesis of glycosaminoglycans by dermal fibroblast cultures derived from normal subjects and progressive systemic sclerosis patients. Arthritis Rheum 28:188–197
Wilder RL, Calandra GB, Garvin AJ, Wright KD, Hansen CT (1982) Strain and sex variation in the susceptibility to streptococcal cell wall-induced polyarthritis in the rat. Arthritis Rheum 25:1064–1072
Wilder RL, Allen JB, Wahl LM, Calandra GB, Wahl SM (1983) The pathogenesis of group A streptococcal cell wall-induced polyarthritis in the rat: comparative studies in arthritis resistant and susceptible inbred rat strains. Arthritis Rheum 26:1442–1451
Williams JD, Lee TH, Lewis RA, Austen F (1985) Intracellular retention of the 5-lipoxgenase pathway product, leukotriene B_4, by human neutrophils activated with unopsonized zymosan. J Immunol 143:2624–2630
Williams JD, Straneva JE, Cooper S, Shadducj RK, Waheed A, Gillis S, Urdal D, Broxmeyer HE (1987) Interactions between purified murine colony-stimulating factors (natural CSF-1, recombinant GM-CSF and recombinant IL-3) on the in vivo proliferation of purified murine granulocyte-macrophage progenitor cells. Exp Hematol 15:1007–1012
Williams LT, Antoniades HN, Goetzl EJ (1983) Platelet-derived growth factor stimulates mouse 3T3 cell mitogenesis and leukocyte chemotaxis through different structural determinants. J Clin Invest 72:1759–1763
Wolpe SD, Davatelis G, Sherry B, Beutler B, Hesse DG, Nguyen HT, Moldawer LL, Nathan CF, Lowry SF, Cerami A (1988) Macrophages secrete a novel heparin-binding protein with inflammatory and neutrophil chemokinetic properties. J Exp Med 167:570–581
Wood DD, Ihrie EJ, Dinarello CA, Cohen PL (1983) Isolation of an interleukin-1-like factor from human joint effusion. Arthritis Rheum 26:975–983
Wright SD, Jong MTC (1986) Adhesion-promoting receptors on human macrophages recognize *E. coli* by binding to lipopolysaccharide. J Exp Med 164:1876–1888
Wright SD, Silverstein SC (1983) Receptors for C3b and C3bi promote phagocytosis but not the release of toxic oxygen from human phagocytes. J Exp Med 158: 2016–2023
Wyler DJ, Stadecker MJ, Dinarello CA, O'Dea JF (1984) Fibroblast stimulation in schistosomiasis. V. Egg granuloma macrophages spontaneously secrete a fibroblast stimulating factor. J Immunol 132:3142–3148
Yocum DE, Allen JB, Wahl SM, Calandra GB, Wilder RL (1986) Inhibition by cyclosporin A of streptococcal cell wall-induced arthritis and hepatic granulomas in rats. Arthritis Rheum 29:262–273
Yoshino S, Cromartie WJ, Schwab JH (1985) Inflammation induced by bacterial cell wall fragments in the rat air pouch. Comparison of rat strains and measurement of arachadonic acid metabolites. Am J Pathol 121:327–336
Zilberstein A, Ruggieri R, Korn JH, Revel M (1986) Structure and expression of cDNA and genes for human interferon-beta-2, a distinct species inducible by growth-stimulating cytokines. EMBO J 5:2529–2537
Zucali JR, Dinarello CA, Oblon DJ, Gross MA, Anderson L, Weiner RS (1986) Interleukin 1 stimulates fibroblasts to produce granulocyte-macrophage colony-stimulating activity and prostaglandin E_2. J Clin Invest 77:1857–1863

CHAPTER 37

Angiogenesis

M. KLAGSBRUN and J. FOLKMAN

A. Introduction

"Angiogenesis" is currently used to describe the growth of new capillary blood vessels by sprouting from established vessels to produce "neovascularization." "Vasculogenesis" denotes the development of blood vessels in the embryo (RISAU and LEMMON 1988; FEINBERG and BEEBE 1983). There is as yet no adequate term to describe the growth of larger vessels such as coronary artery collaterals where actual proliferation of vascular endothelium and smooth muscle lead to enlargement of the lumenal diameter (SCHAPER 1981; PASYK et al. 1982; D'AMORE and THOMPSON 1987). Nor is there a description of endothelial proliferation in the capillaries of hypertrophic heart muscle in the apparent absence of sprouting (HUDLICKA and TYLER 1986). However, the term "non-sprouting angiogenesis" is useful (FOLKMAN 1987).

A systematic study of the mechanism of angiogenesis and the search for angiogenic factors began more than 20 years ago (FOLKMAN 1984, 1985b) after it was demonstrated that angiogenesis was an important process in tumor growth.

This work was initiated by the question of whether the hypervascularity associated with tumor growth was due to dilation of established vessels (the prevailing view in the early 1960s) or to proliferation of new blood vessels. The demonstration that tumor vessels were predominantly new capillaries (CAVALLO et al. 1972; BREM et al. 1972) led to the question of whether this *neo*vascularization was merely a side effect of tumor growth or a key step required for it. Experiments in vivo and in vitro (GIMBRONE et al. 1972; FOLKMAN and HOCHBERG 1973) supported the hypothesis that tumor growth is angiogenesis-dependent (FOLKMAN 1971, 1972). Additional studies with transplantable animal tumors provided evidence that induction of angiogenesis was mediated by soluble tumor-derived factors (FOLKMAN et al. 1971; GIMBRONE et al. 1973b). This evidence was consistent with the results of earlier studies with transparent chambers (ALGIRE 1943; GREENBLATT and SHUBIK 1968; EHRMANN et al. 1968). The first soluble angiogenic factor was isolated from a rat tumor, the Walker carcinosarcoma. Vascularization of the subcutaneous fascia of rats was used as a bioassay, the only available angiogenesis assay at the time (FOLKMAN et al. 1971; TUAN et al. 1973). It soon became apparent that complete purification of an angiogenic factor would be nearly impossible without more sensitive and reproducible bioassays. Therefore, much experimental work in the mid-1970s was devoted to the development of in vitro and in vivo bioassays for angiogenesis.

B. Bioassays for Angiogenesis

I. In Vivo Methods

The most commonly used bioassay for angiogenesis is the chick embryo chorioallantoic membrane (CAM) (AUSPRUNK et al. 1974). The substance to be tested for angiogenic activity is implanted onto the CAM through a window made in the eggshell (FOLKMAN 1974a; FOLKMAN et al. 1976) or in a shell-less embryo in a cultured petri dish (AUERBACH et al. 1974; KLAGSBRUN et al. 1976 FOLKMAN 1985a) or in a paper cup (CASTELLOT et al. 1980). The putative angiogenic factor is applied to the CAM in a carrier of 0.45% methylcellulose (10 μl) which is air dried to form a disk of 2 mm diameter (CRUM et al. 1985). An angiogenic substance may also be applied to the CAM in a polymer pellet (ethylene vinyl acetate copolymer) (LANGER and FOLKMAN 1976) or on a plastic coverslip or a filter paper. Embryos of 6–10 days are commonly used and observed by stereomicroscopy 2–3 days after implantation of the test material. Angiogenic activity is revealed by radial ingrowth of new capillary blood vessels. This neovascularization is confirmed by en face histological microsections of India-ink-injected specimens (HAUDENSCHILD 1980) (Fig. 1).

The CAM assay facilitates the testing of multiple samples and the generation of dose-dilution curves. It has been used to identify almost all of the known angiogenic factors (see below). However, the assay has several drawbacks. False-positive angiogenesis may be induced by any test material which causes cell damage by virtue of abnormal osmolarity, pH, or toxicity. Such substances may induce an inflammatory response or cause focal contraction of the chorioallantoic membrane. Furthermore, inflammatory angiogenesis per se in which infiltrating macrophages or other leukocytes may be the source of angiogenic factors cannot be distinguished from direct angiogenic activity of the test material without detailed histological study and multiple positive and negative controls. Many of these difficulties with the CAM assay have already been discussed by Ryan and Stockley (RYAN and STOCKLEY 1980; RYAN and BARNHILL 1983). Another problem is that angiogenesis may be induced by degradation products of fibrin (DVORAK 1986; THOMPSON et al. 1985) which can leak from CAM vessels in response to injurious test substances. Some of these problems may be overcome by using the yolk sac vessels of the 4-day chick embryo because this system has a markedly reduced inflammatory and immune response (TAYLOR and FOLKMAN 1982; TAYLOR and WEISS 1984).

Several improved techniques for measuring angiogenic activity on the CAM are being developed. SPLAWINSKI et al. (1988) and KREISLE and ERSHLER (1988) have recently reported such improvements.

The cornea micropocket overcomes some of the disadvantages of the CAM assay. In this bioassay, the angiogenic factor is implanted in the cornea

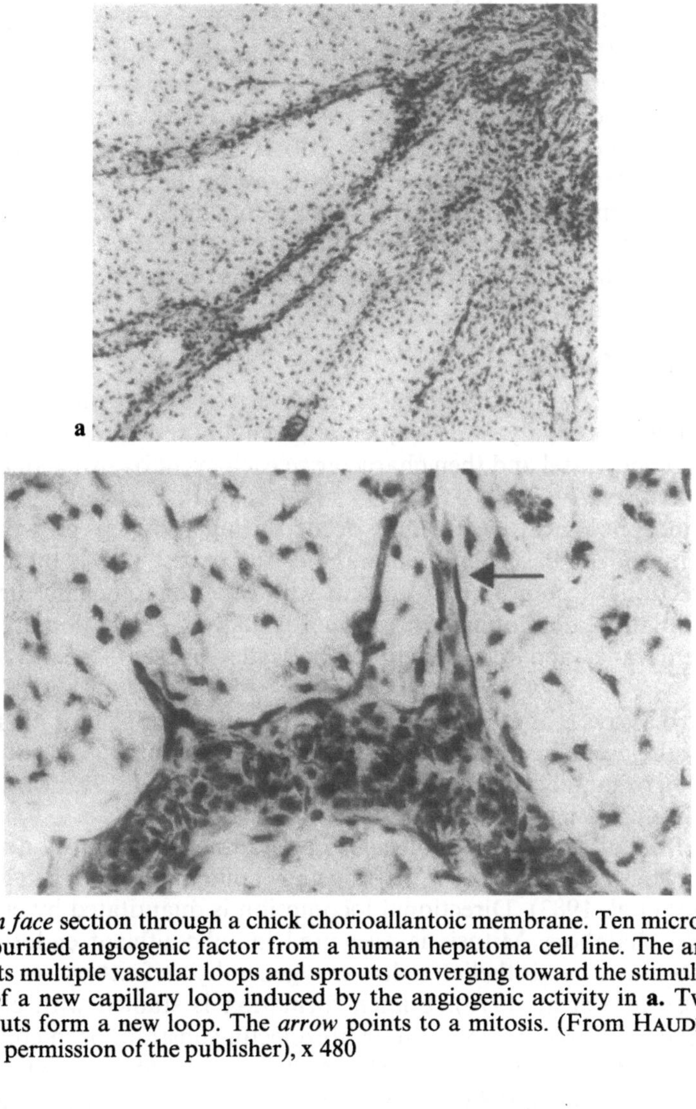

Fig. 1. a *En face* section through a chick chorioallantoic membrane. Ten micrograms of partially purified angiogenic factor from a human hepatoma cell line. The angiogenic pellet elicits multiple vascular loops and sprouts converging toward the stimulus, x 120. **b** Detail of a new capillary loop induced by the angiogenic activity in **a.** Two of the three sprouts form a new loop. The *arrow* points to a mitosis. (From HAUDENSCHILD 1980, with permission of the publisher), x 480

of the rabbit (GIMBRONE et al. 1974), mouse (MUTHUKKARUPPAN and AUERBACH 1979), or rat (GLASER et al. 1980). The test substance is usually implanted in a polymer pellet (LANGER and FOLKMAN 1976). The length and number of new capillaries which enter the avascular cornea can be quantitated periodically by a slit-lamp stereoscope, or at the conclusion of the experiment by image analysis of India-ink-injected specimens (PROIA et al. 1988). The cornea provides the most compelling evidence that *new* capillaries have been induced (BURGER et al. 1983). The major drawback of the cornea is the time and expense required for the use of rabbits, and the fewer substances and concentrations that can be tested.

Other in vivo bioassay systems that have been used include the hamster cheek pouch (SCHREIBER et al. 1986), the subcutaneous space in mice (AUERBACH 1981), and various transparent chambers (HOBBS and CLIFF 1971) implanted subcutaneously.

Recently, biodegradable (THOMPSON et al. 1988) or nonbiodegradable porous sponges (DAVIDSON et al. 1985; THOMAS and GIMENEZ-GALLEGO 1986) have been implanted in animals to study the ingrowth of new blood vessels attracted to the sponge by angiogenic factors contained within it. These techniques facilitate histological and immunocytochemical studies of the new vessels, although the extent of neovascularization is more difficult to quantitate than with other methods.

II. In Vitro Methods

One way to unscramble the problem of how growth is regulated in the vascular system is to analyze the sequential events necessary to develop a new capillary blood vessel and then characterize each event by using an analogous in vitro bioassay.

The morphological events that occur during capillary growth include: degradation of the basement membrane of the parent venule by endothelial cells, directional locomotion in tandem with other endothelial cells, mitosis, lumen formation, development of sprouts and loops, generation of new basement membrane, and recruitment of pericytes (AUSPRUNK and FOLKMAN 1977).

In vitro assays that are quantifiable have been developed for most of these events by the use of capillary (FOLKMAN et al. 1979), aortic (GOSPODAROWICZ et al. 1976), or umbilical vein endothelial cells (JAFFE et al. 1972; MACIAG et al. 1981, 1982a,b; GIMBRONE et al. 1973a).

The effect of angiogenic molecules on endothelial locomotion in vitro is measured by chemokinesis assays using colloidal gold (ZETTER 1980; AZIZKHAN et al. 1983). Directional locomotion is quantitated by chemotaxis in Boyden chambers. Mitosis is measured in subconfluent endothelial cultures. Confluent endothelial cells are generally refractory to mitogens (HAUDENSCHILD et al. 1976). Lumen formation is studied with endothelial cells that are cultured on collagen or fibronectin substrata or in collagen gels (FOLKMAN and HAUDENSCHILD 1980; MACIAG et al. 1982a, b; NICOSIA et al. 1982; MONTESANO et al. 1983; MADRI and WILLIAMS 1983; INGBER et al. 1987).

Angiogenic factors also stimulate increased production of plasminogen activator and collagenase by endothelial cells (RIFKIN et al. 1982; GROSS et al. 1983). Quantitation of these enzymatic activities in vitro provides a novel method of guiding the purification of angiogenic factors (GROSS et al. 1982, 1983). Also, when endothelial cell chemotaxis is stimulated in vitro by angiogenic activity there is a significant increase in cell-associated enzymatic activity capable of degrading both type IV and V collagen (KALEBIC et al. 1983).

It should be emphasized that when in vitro bioassays are employed to guide the purification of an angiogenic factor, results must be confirmed in vivo. It is quite possible for a factor to be angiogenic in vivo but not mitogenic for endothelial cells in vitro (e.g., angiogenin, see below). Conversely, an endothelial mitogen in vitro may not be angiogenic in vivo (e.g., certain low-density lipoproteins).

Assays for angiogenesis are difficult to quantitate in vivo. Nevertheless, with existing techniques it has been possible to identify, characterize, and purify a number of angiogenic factors as follows.

C. Angiogenic Factors

A number of factors have been shown to be angiogenic in vivo. These factors include polypeptides and nonpeptide low molecular weight compounds. In this review we describe primarily the biochemical and biological properties of those polypeptide angiogenic factors that have been purified, sequenced, and cloned (Table 1). A brief summary is given of the low molecular weight non-peptide angiogenic factors which for the most part remain to be fully characterized.

I. Fibroblast Growth Factors

The term fibroblast growth factor (FGF) was first used to describe a cationic polypeptide isolated from pituitary and brain that stimulated cell division in 3T3 cells (GOSPODAROWICZ 1974, 1975). This polypeptide, known as basic FGF (bFGF), has a molecular weight of 17000–18000 and a pI of 9.6 (BOHLEN et al. 1984). Brain was also found to contain another species of FGF, anionic in nature, known as acidic FGF (aFGF) (THOMAS et al. 1980; LEMMON and BRADSHAW 1983; THOMAS 1987). Purified brain aFGF is a polypeptide with a molecular weight of 16000–17000 and a pI of 5 (THOMAS et al. 1984). bFGF and aFGF are structurally related, having a 53% absolute sequence homology (ESCH et al. 1985b). More recent studies indicate that higher molecular weight forms of bFGF exist (SOMMER et al. 1987). In particular, 22000–25000 molecular weight forms of bFGF have been identified in guinea

Table 1. Purified angiogenic factors

Growth factor	Molecular weight
Acidic FGF	16500
Basic FGF	18000
Angiogenin	14100
TGF-α	5500
TGF-β	25000
TNF-α	17000
Platelet-derived ECGF	45000
Angiotropin	4500

pig, rat brain, and rat cells hepatoma cells (MOSCATELLI et al. 1987; PRESTA et al. 1988; PRATS et al. 1989; IBERG et al. 1989). It appears that the translation of these high molecular weight forms is initiated at unusual CTG start codons (PRATS et al. 1989). The biological significance of higher molecular weight bFGF has not been determined yet.

aFGF and bFGF may be part of a family of FGF-related growth factors. Several oncogenes have been described that encode for proteins that have a 40%–50% homology for bFGF and aFGF. The first FGF homolog to be described was *int-2,* which is a product of integration of mammary tumor virus into the host genome (DICKSON and PETERS 1987). The *hst* oncogene was isolated from a human stomach cancer (SAKAMOTO et al. 1986; TAIRA et al. 1987). The same oncogene was later isolated from Kaposi's tumor and named K-*FGF* (DELLI BOVI and BASILICO 1987). KFGF is glycosylated and secreted as a mature protein of 176 amino acids (DELLI BOVI et al. 1988). A third oncogene, *FGF-5,* was isolated from a human bladder tumor (ZHAN et al. 1988).

One of the most interesting and potentially important properties of aFGF and bFGF is their affinity for heparin (SHING et al. 1984; LOBB and FETT 1984; LOBB et al. 1986; KLAGSBRUN and SHING 1985; MACIAG et al. 1984; CONN and HATCHER 1984; D'AMORE and KLAGSBRUN 1984; COURTY et al. 1985; GOSPODAROWICZ et al. 1984; HAUSCHKA et al. 1986; SULLIVAN and KLAGSBRUN 1985). When applied to columns of immobilized heparin, aFGF is eluted with 1 *M* NaCl and bFGF with 1.5 *M* NaCl. In contrast, PDGF, which is as cationic as bFGF, is eluted with 0.5 *M* NaCl and EGF does not bind at all. Heparin-affinity chromatography has become an essential step for the characterization and purification of both types of FGF. The affinity of FGF for heparin might reflect an association that occurs in cells and tissues. bFGF is bound to heparin-like molecules, probably heparan sulfate proteoglycan, in subendothelial cell extracellular matrix (VLODAVSKY et al. 1987a; BAIRD and LING 1987) and in basement membrane (FOLKMAN et al. 1988). Heparin as well as heparan sulfate protect FGF from denaturation and degradation (GOSPODAROWICZ and CHENG 1986; SASKELA et al. 1988). Thus, heparin-like molecules might bind FGF in vivo and form complexes in which FGF is stabilized.

FGF is widely distributed (FOLKMAN and KLAGSBRUN 1987a). aFGF has been found mainly in neural tissue such as brain, hypothalamus, retina and eye, and also in bone. bFGF appears to be more ubiquitous and is found in neural tissue, skeletal tissue, reproductive tissue, glands, organs, white blood cells, and tumors. In fact, it appears that FGF may be one of the most widely detributed growth factors in the body.

FGF is also synthesized by many cells in culture (FOLKMAN and KLAGSBRUN 1987a). Biosynthetic studies indicate clearly that bFGF and aFGF are cellular rather than secreted proteins. The cell association of bFGF was first demonstrated in hepatoma cells (KLAGSBRUN et al. 1986) and subsequently in endothelial cells (VLODAVSKY et al. 1987a, b; SCHWEIGERER et al. 1987b). Most other growth factors are secreted including PDGF, TGF-α,

TGF-β, and TNF-α. The lack of bFGF and aFGF secretion is consistent with the absence of a consensus signal peptide in the open reading frame. Cells transfected with a bFGF cDNA to which an immunoglobulin heavy-chain signal peptide is attached upstream are transformed in culture and produce large, rapidly growing tumors in syngeneic animals (ROGELJ et al. 1988). The lack of bFGF secretion under normal circumstances might be part of a regulatory mechanism in which bFGF, which is abundant in many nongrowing tissues, is prevented from stimulating cell proliferation in either a paracrine or autocrine manner.

bFGF synthesized by endothelial cells is found not only to be associated with these cells but with the subendothelial cell extracellular matrix as well (VLODAVSKY et al. 1987a; BAIRD et al. 1987). In the cornea it can be shown that bFGF is mostly associated with Descemet's membrane, a basement membrane produced by corneal endothelial cells (FOLKMAN et al. 1988). bFGF can be released from extracellular matrix and basement membrane by displacement with heparin or by degradation with heparinases and heparanases (VLODAVSKY et al. 1987a; BAIRD et al. 1987; FOLKMAN et al. 1988; BASHKIN et al. 1989). The association of bFGF with cells, extracellular matrix, and basement membrane has led to the speculation that bFGF may be a stored growth factor that can be potentially released by degradative enzymes. The release of "stored" FGF might be a rapid way to mobilize growth factor in reparative processes such as wound healing.

Although originally described as mitogens for fibroblasts, both bFGF and aFGF were soon found to be potent mitogens for endothelial cells (THOMAS 1987; FOLKMAN and KLAGSBRUN 1987a; GOSPODAROWICZ et al. 1976; DUTHU and SMITH 1980). FGF is also chemotactic for endothelial cells (PRESTA et al. 1986; CONNOLLY et al. 1987). Heparin might play a role in FGF-induced mitogenesis. Heparin potentiates the stimulation of endothelial cell proliferation by aFGF 100-fold (THORNTON et al. 1983; GIMENEZ-GALLEGO et al. 1986). Heparin might potentiate FGF activity by protecting it against denaturation and degradation (GOSPODAROWICZ and CHENG 1986; SASKELA et al. 1988) or by decreasing the K_d of FGF binding (SCHREIBER et al. 1985).

Both bFGF and aFGF are angiogenic factors in the CAM and cornea bioassays at levels as low as 10–100 ng (SHING et al. 1985; FOLKMAN and KLAGSBRUN 1987a; ABRAHAM et al. 1986; THOMAS and GIMENEZ-GALLEGO 1986; ESCH et al. 1985a). bFGF has been shown to induce the infiltration of granulation tissue containing highly dilated blood vessels into polyvinyl sponges that are implanted into rats (DAVIDSON et al. 1985). aFGF complexed to gelatin induces angiogenesis in the neck and peritoneal cavities of the rat (THOMPSON et al. 1988). The angiogenic activity of FGF in vivo is compatible with its properties in vitro. Angiogenesis in vivo is a complex process involving the degradation of capillary basement membrane, the migration and proliferation of endothelial cells, and tube formation. FGF in vitro is mitogenic and chemotactic for endothelial cells, stimulates the production by endothelial cells of collagenase and plasminogen activator, proteases capable of degrading basement membrane (GROSS et al. 1983), and induces capillary endothelial cells to migrate into three-dimensional collagen matrices to form capillary-like tubes (MONTESANO et al. 1986).

In summary, both aFGF and bFGF are potent angiogenesis factors. Since they are both chemotactic and mitogenic for endothelial cells, they may be able to act directly on these cells to stimulate angiogenesis. It is not clear how FGF mediates angiogenesis in vivo. Since it is not a secreted protein, it may have to be released from cells by alternative mechanisms. One of these may be cell lysis. Indeed, FGF might be released from cellular storage sites during injury and then proceed to repair the injury by stimulating connective tissue growth and angiogenesis.

II. Angiogenin

Angiogenin is a single-chain cationic (pI > 9.5) polypeptide containing 123 amino acids and having a molecular weight of 14123 (FETT et al. 1985; STRYDOM et al. 1985; KURACHI et al. 1985; RIORDAN and VALLEE 1988). The angiogenin gene was cloned from a human liver cDNA library. Unlike the fibroblast growth factors, angiogenin contains a signal peptide and is a secreted protein.

One of the interesting properties of angiogenin is its homology to pancreatic RNase. Of the 123 amino acids in human angiogenin, 35% are identical with those at the corresponding positions in human pancreatic RNase. Including conservative replacements, the overall homology is 68%. Despite the homology, pancreatic RNase is not angiogenic. Angiogenin exhibits ribonucleolytic activity but differs distinctly from pancreatic RNase in specific RNA-hydrolyzing properties (RIORDAN and VALLEE 1988; SHAPIRO et al. 1986; RYBAK and VALLEE 1988). For example, angiogenin is not very active toward conventional substrates of RNase, such as wheat germ RNA. Angiogenin does hydrolyze 18S and 28S ribosomal RNA but into relatively large 100- to 500-nucleotide fragments. Compared with pancreatic RNase, about 10^4- to 10^5-fold more angiogenin is needed for ribonucleolytic activity.

Angiogenin is a potent and specific inhibitor of cell-free protein synthesis, at levels considerably lower than pancreatic RNase (ST. CLAIR et al. 1987). The inhibitory effect seems to be a property of the RNase activity of angiogenin on ribosomal RNA. Adding ribosomes to the cell-free protein synthesis system reverses the inhibitory effect. However, no inhibitory effects of angiogenin on protein synthesis have been demonstrated in vivo.

The ribonucleolytic activity of angiogenin appears to be necessary for its angiogenesis activity. Inhibiting RNase activity with bromoacetate abolishes angiogenesis (SHAPIRO et al. 1987b). Human placental RNase inhibitor (PRI) also inhibits angiogenic activity (SHAPIRO and VALLEE 1987). PRI is 60 times more effective against angiogenin than it is against pancreatic RNase. Although ribonucleolytic activity is necessary for angiogenin activity, it is not clear how this activity affects angiogenesis.

Angiogenin was first isolated from the conditioned medium of an established human adenocarcinoma cell line HT-29 which produced about 0.5 μg/ml (FETT et al. 1985). A number of tumor cell lines, including lung carcinoma and hepatoma cells, secrete angiogenin (SHAPIRO et al. 1987a). However, tumors are not the only source of angiogenin. Angiogenin has been found in human plasma (60–150 μg/liter) (BOND and VALLEE 1988) and in bovine plasma (30–80 μg/liter) (RYBAK et al. 1987). Normal cells, e.g., WI-38 produce angiogenin, at levels comparable to those of HT-29 cells. Stimulated lymphocytes synthesize angiogenin and it has been suggested that blood cells may be the source of plasma-derived angiogenin.

Analysis of angiogenin mRNA distribution in tissues suggests that it is made predominantly in adult liver (WEINER et al. 1987). Angiogenin synthesis in rat liver was found to be developmentally regulated, being low in the developing fetus, higher in the neonate, and maximal in the adult. Interestingly, it was found that angiogenin gene expression was not temporally related to vascular development in the rat, a process which is highest in the developing fetus and lowest in the adult.

Angiogenin stimulates extensive blood vessel growth in the CAM and in the rabbit cornea (FETT et al. 1985). Stimulation of the CAM occurs with application of as little as 0.5 ng/egg and 50 ng/eye. Unlike fibroblast growth factors, angiogenin does not appear to be a mitogen for endothelial cells nor does it affect endothelial cell morphology (BICKNELL and VALLEE 1988). However, angiogenin does have at least one effect on cultured aortic, umbilical vein, and capillary endothelial cells. It activates inositol-specific phospholipase C, promoting a transient twofold increase in intracellular 1,2-diacylglycerol, a putative cellular activator of protein kinase C, as well as a 20% increase in inositol triphosphate. The maximum effect occurs with 1 ng/ml angiogenin. However, despite these findings no angiogenin-induced calcium flux was found. The angiogenin effect on inositol triphosphate is very small compared with bradykinin. The relevance of these effects of angiogenin on endothelial cells to its angiogenic activity is unclear.

In summary, angiogenin is a secreted angiogenic factor synthesized in normal and tumor tissue. It is not chemotactic or mitogenic for endothelial cells or any other cell type. Thus, angiogenin probably acts by different mechanisms than does FGF.

III. Transforming Growth Factor-α

Transforming growth factor-α (TGF-α) is a single-chain 50-amino-acid polypeptide with a molecular weight of about 5500 that binds to the epidermal growth factor (EGF) receptor and has a 40% homology to EGF (MARQUADT et al. 1984; DERYNCK et al. 1985). TGF-α was first described as a mitogen secreted by transformed fibroblasts (TODARO et al. 1980; ANZANO et al. 1983). It is also secreted by a variety of tumor cells and by normal cells such as macrophages (MADTES et al. 1988). An important property of TGF-α is the ability to transform normal cells in a reversible manner into cells with a transformed phenotype characterized by the ability to grow as colonies in soft agar and with a lowered serum requirement. Because it is a transforming protein and often found elevated in tumor cells, TGF-α has been suggested as a mediator that stimulates the autocrine growth of tumor cells.

Besides fibroblasts, TGF-α stimulates the proliferation of a number of cell types in culture, including epithelial cells and endothelial cells (DERYNCK et al. 1984; SCHREIBER et al. 1986). TGF-α is an angiogenesis factor. TGF-α injected subcutaneously into the hamster cheek pouch at 0.3–1 μg stimulates capillary proliferation and an increase in the labeling index of endothelial cells (SCHREIBER et al. 1986). EGF is also active in the hamster cheek pouch but is angiogenic only at a dose of 10 μg.

Although very few data are available at present, the synthesis of TGF-α by tumors and macrophages coupled with its angiogenic activity suggests that TGF-α may be an important mediator of tumor vascularization and inflammation.

IV. Transforming Growth Factor-β

Transforming growth factor-β (TGF-β) is a homodimeric polypeptide with a molecular weight of 25000 (SPORN et al. 1987). Each monomer is a 112-amino-acid polypeptide that represents the carboxyl-terminal portion of a 390-amino-acid precursor. Despite the prevalent nomenclature, TGF-β is a protein totally distinct from TGF-α. There are two molecular weight forms of TGF-β, TGF-β 1, which appears to be the most abundant, and TGF-β 2. These two forms are highly homologous but may have different receptors. TGF-β was originally purified from platelets, placenta, and kidney, but has also been found in many other tissues as well, including cartilage, bone, and tumor cells (ASSOIAN et al. 1983, FROLIK et al. 1983; ROBERTS et al. 1981; SEYEDIN et al. 1985, 1986).

An interesting property of TGF-β is that it is secreted in a biologically inactive or latent form. The latent form can be activated by heat, acidification, and proteases (KESKI-OJA et al. 1987; LAWRENCE et al. 1984, 1985). Activation of latent TGF-β in vivo by proteases could be a regulatory mechanism for mediating TGF-β activity.

TGF-β is a multifunctional regulator that both inhibits and stimulates cell growth (SPORN and ROBERTS 1988). TGF-β was originally characterized as a factor that promoted anchorage-independent growth of normal fibroblasts (MOSES et al. 1985). However, it became apparent soon after that TGF-β was a potent inhibitor of the proliferation of cell types such as epithelial cells, endothelial cells, and fibroblasts (TUCKER et al. 1984; BAIRD and DURKIN 1986; FRATER-SCHRODER et al. 1986; HEIMARK et al. 1986; ANZANO et al. 1986). In addition to its effect on cell proliferation, TGF-β strongly enhances the formation of extracellular matrix, a process important in differentiation, by stimulating the secretion of matrix proteins and inhibiting their degradation (IGNOTZ and MASSAGUE 1986).

In vivo, TGF-β is considered to be a mediator of inflammation and repair as well as a mediator of the differentiation of tissues of mesenchymal origin, such as bone, muscle, and blood vessels. There is evidence that TGF-β can also mediate angiogenesis. When injected subcutaneously into the nape of the neck of mice at a dose of 1 μg, TGF-β caused the formation of a highly vascular granulation tissue at the site of injection in 2–3 days (ROBERTS et al. 1986). Neither EGF nor PDGF had a similar effect. On the other hand, TGF-β inhibits angiogenesis in the CAM (DUGAN et al. 1988). The mechanism of angiogenesis stimulation is unclear since TGF-β is a potent inhibitor of capillary and aortic endothelial cell growth in culture whether the substratum is plastic, collagen, or basement membrane (BAIRD and DURKIN 1986; FRATER-SCHRODER et al. 1986; HEIMARK et al. 1986; TAKEHARA et al. 1987; MULLER et al. 1987). One possible explanation is that TGF-β promotes angiogenesis by promoting the differentiation of endothelial cells after their proliferative phase has ended, possibly by inducing the synthesis of matrix. Alternatively, TGF-β-induced neovascularization following subcutaneous injection may be a secondary response occurring in scar formation. TGF-β is highly chemotactic for cells involved in tissue repair, such as fibroblasts and monocytes (WAHL et al. 1987). These secondary cells might be responsible for producing angiogenesis factors that are mitogenic for endothelial cells.

V. Tumor Necrosis Factor

Tumor necrosis factor (TNF-α) is an anionic 154- to 157-amino-acid polypeptide with a molecular weight of 17000 (SHERRY and CERAMI 1988; HARANAKA et al. 1986; BEUTLER et al. 1985). It is synthesized as a prohormone with a precursor extension sequence of 76–80 amino acids. TNF-α is structurally related to lymphotoxin (28% sequence homology), which is produced by mitogen-stimulated lymphocytes and which is known as TNF-β (GRAY et al. 1984). Unlike FGF, TNF-α has a signal sequence compatible with its being a secreted protein.

TNF-α is synthesized primarily by activated macrophages and by some tumor cells. It is a multifunctional polypeptide that is a pleiotropic mediator of inflammation and immunity. Among its properties are cytotoxicity for tumor cells and induction of hemorrhagic necrosis in tumors (CARSWELL et al. 1985). TNF-α induces cachexia and is also known as cachectin. The term cachectin/TNF-α is often used (BEUTLER and CERAMI 1986). Another important property of TNF-α is mediation of endotoxic shock.

TNF-α has a wide range of activities that affect endothelial cells and which may be important in the response of these cells to infection and injury. For example, TNF-α stimulates GM-CSF synthesis, interleukin-1 synthesis, and the induction of ICAM-1, which in turn promotes leukocyte adhesion to endothelial cells (BEVILACQUA et al. 1985; POBER et al. 1986a, b, 1987).

TNF-α has been reported by several groups to be angiogenic (LEIBOVICH et al. 1987; FRATER-SCHRODER et al. 1987), and, in fact, has been claimed to be responsible for the total angiogenic activity of macrophages (LEIBOVICH et al.

1987). In one study, it was demonstrated that TNF-α stimulated angiogenesis in the cornea at 3.5 ng and in the CAM at 1 ng without any evidence of inflammation (LEIBOVICH et al. 1987). TNF-α stimulated angiogenesis at lower concentrations than did FGF, angiogenin, EGF, and TGF-α. However, another group (FRATER-SCHRODER et al. 1987) found that, while TNF-α is indeed angiogenic in the cornea, higher amounts (5 μg) were required compared with FGF (0.5 μg). Besides being angiogenic in vivo, TNF-α stimulates EC chemotaxis and formation of capillary tubes on type I collagen culture in vitro (LEIBOVICH et al. 1987). However, TNF-α is at the same time a potent inhibitor of endothelial cell proliferation in culture (at concentrations as low as 1 ng/ml) (LEIBOVICH et al. 1987; FRATER-SCHRODER et al. 1987; SCHWEIGERER et al. 1987a). Thus, TNF-α is similar in many ways to TGF-β. Both polypeptides promote angiogenesis in vivo and tube formation in vitro, and inhibit endothelial cell proliferation in vitro. It may be that TNF-α, as well as TGF-β, is a promoter of the differentiation phase of endothelial cells, e.g., tube formation and matrix production, rather than the proliferative phase. Alternatively, TNF-α stimulation of angiogenesis might occur by some indirect mechanism, e.g., the stimulation of secondary cells to produce angiogenic factors.

VI. Platelet-Derived Endothelial Cell Growth Factor

The platelet-derived endothelial cell growth factor (PD-ECGF) is a newly described growth factor that has recently been purified to homogeneity (MIYAZONO et al. 1987) and whose gene has been cloned (ISHIKAWA et al. 1989). PD-ECGF is an acidic single-chain polypeptide with a molecular weight of 45000. It is heat and acid labile but resistant to reducing agents. Unlike FGF it does not bind to heparin nor does heparin potentiate its activity. PD-ECGF has two interesting properties. First of all, it is the only endothelial cell growth factor in platelets. Previously, platelets have been shown to contain PDGF, which is not mitogenic for endothelial cells, TGF-β, which inhibits endothelial cell proliferation, and EGF which at best is a poor mitogen for endothelial cells. Secondly, PD-ECGF stimulates porcine and human endothelial cells but not human fibroblasts. Thus, if these results are borne out for other cell targets, PD-ECGF might have the distinct property of being a specific endothelial cell mitogen. Besides being an endothelial cell mitogen, PD-ECGF ECGF is also angiogenic. However, at this writing, it is too early to assess the role of the newly described PD-ECGF as a mediator of angiogenesis in vivo.

VII. Angiotropin

Angiotropin has been isolated from cultures of peripheral porcine monocytes activated by concanavalin A (HOCKEL et al. 1987, 1988). It is a copper-containing polyribonucleopolypeptide with a molecular weight of 4500. No angiotropin amino acid sequence data or cDNA is available at present. Angiotropin is not an endothelial cell mitogen but does stimulate endothelial cell migration and appears to differentiate capillary endothelial cells in culture by changing their phenotype from contact-inhibited epithelioid cells into migrating fibroblastoid cells. The effects of angiotropin on endothelial cells may be specific since this factor does not stimulate 3T3 cells to migrate, proliferate, or differentiate.

Angiotropin at a dose of 2500 pg induces angiogenesis in the CAM and cornea bioassays. When injected into the dorsal skin of a rabbit ear lobe, angiotropin induces phenotypic changes in capillary and postcapillary venule endothelial cells, vascular dilation, and angiogenesis. These events occur in two stages. Two days post-injection, capillaries are highly dilated. By day 8 blood vessels are reduced to normal size but the number of microvessels is significantly increased. In addition, the angiogenesis is associated with epidermal and stromal proliferation but not with tissue necrosis or scar formation. Thus, angiotropin may be considered to be a chemical mediator produced by macrophages that initiates a cascade of inflammatory and wound-healing events including angiogenesis. As with the other angiogenic factors that are not endothelial cell mitogens, e.g., TGF-β and TNF-α, the mechanism of angiotropin-mediated angiogenesis is not clear.

VIII. Low Molecular Weight Nonpeptide Angiogenesis Factors

A number of low molecular weight compounds have been reported to be angiogenic. They have been isolated from tumors, macrophages, serum, and fat, to mention just a few sources. These factors include-well-known metabolites, such as prostaglandins and nicotinamide as well as poorly characterized molecules. In general, these compounds are angiogenic in CAM or cornea bioassays but are not mitogenic for endothelial cells in culture. It has been very difficult to ascertain the role of low molecular weight nonpeptides as mediators of angiogenesis. One problem has been the question of whether these compounds are actually angiogenic factors as opposed to cofactors and nutrients that promote or enhance the angiogenic response. Another major problem has been the lack of purity of many of the factors. The following is a brief summary of the better-characterized low molecular weight angiogenic factors.

Several lipids have been reported to be angiogenic. Prostaglandins are elevated in tumors, wounds, and inflammatory exudates, processes characterized by angiogenic responses. The prostaglandins PGE_1 and PGE_2, but not PGA or PGF, have been shown to be angiogenic in the CAM and cornea assays at doses of 20 ng to 1 μg (Form and Auerbach 1983; Ziche et al. 1982; Ben Ezra 1978). 3T3 cells that have differentiated into adipocytes secrete as yet uncharacterized polar lipids that are angiogenic in the CAM bioassay (Castellot et al. 1982; Dobson et al. 1985; Castellot et al. 1980). Another as yet uncharacterized lipid angiogenic factor has been isolated from omentum (Goldsmith et al. 1984). The presence of angiogenic factors in adipocytes and omentum might be relevant to the mechanism of fat mobilization considering that adipose tissue is highly vascularized.

Adenosine and adenosine diphosphate (FRASER et al. 1979; DUSSEAU et al. 1986) are angiogenic in the CAM and cornea assays. These compounds are vasodilators that accumulate in response to tissue hypoxia. It is possible that they can augment tissue oxygen tension by inducing a vasoproliferative as well as a vasodilatory response. Tumor-derived nicotinamide has also been found to be angiogenic (KULL et al. 1987). The physiological significance of nicotinamide-mediated angiogenesis is unclear but may be related to the observation that related compounds such as nicotinate are vasodilatory.

A series of angiogenic factors have been isolated from tumors, wound fluid, synovial fluid, serum, and retina (TUAN et al. 1973; BROWN et al. 1980; AUERBACH et al. 1981 KNIGHTON et al. 1983; KISSUN et al. 1982; FOLKMAN 1974b; ODEDRA and WEISS 1987; FENSELAU et al. 1981; BANDA et al. 1982). These factors have been difficult to purify or even characterize. It is not clear whether they are novel compounds or related to well-characterized angiogenic factors as described above.

IX. Mechanisms of Angiogenesis Factor Action

The angiogenic factors described above vary greatly in their biological properties with respect to being endothelial cell mitogens, endothelial cell chemotactic factors, and stimulators of tube formation (Table 2). They also differ in their target cell specificity, with PD-ECGF and angiotropin being possibly specific for endothelial cells while the other factors are multipotential. Another difference is that all of the angiogenic factors except FGF and PD-ECGF are secreted by cells. This observation raises the question of how FGF and PD-ECGF can act in a paracrine manner. One possibility is that they are released from lysed cells as a response to tissue damage.

The differences in biological properties suggest that different angiogenic factors must act by different mechanisms. Some angiogenic factors (FGF, TGF-α and PD-ECGF) are endothelial cell mitogens while others are not mitogenic (angiogenin) or actually inhibitory (TGF-β, TNF-α, and angiotropin). Although TGF-β and TNF-α inhibit endothelial cell proliferation, they do stimulate tube formation, suggesting that these polypeptides might be ac-

Table 2. Biological properties of polypeptide angiogenic factors

	FGF	Angiogenin	TGF-α	TGF-β	TNF-α	PD-ECGF	Angiotropin
Angiogenesis	+	+	+	+	+	+	+
EC mitogenicity	+	−	+	INH	INH	+	INH
EC chemotaxis	+	ND	ND	ND	+	+	+
EC tube formation	+	ND	ND	+	+	ND	ND
EC specificity	−	ND	−	−	−	+	+
Secretion	−	+	+	+	+	−	+

EC, endothelial cells; ND, no data available; INH, inhibitor of proliferation.

tive in endothelial cell differentiation. These observations suggest that angiogenesis might represent a multiple event requiring multiple factors. One speculation is that there are two classes of angiogenesis factor, those that stimulate proliferation and those that stimulate differentiation. FGF, TGF-α, and PD-ECGF might be endothelial cell proliferative factors while TGF-β, TNF-α, angiogenin, and angiotropin might belong to the class of endothelial cell differentiation factors. One might draw an analogy between blood vessel formation and cell division. Cell division requires competence factors that allow cells to enter the cell cycle and progression factors that allow cells to progress through the cycle and divide (PLEDGER et al. 1978). Perhaps, the proliferative endothelial cell factors make endothelial cells competent for other angiogenic factors which in turn stimulate blood vessel differentiation and development.

An alternative mechanism that would explain the different biological properties of the various angiogenic factors is that some of them might act directly on endothelial cells and stimulate them to move, divide, or form tubes. However, other angiogenic factors might act indirectly to trigger secondary cells to produce angiogenic factors that then directly stimulate endothelial cells. A key cell in this scenario would be the macrophage. The macrophage produces at least four angiogenic factors, i.e., FGF, TGF-α, TGF-β, and TNF-α. It is conceivable that many of the angiogenic factors, particularly those low molecular weight factors associated with inflammation, act indirectly via the macrophage to stimulate angiogenesis.

Finally, it is worth considering that biological activities measured in vitro do not necessarily recapitulate what occurs in vivo. Vascular endothelial cells are bipolar with a luminal side and a basement membrane side. Angiogenic factors added in vitro interact with the luminal side. However, in vivo these same angiogenic factors might act on the extravascular basement membrane side as well as on the intravascular luminal side. It is conceivable that the effects of angiogenic factors on endothelium in vivo are quite dependent on the site of factor attachment (FOLKMAN and KLAGSBRUN 1987b).

D. Physiological Regulation of Angiogenic Molecules

With several purified angiogenic factors in hand, it is now necessary to understand how they operate to form a whole capillary network.

A number of questions arise. For example, why is the proliferation of vascular endothelial cells in most normal tissues so extremely slow, with turnover times measured in years, when in fact angiogenic factors have been isolated from almost all normal tissues (DENEKAMP 1984; ENGERMAN et al. 1967)? How are these potent angiogenic factors maintained in a functionally inactive state, yet with the potential to become active mitogens so that during angiogenesis in ovulation or wound healing endothelial cells can proliferate rapidly with turnover times measured in days (DENEKAMP 1984)? How does the mitogenic activity of endothelial cell growth factors such as FGF account for the capacity of endothelial cells to form complex capillary networks instead of just a pile of endothelial cells?

These questions suggest that angiogenic factors do not operate alone, but that their final effect upon a target cell will be determined by a variety of neighborhood conditions. Sporn and Roberts have suggested that the action of peptide growth factors depends upon the context of other soluble regulatory molecules present (SPORN and ROBERTS 1988). The studies of INGBER et al. (1987) indicate that the action of FGF, for example, may depend not only upon other soluble regulatory factors, but upon the presence of insoluble complexes such as heparan sulfate, and also upon the mechanical context of the target cell upon which it acts. Some of these modulating factors are outlined below.

I. Role of Extracellular Matrix in Modulating Angiogenic Factors

Of the angiogenic factors identified so far, FGF has been studied in sufficient detail to reveal that its effects on endothelial cells are very dependent upon extracellular matrix. A compelling experiment suggests that, in the presence of a constant concentration of FGF, capillary endothelial cells may switch between phases of growth, quiescence, tube formation (differentiation), or involution, depending upon the adhesivity or mechanical integrity of their extracellular matrix (INGBER et al. 1987). Endothelial cell spreading can be precisely controlled by culturing the cells in FGF-containing chemically defined medium on bacteriological dishes coated with varying concentrations of fibronectin, a major extracellular matrix component. High fibronectin concentrations provide a strongly adhesive substratum which promotes extensive cell spreading and proliferation. Low coating densities of fibronectin lead to cell rounding and loss of viability. Intermediate fibronectin coating densities promote partial cell extension, but the substratum in this case is not sufficiently adhesive to prevent intermittent retraction of cell membrane. Partial retraction of multicellular aggregates results in cell shortening, cessation of growth, and formation of branching tubular networks within 24–48 h. None of these events occur in the absence of FGF. When these findings are taken together with studies of other substrata which effect cell growth by changes in cell spreading (EMERMAN and PITELKA 1977; SCHOR et al. 1979; HARRIS et al. 1980; Ingber and JAMIESON 1985), they suggest that matrix proteins may modulate responsiveness to soluble growth factors by altering adhesive interactions and subsequent tensile forces generated within the intracellular cytoskeleton.

This conceptual framework has special relevance for a capillary developing under the stimulus of an angiogenic factor, where endothelial cells oriented in tandem exist in different phases of mitosis, quiescence, and lumen formation (AUSPRUNK and FOLKMAN 1977).

II. Mast Cells and Heparin as Potentiators of Angiogenesis

There is increasing evidence that tissue mast cells augment angiogenesis once it has been initiated by an angiogenic factor, and that this potentiation may be mediated by heparin.

Mast cells accumulate in the most highly vascularized areas of certain tumors (SELYE 1965) as well as in tissue neovascularized by inflammatory or immune reactions. On the basis of these and other observations that during immune angiogenesis mast cell infiltration preceded neovascularization it was postulated that mast cells might enhance angiogenesis (RYAN 1970; SMITH and BASU 1970). During experimental tumor-induced angiogenesis, mast cell influx not only preceded capillary ingrowth, but the number of mast cells significantly exceeded the ratio of mast cells to capillaries found in normal tissues. Mast cell density around the tumor implant was approximately 40 times the normal mast cell count (KESSLER et al. 1976). Conditioned medium from mast cells isolated from the rat peritoneum were found to stimulate chemokinesis of capillary endothelial cells in vitro (AZIZKHAN et al. 1980). Furthermore, a variety of tumors were found to be chemotactic for mast cells (ZETTER 1980). Mast-cell-deficient mice exhibit a decreased rate of tumor angiogenesis (STARKEY et al. 1988). Experimental evidence suggests that the effect of mast cells on angiogenesis may be mediated through mast cell heparin. The factor in mast-conditioned medium which stimulated chemokinesis in endothelial cells was found to be heparin (AZIZKHAN et al. 1980) and heparin (including nonanticoagulant fractions) also potentiated tumor-induced angiogenesis in vivo (TAYLOR and FOLKMAN 1982; CASTELLOT et al. 1982). Mast cell granules caused proliferation of microvascular endothelial cells (MARKS et al. 1986). Heparin was found to potentiate the proliferation of vascular endothelial cells in culture in the presence of a growth factor now known to be acidic FGF (THORNTON et al. 1983) and heparin also potentiated the chemotactic activity of this growth factor for endothelial cells (TERRANOVA et al. 1985). Taken together, these findings indicate that mast cells may augment capillary growth once it has been initiated by an angiogenic molecule, and that mast cell heparin may be one of the mediators of this potentiation. Another mechanism of mast cell potentiation may be based on release of stored basic FGF from basement membranes (see below), either by heparin displacement or through activation of local collagenases by mast cell proteases.

III. Storage of Basic FGF in Basement Membrane – Role of Heparan Sulfate

Recent data show that extracellular matrix and basement membranes are storage sites for basic FGF. bFGF can be extracted from extracellular matrix of vascular and corneal endothelial cells in vitro (VLODAVSKY et al. 1986, 1987b) and from corneal basement membranes in vivo (FOLKMAN et al. 1988). bFGF appears to be bound to heparan sulfate, which is a major component of basement membrane (KRAMER 1971) and can be rapidly and almost completely released from basement membranes in an active form (i.e., mitogenic to capillary endothelial cells) by extraction with heparin, heparan sulfate, but not by other glycosaminoglycans. Enzymatic degradation of the basement membrane by collagenase or heparanase also releases active bFGF (FOLKMAN et al. 1988). Heparan sulfate could protect bound basic FGF from enzymatic or physical degradation because heparin has been shown to protect both basic and acidic FGF from denaturation and degradation (SCHREIBER et al. 1985; GOSPODAROWICZ and CHENG 1986). It is possible that storage of basic FGF as a complex with heparan sulfate maintains the angiogenic molecule in a stable yet potentially active form. The bFGF could be released by mast cell heparin or by enzymatic degradation of the basement membrane (BAIRD and LING 1987; KRAMER et al. 1982; BASHKIN et al. 1989). This storage capacity of the extracellular matrix may be another mechanism of restricting access of potent angiogenic factors to vascular endothelial cells until there is a wound deep enough to injure extracellular matrix.

IV. Regulation of Angiogenic Factors by Pericytes

Established capillaries are surrounded by pericytes which are enveloped in the capillary basement membrane and make close contact with capillary endothelial cells (MAZANET and FRANZINI-ARMSTRONG 1982). Pericytes are sparse or absent at the tips of growing capillaries and reappear in mature, nongrowing capillaries (AUSPRUNK and FOLKMAN 1977). While there are fewer pericytes than endothelial cells in most microvascular beds (TILTON et al. 1979), the highest ratio of pericytes to endothelial cells is in the retina where endothelial turnover is the lowest (TILTON et al. 1985; CUTHERBERTSON and MANDEL 1986). Furthermore, pericyte "dropout" correlates with the onset of retinal neovascularization in diabetics (D'OLIVEIRA 1966). These observations suggest that pericytes may downregulate endothelial proliferation. To elucidate the mechanism of this regulation, D'Amore and her associates recently isolated pure cultures of pericytes from the retina using specific immunocytochemical markers (HERMAN and D'AMORE 1985). In coculture systems, contact between pericytes and endothelial cells was found to be necessary for suppression of endothelial proliferation by pericytes. Furthermore, a single pericyte could contact and inhibit the growth of up to ten endothelial cells (ORLIDGE and D'AMORE 1986). Both endothelial cells and pericytes secreted latent TGF-β into the medium. However, only after contact between the two call types did active TGF-β appear in the medium (ORLIDGE and D'AMORE 1988). Furthermore, the addition of exogenous FGF could overcome the inhibitory effect of pericytes. These experiments suggest that capillary growth may be modulated by a balance of FGF derived from endothelial cells and TGF-β derived from local pericytes.

V. Endocrine Regulation of Angiogenesis

The female reproductive system is the most active site of physiological angiogenesis. Angiogenesis occurs at regular intervals in the ovary and endometrium, and also plays a major role in the formation of the placenta (FINDLAY 1986).

1. Ovary

As the ovarian follicle develops, new capillary blood vessels appear in its wall. After the follicle ruptures, leading to the formation of the corpus luteum, neovascularization accelerates. Ultimately, each luteal cell is in close approximation to a capillary blood vessel (BASSET 1943; JAKOB et al. 1977; DIZEREGA and HODGEN 1980; ZELEZNIK et al. 1981). Angiogenesis induced by rat and rabbit corpus lutea seems to be gonadotropin responsive (GOSPODAROWICZ and THAKRAL 1978; KOOS and LEMAIRE 1983). The origin of angiogenic activity in the ovary is not clear. However, the fluid in human ovarian follicles has recently been found to be highly angiogenic (FREDERICK et al. 1984) and GOSPODAROWICZ et al. showed that the angiogenic activity in corpus luteum is closely related to FGF (GOSPODAROWICZ et al. 1985). Extracts of ovarian stroma are angiogenic even in the absence of a follicle or a corpus luteum (MAKRIS et al. 1984). It is possible that all of these angiogenic activities in different components of the ovary may be mediated by FGF (MORMEDE et al. 1985).

An unresolved problem is how FGF-mediated angiogenesis is hormonally regulated in the ovary. An increase in FGF activity may be partly induced by gonadotropin (BAIRD et al. 1986). FGF itself may also participate in a feedback loop which leads to involution of the ovarian follicle, by virtue of the fact that FGF can inhibit FSH-induced aromatase activity in granulosa cells and thus suppress estradiol formation (BAIRD et al. 1986).

Endothelial cells in vitro have binding sites for estrogen. However, no direct link has been established between estrogen and a regulatory pathway for blood vessel growth or regression (COLBURN and BUONASSISSI 1978).

2. Endometrium

The endometrium also appears to induce neovascularization on a cyclic schedule. This angiogenic activity has been demonstrated by the transplantation of primate endometrium into the hamster cheek pouch. The extent of new *capillary* growth and bleeding was under the influence of ovarian steroids (ABEL 1985). Of interest is that endometrial neovascularization during the menstrual cycle of women and nonhuman primates involves not only capillary growth but arterial growth. DNA synthesis increases in the vascular endothelial cells of the small arteries of the endometrium, mainly in the midsecretory phase of the menstrual cycle (FERENCZY et al. 1979). It is not known whether capillary and arterial proliferation are under the control of the same or different angiogenic factors.

If FGF is considered as the principal mediator of angiogenesis in the ovary and in endometrium, heparin could be acting to potentiate angiogenesis in these tissues (FINDLAY 1986). Follicular fluid (STANGROOM and WEAVERS 1962) and endometrial fluids (FOLEY et al. 1978) both contain heparin-like anticoagulant activity (STANGROOM and WEAVERS 1962; FOLEY et al. 1978).

3. Placenta

Angiogenesis is also very extensive in the early stages of implantation of the fertilized ovum and development of the placenta in women (FINDLAY 1986; EDWARDS 1980) and in other species. In women the vascular supply in the fetal membranes is established before the point at which the first menstrual period is missed (EDWARDS 1980; FINDLAY 1986). However, the pathway by which this angiogenesis is turned on is not known. It is known, however, that placenta produces bFGF (SOMMER et al. 1987). A central problem in endocrinology is to understand how angiogenesis in various tissues of the reproductive system is so tightly regulated and to learn how it is governed hormonally.

VI. Role of Hypoxia in Regulating Angiogenic Factors

Certain angiogenic peptides may be released by reduced oxygen tension in tissue. This concept has developed mainly from the study of angiogenesis in wound healing. The biology of wound healing is a large and complex field and cannot be reviewed here (see for example: HUNT 1980; WHALEN and ZETTER 1989). However, one aspect of wound healing, the induction of angiogenesis by macrophages, provides an interesting model for what may be a general regulatory principle in angiogenesis.

The influx of macrophages into a wound is an early and critical rate-limiting event in wound healing (LEIBOVICH and ROSS 1975). Macrophages release angiogenic activity (POLVERINI et al. 1977; THAKRAL et al. 1979), and those macrophages isolated from wounds release a non-FGF low molecular weight factor (BANDA et al. 1982) as well as bFGF (BAIRD et al. 1985) and TNF-α (LEIBOVICH et al. 1987; FRATER-SCHRODER et al. 1986). It has been found that angiogenic activity from macrophages varies inversely with tissue oxygen tension. When oxygen tension is low, as in the depths of a wound, macrophage angiogenic activity is at its peak. When new capillaries enter the wound and oxygen tension rises, macrophage angiogenic activity is significantly reduced or turned off (KNIGHTON et al. 1983). The mechanism by which changes in local oxygen tension effect macrophage production of angiogenic activity is still not clear. However, recent evidence suggests that the hypoxic stimulus for macrophage production of angiogenic activity is mediated directly by pH, principally by lactate production. Thus, at lowest oxygen tensions of approximately 15 mmHg, where tissue pH is approximately 6.5, macrophages produce the highest angiogenic activity (KNIGHTON et al. 1987).

E. Pathological Angiogenesis

Physiological angiogenesis is usually brief, tightly regulated and self-limited. In contrast, pathological angiogenesis persists, often indefinitely. A review of angiogenic factors would be incomplete without a brief discussion of the pathology that may result when angiogenic activity persists, or is activated at the wrong time or in the wrong tissue. Diabetic retinopathy, neovascular glaucoma, and retrolental fibroplasia are just a few examples of diseases dominated by pathological neovascularization. In fact, pathological neovascularization in each compartment of the eye leads to such extensive damage that neovascularization per se is the leading cause of blindness worldwide (GARNER 1986). Other examples are listed in Table 3, although the list is not meant to be all-inclusive. These diseases may be called "angiogenic-dependent diseases," because they share at least two characteristics in common: (a) an abnormality of capillary blood vessel growth is a principal pathological feature and (b) the possibility that therapeutic control of the abnormal capillary growth would ameliorate or eliminate other manifestations of the disease. These diseases occur in many branches of medicine and surgery, and until recently have been thought to be unrelated.

While it is beyond the scope of this chapter to discuss angiogenic diseases in any detail, from the perspective of angiogenic factors, rheumatoid arthritis may be used as an illustration. The affected joint goes mainly through two stages of pathology. At first the synovial lining of the joint becomes inflamed with a cellular infiltrate, mostly mononuclear cells and macrophages. In the second phase, new capillary blood vessels grow in the synovial membrane and eventually invade the lateral aspects of the cartilage. This leads to cartilage destruction and increasing loss of joint function. The *neo*vascularization component of the disease has at least two roles: (a) it is thought to be the basis of irreversible destruction of cartilage (i.e., chondrolysis) which almost never oc-

Table 3. Angiogenesis-dependent diseases

Ophthalmology	Neurology
Diabetic retinopathy	Osler-Weber Syndrome
Corneal graft neovascularization	Orthopedics
Neovascular glaucoma	Nonunion fractures
Trachoma	Radiology
Retrolental fibroplasia	Arteriovenous malformations
Dermatology	Oncology
Psoriasis	Solid tumors
Pyogenic granuloma	Surgery
Cardiology	Granulations – burns
Atherosclerotic plaques	Vascular adhesions
Pediatrics	Hypertrophic scars
Hemangioma	Delayed wound healing
Angiofibroma	Internal medicine
Hemophilic joints	Arthritis
	Scleroderma

curs in the absence of neovascularization; (b) it provides an additional conduit for the delivery of mononuclear cells to the joint cavity (MATSUBARA and ZIFF 1987). It is likely that one or more macrophage-derived angiogenic factors are responsible for the neovascularization of the joint (KOCH et al. 1986). Inhibition of angiogenesis at a sufficiently early stage could possibly prevent cartilage destruction. Furthermore, there is a marked increase in mast cells in many rheumatoid arthritic joints (CRISP et al. 1984) and they may augment neovascularization of the joint.

The most persistent and intense angiogenesis is usually associated with tumor growth, and progressive tumor growth has been shown to be angiogenesis dependent (FOLKMAN 1985b). Tumors also appear to be capable of considerable redundancy in stimulating angiogenesis, and may deploy more than one angiogenic factor in the induction of neovascularization. For example, in addition to producing their own angiogenic factors (e.g., bFGF, angiogenin, or TGF-α), certain tumors can also attract macrophages and activate them to release angiogenic factors (POLVERINI and LEIBOVICH 1984). Tumors may release angiogenic factors stored in the extracellular matrix by producing specific collagenases and heparanases (VLODAVSKY et al. 1983). Tumors may release a vascular permeability factor that causes leakage of fibrinogen from postcapillary venules (SENGER et al. 1983). The fibrin that subsequently forms in the extravascular space may facilitate formation of a new capillary network (DVORAK 1986) because fibrin can stimulate endothelial locomotion (KADISH et al. 1979) and can recruit new macrophages. It is possible that fibrin degradation products may activate macrophages to secrete angiogenic peptides.

Most experimental evidence reveals that once angiogenic activity is "turned on" in a population of tumor cells it is rarely, if ever, turned off. However, there is very little understanding about when and how angiogenic activity is first expressed during the progression from a preneoplastic state to a malignant tumor.

F. Angiogenesis Inhibitors

A strategy for inhibiting pathological angiogenesis could be directed at two possible targets: (a) angiogenic factors or (b) endothelial cells. The first strategy would be based upon blocking expression or production of angiogenic factors, or neutralizing their activity. The second strategy would be to block capillary endothelial cells from responding to any angiogenic factor. All of the angiogenic inhibitors that have been discovered so far operate on this latter basis.

The first angiogenesis inhibitor was found in cartilage (EISENSTEIN et al. 1973; BREM and FOLKMAN 1975). When the partially purified inhibitor was infused intraarterially into the vascular bed of tumors in the mouse and rabbit eye, tumor growth was suppressed (LANGER et al. 1980). Considerable progress has been made in the purification and partial sequencing of this inhibitor but sufficient material is not yet available for systematic administration to tumor-bearing mice. Protamine was also found to inhibit angiogenesis (TAYLOR and FOLKMAN 1982) but the cumulative toxicity from prolonged administration made it of no use for further animal or human study. The main toxic side effects of protamine, i.e., hypocalcemia (POTTS et al. 1984) and hypotension, are unrelated to inhibition of capillary growth. In fact, the side effects, if any, of prolonged inhibition of capillary growth in a normal adult animal are still unknown.

Stronger inhibition of angiogenesis was obtained with mixtures of cortisone (or hydrocortisone) and heparin, although neither drug alone was an effective angiogenesis inhibitor (FOLKMAN et al. 1983). The antiangiogenic property of the corticosteroids was subsequently found to be dissociable from their glucocorticoid (antiinflammatory) and mineralocorticoid (salt-retaining) properties (CRUM et al. 1985). A small family of "angiostatic steroids" was identified with common structure-activity relationships. Many members of this family comprise the natural metabolites of cortisone, e.g., tetrahydrocortisol, and have no other known activity except for their capacity to inhibit angiogenesis and bring about regression of growing capillaries (CRUM et al. 1985; FOLKMAN and INGBER 1987). The angiostatic potency of the natural steroids has been greatly increased by synthetic derivatives, for example, those based on the structure of dexamethasone (CRUM et al. 1985).

It is a curious and unexplained fact that heparin is required for the manifestation of antiangiogenic activity of every angiostatic steroid discovered so far, and, for some of these steroids, heparin is synergistic. We can only speculate that the role of heparin or its analogs in this "pair effect" may be to transport steroids into endothelial cells. It is not known whether classic receptors exist on endothelial cells for several angiostatic steroids, such as tetrahydrocortisol.

Whatever the mechanism by which heparin is synergistic with angiostatic steroids, this new property of heparin is independent of anticoagulant activity. Nonanticoagulant fragments of heparin produced by enzymatic cleavage (FOLKMAN et al. 1983) or by chemical synthesis (CHOAY et al. 1983) are as good or better than heparin itself as potentiators of angiostatic steroids. However, this potentiation varies with heparins from different manufacturers and even from batch to batch (FOLKMAN 1986). Heparin preparations are nonuniform and heterogeneous in composition, molecular size, structure, position of substituents (*N*-sulfate, *O*-sulfate and glucuronic acid, and sequence) (LINHARDT et al. 1982).

This heterogeneity is believed to be the major cause of the observed variability in antiangiogenic activity of different heparin preparations reported by us (FOLKMAN et al. 1983) and by others (for review see FOLKMAN et al. 1989a). Recently, this problem has been overcome by the discovery of two simple substitutes of heparin in the pair effect with angiostatic steroids. INOUE et al. (1988) reported a bacterial-derived peptidoglycan that was highly antiangiogenic when administered with tetrahydrocortisol.

In another study, relatively simple molecules which mimic heparin in the "pair effect" with angiostatic steroids were synthesized (FOLKMAN et al. 1989a). These compounds belong to the family of hydrophilic cycloamyloses, of which β-cyclodextrin tetradecasulfate is the most potent. Nonsulfated cyclodextrins are inactive. In the chick embryo chorioallantoic membrane assay β-cyclodextrin tetradecasulfate had more than 100 times the activity of heparin (on a weight basis). In the rabbit cornea it has the advantage that it could be administered topically as eyedrops. These compounds may replace heparin as potentiators of angiostatic steroids.

While the mechanism by which heparin or the saccharides which mimic it potentiate angiostatic steroids is unclear, the cellular basis of regression of capillary blood vessel growth induced by steroid-heparin pairs has been elucidated. Immunofluorescent studies using antibodies to fibronectin, laminin, and collagen, the major components of the vascular basement membrane, show that during capillary regression under the influence of steroid-heparin combinations on the chorioallantoic membrane there is rapid breakdown and fragmentation of the basement membrane of growing capillaries (INGBER et al. 1986). This degradation can be detected within 24–48 h. Capillary endothelial cells become rounded and many separate from their extracellular matrix, leading to retraction of their capillary sprouts. These results are consistent with our previous reports that alterations in substratum (FOLKMAN and MOSCONA 1978) or in extracellular matrix (INGBER et al. 1987) prohibit cell spreading or cause cell rounding and also inhibit the growth of capillary endothelial cells in vitro.

Angiostatic steroids do not induce degradation in nongrowing capillaries or in larger veins or arteries, at least within the time period of these experiments. Also, basement membranes of epithelium only microns away from these regressing capillaries do not undergo dissolution.

The specificity of angiostatic steroids for growing capillaries is unparalleled. This specificity may be based on the relatively higher rate of basement membrane turnover that is part of normal capillary growth. It is also possible that larger vessels and nongrowing capillaries are invested by a more dense basement membrane. For example, basement membranes of larger vessels stained more heavily with fibronectin antibodies in our studies than did fine, nongrowing capillaries.

On the basis of these findings, other molecules which interfere with synthesis or deposition of basement membrane components are also being examined for antiangiogenic activity. Recently, proline analogs such as *cis*-hydroxyproline and l-azetidine carboxylic acid were found to be potent angiogenesis inhibitors in the chick embryo (INGBER and FOLKMAN 1988). However, it remains to be seen if this mechanism of action will serve as a general guideline for the discovery of novel angiogenesis inhibitors.

A central question in the field of angiogenesis inhibition is whether it will ever be possible to turn off the expression of angiogenic activity at its source, either in tumors or in a given normal tissue. Until that time, it would seem prudent to continue to identify angiogenesis inhibitors which block endothelial cells from responding to various angiogenic factors.

G. Future Directions

Much progress has been made in recent years in understanding the biology of angiogenesis and the factors that stimulate and inhibit angiogenesis. A number of important questions need to be addressed as follows:

1. Are there other as yet unidentified angiogenic factors? To date, eight angiogenic factors have been purified, most of which have been sequenced and cloned. However, other angiogenic factors are poorly characterized and remain to be purified. It is highly probable that novel angiogenic factors exist which remain to be elucidated.
2. Are there natural angiogeneic inhibitors in the body? Several synthetic inhibitors, e.g., angiostatic steroids, have been shown to inhibit angiogenesis. Yet very little is known about the existence of naturally occurring angiogenic inhibitors in tissues. Since angiogenesis is such a rare event, it is plausible that inhibitors exist. One possible source of a naturally occurring inhibitor might be the pericyte which has been shown to produce inhibitors of endothelial cell growth (ORLIDGE and D'AMORE 1988). Other normal cells may secrete inhibitors of angiogenesis, as for example the recent report by RASTINEJAD (1989).
3. What is the normal physiological role of angiogenic factors? Most angiogenic factors, e.g., FGF, have been tested in model systems such as the CAM, the cornea, and wound-healing models. It would be important to know if these angiogenic factors are involved in physiological vascularization. The female reproductive system, which is characterized by vascularization of ovary, endometrium, and placenta, would be an excellent system for the analysis of angiogenic factor function.
4. How is angiogenesis regulated in normal tissue? The presence of multiple angiogenic factors coupled with the low amount of angiogenesis that occurs in the body suggests tight regulation. Mechanisms might include storage of factors so they do not interact with their receptors (FGF), activation of latent forms (TGF-β), and the presence in tissue of angiogenic inhibitors. Angiogenic factors might be activated only when needed, e.g., wound healing and monthly vascularization of the ovary.
5. How is angiogenesis induced in tumors? Tumors may be dormant for long periods, e.g., cervical carcinoma, and then become vascularized. It would be important to know how tumors switch from the avascular to the vascular state. A pertinent model for the programmed vascularization of pancreatic tumors in transgenic mice has been proposed by FOLKMAN et al. (1989b).
6. How are large blood vessels formed? Most of the work in the angiogenesis field involves capillary formation where only one cell is involved. However, larger blood vessels are more complex, since they contain endothelial cells and smooth muscle cells in well-defined ratios. How these larger assemblies are formed is unknown. Endothelial and smooth muscle mitogens working in a paracrine fashion might be important for vein and artery formation.

An exciting and challenging field is developing based on blood vessel growth and regression. Hopefully, a detailed elucidation of the various molecules which control this process will enlarge our understanding of physiological and pathological angiogenesis.

References

Abel MH (1985) Prostanoids and menstruation. In: Baird DT, Michie EA (eds) Mechanism of menstrual bleeding. Raven, New York, pp 139–156
Abraham JA, Mergia A, Whang JL, Tumolo A, Friedman J, Hjerrild KA, Gospodarowicz D, Fiddes JC (1986) Nucleotide sequence of a bovine clone encoding the angiogenic protein, basic fibroblast growth factor. Science 233:545–548
Algire GH (1943) Microscopic studies of the early growth of a transplantable melanoma of the mouse, using the transparent-chamber technique. JNCI 4:13–20
Anzano MA, Roberts AB, Smith JM, Sporn MB, DeLarco JE (1983) Sarcoma growth factor from conditioned medium of virally transformed cells is composed of both type α and type β transforming growth factors. Proc Natl Acad Sci USA 80:6264–6268
Anzano MA, Roberts AB, Sporn MB (1986) Anchorage independent growth of primary rat embryo cells is induced by platelet-derived growth factor and inhibited by type-beta transforming growth factor. J Cell Physiol 126:312–318
Assoian RK, Komoriya A, Meyers CA, Miller DM, Sporn MB (1983) Transforming growth factor-beta in human platelets. J Biol Chem 258:7155–7160
Auerbach R (1981) Angiogenesis-inducing factors: a review. In: Pick E (ed) Lymphokines. Academic, London, pp 69–88
Auerbach R, Kubai L, Knighton D, Folkman J (1974) A simple procedure for the long-term cultivation of chicken embryos. Dev Biol 41:391–394
Ausprunk DH, Folkman J (1977) Migration and proliferation of endothelial cells in preformed and newly formed blood vessels during tumor angiogenesis. Microvasc Res 14:53–65
Ausprunk DH, Knighton DR, Folkman J (1974) Differentiation of vascular endothelium in the chick chorioallantois: a structural and autoradiographic study. Dev Biol 38:237–248
Azizkhan R, Azizkhan J, Zetter B, Folkman J (1980) Mast cell heparin stimulates migration of capillary endothelial cells in vitro. J Exp Med 152:931–944
Azizkhan JC, Sullivan R, Azizkhan R, Zetter B, Klagsbrun M (1983) The stimulation of capillary endothelial cell migration by chondrosarcoma-derived growth factors. Cancer Res 43:3281–3286
Baird A, Durkin T (1986) Inhibition of endothelial cell proliferation by type-beta transforming growth factor: interactions with acidic and basic fibroblast growth factors. Biochem Biophys Res Commun 138:476–482
Baird A, Ling N (1987) Fibroblast growth factors are present in the extracellular matrix produced by endothelial cells in vitro: implications for a role of heparinase-like enzymes in the neovascular response. Biochem Biophys Res Commun 142:428–435
Baird A, Mormede P, Bohlen P (1985) Immunoreactive fibroblast growth factor in cells of peritoneal exudate suggests its identity with macrophage-derived growth factor. Biochem Biophys Res Commun 126:358–364
Baird A, Esch F, Mormede P, Ueno N, Ling N, Bohlen P, Ying SY, Wehrenberg W, Guillemin R (1986) Molecular characterization of fibroblast growth factor: distribution and biological activities in various tissues. In. Greep RO (ed) Recent progress in hormone research, vol 42. Academic, New York, pp 143:205
Banda MJ, Knighton DR, Hunt TK, Werb Z (1982) Isolation of a nonmitogenic angiogenesis factor from wound fluid. Proc Natl Acad Sci USA 79:7773–7777
Bashkin P, Doctrow S, Klagsbrun M, Svahn CM, Folkman J, Vlodavsky I (1989) Basic fibroblast growth factor binds to subendothelial extracellular matrix and is released by heparanase and heparin-like molecules. Biochemistry 28:1737–1743
Basset D (1943) The changes in the vascular pattern of the ovary of the albino rat during the estrous cycle. Am J Anat 73:251–291

Ben Ezra D (1978) Neovasculogenic ability of prostaglandins, growth factors and synthetic chemoattractants. Am J Ophthalmol 86: 455–461
Beutler B, Cerami A (1986) Cachetin and tumour necrosis factor as two sides of the same biological coin. Nature 320:584–588
Beutler B, Mahoney J, LeTrang N, Pekala P, Cerami A (1985) Purification of cachectin, a lipoprotein lipase suppressing hormone secreted by endotoxin-induced RAW 264.7 cells. J Exp Med 161:984–995
Bevilacqua MP, Pober JS, Wheeler ME, Cotran RS, Gimbrone MA Jr (1985) Interleukin 1 acts on cultured human vascular endothelium to increase the adhesion of polymorphonuclear leukocytes, monocytes, and related leukocyte cell lines. J Clin Invest 76:2003–2011
Bicknell R, Vallee BL (1988) Angiogenin activates endothelial cell phospholipase C. Proc Natl Acad Sci USA 85:5961–5965
Bohlen P, Baird A, Esch F, Ling N, Gospodarowicz D (1984) Isolation and partial molecular characterization of pituitary fibroblast growth factor. Proc Natl Acad Sci USA 81:5364–5368
Bond MD, Vallee BL (1988) Isolation of bovine angiogenin using a placental ribonuclease inhibitor binding assay. Biochemistry 27:6282
Brem S, Folkman J (1975) Inhibition of tumor angiogenesis mediated by cartilage. J Exp Med 141:427–438
Brem S, Cotran R, Folkman J (1972) Tumor angiogenesis: a quantitative method for histologic grading. JNCI 48:347–356
Brown RA, Weis JB, Tomlinson IW, Philipps P, Kumar S (1980) Angiogenic factor from synovial fluid resembling that from tumors. Lancet 8170:682–685
Burger PC, Chandler DB, Klintworth GK (1983) Corneal neovascularization as studied by scanning electron microscopy of vascular casts. Lab Invest 48:169–180
Carswell EA, Old LJ, Kassel RL, Green S, Fiore N, Williamson B (1975) An endotoxin-induced serum factor that causes tumor necrosis. Proc Natl Acad Sci USA 72:3666–3670
Castellot JJ, Karnovsky MJ, Spiegelman BM (1980) Potent stimulation of vascular endothelial cell growth by differentiated 3T3 adipocytes. Proc Natl Acad Sci USA 77:6007–6011
Castellot JJ Jr, Karnovsky MJ, Spiegelman BM (1982) Differentiation-dependent stimulation of neovascularization and endothelial cell chemotaxis by 3T3 adipocytes. Proc Natl Acad Sci USA 79:5597–5601
Cavallo T, Sade R, Folkman J, Cotran RS (1972) Tumor angiogenesis: rapid induction of endothelial mitoses demonstrated by autoradiography. J Cell Biol 54:408–420
Choay J, Petitou M, Lormeau JC, Sinay P, Casu B, Gatti G (1983) Structure-activity relationship of heparin: a synthetic pentasaccharide with high affinity for antithrombin III and eliciting high anti-factor Xa activity. Biochem Biophys Res Commun 116:492–499
Colburn P, Buonassissi V (1978) Estrogen-binding sites in endothelial cell cultures. Science 201:817–819
Conn G, Hatcher VB (1984) The isolation and purification of two anionic endothelial cell growth factors from human brain. Biochem Biophys Res Commun 124:262–268
Connolly DT, Stoddard BL, Harakas NK, Feder J (1987) Human fibroblast-derived growth factor is a mitogen and chemoattractant for endothelial cells. Biochem Biophys Res Commun 144:705–712
Courty J, Loret C, Moenner M, Chevallier B, Lagente O, Courtois Y, Barritault D (1985) Bovine retina contains three growth factor activities with different affinity for heparin: eye-derived growth factor I, II, and III. Biochimie 67:265–269
Crisp A, Chapman CM, Kirkham SE, Schiller AL, Krane SM (1984) Articular mastrocytosis in rheumatoid arthritis. Arthritis Rheum 27:845–851
Crum R, Szabo S, Folkman J (1985) A new class of steroids inhibits angiogenesis in the presence of heparin or a heparin fragment. Science 230:1375–1378
Cutherbertson RA, Mandel TE (1986) Anatomy of the mouse retina. Endothelial cell-pericyte ratio and capillary distribution. Invest Ophthalmol Vis Sci 26:68–73

D'Amore PA, Klagsbrun M (1984) Endothelial cell mitogens derived from retina and hypothalamus: biochemical and biological similarities. J Cell Biol 99:1545–1549
D'Amore PA, Thompson RW (1987) Collateralization in peripheral vascular disease. In: Strandness D, Didsheim P, Clowes A, Watson J (eds) Vascular diseases. Grune and Stratton, Orlando, pp 319–333
Davidson JM, Klagsbrun M, Hill KE, Buckley A, Sullivan R, Brewer S, Woodward SC (1985) Accelerated wound repair, cell proliferation, and collagen accumulation are produced by a cartilage-derived growth factor. J Cell Biol 100:1219–1227
Delli Bovi PD, Basilico C (1987) Homology between fibroblast growth factor and a transforming gene from Kaposi's sarcoma. Proc Natl Acad Sci USA 84:5660–5664
Delli Bovi P, Curatola AM, Newman KM, Sato Y, Moscatelli D, Hewick RM, Rifkin DB, Basilico C (1988) Processing, secretion and biological properties of a novel growth factor of the fibroblast growth family with oncogenic potential. Mol Cell Biol 8:2933–2941
Denekamp J (1984) Vasculature as a target for tumour therapy. In: Hammersen F, Hudlicka O (eds) Progress in applied microcirculation. Karger, Basel, pp 28–38
Derynck R, Roberts AB, Winkler ME, Chen EY, Goeddel DV (1984) Human transforming growth factor-α: precursor structure and expression in *E.coli*. Cell 38:287–297
Derynck R, Roberts AB, Eaton DH, Winkler ME, Goeddel DV (1985) Human transforming growth factor-alpha: precursor sequence, gene structure and heterologous expression. In: Feramisco J, Ozanne B, Stiles C (eds) Cancer cells, vol 3. Growth factors and transformation. Cold Spring Harbor Laboratory, Cold Spring Harbor, pp 79–86
Dickson C, Peters G (1987) Potential oncogene product related to growth factors. Nature 326:833
DiZerega G, Hodgen G (1980) Fluorescence localization of luteinizing hormone/human chorionic gonadotropin uptake in the primate ovary. II. Changing distribution during selection of dominant follicle. J Clin Endocrinol Metab 51:903–907
Dobson DE, Castellot JJ, Spiegelman BM (1985) Angiogenesis stimulated by 3T3-adipocytes is mediated by prostanoid lipids. J Cell Biol 101:109a
D'Oliveira F (1966) Pericytes and diabetic retinopathy. Br J Ophthalmol 50:134–143
Dugan JD Jr, Roberts AB, Sporn MB, Glaser BM (1988) Transforming growth factor beta (TGFβ) inhibits neovascularization in vivo. J Cell Biol 107:579a
Dusseau JW, Hutchins PM, Malbasa DS (1986) Stimulation of angiogenesis by adenosine on the chick chorioallantoic membrane. Circ Res 59:163–170
Duthu GS, Smith JR (1980) In vitro proliferation and lifespan of bovine aorta endothelial cells: effect of culture conditions and fibroblast growth factor. J Cell Physiol 103:385–392
Dvorak HF (1986) Tumors: wounds that do not heal. Similarities between tumor stroma generation and wound healing. N Engl J Med 315:1650–1659
Edwards RG (1980) Conception in the human female. Academic, London
Ehrmann RL, Knoth M (1968) Choriocarcinoma: transfilter stimulation of vasoproliferation in the hamster cheek pouch – studied by light and electron microscopy. JNCI 41:1329–1341
Eisenstein R, Sorgente N, Soble L, Miller A, Kuettner KE (1973) The resistance of certain tissues to invasion: penetrability of explanted tissues by vascularized mesenchyme. Am J Pathol 73:765–774
Emerman JT, Pitelka DR (1977) Maintenance and induction of morphological differentiation in dissociated mammary epithelium of floating collagen membranes. In Vitro 13:316–328
Engerman RL, Pfaffenbach D, Davis MD (1967) Cell turnover of capillaries. Lab Invest 17:738–743
Esch F, Baird A, Ling N, Ueno N, Hill F, Denoroy L, Klepper R, Gospodarowicz D, Bohlen P, Guillemin R (1985a) Primary structure of bovine pituitary basic fibroblast growth factor (FGF) and comparison with the amino-terminal sequence of bovine acidic FGF. Proc Natl Acad Sci USA 82:6507–6511

Esch F, Ueno N, Baird A, Hill F, Denoroy L, Ling N, Gospodarowicz D, Guillemin R (1985b) Primary structure of bovine brain acidic fibroblast growth factor (FGF). Biochem Biophys Res Commun 133:554–562

Feinberg RN, Beebe DC (1983) Hyaluronate in vasculogenesis. Science 220:1177–1179

Fenselau A, Watt S, Mello RJ (1981) Tumor angiogenic factor: purification from the Walker 256 rat tumor. J Biol Chem 256:9605–9611

Ferenczy A, Bertrand G, Gelfand MM (1979) Proliferation kinetics of human endometrium during the normal menstrual cycle. Am J Obstet Gynecol 133:859–867

Fett JW, Strydom DJ, Lobb RF, Alderman EM, Bethune JL, Riordan JF, Vallee BL (1985) Isolation and characterization of angiogenin, an angiogenic protein from human carcinoma cells. Biochemistry 24:480–5486

Findlay JK (1986) Angiogenesis in reproductive tissues. J Endocrinol 111:357–366

Foley MF, Griffin BD, Zuzel M, Aparicio SR, Bradbury K, Bird CC, Clayton JK, Jenkins DM, Scott JS, Rajah CM, McNicol GP (1978) Heparin-like activity in uterine fluid. Br Med J II:322–324

Folkman J (1971) Tumor angiogenesis: therapeutic implications. N Engl J Med 285:1182–1186

Folkman J (1972) Anti-angiogenesis: new concept for therapy of solid tumors. Ann Surg 175:409–416

Folkman J (1974a) Tumor angiogenesis factor. Cancer Res 34:2109–2113

Folkman J (1974b) Tumor angiogenesis. Adv Cancer Res 19:331–358

Folkman J (1984) Angiogenesis. In: Jaffe EA (ed) Biology of endothelial cells. Martinus Nijhoff, Boston, pp 412–428

Folkman J (1985a) Angiogenesis and its inhibitors. In: DeVita VT Jr, Hellman S, Rosenberg SA (eds) Important advances in onoclogy. Lippincott, Philadelphia, pp 42–62

Folkman J (1985b) Tumor angiogenesis. In: Klein G, Weinhouse S (eds) Advances in Cancer Research. Academic, New York, pp 175–203

Folkman J (1986) How is blood vessel growth regulated in normal and neoplastic tissue? G.H.A. Clowes Memorial Award Lecture. Cancer Res 46:467–473

Folkman J (1987) Angiogenesis. In: Verstraete M, Vermylen J, Lijnan R, Arnout J (eds) Thrombosis and haemostasis. Leuven University Press, Leuven, pp 583–596

Folkman J, Haudenschild C (1980) Angiogenesis in vitro. Nature 288:551–556

Folkman J, Hochberg M (1973) Self-regulation of growth in three dimensions. J Exp Med 138:745–753

Folkman J, Ingber DE (1987) Angiostatic steroids: method of discovery and mechanism of action. Ann Surg 206:374–384

Folkman J, Klagsbrun M (1987a) Angiogenic factors. Science 235:442–447

Folkman J, Klagsbrun M (1987b) A family of angiogenic peptides. Nature 329:671–672

Folkman J, Moscona A (1978) Role of cell shape in growth control. Nature 273:346–349

Folkman J, Merler E, Abernathy C, Williams C (1971) Isolation of a tumor factor responsible for angiogenesis. J Exp Med 133:275–288

Folkman J, Knighton D, Klagsbrun M (1976) Tumor angiogenesis activity in cells grown in tissue culture. Cancer Res 36:110–114

Folkman J, Haudenschild C, Zetter BR (1979) Long-term culture of capillary endothelial cells. Proc Natl Acad Sci USA 76:5217–5221

Folkman J, Langer R, Linhardt R, Haudenschild C, Taylor S (1983) Angiogenesis inhibition and tumor regression caused by heparin or a heparin fragment in the presence of cortisone. Science 221:719–725

Folkman J, Klagsbrun M, Sasse J, Wadzinski M, Ingber D, Vlodavsky I (1988) Heparin-binding angiogenic protein – basic fibroblast growth factor – is stored within basement membrane. Am J Pathol 130:393–400

Folkman J, Weisz P, Joullie M, Li W, Ewing W (1989a) Control of angiogenesis with synthetic heparin substitutes. Science 243:1490–1493

Folkman J, Watson K, Ingber D, Hanahan D (1989) Induction of angiogenesis during the transition from hyperplasia to neoplasia. Nature 339:58–61

Form DM, Auerbach R (1983) PGE2 and angiogenesis. Proc Soc Exp Biol Med 172:214–218
Fraser RA, Ellis M, Stalker AL (1979) Experimental angiogenesis in the chorioallantoic membrane. In: Lewis DH (ed) Current advances in basic and clinical microcirculatory research. Karger, Basel, p 25
Frater-Schroder M, Muller G, Birchmeier W, Bohlen P (1986) Transforming growth factor-beta inhibits endothelial cell proliferation. Biochem Biophys Res Commun 137:295–302
Frater-Schroder M, Risau W, Hallmann R, Gautschi R, Bohlen P (1987) Tumor necrosis factor type-α, a potent inhibitor of endothelial cell growth in vitro, is angiogenic in vivo. Proc Natl Acad Sci USA 84:5277–5281
Frederick JL, Shimanuki T, DiZerega GS (1984) Initiation of angiogenesis by human follicular fluid. Science 224:389–390
Frolik CA, Dart LL, Meyers CA, Smith DM, Sporn MB (1983) Purification and initial characterization of a type beta transforming growth factor from human placenta. Proc Natl Acad Sci USA 80:3676–3680
Garner A (1986) Ocular angiogenesis. Int J Exp Pathol 28:249–309
Gimbrone MA Jr, Leapman S, Cotran RS, Folkman J (1972) Tumor dormancy in vivo by prevention of neovascularization. J Exp Med 136:261–276
Gimbrone MA Jr, Cotran RS, Folkman J (1973a) Endothelial regeneration and turnover. Studies with human endothelial cell cultures. Ser Haematol 6:453–455
Gimbrone MA Jr, Leapman S, Cotran R, Folkman J (1973b) Tumor angiogenesis: iris neovascularization at a distance from experimental intraocular tumors. JNCI 50:219–228
Gimbrone MA Jr, Cotran RS, Folkman J (1974) Tumor growth neovascularization: an experimental model using rabbit cornea. JNCI 52:413–427
Gimenez-Gallego G, Conn G, Hatcher VB, Thomas KA (1986) Human brain-derived acidic and basic fibroblast growth factors: amino terminal sequences and specific mitogenic activities. Biochem Biophys Res Commun 135:541–548
Glaser BM, D'Amore PA, Seppa H, Seppa S, Schiffmann E (1980) Adult tissues contain chemoattractants for vascular endothelial cells. Nature 288:483–484
Goldsmith HS, Griffith AL, Kupferman A, Catsimpoolas N (1984) Lipid angiogenic factor from omentum. JAMA 252:2034–2036
Gospodarowicz D (1974) Localization of a fibroblast growth factor and its effect alone and with hydrocortisone on 3T3 cell growth. Nature 249:123–129
Gospodarowicz D (1975) Purification of a fibroblast growth factor from bovine pituitary. J Biol Chem 250:2515–2520
Gospodarowicz D, Cheng J (1986) Heparin protects basic and acidic FGF from inactivation. J Cell Physiol 128:475–484
Gospodarowicz D, Thakral K (1978) Production of a corpus luteum angiogenic factor responsible for proliferation of capillaries and neovascularization of the corpus luteum. Proc Natl Acad Sci USA 75:847–851
Gospodarowicz D, Moran J, Braun D, Birdwell CR (1976) Clonal growth of bovine endothelial cells in culture: fibroblast growth factor as a survival factor. Proc Natl Acad Sci USA 73:4120–4124
Gospodarowicz D, Cheng J, Lui GM, Baird A, Bohlen P (1984) Isolation by heparin-Sepharose affinity chromatography of brain fibroblast growth factor: identity with pituitary fibroblast growth factor. Proc Natl Acad Sci USA 81:6963–6967
Gospodarowicz D, Cheng J, Lui G, Baird A, Esch F, Bohlen P (1985) Corpus luteum angiogenic factor is related to fibroblast growth factor. Endocrinology 117: 2383–2391
Gray PW, Aggarwal BB, Benten CV, Bringman TS, Henzel WJ, Jarrett JA, Leung DW, Moffat B, Ng P, Sverdersky LP, Palladino MA, Nedwin GE (1984) Cloning and expression of cDNA for human lymphotoxin, a lymphokine with tumor necrosis activity. Nature 320:584–588

Greenblatt M, Shubik P (1968) Tumor angiogenesis: transfilter diffusion studies in the hamster by the transparent chamber technique. JNCI 41:111–124

Gross JL, Moscatelli D, Jaffe EA, Rifkin DB (1982) Plasminogen activator and collagenase production by cultured capillary endothelial cells. J Cell Biol 95:974–981

Gross JL, Moscatelli D, Rifkin DB (1983) Increased capillary endothelial cell protease activity in response to angiogenic stimuli in vitro. Proc Natl Acad Sci USA 80: 2623–2627

Haranaka KE, Carswell A, Williamson BD, Prendergast JS, Satomi N, Old LJ (1986) Purification, characterization, and antitumor activity of nonrecombinant mouse tumor necrosis factor. Proc Natl Acad Sci USA 83:3949–3953

Harris AK, Wild P, Stopak D (1980) Silicone rubber substrata: a new wrinkle in the study of cell locomotion. Science 208:177–179

Haudenschild CC (1980) Growth control of endothelial cells in atherogenesis and tumor angiogenesis. In: Altura BM, Davis E, Harders H (eds) Vascular endothelium and basement membranes. Karger, Basel, pp 226–251 (Advances in microcirculation, vol 9)

Haudenschild CC, Zahniser D, Folkman J, Klagsbrun M (1976) Human endothelial cells in culture. Lack of response to serum growth factors. Exp Cell Res 98:175–183

Hauschka PV, Iafrati TA, Doleman SD, Klagsbrun M (1986) Growth factors in bone matrix: isolation of multiple types by affinity chromatography on heparin-Sepharose. J Biol Chem 261:12665–12674

Heimark RL, Twardzik DR, Schwartz SM (1986) Inhibition of endothelial regeneration by type-beta transforming growth factor from platelets. Science 233:1078–1080

Herman IM, D'Amore PA (1985) Microvascular pericytes contain muscle and nonmuscle actin. J Cell Biol 101:43–52

Hobbs JF, Cliff WJ (1971) Observations on tissue grafts established in rabbit ear chambers. A combined light and electron microscopy study. J Exp Med 134:963–971

Hockel M, Sasse J, Wissler JH (1987) Purified monocyte-derived angiogenic substance (angiotropin) stimulates migration, phenotypic changes, and "tube formation" but not proliferation of capillary endothelial cells in vitro. J Cell Physiol 133:1–13

Hockel M, Jung W, Vaupel P, Rabes H, Khaledpour C, Wissler JH (1988) Purified monocyte-derived angiogenic substance (angiotropin) induces controlled angiogenesis associated with regulated tissue proliferation in rabbit skin. J Clin Invest 82: 1075–1090

Hudlicka O, Tyler KR (1986) Angiogenesis: the growth of the vascular system. Academic press, London, pp 101–120

Hunt TK (1980) Wound healing and wound infection: theory and surgical practice. Appleton-Century-Crofts, New York

Iberg N, Rogelt S, Fanning P, Klagsbrun M (1989) Purification of 18- and 22 kDa forms of basic fibroblast growth factor from rat cells transformed by the *ras* oncogene. J Biol Chem 264:19951–19955

Ignotz R, Massague J (1986) Transforming growth factor-beta stimulates the expression of fibronectin and collagen and their incorporation into the extracellular matrix. J Biol Chem 261:4337–4345

Ingber DE, Folkman J (1988) Inhibition of angiogenesis through modulation of collagen metabolism. Lab Invest 59:44–51

Ingber DE, Jamieson JD (1985) Cell as tensegrity structures: architectural regulation of histodifferentiation by physical forces transduced over basement membrane. In: Anderson LC, Gahmberg CG, Ekblom P (eds) Gene expression during normal and malignant differentiation. Academic, Orlando, pp 13–32

Ingber DE, Madri JA, Folkman J (1986) A possible mechanism for inhibition of angiogenesis by angiostatic steroids: induction of capillary basement membrane dissolution. Endocrinology 119:1768–1775

Ingber DE, Madri JA, Folkman J (1987) Endothelial growth factors and extracellular matrix regulate DNA synthesis through modulation of cell and nuclear expansion. In Vitro Cell Dev Biol 23:387–394

Ishikawa F, Miyazono K, Hellman U, Wernstedt C, Hagiwara K, Usuki K, Takaku F, Heldin CH (1989) Identification of angiogenic activity and the cloning and expression of platelet-derived endothelial cell growth factor. Nature 338:557–562
Jaffe EA, Nachman RL, Becker CG, Minick CR (1972) Culture of human endothelial cells derived from umbilical veins: identification by morphologic and immunologic criteria. J Clin Invest 52:2745–2756
Jakob W, Jentzsch KD, Mauersberger B, Oehme P (1977) Demonstration of angiogenesis activity in the corpus luteum of cattle. Exp Pathol 13:231–236
Kadish JL, Butterfield CE, Folkman J (1979) The effect of fibrin on cultured vascular endothelial cells. Tissue Cell 11:99–108
Kalebic T, Garbisa S, Glaser B, Liotta LA (1983) Basement membrane collagen: degradation by migrating endothelial cells. Science 221:281–283
Keski-Oja J, Lyons RM, Moses HL (1987) Inactive secreted form(s) of transforming growth factor-beta: activation by proteolysis. J Cell Biochem [Suppl] 11 a:60
Kessler D, Langer R, Pless N, Folkman J (1976) Mast cells and tumor angiogenesis. Int J Canc 18:703–709
Kissun RD, Hill CR, Garner A, Phillips P, Kumar S, Weiss JB (1982) A low-molecular-weight angiogenic factor in cat retina. Br J Ophthalmol 66:165–169
Klagsbrun M, Shing Y (1985) Heparin affinity of anionic and cationic capillary endothelial cell growth factors: analysis of hypothalamus-derived growth factors and fibroblast growth factors. Proc Natl Acad Sci USA 82:805–809
Klagsbrun M, Knighton D, Folkman J (1976) Tumor angiogenesis in cells grown in tissue culture. Cancer Res 36:110–114
Klagsbrun M, Sasse J, Sullivan R, Smith JA (1986) Human tumor cells synthesize an endothelial cell growth factor that is structurally related to basic fibroblast growth factor. Proc Natl Acad Sci USA 83:2448–2452
Knighton DR, Hunt TK, Scheuenstahl H, Halliday BJ, Werb Z, Banda MJ (1983) Oxygen tension regulates the expression of angiogenesis factor by macrophages. Science 221:1283–1285
Knighton D, Schumerth S, Fiegel V (1987) Environmental regulation of macrophage angiogenesis. In: Rifkin DB, Klagsbrun M (eds) Current communications in molecular biology. Cold Spring Harbor Laboratory, Cold Spring Harbor, pp 150–154
Koch AE, Polverini PJ, Leibovich SJ (1986) Stimulation of neovascularization by human rheumatoid synovial tissue macrophages, Arthritis Rheum 29:471–479
Koos R, Lemaire W (1983) Evidence for an angiogenic factor from rat follicles. In: Greenwald GS, Terranova PF (eds) Factors regulating ovarian function. Raven, New York, pp 191–195
Kramer P (1971) Heparan-sulfates of cultured cells: I. Membrane-associated in cell-sap species in Chinese hamster cells. Biochemistry 10:1443–1445
Kramer RH, Vogel GL, Nicolson GL (1982) Solubilization and degradation of subendothelial matrix glycoproteins and proteoglycans by metastatic tumor cells. J Biol Chem 257:2678–2686
Kreisle RA, Ershler WB (1988) Investigation of tumor angiogenesis in an Id mouse model: role of host-tumor interactions. JNCI 80:849–854
Kull FC Jr, Brent DA, Parikh I, Cuatrecasas P (1987) Chemical identification of a tumor-derived angiogenic factor. Science 236:843–845
Kurachi K, Davie EW, Strydom DJ, Riordan JF, Vallee BL (1985) Sequence of the cDNA and gene for angiogenin, a human angiogenesis factor. Biochemistry 24: 5494–5499
Langer R, Folkman J (1976) Polymers for the sustained release of proteins and other macromolecules. Nature 263:797–800
Langer RS, Conn H, Vacanti JP, Haudenschild C, Folkman J (1980) Control of tumor growth in animals by infusion of an angiogenesis inhibitor. Proc Natl Acad Sci USA 77:4431–4335

Lawrence DA, Pircher R, Kryceve-Martinerie C, Jullien P (1984) Normal embryo fibroblasts release transforming growth factors in a latent form. J Cell Physiol 121:184–188

Lawrence DA, Pircher R, Jullien P (1985) Conversion of a high molecular weight latent beta-TGF from chicken embryo fibroblasts into a low molecular weight active beta-TGF under acidic conditions. Biochem Biophys Res Commun 133:1026–1034

Leibovich SJ, Ross R (1975) The role of macrophages in wound repair: a study with hydrocortisone and antimacrophage serum. Am J Pathol 78:71–100

Leibovich SJ, Polverini PJ, Shepard HM, Wiseman DM, Shively V, Nuseir N (1987) Macrophage-induced angiogenesis is mediated by tumour necrosis factor-α. Nature 329:630–632

Lemmon SK, Bradshaw R (1983) Purification and partial characterization of bovine pituitary fibroblast growth factor. J Cell Biochem 21:195–208

Linhardt RJ, Grant A, Coonery CL, Langer R (1982) Differential anticoagulant activity of heparin fragments prepared using microbial heparinase. J Biol Chem 257:7310–7313

Lobb RR, Fett JW (1984) Purification of two distinct growth factors from bovine neural tissue by heparin affinity chromatography. Biochemistry 23:6295–6299

Lobb RR, Sasse J, Shing Y, D'Amore PA, Sullivan R, Jacobs J, Klagsbrun M (1986) Purification and characterization of heparin-binding growth factors. J Biol Chem 261:1924–1928

Maciag T, Hoover GA, Stemerman MB, Weinstein R (1981) Serial propagation of endothelial cells in vitro. J Cell Biol 91:420–426

Maciag T, Hoover GA, van der Spek J, Stemerman MB, Weinstein R (1982a) Growth and differentiation of human umbilical-vein endothelial cells in culture. In: Book A, Sato GH, Pardee AB, Sirbasku DA (eds) Growth of cells in hormonally defined media. Cold Spring Harbor Laboratory, Cold Spring Harbor, pp 525–538

Maciag T, Kadish J, Wilkins L, Stemerman MB, Weinstein R (1982b) Organization behavior of human umbilical vein endothelial cells. J Cell Biol 94:511–520

Maciag T, Mehlman T, Friesel R, Schreiber AB (1984) Heparin binds endothelial cell growth factor, the principal cell mitogen in bovine brain. Science 225:932–935

Madri J, Williams SK (1983) Capillary endothelial cell cultures: phenotypic modulation by matrix components. J Cell Biol 97:153–165

Madtes DK, Raines EW, Sakariassen KS, Assoian RK, Sporn MB, Bell GI, Ross R (1988) Induction of transforming growth factor-α in activated human alveolar macrophages. Cell 53:285–293

Makris A, Ryan KJ, Yasumizu T, Hill CL, Zetter BR (1984) The non-luteal porcine ovary as a source of angiogenic activity. Endocrinology 15:1672–1677

Marks RM, Roche WR, Czerniecki M, Penny R, Nelson DS (1986) Mast cell granules cause proliferation of human microvascular endothelial cells. Lab Invest 55: 289–294

Marquardt H, Hunkapiller MW, Hood LE, Todaro GJ (1984) Rat transforming growth factor type I: structure and relationship to epidermal growth factor. Science 223:1079–1082

Matsubara T, Ziff M (1987) Inhibition of human endothelial cell proliferation by gold compounds. J Clin Invest 79:1440–1446

Mazanet R, Franzini-Armstrong C (1982) Scanning electron microscopy of pericytes in rat red muscle. Microvasc Res 23:361–369

Miyazono K, Okabe T, Urabe A, Takaku F, Heldin C-H (1987) Purification and properties of an endothelial cell growth factor from human platelets. J Biol Chem 262:4098–4103

Montesano R, Orci L, Vassalli P (1983) In vitro rapid organization of endothelial cells into capillary-like networks is promoted by collagen matrices. J Cell Biol 97: 1648–1652

Montesano R, Vassali JD, Baird A, Guillemin R, Orci L (1986) Basic fibroblast growth factor induces angiogenesis in vitro. Proc Natl Acad Sci USA 83:7297–7301

Mormede P, Baird A, Pigeon P (1985) Immunoreactive fibroblast growth factor (FGF) in rat tissues: molecular weight forms and the effects of hypophysectomy. Biochem Biophys Res Commun 120:1108–1113

Moscatelli D, Silverstein J, Manejias R, Rifkin DB (1987) M_r 25000 heparin binding protein from guinea pig brain is a high molecular weight form of basic fibroblast growth factor. Proc Natl Acad Sci USA 84:5778–5782

Moses HL, Tucker RF, Leof EB, Coffey RJ, Halper J, Shipley GD (1985) Type beta transforming growth factor is a growth stimulator and a growth inhibitor. Cancer Cells (Cold Spring Harbor) 3:65–71

Moses MA, Sudhalter J, Langer R (1988) A Cartilage-derived Collagenase inhibitor of capillary all proliferation. Invest opth 29:95

Muller G, Behrens J, Nussbaumer U, Bohlen P, Birchmeier W (1987) Inhibitory action of transforming growth factor-β on endothelial cells. Proc Natl Acad Sci USA 84:5600–5604

Muthukkarauppan VR, Auerbach R (1979) Angiogenesis in the mouse cornea. Science 205:1416–1417

Nicosia RF, Tchao R, Leighton J (1982) Histotypic angiogenesis in vitro: light microscopic, ultrastructural and radioautographic. In Vitro 18:538–549

Odedra R, Weiss JB (1987) A synergistic effect on microvessel cell proliferation between basic fibroblast growth factor (FGFb) and endothelial cell stimulating angiogenesis factor (ESAF). Biochem Biophys Res Commun 143:947–953

Orlidge A, D'Amore P (1986) Pericyte and smooth muscle cell modulation of endothelial cell proliferation. J Cell Biol 103:471 a

Orlidge A, D'Amore P (1988) Endothelial cell-pericyte cocultures produce activated TGF-β which inhibits endothelial cell growth. Invest Ophthalmol Vis Sci 29:109

Pasyk S, Schaper W, Schaper J, Pasyk K, Miskiewicz G, Strinseifer B (1982) DNA synthesis in coronary collaterals after coronary occlusion in the conscious dog. Am J Physiol 242 (Heart Circ Physiol 11):H1031–H1037

Pledger WJ, Stiles CD, Antoniades HN, Scher CD (1978) An ordered sequence of events is required before BALB/c-3T3 cells become committed to DNA synthesis. Proc Natl Acad Sci USA 75:2839–2843

Pober JS, Bevilacqua MP, Mendrick DL, LaPierre LA, Fiers W, Gimbrone MA Jr (1986a) Two distinct monokines, interleukin 1 and tumor necrosis factor, each independently induce biosynthesis and transient expression of the same antigen on the surface of cultured human vascular endothelial cells. J Immunol 137:1680–1687

Pober JS, Gimbrone MA Jr, LaPierre LA, Mendrick DL, Fiers W, Rothlein R, Springer TA (1986b) Overlapping patterns of activation of human endothelial cells by interleukin 1, tumor necrosis factor, and immune interferon. J Immunol 137:1893–1896

Pober JS, LaPierre LA, Stophen AH, Brock TA, Springer TA, Fiers W, Bevilacqua MP, Mendrick DL, Gimbrone MA Jr (1987) Activation of cultured human endothelial cells by recombinant lymphotoxin: comparison with tumor necrosis factor and interleukin 1 species. J Immunol 138:3319–3324

Polverini P, Cotran R, Gimbrone M Jr, Unanue E (1977) Activated macrophages induce vascular proliferation. Nature 269:804–806

Polverini P, Leibovich S (1984) Induction of neovascularization in vivo and endothelial proliferation in vitro by tumor-associated macrophages. Lab Invest 51:635–642

Potts M, Dopplet S, Taylor S, Folkman J, Neer R, Potts JT Jr (1984) Protamine: a powerful in vivo inhibitor of bone resorption. Calcif Tissue Int 36:189–193

Prats H, Kaghad M, Prats AC, Klagsbrun M, Lelias JM, Liauzun P, Chalon P, Tauber JP, Amalric F, Smith JA, Caput D (1989) High molecular weight forms of basic fibroblast growth factor are initiated by alternative CUG codons. Proc Natl Acad Sci USA (86:1836–1840)

Presta M, Moscatelli D, Silverstein JJ, Rifkin DB (1986) Purification from a human hepatoma cell line of a basic FGF like molecule that stimulates capillary endothelial cell plasminogen activator production, DNA synthesis and migration. Mol Cell Biol 6:4060–4066

Presta M, Rusnati M, Maier JAM, Ragnotti G (1988) Purification of basic fibroblast growth factor from rat brain: identification of a Mr 22,000 immunoreactive form. Biochem Biophys Res Commun 155:1161
Proia AD, Chandler MB, Haynes WL, Smith CS, Suvarnamani C, Erkel F, Klintworth GK (1988) Quantitation of corneal neovascularization using computerized image analysis. Lab Invest 58:473–479
Rastinejad F, Polverini PJ and Bouck NP (1989) Regulation of the activity of a new inhibitor of angiogenesis by a cancer suppressor gene. CELL 56:345–355
Rifkin DB, Gross JL, Moscatelli D, Jaffe E (1982) Proteases and angiogenesis: production of plasminogen activator and collagenase by endothelial cells. In: Nossel H, Vogel HJ (eds) Pathobiology of the endothelial cell. Academic, New York, pp 191–197
Riordan JF, Vallee BL (1988) Human angiogenin, an organogenic protein. Br J Cancer 57:587–590
Risau W, Lemmon V (1988) Changes in the vascular extracellular matrix during embryonic vasculogenesis and angiogenesis. Dev Biol 125:441–450
Roberts AB, Anzano MA, Lamb LC, Smith JM, Sporn MB (1981) New class of transforming growth factors potentiated by epidermal growth factor: isolation from non-neoplastic tissues. Proc Natl Acad Sci USA 78:5339–5343
Roberts AB, Sporn MB, Assoian RK, Smith JM, Roche NS, Wakefield LM, Heine UI, Liotta LA, Falanga V, Kehrl JH, Fauci AS (1986) Transforming growth factor type-beta: rapid induction of fibrosis and angiogenesis in vivo and stimulation of collagen formation in vitro. Proc Natl Acad Sci USA 83:4167–4171
Rogelj S, Weinberg RA, Fanning P, Klagsbrun M (1988) Basic fibroblast growth factor fused to a signal peptide transforms cells. Nature 331:173–175
Ryan T (1970) Factors influencing the growth of vascular endothelium in the skin. Br J Dermatol 82:99–111
Ryan TJ, Barnhill RL (1983) Physical factors and angiogenesis. Ciba Found Symp: 80–94
Ryan TJ, Stockley AT (1980) Mechanical versus biochemical factors in angiogenesis. Microvasc Res 20:258–259
Rybak SM, Vallee BL (1988) Base cleavage specificity of angiogenin with *Saccharomyces cerevisiae* and *Escherichia coli* 5S RNAs. Biochemistry 27:2288–2294
Rybak SM, Fett JW, Yao QZ, Vallee BL (1987) Angiogenin mRNA in human tumor and normal cells. Biochem Biophys Res Commun 146:1240–1248
Sakamoto H, Mori M, Taira M, Yoshida T, Matsukawa S, Shimizu K, Sekiguchi M, Terada M, Sugimura T (1986) Transforming gene from human stomach cancers and a non-cancerous portion of stomach mucosa. Proc Natl Acad Sci USA 83:3997–4001
Saksela O, Moscatelli D, Sommer A, Rifkin DB (1988) Endothelial cell-derived heparan sulfate binds basic fibroblast growth factor and protects it from proteolytic degradation. J Cell Biol 107:743–751
Schaper W (1981) The collateral circulation of the heart. American Elsevier, New York
Schaper W, DeBrabander M, Lewi P (1971) DNA synthesis and mitoses in coronary collateral vessels of the dog. Circ Res 28:671–679
Schor AM, Schor SL, Kumar S (1979) Importance of a collagen substratum for stimulation of capillary endothelial cell proliferation by tumor angiogenesis factor. Int J Cancer 24:225–234
Schreiber AB, Kenney J, Kowalski WJ, Friesel R, Mehlman T, Maciag T (1985) Interaction of endothelial cell growth factor with heparin: characterization of receptor and antibody recognition. Proc Natl Acad Sci USA 82:6138–6142
Schreiber AB, Winkler ME, Derynck R (1986) Transforming growth factor-alpha: a more potent angiogenic mediator than epidermal growth factor. Science 232: 1250–1253
Schweigerer L, Malerstein B, Gospodarowicz D (1987a) Tumor necrosis factor inhibits the proliferation of cultured capillary endothelial cells. Biochem Biophys Res Commun 143:997–1004

Schweigerer L, Neufeld G, Friedman J, Abrahan JA, Fiddes JC, Gospodarowicz D (1987b) Capillary endothelial cells express basic fibroblast growth factor. Nature 325:257–259

Selye H (1965) The mast cells. Butterworth, Washington, p 293

Senger DR, Galli SJ, Dvorak AM, Perruzzi CA, Harvey VS, Dvorak HF (1983) Tumor cells secrete a vascular permeability factor that promotes accumulation of ascites fluid. Science 219:983–985

Seyedin SM, Thomas TC, Thompson AY, Rosen DM, Piez KA (1985) Purification and characterization of two cartilage-inducing factors from bovine demineralized bone. Proc Natl Acad Sci USA 82:2267–2271

Seyedin SM, Thompson AY, Bentz H, Rosen DM, McPherson JM, Conti A, Siegel NR, Galluppi GR, Piez KA (1986) Cartilage-inducing factor-A. J Biol Chem 261:5693–5695

Shapiro R, Vallee BL (1987) Human placental ribonuclease inhibitor abolishes both angiogenic and ribonucleolytic activities of angiogenin. Proc Natl Acad Sci USA 84:2238–2241

Shapiro R, Riordan JF, Vallee BL (1986) Characteristic ribonucleolytic activity of human angiogenin. Biochemistry 25:3527–3532

Shapiro R, Strydom DJ, Olson KA, Vallee BL (1987a) Isolation of angiogenin from normal human plasma. Biochemistry 26:5141–5146

Shapiro R, Weremowicz S, Riordan JF, Vallee BL (1987b) Ribonucleolytic activity of angiogenin: essential histidine, lysine, and arginine residues. Proc Natl Acad Sci USA 84:8783–8787

Sherry B, Cerami A (1988) Cachectin/tumor necrosis factor exerts endocrine, paracrine, and autocrine control of inflammatory responses. J Cell Biol 107:1269–1277

Shing Y, Folkman J, Sullivan R, Butterfield C, Murray J, Klagsbrun M (1984) Heparin affinity: purification of a tumor-derived capillary endothelial cell growth factor. Science 223:1296–1299

Shing Y, Folkman J, Haudenschild C, Lund D, Crum R, Klagsbrun M (1985) Angiogenesis is stimulated by a tumor-derived endothelial cell growth factor. J Cell Biochem 29:275–287

Smith S, Basu P (1970) Mast cells in corneal immune reaction. Can J Ophthalmol 5:175–183

Sommer A, Brewer MT, Thompson RC, Moscatelli D, Presta M, Rifkin DB (1987) A form of human fibroblast growth factor with an extended amino terminus. Biochem Biophys Res Commun 42:543–550

Splawinski J, Michna M, Palczak R, Konturek S, Splawinski B (1988) Angiogenesis: quantitative assessment by the chick chorioallantoic membrane assay. Meth Find Exp Clin Pharmacol 10:221–226

Sporn MB, Roberts AB (1988) Peptide growth factors are multifunctional. Nature 332:217–219

Sporn MB, Roberts AB, Wakefield LM, de Crombrugghe B (1987) Some recent advances in the chemistry and biology of transforming growth factor-beta. J Cell Biol 105:1039–1045

St. Clair DK, Rybak SM, Riordan JF, Vallee BL (1987) Angiogenin abolishes cell-free protein synthesis by specific ribonucleolytic inactivation of ribosomes. Proc Natl Acad Sci USA 84:8330–8334

Stangroom JE, Weavers R (1962) Anticoagulant activity of equine follicular fluid. J Reprod Fertil 3:269–282

Starkey JR, Crowle PK, Taubenberger S (1988) Mast-cell deficient W/W^v mice exhibit a decreased rate of tumor angiogenesis. Int J Cancer 42:48–52

Strydom DJ, Fett JW, Lobb RR, Alderman EM, Bethune JL, Riordan JF, Vallee BL (1985) Amino acid sequence of human tumor-derived angiogenin. Biochemistry 24:5486–5494

Sullivan R, Klagsbrun M (1985) Purification of cartilage-derived growth factor by heparin affinity chromatography. J Biol Chem 260:2399–2401

Taira M, Yoshida T, Miyagawa K, Sakamoto H, Terada M, Sugimura T (1987) cDNA sequence of human transforming gene *hst* and identification of the coding sequence required for transforming activities. Proc Natl Acad Sci USA 84:2980–2984
Takehara K, LeRoy EC, Grotendorst GR (1987) TGF-β inhibition of endothelial cell proliferation: alteration of EGF binding and EGF-induced growth-regulatory (competence) gene expression. Cell 49:415–422
Taylor CM, Weiss JB (1984) The chick vitelline membrane as a new test system for angiogenesis and antiangiogenesis. Int J Microcirc Clin Exp 3:337
Taylor S, Folkman J (1982) Protamine is an inhibitor of angiogenesis. Nature 297: 307–312
Terranova VP, DiFlorio R, Lyall RM, Hic S, Friesel R, Maciag T (1985) Human endothelial cells are chemotactic to endothelial cell growth factor and heparin. J Cell Biol 101:2330–2334
Thakral K, Goodson W, Hunt T (1979) Stimulation of wound blood vessel growth by wound macrophages. J Surg Res 26:430
Thomas KA (1987) Fibroblast growth factors. FASEB J 1:434–440
Thomas K, Gimemez-Gallego G (1986) Fibroblast growth factors: broad spectrum mitogens with potent angiogenic activity. Trends Biochem Sci 11:81–84
Thomas KA, Rios-Candelore M, Fitzpatrick S (1984) Purification and characterization of acidic fibroblast growth factor from bovine brain. Proc Natl Acad Sci USA 81:357–361
Thomas KA, Riley MC, Lemmon SK, Baglan NC, Bradshaw RA (1980) Brain fibroblast growth factor. J Biol Chem 255:5517–5520
Thompson JA, Anderson KD, DiPietro JM, Zweibel JA, Zametta M, Anderson WF, Maciag T (1988) Site-directed neovessel formation in vivo. Science 241:1349–1352
Thompson WD, Campbell R, Evans T (1985) Fibrin degradation response in the chick embryo chorioallantoic membrane. J Pathol 145:27–37
Thornton S, Mueller S, Levine E (1983) Human endothelial cells: use of heparin in cloning and long-term serial cultivation. Science 222:623–625
Tilton RG, Kilo C, Williamson JR (1979) Pericyte-endothelial relationships in cardiac and skeletal muscle capillaries. Microvasc Res 18:325–335
Tilton RG, Miller EJ, Kilo C, Williamson JR (1985) Pericyte form and distribution in rat retinal and uveal capillaries. Invest Ophthalmol Vis Sci 26:68–73
Todaro GJ, Fryling C, DeLarco JE (1980) Transforming growth factors produced by certain human tumor cells: polypeptides that interact with epidermal growth factor receptors. Proc Natl Acad Sci USA 77:5258–5262
Tuan D, Smith S, Folkman J, Merler E (1973) Isolation of the non-histone proteins of rat Walker carcinoma and their association with tumor angiogenesis. Biochemistry 12:3159–3165
Tucker RF, Shipley GD, Moses HL, Holley RW (1984) Growth inhibitor from BSC-1 cells closely related to platelet type beta transforming growth factor. Science 226:705–707
Vlodavsky I, Fuks Z, Bar-Ner M, Ariav Y, Schirrmacher V (1983) Lymphoma cell mediated degradation of sulfated proteoglycans in the subendothelial cell extracellular matrix: relationship to tumor metastasis. Cancer Res 43:2704–2711
Vlodavsky I, Sullivan R, Fridman R, Sasse J, Folkman J, Klagsbrun M (1986) Heparin-binding endothelial cell growth factor produced by endothelial cells and sequestered by the endothelial extracellular matrix. J Cell Biol 103 (2):98 a
Vlodavsky I, Folkman J, Sullivan R, Fridman R, Ishai-Michaeli R, Sasse J, Klagsbrun M (1987a) Endothelial cell-derived basic fibroblast growth factor; synthesis and deposition into subendothelial extracellular matrix. Proc Natl Acad Sci USA 84:2292–2296
Vlodavsky I, Friedman R, Sullivan R, Sasse J, Klagsbrun M (1987b) Aortic endothelial cells synthesize basic fibroblast growth factor which remains cell-associated and platelet-derived growth factor-like protein which is secreted. J Cell Physiol 131:402–408

Wahl SM, Hunt DA Wakefield LM, McCartney-Francis N, Wahl LM, Roberts AB, Sporn MB (1987) Transforming growth-factor beta (TGF-beta) induces monocyte chemotaxis and growth factor production. Proc Natl Acad Sci USA 84:5788–5792
Weiner HL, Weiner LH, Swain JL (1987) Tissue distribution and developmental expression of the messenger RNA encoding angiogenin. Science 237:280–282
Whalen GF, Zetter BR (1989) Angiogenesis and wound healing. In: Diegelman RF, Cohen K, Lindblatt WJ (eds) Wound healing: Biochemical and clinical aspects. Saunders, Philadelphia (in press)
Zeleznik A, Schuller H, Reichert L Jr (1981) Gonadatotropin-binding sites in the rhesus monkey ovary: role of the vasculature in the selective distribution of human chorionic gonadotropin to the preovulatory follicle. Endocrinology 109:356–362
Zetter BR (1980) Migration of capillary endothelial cells is stimulated by tumor-derived factors. Nature 285:41–43
Zhan X, Bates B, Hu X, Goldfarb M (1988) The human FGF-5 oncogene encodes a novel protein related to fibroblast growth factors. Mol Cell Biol 8:3487
Ziche M, Jones J, Gullino P (1982) Role of prostaglandin E1 and copper in angiogenesis. JNCI 69:475–482

CHAPTER 38

Metastasis

E. SCHIFFMANN, M. L. STRACKE, and L. A. LIOTTA

A. Introduction

Metastasis is the major cause of morbidity and death for cancer patients (SUGARBAKER et al. 1982). Treatment modalities such as surgery, chemotherapy, and radiotherapy can now cure approximately 50% of the patients who develop a malignant tumor. The majority of the patients in the treatment failure group succumb to the direct effects of the metastases or to complications associated with treatment of metastases. The dispersed anatomical location of metastases and their heterogeneous cell composition prevent surgical removal and limit the response to systemic anticancer agents. Consequently, a major challenge to cancer scientists is the development of improved methods to predict the metastatic aggressiveness of a patient's individual tumor, prevent local invasion, and identify and treat clinically silent micrometastases. The approach to this problem has been to understand the biochemical basis of invasion and metastasis in pursuit of identifying tumor-specific molecules which may suggest diagnostic and therapeutic strategies (FIDLER and HART 1982; POSTE 1983). Over the past several years, significant progress has been made toward this goal. Recently, a new class of cytokines has been identified which may play an important role in tumor cell motility and invasion (LIOTTA et al. 1986; ATNIP et al. 1987; STRACKE et al. 1987). We will review this rapidly moving field.

B. Invasion as an Active Process

Cancer invasion is an active process requiring tumor cell locomotion. This simple fact was not appreciated until recently due to the dominance of the mechanical theory of invasion (TYZZER 1913, COMAN 1973; EAVES 1973). The mechanical theory held that tumor invasion is a passive process caused by simple growth pressure. The expanding tumor mass was thought to compress and passively destroy surrounding host tissue by blood vessel compression and resulting anoxia. This same mechanical growth pressure could allow the tumor cells to break through blood vessel walls. Extravasation could be explained based on colony formation and expansion within the blood vessel lumen with the resultant rupture of the vessel wall. Under the passive mechanism, tumor cells flaked off the primary tumor and became distributed in the host tissue. Reduced cell-to-cell cohesiveness of tumor cells as emphasized by COMAN (1973) contributed to detachment. Thus, until 1970, tumor invasion and metastases were thought to be the result of passive growth pressure coupled with low tumor cell cohesiveness (EAVES 1973). It was not until the 1960s and 1970s that active mechanisms were seriously investigated.

The simple mechanical theory of invasion could not account for the difference in invasive behavior between many rapidly proliferating benign and malignant neoplasms. Leiomyomas of the uterus frequently generate significant growth pressure, which can be readily appreciated on gross section. Nevertheless, these tumors compress (but do not damage) the surrounding host muscle and never invade or metastasize. Fibroadenomas of the breast can proliferate rapidly and expand, but they also fail to invade and metastasize. Mechanical pressure alone could not explain how the tumor cells can traverse host cellular, connective tissue, and vascular wall mechanical barriers. The theoretical framework therefore switched to a search for active mechanisms involved in facilitating the movement of tumor cells through host barriers. Possible mechanisms include tumor cell ameboid motility, lytic enzyme secretion, and secretions of other factors by invasive tumor cells. Direct experimental evidence using in vitro invasion assays indicated that growth pressure alone was not sufficient for invasion (MEYVISCH et al. 1983; THORGEIRSSON et al. 1984). Vascularization was also not required for invasion and host tissue destruction in vitro (MEYVISCH et al. 1983; THORGEIRSSON et al. 1984). Inhibitors of DNA synthesis did not block invasion (THORGEIRSSON et al. 1984), whereas inhibitors of protein synthesis or microtubule formation did block invasion (MEYVISCH et al. 1983). In parallel with these theoretical developments, there has been a revolution in the appreciation of tumor heterogeneity (FIDLER and HART 1982) and the roles of the cell surface (POSTE 1983), the host immune system (HANNA and KEY 1982; FROST and KERBEL 1983), and endothelial interactions (FOLKMAN 1985) as modulators of invasion and metastases. Thus, the theories of the mechanisms that play a role in tumor invasion and metastases have progressed from those of a simple passive mechanical explanation to one of a highly complex cascade of active biochemical and cellular factors.

C. Interaction of Tumor Cells with the Extracellular Matrix

The mammalian organism is composed of a series of tissue compartments separated from one another by two types of extracellular matrix: basement membranes and interstitial stroma (HAY 1982; YAMADA et al. 1985).

The extracellular matrix is a complex meshwork of collagen and elastin, embedded in a viscoelastic ground substance composed of proteoglycans and glycoproteins (HAY 1982; YAMADA et al. 1985). The matrix exists as a three-dimensional supporting scaffold that isolates tissue compartments, mediates cell attachment, and determines tissue architecture. The matrix functions as a selective macromolecular filter and also influences mitogenesis, morphogenesis, and cytodifferentiation. The matrix is postulated to exert chemical and mechanical influences on the shape and biochemical interactions between normal cells and the matrix. This may be altered in neoplasia and may influence tumor proliferation and invasion.

For a given type of extracellular matrix, the molecular composition is tissue specific. Molecules such as collagen type IV, laminin, entactin, and basement membrane, proteoglycan are uniquely localized in the basement membrane, which is formed by epithelium, endothelium, Schwann's cells, and myocytes. The interstitial matrix associated with stromal fibroblasts and myofibroblasts contains collagen types I and III, fibronectin, and specific types of proteoglycans and glycoproteins (HAY 1982). The matrix components produced by a cell generally reflect its tissue of origin.

During the transition from in situ to invasive carcinoma (VRACKO 1974; WICHA et al. 1980), tumor cells penetrate the epithelial basement membrane and enter the underlying interstitial stroma. Once the tumor cells enter the stroma, they gain access to lymphatics and blood vessels for further dissemination. Fibrosarcomas and angiosarcomas, developing from stromal cells, invade surrounding muscle basement membrane and destroy myocytes. Tumor cells must cross basement membranes to invade nerve and most types of organ parenchyma. During intravasation or extravasation, the tumor cells of any histological origin must penetrate the subendothelial basement membrane. In the distant organ where metastatic colonies are initiated, extravasated tumor cells must migrate through the perivascular intestinal stroma before tumor colony growth occurs in the organ parenchyma. Therefore, tumor cell interaction with the extracellular matrix occurs at multiple stages in the metastatic cascade (YAMADA et al. 1985).

D. Three Stages in Invasion

A three-stage hypothesis has been proposed to describe the sequence of biochemical events that occurs during tumor cell invasion of extracellular matrix (FOLKMAN 1985). The first stage is tumor cell attachment to the matrix. Attachment may be mediated through specific glycoproteins such as laminin or fibronectin that bind to tumor cell plasma membrane receptors. Following attachment, the tumor cell secretes hydrolytic enzymes (or induces host cells to secrete enzymes) that can locally degrade the matrix (including degradation of the attachment glycoproteins). Matrix lysis most likely takes place in a highly localized region close to the tumor cell surface, where the amount of active enzyme exceeds the natural protease inhibitors present in the serum and in the matrix itself (LIOTTA et al. 1979, 1982; STRAULI 1980; WOOLLEY et al. 1980; NAKAJIMA et al. 1984). In contrast to the invasive tumor cell, when the normal cell or benign tumor cell attaches to the matrix, it may respond by shifting into a resting or differentiated state. The third stage is tumor cell locomotion into the region of the matrix modified by proteolysis. The major emphasis of this review is the role of tumor cell cytokines in the regulation of locomotion as a necessary component of this third step. The direction of the locomotion may be influenced by chemotactic factors and autocrine motility factors. The chemotactic factors derived from serum, organ parenchyma, or the matrix itself may influence the organ specificity of metastases. Continued invasion of the matrix may take place by cyclic repetition of these three stages.

E. Agents Inducing Migration: Autocrine Motility Factors

Cell motility is necessary for tumor cells to traverse many stages in the complex cascade of invasion. Such stages could include the detachment and subsequent infiltration of cells from the primary tumor into adjacent tissue, the migration of the cells through the vascular wall into the circulation (intravasation), and the extravasation of the cells to a secondary site. The movement of cells through biological barriers such as the endothelial basement membranes of the vasculature may well occur by means of chemotactic mechanisms. Indeed, studies on in vitro chemotaxis of some tumor cells report that a variety of compounds such as complement-derived materials, collagen peptides, formyl peptides, and certain connective tissue components can act as chemoattractants (LAM et al. 1981; MCCARTHY et al. 1985). While these agents may well contribute to the directional aspects of a motile response, they are not sufficient to initiate the intrinsic locomotion of tumor cells. The availability of soluble attractants to the tumor cell is greatly dependent upon the host even in those cases in which the production of attractants is the result of tumor cell-host tissue interaction. At best, it seems that the cell would have access to such motility stimuli at sporadic and irregular intervals, conditions unfavorable to a sustained migration of highly invasive cells. With these considerations in mind and stimulated by the studies of Todaro, Sporn, and coworkers (TODARO et al. 1980; ANZANO et al. 1983), in which they demon-

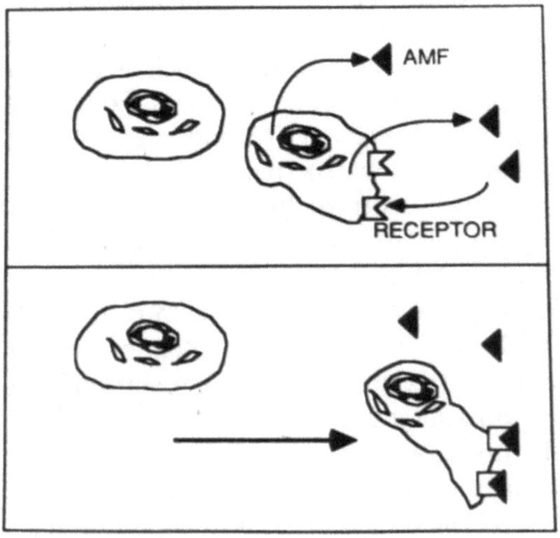

Fig. 1. AMF hypothesis: cells produce a protein factor (*AMF*) which binds to specific cell surface receptors resulting in the triggering of an intrinsic motile response. The factor or factors can also recruit neighboring cells via a paracrine mechanism

strated autocrine growth factors for transformed cells, we investigated the possibility that such cells could elaborate autocrine motility factors. The action of these substances might, in part, explain both the markedly invasive character and the metastatic property of malignant cells. Thus, under the influence of such an autocrine material, a tumor cell might move out into the surrounding host tissue and also exert a "recruiting" effect on adjacent tumor cells in the presence of a gradient of attractant (Fig. 1). Conceivably, such factors might also attract fibroblastic cells of the host, resulting in the phenomenon of desmoplasia, characteristic of invasive tumors.

F. Melanoma Autocrine Motility Factor

We have found that the human melanoma cell line A2058 produces in culture a material that markedly stimulates its own motility. These cells respond in a dose-dependent manner to various concentrations of conditioned medium obtained by incubating confluent cells in serum-free medium, an indication that the motility factor is derived from the cell. Motility was measured by the modified Boyden chamber procedure (Fig. 2). Using this assay and the

A 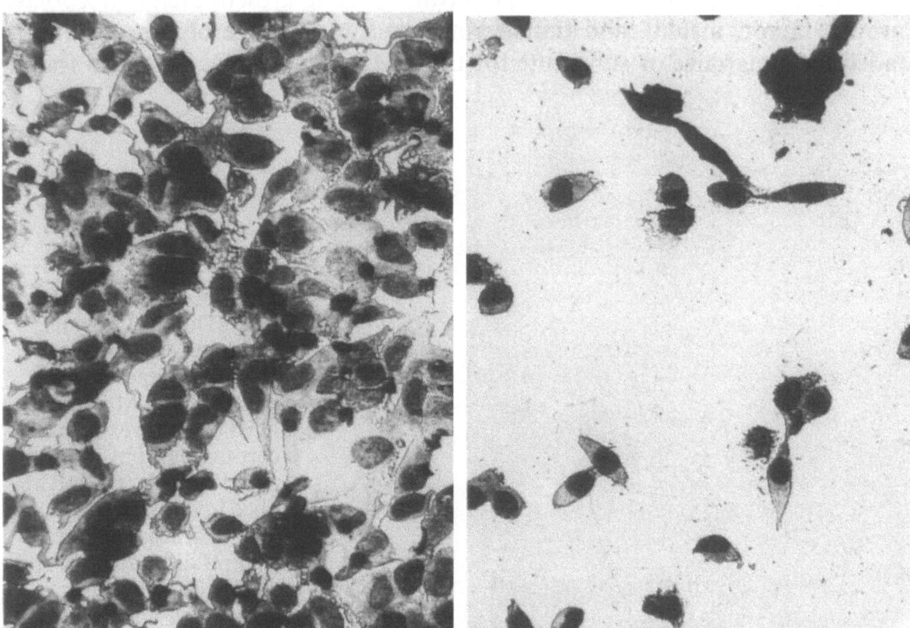 B

Fig. 2 A, B. AMF stimulated migration of human melanoma cells. A2058 human melanoma cells were placed into the top chamber of a modified Boyden chamber multiwell. The cells were allowed to adhere to the surface of a Nucleopore polycarbonate filter containing an 8-μm-diameter pore. The tumor cells migrating after 4 h to the opposite face of the filter are shown for a high-power field with **A** or without **B** the addition of approximately 1 n*M* AMF "checkerboard" analysis (ZIGMOND and HIRSCH 1973), we have also found that the conditioned medium factor has both chemotactic (directional) and chemokinetic (randomly motile) properties. That is, the cells respond to positive gradients of the motility factor as well as to high uniform concentrations of attractant. In fact, the chemokinetic effect was stronger than the chemotactic one. These observations accord with our hypothesis that an autocrine motility factor (AMF) contributes to invasion and metastasis of malignant cells. Since a variety of chemotactic factors stimulate adherence and spreading as well as motility, we determined that AMF unequivocally stimulated motility. Cells were preattached for 1 h to filters in serum-free medium. The preattached cells were then inverted and the cells were allowed to respond during a 3-h incubation period by migrating against gravity to AMF or control medium in the upper well of the Boyden chamber. A mean of 65 cells/field (× 500) migrated from the bottom to the top of the filter in response to AMF compared with 14 cells/field in the control medium. These results indicate that AMF stimulates motility distinctly from initial adherence of the cells to the substratum. A similar AMF was produced by human colon and breast carcinoma cells.

Formylated peptides, fibronectin, and whole laminin neither stimulated motility nor inhibited AMF-stimulated motility. Epidermal growth factor, platelet-derived growth factor, type β transforming growth factor, fibroblast growth factor, insulin, and transferrin over a wide range of concentrations failed to antagonize or substitute for the AMF activity. We have also found

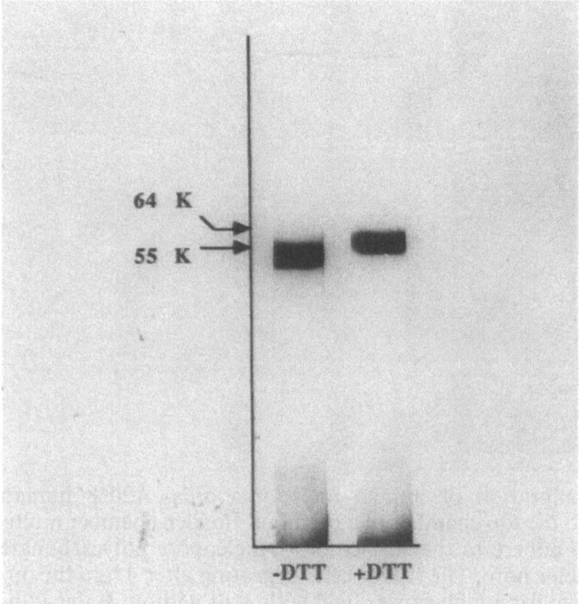

Fig. 3. Analysis of AMF by $NaDodSO_4$/polyacrylamide gel electrophoresis (5%–15% gradient separating gel). Material isolated by anion exchange chromatography was iodinated with $Na^{125}I$ and applied in the presence (*lane 2*) or absence (*lane 1*) of 0.5 *mM* dithiothreitol and was detected by autoradiography. The purified AMF factor binds to cells with an affinity of 0.5 n*M* and approximately 30000 receptor/cell. Aminoterminal sequence analysis of AMF reveals that it is a unique protein

(data not shown) that polyclonal antibodies to laminin, fibronectin, or plasminogen activator did not immunoprecipitate AMF. Using another highly motile cell type, we found that AMF did not stimulate motility in rabbit neutrophils.

I. Isolation and Characterization

Conditioned protein-free medium which elicited both large (10%–15% of the cells migrated) dose-dependent chemotactic and randomly motile responses (checkerboard analysis) was used to isolate AMF. The conditioned medium, after concentration (Amicon), was subjected to molecular sieve chromatography. AMF emerged as a broad major peak between 40 and 65 kDa (data not shown). The AMF was further isolated by fast-performance liquid chromatography.

The AMF activity was iodinated and found to comprise a single major component (electrophoresis) of ≈55 kDa without reduction of disulfide bonds. Upon reduction with 5 m*M* dithiothreitol, the migration of this component on the gel became slower, indicating the existence of interchain disulfide bonds (Fig. 3). A maximal chemotactic response to the purified AMF was elicited at a concentration of 10 n*M*. Amino acid analysis of AMF revealed a high content of glycine, serine, glutamic acid, and aspartic acid residues. Both tyrosine and cysteine are present, the latter concordant with the existence of interchain disulfide bonds as indicated by the altered electrophoretic mobility of AMF before and after treatment with dithiothreitol.

II. Some Chemical Properties of the Protein

AMF activity was markedly decreased after incubation with bacterial proteinase K but not with chymotrypsin, DNase, or RNAse. AMF was inactivated by heating at 100° C. Exposure to dithiothreitol virtually eliminated activity, while $PhMeSO_2F$ treatment had no effect. These results suggest that AMF is a protein and that its active conformation is stabilized by disulfide bonds. Furthermore, the activity did not seem to depend on a serine esterase activity. The activity was stable over a broad pH range (pH 4–11) but became inactivated below pH 4.

III. Signal Transduction in Tumor Cells

Because some cells require ongoing protein synthesis to develop a motile response, we determined whether inhibition of protein synthesis affected the response of the melanoma cells to its autocrine factor. We found that concentrations of cycloheximide that eliminated de novo protein synthesis had no effect on stimulated cell motility. Therefore, the cell protein components required for developing a motile response appear to be stable for the duration of migration (4 h).

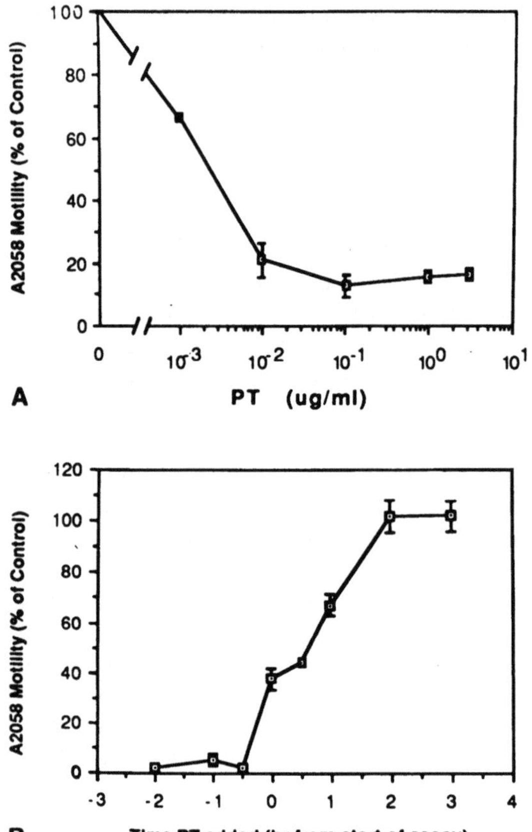

Fig. 4. A Effect of different doses of pertussis toxin on A2058 motility. Pertussis toxin was added to the cells at the indicated concentrations for 2 h prior to and during the assay. **B** Time course of the pertussis toxin effect on A2058 motility. Pertussis toxin (0.5 μg/ml) was added to the cells for various periods of preincubation or after the start of the assay. All data are expressed as means ±SEM

Studies with leukocytes (BOKOCH and GILMAN 1984; SMITH et al. 1986) have implicated a guanine nucleotide protein (G protein) in the receptor-mediated initiation of a motile response in these cells. The evidence is convincing that the locomotion of certain tumor cells also directly involves a G protein (STRACKE et al. 1987). Pertussis toxin, known to inhibit action of the Gi protein of the adenylate cyclase pathway (KATADA and UI 1982) profoundly and rapidly inhibited the AMF-stimulated migration in vitro of A2058 melanoma cells (STRACKE et al. 1987) and two breast cancer cell lines (GUIRGUIS et al. 1987). In the melanoma cell line 0.5 μg/ml pertussis toxin (Fig. 4) completely blocked motility without affecting growth in culture (M. L. STRACKE, unpublished results). However, the adenylate cyclase pathway does not appear to be directly involved in the motility response since agents which selectively modulate or have a role in this pathway, e.g., cholera toxin, forskolin, the cyclic AMP analog 8-bromoadenosine 3′:5′-cyclic monophosphate, and the cyclase inhibitor 2′,5′-dideoxyadenosine, all had minor effects on cell migration. We have also found that agents which stimulate (phorbol ester, oleyl acetyl glycerol) and inhibit (trifluoroperazine) protein kinase C had no significant effect upon motility (M. L. STRACKE, unpublished results). It is likely, then, that effector systems other than that of adenylate cyclase are mediated by a G protein in producing tumor cell motility. G proteins have been shown to act in a variety of second messenger pathways, including phospholipase A2 (OKAJIMA and UI 1984), phospholipase C (KIKUCHI et al. 1986), and activation of calcium channels (HESCHELER et al. 1987). Specifically, in the neutrophil, pertussis toxin inhibits both lipase enzymes as well as cell motility (MOLSKI et al. 1984; LAD et al. 1985). Evidence that suggests a role for phospholipase A2 in tumor cell locomotion has been obtained with the melanoma cell. Quinacrine, an agent that inhibits phospholipase A2, markedly reduced AMF-stimulated migration (LIOTTA et al. 1986). Additionally, deaza-adenosine, an inhibitor of biological methylation (GURANOWSKI et al. 1981), was found to reduce markedly both membrane phospholipid methylation and AMF-stimulated motility, whereas AMF itself caused a sustained increase in the methylation of phosphatidyl choline (Ptd Cho) in melanoma cells (Fig. 5, legend). Since Ptd Cho is the major substrate for phospholipase A2, these findings are consistent with a role for this enzyme in tumor cell motility. Studies with a murine tumor cell line, BU-L, suggest that metabolism of arachidonic acid, a product of the lipase reaction, may play a role in tumor cell motility (BOIKE et al. 1987). Lipoxygenase inhibitors such as quercetin, nordihydroguaretic acid, and nafazatrom significantly reduced stimulated motility, but indomethacin, a cyclo-oxygenase-blocking agent, had no effect. Calmidazolium also substantially inhibited motility. Collectively, these results are in accord with both the lipoxygenase pathway for arachidonate metabolism and a calmodulin-mediated mobilization of calcium participating in migration of certain tumor cells.

However, a role for the cyclo-oxygenase pathway cannot be ruled out. It has been reported that phorbol myristate acetate and laminin-stimulated motility in murine fibrosarcoma cells are inhibited by prostaglandins of the E series (HE et al. 1986). Preliminary studies (M. L. STRACKE, unpublished material) with human melanoma cells indicate that calcium channel blocking agents inhibit AMF-stimulated motility. On the other hand, calcium ionophores were found to stimulate motility. AMF stimulates the formation of inositol trisphosphate (IP_3) in melanoma cells. Also, the stimulated motility of the cells was markedly reduced by lithium, an inhibitor of the phosphoinositide pathway (KOHN et al. 1989). These results are consistent with the participation of phospholipase C (PLC, Fig. 6) in the generation of motility (SMITH et al. 1986) in melanoma cells.

Still other membrane-associated enzymes may be directly involved in locomotion. It has recently been shown that cathepsin B in highly metastatic variants of a murine melanoma cell line (SLOANE et al. 1986) is located chiefly in the cell membrane, whereas it is mainly in the cytosol in weakly metastatic cells. Protein inhibitors of the enzyme which were derived from normal liver and tumor tissue were found to inhibit AMF-induced motility in both the human A2058 melanoma cell and the murine Walker carcinoma cell (BOIKE et al. 1986). These studies suggest a role for cathepsin B in the locomotory

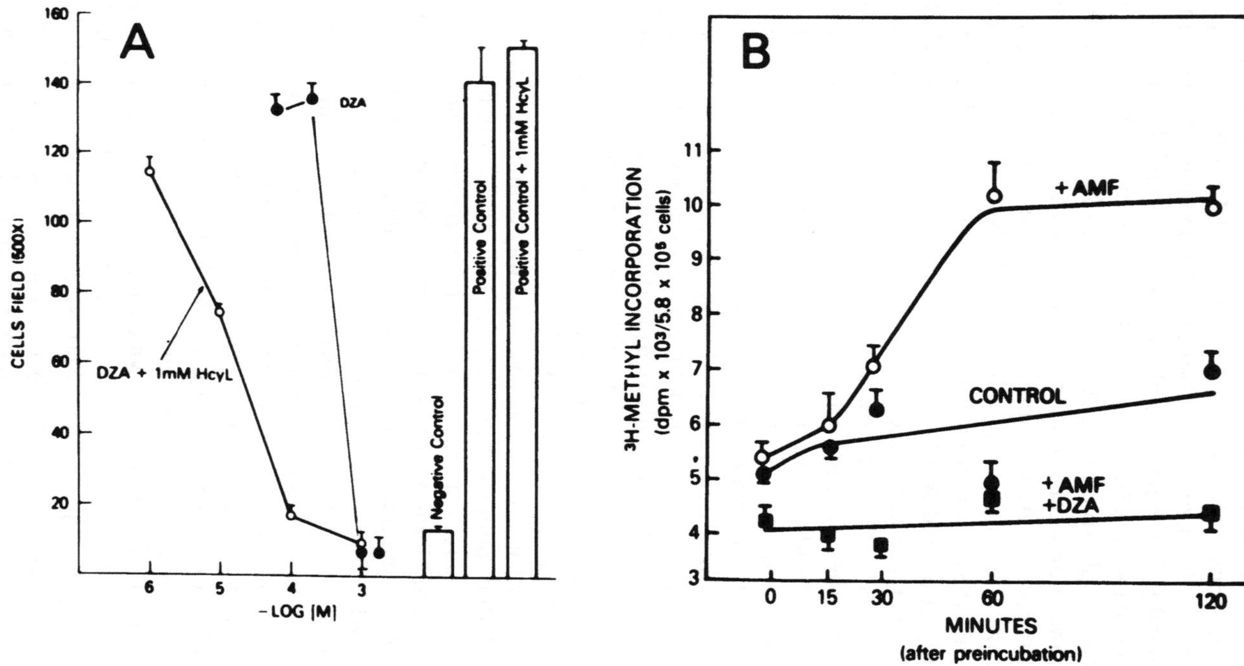

Fig. 5. A Inhibition of chemotaxis of melanoma cells to AMF by cAdo alone (●) and by cAdo and 1 m*M* homocysteine thiolactone (○). *Bars: a,* negative control; *b,* positive control; *c,* positive control plus 1 m*M* homocysteine thiolactone. **B** AMF-stimulated methylation of membrane phospholipids in A2058 tumor cells. cAdo was added at a concentration of 1 m*M*. ○, With AMF; ●, control; ■, with AMF and cAdo. The rate of formation of phosphatidyl choline, Ptd Cho, via methylation was the most rapid compared with the other methylated phospholipids (not shown)

process itself since the in vitro assay conditions for motility do not present the cell with a physiological matrix barrier through which it must pass as would be the case in vivo. Under the latter circumstances, it has been amply demonstrated that malignant cells have activated membrane-associated type IV collagenase and also the ability to secrete this matrix-degrading enzyme during their invasive course.

From these considerations, it is likely that the generation of a motile response in tumor cells initially involves a direct role for a G protein which interacts with an activated receptor and then transduces the signal to an effector system such as the phosphodiesterase IP_3 pathway and phospholipase A2 (Fig. 6). The subsequent production of arachidonate and its metabolism via

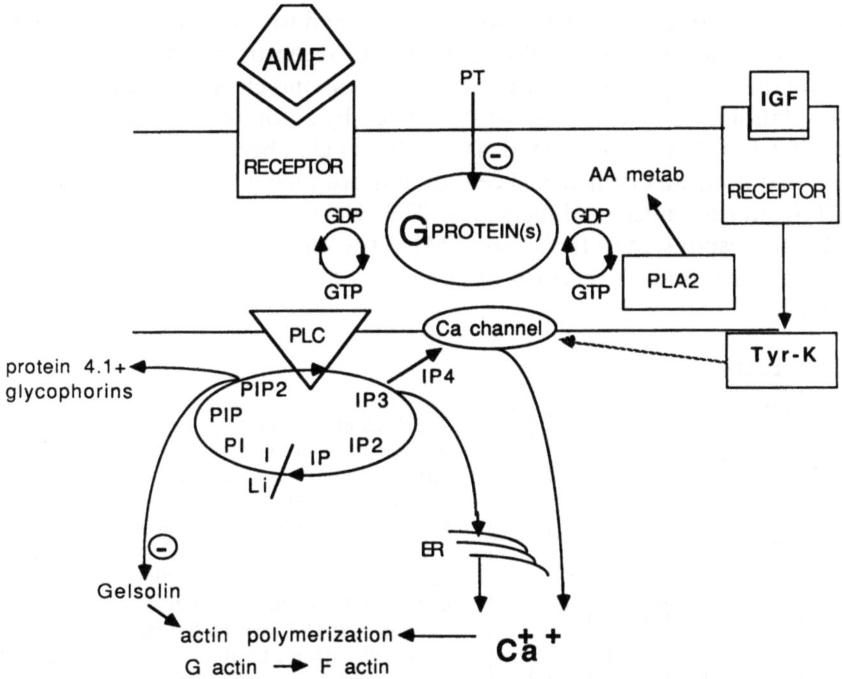

Fig. 6. Signal transduction pathways of tumor cell motility. Two pathways are depicted: the activation of phospholipases (PLC, PLA_2) by AMF acting through its receptor, and the activation of a tyrosine kinase by insulin-like growth factors acting through the IGF, receptor (vide infra). All evidence to date indicates that both the motile response to AMF and the latter's stimulation of the PLC pathways are mediated by a pertussis (PT)-sensitive G protein. The increased levels of inositol tris phosphate (IP_3) are considered to mobilize intracellular Ca^{2+}, which activates the cytoskeleton to generate a motile response. The adenyl cyclase pathway probably does not play a direct role lin signal transduction. Additionally, phospholipase hydrolysis may locally alter membrane fluidity with effects on receptor aggregation and action assembly. The putative tyrosine kinase pathway shown here to be stimulated by the IGFs does not appear to be sensitive to pertussis toxin since PT did not inhibit IGF-stimulated motility. Alternate pathways of response to different stimuli of motility would be advantageous to the malignant cell's dissemination in the host

lipoxygenase may contribute to the mobilization of calcium by IP_3 and DAG, which could also be required for changes in the cytoskeleton that are essential for locomotion. With respect to a role for cathepsin B, it is conceivable that AMF may stimulate its activity within the membrane to cause a specific cleavage of a proenzyme whose active form, e.g., protein kinase C (PONTREMOLI et al. 1986), is required for the motile response.

G. Unique Features of Tumor Cell Motility

The molecular mechanisms of signal transduction just discussed do not yet provide a basis for concluding that migrating tumor cells employ a uniquely different set of reactions from nontransformed cells. However, there are some features of tumor cell components that would appear to be unique. It is likely that there are at least two gene products required for motility that are characteristic of metastatic tumor cells: AMF and its receptor. Supporting this assertion are the findings that neutrophils do not respond chemotactically to AMF, and melanoma cell motility is not stimulated by a potent leukoattractant such as F Met-Leu-Phe (LIOTTA et al. 1986). It has also been shown (LIOTTA et al. 1986) that nontransformed 3T3 cells do not produce AMF but can respond to AMF from *ras*-transfected metastatic 3T3 cells (Fig. 7). The latter, moreover, are highly responsive to their own AMF. It would be of interest to determine whether AMF and its receptor are products of new or known activated oncogenes.

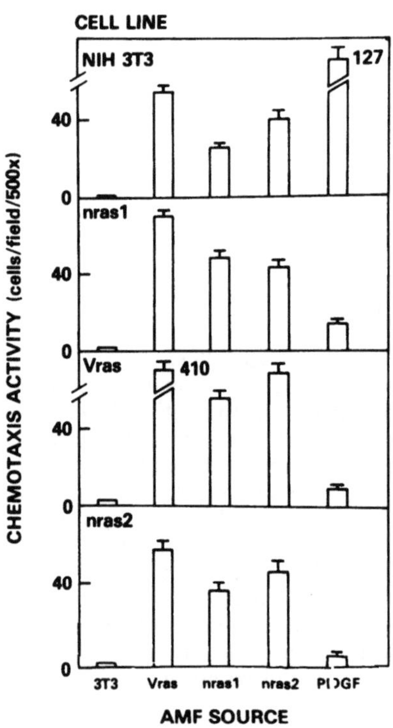

Fig. 7. Chemotactic response to NIH 3T3 cells or their *ras*-oncogene-transformed counterparts to their own conditioned medium. Purified platelet-derived growth factor (PDGF) (10–100 ng/ml) was used as a control. The label on the abscissa refers to the source of conditioned medium: *3T3*, NIH 3T3 parent cells; *VRAS*, tertiary transfectant v-*ras* oncogene; *NRAS1*, *NRAS2*, n-*ras* transfectants from two different human tumor sources. The chemotactic responses (mean ± SEM; $n = 9$) for four different test cell lines are shown. Unstimulated random migration was < 5% of directed migration. The NIH 3T3 cells did not produce AMF (not detected even in × 10 concentrated medium). In contrast, all the transformed lines produced a potent chemoattractant for both the NIH 3T3 cells and all the other transformed lines. The motility factor produced by NIH 3T3 cells was inhibited by antibodies made against purified human tumor cell AMF. The maximum PDGF response over a range of doses is shown

H. Growth Factors as Motility Stimulants

A metastatic tumor cell may, in addition to its responsiveness to AMF, have a variety of receptors for other motility-stimulating agents. These are generally less effective than the AMF system in stimulating migration, but their presence obviously gives these cells enhanced invasiveness.

Melanoma cells contain unoccupied laminin receptors (WEWER et al. 1987) and have been found to respond chemotactically and haptotactically to laminin (MCCARTHY et al. 1983; MCCARTHY and FURCHT 1984). Since laminin is a major constituent of basement membranes, this property of the cell should facilitate its passage through the membrane during metastasis. As laminin is synthesized by melanoma cells, this glycoprotein may be an example of an AMF. However, its potency in the human melanoma cell is considerably less than the 55 kDa protein secreted by these cells. Moreover, the cells do not secrete laminin appreciably.

I. Thrombospondin

Thrombospondin is another well-characterized protein that melanoma cells secrete (ROBERTS et al. 1987) and respond to with locomotion (TARABOLETTI et al. 1987). The motility response has been investigated recently in some detail (TARABOLETTI et al. 1987). Thrombospondin was shown to stimulate both chemotaxis and haptotaxis strongly (migration to a gradient of matrix-bound attractant) at concentrations as low as 0.1 μM. With the aid of antibodies which bound to specific sites on the molecule, it was determined that the amino-terminal domain was essential for chemotaxis while the carboxy-terminal region was necessary for haptotaxis. The internal core region of thrombospondin did not appear to be required for either type of motility. The synthetic peptide GRDGS blocked fibronectin-stimulated haptotaxis but had no effect upon that induced by thrombospondin or laminin. These studies indicate that thrombospondin stimulates motility in malignant tumor cells in a highly specific manner. The type of migratioin evoked depends upon a specific domain of the attractant. This property of thrombospondin may be quite relevant to the metastatic process. At sites of endothelial injury, the accumulation of activated platelets would release levels of thrombospondin that would be sufficient to stimulate motility in tumor cells in both haptotactic and chemotactic ways. Malignant cells in the circulation could then be arrested at such sites in haptotactic fashion and, subsequently, in response to soluble stimuli, invade the endothelial basement membrane in their course of establishing secondary metastatic foci.

II. Bombesin

Bombesin, a neuropeptide, is synthesized and secreted by small cells of lung carcinoma (SCCL) cells in culture (MOODY et al. 1981). This peptide is both a potent mitogen (CUTTITTA et al. 1985) and chemoattractant (RUFF et al. 1985) for these cells. A role for bombesin in the metastasis of the SCCL cells may involve the dispersal of an original locus of cells to enter the vasculature in response to secreted bombesin as well as exiting the vasculature opposite a secondary target organ such as the brain which may emit bombesin as a chemical signal. Another example of "recruitment" of tumor cells by an organ-derived attractant may be the finding that nerve growth factor stimulates migration in murine carcinoma cells (KAHAN and KRAMP 1987).

III. Insulin-Like Growth Factors

Recently, insulin-like growth factors (IGFs) have been shown to be potent stimulants of human melanoma cell motility (STRACKE et al. 1988). The order of potencies is: IGF-I > IGF-II ~ insulin with ED_{50}s of 5 nM and 150–200 nM respectively (Fig. 8). A high-affinity receptor for IGF-I was identified on the cells with a $K_D \sim 3 \times 10^{-10}$ M and an estimated 5000 receptors/cell (Fig. 9). Although this is a small number, studies with leukocytes have indicated that only a small number of receptors need be occupied to produce a motile response. Labeled insulin, however, did not bind appreciably to the melanoma cells. This suggested that insulin acted at the IGF-I receptor to stimulate motility, a result consistent with its demonstrated growth-promoting activity in cells lacking an insulin receptor (RODECK et al. 1987). IGF-I produced a maximal response that was equivalent to that evoked by AMF, an exceptionally potent motility stimulant for melanoma cells. However, the insulin peptides produced effects which differed from those induced by AMF in two respects: the peptide-stimulated migration was predominantly chemotactic, whereas the response to AMF has a greater chemokinetic component; the peptide-stimulated motility was not inhibited by pertussis toxin treatment of cells, while the AMF-induced migration was practically eliminated by toxin treatment. At the biochemical level there ap-

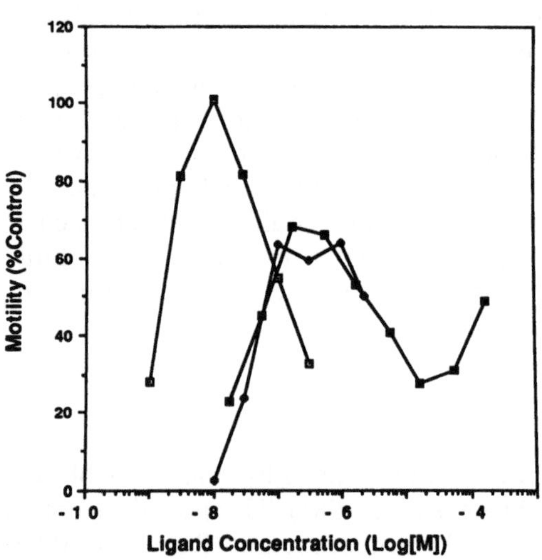

Fig. 8. Motile response of melanoma cells as a function of concentrations of insulin peptides: –■–, insulin; –□–, IGF-I; –◆–, IGF-II. Autocrine motility factor was a positive control with results expressed as percentage of control minus background

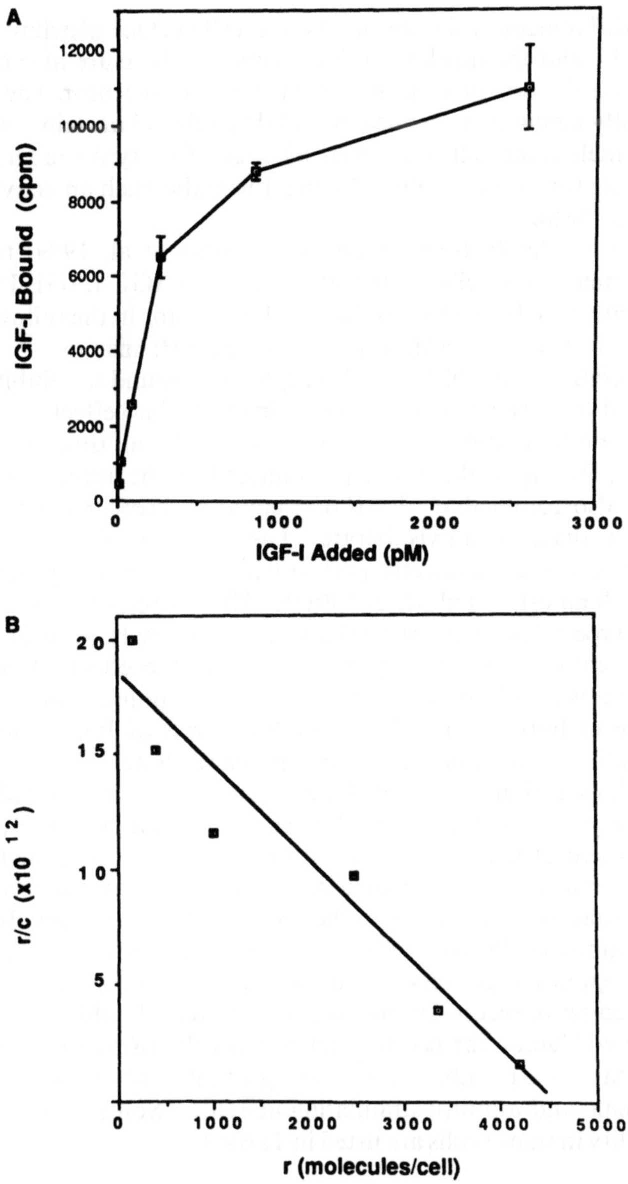

Fig. 9. A Binding of ^{125}I-labeled IGF-I to melanoma cells. Varying concentrations of labeled peptide were added to cells for 2 h at 15° C. Nonspecific binding was determined by adding 100-fold excess of unlabeled peptide. Specific binding (shown) is the total minus the nonspecific binding. **B** Scatchard analysis shows a $K_d \sim 3 \times 10^{-10}$ *M* with an estimated 5000 receptors/cell

pear to be differences in the specificity and cell surface distribution of receptors for AMF and the insulin peptide receptors; there are also differences in the signal transduction mechanism for each type of receptor. The AMF receptor apparently mediates a chemical signal through a G protein, while the IGF_1 receptor in melanoma cells may exert its effect via a tyrosine kinase since the activated receptor in other cells is known to exhibit such an activity (MÄKELÄ and ALITALA 1986).

Evidence has lately been obtained (STRACKE et al. 1989) that the IGF receptor on melanoma cells is a true type I protein. IGF-I, IGF-II, and insulin competed with ^{125}I-IGF-I for binding to the receptor in the following order of their IC_{50}s: IGF-I, ~ 2 n*M*; IGF-II, ~ 150 n*M*; insulin, ~ 300 n*M*. An antibody specific for the IGF type I receptor was found to inhibit the binding of ^{125}I-IGF-I to cells, while a control antibody had no effect. In addition, the receptor antibody inhibited the motility induced by all three insulin peptides but had no effect upon the motility stimulated by the autocrine motility factor. It was also found that ^{125}I-IGF-I could be crosslinked to a cell component with disuccinimidylsuberate. The crosslinked product could be detected on gel electrophoresis performed under reducing conditions and migrated with an estimated M_r of 130000. This corresponds to a known subunit of the type I IGF-I receptor (KASUGA et al. 1981). Additional evidence that this receptor does not appear to mediate the effect of the autocrine motility factor is the observation that the motile response of melanoma cells in the presence of both maximally stimulating levels of IGF-I and AMF was greater than the response induced by either ligand alone.

The biological significance of these studies may be that highly malignant tumor cells possess a variety of modalities to mount a motile response. These could be present at the receptor and transducer levels, as stated above. The ability of the cell to respond in such diverse ways would confer an advantage in their metastatic dissemination in the host: AMF could be involved in the initial dissociation of cells from the primary tumor and paracrine factors such as the insulin peptides might serve as site-specific attractants in signaling cells to invade and colonize secondary metastatic foci. IGF-I, additionally, is known to be a growth factor for certain melanoma cells (RODECK et al. 1987). It, therefore, may be a member of a growing number of cytokines which have both mitogenic and motility-stimulating activities. Some growth factors that induce motility in tumor cells are listed in Table 1.

Table 1. Growth factors stimulating motility in tumor cells

Factor	Responding cell	References
Bombesin	Small cells of lung carcinoma	RUFF et al. (1985)
Thrombospondin	A2058 human melanoma cells	TARABOLETTI et al. (1987)
Insulin-like growth factor	A2058 human melanoma cells	STRACKE et al. (1988)
Nerve growth factor	Mouse embryonal carcinoma cells	KAHAN and KRAMP (1987)

There is also some evidence to suggest that in certain cases metastasis may be aided by the release of specific organ-derived attractants for tumor cells into the circulation. It has been found that extracts of lung and liver, e.g., stimulate motility selectively in those tumor cells which colonize correspondingly the lung and liver in nude mice (HUJANEN and TERRANOVA 1985). These substances have not yet been characterized.

I. Autocrine Motility Responses in Nontransformed Cells

The cellular slime mold at a stage of its morphogenetic cycle both produces and responds chemotactically to cyclic AMP (BONNER et al. 1969). Another species of slime mold, *Polysphondylium,* both synthesizes and responds to a dipeptide of glutamic acid and ornithine, blocked at both N-terminal and C-terminal positions (SHIMOMURA et al. 1982). The role of these autocrine factors is to promote the aggregation of each of the respective slime mold cells to a multicellular stage of its developmental program.

Leukocytes such as polymorphonuclear neutrophils (PMNs) generate and release leukoattractants in response to external stimuli. PMNs exposed to formyl peptides produce chemotactic factors derived from arachidonate such as leukotriene B4 (GOETZL 1980). When these cells are presented with a suspension of urate crystals, a potent protein attractant is formed (SPILBERG et al. 1982). Such behavior may function in the amplification of a motile response which could bring the phagocyte more rapidly to a site of injurious agents in the host. However, the PMNs must first be activated to produce leukoattractants, whereas tumor cells, at least in culture, produce AMF apparently in the course of growth.

Some differentiated mammalian cells capable of replication can respond to autocrine substances. Fibroblasts, for example, synthesize fibronectin YAMADA and WESTON 1974) and can respond chemotactically and haptotactically to this matrix component (GAUSS-MÜLLER et al. 1980). It has also been reported that collagen and its degradation products stimulated motility in fibroblasts (POSTLETHWAITE et al. 1978). Such factors could play a role in wound healing.

STOKER et al. (1987) have reported that a "scatter" factor is produced by embryonic fibroblasts and fibroblastic cells from certain organs. However, the target cells are epithelial cells. The factor therefore has paracrine features. A2058 melanoma cells neither produce nor respond to this material. However, "scatter" factor and AMF from A2058 cells are both heat-labile proteins with similar molecular weights and may represent a new class of cytokines that induce cell migration in both normal and pathological processes such as development, metastasis, and wound healing.

Early events in migration may involve pseudopodia protrusion. During the course of invasion, the same tumor cell must interact with a variety of extracellular matrix proteins as it traverses each tissue barrier. For example, the tumor cell encounters laminin and type IV collagen when it penetrates the basement, and type I collagen and fibronectin when it crosses the interstitial stroma. It has recently been shown that cells express specific cell surface receptors which recognize extracellular matrix proteins. The first example of such a receptor is the laminin receptor, which binds to laminin with nanomolar affinity (WEWER et al. 1986). Laminin receptors have been shown to be augmented in actively invading tumor cells, and may play an important role in tumor cell interaction with the basement membrane. RGD recognition receptors are another class of cell surface proteins whicn bind extracellular matrix proteins which in turn contain the protein sequence Arg-Gly-Asp (DUBAND et al. 1986). Such proteins include fibronectin, collagen type I, and vitronectin. The process of cell migration undoubtedly requires a series of adhesion and detachment steps resulting in traction and propulsion. Studies using the AMF-stimulated motility as model system have revealed an important function of pseudopodia protrusion in this process. AMF stimulates motility on a variety of different substrata. Therefore, its action is independent of the mechanism of attachment. Furthermore, AMF induces the rapid protrusion of pseudopodia in a time- and dose-dependent manner (GUIRGUIS et al. 1987). Isolation of the induced pseudopodia reveals that they are highly enriched in their content of laminin and fibronectin matrix receptors. Since cell pseudopodia formation is known to be a prominent feature of actively motile cells, we can now set forth a working hypothesis to explain the early events in cell motility (Fig. 10). Cytokines such as AMF which stimulate intrinsic motility may induce exploratory pseudopodia prior to cell translocation. Such pseudopodia may express augmented levels of matrix receptors

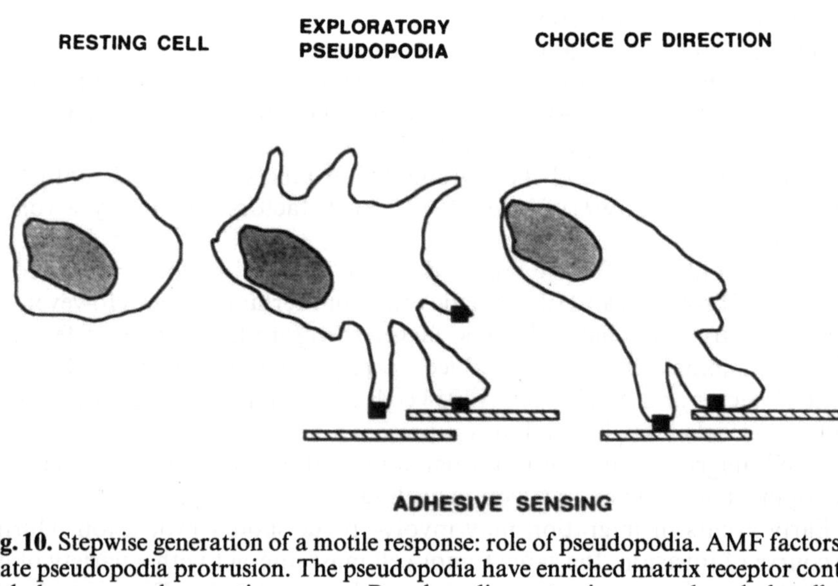

Fig. 10. Stepwise generation of a motile response: role of pseudopodia. AMF factors initiate pseudopodia protrusion. The pseudopodia have enriched matrix receptor content and also serve as locomotion organs. Pseudopodia protrusion precedes whole cell vectorial translocation

(and possibly proteinases). The protruding pseudopodia may serve multiple functions including acting as "sense organs" to interact with the extracellular matrix proteins and thereby locate directional cues, providing propulsive traction for locomotion, and even inducing local matrix proteolysis to assist the penetration of the matrix.

J. Autocrine Motility Factors as Markers of Malignancy

Recently, some clinically related studies have demonstrated a correlation between cell motility and malignancy. MOHLER et al. (1987) observed that highly metastatic prostatic adenocarcinoma cells showed a significantly greater rate of translocation than did prostatic cells of lower malignancy. CHODAK et al. (1985) reported the presence of a protein from the urine of a bladder cancer patient that stimulated phagokinesis in 3T3 cells. In an application of the studies on the AMF from melanoma cells (LIOTTA et al. 1986), GUIRGUIS et al. (1988) determined the relative stimulation of motility in tumor cell lines by levels of AMF that were present in urine preparations from 51 subjects, which included bladder cancer patients, those with benign bladder conditions, and normals. A high correlation was found between tumor grade and motility (Table 2). In the same study, it was found that in some follow-up procedures with patients who had been treated for bladder cancer the presence of a given level of AMF in urine accurately forecast a recurrence of the cancer prior to the actual pathological diagnosis. Since the analysis of urine for AMF involves noninvasive techniques, this approach could prove to be a powerful adjunct to diagnoses of malignancies using easily accessible body fluids.

Table 2. Mean motility values for urine samples from patients with different grades of transitional cell carcinoma of the bladder (GUIRGUIS et al. 1988)

Grade	Mean motility values[a, b]	No. of cases
Benign	47.3 ± 33.6	27
I	105.5 ± 70.0	4
II	186.4 ± 47.8	7
III	229.9 ± 63.2	13

[a] Motility values were determined by laser densitometry. A breast tumor cell line (MDA-435s) was used as the responder cell in a chemotaxis assay.
[b] P values for each grade were less than 0.06.

References

Anzano MA, Roberts AB, Smith JM, Sporn MB, De Larco JE (1983) Sarcoma growth factors from conditioned media of virally transformed cells composed of both type α and type β growth factors. Proc Natl Acad Sci USA 80:6264–6268

Atnip KD, Haney L, Nicolson GL, Dabbous MK (1987) Chemotactic response of rat mammary adenocarcinoma cell clones to tumor-derived cytokines. Biochem Biophys Res Commun 146:996–1002

Boike GM, Honn KV, Lah TT, Guirguis RA, Schiffmann E, Liotta LA, Sloane BF (1986) Cysteine proteinases and chemotaxis in Walker 256 carcinosarcoma. J Cell Biol 103:199

Boike GM, Sloane BF, Deppe G, Stracke M, Schiffmann E, Liotta LA, Honn KV (1987) The role of calcium and arachidonic acid metabolism in the chemotaxis of a new murine tumor line. Am Assoc Cancer Res 28:82

Bokoch GM, Gilman AG (1984) Inhibition of receptor-mediated release of arachidonic acid by pertussis toxin. Cell 39:301–308

Bonner JT, Barkley DS, Hall EM, Knign TM, Mason JW, O'Keefe G, 3rd, Wolfe PB (1969) Acrasin, acrasinase, and the sensitivity to acrasin in dictyostelium discoideum. Dev Biol 20:72-87

Chodak GW, Klagsbrun M, Shing YW (1985) Partial characterization of a cell motility factor from human urine. Cancer Res 45:690–694

Coman DR (1973) Mechanism responsible for the origin and distribution of blood-borne tumor metastases. Cancer Res 13:397–405

Cuttitta F, Carney DN, Mulshine J, Moody TW, Fedorko J, Fischler A, Minna JD (1985) Bombesin-like peptides can function as autocrine growth factors in human small-cell lung cancer. Nature 316:823–826

Duband JL, Rocher S, Chen WT, Yamada KM, Thiery JP (1986) Cell adhesion and migration in the early vertebrate embryo: location and possible role of the putative fibronectin receptor complex. J Cell Biol 102:160–178

Eaves G (1973) The invasiveness growth of malignant tumors as a purely mechanical process. J Pathol 109:233–237

Fidler IJ, Hart IR (1982) Biologic diversity in metastatic neoplasms: origins and implications. Science 217:998–1003

Folkman J (1985) Tumor angiogenesis. Adv Cancer Res 43:175–203

Frost P, Kerbel RS (1983) Immunology of metastasis. Can the immune response cope with dissemination tumor? Cancer Metastasis Rev 2:239–256

Gauss-Muller V, Kleinman HK, Martin GR, Schiffmann E (1980) Role of attachment factors and attractants in fibroblast chemotaxis. J Lab Clin Med 96:1071–1080

Goetzl EJ (1980) A role for endogeneous mono-hydroxy-eicosatetraenoic acids (HETEs) in the regulation of human neutrophil migration. Immunology 40:709–719

Guirguis R, Margulies I, Taraboletti G, Schiffman E, Liotta LA (1987) Cytokine-induced pseudopodial protrusion is coupled to tumor cell migration. Nature 329:261–263

Guirguis R, Schiffmann E, Liu B, Birkbeck D, Engel J, Liotta L (1988) Detection of autocrine motility factors in urine as markers of bladder cancer. JNCI 80:1203–1211

Guranowski A, Montgomery JA, Cantoni GL, Chiang PK (1981) Adenosine analogues as substrates and inhibitors of *S*-adenosylhomocysteine hydrolase. Biochemistry 20:110–115

Hanna MG, Key ME (1982) Immunotherapy of metastases enhances subsequent chemotherapy. Science 217:367–369

Hay ED (1982) Cell biology of extracellular matrix. Plenum, New York

He XM, Fligiel SE, Varani J (1986) Modulation of tumor cell motility by prostaglandins and inhibitors of prostaglandin synthesis. Exp Cell Biol 54:128–137

Hescheler J, Rosenthal W, Trautwein W, Schultz G (1987) The GPT-binding protein. Go, regulates neuronal calcium channels. Nature 325:445–447
Hujanen ES, Terranova VP (1985) Migration of tumor cells to organ-derived chemoattractants. Cancer Res 45:3517–3521
Kahan BW, Kramp DC (1987) Nerve growth factor stimulation of mouse embryonal cell migration. Cancer Res 47:6324–6328
Kasuga M, Van Obberghen E, Nissley SP, Rechler MM (1981) Demonstration of two subtypes of insulin-like growth factor receptors by affinity cross-linking. J Biol-Chem 256:5305–5308
Katada T, Ui M (1982) Direct modification of the membrane adenylate cyclase system by islet-activating protein due to ADP-ribosylation of a membrane protein. Proc Natl Acad Sci USA 79:3129–3133
Kikuchi A, Kozawa O, Kaibuchi K, Katada T, Ui M, Takai Y (1986) Direct evidence for involvement of a guanine nucleotide-binding protein in chemotactic peptide-stimulated formation of inositol bisphosphate and trisphosphate in differentiated human leukemia (HL-60) cells. Reconstitution with Gi or Go of the plasma membranes ADP-ribosylated by pertussis toxin. J Biol Chem 261:11558–11562
Kohn EC, Liotta LA, Schiffmann E (1990) Autocrine motility factor stimulates phosphatidyl inositol turnover in human melanoma cells. Biochim Biophys Res Comm 166:757–764
Lad PM, Olson CV, Grewal IS, Scott SJ (1985) A pertussis toxin-sensitive GTP-binding protein in the human neutrophil regulates multiple receptors, calcium mobilization, and lectin-induced capping. Proc Natl Acad Sci USA 82:8643–8647
Lam WC, Delikatny JE, Orr FW, Wass J, Varani J, Ward PA (1981) The chemotactic response of tumor cells: a model for cancer metastasis. Am J Pathol 104:69–76
Liotta LA, Abe S, Gehron P, Martin GR (1979) Preferential digestion of basement membrane collagen by an enzyme derived from a metastatic murine tumor. Proc Natl Acad Sci USA 76:226
Liotta LA, Thorgeirsson UP, Garbisa S (1982) Role of collagenases in tumor cell invasion. Cancer Metastasis Rev 1:277–297
Liotta LA, Mandler R, Murano G, Katz DA, Gordon RK, Chiang PK, Schiffmann E (1986) Tumor cell autocrine motility factor. Proc Natl Acad Sci USA 83:3302–3306
Mäkelä TP, Alitalo K (1986) Tyrosine kinases in control of cell growth and transformation. Med Biol 64:325–330
McCarthy JB, Furcht LT (1984) Laminin and fibronectin promote the haptotactic migration of B16 melanoma cells in vitro. J Cell Biol 98:1474–1480
McCarthy JB, Palm SL, Furcht LT (1983) Migration by haptotaxis of a Schwann cell tumor line to the basement membrane glycoprotein laminin. J Cell Biol 97:772–777
McCarthy JB, Basara ML, Palm SL, Sas DR, Furcht LT (1985) Stimulation of haptotaxis and migration of tumor cells by serum spreading factor. Cancer Metastasis Rev 4:125–152
Meyvisch C, Storme G, Bruyneel E, Mareel M (1983) Invasiveness and tumorigenicity of MO_4 mouse fibrosarcoma cells pretreated with microtubule inhibitors. Clin Exp Metab 1:17–28
Mohler JL, Partin AW, Coffey DS (1987) Prediction of metastatic potential by a new grading system of cell motility. J Urol 138:168–170
Molski TF, Naccache PH, Marsh ML, Kermode J, Becker EL, Sha'afi RI (1984) Pertussis toxin inhibits the rise in the intracellular concentration of free calcium that is induced by chemotactic factors in rabbit neutrophils: possible role of the "G proteins" in calcium mobilization. Biochem Biophys Res Commun 124:644–650
Moody TW, Pert CB, Gazdar AF, Carney DN, Minna JD (1981) High levels of intracellular bombesin characterize human small-cell lung carcinoma. Science 214:1246–1248
Nakajima M, Custead SE, Welch DR, Nicolson GL (1984) Type IV collagenolysis: relation to metastatic properties of rat 13762 mammary adenocarcinoma metastatic cell clones. Am Assoc Cancer Res Proc 244:162

Okajima F, Ui M (1984) ADP-ribosylations of the specific membrane protein by islet-activating protein, pertussis toxin, associated with inhibition of a chemotactic peptide-induced arachidonate release in neutrophils. A possible role of the toxin substrate in Ca^{2+}-mobilizing biosignaling. J BiolChem 259:13863–13871

Pontremoli S, Melloni E, Michetti M, Sacco O, Salmino F, Separatore B, Horecker BL (1986) Biochemical responses in activated human neutrophils mediated by protein kinase C and Ca^{2+}-requiring proteinase. J Biol Chem 261:8309–8313

Poste G (1983) Experimental systems for analysis of the surface properties of metastatic tumor cells. In: Nowotny A (ed) Biomembranes, vol 2. Plenum, New York, pp 341–364

Postlethwaite AE, Seyer JM, Kang AH (1978) Chemotactic attraction of human fibroblasts to type I, II, and III collagens and collagen-derived peptides. Proc Natl Acad Sci USA 75:871–875

Roberts DD, Sherwood JA, Ginsburg V (1987) Platelet thrombospondin mediates attachment and spreading of human melanoma cells. J Cell Biol 104:131–139

Rodeck U, Herlyn M, Menssen HD, Furlanetto RW, Koprowski H (1987) Metastatic but not primary melanoma cell lines grow in vitro independently of exogenous growth factors. Int J Cancer 40:687–690

Ruff M, Schiffmann E, Terranova V, Pert CB (1985) Neuropeptides are chemoattractants for human tumor cells and monocytes: a possible mechanism for metastasis. Clin Immunol Immunopathol 37:387–396

Shimomura O, Suthers HLB, Bonner JT (1982) Chemical identity of the acrasin of the cellular slime mold, *Polysphondylium violaceum*. Proc Natl Acad Sci USA 79:7376–7379

Sloane BF, Rozhin J, Johnson K, Taylor H, Crissman JD, Honn KV (1986) Cathepsin B association with plasma membrane in metastatic tumors. Proc Natl Acad Sci USA 83:2483–2487

Smith CD, Cox CC, Snyderman R (1986) Receptor-coupled activation of phosphoinositide-specific phospholipase C by an N protein. Science 232:97–100

Spilberg I, Mehta J, Simchowtiz L (1982) Induction of a chemotactic factor from human neutrophils by diverse crystals. J Lab Clin Med 100:399–404

Stoker M, Gherardi E, Perryman M, Gray J (1987) Scatter factor is a fibroblast-derived modulator of epithelial cell mobility. Nature 327:239–242

Stracke ML, Guirguis R, Liotta LA, Schiffmann E (1987) Pertussis toxin inhibits stimulated motility independently of the adenylate cyclase pathway in human melanoma cells. Biochem Biophys Res Commun 146:339–345

Stracke ML, Kohn EC, Aznavoorian SA, Wilson LL, Salomon D, Krutzsch HC, Liotta LA, Schiffmann E (1988) Insulin-like growth factors stimulate chemotaxis in human melanoma cells. Biochem Biophys Res Commun 153:1076–1083

Stracke ML, Engel JD, Rechler MM, Liotta LA, Schiffmann E (1989) Insulin-like growth factor-I and autocrine motility factor stimulate chemotaxis in human melanoma cells by different receptors and biochemical pathways. J Cell Biol 107:17a

Strauli P (1980) Proteinases and tumor invasion. In: Strauli P, Barrett AJ, Baici A (eds) Proteinases and tumor invasion. Raven, New York, pp 215–222

Sugarbaker EV, Weingrad DN, Roseman JM (1982) Observations on cancer metastases. In: Liotta LA, Hart IR (eds) Tumor invasion and metastasis. Martinus Nijhoff, The Hague, pp 427–465

Taraboletti G, Roberts DD, Liotta LA (1987) Thrombospondin-induced tumor cell migration: haptotaxis and chemotaxis are mediated by different molecular domains. J Cell Biol 105:2409–2415

Thorgeirsson UP, Turpeenniemi-Hujanen T, Neckers LM, Johnson DW, Liotta LA (1984) Protein synthesis but not DNA synthesis is required for tumor cell invasion in vitro. Invasion Metastasis 4:73–84

Todaro GJ, Fryling C, De Larco JE (1980) Transforming growth factors produced by certain human tumor cells. Proc Natl Acad Sci USA 77:5258–5262

Tyzzer EE (1913) Factors in the production and growth of tumor metastases. J Med Res 28:309–330
Vracko R (1974) Basal lamina scaffold-anatomy and significance for maintenance of orderly tissue structures. Am J Pathol 77:313–338
Wewer UM, Liotta LA, Jaye M, Ricca GA, Drohan WN, Claysmith AP, Rao CN, Wirth P, Coligan JE, Albrechtsen R, Mudryj M, Sobel ME (1986) Altered levels of laminin receptor mRNA in various human carcinoma cells that have different abilities to bind laminin. Proc Natl Acad Sci USA 83:7137–7141
Wewer U, Taraboletti G, Albrechtsen R, Sobel M, Liotta LA (1987) Role for laminin receptor in tumor migration. Cancer Res 47:5691–5698
Wicha MS, Liotta LA, Garbisa S, Kidwell WR (1980) Basement membrane collagen requirements for attachment and growth of mammary epithelium. Exp Cell Res 124:181–190
Woolley DE, Tetlow LC, Mooney CJ, Evanson JM (1980) Human collagenase and its extracellular inhibitors in relation to tumor invasiveness. In: Strauli P, Barrett AJ, Baici A (eds) Proteinases and tumor invasion. Raven, New York, pp 97–113
Yamada KM, Weston JA (1974) Isolation of a major cell surface glycoprotein from fibroblasts. Proc Natl Acad Sci USA 71:3492–3496
Yamada KM, Akiyama SK, Hasegawa T, Hasegawa E, Humphries MJ, Kennedy DW, Nagata K, Urushihara H, Olden K, Chen WT (1985) Recent advances in research on fibronectin and other cell attachment proteins. J Cell Biochem 28:79–97
Zigmond SH, Hirsch JG (1973) Leukocyte locomotion and chemotaxis. New methods for evaluation, and demonstration of a cell-derived chemotactic factor. J Exp Med

CHAPTER 39

Expression of Growth Factors and Their Receptors in Development

D. C. Lee and K. M. Han

A. Introduction

The discovery and characterization of growth factors and their receptors has led to considerable interest in the roles these molecules play in early development. The relatively recent cloning of cDNAs for many of these molecules has meant that efforts to understand the developmental functions of growth factors need no longer be limited to in vitro studies. Information gained from these analyses can now be weighed in the light of detailed knowledge of their developmental expression. Thus, our aim in writing this review has been to integrate data gleaned from studies of the effects of various growth factors on fetal tissues and cells in vitro with those aimed to analyzing their expression in vivo. The inevitable limitations of space have prompted us to concentrate on the latter.

Development is a complex process in which a single cell, the zygote, grows and differentiates into a multicellular organism. This process is dependent on both genetic and epigenetic influences on the developing embryo. In higher animals, development occurs in utero where the fetal and maternal circulatory systems are linked by a specialized organ, the placenta. Thus, development may not be regulated only by fetal gene products, but may also be influenced by humoral factors derived from maternal tissues and/or placenta. However, it is important to recognize that development does not end with parturition. Most, if not all, tissues and organs continue to develop in postnatal life. Thus our review will emphasize, but will not be restricted to, the role of growth factors during embryonic and fetal development.

A number of themes will hopefully emerge from what follows. An important one is that growth factors should not be thought of simply as "mitogens." Many growth factors have both growth-promoting and growth-inhibiting activities, depending on the biological context. In addition, growth factors can influence the differentiation of animal cells in culture, and there is increasing evidence from studies of lower animals that growth factors (or growth factor-related molecules) can act as inductive signals in early differentiation events. Examples include the findings that: (a) segregation of products to the animal and vegetal poles is essential for normal polarization of the embryo, and a transforming growth factor (TGF-β)-related molecule, Vg-1, is one of the products localized to the amphibian vegetal pole (Weeks and Melton 1987); (b) two mammalian growth factors, fibroblast growth factor

(FGF) and embryonal carcinoma-derived growth factor, will mimic amphibian mesoderm induction signals (SLACK et al. 1987); (c) the Notch locus, expression of which is essential for neurulation of the *Drosophila* embryo, encodes a molecule related to the epidermal growth factor (EGF) precursor (WHARTON et al. 1985); and (d) expression of lin-12, which also encodes an EGF-related molecule, is required for normal cell fate switching in the development of the nematode reproductive system (GREENWALD 1985). Thus, the classical definition of "growth factor" may be too narrow, and this confounds attempts to interpret function simply based on the location and ontogeny of expression.

A second major theme relates to the distinction that growth factors, unlike the classical endocrine hormones, are produced by a wide range of separate and anatomically distinct cell types. Furthermore, many of these same cells also produce the appropriate receptor. This has given rise to the notion that growth factors are not limited to endocrine actions, but can also act through paracrine and autocrine mechanisms (SPORN and TODARO 1980). Indeed, given the potency of growth factors and the ubiquitous distribution of many of the growth factor receptors, it seems likely that mechanisms must exist whereby the actions of these molecules can be highly localized – for example, during development when inductive signals are passed between relatively small numbers of cells. Hence, it may be significant that a number of growth factors including all members of the EGF family, CSF-1, and possibly a developmental form of the platelet-derived growth factor (PDGF) A-chain are synthesized as integral membrane precursors. Assuming that these membrane-anchored "precursors" are biologically active (see the discussion of TGF-α below), this might provide a form of cell-cell interaction in which paracrine actions are limited to contiguous cells. It should also be considered that the biological activities of these larger, transmembrane molecules may differ from those of the smaller secreted growth factors.

We are, undoubtedly, entering an exciting period in terms of efforts to understand the role of growth factors in development. The realization that analogous molecules are expressed during the development of lower eukaryotes allows for greater manipulation, including the use of genetics. In addition, a number of new and improving technologies hold great promise. For example, polymerase chain reaction (PCR) amplification now allows the detection of specific transcripts in single cells, and this has already been applied to the study of growth factor expression in preimplantation embryos (RAPPOLEE et al. 1988). Methodologies for localizing growth factor expression in embryos by in situ hybridization and immunocytochemistry are improving, and the ability to study in transgenic animals expression of heterologous genes which have been divorced from their normal regulatory elements may provide new insights. Finally, a new approach on the horizon, namely that of inactivating genes through homologous recombination, may ultimately provide a powerful quasigenetic means by which to address the function of growth factors in development.

What follows is an attempt to summarize the current state of knowledge concerning those growth factors for which a significant body of development work exists.

B. The EGF/TGF-α Family of Growth Factors

I. Epidermal Growth Factor

1. Introduction

Epidermal growth factor (EGF) is a potent mitogen for epithelial and mesenchymal cells. The mature, 53-amino-acid polypeptide is cleaved from the extracellular domain of a large integral membrane glycoprotein precursor (SCOTT et al. 1983a; GRAY et al. 1983). This approximately 1200-amino-acid prepro-EGF contains, in addition to the mature growth factor, seven additional EGF-like sequences in the extracellular domain. It is not yet known whether the integral membrane precursor is biologically active, or whether any of the additional EGF-like repeats are proteolytically released as bioactive molecules.

Prepro-EGF is encoded by a 4.7-kb mRNA. In adult mice, the highest levels of expression of this mRNA are found in the kidney and in the male submaxillary gland. In addition, lower levels of expression in the mammary gland, pancreas, duodenum, pituitary gland, lung, spleen, brain, and uterus have been reported (RALL et al. 1985).

2. Developmental Expression of the EGF Receptor

EGF binds to a 170-kDa transmembrane glycoprotein, the EGF receptor. Binding activates the receptor's intrinsic tyrosine kinase activity leading to its autophosphorylation, internalization, and downregulation. The rat EGF receptor is encoded by transcripts of 9.6, 6.5, 5.0, and 2.8 kb. The last of these is expressed in a variety of adult tissues, as well as in the fetus, and may encode a smaller secreted form of the receptor (L. A. PETCH and H. S. EARP, unpublished observations). As detected by Northern analysis, the levels of EGF receptor mRNA in the whole rat fetus increase significantly between days 11 and 14 of gestation (D. C. LEE, unpublished observations), while receptor-mediated tyrosine kinase activity has been observed in mouse fetuses as early as day 10 (HORTSCH et al. 1983). Moreover, EGF-binding activity has been demonstrated in a variety of fetal murine tissues (NEXO et al. 1980; ADAMSON et al. 1981) and in blastocyst outgrowths (ADAMSON and MEEK 1984). By day 13, EGF receptor tyrosine kinase activity is present in skin, developing skeletal muscle, and various internal organs, but notably missing from brain and liver (HORTSCH et al. 1983). The liver does not acquire receptor tyrosine kinase activity until around day 18. That the fetal receptor is functional is further suggested by the fact that binding of EGF stimulates DNA synthesis in some fetal tissues in vitro (ADAMSON et al. 1981). Finally, in both rodents and humans, the uterus, decidua, and placenta also contain relatively high levels of EGF receptor and/or its mRNA (O'KEEFE et al. 1974; D. C. LEE, unpublished observations).

A *Drosophila* homolog related to the mammalian EGF receptor and the neu protooncogene has recently been identified (WADSWORTH et al. 1985; LIVNEH et al. 1985). Two, alternatively spliced, transcripts from this gene are present at low abundance in the maternal RNA stored in unfertilized eggs and in 2-h-old embryos (LEV et al. 1985). The abundance of both transcripts rises sharply between 2 and 5 h after egg deposition, remains high throughout embryogenesis, and then declines to lower levels in the larval and pupal stages. An in situ hybridization analysis has shown that the transcripts are uniformly distributed in embryos, but localize to proliferating tissues of the imaginal disks and brain cortex in larvae (SCHEJTER et al. 1986). Adult transcripts are found mainly in the brain and ganglia. The ligand for this *Drosophila* receptor homolog has not yet been identified.

3. Biological Actions of Exogenous EGF

A plethora of studies have described the effects of EGF on developing tissues and cell types. These date back to the early studies of COHEN, which demonstrated the ability of EGF to stimulate epidermal growth (COHEN 1965) and induce premature eyelid opening and incisor eruption in newborn mice (COHEN 1962). Later studies have detailed the in vivo and in vitro effects of EGF on the proliferation and differentiation of various tissues and organs during fetal and neonatal development. For example, injection of EGF into fetal lambs (SUNDELL et al. 1980) and rabbits (CATTERTON et al. 1979) accelerates the morphological maturation of the lungs, and the addition of EGF to chick embryo tracheal and bronchial tree lung cultures enhances the rate of DNA synthesis (GOLDIN and OPPERMAN 1980). In addition, EGF may also stimulate the biosynthesis of surfactant precursors (GROSS et al. 1986), although this activity has been disputed (LEHEUP et al. 1983; KELLER and LADDA 1981).

Investigations by PRATT and colleagues have suggested that EGF, acting either alone or in combination with other agents, can influence craniofacial development, particularly that of the palate. The latter has been shown to contain EGF-binding activity by day 13 of mouse development (NEXO et al. 1980). Mesenchymal cells from the human embryonic palate are highly responsive to EGF (YONEDA and PRATT 1981), and EGF has been shown to stimulate palatal mesenchymal synthesis of collagen, hyaluronic acid, and possibly fibronectin (SILVER et al. 1984). Furthermore, in cell and organ cultures of mouse and rat embryonic palatal shelves, EGF promotes continued DNA synthesis, and inhibits the normal breakdown of the medial epithelium (GROVE and PRATT 1984), a process described as programmed cell death. This breakdown concludes the normal sequence of events whereby the primary palatal shelves fuse and the nasal and oral cavities are separated. Inhibition of this process can yield palatal deformities including cleft palate. Thus, in studies of human palatal shelves, EGF has been shown to potentiate the action of a known cleft palate teratogen, retinoic acid (ABBOTT and PRATT 1987). These studies suggest that if EGF functions in palatal development it would be to promote the continued proliferation of the medial epithelium until the time of fusion, at which point either the availability or efficacy of the growth factor must be altered. Finally, changes in the pattern of ^{125}I-EGF binding have been interpreted to suggest that EGF may play a role in early tooth formation (PARTANEN and THESLEFF 1987).

Functional EGF receptors have been localized to the human fetal small intestine and colonic epithelium as early as 12 weeks of gestation (POTHIER and MENARD 1988). In addition, GALLO-PAYET et al. (1987) demonstrated the presence of EGF receptors along the entire length of the mouse small intestine. The binding activity was minimal at birth but increased throughout postnatal life and reached adult values by the 26th day. Similarly, EGF-binding activity has been demonstrated in both the proximal and distal colon of suckling mice (MENARD et al. 1987). The presence of EGF receptors in the gastrointestinal epithelium, together with the demonstration that milk contains considerable quantities of EGF (CARPENTER 1980) that is readily absorbed by neonates (THORNBURG et al. 1984), has prompted investigations into the biological effects of EGF on the developing gut. The results have been inconsistent and an obvious pattern has not emerged. While the addition of EGF to organ cultures of 8-day-old mouse small intestine had no effect on either brush border enzymatic activities or DNA synthesis, the same investigators found that when EGF was injected into mice from 8 to 11 days of age it induced the precocious appearance of sucrase and enhanced the activities of a number of brush border enzymes (MALO and MENARD 1982). The addition of EGF to explants of fetal mouse duodenum increased the levels of alkaline phosphatase, maltase, and trehalase activities (BEAULIEU et al. 1985), but when EGF was added to organ cultures of human fetal jejunum, it produced a significant increase in lactase activity, inhibited the usual increase in sucrase, trehalase, and glucoamylase activities, and reduced the labeling index of the epithelial cells (MENARD et al. 1988). Finally, several studies have examined the effects of EGF on the development of the stomach. Injection of EGF into suckling rats stimulated mucosal growth (as measured by DNA, RNA, and protein content) (JOHNSON and GUTHRIE 1980; DEMBINSKI and JOHNSON 1985) and ornithine decarboxylase expression (FELDMAN et al. 1978), but did not lead to precocious maturation or functional development as measured by antral or serum gastrin levels.

The results of these studies on the developing gut suggest that the response to EGF can vary with the species, stage of development, and the method of study, i.e., in vitro versus in vivo. Other in vivo studies suggest that the biological response can also depend on the route of administration. For example, orogastric administration of EGF to neonatal rat pups induced precocious maturation of intestinal brush border disaccharidase activities but had no effect on pancreatic amylase (O'LOUGHLIN et al. 1985). However, when EGF was administered intraperitoneally the opposite response was observed. To summarize, despite a large body of work describing the effects of EGF on the developing gut, the evidence for a biological role is not convincing. By and large, the effects noted are rather modest and inconsistent, and there is no evidence for EGF's involvement in the dramatic changes that occur around the 3rd week of postnatal life when cell proliferation increases and enzyme patterns change from those of the suckling animal to those of the adult. Moreover, the role of milk-borne EGF is uncertain since the development and maturation of the gut in mice fed an EGF-free artificial diet was not obviously altered by the addition of EGF.

In the nervous system, EGF has been shown to bind to, and stimulate DNA synthesis in, mouse cerebellar astrocytes (LEUTZ and SCHACHNER 1981). Similarly, EGF increased the enzyme activities characteristic of glial cells (HONEGGER and GUENTERT-LAUBER 1983). However, a role for EGF in the development of the brain is questionable since EGF receptor is not detectable in that tissue until at least several days after birth (HORTSCH et al. 1983; GOMEZ-PINILLA et al. 1988). In addition, PROBSTMEIER and SCHACHNER (1986), using a sensitive double-site enzyme immunoassay, have failed to detect any EGF immunoreactivity in the developing rodent brain, and EGF presumably does not cross the blood-brain barrier.

Skin development in fetal lambs in vivo appears to be accelerated by the infusion of EGF, which causes hypertrophy of the sebaceous and sweat glands, and thickening of the epidermis (DOLLING et al. 1983). Studies in vitro have shown that EGF can stimulate proliferation and keratinization in chick embryonic epidermis (COHEN 1965; BERTSCH and MARKS 1974). And, as already noted, EGF receptors have been localized to the skin of the rodent fetuses by tyrosine kinase activity (HORSTCH et al. 1983) and by iodinated EGF binding (GREEN et al. 1984). The latter group has also shown that the binding activity increases significantly between 16 and 17 days of gestation, coincident with the period of maximal epidermal growth (STERN et al. 1971), and that it becomes restricted to the basal epidermal cells. Interestingly, between 17 and 20 days of gestation, EGF receptor activity appears to be depleted from basal cells overlying the dermal condensates that mark the first stage of hair follicle formation. The possibility that this receptor population is already occupied or downregulated was interpreted by the authors to suggest a role for either EGF, or an EGF-like molecule, in the development of hair follicles.

4. Developmental Expression of EGF

The previously described findings clearly establish that EGF receptor is widely expressed in fetal organs, and that exogenous EGF can modulate the growth and biochemical characteristics of developing tissues. They do not, however, establish a normal physiological role for EGF during development. And, despite the attractiveness of presuming a developmental role for EGF, there is still surprisingly little evidence of its expression in the fetus. Although the data were not presented, KASSELBERG et al. (1985) have reported the immunocytochemical localization of EGF in several human fetal tissues during the early second trimester. These include the placenta, salivary glands, duodenum (Brunner's glands), stomach, and pituitary glands. An early study by NEXO et al. (1980) used both a radioimmunoassay and a radioreceptor assay to quantitate EGF-like activity in fetal mice from 11.5 to 17.5 days of gestation. Whereas the radioreceptor assay detected an EGF-like substance at 11.5 days, with the levels of this activity increasing significantly between 15.5 and 17.5 days, immunoreactive EGF was barely detectable at 14.5 days. Furthermore, the immunoassay consistently yielded values five- to tenfold lower than the radioreceptor assay and showed a much less significant increase between 15.5 and 17.5 days. This contrasted with maternal EGF, which yielded similar values from the two assays. These data were interpreted to suggest that the fetal form of EGF differs from that of the adult, and may correspond to the then newly discovered TGF-α (see below). More recently, POPLIKER et al. (1987) used a cloned cDNA probe to analyze mouse fetuses and neonates for the presence of EGF mRNA. They were unable to detect EGF mRNA by either dot blot or Northern analysis in whole or partially dissected fetuses, fetal membranes, or placentae from day 9 of gestation through the early neonatal period. Furthermore, EGF mRNA was not detected in kidney samples until approximately 2 weeks after birth, and was still absent from submaxillary gland samples at weaning. Finally, RAPPOLEE et al. (1988) failed to detect EGF transcripts in preimplantation mouse embryos using PCR amplification of cDNAs – a protocol that did detect mRNAs for other growth factors. Thus, the evidence for fetal expression of EGF is either lacking or negative. Unfortunately, localized in situ analyses of EGF mRNA expression in fetuses have not yet been reported for any species.

5. Transplacental Transport of Maternal EGF

The inability to detect fetal expression of EGF has prompted investigations into whether maternal sources could contribute EGF to the embryo. CALVERT et al. (1982) injected pregnant mice at 15, 16, and 17 days of gestation with EGF, and examined fetuses at 18 days of gestation. They found that EGF treatment had no effect on the weight of the fetus or the length of the small intestine, and there was no change in the protein or DNA content. However, brush border alkaline phosphatase and trehalase activities were augmented. LEUNG et al. (1986) injected [^{125}I]EGF into term-pregnant rats and found that, while radioactive material localized to placenta, none of the fetal organs were labeled. In contrast, when POPLIKER et al. (1987) injected 2×10^6 cpm into the uterine arteries, 6000 cpm localized to the fetus. This corresponds to 0.3% of the total injected cpm, or 18% of the counts extracted from the entire uterine complex. When the EGF was, instead, injected into the placenta, 27% of the uterine complex radioactivity was extracted from the fetus. For comparison, placental injections of [^{125}]insulin yielded a tenfold lower recovery of cpm from the fetus. Unfortunately, the significance of these results is not entirely clear since the nature of the radioactive material, and hence its identity with intact EGF, was not established. In addition, given the difficulty of the placental injections, the insulin versus EGF comparison might have been more convincing had the number of animals studied been made clear. Given the uncertain nature of the above findings, the possibility of transplacental transport of EGF, and EGF-like peptides, from maternal sources remains open.

II. Transforming Growth Factor-α

1. Introduction

The mature form of transforming growth factor-α (TGF-α) is a single polypeptide of 50 amino acids that is structurally and functionally related to EGF (see Chap. 4). TGF-α binds the EGF receptor with an affinity comparable to that of EGF, and binding similarly leads to activation of the receptor's tyrosine kinase activity. Although TGF-α mRNA and protein have been detected in a number of normal tissues, expression of this mitogen is distinguished from that of EGF by the fact that it is frequently enhanced or induced by neoplastic transformation. Thus, TGF-α synthesis is most prevalent or abundant in tumor cells and cells transformed by retroviruses, oncogenes, and chemicals.

Like EGF, the mature, 5-kDa, TGF-α is also cleaved from a conserved integral membrane glycoprotein. The human (Derynck et al. 1984) and rat (Lee et al. 1985a) precursors are 160 and 159 amino acids in length, respectively. The proteolytic processing of this precursor by an elastase-like enzyme appears to be inefficient in many cell types. This leads to the secretion of larger biologically active forms of TGF-α and, in some cases, accumulation of the precursor in the plasma membrane. That the integral membrane precursor may be biologically active is suggested by studies of mutant precursor that cannot be cleaved to release mature growth factor. When cells expressing the recombinant mutant precursor are cocultured with A431 cells, EGF receptor on the surface of the latter cells is autophosphorylated and there is a rapid rise on the surface of the latter cells is autophosphorylated and there is a rapid rise in the intracellular A431 calcium levels (S. Wong, L. Winchell, and D. C. Lee, unpublished observations). These data suggest that the actions of growth factors can, by anchoring precursors to the membrane, remain highly localized – a suggestion that may have relevance to other growth factors which are similarly processed from transmembrane molecules. These include EGF, CSF-1, and, possibly, a developmental form of the PDGF A-chain (see below). In addition, it may also be relevant to a consideration of molecules like Notch and lin-12 which, though not yet shown to be growth factors, are integral membrane proteins with EGF-like repeats in their extracellular domains (see below).

2. Developmental Expression of TGF-α

Injected synthetic TGF-α has been shown to have essentially the same physiological effects on newborn mice as EGF (Smith et al. 1985; Tam 1985). That TGF-α might have a role as a developmental growth factor was first suggested by NEXO et al. (1980), who found that mouse embryos between days 11.5 and 17.5 of gestation contained more EGF receptor-binding activity than could be accounted for by an EGF-specific radioimmunoassay. Whereas immunoreactive species were barely detectable before day 14.5, appreciable amounts of EGF-receptor-competing activity could be detected as early as 11.5 days. Both assays detected increasing levels of activity between days 15.5 and 17.5 of gestation. Since the authors found maternal EGF to be equireactive in the radioreceptor and immunoassays, they hypothesized the existence of a fetal growth factor related to, but distinct from, EGF. They further suggested that this EGF-like activity might correspond to the then newly discovered TGF-α. Subsequently, TWARDZIK et al. (1982a), PROPER et al. (1982), and MATRISIAN et al. (1982) reported the extraction from 12- through 19-day-old mouse embryos of activities that promoted growth of normal fibroblasts in soft agar, stimulated DNA synthesis, and competed with EGF for binding to receptor. These activities were ascribed to "transforming growth factors" because of the contemporary (but incorrect) assumption that stimulation of anchorage-independent growth was a property unique to TGF-α, and not shared by EGF.

Additional support for developmental expression of TGF-α came from TWARDZIK (1985), who reported that the EGF-competing activity isolated from mouse embryos was immunologically distinct from EGF, but did compete in radioimmunoassay (RIA) with a synthetic peptide whose sequence corresponds to the carboxy-17 amino acid residues of rat TGF-α. The highest levels of activity were found in day-7 embryos, with a considerably smaller peak at day 13. Simultaneously, LEE et al. (1985b) described the detection of the 4.5-kb TGF-α mRNA in rat fetuses from days 8–10 of gestation using a cloned rat TGF-α cDNA as a hybridization probe. This transcript was not detected in fetuses from days 13–18 of gestation. In a subsequent report, however, LEE and colleagues used Northern and in situ hybridization analyses to localize TGF-α transcripts to the maternal decidua (HAN et al. 1987a). Since they did not detect expression in older (day 10) fetuses by Northern analysis, or in younger embryos by in situ analysis of selected sections, they suggested that during early gestation the predominant expression of this growth factor occurs in the decidua. They further suggested that their previous detection of transcripts in the young fetus was due to decidual contamination. However, RAPPOLEE et al. (1988) have recently detected TGF-α transcripts in preimplantation mouse blastocysts using PCR amplification of reverse transcriptase products. They also detected TGF-α antigen which they found to be concentrated in punctate structures in the perinuclear area of all cells of the blastocyst. In addition, WILCOX and DERYNCK (1988) detected TGF-α transcripts by in situ hybridization analysis in the mouse fetus at days 9 and 10 of gestation. TGF-α transcripts were detected in the otic vesicle, oral cavity, pharyngeal pouch, first and second branchial arches, and the developing mesonephric tubules of the kidneys only at days 9 and 10 of gestation, and were absent from older fetuses. Judging from the lengths of time the fetal sections were exposed to emulsion (4–8 weeks for ^{35}S probes), it appears that, even at days 9 and 10, TGF-α mRNA was expressed at low levels in these tissues. This might explain the failure of HAN et al. (1987a) to detect these transcripts by Northern analysis in rat fetuses of a comparable age. Alternatively, there may be differences in the developmental expression of the mouse and rat TGF-α genes. Finally, FREEMARK and COMER (1987) have reported the extraction of an activity from ovine kidney at days 123–137 of gestation that competes for binding to EGF receptor, stimulates the soft agar growth of NRK cells, but is not recognized by an EGF RIA. Although it is difficult to relate the gestational ages of mice and sheep, and the identity of this activity with TGF-α has not been verified, these results may be consistent with the detection of TGF-α mRNA in the developing mouse kidney as described above.

As previously noted, HAN et al. (1987a) detected TGF-α transcripts in the proliferating rat decidua. Expression was not observed in the nonpregnant uterus, or in the uterus of pregnant animals prior to the onset of decidualization at about day 5.5. Transcript levels in the decidua were highest at day 8 of gestation, and then declined as the decidua was resorbed. An in situ hybridization analysis using TGF-α-specific cRNA probes revealed a gradient of expression across the decidua with the highest levels of TGF-α mRNA found in those regions closest to the embryo. The significance of this gradient is presently unknown. However, it does not appear to imply that induction of TGF-α mRNA is established by implantation of the embryo; expression can be observed when decidualization is induced by artificial stimulation of estrogen-primed endometrium (V. K. M. HAN, D. C. LEE, and T. KENNEDY, unpublished results).

The function of TGF-α expressed by the rodent decidua is presently unknown. Formation of the decidua (decidualization) is a poorly understood phenomenon involving cell proliferation and migration that is coordinated with cellular growth in the uterine epithelium. It apparently results from differentiation and proliferation of stem cells in the endometrial stroma that may have originally derived from the bone marrow (KEARNS and LALA 1982). Since both decidua and endometrium express EGF receptor mRNA (D. C. LEE, unpublished observations) and contain EGF-binding sites (RAO et al. 1984; CHEGINI and RAO 1985), it seems most likely that TGF-α acts to initiate or potentiate the massive proliferation of these cells through paracrine or autocrine mechanisms. In addition, since EGF can induce prolactin synthesis in the GH4 rat pituitary cell line (MURDOCH et al. 1982), and decidual cells secrete a prolactin-like protein (GOLANDER et al. 1979), TGF-α may act to regulate hormone production in these tissues. However, it is also possible (as noted above) that some of the TGF-α produced in the decidua and/or placenta is transported to the fetus. Finally, it is worth noting that the urine of pregnant women has been reported to contain elevated levels of TGF-α-like activity (TWARDZIK et al. 1982b).

To summarize, the aforementioned data indicate that TGF-α transcripts are expressed during early development, and suggest that the results of studies describing the actions of EGF on developing tissues may actually be more relevant to TGF-α. However, no study to date has yet described the expression of TGF-α transcripts in fetal rats or mice beyond day 10 of gestation, despite the fact that TGF-α-like activity has been extracted from fetuses between days 12 and 17 of gestation (NEXO et al. 1980; PROPER et al. 1982; TWARDZIK et al. 1982a; TWARDZIK 1985). This discrepancy may be explained, in part, by differences in the stability of TGF-α mRNA and protein. Alternatively, it could be the result of expression of a distinct EGF-like growth factor. Finally, it is also possible that TGF-α is transported to the embryo from the decidua which appears to express TGF-α mRNA at considerably higher levels than the fetus. However, transplacental transport of TGF-α has not been specifically examined and, as described above, studies with EGF have been inconclusive. Clearly, fetal expression of this growth factor requires further investigation and verification.

III. Link Between EGF-Related Growth Factors and Homeotic Loci

The structure and integral membrane properties of the precursors to the EGF family of growth factors [EGF, TGF-α, and vaccinia virus growth factor (VVFG)] suggest that they belong to a growing family of secreted and transmembrane proteins that share the presence of EGF-like sequences. This diverse group of proteins includes a variety of extracellular proteases (blood clotting factors IX and X, tissue plasminogen activator), the LDL receptor, and the products of two homeotic loci, Notch and lin-12, of *Drosophila* and nematodes, respectively. The latter gene products are of obvious developmental interest. The Notch locus was originally named for the subtle dominant phenotype of most alleles – incisions in wing margins of heterozygous flies. However, the recessive null phenotype is considerably more dramatic, and homozygous embryos display abnormal cell-fate switching. Shortly after gastrulation in wild-type embryos, cells move up from the ventral ectoderm to become neuroblasts. Three waves of recruitment ultimately result in a fourth of the ectodermal cells becoming neuroblasts. However, with Notch mutants, most or all of the ectodermal cells become neuronal. The resulting embryos have been described as "all brains and no skin," and die young. A variety of studies suggest that cell-cell interactions play an important role in the determinative stages leading to dermoblast and neuroblast differentiation (CAMPOS-ORTEGA 1985; DOE and GOODMAN 1985a, b).

The Notch locus was cloned in 1983 (ARTAVANIS-TSAKONAS et al. 1983), and shown to encode a 10.1-kb transcript derived from 40 kb of genomic DNA. Sequence analysis of this transcript revealed a large open reading frame of 2703 amino acids that includes potential signal peptide and transmembrane domains (WHARTON et al. 1985). The amino-terminal (extracellular) portion of this presumed integral membrane protein is composed of 36 tandem repeats of a 40-amino-acid-long sequence which bears striking homology to EGF. Although it is not yet known whether this molecule is cleaved to release one or more EGF-like units, small patches of mutant Notch tissue in a wild-type animal still express the mutant phenotype. This suggests that the Notch product(s) does not diffuse far. However, as noted above, we have recently shown that a mutant integral membrane TGF-α precursor which cannot be cleaved to release mature growth factor can activate EGF receptors on the surface of adjacent cells. This raises the possibility that the intact Notch protein, which is much larger and contains multipole EGF-like units, may be able to activate EGF receptors either in the same membrane or on the surface of adjacent cells. Other features of the Notch sequence suggest that it may function as a signal transducer. The putative cytoplasmic domain contains sequences reminiscent of a nucleotide-binding site and, as already noted, Notch is known to be cell autonomous. Ultimately, any models proposed to explain how Notch might control cell lineage must account for the observation that Notch expression in the embryo appears to be widespread and not limited to just those cells participating in neurogenesis (HARTLEY et al. 1987). Genetic analyses suggest that the Notch locus is a complex one, and the possibilities of differential RNA splicing or protein processing should not be ignored.

The lin-12 ("cell lineage abnormal") locus of *Caenorhabditis elegans* is a homeotic gene that controls so-called binary decisions during development. The cell lineage of *C. elegans* includes many examples of repeating patterns in which different groups of cells proceed through the same set of cell divisions and generate analogous body parts. However, in some cases, homologous cells in two parallel lineages will have distinctive fates, and it is these distinctive cells which are often affected by lin-12 mutations. Recessive mutations which apparently inactivate the gene cause both cells to adopt one of two distinctive fates. Dominant mutations which increase the activity of the gene cause both cells to adopt the other distinctive fate. Thus, mutations in the lin-12 locus affect many structures in the worm. GREENWALD (1985) recently cloned the lin-12 sequence by a clever approach in which the nematode mobile element Tc1 was used to "tag" the gene. Although the complete nucleotide sequence of this gene has not yet been reported, the sequence of a 3-kb restriction fragment revealed three complete open reading frames that are presumed to correspond to exons. Given this assumption, the resulting peptide would contain 11 tandem repeats of a 38-amino-acid sequence that is homologous to EGF. Whether the lin-12 product is a secretory or integral membrane protein is presently unknown.

The linking of homeotic loci and the EGF family of growth factors is a potentially exciting development that will undoubtedly encourage the testing of new hypotheses. While it is not yet known whether the Notch and lin-12 products function as growth factors, it is also true that we know virtually nothing about the functions and biological activities of the integral membrane EGF and TGFα precursors they resemble. The possibility that some homeotic loci encode growth factors suggests that the critical role these genes play could be as fundamental as arresting or releasing the cell cycle.

IV. Developmental Expression of the Neu Oncogene

The neu oncogene (also referred to as c-erbB-2 and Her2) encodes a 185-kDa transmembrane glycoprotein with tyrosine kinase activity that is structurally related to the EGF receptor (Bargmann et al. 1986). It is thought that this protein is the receptor for an as yet unidentified growth factor. Kokai et al. (1987) have shown that the neu gene is expressed in mid-gestation rat fetuses (days 14–18) in a variety of tissues including nervous system, connective tissue, and secretory epithelium, but not lymphoid tissue.

C. β-Type TGFs

I. Introduction

Transforming growth factor-β (TGF-β) was originally discovered by virtue of its ability to promote the TGF-α-induced neoplastic transformation of fibroblasts in culture (see Chap. 8). However, it is now known to have pleiotropic actions, influencing migration, proliferation, and differentiation of a wide variety of cell types. Moreover, TGF-β is a bifunctional molecule: it stimulates the soft-agar growth of mesenchymal cells but inhibits the growth of most epithelial cells. TGF-β is found in many tissues with especially high concentrations in bone (Seyedin et al. 1985) and platelets (Assoian et al. 1983). This suggests a possible role in tissue repair and bone formation/remodeling. Perhaps consistent with this speculation are the findings that TGF-β (a) has a direct organizational action on capillary endothelial cells in vitro (Madri et al. 1988) and (b) acts to increase formation of connective tissue proteins including collagens, fibronectin, and proteoglycans (Ignotz and Massague 1986; Roberts et al. 1986). These latter increases appear to be the result, in part, of induced synthesis of the respective mRNAs (Ignotz et al. 1987). In addition, TGF-β may indirectly inhibit the degradation of these molecules by inducing the expression of protease inhibitors (Laiho et al. 1986). Other reported actions of TGF-β include the inhibition of certain growth-related genes in keratinocytes (Coffey et al. 1988) and modulation of the number of cell surface EGF receptors (Assoian et al. 1984). Finally, TGF-β may function as a negative paracrine regulator of liver regeneration since the levels of its mRNA in nonparenchymal cells were found to peak following the major wave of DNA synthesis and mitosis (Braun et al. 1988).

TGF-β1 is a homodimer of a 112-amino-acid polypeptide that is cleaved from the carboxy terminus of a 390-amino-acid precursor (Derynck et al. 1985). Secreted in a latent form, TGF-β1 can be activated by proteases, acid pH, and denaturing agents including urea (Gentry et al. 1987). Recently, a second minor form of TGF-β (TGF-β2) has been described. Although the activities of the two TGF-β molecules are similar, they can be distinguished: Whereas TGF-β1 is a potent inhibitor of hematopoietic progenitor cell proliferation, TGF-β2 is not (Ohta et al. 1987). Both species bind to ubiquitous cell surface receptors with high affinity. The largest of these, as yet, ill-defined receptor species has an apparent native molecular weight of 550–600 kDa, or 280–300 kDa after reduction. Lower molecular weight forms of the receptor have also been described. Whether there are separate receptors for the different TGF-β forms is not yet known, though there is apparent crossreactivity (Cheifetz et al. 1987).

TGF-β1 and -β2 are highly conserved molecules that appear to be part of a larger family of polypeptides. Members of this family include inhibin, Mullerian-inhibiting substance, and a product of the *Drosophila* decapentaplegic gene complex which controls dorsoventral patterning in the fly embryo (PADGETT et al. 1987). Inhibin regulates the production of follicle-stimulating hormone by the anterior pituitary gland, and may have intragonadal actions as well. Recent evidence suggests that, in addition, it may have a role in the placenta (PETRAGLIA et al. 1987). Inhibin-like immunoreactivity is present in the cytotrophoblast layer of the human term placenta and in cultures of human trophoblasts, and the secretion of inhibin in these latter cells can be induced by human chorionic gonadotropin (hCG). As the name suggests, Mullerian-inhibiting substance (MIS) is responsible for the developmental regression of Mullerian ducts, the primordia of fallopian tubes, uterus, and the upper part of the vagina. During fetal life, MIS production is maximal during the period of physiological regression of Mullerian ducts in vivo (VIGIER et al. 1984) and then decreases, so that in the rat it is undetectable during the 2nd week of life (DONAHOE et al. 1976).

The finding that TGF-β is related to molecules of developmental importance, in conjunction with its profound effects on cellular function and presumed function in tissue repair, has led numerous investigators to suggest that it must play a critical role in embryogenesis. This postulate has been supported by the findings that (a) TGF-β inhibits myogenic (MASSAGUE et al. 1986; OLSON et al. 1986) and adipogenic (IGNOTZ and MASSAGUE et al. 1985) differentiation in culture; (b) muscle differentiation is accompanied by a loss of TGF-β receptors (EWTON et al. 1988); (c) TGF-β stimulates parameters of bone growth in fetal rat calvariae, which are also a source for the activity (CENTRELLA et al. 1986); and (d) the differentiation of murine embryonal carcinoma cells is accompanied by the appearance of high-affinity receptors for TGF-β (RIZZINO 1987).

II. Developmental Expression of TGF-β

The first evidence for developmental expression came from the findings that TGF-β could be extracted from whole mouse fetuses (PROPER et al. 1982) and term placenta (FROLIK et al. 1983). Subsequent immunohistochemical studies by ELLINGSWORTH et al. (1986) utilized an antibody directed against residues 1–30 of cartilage-inducing factor A (CIF-A), a molecule now believed to be identical to TGF-β1. Marked staining with this antibody suggested the presence of TGF-β1 in osteocytes and chondrocytes of bovine fetuses at 6 months gestation. Additional staining was found in the liver, where clusters of hematopoietic stems cells stained intensely, but hepatocytes and stromal cells did not. Hematopoietic stem cells also stained in the bone marrow, along with megakaryocytes and mononuclear cells. In the kidney, staining was limited to epithelial cells lining the calyses, and, in the thymus, to Hassall's corpuscles and some medullary thymocytes. ROBEY et al. (1987) detected TGF-β mRNA by Northern analysis in both fetal rat bone, and fetal bovine bone-forming cells. They further described the synthesis and secretion of TGF-β by fetal bovine bone cells and demonstrated the presence of high-affinity TGF-β receptors on these cells. Finally, they found barely detectable levels of TGF-β mRNA in most fetal rat and bovine tissues (with the exception of the ameloblast layer of the developing tooth which expressed high levels of this mRNA; see below), but significantly higher levels in primary cell cultures prepared from these tissues.

An extensive study by HEINE et al. (1987) found continuous high expression of TGF-β1 mRNA in day 8–18 mouse fetuses. In addition, using the same antibody developed by ELLINGSWORTH et al. (1986), they found specific immunohistochemical staining for TGF-β in a wide variety of fetal mouse tissues. This staining was closely associated with either mesenchyme per se or tissues derived from it including connective tissue, cartilage, and bone. TGF-β staining was particularly conspicuous in tissues derived from neural crest mesenchyme. These include the developing palate, larynx, facial mesenchyme, nasal sinuses, meninges, and teeth. Moreover, TGF-β expression was found in tissues in which critical mesenchymal-epithelial interactions occur, and particularly intense staining correlated with remodeling of the mesenchyme. Examples of the latter include the formation of digits from limb buds, formation of the palate, and formation of heart valves. Specific examples of staining are best characterized in terms of potential functions for TGF-β. These include: (a) the formation/remodeling of cartilage and bone; (b) early angiogenesis and the subsequent development of cardiovascular structures; and (c) the differentiation of specialized structures which require interactions between mesenchyme and adjacent epithelial cells.

With respect to cartilage and bone, HEINE et al. (1987) found marked TGF-β immunoreactivity in areas of active capillary growth and bone matrix formation, in areas of joint formation, and in the cytoplasm of osteoblasts per se. At day 15, strong staining was observed in areas of intramembranous ossification of flat bones as in the calvarium. The notochord showed staining in surrounding collagen sheets, and in small clusters of centrally located cells which coincided with the future localization of intervertebral disks. The development of the centra of the future definitive vertebrae from individual sclerotomal segments was preceded by an interesting change in TGF-β distribution: The staining became intensified in the posterior half of one segment and in the anterior half of the following segment, thus reflecting the manner in which these sclerotomal segments fuse to form vertebrae.

Conspicuous staining for TGF-β correlated with marked angiogenic activity, i.e., in the marginal blood sinuses of limbs between days 13 and 15. In addition, apparent expression of TGF-β in the heart between days 11 and 13 correlated with a time of cytodifferentiation and remodeling in this organ. Staining was particularly intense in the mesenchymal cushion tissue – a mass of connective tissue composed of collagens, fibronectin, and proteoglycans. At later times (i.e., day 15), there was still strong TGF-β staining in the mesenchyme of the valves. Interestingly, there was no apparent staining of either the endothelial or smooth muscle cells of the developing vascular system.

The correlation between TGF-β staining and the differentiation of specific structures in which an interaction between mesenchyme and epithelium is known to be important was particularly interesting. For example, the staining pattern of TGF-β in the mesenchyme surrounding the developing follicle seemed to correlate with stages in development. Thus, in the case of hair follicles of the snout, staining in the mesenchyme was weak prior to the actual appearance of hair germ cells. Staining in the mesenchyme intensified as the development of the hair follicles accelerated (around day 13), and then disappeared after hair follicle formation was completed. And, mesenchyme underlying those regions of skin that do not participate in hair follicle formation did not stain for TGF-β. Development of the teeth provided another example of apparent TGF-β expression correlating with particular stages in the development of a structure. Thus, staining became locally intense around day 15 in cells of neural crest origin as they became incorporated into the dental papilla. Staining was subsequently observed between the inner and outer dental epithelium, and in tissue surrounding the capillaries that infiltrate the developing teeth. This latter staining is consistent with the finding of high levels of TGF-β mRNA in cells from the ameloblastic layer of fetal bovine teeth (ROBEY et al. 1987).

WILCOX and DERYNCK (1988) recently described an in situ hybridization analysis of TGF-β1 mRNA expression in fetal mouse tissues. Using a cRNA probe derived from a cloned TGF-β1 cDNA, they found TGF-β1-mRNA-expressing cells scattered throughout the blood islands and capillaries of day-9 to -10 fetuses. Staining for endogenous peroxidase on adjacent sections suggested that, consistent with their appearance, most of the TGF-β1-mRNA-positive cells were erythroid in nature. The only exception was the appearance of TGF-β1 mRNA in the innermost layer of cells in developing ventricles. Between days 10 and 12, the budding liver became the primary source of TGF-β1 mRNA expression. This expression was limited to clusters of hematopoietic cells by day 14 and was nondetectable by day 16. Finally, RAPPOLEE et al. (1988) have used PCR amplification of cDNAs to demonstrate the presence of TGF-β1 transcripts in preimplantation blastocysts.

Taken together, these various studies support a function for TGF-β in development. However, there are certain inconsistencies, especially when comparing the recent studies of TGF-β expression during fetal mouse development. The in situ hybridization analysis by WILCOX and DERYNCK (1988) suggests a more limited expression than does the immunohistochemical analysis of HEINE et al. (1987). Whether these discrepancies are the result of differences in either assay specificities (i.e., the antibody might recognize other related molecules, whereas the cRNA probe is specific for TGF-β1) or sensitivities (the relative abundance of mRNA versus protein might be significantly different), or have yet other explanations – for example, the staining is detecting sites of localization, not synthesis – needs to be clarified. Nevertheless, there is an emerging picture of TGF-β expression in bone development, and in hematopoiesis. A potential role for this molecule in the former is consistent with its known biological actions. What the action of

TGF-β would be in the function and development of hematopoietic cells is less clear, though, as noted above, TGF-β1 inhibits hematopoietic progenitor cell proliferation. In addition, the notion, suggested by the data of HEINE et al. (1987), that TGF-β functions in the remodeling of embryonic tissues, is attractive given both its presumed role in regulating repair of adult tissues and its ability to regulate proteases like plasminogen activator. Clearly our view is still fragmentary, not the least because virtually nothing is known about fetal expression of the TGF-β receptor(s).

III. Role for TGF-β in Amphibian Development

Finally, an exciting new discovery stems from recent studies which implicate TGF-β2 and related molecules in early amphibian development. When the amphibian egg is laid it has an obvious pigmented animal pole and a yolky vegetal pole. Cells from the animal pole subsequently develop into ectoderm, and cells from the vegetal pole into endoderm. Moreover, there are interactions between these cells in the blastula, with the vegetal cells inducing neighboring animal cells to become mesoderm. It has long been presumed that the different properties of animal and vegetal cells are established by "determinants" that are differentially localized in the egg. The nature of some of these determinants is suggested by the studies of MELTON and colleagues, who described the isolation of cDNA clones corresponding to mRNAs that are localized in unfertilized eggs (REBAGLIATI et al. 1985). One that is localized to the vegetal pole, Vg1, has been sequenced and shown to encode a protein that is 38% homologous to mature TGF-β1 (WEEKS and MELTON 1985). What makes this observation all the more significant is the finding of KIMELMAN and KIRSCHNER (1987) that the mesoderm induction described above can be mimicked by FGF and TGF-β. In other words, TGF-β will potentiate the FGF-induced synthesis of muscle actin (a mesoderm marker) in animal hemisphere cells. That they were also able to identify an mRNA that codes for an FGF-like molecule in the early *Xenopus* embryo is consistent with the postulated function for these growth factors. ROSA et al. (1988) have subsequently shown that TGF-β2, but not TGF-β1, will induce α-actin mRNA in amphibian animal cap cells, and that the inducing activity secreted by *Xenopus* XTC cells is inhibited by antibodies to the former and not the latter. These results are exciting, not just because they indicate a role for TGF-β in the early embryo, but because of the profound implication that growth factors might act as inductive signals during vertebrate development.

D. Insulin-Like Growth Factors/Somatomedins

I. Introduction

Insulin-like growth factors (IGFs) or somatomedins (Sms) are single-chain polypeptides with chemical structures similar to proinsulin (see Chap. 6). Two distinct forms of IGF have been identified: IGF I (Sm-C) and IGF II [also known as multiplication-stimulating activity (MSA)]. Human IGF I is a basic polypeptide of 70 amino acids that is derived from a 130-amino-acid precursor (JANSEN et al. 1983; LE BOUC et al. 1986). IGF II, on the other hand, is a slightly acidic polypeptide of 67 amino acids that is cleaved from a precursor of 180 amino acids (BELL et al. 1984; DULL et al. 1984; JANSEN et al. 1985). The mature forms of human IGFs I and II share approximately 60% amino acid sequence identity, and display greater than 90% homology with their rodent counterparts (WHITFIELD et al. 1984). Additional variant forms of IGFs have been identified in human and rat tissues. The latter are believed to represent differentially processed species derived from the same precursor peptides.

Studies of IGFs provided early support for the tenet that growth factor actions are not limited to mitogenesis. In fact, one of the more intriguing facets of IGFs is their ability to promote differentiation. Thus, IGFs have been shown to facilitate the differentiation of (a) L6 myoblasts into myotubes (EWTON and FLORINI 1981); (b) lens epithelium (BEEBE et al. 1986), and (c) ovarian granulosa cells (ADASHI et al. 1984; DAVOREN et al. 1985).

II. Expression of IGF Receptors and Binding Proteins

IGFs bind with high affinity to two distinct types of "receptors." The type I molecule, which structurally resembles the insulin receptor, is a heterodimer of two 130-kDa and two 95-kDa subunits. It preferentially binds IGF I, and also shows weak affinity for insulin. The type II "receptor," on the other hand, is a single polypeptide of 250 kDa that binds IGF II with high affinity, but does not bind either IGF I or insulin. As pointed out in a recent review by ROTH et al. (1988), the term "receptor" should be used cautiously with respect to the type II molecule. Whereas the type I receptor has an associated tyrosine kinase activity, the type II receptor does not. Indeed, it has yet to be shown that this molecule propagates a signal in response to IGF II, and binding to the type II receptor molecule has not been convincingly linked to a biological action other than internalization. Furthermore, the type II receptor has recently been shown to be identical to a receptor for mannose-6-phosphate. Since the latter has been implicated in the targeting of lysosomal enzymes, the significance of IGF II binding to this species is presently unclear. Quite possibly, it is the binding of IGF II to the type I receptor that is significant vis-a'-vis the biological actions of this growth factor. Thus, although human fetal fibroblasts express the type II receptor, IGF II stimulation of amino acid transport and DNA synthesis in these cells appears to be mediated through the type I molecule (CONOVER et al. 1986, 1987). Finally, in this regard, it is worth noting that a recent study suggests that IGF I and IGF II bind to different sites on the type II receptor (CASELLA et al. 1986).

Both types of IGF receptor have been identified in membrane preparations and/or cultured cells from virtually every fetal tissue (reviewed by NISSLEY and RECHLER 1984a, b), as well as from the placenta (BHAUMICK et al. 1981; DAUGHADAY et al. 1981; MASSAGUE and CZECH 1982). Both type I and type II receptors are detectable in whole mouse embryos as early as day 9 of gestation – and in apparent greater abundance than insulin receptors (SMITH et al. 1987). Similarly, whereas insulin receptors are barely detectable in chick embryos before days 3–4 (DOETSCHMAN et al. 1975), and do not become prominent until the 2nd week of development (HENDRICKS et al. 1984), IGF receptors are demonstrable as early as day 2 (DEPABLO et al. 1985). Whether IGF receptors are expressed earlier in chick embryogenesis is less clear. While MATTSON et al. (1988) reported insulin and IGF binding to mouse oocytes and preimplantation embryos by light microscopic autoradiography, SMITH et al. (1987) failed to detect either type I or type II receptors in mouse blastocysts by affinity labeling. Finally, KIESS et al. (1987) have recently described a soluble form of the type II IGF receptor that is present at higher levels in fetal as opposed to adult rat serum. For the reasons described above, the significance of this latter finding is presently unclear.

One of the features that distinguishes IGF physiology from that of other growth factors is the existence of specific, high-affinity binding proteins. The vast majority of IGF in adult serum is bound to a growth-hormone-dependent complex that contains a 50-kDa binding subunit, as well as a 100-kDa nonbinding polypeptide. In addition, non-growth-hormone-dependent binding proteins of approximately 26–40 kDa are also present, though these smaller species are usually not saturated. The larger binding complex has not been detected during development (SMITH et al. 1987). However, the smaller forms are present at relatively high concentrations in human amniotic fluid (DROP et al. 1979; D'ERCOLE et al. 1985) and have been extracted from whole mouse embryos from days 9–12 of gestation (SMITH et al. 1987). In addition, immunohistochemical staining using antibodies directed against the binding proteins present in amniotic fluid suggests that the latter colocalize with IGFs in human fetal tissues (HILL et al. 1989). The functional significance of these binding proteins and, in particular, whether they facilitate or inhibit IGF action is controversial (DROP et al. 1979; ELGIN et al. 1987).

III. IGFs in Fetal Tissues and Fluids

Both IGF I and IGF II are detected in human fetal plasma as early as 15 weeks gestation (ASHTON et al. 1988) and IGF I is present in human fetal tissue extracts from 12 weeks gestation (D'ERCOLE et al. 1986). Although the levels of both IGF I and IGF II may increase as much as twofold from 32 weeks to term (GLUCKMAN and BRINSMEAD 1976; ASHTON and VESEY 1978; ZAPF et al. 1981; BENNETT et al. 1983), the concentrations in human fetal sera and newborn cord blood remain low compared with adult or maternal sera (D'ERCOLE et al. 1980b; D'ERCOLE and UNDERWOOD 1981). This contrasts with other animal species in which IGF I levels steadily rise to adult concentrations through gestation, and IGF II levels in late pregnancy are high compared with adult values. A particularly striking example is that of fetal rat serum which, in late gestation, contains 20- to 100-fold more immunoreactive IGF II than does maternal serum (MOSES et al. 1980; DAUGHADAY et al. 1982). Thereafter, IGF II levels fall to adult values by the time of weaning. Similarly in lambs and guinea pigs, IGF II levels in the fetal serum rise to two- to four-fold higher than those in the maternal blood, and decline to adult values at term (GLUCKMAN and BUTLER 1983; DAUGHADAY et al. 1986).

IGFs have been shown to be secreted by a large variety of fetal cells in culture. These include fetal chick, mouse, rat and human hepatocytes (HASELBACHER et al. 1980; RICHMAN et al. 1985; STRAIN et al. 1987), human and rat embryo fibroblasts (ADAMS et al. 1983; ATKINSON et al. 1980; CLEMMONS et al. 1981), and human fetal pancreas cells (HILL et al. 1987). In addition, IGF II has also been shown to be produced by murine Dif 5 cells (NAGARAJAN et al. 1985). The latter are differentiated endoderm-like cells derived from F9 embryonal carcinoma cells in response to retinoic acid. These cells possess properties similar to mouse embryonic stem cells, and are capable of proliferating in the absence of serum or other hormonal supplements. In contrast, undifferentiated embryonal carcinoma cell lines produce negligible IGF II. These data have been interpreted to suggest that visceral endoderm may be a source for IGF II in the developing embryo.

The aforementioned data are consistent with roles for IGFs I and II during embryonic and fetal development. Indeed, the size of the fetus and newborn have been positively correlated with serum IGF I levels (GLUCKMAN and BRINSMEAD 1976; ASHTON and VESEY 1978; D'ERCOLE and UNDERWOOD 1981; ASHTON et al. 1988), and cord blood IGF I concentrations are depressed in small-for-gestational-age infants (FOLEY et al. 1980). Moreover, studies utilizing animal models of intrauterine growth retardation appear to suggest that birth size is directly correlated to the serum or tissue levels of IGF I (VILEISIS and D'ERCOLE 1986; BASSETT and GLUCKMAN 1987). In contrast, none of these studies suggest that IGF II levels correlate with either fetal or newborn weight and length.

IV. Developmental Expression of IGF Genes

Numerous reports describe the detection of IGF I and IGF II transcripts in embryonic and fetal tissues. A number of generalities emerge from these studies. However, before listing these conclusions, it should be noted that in rodents and humans multiple mRNAs have been detected for both IGF I and IGF II. These transcripts may be the result of multiple transcription initiation and termination events (FRUNZIO et al. 1986; GRAY et al. 1987; DE PAGTER-HOLTHUIZEN et al. 1987, 1988), as well as alternative splicing (ROTWEIN et al. 1987a). Certain of these variations are of developmental interest. For example, human IGF II transcripts appear to derive from three distinct promoters, two of which are utilized only during fetal life (DE PAGTER-HOLTHUIZEN et al. 1987). However, the biological significance of these various transcripts is presently unclear, and in one study of IGF II expression in fetal rats, probes representing the coding region detected fewer transcripts than a probe derived from the 5' untranslated portion (GRAY et al. 1987). Hence, in the following summary, we will describe the observation of IGF transcripts without detailed reference to the sizes of the transcripts. Conclusions which emerge from the various studies are as follows:

1. Messenger RNAs encoding both IGF I and IGF II are expressed in a variety of embryonic and fetal tissues relatively early in gestation. For example, using an RNase protection assay, ROTWEIN et al. (1987 b) demonstrated the presence of IGF I mRNA in whole rat embryos as early as day 11 of gestation. The level of IGF I mRNA increased almost tenfold between days 11 and 13 of gestation, and then remained relatively constant thereafter. This assay also detected IGF II transcripts, the level of which remained unchanged from day 13 of gestation through term. Similarly, GRAY et al. (1987) found that expression of IGF II mRNAs in fetal liver and placenta was low at day 11 of gestation, increased dramatically at day 12, and then remained essentially unchanged through term. IGF I transcripts have been detected in fetal liver, lung, brain, and intestine (LUND et al. 1986). BROWN et al. (1986) found IGF II transcripts in 11 of 13 fetal tissues examined, with the highest levels in muscle, skin, lung, liver, intestine, and thymus. Lower levels were observed in heart, brain stem, cerebral cortex, kidney, and hypothalamus. In this study, IGF II transcripts were not detected in the spleen and pancreas.

Analogous results have also been reported for expression of human IGF genes. For example, HAN et al. (1988) found low levels of IGF I expression in every fetal tissue examined from 16 through 20 weeks of gestation, with the highest levels in the placenta and stomach. These investigators also detected IGF II mRNA in all tissues examined except cerebral cortex and hypothalamus. Both IGF I and IGF II transcripts were detected as early as 8 weeks of gestation in human liver and heart (V. K. M. HAN, unpublished observations). Others have described the expression of IGF II mRNAs in human fetal muscle, tongue, liver, kidney, lung, and adrenal gland, as well as in the placenta (SCOTT et al. 1985; GRAY et al. 1987).

2. Transcripts for IGF II are more abundant than those encoding IGF I in most fetal tissues. For example, in the human fetus, the ratio of IGF II to IGF I transcripts varied from 2 in the thymus and spleen to 650 in the liver (HAN et al. 1988). While this is generally consistent with the finding that serum levels of IGF II are higher than those of IGF I (MOSES et al. 1980), the ratio of IGF II to IGF I in the serum is considerably lower than that suggested by the relative levels of the two mRNAs in many tissues. A corollary of this observation worth noting is that the levels of the two growth factors in serum may not be indicative of the levels found locally in particular tissues.

In the postnatal rat, IGF I transcripts continue to be expressed at relatively high levels in the liver, whereas IGF II mRNA is largely confined to neural tissues (LUND et al. 1986; BROWN et al. 1986; MURPHY et al. 1987; GRAY et al. 1987). Likewise in humans, significant IGF II expression is observed only in skin, nerve, and muscle (GRAY et al. 1987).

3. In both rats and humans, mesenchymal tissues predominate in the expression of IGF I and, especially, IGF II transcripts. For example, HAN et al. (1987 c) found transcripts for both growth factors in the perisinusoidal cells of liver, perichondrium of cartilage, the sclera of the eye, as well as connective tissue layers and the sheaths, septa, and capsules of each organ and tissue. All of the hybridizing regions comprised predominantly fibroblasts or other cells

of mesenchymal origin – a finding that is consistent with earlier reports that fibroblasts (including rat embryo fibroblasts) secrete both IGF I (ATKINSON et al. 1980; CLEMMONS et al. 1981) and IGF II (ADAMS et al. 1983). HAN et al. (1987c) suggested on the basis of these results that, since fibroblasts are both widely distributed and anatomically integrated into tissues and organs, they are ideally suited for the production of IGFs which might then exert paracrine effects on surrounding cells. Perhaps consistent with this speculation is their finding that in many fetal tissues immunoreactive IGFs are localized to epithelial cells and not the mesenchymal cells that appear to express the mRNAs (HAN et al. 1987b). For example, in the human fetal pancreas, IGFs are localized to the B cells of islets (HILL et al. 1987), whereas IGF transcripts are found in adjacent fibrous tissue. The authors proposed that in these cases they were detecting sites of binding rather than synthesis – an interpretation that was supported by the apparent colocalization of immunoreactive IGF-binding proteins (HILL et al. 1989). (As an aside, it should be noted that these results suggest that analyses of growth factor expression that are limited solely to immunoreactive localization should be interpreted with caution.)

Similar results have been reported for in situ hybridization analyses of fetal rats. For example, BECK et al. (1987) localized significant IGF II expression to the yolk sac, hepatic buds, dermatomyotome, sclerotome, branchial arch mesoderm, and septum transversum between days 10.5 and 14.5 of gestation. At later stages (between days 15.5 and 21.5), relatively intense hybridization was observed in the yolk sac, liver, muscle, precartilaginous mesenchymal condensations, perichondrium and immature chondrocytes, and periosteum and centers of intramembranous ossification and developing sclera. STYLIANOPOULOU et al. (1988) obtained similar results, and in noting the significant overlap between IGF II mRNA distribution and TGF-β immunostaining suggested a possible functional relationship between the two growth factors during rat embryogenesis. In the latter study, cells of other germ layers that appeared to express IGF II mRNA included liver and bronchial epithelium (endoderm), as well as choroid plexus, pituitary rudiment, and auditory placode (ectoderm).

The finding that (a) expression of fetal IGFs is widespread and (b) serum IGF levels do not reflect localized tissue levels supports the contention of HAN et al. (1987c) that IGFs can act through autocrine and/or paracrine mechanisms. It should be noted that this represents a change in emphasis from the early studies which largely focused on GH-dependent serum IGF levels and the implied endocrine actions of these growth factors.

Finally, RAPPOLLEE and colleagues have recently utilized PCR amplification to study the expression of IGFs and their receptors in mouse blastocysts from the 2- to the 64-cell-stage (corresponding to about day 3.5–4.5 of gestation). They detected transcripts for the insulin, as well as types I and II IGF receptors beginning at the eight-cell stage when the embryo compacts (D. RAPPOLLEE, unpublished observations). Interestingly, this was preceded by the expression of IGF II transcripts which could be detected in the two-cell embryo and were found throughout gestation. In contrast, IGF I transcripts were not identified until day 8.5 of gestation. Taken together with the finding that in the fetus (but not the adult) IGF II expression is generally higher than that of IGF I, these data suggest a significant role for IGF II during development.

E. Platelet-Derived Growth Factor

I. Introduction

Platelet-derived growth factor (PDGF) is the major mitogen found in serum (see Chap. 5). It stimulates the proliferation and chemotaxis of fibroblasts and smooth muscle cells, and stimulates collagen, glycosaminoglycan, and collagenase production by fibroblasts. These activities, together with the fact that PDGF is found in the α-granules of circulating platelets and is released into serum during blood clotting, has suggested that it plays a role in the initiation of wound healing.

In humans, PDGF comprises two related chains (A and B) which form either homo- (AA, BB) or heterodimers (AB). The A- and B-chains are the products of separate genes; the B gene is the human homolog of the simian sarcoma virus v-sis oncogene. Since some tumor cells express this product along with the PDGF receptor, it may function as an autocrine factor in the development of malignancy. The A-chain is also expressed by human tumor cells, and PDGF-AA has been identified as a product of melanoma, osteosarcoma, and glioblastoma cells. The extent to which the homo- and heterodimers of PDGF have distinct activities and, possibly, separate receptors has not yet been clarified.

II. Developmental Expression of PDGF

A developmental role for PDGF is suggested by several recent findings. MERCOLA et al. (1988) described the cloning from *Xenopus* oocytes and embryos of cDNAs which are homologous to a human A-chain cDNA, and whose sequence predicts two different forms of the A-chain. These two forms appear to result from alternative splicing and contain different COOH-termini. The larger of these species differs in a significant way from the human sequence only by the insertion of a largely hydrophobic stretch of 11 amino acids into the normally hydrophilic COOH-terminus. This raises the interesting possibility that this form of the A-chain may be membrane associated. Northern blots probed with the cloned *Xenopus* cDNA revealed the presence in oocytes of two maternally derived A-chain transcripts of approximately 3 kb. These transcripts decreased in abundance during development, but then reappeared in the late gastrula stage after the onset of embryonic transcription. At these later times, a larger transcript of approximately 7 kb predominated. The relationship of this latter transcript to the smaller species has not yet been elucidated. Finally, although a *Xenopus* B-chain gene was detected by

Southern blot analysis, expression of the corresponding transcript was not apparent at any stage during development.

That PDGF might also be involved in early mammalian development is suggested by a variety of evidence. First, SEIFERT et al. (1984) found activity that bound to PDGF receptor was secreted by cultured rat aortic smooth muscle cells (SMCs) isolated from 13- to 18-day-old pups but not from adult animals. MAJESKY et al. (1988) further showed that PDGF B-chain transcripts accumulated in passaged newborn rat SMC but not adult rat SMC, whereas PDGF A-chain mRNA was found at comparable levels in SMC from both age groups. In addition, B-chain transcripts appear to be relatively short-lived compared with A-chain transcripts. Although these results may arise from culture artifacts, they suggest differential regulation of the two chains during development. Second, RIZZINO and BOWEN-POPE (1985) found that media conditioned by murine embryonal carcinoma (EC) cells (which resemble derivatives of the inner cell mass of the mouse blastocyst) contained a PDGF-like activity. Third, reminiscent of the aforementioned *Xenopus* studies, RAPPOLEE et al. (1988) have recently used coupled reverse transcription and PCR amplification to demonstrate the presence of PDGF A-transcripts in early cavitation mouse blastocysts (32- to 64-cell stage). Lower levels of this transcript were also detected in unfertilized ovulated oocytes as a maternal product that disappeared between the two- and eight-cell stages. These investigators also detected PDGF antigens localized to punctate structures in the perinuclear area of all cells of the blastocyst.

There is currently little direct information concerning the developmental expression of the PDGF receptor. However, VU et al. (1989) have recently detected a novel PDGF receptor transcript in mouse embryonal carcinoma cells, as well as in embryonic stem cells. This novel transcript is missing approximately 1100 bps from the 5' end of the normal mRNA, thus resulting in the production of a truncated receptor that lacks most of the extracellular domain. They have further shown that the in vitro differentiation of the embryonal carcinoma cells leads to increased production of the normal, full-length transcript. These data raise the possibility of a developmental form of the PDGF receptor whose activity (in a manner analogous to that of erbB) may be ligand independent. These investigators are presently examining embryos and fetuses for the expression of this novel transcript.

F. Fibroblast Growth Factor and Related Molecules

I. Introduction

Fibroblast growth factor (FGF) has been shown to influence the growth of many different cell types (see Chap. 7). To date, two forms of FGF have been described. The acidic form, which was originally isolated from brain, appears to be primarily associated with neural tissue and may promote the outgrowth of neuronal processes (LIPTON et al. 1988). In contrast, basic FGF has been detected in many tissues and cell types, including pituitary, placenta, endothelial, and tumor cells. Basic FGF shares 55% sequence homology with the acidic molecule and, in addition, exhibits 40% amino acid sequence identity with two newly discovered oncogenes, hst and int-2. Both acidic and basic FGF appear to bind to the same high-affinity receptor.

II. Developmental Expression of FGF

Although its physiological roles have not yet been defined, FGF has been implicated in endothelial cell proliferation and migration during angiogenesis. Capillary endothelial cells have been shown to express, and proliferate in response to basic FGF (SCHWEIGERER et al. 1987). In addition, both acidic and basic FGF have been shown to induce neutrophic factors and promote the survival and differentiation of embryonic and neonatal neurons in culture (UNSICKER et al. 1987). Finally, both forms of FGF also repress the onset of skeletal muscle differentiation, and thereby prevent (a) the acquisition of a permanent postmitotic phenotype; (b) the transcriptional activation of skeletal muscle specific genes; and (c) cell fusion (CLEGG et al. 1987). Once FGF is removed, greater than 95% of the cells become postmitotic within 12 h, and surface FGF receptors are undetectable by 24 h. Once committed, the cells are refractory to FGF – even prior to the decline in surface receptor levels (OLWIN and HAUSCHKA 1988). These various findings suggest a possible role for FGF during development, and this suggestion is supported by studies of EC cell lines which may mimic early stages of mammalian development. Thus, RIZZINO et al. (1988) found that three different EC cell lines, including a multipotent human EC cell line, produce a heat-labile, heparin-binding factor that competes with FGF for binding to membrane receptors, and is immunologically related to basic FGF. Retinoic-acid-induced differentiation of the EC cell lines leads to a significant decrease in the production of this activity and is accompanied by a tenfold increase in the number of FGF receptors. Moreover, the growth of the differentiated cells is stimulated by FGF, whereas that of the undifferentiated EC cells is not. Whether these results suggest that the differentiation of EC cells is accompanied by a switch from the autocrine production of FGF to a dependence on paracrine or endocrine sources or, instead, has an alternative explanation is unclear. In addition, the identity with FGF of the material produced by the parental cells has not been fully established.

An activity which stimulates the growth of a mouse muscle cell line, competes for binding to the FGF receptor on 3T3 cells, and is retained on heparin columns has recently been extracted from whole chick embryos, developing limb buds, and unfertilized egg yolk and white (SEED et al. 1988). The levels of this activity are constant between days 2.5 and 6 when organogenesis is occurring, and then increase dramatically between days 9 and 13. Concomitant with this increase is a decline in the relative levels extracted from the limb bud (compared with whole body). In addition, it is also accompanied by a decline in the apparent levels of FGF receptor, so that by day 17 no receptor activity is detected in muscle or heart, and only low levels are present in liver, eye, and brain (B. OLWIN, unpublished observations).

Further corroboration of a developmental role for FGF comes, again, from studies of *Xenopus*. SLACK et al. (1987) found that the heparin-binding growth factors, basic FGF, and embryonal-carcinoma-derived growth factor (ECDGF) can mimic natural signals and induce the appearance of mesoderm in developing early frog embryos. No other growth factors tested were active. These findings have been confirmed by KIMELMAN and KIRSCHNER (1987), who found that FGF will induce the appearance of embryonic muscle actin mRNA in animal cap cells, and that this effect is potentiated by TGF-β. The physiological significance of this finding is further suggested by the fact that these authors were able to isolate from an oocyte cDNA library a clone which is homologous to basic FGF. The carboxy-terminus of the predicted *Xenopus* protein shares 54 out of 61 amino acids with basic FGF.

Finally, a recent study which utilized coupled reverse transcription and PCR amplification failed to detect expression of basic FGF mRNA in preimplantation mouse embryos (RAPPOLEE et al. 1988). Since this same methodology did allow detection of TGF-α, PDGF, and TGF-β transcripts, this finding suggests that if basic FGF is expressed in the embryo it must be postimplantation.

III. Developmental Expression of Related Molecules

As noted above, a growth factor (ECDGF) purified from the culture medium of PC13 murine EC cells (HEATH and ISACKE 1984) shares with FGF the properties of binding to heparin and ability to induce mesoderm in *Xenopus* embryos. However, molecular studies characterizing the possible expression of this molecule during development have not yet been reported.

A newly discovered oncogene, int-2, has recently been shown to encode a protein with 40% amino acid sequence homology to basic FGF (DICKSON and PETERS 1987). int-2 was first recognized as a site with frequent proximity to integrated mouse mammary tumor virus (MMTV). Expression of int-2 is highly restricted in normal cells: int-2 transcripts are found in many mammary tumors but have not been detected in normal adult tissues. However, JAKOBOVITS et al. (1986) have detected int-2 expression in differentiated mouse EC cells and in day-7.5 mouse egg cylinders by Northern blot analysis. These transcripts were not detected in early somite-stage embryos 24 h later. Based on this observation and a variety of indirect evidence, it was suggested that the expression at day 7.5 is probably confined to the extraembryonic endoderm that surrounds the embryonic component of the egg cylinder. WILKINSON et al. (1988) used in situ hybridization analysis to corroborate the increased expression in differentiated versus and undifferentiated EC cells and, through comparison with sections of embryo, to confirm that expression occurs in parietal, but not visceral, endoderm. int-2 transcripts were also detected between days 7.5 and 9.5 in the embryo proper. For example, expression was observed in those cells that migrate through the primitive streak to form mesodermal structures. These include cells that give rise to the extraembryonic mesoderm that surrounds the entire exocoelomic cavity, as well as those cells that ultimately organize to form somites. Interestingly, in both cases, expression appears to be limited to migrating cells and is not detected in the more organized derivatives. Finally, high levels of regionalized int-2 expression were also observed in nonmesodermal cells in day-8.5 through -9.5 embryos. Expression was detected on the lower surface of the endodermally derived pharyngeal pouches and in neuroepithelial cells of the hindbrain adjacent to the otocyst. While the role of int-2 in these various contexts is uncertain, the pattern of expression is not inconsistent with the known properties of FGF to stimulate cell migration and act as an induction signal. Since the neuroepithelium adjacent to the developing otocyst is believed to be the source of an induction signal, an intriguing possibility is that int-2 serves as an inducer.

G. Hematopoietic Growth Factors

I. Colony-Stimulating Factor 1 and Its Receptor (c-*fms*)

1. Introduction

Colony-stimulating factor 1 (CSF-1) is a homodimeric glycoprotein growth factor that has been shown to regulate the survival, proliferation, and differentiation of mononuclear phagocytes (see Chap. 15). Cells of this lineage express a high-affinity, 165-kDa, glycoprotein receptor which has an associated tyrosine kinase activity and is either closely related, or identical, to the c-*fms* protooncogene product. As in the case of the EGF family of growth factors, the secreted form of CSF-1 appears to be cleaved from an integral membrane precursor.

A role for CSF-1 in development was first suggested by Bradley et al. (1971), who found that extracts of mouse placenta, uterus, and fetal membranes had 10- to 50-fold higher levels of CSF-like activity than other maternal tissues. Subsequently, Bartocci et al. (1986) used a specific radioimmunoassay to demonstrate a pregnancy-associated increase of 1000-fold in the levels of CSF-1 in the uterus. They further showed that uterine CSF-1 could be induced in normal females by chorionic gonadotropin (CG), and that the inductive effect of CG was largely blocked in ovariectomized mice. These results suggested that, during pregnancy, CSF-1 expression in the uterus is regulated by CG via the latter's influence on ovarian function. Pollard et al. (1987) extended these findings by demonstrating the presence of CSF-1 mRNA in pregnant uterus, where it was preferentially localized to columnar cells in the luminal and glandular secretory epithelium of the endometrium. In contrast, CSF-1 mRNA was not detected in nonpregnant uterus, and low levels were found in the placenta. Interestingly, the comparative ratio of the two normal CSF-1 transcripts was reversed. Thus, whereas in mouse L cells a 4.6-kb transcript predominates, in the pregnant uterus a 2.3-kb transcript is most abundant. Since the two transcripts differ only in their 3' untranslated sequence, the changing ratio suggests developmental-specific differences in mRNA splicing. The end result may be increased stability of the uterine transcript.

POLLARD et al. (1987) also examined hormonal influences on CSF-1 expression in the uterus of ovariectomized mice using a protocol that simulates implantation and decidualization. They found that, while a combined treatment with estrogen and progesterone resulted in a fourfold increase in uterine CSF-1, sequential treatment with hormones followed by arachis oil (which stimulates decidualization) led to a 15-fold induction. Since arachis oil alone was ineffective, these data suggest that CSF-1 production is not associated with decidual cell proliferation per se, but is potentiated by decidualization. Extending the results of BARTOCCI et al. (1986), they also showed that the antiprogestin RU 486 abrogates the CG induction of uterine CSF-1. This further supports the interpretation that CG regulates CSF-1 production in the pregnant uterus indirectly by stimulating ovarian synthesis of estrogen and progesterone. Finally, AZOULAY et al. (1987) found high levels of CSF-1-like bioactivity in amniotic fluid: Activity was already high at day 9 of gestation, increased to a maximum at days 11 and 12, and then decreased gradually till birth. The next highest levels of activity were found in samples of yolk sac and amnion, with extracts of whole fetuses or placenta having little or no detectable activity (though media conditioned by dissociated fetal cells did appear to contain some CSF-1-like activity). These investigators also detected CSF-1 mRNA in a preparation of extraembryonic tissues, but not in whole or partially dissected fetuses. Although they did not apparently examine the uterus directly, it may have contaminated the extraembryonic tissues.

2. Developmental Expression of c-*fms*

Expression of c-*fms* (the putative CSF-1 receptor) during rodent development was first characterized by VERMA and colleagues. For example, MULLER et al. (1983) found significant expression of the approximately 4-kb c-*fms* transcript in placenta and extraembryonic membranes (yolk sac and amnion), but did not detect c-*fms* mRNA in the fetus at any age. Expression was highest in the placenta, where it increased 15-fold from day 7 through day 15 of gestation before leveling off. Similarly, expression in the extraembryonic membranes, which was two- to threefold lower at all time points, increased sixfold from days 12 through 18 of gestation. More recently, RETTENMIER et al. (1986) detected c-*fms* mRNA in choriocarcinoma cell lines that are derived from placental trophoblasts. They further demonstrated the immunoprecipitation from these cells of glycoproteins that had associated tyrosine kinase activity and whose tryptic peptides were identical to those of c-*fms* immunoprecipitated from peripheral blood mononuclear cells. These species appear to be surface localized and functional since choriocarcinoma cells were shown to bind CSF-1 with high affinity.

Considered together, the aforementioned data concerning the expression of CSF-1 and c-*fms* suggest that CSF-1 synthesized in the uterus in response to progesterone and estrogen regulates placental trophoblast proliferation and differentiation. This interpretation is supported by the following findings: (a) CSF-1 synthesis is localized to luminal and glandular secretory epithelial cells whose mitotic index and secretory activity is regulated by estrogen and progesterone; (b) cells derived from placental trophoblasts express functional CSF-1 receptors; and (c) CSF-1 is a mitogen for fetally derived trophoblasts (ATHANASSAKIS et al. 1987). As pointed out by POLLARD et al. (1987), the close proximity of placental trophoblasts to uterine secretory epithelium would permit their stimulation by secreted growth factor, and even raises the possibility that the transmembrane precursor may be biologically active. In addition, uterine CSF-1 may cause local proliferation and differentiation of cells of the mononuclear phagocytic lineage giving rise to mature macrophages. The uterus is populated by cells which are thought to have originated in the bone marrow (KEARNS and LALA 1982), and macrophages have been identified on the maternal side of the placenta's fetal/maternal border. Although their role is unknown, it may be to contribute to immunosuppression of the maternal response to the allogeneic fetus.

II. Related Growth Factors

AZOULAY et al. (1987) did not detect IL-3 or GM-CSF transcripts in placenta, fetal membranes, or whole fetuses by Northern analysis. Although they did detect low levels of G-CSF mRNA in samples of fetal membranes (yolk sac plus amnion), the transcript was abnormally large (2.8 versus the usual 1.6 kb), and fetal membrane extracts (which contained CSF-1 activity) did not contain G-CSF activity. RAPPOLEE et al. (1988) failed to detect G-CSF transcripts in preimplantation mouse embryos, despite the fact that their coupled reverse transcription/PCR amplification protocol detected mRNAs for other growth factors.

III. Interleukins-2 and -4

Relatively little data are presently available concerning the developmental role/expression of the various interleukins. RAULET (1985) found that fetal mouse thymocytes express receptors for interleukin 2 (IL-2) – the T-cell growth factor – without in vitro induction, and that they proliferate vigorously in an IL-2-dependent fashion if provided a costimulating mitogen. SIDERAS et al. (1988) examined the expression of IL-4 in fetal thymocytes. IL-4 induces B lymphocytes to switch form IgM to IgG1 and IgE, increases the level of Ia molecules on the surface of resting B lymphocytes, supports the proliferation of some IL-2-dependent cell lines and mast cells, promotes the growth of early lymphoid precursors from bone marrow, and induces growth and differentiation of normal intrathymic T-cell precursors from fetal mice in vitro. These investigators found that at day 14 of gestation roughly 10% of fetal thymocytes express mRNA for IL-4 after stimulation by phorbol 12-myristate 13-acetate (PMA) and the calcium ionophore, ionomycin. The frequency of thymocytes expressing IL-4 mRNA decreased significantly by day 16 of gestation, with very few expressing cells by day 18. The authors interpreted these findings to suggest that the number of cells with such potential decreases as they differentiate in the thymus. They further suggest that IL-4 plays an essential role in the early stages of T-cell development in the thymus, and that it promotes the growth and differentiation of the T-cell progenitors.

H. Nerve Growth Factor

I. Introduction

Nerve growth factor (NGF) was the first "growth factor" to be isolated and characterized (COHEN et al. 1960). Ironically, however, it is not a "growth factor" in the classic sense of promoting mitogenesis. Rather, its primary function is to mediate differentiation and promote the survival of certain cells of the developing nervous system, as well as neural crest derivatives. Since NGF has primarily been studied in the context of the developing nervous system, and this topic is, therefore, reviewed extensively in Chap. 23, we will limit our review to recent studies that describe the expression of genes encoding NGF and its receptor during embryogenesis.

NGF is a dimer of two identical 118-amino-acid polypeptides. The latter are cleaved from a precursor of 307 amino acids (SCOTT et al. 1983 b; ULLRICH et al. 1983), which may also give rise to additional polypeptides of unknown function. NGF has been isolated both as the dimer (BOCCINI and ANGELETTI 1969) and as a high-molecular-weight complex consisting of two additional proteins, one of which may be involved in processing the NGF precursor. NGF binds to two types of receptors which display either high affinity and low capacity, or low affinity and high capacity (VALE and SHOOTER 1984). These are referred to as "slow" and "fast" receptors, respectively, based on the rates of dissociation. Pharmacological dose-response curves indicate that the slow receptors are responsible for receptor-mediated internalization, as well as long-term responses to NGF which include enhanced survival of neurons and promotion of neurite outgrowth. In contrast, the larger population of fast receptors mediates at least some of the more rapid effects of NGF including the stimulation of amino acid uptake (MISKO et al. 1987).

A large body of literature describes the action of NGF in the developing peripheral nervous system (PNS). The accumulated knowledge can be summarized as follows: (a) NGF is produced in limited amounts by peripheral target tissues of the sympathetic and sensory neurons; (b) the growth factor is secreted into the synaptic clefts where it binds to both classes of receptors, but most importantly to the high-affinity slow receptors; and (c) the growth factor is taken up into the nerve terminal by receptor-mediated internalization and is transported retrograde along axons to the neuronal cell bodies; the cascade of reactions that lead to neuronal survival and maintenance are initiated by signal transduction somewhere along this pathway.

The evidence supporting NGF action in the development of the PNS is irrefutable. Though still controversial, accumulating data suggest that NGF may also promote the development of the CNS. For example, cultures of dissociated neurons from the forebrain or septal region of the developing rat brain synthesize higher levels of choline acetyltransferase (ChAT) in response to NGF (HEFTI et al. 1985). However, neither the survivability nor degree of neurite extension of cholinergic neurons were affected, and anti-NGF antisera did not affect any of the measured parameters. On the other hand, MARTINEZ et al. (1985) showed that treatment of fetal rat striatal organotypic cultures with NGF produced a 5- to 12-fold increase in the specific activity of ChAT in a dose-dependent fashion, and this effect was specifically blocked by NGF antisera. Moreover, JOHNSTON et al. (1987) showed increased ChAT activity in the cell bodies and fibers of developing neurons of the septum, hippocampus, and caudate putamen of 2-day-old rats following intraventricular injection of NGF. The response appeared to be regionally and developmentally specific, thus providing additional evidence in support of the hypothesis that NGF acts as a trophic factor during CNS development. This hypothesis is further supported by recent studies that describe the cellular localization of NGF and its mRNA.

II. Localization of NGF and Its Receptor

FINN et al. (1987) have described the distribution of NGF immunoreactivity in the developing mouse nervous system. The intensity of staining was greatest at days 15 and 16 of gestation, with the antigen appearing to be widespread and present in high concentrations in both the PNS and CNS. The most intense immunostaining was found in cranial nerve tracts, as well as the hippocampus, developing white matter of the spinal cord, and tegmentum. Less intense staining was seen in the diencephalic regions, spinal cord gray matter, medullary fiber tracts, and cerebellum. Correspondingly, YAN and JOHNSON (1987) showed that NGF receptor was present in whole mouse embryo homogenates as early as day 10 of gestation. The receptor levels increased approximately threefold from day 11 through day 18, and then declined slightly immediately prior to term. Subsequently, NGF receptor content in sciatic nerve homogenates declined 23-fold from newborn to adulthood. These data raise the possibility that changes in the levels of NGF receptor are, in part, responsible for changes in tissue responsiveness during development. Using ^{125}I-NGF-binding autoradiography, RAIVICH et al. (1987) mapped the expression of NGF receptors during chicken cranial development. Intense autoradiographic labeling was observed in the classical target sites, the proximal cranial sensory ganglia, and the sympathetic superior cervical ganglia throughout development and after hatching. Of interest is the specific binding observed in the various brain regions during early brain development, as well as certain nonnervous tissues such as parts of the otic vesicles epithelium and skeletal muscle anlagen of the head. These various findings are consistent with a role for NGF in CNS development. In addition, given that human NGF receptor has been identified on cells derived from all three germ layers, they also suggest that NGF may also function in the differentiation of nonnervous tissues (THOMSON et al. 1988).

The availability of cloned genes encoding both NGF and its receptors has made possible analyses of the expression of these genes in developing tissues. For example, WHITTEMORE et al. (1986) showed that the levels of NGF mRNA in the developing rat brain increase until 3 or 4 weeks after birth, at which time they plateau. Analogous results were obtained by SELBY et al. (1987), who used an S1 nuclease assay to show that NGF transcripts increase in the cerebral cortex during fetal life and the immediate postnatal period, reach a peak at 20 days postbirth, and then stabilize at a slightly lower level. Although there are minor differences, roughly similar patterns of NGF mRNA expression are seen in the eye and heart. These results should be viewed in light of the fact that, although in the CNS and PNS the majority of ingrowing nerve terminals arrive at their target organs before birth, the connections continue to mature physiologically for weeks. This maturation of synaptic connections thus coincides with the increased level of expression of NGF mRNA. That innervation per se does not induce NGF expression was established in early deprivation studies in which whisker pads were removed (SELBY et al. 1987).

LARGE et al. (1986) compared the distribution of NGF protein and mRNA in newborn and postnatal rats. NGF protein was assayed by radioimmunoassay of tissue extracts, and mRNA levels by Northern analysis. They found that in adult rats the hippocampus and neocortex contained the highest concentrations of NGF mRNA, but relatively low levels of NGF protein. In contrast, the basal forebrain contained high levels of NGF protein but low levels of NGF transcript. These data are consistent with the retrograde transport of NGF protein to this region from the neocortex and hippocampus.

EBENDAL et al. (1986) examined the expression of NGF mRNA in various tissues of the adult chicken, as well as in the developing chick embryo. They observed high levels of expression in the heart and brain of 10-week-old roosters, and lower levels in the spleen, liver, and skeletal muscle. These findings are consistent with a correlation between NGF expression and the density of sympathetic innervation in peripheral organs. In the adult avian brain, NGF mRNA is found at higher concentrations in the optic tectum and cerebellum than in the cortex and hippocampus. The fact that this distribution differs from that described for adult rat brain may be explained, in part, by the diverging evolution of the avian and mammalian forebrain. Finally, in the later stages of development, NGF mRNA was expressed in the heart and brain, but at lower levels than in the adult.

Finally, BUCK et al. (1987) examined the expression of NGF receptor mRNA in the CNS and PNS of the developing rat. They detected transcripts in the sympathetic and sensory ganglia, as well as in the septum-basal forebrain of the neonatal rat. In postnatal life, the level of NGF receptor mRNA was markedly reduced in the sensory ganglia, but increased two- to fourfold in the basal forebrain and sympathetic ganglia. This suggests that NGF receptor expression is developmentally regulated in specific areas of the nervous system in a differential fashion. Immunocytochemical staining for the NGF receptor in developing nonmyelinated nerves from human fetuses was intense in the Schwann cells around 13–14 weeks of gestation, but declined beginning around 16–18 weeks. This change in staining intensity correlates with the onset of myelination. In the adult peripheral nerve, weak staining was confined to a few endoneural cells and neural-crest-derived perineurium. This apparent change in the expression of NGF receptor is consistent with similar studies of rat sciatic nerve (YAN and JOHNSON 1987), and with the findings of ZIMMERMAN and SUTTER (1983), who observed a decrease in NGF receptor expression in chick sensory ganglia Schwann cells following myelination.

I. Conclusions

Perhaps the most significant conclusion to be drawn from the preceding summary is that (with the conspicuous exception of EGF) many of the growth factors and their receptors are expressed early in development. Indeed, the recent studies of RAPPOLLEE and colleagues suggest that, in some cases, the transcripts that encode these molecules are present, and probably translated, in preimplantation blastocysts – as early as the two-cell stage. This raises the intriguing possibility that, prior to implantation and the formation of the placenta, the developing embryo is influenced not only by its own production of growth factors, but also by maternal growth factors that are present in the uterine fluids. In any event, these findings are consistent with the hypothesis that growth factors play vital roles in early mammalian development. Given the multifaceted nature of their actions, and the precedent that growth factor homologs act as inducing signals in amphibian development, it seems unlikely that these roles are limited to stimulating mitogenesis. However, precisely because of their diverse and sometimes bifunctional actions, and the fact that growth factors may act through autocrine, paracrine, and endocrine mechanisms, achieving an understanding of these roles will be a rather more difficult task than determining when and where they are expressed. Unraveling these functions will be the real challenge of the next decade.

Acknowledgements. We are especially grateful to Shelley Earp, Joe D'Ercole, Noreen Luetteke, Brad Olwin, Thienu Vu, Rusty Williams, and Dan Rappollee for helpful discussions, and for sharing unpublished data. We also thank Angie Boudwin and Pat McCallum for assistance in preparing the manuscript.

References

Abbott BD, Pratt RM (1987) Human embryonic palatal epithelial differentiation is altered by retinoic acid and epidermal growth factor in organ culture. J Craniofac Genet Dev Biol 7:241–265

Adams SO, Nissley SP, Handwerger S, Rechler MM (1983) Developmental patterns of insulin-like growth factor I and II synthesis and regulation in rat fibroblasts. Nature 302:150–153

Adamson ED, Meek J (1984) The ontogeny of epidermal growth factor receptors during mouse development. Dev Biol 103:62–70

Adamson ED, Deller MJ, Warshaw JB (1981) Functional EGF-receptors are present on mouse embryo tissues. Nature 291:656–659

Adashi EY, Resnick CE, Svoboda ME, Van Wyk JJ (1984) A novel role for somatomedin-C in the cytodifferentiation of the ovarian granulosa cell. Endocrinology 115:1227–1231

Artavanis-Tsakonas S, Muskavitch MAT, Yedvobnick B (1983) Molecular cloning of Notch, a locus affecting neurogenesis in *Drosophila melanogaster*. Proc Natl Acad Sci USA 80:1977–1981

Ashton IK, Vesey J (1978) Somatomedin activity in human cord plasma and relationship to birth size, insulin, growth hormone, and prolactin. Early Hum Dev 2:115–222

Ashton IK, Zapf J, Einschenk I, MacKenzie IZ (1988) Insulin-like growth factor (IGF) 1 and 2 in human foetal plasma and relationship to gestational age and foetal size during mid-pregnancy. Acta Endocrinol 110:558–563

Assoian RK, Komoriya A, Meyers CA, Miller DM, Sporn MB (1983) Transforming growth factor-beta in human platelets. J Biol Chem 258:7155–7160

Assoian RK, Frolik CA, Roberts AB, Miller DM, Sporn MB (1984) Transforming growth factor-beta controls receptor levels for epidermal growth factor in NRK fibroblasts. Cell 36:35–41

Athanassakis I, Bleackley RC, Peatkau V, Guilbert L, Barr PJ, Wegmann TG (1987) The immunostimulatory effect of T cell and T cell lymphokines on murine fetally derived placental cells. J Immunol 138:37–44

Atkinson PR, Weidman ER, Bhaumick B, Bala RM (1980) Release of somatomedin-like activity by cultured WI-38 human fibroblasts. Endocrinology 106:2006–2012

Azoulay M, Webb CG, Sachs L (1987) Control of hematopoietic cell growth regulators during mouse fetal development. Mol Cell Biol 7:3361–3364

Bargmann CI, Hung MC, Weinberg RA (1986) The neu oncogene encodes an epidermal growth factor receptor-related protein. Nature 319:226–230

Bartocci A, Pollard JW, Stanley ER (1986) Regulation of colony-stimulating factor I during pregnancy. J Exp Med 86:956–961

Bassett NS, Gluckman PD (1987) Insulin-like growth factors (IGFs) in experimental fetal growth retardation in the sheep. Abstracts Annual Meeting, Endocrine Society, Indianapolis, p 853

Beaulieu JF, Menard D, Calvert R (1985) Influence of epidermal growth factor on the maturation of the fetal mouse duodenum in organ culture. J Pediatr Gastoenterol Nutr 4:476–481

Beck F, Samani NJ, Penschow JD, Thorley B, Tregear GW, Coghlan JP (1987) Histochemical localization of IGF-I and -II mRNA in the developing rat embryo. Development 101:175–184

Beebe DC, Snellings K, Silver N, Van Wyk JJ (1986) Control of lens cell differentiation and ion fluxes by growth factors. Prog Clin Biol Res 217 A:365–369

Bell GI, Merryweather JP, Sanchez-Rescador R, Stempien MM, Priestley L, Scott J, Rall LB (1984) Sequence of a cDNA clone encoding human preproinsulin-like growth factor II. Nature 310:775–777

Bennett A, Wilson DM, Liu F, Nagashima R, Rosenfeld RG, Hintz RL (1983) Levels of insulin-like growth factors I and II in human cord blood. J Clin Endocrinol Metab 57:609–612

Bertsch S, Marks F (1974) Effect of foetal calf serum and epidermal growth factor on DNA synthesis in explants of chick embryo epidermis. Nature 251:517–519

Bhaumick B, Bala RM, Hollenberg MD (1981) Somatomedin receptor of human placenta: solubilization, photolabelling, partial purification and comparison with insulin receptor. Proc Natl Acad Sci USA 78:4278–7283

Bocchini V, Angeletti PU (1969) The nerve growth factor: purification as a 30000-molecular-weight protein. Proc Natl Acad Sci USA 64:787–794

Bradley TR, Stanley ER, Sumner MA (1971) Factors from mouse tissues stimulating colony growth of mouse bone marrow cells in vitro. Aust J Exp Biol Med Sci 49:595–603

Braun L, Mead JE, Panzica M, Mikumo R, Bell GI, Fausto N (1988) Transforming growth factor β mRNA increases during liver regeneration: a possible paracrine mechanism of growth regulation. Proc Natl Acad Sci USA 85:1539–1543

Brown AL, Graham DE, Nissley SP, Hill DJ, Strain AJ, Rechler MM (1986) Developmental regulation of insulin-like growth factor II mRNA in different rat tissues. J Biol Chem 261:13144–13150

Buck CR, Martinez HJ, Black IB, Chao MV (1987) Developmentally regulated expression of the nerve growth factor receptor gene in the periphery and brain. Proc Natl Acad Sci USA 84:3060–3063

Calvert R, Beaulieu JF, Menard D (1982) Epidermal growth factor (EGF) accelerates the maturation of fetal mouse intestinal mucosa in utero. Experientia 38:1096–1097

Campos-Ortega JA (1985) Genetics of early neurogenesis in *Drosophila melanogaster*. TINS 84:245–250

Carpenter G (1980) Genetics growth factor is a major growth promoting agent in human milk. Science 210:198–199

Casella SJ, Han VKM, Svoboda ME, D' Ercole AJ, Van Wyk JJ (1986) Insulin-like growth factor II binding to the type II somatomedin receptor: evidence for a separate high affinity binding site on type I receptors. J Biol Chem 261:9268–9272

Catterton WZ, Escobedo MB, Sexson WR, Gray ME, Sundell HW, Stahlman MT (1979) Effect of epidermal growth factor on lung maturation in fetal rabbits. Pediatr Res 13:104–108

Centrella M, Massagué J, Canalis E (1986) Human platelet-derived transforming growth factor-β stimulates parameters of bone growth in fetal rat calvariae. Endocrinology 119:2306–2312

Chegini N, Rao CV (1985) Epidermal growth factor binding to the human amnion, chorion, decidua, and placenta from mid- and term pregnancy: quantitative light microscopic autoradiographic studies. J Clin Endocrinol Metab 61:529–535

Cheifetz S, Weatherbee JA, Tsang MLS, Anderson JK, Mole JE, Lucas R, Massagué J (1987) The transforming growth factor-beta system, a complex pattern of cross-reactive ligands and receptors. Cell 48:409–415

Clegg CH, Linkhart TA, Olwin RB, Hauschka SD (1987) Growth factor control of skeletal muscle differentiation: commitment to terminal differentiation occurs in G1 phase and is repressed by fibroblast growth factor. J Cell Biol 105:949–956

Clemmons DR, Underwood Le, Van Wyk JJ (1981) Hormonal control of immunoreactive somatomedin production by cultured human fibroblasts. J Clin Invest 67:10–17

Coffey RJ, Bascom CC, Sipes NJ, Graves-Deal R, Weissman BE, Moses HL (1988) Selective inhibition of growth-related gene expression in murine keratinocytes by transforming growth factor β. Mol Cell Biol 8:3088–3093

Cohen S (1960) Purification of a nerve-growth promoting protein from the mouse salivary gland and its neurocytotoxic antiserum. Proc Natl Acad Sci USA 46:302–311

Cohen S (1962) Isolation of a mouse submaxillary gland protein accelerating incisor eruption and eyelid opening in the newborn animal. J Biol Chem 237:1555–1562

Cohen S (1965) The stimulation of epidermal proliferation by a specific protein (EGF). Dev Biol 12:394–407

Conover CA, Rosenfeld RG, Hintz RL (1986) Hormonal control of the replication of human fetal fibroblasts: role of somatomedin-C/insulin-like growth factor I. J Cell Physiol 128:47–54

Conover CA, Rosenfeld RG, Hintz RL (1987) Insulin-like growth factor II binding and action in human fetal fibroblasts. J Cell Physiol 133:560–566

Daughaday WH, Mariz IK, Trivedi B (1981) A preferential binding site for insulin-like growth factor II in human and rat placental membranes. J Clin Endocrinol Metab 53:282–288

Daughaday WH, Parker KA, Borowsky S, Trivedi B, Kapadia M (1982) Measurement of somatomedin-related peptides in fetal, neonatal, and maternal rat serum by IGF-

I radioimmunoassay, IGF II radioreceptor assay (RRA) and multiplication stimulating activity RRA after acid-ethanol extraction. Endocrinology 110:575–581
Daughaday WH, Yanow CE, Kapadia M (1986) Insulin-like growth factors I and II in maternal and fetal guinea pig serum. Endocrinology 119:490–494
Davoren JB, Hsueh AJW, Li CH (1985) Somatomedin-C augments FSH-induced differentiation of cultured rat granulosa cells. Am J Physiol 12:E 26
Dembinski AB, Johnson LR (1985) Effect of epidermal growth factor on the development of rat gastric mucosa. Endocrinology 116:90–94
DePablo F, Herrandez E, Collia F, Gomez JA (1985) Untoward effects of pharmacological doses of insulin in early chick embryos. Through which receptors are they mediated? Diabetologia 28:308–315
de Pagter-Holthuizen P, Jansen M, van Schaik FMA, van der Kammen R, Oosterwijk C, Van den Brande JL, Sussenbach JS (1987) The human insulin-like growth factor II gene contains two development-specific promoters. FEBS Lett 214:259–264
de Pagter-Holthuizen P, Jansen M, van der Kammen RA, van Schaik FMA, Sussenback JS (1988) Differential expression of the human insulin-like growth factor II gene: characterization of the IGF II mRNAs and an mRNA encoding a putative IGF-II associated protein. Biochim Biophys Acta 950:282–295
D'Ercole AJ, Underwood LE (1981) Growth factors in fetal growth and development. In: Novy MJ, Resko JA (eds) Fetal endocrinology: ORPC symposia on reproductive biology, vol 1. Academic, New York, pp 155–182
D'Ercole AJ, Applewhite GT, Underwood LE (1980a) Evidence that somatomedin is synthesized by multiple tissues in the fetus. Dev Biol 75:315–328
D'Ercole AJ, Wilson DF, Underwood LE (1980b) Changes in the circulating form of serum somatomedin-C during fetal life. J Clin Endocrinol Metab 51:674–676
D'Ercole AJ, Drop SLS, Kortleve DJ (1985) Somatomedin-C/insulin-like-growth-factor-I-binding proteins in human amniotic fluid and in fetal and postnatal blood: evidence of immunological homology. J Clin Endocrinol Metab 61:612–617
D'Ercole AJ, Hill DJ, Strain AJ, Underwood LE (1986) Tissue and plasma somatomedin-C/insulin-like growth factor I (Sm-C/IGF I) concentrations in the human fetus during the first half of gestation. Pediatr Res 20:253–255
Derynck R, Roberts AB, Winkler ME, Chen EY, Goeddel DV (1984) Human transforming growth factor-alpha: precursor structure and expression in *E. coli.* Cell 38:287–297
Derynck R, Jarrett JA, Chen EY, Eaton DH, Bell JR, Assoian RK, Roberts AB, Sporn MB, Goeddel DV (1985) Human transforming growth factor-β cDNA sequence and expression in human tumor cells. Nature 316:701–705
Dickson C, Peters G (1987) Potential oncogene product related to growth factors. Nature 326:833
Doe CQ, Goodman CS (1985a) Early events in insect neurogenesis. I. Development and segmental differences in the pattern of neuronal precursor cells. Dev Biol 3:193–205
Doe CQ, Goodman CS (1985b) Early events in insect neurogenesis. II. The role of cell interactions and cell lineage in the determination of neuronal precursor cells. Dev Biol 3:206–219
Doetschman TC, Havaranis AS, Herrmann H (1975) Insulin binding to cells of several tissues of the early chick embryo. Dev Biol 47:228–236
Dolling M, Thorburn GD, Young IR (1983) Effects of epidermal growth factor on the skin of the fetal lamb. J Anat 136:656
Donahoe PK, Ito Y, Marfatia S, Hendren WH (1976) The production of Mullerian inhibiting substance by the fetal, neonatal and adult rat. Biol Reprod 15:329–334
Drop SLS, Valiquette G, Guyda HJ, Corvol MT, Posner BI (1979) Partial purification and characterization of a binding protein for insulin-like activity (ILAs) in human amniotic fluid: a possible inhibitor of insulin-like activity. Acta Endocrinol 90: 505–518

Dull TJ, Gray A, Hayflick JS, Ullrich A (1984) Insulin-like growth factor II precursor gene organization in relation to the insulin gene family. Nature 310:777–781

Ebendal T, Larhammar D, Persson H (1986) Structure and expression of the chicken β nerve factor. EMBO J 5:1483–1487

Elgin RG, Busby WH Jr, Clemmons DR (1987) An insulin-like growth factor (IGF) binding protein enhances the biologic response to IGF I. Proc Natl Acad Sci USA 84:3254–3258

Ellingsworth LR, Brennan JE, Fok K, Rosen DM, Bentz H, Piez KA, Seyedin SM (1986) Antibodies to the N-terminal portion of cartilage-inducing factor A and transforming growth factor β. J Biol Chem 261:12362–12367

Ewton DZ, Florini JR (1981) Effects of the somatomedins and insulin on myoblast differentiation in vitro. Dev Biol 86:31–39

Ewton DZ, Spizz G, Olson EN, Florini JR (1988) Decrease in transforming growth factor-β binding and action during differentiation in muscle cells. J Biol Chem 263:4029–4032

Feldman E, Aures D, Grossman M (1978) Epidermal growth factor stimulates ornithine decarboxylase activity in the digestive tract of mouse. Proc Soc Exp Biol Med 159:400–402

Finn PJ, Ferguson IA, Wilson PA, Vahaviolos J, Rush RA (1987) Immunohistochemical evidence for the distribution of nerve growth factor in the embryonic mouse. J Neurocytol 16:639–647

Foley TD, DePhillip R, Perricelli A, Miller A (1980) Low somatomedin activity in cord serum from infants with intrauterine growth retardation. J Pediatr 96:603–610

Freemark M, Comer M (1987) Epidermal growth factor (EGF)-like transforming growth factor (TGF) activity and EGF receptors in ovine fetal tissues: possible role for TGF in ovine fetal development. Pediatr Res 22:609–615

Frolik CA, Dart LL, Meyers CA, Smith DM, Sporn MB (1983) Purification and initial characterization of a type-β transforming growth factor from human placenta. Proc Natl Acad Sci USA 80:3676–3680

Frunzio R, Chiariotti L, Brown AL, Graham DE, Rechler MM, Bruni CB (1986) Structure and expression of the rat insulin-like growth factor II (rIGF-II) gene. J Biol Chem 261:17138–17149

Gallo-Payet N, Pothier P, Hugon JS (1987) Ontogeny of EGF receptors during postnatal development of mouse small intestine. J Pediatr Res 6:114–120

Gentry LE, Webb NR, Lim GJ, Brunner AM, Ranchalis JE, Twardzik DR, Lioubin MN, Marquardt H, Purchio AF (1987) Type 1 transforming growth factor beta: amplified expression and secretion of mature and precursor polypeptides in Chinese hamster ovary cells. Mol Cell Biol 7:3418–3427

Gluckman PD, Brinsmead MW (1976) Somatomedin in cord blood: relationship to gestational age and birth size. J Clin Endocrinol Metab 43:1378–1381

Gluckman PD, Butler JH (1983) Parturition-related changes in insulin-like growth factors I and II in the perinatal lamb. J Endocrinol 99:223–232

Golander A, Hurley T, Barrett J, Handwerger S (1979) Synthesis of human prolactin by decidua in vitro. J Endocrinol 82:263–268

Goldin GV, Opperman LA (1980) Induction of supernumerary tracheal buds and the stimulation of DNA synthesis in the embryonic chick lung and trachea by epidermal growth factor. J Embryol Exp Morphol 60:235–243

Gomez-Pinilla F, Knauer DJ, Nieto-Sampedro M (1988) Epidermal growth factor receptor immunoreactivity in rat brain. Development and cellular localization. Brain Res 438:385–390

Gray A, Dull TJ, Ullrich A (1983) Nucleotide sequence of epidermal growth factor cDNA predicts a 128000-molecular weight protein precursor. Nature 303:722–725

Gray A, Tam AW, Dull TJ, Hayflick J, Pintar J, Cavenee WK, Koufos A, Ullrich A (1987) Tissue-specific and developmentally regulated transcription of the insulin-like growth factor 2 gene. DNA 6:283–295

Green MR, Phil D, Couchman JR (1984) Distribution of epidermal growth factor receptors in rat tissues during embryonic skin development, hair formation, and the adult hair growth cycle. J Invest Dermatol 83:118–123
Greenwald I (1985) lin-12, A nematode homeotic gene, is homologous to a set of mammalian proteins that includes epidermal growth factor. Cell 43:583–590
Gross I, Dynia DW, Rooney SA, Smart DA, Warshaw JB, Sissom JF, Hoath SB (1986) Influence of epidermal growth factor on fetal rat lung development in vitro. Pediatr Res 20:473–477
Grove RI, Pratt RM (1984) Influence of epidermal growth factor and cyclic AMP on growth and differentiation of palatal epithelial cells in culture. Dev Biol 106: 427–437
Han VKM, Hunter ES, Pratt RM, Zendegui JG, Lee DC (1987a) Expression of rat transforming growth factor alpha mRNA during development occurs predominantly in the maternal decidua. Mol Cell Biol 7:2335–2343
Han VKM, Hill DJ, Strain AJ, Towle AC, Lauder JM, Underwood LE, D'Ercole AJ (1987b) Identification of somatomedin-insulin-like growth factor immunoreactive cells in the human fetus. Pediatr Res 22:245–249
Han VKM, D'Ercole AJ, Lund PK (1987c) Cellular localization of somatomedin (insulin-like growth factor) messenger RNA in the human fetus. Science 236: 193–197
Han VKM, Lund PK, Lee DC, D'Ercole AJ (1988) Expression of somatomedin/-insulin-like growth factor messenger ribonucleic acids in the human fetus: identification, characterization and tissue distribution. J Clin Endocrinol Metab 66:422–429
Hartley DA, Xu T, Artavanis-Tsakonas S (1987) The embryonic expression of the Notch locus of *Drosophila melanogaster* and the implications of point mutations in the extracellular EGF-like domain of the predicted protein. EMBO J 6:3407–3417
Haselbacher GK, Andres RY, Humbel RE (1980) Evidence for the synthesis of a somatomedin similar to insulin-like growth factor I by chick embryo liver cells. Eur J Biochem 111:245–250
Heath JK, Isacke CM (1984) PC13 embryonal carcinoma-derived growth factor. EMBO J 3:2957–2962
Hefti F, Hartikka J, Eckentstein F, Gnahn H, Heumann R, Schwab M (1985) Nerve growth factor increases choline acetyltransferase but not survival or fiber outgrowth of cultured fetal septal cholinergic neurons. Neuroscience 14:55–61
Heine UI, Munoz EF, Flanders KC, Ellingsworth LR, Lam HYP, Thompson NL, Roberts AB, Sporn MB (1987) Role of transforming growth factor-β in the development of the mouse embryo. J Cell Biol 105:2861–2876
Hendricks SA, De Pablo F, Roth J (1984) Early development and tissue-specific patterns of insulin binding in chick embryo. Endocrinology 115:1315–1321
Hill DJ, Frazer A, Swenne I, Wirdnam PK, Milner RDG (1987) Somatomedin-C in human fetal pancreas. Cellular localization and release during organ culture. Diabetes 36:465–470
Hill DJ, Clemmons DR, Wilson S, Han VKM, Strain AJ, Milner RDG (1989) Immunological distribution of one form of insulin-like growth factor (IGF) binding protein and IGF peptide in human fetal tissues. J Mol Endocrinol 2:31–38
Honegger P, Guentert-Lauber B (1983) Epidermal growth factor (EGF) stimulation of cultured brain cells. I. Enhancement of the developmental increase in glial enzymatic activity. Dev Brain Res 11:245–252
Hortsch M, Schlessinger J, Gootwine E, Webb C (1983) Appearance of functional EGF-receptor kinase during rodent embryogenesis. EMBO J 2:1937–1941
Ignotz RA, Massagué J (1985) Type β transforming growth factor controls the adipogenic differentiation of 3T3 fibroblasts. Proc Natl Acad Sci USA 82:8530–8534
Ignotz RA, Massagué J (1986) Transforming growth factor-beta stimulates the expression of fibronectin and collagen and their incorporation into the extracellular matrix. J Biol Chem 261:4337–4345

Ignotz RA, Endo T, Massagué J (1987) Regulation of fibronectin and type 1 collagen mRNA levels by transforming growth factor-beta. J Biol Chem 262:6443–6446

Jakobovits A, Shackleford GM, Varmus HE, Martin GR (1986) Two proto-oncogenes implicated in mammary carcinogenesis, int-1 and int-2, are independently regulated during mouse development. Proc Natl Acad Sci USA 83:7806–7810

Jansen M, van Schaik FMA, Ricker AT, Bullock B, Woods DE, Gabbay KH, Nussbaum AL, Sussenbach JS, Van den Brande JL (1983) Sequence of cDNA encoding human insulin-like growth factor I precursor. Nature 306:609–611

Jansen M, van Schaik FMA, van Tol H, Van den Brande JL, Sussenbach JS (1985) Nucleotide sequences of cDNAs encoding precursors of human insulin-like growth factor II (IGF-II) and an IGF-II variant. FEBS Lett 179:243–246

Johnson LR, Guthrie PD (1980) Stimulation of rat oxyntic gland mucosal gland mucosal growth by epidermal growth factor. Am Soc Physiol 238:45

Johnson D, Lanahan A, Buck CR, Sengel A, Morgan C, Mercer E, Bothwell M, Chao M (1986) Expression and structure of the human NGF receptor. Cell 47:545–554

Johnston MV, Rutkowski JL, Wainer B, Long JB, Mobley WC (1987) NGF effects on developing forebrain cholinergic neurons are regionally specific. Neurochem Res 12:985–994

Kasselberg AG, Orth DN, Gray ME, Stahlman MT (1985) Immunocytochemical localization of human epidermal growth factor/urogastrone in several human tissues. J Histochem Cytochem 33:315–322

Kearns M, Lala PK (1982) Bone marrow origin of decidual cell precursors in the pseudopregnant mouse uterus. J Exp Med 155:1537–1554

Keller GH, Ladda RL (1981) Correlation between phosphatidylcholine labelling and hormone receptor levels in alveolar type II epithelial cells: effects of dexamethasone and epidermal growth factor. Arch Biochem Biophys 211:321–326

Kiess W, Greenstein LA, White RM, Lee L, Rechler MM, Nissley SP (1987) Type II insulin-like growth factor receptor is present in rat serum. Proc Natl Acad Sci USA 84:7720–7724

Kimelman D, Kirschner M (1987) Synergistic induction of mesoderm by FGF and TGFβ and the identification of an mRNA coding for FGF in the early *Xenopus* embryo. Cell 51:869–877

Kokai Y, Cohen JA, Drebin JA, Greene MI (1987) Stage- and tissue-specific expression of the neu oncogene in rat development. Proc Natl Acad Sci USA 84:8498–8501

Laiho M, Saksela O, Andreasen PA, Keski-Oja J (1986) Enhanced production and extracellular deposition of the endothelial-type plasminogen activator inhibitor in cultured human lung fibroblasts by transforming growth factor-beta. J Cell Biol 103:2403–2410

Large T, Bodary S, Clegg D, Weskamp G, Otten U, Reichardt L (1986) Nerve growth factor gene expression in the developing rat brain. Science 234:352–355

Le Bouc Y, Dreyer D, Jaeger F, Binoux M, Sondermeyer P (1986) Complete characterization of the human IGF-I nucleotide sequence isolated from a newly constructed adult liver cDNA library. FEBS Lett 196:108–112

Lee DC, Rose TM, Webb NR, Todaro GJ (1985a) Cloning and sequence analysis of a cDNA for rat transforming growth factor-alpha. Nature 313:489–491

Lee DC, Rochford R, Todaro GJ, Villarreal LP (1985b) Developmental expression of rat transforming growth factor-alpha mRNA. Mol Cell Biol 5:3644–3646

Leheup BP, Gray ME, Stahlman MT, LeQuire VS (1983) Synergistic effect of epidermal growth factor and retinoic acid on lung phospholipid synthesis. Pediatr Res 17:381A

Lenoir D, Honegger P (1983) Insulin-like growth factor I (IGF I) stimulates DNA synthesis in fetal rat brain cultures. Dev Brain Res 7:205–213

Leung YK, Elitsur Y, Lee PC, Lebenthal E (1986) Binding of epidermal growth factor (EGF) to pancreatic acini of fetal and postnatal rats and possible transfer of EGF through the placenta. Gastroenterology 90:1518

Leutz A, Schachner M (1981) Epidermal growth factor stimulates DNA-synthesis of astrocytes in primary cerebellar cultures. Cell Tissue Res 220:393–404

Lev Z, Shilo BZ, Kimchie Z (1985) Developmental changes in expression of the *Drosophila melanogaster* epidermal growth factor receptor gene. Dev Biol 110:499–502

Lipton SA, Wagner JA, Madison RD, D'Amore PA (1988) Acidic fibroblast growth factor enhances regeneration of processes by postnatal mammalian retinal ganglion cells in culture. Proc Natl Acad Sci USA 85:2388–2392

Livneh E, Glazer L, Segal D, Schlessinger J, Shilo BZ (1985) The *Drosophila* EGF receptor gene homolog: conservation of both hormone binding and kinase domains. Cell 40:599–607

Lund PK, Moat-Staats BM, Hynes MA, Simmons JG, Jansen M, D'Ercole AJ, Van Wyk JJ (1986) Somatomedin-C/insulin-like growth factor-I and insulin-like growth factor II mRNAs in fetal rat and adult tissues. J Biol Chem 261:14539–14544

Madri JA, Pratt BM, Tucker AM (1988) Phenotypic modulation of endothelial cells by transforming growth factor-β depends on the composition and organization of the extracellular matrix. J Cell Biol 106:1375–1384

Majesky MW, Benditt EP, Schwartz SM (1988) Expression and developmental control of platelet-derived growth factor A-chain and B-chain/sis genes in rat aortic smooth muscle cells. Proc Natl Acad Sci USA 85:1524–1528

Malo C, Menard D (1982) Influence of epidermal growth factor on the development of suckling mouse intestinal mucosa. Gastroenterology 83:28–35

Martinez HJ, Dreyfuss CF, Jonakait GM, Black IB (1985) Nerve growth factor promotes cholinergic development in brain striatal cultures. Proc Natl Acad Sci USA 82:7777–7781

Massague J, Czech M (1982) The subunit structure of two distinct receptors for insulin-like growth factors I and II and their relationship to insulin receptor. J Biol Chem 257:5038–5045

Massague J, Cheifetz S, Endo T, Nadal-Ginard B (1986) Type β transforming growth factor is an inhibitor of myogenic differentiation. Proc Natl Acad Sci USA 83:8206–8210

Matrisian LM, Pathak M, Magun B (1982) Identification of an epidermal growth factor-related transforming growth factor from rat fetuses. Biochem Biophys Res Commun 107:761–769

Mattson BA, Rosenblum IY, Smith RM, Heyner S (1988) Autoradiographic evidence for insulin and insulin-like growth factor binding to early mouse embryos. Diabetes 37:585–589

Menard D, Pothier P, Gallo-Payet N (1987) Epidermal growth factor receptors during postnatal development of the mouse colon. Endocrinology 121:1548–1554

Menard D, Arsenault P, Pothier P (1988) Biologic effects of epidermal growth factor in human fetal jejunum. Gastroenterology 94:656–663

Mercola M, Melton DA, Stiles CD (1988) Platelet-derived growth factor A chain is maternally encoded by *Xenopus* embryos. Science 241:1223–1225

Misko TP, Radeke MJ, Shooter EM (1987) Nerve growth factor in neuronal development and maintenance. J Exp Biol 132:177–190

Moses AC, Nissley SP, Short PA, Rechler MM, White RM, Knight AB, Higa OZ (1980) Elevated levels of multiplication-stimulating activity, an insulin-like growth factor, in fetal rat serum. Proc Natl Acad Sci USA 77:3649–3653

Muller R, Slamon DJ, Adamson ED, Tremblay JM, Muller D, Cline MJ, Verma IM (1983) Transcription of c-onc genes c-ras^{ki} and c-fms during mouse development. Mol Cell Biol 3:1062–1069

Murdoch GH, Potter E, Nicolaisen AK, Evans RM, Rosenfeld MG (1982) Epidermal growth factor rapidly stimulates prolactin gene transcription. Nature 300:192–194

Murphy LJ, Bell GI, Friesen HG (1987) Tissue distribution of insulin-like growth factor I and II messenger ribonucleic acid in the adult rat. Endocrinology 120:1279–1282

Nagarajan L, Anderson WB, Nissley SP, Rechler MM, Jetten AM (1985) Production of insulin-like growth factor II (MSA) by endoderm-like cells derived from embryonal carcinoma cells: possible mediator of embryonal cell growth. J Cell Physiol 124:199–206

Nexo E, Hollenberg MD, Figueora A, Pratt RM (1980) Detection of epidermal growth factor-urogastrone and its receptor during fetal mouse development. Proc Natl Acad Sci USA 77:2782–2785
Nissley SP, Rechler MM (1984a) Insulin-like growth factors: biosynthesis, receptors, and carrier proteins. In: Li CH (ed) Hormonal proteins and peptides, vol 12. Academic, New York, pp 127–203
Nissley SP, Rechler MM (1984b) Somatomedin/insulin-like growth factor tissue receptors. Clin Endocrinol Metab 13:43–67
Ohta M, Greenberger JS, Anklesaria P, Bassols A, Massagué J (1987) Two forms of transforming growth factor-β distinguished by multipotential haematopoietic progenitor cells. Nature 329:539–541
O'Keefe E, Hollenberg MD, Cuatrecasas P (1974) Epidermal growth factor characteristics of specific binding in membranes from liver, placenta, and other target tissues. Arch Biochem Biophys 164:518–526
O'Loughlin EV, Chung M, Hollenberg M, Hayden J, Zahavi I, Gall DG (1985) Effect of epidermal growth factor on ontogeny of the gastrointestinal tract. Am J Physiol 249:G 674–678
Olson EN, Sternberg E, Hu JS, Spizz G, Wilcox C (1986) Regulation of myogenic differentiation by type β transforming growth factor. J Cell Biol 103:1799–1805
Olwin BB, Hauschka SD (1988) Cell surface fibroblast growth factor and epidermal growth factor receptors are permanently lost during skeletal muscle terminal differentiation in culture. J Cell Biol 107:761–769
Padgett RW, St. Johnston RD, Gelbart WM (1987) A transcript from a *Drosophila* pattern gene predicts a protein homologous to the transforming growth factor-β family. Nature 325:81–84
Partanen AM, Thesleff I (1987) Localization and quantitation of ^{125}I-epidermal growth factor binding on mouse embryonic tooth and other embryonic tissues at different developmental stages. Dev Biol 120:186–197
Petraglia F, Sawchenko P, Lim ATW, Rivier J, Vale W (1987) Localization, secretion and action of inhibin in human placenta. Science 237:187–189
Pollard JW, Bartocci A, Arceci R, Orlofsky A, Ladner MB, Stanley ER (1987) Apparent role of the macrophage growth factor, CSF-1, in placental development. Nature 330:484–486
Popliker M, Shatz A, Avivi A, Ullrich A, Schlessinger J, Webb CG (1987) Onset of endogenous synthesis of epidermal growth factor in neonatal mice. Dev Biol 119:38–44
Pothier P, Menard D (1988) Presence and characteristics of epidermal growth factor receptors in human fetal small intestine and colon. FEBS Lett 228:113–117
Probstmeier R, Schachner M (1986) Epidermal growth factor is not detectable in developing and adult rodent brain by a sensitive double-site enzyme immunoassay. Neurosci Lett 63:290–294
Proper JA, Bjornson CL, Moses HL (1982) Mouse embryos contain polypeptide growth factor(s) capable of inducing a reversible neoplastic phenotype in nontransformed cells in culture. J Cell Physiol 110:169–174
Raivich G, Zimmerman A, Sutter A (1987) Nerve growth factor (NGF) receptor expression in chicken cranial development. J Comp Neurol 256:229–245
Rall LB, Scott J, Bell GI, Crawford RJ, Penschow JD, Niall HD, Coghlan JP (1985) Mouse prepro-epidermal growth factor synthesis by the kidney and other tissues. Nature 313:228–231
Rao CV, Carman FR, Chegini N, Schultz GS (1984) Binding sites for epidermal growth factor in human fetal membranes. J Clin Endocrinol Metab 58:1034–1042
Rappolee DA, Brenner CA, Schultz R, Mark D, Werb Z (1988) Developmental expression of PDGF, TGF-α, and TGF-β genes in preimplantation mouse embryos. Nature 241:1823–1825
Raulet DH (1985) Expression and function of interleukin-2 receptors on immature thymocytes. Nature 314:101–103

Rebagliati MR, Weeks DL, Harvey RP, Melton DA (1985) Identification and cloning of localized maternal RNAs from *Xenopus* eggs. Cell 42:769–777
Rettenmier CW, Sacca R, Furman WL, Roussel MF, Holt JT, Nienhuis AW, Stanley ER, Scherr CJ (1986) Expression of the human c-fms proto-oncogene product (colony-stimulating factor-1 receptor) on peripheral blood mononuclear cells and choriocarcinoma cell lines. J Clin Invest 77:1740–1746
Richman RA, Beredict MR, Clorini JR, Toly BA (1985) Hormonal regulation of somatomedin secretion by fetal rat hepatocytes in primary culture. Endocrinology 116:180–185
Rizzino A (1987) Appearance of high affinity receptors for type β transforming growth factor during differentiation of murine embryonal carcinoma cells. Cancer Res 47:4386–4390
Rizzino A, Bowen-Pope DF (1985) Production of PDGF-like growth factors by embryonal carcinoma cells and binding of PDGF to their endoderm-like differentiated cells. Dev Biol 110:15–22
Rizzino A, Kuszynski C, Ruff E, Tiesman J (1988) Production and utilization of growth factors related to fibroblast growth factor by embryonal carcinoma cells and their differentiated cells. Dev Biol 129:61–71
Roberts AB, Sporn MB, Assoian RK, Smith JM, Roche NS, Wakefield LM, Heine UI, Liotta LA, Falanga V, Kehrl JH, Fauci AS (1986) Transforming growth factor type-beta: rapid induction of fibrolysis and angiogenesis in vivo and stimulation of collagen formation in vitro. Proc Natl Acad Sci USA 83:4167–4171
Robey PG, Young MF, Flanders KC, Roche NS, Kondaiah P, Reddi AH, Termine JD, Sporn MB, Roberts AB (1987) Osteoblasts synthesize and respond to TGF-beta in vitro. J Cell Biol 105:457–463
Rosa F, Roberts AB, Danielpour D, Dart LL, Sporn MB, Dawid IB (1988) Mesoderm induction in amphibians: the role of TGF-β2-like factors. Science 239:783–785
Roth RA (1988) Structure of the receptor for insulin-like growth factor II: the puzzle amplified. Science 239:1269–1271
Rotwein P, Pollock KM, Didier DK, Drivi GG (1987a) Organization and sequence of the human insulin-like growth factor I gene. Alternative RNA processing produces two insulin-like growth factor precursor peptides. J Biol Chem 261:4828–4832
Rotwein P, Pollock KM, Watson M, Milbrandt JD (1987b) Insulin-like growth factor gene expression during rat embryonic development. Endocrinology 121:2141–2144
Schejter ED, Segal D, Glazer L, Shilo BZ (1986) Alternative 5' exons and tissue-specific expression of the *Drosophila* EGF receptor homolog transcripts. Cell 46:1091–1101
Schweigerer L, Neufeld G, Friedman J, Abraham JA, Fiddes JC, Gospodarowicz D (1987) Capillary endothelial cells express basic fibroblast growth factor, a mitogen that promotes their own growth. Nature 325:257–259
Scott J, Urdea M, Quiroga M, Sanchez-Pescador R, Fong N, Selby M, Rutter WJ, Bell GI (1983a) Structure of a mouse submaxillary messenger RNA encoding epidermal growth factor and seven related proteins. Science 221:236–240
Scott J, Selby M, Urdea M, Quiroga M, Bell GI, Rutter WJ (1983b) Isolation and nucleotide sequence of a cDNA encoding the precursor of mouse nerve growth factor. Nature 302:538–540
Scott J, Cowell J, Robertson ME, Priestley LM, Wadley R, Hopkins B, Pritchard J, Bell GI, Rall LB, Graham CF, Knott TJ (1985) Insulin-like growth factor II gene expression in Wilm's tumour and embryonic tissues. Nature 317:260–264
Seed J, Olwin BB, Hauschka SD (1988) Fibroblast growth factor levels in the whole embryo and limb bud during chick development. Dev Biol 128:50–57
Seifert RA, Schwartz SM, Bowen-Pope DF (1984) Developmentally regulated production of platelet-derived growth factor-like molecules. Nature 311:669–671
Selby MJ, Edwards R, Sharp F, Rutter WJ (1987) Mouse nerve growth factor gene: structure and expression. Mol Cell Biol 7:3057–3064
Seyedin SM, Thomas TC, Thompson AY, Rosen DM, Piez KA (1985) Purification and characterization of two cartilage-inducing factors from bovine demineralized bone. Proc Natl Acad Sci USA 82:2267–2271

Sideras P, Funa K, Zalcberg-Quintana I, Xanthopoulos KG, Kisielow P, Palacios R (1988) Analysis by in situ hybridization of cells expressing mRNA for interleukin 4 in the developing thymus and in peripheral lymphocytes from mice. Proc Natl Acad Sci USA 85:218–221

Silver MH, Murray JC, Pratt RM (1984) Epidermal growth factor stimulates type-V collagen synthesis in cultured murine palatal shelves. Differentiation 27:205–208

Slack JMW, Darlington BG, Heath JK, Godsave SF (1987) Mesoderm induction in early *Xenopus* embryos by heparin-binding growth factors. Nature 326:197–200

Smith JM, Sporn MB, Roberts AB, Derynck R, Winkler ME, Gregory H (1985) Human transforming growth factor-α causes precocious eyelid opening in newborn mice. Nature 315:515–516

Smith EP, Sadler TW, D'Ercole (1987) Somatomedins/insulin-like growth factors, their receptors and binding proteins are present during mouse embryogenesis. Development 101:73–82

Soares MB, Ishii DN, Efstratiadis A (1985) Developmental and tissue specific expression of a family of transcripts related to rat insulin-like growth factor II mRNA. Nucleic Acids Res 13:1119–1134

Sporn MB, Todaro GJ (1980) Autocrine secretion and malignant transformation of cells. N Engl J Med 303:878–880

Stern IB, Dayton L, Duecy J (1971) The uptake of tritiated thymidine in the dorsal epidermis of the fetal and newborn rat. Anat Rec 170:225–234

Strain AJ, Hill DJ, Swenne I, Milner RDG (1987) Stimulation of DNA synthesis in isolated human foetal hepatocytes by insulin-like growth factor I, placental lactogen and growth hormone. J Cell Physiol 132:33–40

Stylianopoulou F, Efstratiadis A, Herbert J, Pintar J (1988) Pattern of the insulin-like growth factor II gene expression during rat embryogenesis. Development 103:497–506

Sundell HW, Gray ME, Serenius FS, Escobedo MB, Stahlman MT (1980) Effects of epidermal growth factor on lung maturation in fetal lambs. Am J Pathol 100:707–726

Tam JP (1985) Physiological effects of transforming growth factor in the newborn mouse. Science 229:673–675

Thomson TM, Rettig WJ, Chesa PG, Green SH, Mena AC, Old LJ (1988) Expression of human nerve growth factor receptor on cells derived from all three germ layers. Exp Cell Res 174:533–539

Thornburg W, Matrisian L, Magun B, Koldovsky D (1984) Gastrointestinal absorption of epidermal growth factor in suckling rats. Am J Physiol 246:G 80–85

Twardzik DR (1985) Differential expression of transforming growth factor-alpha during prenatal development of the mouse. Cancer Res 45:5413–5416

Twardzik DR, Ranchalis JE, Todaro GJ (1982a) Mouse embryonic transforming growth factors related to those isolated from tumor cells. Cancer Res 42:590–593

Twardzik DR, Sherwin SA, Ranchalis JE, Todaro GJ (1982b) Transforming growth factors in the urine of normal, pregnant and tumor-bearing humans. JNCI 69:793–798

Ullrich A, Gray A, Berman C, Dull JJ (1983) Human β-nerve growth factor gene sequence highly homologous to that of mouse. Nature 303:821–825

Unsicker K, Reichert-Preibsch H, Schmidt R, Pettmann B, Labourdette G, Sensenbrenner M (1987) Astroglial and fibroblast growth factors have neurotrophic functions for cultured peripheral and central nervous system neurons. Proc Natl Acad Sci USA 84:5459–5463

Vale RD, Shooter EM (1984) Assaying binding of nerve growth factor to cell surface receptors. Methods Enzymol 109:21–39

Vigier B, Picard JY, Tran D, Legeai L, Josso N (1984) Use of monoclonal antibody against bovine anti-Mullerian hormone. J Reprod Fertil 69:207–214

Vileisis RA, D'Ercole AJ (1986) Tissue and serum somatomedin-C/insulin-like growth factor I in fetal rats made growth retarded by uterine artery ligation. Pediatr Res 20:126–130

Vu TH, Martin GR, Lee P, Mark D, Wang A, Williams LT (1989) Developmentally regulated use of alternative promoters creates a novel PDGF receptor transcript in mouse teratocarcinoma and embryonic stem cells. Mol Cell Biol (in press)
Wadsworth SC, Vincent WS, BilodeWadsworth SC, Vincent WS, Bilodeau-Wentworth D (1985) A *Drosophila* genomic sequence with homology to human epidermal growth factor receptor. Nature 314:178–180
Weeks DL, Melton DA (1987) A maternal mRNA localized to the vegetal hemisphere in *Xenopus* eggs codes for a growth factor related to TGFβ. Cell 51:861–867
Wharton KA, Johansen KM, Xu T, Artavanis-Tsakonas S (1985) Nucleotide sequence from the neurogenic locus *Notch* implies a gene product that shares homology with proteins containing EGF-like repeats. Cell 43:567–581
Whitfield HJ, Bruni CB, Frunzio R, Terrell JE, Nissley SP, Rechler MM (1984) Isolation of a cDNA clone encoding rat insulin-like growth factor II precursor. Nature 312:277–280
Whittemore SR, Ebendal T, Larkfors L, Olson L, Seiger A, Stromberg I, Perrson H (1986) Developmental and regional expression of beta NGF mRNA and (NGF) receptor in rats. Dev Biol 121:139–148
Wilcox JN, Derynck R (1988) Developmental expression of transforming growth factors alpha and beta in mouse fetus. Mol Cell Biol 8:3415–3422
Wilkinson DG, Peters G, Dickson C, McMahon AP (1988) Expression of the FGF-related proto-oncogene int-2 during gastrulation and neurulation in the mouse. EMBO J 7:691–695
Yan Q, Johnson EM Jr (1987) A quantitative study of the developmental expression of nerve growth factor (NGF) receptor in rats. Dev Biol 121:139–148
Yoneda T, Pratt RM (1981) Mesenchymal cells from the human embryonic palate are highly responsive to epidermal growth factor. Science 213:563–565
Zapf J, Walters H, Froesch EWR (1981) Radioimmunological determination of insulin-like growth factors I and II in normal subjects and in patients with growth disorders and extrapancreatic tumor hypoglycemia. J Clin Invest 68:1321–1330
Zimmerman A, Sutter A (1983) Nerve growth factor receptors on glial cells. Cell-cell interaction between neurons and Schwann cells in cultures of chick sensory ganglia. EMBO J 2:879–885

CHAPTER 40

Relationships Between Oncogenes and Growth Control

A. LEUTZ and T. GRAF

A. Introduction

In recent years we have witnessed a revolution in the understanding of the mechanisms which regulate mammalian cell growth. This applies both to normal cells, in which growth is tightly controlled, as well as to cancer cells, which divide in an uncontrolled fashion. With the advent of molecular biology a number of genes have been identified whose products are involved in regulating normal cell growth. In parallel research 40–60 genes that are capable of inducing a transformed phenotype and that have been termed "oncogenes" have been identified. It is now clear that growth control genes on the one hand and oncogenes on the other are largely one and the same. Although unequivocal evidence for this notion was obtained only recently, the concept itself emerged gradually beginning with the discovery of viral oncogenes (v-*onc* genes) and the fact that these genes represent cell-derived sequences (termed protooncogenes, cellular oncogenes, or c-*onc* genes).

Since the discovery of oncogenes about 13 years ago this field has undergone a dramatic explosion covered by more than 4500 publication on oncogenes and 12000 on growth factors, numerous reviews, and several books. It is therefore impossible to cover the whole topic in a relatively short article. Nevertheless, we have attempted to review in this article the relationships that exist between growth factors, their receptors, and other proteins involved in growth control on the one hand, and transforming proteins encoded by oncogenes on the other. These relationships, schematically summarized in Fig. 1, also provide the framework for the sections that follow, where we review the roles of individual oncogenes acting at the levels of growth factors, signal transducers, and nuclear proteins. In particular, we will discuss their origin, mode of oncogenic activation, relationship to growth control, and involvement in human neoplasia. Since much of the more exciting recent progress has been made in the field of nuclear oncogenes, they are discussed in slightly more detail than, e.g., growth factor receptor related oncogenes, most of which have also been reviewed elsewhere in this book. The review then ends with a discussion of how oncogenes can cooperate to induce neoplastic growth, and, in particular, leukemia.

The relationship of individual oncogenes to growth control is fully established for only a few but others can still be usefully grouped based on sequence similarities as well as biochemical properties and subcellular localiza-

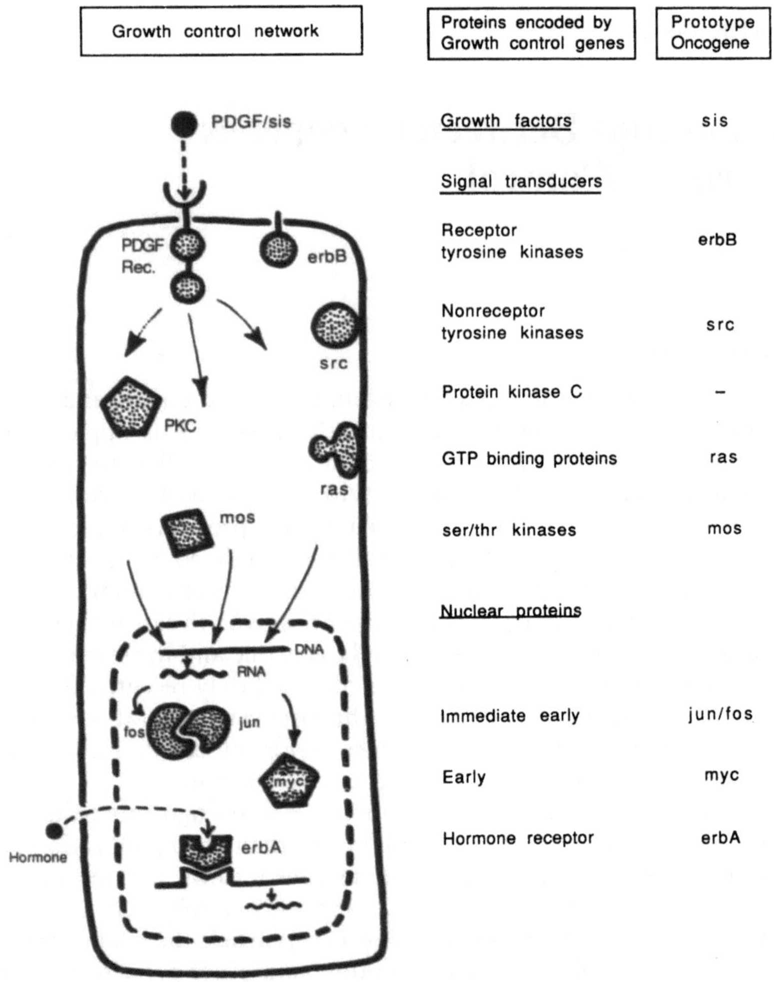

Fig. 1. Growth control network and levels at which oncogenes can act. The scheme illustrates a cell which is stimulated by a growth factor (with platelet-derived growth factor as an example) and proteins which act at different levels of the growth control network. In the *right column* and in the *drawing* examples of oncogenes whose products correspond to growth factors, signal transducers, and nuclear proteins are indicated

tion of the products. It should be stressed, however, that there are several oncogenes which cannot easily be linked to growth control, such as the *mas* oncogene, which is related to the angiotensin receptor gene (Jackson et al. 1988), or the v-*crk* oncogene, whose predicted product is distantly related to phospholipase C (Mayer et al. 1988). There are also several oncogenes, such as v-*ski* and v-*rel*, whose functions are obscure and cannot be placed into the scheme outlined in Fig. 1. They will not be further discussed in this review. The interested reader can find a comprehensive description of most known oncogenes in a recently published book (Reddy et al. 1988).

It should be emphasized that this review is strictly organized on the basis of the growth control "cascade" or "network." This concept implies that binding of a growth factor to its receptor has pleiotropic effects which result in the activation of second messengers as well as in posttranslational modifications such as phosphorylation of various proteins involved in signal transduction. Ultimately the signals must reach the nucleus, where they activate transcriptional regulation and cell division. For simplicity we have represented this sequel of events as a "cascade," but it is well established that proteins involved in signal transduction can interact with others which are at the same level or even further upstream, resulting in a growth control "network." A further complication is that growth stimulation may result in the activation of genes, such as those encoding growth factors themselves, which could act back on the same cell and which might initiate another cycle of growth stimulation. Finally we would like to point out that the action of oncogenes within the framework of the growth control network implies that their transforming effects ultimately result in transcriptional regulation of other genes. It is, known however, that certain oncogene proteins acting on the cell surface, such as the v-*src* protein, have profound effects on cell morphology. This has been shown with fibroblasts transformed with a temperature-sensitive mutant of Rous sarcoma virus whose morphology can be rapidly changed by simple temperature shifts even if the cells are enucleated (Beug et al., 1978). The relative contribution of these direct transforming effects of oncogene proteins to the cell's tumorigenicity is not known.

B. Growth Factor Genes

Based on their biological properties and on their sequence, more than 30 different growth factors have been described to date, most of which are covered by articles in this book. A number of these factors, if overexpressed or aberrantly expressed, have turned out to have oncogenic potential. In addition, several oncogene proteins were found to be related to growth factors. They are the subject of the following section, summarized in Tables 1 and 2.

Table 1. Oncogenes related to growth factor genes

Oncogene	Source	Mode of activation	Relationship to growth factor
v-sis	Simian retrovirus	Retroviral transduction	Corresponds to PDGF B-chain gene
int-1	Murine mammary tumors	Promoter insertion	Not established; homology to *Drosophila* wingless gene
int-2	Murine mammary tumors	Promoter insertion	Similarity to b-FGF gene
hst/K-FGF	Human tumors	Rearrangement/ amplification	Similarity to b-FGF gene

Table 2. Growth factor genes experimentally turned into oncogenes

Growth factor gene	Experimental manipulation of gene	In vitro properties of cells expressing the growth factor	In vivo properties of cells expressing the growth factor
PDGF-B	Linkage to retroviral promoter	Transformation of NIH 3T3 cells	ND
FGF-5	Recombination with retroviral promoter	Transformation of NIH 3T3 cells	ND
b-FGF	Linkage to retroviral promoter and fusion to secretory signal sequence	Transformation of NIH 3T3 cells	Tumors in nude mice
TGF-α	Construction of a TGF-α expressing retrovirus	Transformation of NRK cells	ND
	Linkage to SV40 promoter	Transformation of Rat-1 cells	Tumors in nude mice
GM-CSF	Construction of a GM-CSF-expressing retrovirus	Induction of factor independence of FDC-P1 cells	Tumors in syngenic mice
IL-3	Construction of an IL-3-expressing retrovirus	Induction of factor independence of 32D and FDC-P1 cells	Tumors in nude mice
IL-2	Construction of an IL-2-expressing retrovirus	Induction of factor independence of CTLL-2 cells	Tumors in nude and syngenic mice

ND, not determined.

I. Growth Factor-Type Oncogenes

1. The v-*sis* Oncogene

The v-*sis* gene was first discovered as the oncogene of the simian sarcoma virus (ROBBINS et al. 1982a, b) and later as that of the Parodi-Irgens feline sarcoma virus (BESMER et al. 1983). The sequence of its predicted product is very similar to that of the B-chain gene of platelet-derived growth factor (PDGF; DOOLITTLE et al. 1983; WATERFIELD et al. 1983; ROBBINS et al. 1983; JOHNSSON et al. 1984). The divergence between the two protein sequences (around 10%) is probably mostly due to species differences since v-*sis* was derived from a New World monkey while the PDGF sequence used for comparison was of human origin.

There is evidence which indicates that the products of v-*sis* and c-*sis* are functionally similar if not identical (GAZIT et al. 1984). Both proteins are mitogenic for fibroblasts, bind to PDGF receptors, and induce their phosphorylation on tyrosine (LEAL et al. 1985; JOHNSSON et al. 1985a). This suggests that mammalian fibroblasts transformed by v-*sis* grow in an autocrine fashion. The concept of autocrine growth stimulation, proposed first by SPORN and TODARO (1980) for transforming growth factors, postulates that a given growth factor binds to the cell's own receptors and thus elicits a mitogenic signal. In accordance with this notion are the findings that the v-*sis* gene product induces DNA synthesis in uninfected quiescent fibroblasts and that only cells that express PDGF receptors can be transformed by the oncogene (LEAL et al. 1985). Another prediction derived from the autocrine hypothesis is that it should be possible to block the self-stimulation of growth by the v-*sis* gene product using antibodies specific to PDGF/sis protein or by treatment with suramin. However, while some authors were able to interrupt the autocrine loop (JOHNSSON et al. 1985b; BETSHOLTZ et al. 1985) others could not or could do so only partially (HUANG et al. 1984; JOSEPHS et al. 1984; ROBBINS et al. 1985; HANNINK and DONOGHUE 1986). In addition, chronic exposure of normal fibroblasts to PDGF does not lead to cell transformation (ASSOIAN et al. 1984). An explanation for this apparent discrepancy might be provided by differences in the cell systems used: those authors who could interrupt the autocrine loop worked with primary human fibroblast cultures while those who failed worked with established rodent fibroblast cell lines such as NRK or NP1. Based on their failure to efficiently block the autocrine loop, HUANG et al. (1984) formulated the "internal autocrine hypothesis" where the $p28^{v\text{-}sis}$ protein is thought to activate the PDGF receptor during its processing in the endoplasmic reticulum and in the Golgi apparatus. The hypothesis has recently received experimental support from experiments performed by KEATING and WILLIAMS (1988). Their studies indicate that in NRK cells transformed by v-*sis* mature receptor molecules fail to accumulate because the receptor precursors rapidly enter the degradative pathway. In contrast, normal NRK cells express the mature form of the receptor which is stable and not degraded in lysosomes even after stimulation with PDGF. These data indicate that autocrine activation of the PDGF receptor may occur in intracellular compartments of certain cell types transformed by v-*sis*. v-*sis* is frequently expressed in human tumor cell lines but there is no clear evidence suggesting its involvement in human neoplasia (for a review see HELDIN et al. 1987).

2. The *int*-1 and *int*-2 Oncogenes

These two genes were originally identified as retroviral integration sites. Mammary tumors induced by mouse mammary tumor virus (MMTV) in C3H mice have viral integration sites clustered predominantly within 15 kb of the *int*-1 locus or within 30 kb of the *int*-2 locus (NUSSE and VARMUS 1982; PETERS et al. 1983). The observation that, in contrast to normal mammary tissues, these tumors express significant levels of the corresponding mRNAs (FUNG et al. 1985; DICKSON et al. 1984) suggested that *int*-1 and *int*-2 represent cellular oncogenes whose transcriptional activation (by promoter/enhancer insertion through the retroviral LTR sequences) contributes to oncogenesis. This hypothesis was directly confirmed by the demonstration that the *int*-1 gene can transform a mammary epithelial cell line (BROWN et al. 1986, RIJSEWIJK et al. 1987b). Sequencing of *int*-1 revealed the typical structure of a secreted or membrane-bound protein, possibly a growth factor containing a hydrophobic aminoterminus with potential signal peptide function, N-glycosylation sites, and a cysteine-rich carboxy-terminus (VAN OOYEN and NUSSE 1984; VAN OOYEN et al. 1985). In support of this notion is the finding that the *int*-1 gene product enters the secretory pathway and can also be found in low amounts in cell supernatants (PAPKOFF et al. 1987; BROWN et al. 1987). It now appears that most of the protein is attached to the extracellular matrix (A. BROWN, personal communication). Experiments performed with the *Drosophila int*-1 homolog, the *wingless* gene (RIJSEWIJK et al. 1987a), indicate that cells expressing this gene are involved in segmentation polarity control and influence the fate of neighboring cells, apparently via a diffusible factor (CABRERA et al. 1987; WIESCHAUS and RIGGLEMAN 1987).

Although direct proof of its capacity to transform cells and to encode a growth factor is still lacking, the *int*-2 gene was found to be related to basic fibroblast growth factor (FGF) based on amino acid sequence similarities (DICKSON and PETERS 1987). Studies on the distribution of expression of *int*-1 and *int*-2 in mice at different stages of development revealed that these genes are predominantly expressed in specific areas of the embryo (SHACKLEFORD and VARMUS 1987; WILKINSON et al. 1987, 1988), suggesting a role in development. These studies also indicate that the products of these genes act as morphogens rather than as mitogens and possibly have no role in normal mammary gland development (for review see NUSSE 1989). The detection of *int* gene activation in mammary tumors is, not restricted to mice since in a recent study the *int*-2 gene has been found to be amplified in about 15% of human breast tumors examined (LIDEREAU et al. 1988).

3. The *hst* Oncogene

The recently discovered human gene *hst* (for human stomach cancer) or K-FGF gene (for Kaposi-FGF) is one of the best examples of an oncogene corresponding to a growth factor gene. This gene, which is related to the FGF gene family, was discovered independently in different laboratories searching for new oncogenes in human stomach cancers (YOSHIDA et al. 1987; SAKAMOTO et al. 1988; KODA et al. 1987) or Kaposi sarcomas (DELLI-BOVI et al. 1987). Both the originally isolated oncogenes as well as their subsequently cloned cellular homologs induce transformation and growth factor independence when overexpressed in fibroblasts or endothelial cells (TAIRA et al. 1987, SAKAMOTO et al. 1986; DELLI-BOVI et al. 1988). This suggests that oncogenic activation of the *hst* gene does not involve alterations in its coding domain but that gene rearrangements or amplifications lead to its overexpression (YOSHIDA et al. 1988).

II. Growth Factor Genes Experimentally Shown to be Capable of Acting as Oncogenes

FGF-5 is a member of the fibroblast growth factor family, exhibiting a 40%–50% sequence homology to both basic and acidic FGF. The FGF-5 gene has been shown to transform NIH 3T3 cells (a mouse fibroblast cell line) following a recombinational event leading to its overexpression (ZHAN et al. 1988). Transformation of NIH 3T3 cells was also obtained with the basic

FGF gene itself but only when it was fused to a secretory signal sequence (ROGELJ et al. 1988). This suggests that entrance into the secretion pathway with subsequent binding of the receptor is a prerequisite for the capacity of basic FGF to act in an autocrine fashion. Another example of a growth factor gene with transforming properties is transforming growth factor-α (TGF-α). This gene, when placed under the control of the SV40 promoter, transforms Rat-1 cells and induces their tumorigenicity in nude mice (ROSENTHAL et al. 1986). Likewise, an artificially synthesized TGF-α gene (TAM et al. 1984) expressed from a retroviral vector is capable of transforming NRK cells (WATANABE et al. 1987). Similar results were obtained with a synthetic EGF gene (STERN et al. 1987). These observations on FGFs and TGF-α extend the conclusions already drawn from the v-*sis*/PDGF experiments: that growth factor genes, if overexpressed or ectopically expressed, can behave as oncogenes. Therefore, for this category of oncogenes the distinction from protooncogenes is only meaningful in terms of their respective operational definitions.

The experiments discussed so far concerning the transforming ability of growth factors were all performed in fibroblasts or endothelial cells. Transformation of these cells is usually scored as their ability to form foci of densely packed cells, to exhibit morphological alterations, or to develop colonies in agar. In contrast to these cell types, hematopoietic cells or cell lines generally die within hours or days after withdrawal of their cognate growth factor. Most hematopoietic cell lines also fail to form tumors in syngenic or nude mice. Such cell lines have been extensively used to study the effects of growth factor genes which had not previously been associated with oncogenesis. A well-known example is the IL-3/GM-CSF-dependent mouse myeloid cell line FDC-P1 which, following introduction of a retrovirus expressing the GM-CSF gene, becomes factor independent and tumorigenic (LANG et al. 1985). Interestingly, however, attempts to interrupt the presumptive autocrine loop by treating the cells with specific GM-CSF antibodies were unsuccessful (LANG et al. 1985), again raising the possibility of an internal mechanism of receptor stimulation by GM-CSF such as discussed for the v-*sis* protein. Similar experiments were reported with the IL-3 gene inserted into a retroviral vector. FCD-P1 cells infected with the IL-3 virus become growth factor independent and form tumors in nude mice. In addition, infection of mouse hematopoietic cells with the IL-3 virus induces the development of normal hematopoetic colonies in the absence of IL-3 (WONG et al. 1987). Spontaneous mutants of the FDC-P1 cell line which are factor independent and capable of forming tumors have also been isolated. An analysis of a number of such factor-independent derivatives revealed that they contained either a rearranged GM-CSF gene or a rearranged IL-3 gene and also produced the corresponding factor (STOCKING et al. 1988). The concept emerging from these studies is that autocrine growth is an important step in leukemogenesis. This applies not only to hematopoietic neoplasms but also to lymphoid neoplasms since introduction of the IL-2 gene into the human T-cell line CTLL-2 renders the cells factor independent and tumorigenic in nude as well as well as in syngenic mice (YAMADA et al. 1987).

C. Signal Transducer Genes

The transmission of the signal from the cell surface to the cell nucleus is probably the most complicated and least understood area within the growth control field. Largely through the study of oncogenes, some of the proteins which play a role in this process have been identified. In this section we discuss the oncogenes with a recognized or suspected role in signal transduction. We have included the receptor tyrosine kinase-type oncogenes in this section because the corresponding receptors consist of a "receiving" extracellular and a "transmitting" intracellular domain. The "receiving" domain corresponds to the N-terminus of the molecule, while the "transmitting" domain, encoding the tyrosine kinase, corresponds to the C-terminus. It has been suggested (YARDEN and ULLRICH 1988) that tyrosine kinase receptors have originally evolved through fusion of receptors lacking a significant cytoplasmic domain (such as the IL-2 receptor) with tyrosine kinases which lack a ligand-binding domain (such as the *src* protein). The receptor tyrosine kinase-type oncogenes are listed in Table 3. All other oncogenes besides receptors which fall in the class of suspected signal transducers are summarized in Table 4.

I. Receptor Tyrosine Kinase-Type Oncogenes

1. The v-*erb*B Oncogene

This oncogene was originally found in the strain R (or ES4) of avian erythroblastosis virus (AEV; ROUSSEL et al. 1979; VENNSTRÖM and BISHOP 1982) and later in the H strain of AEV (YAMAMOTO et al. 1983a). Com-

Table 3. Growth factor receptor/tyrosine kinase-type oncogenes

Oncogene	Corresponding receptor gene	Mode of activation	Original source
v.erbB	EGF receptor	Truncation, point mutations	Avian retroviruses
neu cells (erbB-2)	Similarity to EGF receptor	Point mutation	Rat neuroblastoma
v-fms	CSF-1 receptor	Point mutations; truncation	Feline retrovirus
v-kit	Distant similarity to PDGF and CSF-1 receptors	Probably truncation and mutation	Feline retrovirus
v-ros	Similarity to insulin receptor	Probably truncation	Avian retrovirus
met	Similarity to insulin receptor	Probably truncation	Human osteosarcoma
v-sea	Similarity to insulin receptor	Probably truncation	Avian retrovirus
trk	?	Probably truncation	Human coloncarcinoma
ret	?	Probably truncation	Human monocytic cells

Table 4. Signal transducer-type oncogenes other than receptors

Oncogene	Function of product	Original source of gene	Subcellular localization of product
v-src	Tyrosine kinase	Avian retroviruses	Inner plasma membrane, cytoskeleton
lck[a]	Tyrosine kinase	Nucleic acid sequence homology to src	Cytoskeleton
v-yes	Tyrosine kinase	Avian retrovirus	Inner plasma membrane, cytoskeleton
v-fgr	Tyrosine kinase	Feline retrovirus	Cytoplasm
hck[a]	Tyrosine kinase	Nucleic acid sequence homology to src and lck	
lyn[a]	Tyrosine kinase	Nucleic acid sequence homology to yes	
fyn[a]	Tyrosine kinase	Nucleic acid sequence homology to src and yes	
v-abl	Tyrosine kinase	Murine retrovirus	Inner plasma membrane, cytoplasm
v-fps/fes	Tyrosine kinase	Avian/feline retroviruses	Inner plasma membrane, cytoplasm
v-H-ras	GTP-binding protein	Murine retrovirus	Inner plasma membrane
v-K-ras	GTP-binding protein	Murine retrovirus	Inner plasma membrane
N-ras	GTP-binding protein	Human neuroblastoma	Inner plasma membrane
v-mil/raf	Serine/threonine kinase	Avian/murine retroviruses	Cytoplasm
v-mos	Serine/threonine kinase	Murine retrovirus	Cytoplasm
pim-1	Serine/threonine kinase[b]	Murine leukemia cells	?

[a] Not yet shown to be activatable as an oncogene.
[b] Deduced from sequence comparisons.

parisons between the predicted protein sequence of the v-*erb*B oncogene of the H strain and that of the human epidermal growth factor (EGF)-receptor revealed a high degree of similarity (Downward et al. 1984b). It is now accepted that v-*erb*B has evolved from the avian c-*erb*B/EGF receptor gene (Lax et al. 1985; Nilsen et al. 1985). v-*erb*B exhibits a large N-terminal truncation and thus lacks almost the complete extracellular portion responsible for ligand binding. In addition, it lacks variable portions of the C-terminus including one or two out of the three tyrosine autophosphorylation sites. Finally it contains several point mutations relative to c-*erb*B (Nilsen et al. 1985; Lax et al. 1985; Downward et al. 1984a, b; Ullrich et al. 1984; Yamamoto et al. 1983a, b). These observations led to the hypothesis that v-*erb*B transforms cells by continuously emitting mitogenic signals even in the absence of ligand (Downward et al. 1984a, b; Ullrich et al. 1984). This hypothesis was confirmed by the observation that the v-*erb*B protein, unlike the EGF receptor, displays a constitutively active tyrosine kinase activity and phosphorylates itself on tyrosine (Ushiro and Cohen 1980; Kris et al. 1985).

Which of the three structural modifications, N-terminal truncation, C-terminal truncation, or point mutations, are essential for the activation of the oncogenic potential of the EGF receptor? Overexpression of the complete human EGF receptor in NIH 3T3 cells or chick fibroblasts leads to ligand-dependent cell transformations as demonstrated by the formation of foci and colonies in soft agar (VELU et al. 1987; DI FIORE et al. 1987; KHAZAIE et al. 1988). Studies with chick cells showed that the receptor with an N-terminal truncation causes ligand-independent fibroblast transformation which can be further enhanced by an additional C-terminal truncation. In contrast, the N-terminal truncation is sufficient to activate the full erythroblast-transforming capacity of the EGF receptor (NILSEN et al. 1985; KHAZAIE et al. 1988). This suggests that the extreme C-terminus of the receptor contains a regulatory domain that is active in fibroblasts but not in erythroblasts. Since normal erythroblasts are not known to express the EGF receptor gene, this suggests that they lack regulatory circuits which are normally coupled to the EGF response in fibroblasts.

Ligand-induced activation of receptor kinase and of mitogencity is thought to require the dimerization of EGF receptors. Compatible with this notion is the finding that bivalent but not monovalent receptor antibodies can act as agonists (YARDEN and SCHLESSINGER 1987; GILL et al. 1984; BISWAS et al. 1985). Whether dimerization of v-*erb*B protein is a prerequisite of its transforming capacity is not known, although cell transformation correlates with both cell surface expression (BEUG and HAYMAN 1984) and tyrosine kinase activity of the protein (HAYMAN et al. 1986). Extensive studies have been performed in search for cellular substrates of the v-*erb*B protein/EGF receptor and a number of candidate proteins, located either in the cytoplasm or in the cytoskeleton, have been proposed (see for example review by MARTIN 1985). However, it is still not clear whether any of them are physiologically relevant and if the substrates differ between the normal receptor and its oncogenic derivative.

Several human malignancies have been reported, most notably squamous cell carcinomas and glioblastomas, in which the EGF-receptor gene is amplified and/or overexpressed (YU et al. 1984; MERLINO et al. 1984; KING et al. 1985b; LIBERMANN et al. 1985). Whether or not these alterations in EGF receptor expression play a causal role in tumor development remains to be seen.

2. The v-*fms* Oncogene

This gene was originally discovered as the oncogene of the SM strain of feline sarcoma virus (FeSV; FRANKEL et al. 1976; DONNER et al. 1982) and later as that of another feline sarcoma virus isolate (BESMER et al. 1986b). It encodes an integral transmembrane glycoprotein with a tyrosine protein kinase domain (HAMPE et al. 1984; COUSSENS et al. 1986) whose cellular homolog is almost exclusively expressed in monocytes and macrophages. These observations led to the discovery that the c-*fms* gene product corresponds to the receptor of the macrophage colony-stimulating factor (M-CSF or CSF-1, SHERR et al. 1985). As for v-*erb*B, the v-*fms* protein must be expressed at the cell surface to induce cell transformation (ROUSSEL et al. 1984, LYMAN et al. 1987) and becomes autophosphorylated at tyrosine in the presence of added ligand (DOWNING et al. 1988; TAMURA et al. 1986). However, in contrast to the v-*erb*B protein, it can still bind CSF-1, due to the presence of a complete ligand-binding domain (SACCA et al. 1986). In analogy of the v-*erb*B gene product, v-*fms* protein exhibits a C-terminal truncation which eliminates tyrosine 969, one of the three potential tyrosine autophosphorylation sites (HAMPE et al. 1984). Replacement of C-terminal v-*fms* sequences with c-*fms* sequences reduces its transforming potential while mutation of the tyrosine 969 to phenylalanine again restores the transforming potential (BROWNING et al. 1986). Two recent studies have now demonstrated that both the transforming potential and tyrosine autophosphorylation of c-*fms* can be activated by a point mutation in position 301 (located within the ligand-binding domain) and that its transforming capacity can be further enhanced by a mutation in tyrosine 969 (WOOLFORD et al. 1988; ROUSSEL et al. 1988). Thus, in a fashion similar to EGF-receptor activation, truncation or alteration of the C-terminus of the CSF-1 receptor enhances its transforming potential, probably by disrupting inhibitory feedback circuits.

Although v-*fms* effectively transforms fibroblasts it does not transform myelomonocytic cells. This is surprising in view of the fact that only the latter cell type expresses the CSF-1 receptor and responds to signaling by the activated CSF-1 receptor. However, v-*fms* does induce factor independence and tumorigenicity in a CSF-1-dependent murine macrophage cell line, presumably by a nonautocrine mechanism (WHEELER et al. 1986; SHERR 1987). These observations raise the possibility that the signals emitted by the mutated CSF-1 receptor are different from those emitted by the receptor activated by ligand. Another possibility is that signaling by CSF-1 in normal cells does not induce the growth of myeloid cells but only supports their survival, an effect which might allow the cells to divide in response to internal or external signals of unknown nature.

At present no clear-cut evidence exists which points to the involvement of the c-*fms*/CSF-1 receptor gene in human malignancies. In certain cases of human acute myelogenous leukemia c-*fms* appears to be expressed in cells which lack other markers of myelomonocytic differentiation (see review by Sherr 1989). However, the significance of this seemingly aberrant expression is still unclear.

3. Other Receptor-Type Tyrosine Kinase Oncogenes

Besides v-*erb*B and v-*fms* at least six additional oncogenes exist whose structures suggest that they correspond to receptor tyrosine kinases. The *neu* gene (also called *erb*B2; KING et al. 1985a; SEMBA et al. 1985; or HER-2, COUSSENS et al. 1985) was initially isolated from a rat neuroblastoma induced by ethylnitrosourea (SHIH et al. 1981). It encodes a 185-kDa protein that closely resembles the EGF receptor (PADHY et al. 1982; SCHECHTER et al. 1984; BARGMANN et al. 1986a) but for which a ligand has not yet been identified. Interestingly, the transforming potential of the *neu* gene is activated by a single amino acid substitution in the transmembrane domain of the protein (BARGMANN et al. 1986b). This suggests that the receptor encoded by *neu* must undergo a conformational change to activate its signaling function. The *neu* gene hase been found to be amplified in human mammary carcinomas (KING et al. 1985a; SLAMON et al. 1987; KRAUS et al. 1987) and in human salivary gland carcinomas (SEMBA et al. 1985), although it is not clear whether the amplified genes carry an activating mutation. It was recently found that overexpression of the mutated *neu* gene in mammary epithelia of transgenic mice leads to the rapid development of mammary carcinomas (MULLER et al. 1988).

The v-*kit* oncogene was discovered as the transforming gene of the HZ strain of feline sarcoma virus (BESMER et al. 1986a). It encodes a receptor-like molecule which, like the PDGF and CSF-1 receptors, has a tyrosine kinase domain interrupted by a stretch a amino acids of unknown function (BESMER et al. 1986a). This insert exhibits a significant sequence similarity to that of the PDGF receptor (YARDEN et al. 1986). Even though the ligand of the c-*kit* receptor is not yet known, recent experiments suggest that it plays a role in certain type of stem cells. Thus, mice mutated in the *W* locus, which are characterized by the abnormal development of germ cells, pigment cells, hematopoietic progenitors, and mast cells, exhibit rearrangements in the c-*kit* locus (GEISSLER et al. 1988).

The v-*ros* oncogene is responsible for the sarcoma-inducing potential of the UR2 avian sarcoma virus (BALDUZZI et al. 1981). Its predicted protein sequence resembles that of the β-chain of the insulin receptor (NECKAMEYER et al. 1986). It is not clear, however, whether in an analogy to the insulin receptor it can associate with another polypeptide to form a functional receptor. The high expression of c-*ros* in kidneys (NECKAMEYER et al. 1986; SHIBUYA et al. 1982; BIRCHMEIER et al. 1986) suggests that the corresponding putative receptor functions in this organ.

Two further oncogenes whose products exhibit distant similarities to the insulin receptor β-chain and which perhaps correspond to growth factor receptors are the *met* and v-*sea* oncogenes. The *met* gene was detected following transfection of DNA from human osteosarcoma cells into NIH 3T3 cells (COOPER et al. 1984) while v-*sea* was identified as the transforming gene of the avian S13 erythroblastosis/sarcoma virus (BENEDICT et al. 1985). Both the *met*- and v-*sea*-encoded proteins are truncated in their N-termini and fused to other sequences (PARK et al. 1986; TEMPEST et al. 1986; HAYMAN et al. 1985). A naturally occurring variant of the S13 virus carries a temperature-sensitive v-*sea* oncogene which synthesizes an active tyrosine kinase at 37° C but not at 42° C (KNIGHT et al. 1988). This mutant has become very useful in studies on the mechanism of erythroid cell transformation (KNIGHT et al. 1988).

The *trk* oncogene was discovered following transfection of NIH 3T3 cells with DNA of a human colon carcinoma (MARTIN-ZANCA et al. 1986). It was found to contain a tyrosine kinase and a membrane-spanning domain, with its

N-terminal portion fused to nonmuscle tropomyosin (MARTIN-ZANCA et al. 1986). In another series of transfection experiments using NIH 3T3 cells the *trk* oncogene was reisolated now exhibiting a truncation in its N-terminal domain but without fusion to the tropomyosin molecule (KOZMA et al. 1988). These experiments suggest that the *trk* gene is frequently rearranged and activated during transfection and that the tropomyosin domain of the originally described oncogene plays no direct role in transformation.

The *ret* oncogene was probably also activated during transfection experiments with NIH 3T3 cells (TAKAHASHI et al. 1987) using DNA from a human monocytic cell line. The predicted protein sequence contains two hydrophobic domains in the N-terminal portion, suggesting that the corresponding putative receptor molecule traverses the membrane twice (TAKAHASHI et al. 1987).

II. Tyrosine Kinase-Type Oncogenes Lacking a Transmembrane Domain

Since the discovery of the v-*src* oncogene, which encodes a membrane-associated tyrosine kinase lacking a transmembrane domain, at least another six genes have been found (v-*yes*, v-*fgr*, *hck*, *lyn*, *fyn*, and *lck*) which are highly similar to v-*src* in a region spanning 469 amino acids. Two additional genes, namely v-*abl* and v-*fps*/v-*fes*, are more distantly related (for a recent review see COOPER 1989). Five of these genes, v-*src*, v-*yes*, v-*fgr*, v-*abl*, and v-*fps*/*fes* were found as transforming genes in avian and mammalian retroviruses. *lck*, *lyn*, and *fyn* were identified through their similarities to known tyrosine kinases on the nudeic acid level while *lck* was found through the tyrosine kinase activity of its protein product. Even though transformation of cells by these latter genes has not yet been demonstrated, it is likely that they can be oncogenically activated. It is still not clear whether tyrosine kinases without receptor function play a part in normal growth control. Fibroblasts transformed by v-*src* exhibit properties which are very similar to cells transformed by receptor tyrosine kinase type oncogenes (ROYER-POKORA et al. 1978). The cellular homolog of v-*src*, the c-*src* gene, encodes a phosphoprotein of 60 kDa ($pp60^{c\text{-}src}$) which is highly expressed in brain (COTTON et al. 1983), platelets (GOLDEN et al. 1986), and chromaffin cells (PARSONS and CREUTZ 1986), suggesting a role in differentiation or secretion. A possible link to growth control comes from the recent observations that , in quiescent fibroblasts exposed to PDGF, $pp60^{c\text{-}src}$ becomes phosphorylated at serine and tyrosine residues with a concomitant activation of its protein tyrosine kinase activity (GOULD and HUNTER 1988). This raises the possibility that signals emitted by the activated PDGF-receptor tyrosine kinase are amplified by the cellular *src* protein. It has also been suggested that the c-*src* protein or other members of the *src* family act as signal-transducing effectors which associate with certain growth factor receptors that lack a tyrosine kinase domain, such as the IL-2 receptor (for review see YARDEN and ULLRICH 1988). However, no evidence has yet been reported which supports this notion.

The mechanisms by which c-*src* becomes activated as an oncogene have been studied in great detail and are complex. Simple overexpression of c-*src* does not lead to overt cell transformation and the protein's tyrosine kinase activity remains barely detectable (IBA et al. 1984; SHALLOWAY et al. 1984). Deletions of the C-terminal end which include the tyrosine at position 527, or a change of this residue into another amino acid, activate the transforming capacity of the gene and lead to a constitutive expression of tyrosine kinase acactivity (CARTWRIGHT et al. 1987; REYNOLDS et al., submitted; KMIECIK and SHALLOWAY 1987; PIWNICA-WORMS et al. 1987). The specific activity of the tyrosine kinase and the transforming potential of $pp60^{c\text{-}src}$ can also be activated by the polyoma virus middle T antigen. The middle T protein inhibits the autophosphorylation of $pp60^{c\text{-}src}$ at tyrosine residues, probably including residue 527, through the formation of a protein complex (COURTNEIDGE and SMITH 1983; BOLEN et al. 1984; CARTWRIGHT et al. 1986; COURTNEIDGE 1985). Thus, as already discussed for the EGF and CSF-1 receptors, a specific C-terminal tyrosine of $pp60^{c\text{-}src}$ has a negative regulatory effect on the transforming ability and kinase activity of the protein. In addition to mutation in the C-terminus, point mutations in the N-terminal one-third of the protein have been reported to activate the transforming capacity of $pp60^{c\text{-}src}$ (KATO et al. 1986).

Much less is known about the function and mechanisms of oncogenic activation of most other genes in this family. An interesting gene with a remarkably potent capacity to transform murine B-cells is the v-*abl* oncogene, first discovered in the Abelson murine leukemia virus (for review see WITTE 1986). c-*abl*, the cellular counterpart of this gene, almost certainly contributes to the formation of human neoplasms since it has been found to be rearranged in 90%–95% of hematopoietic cells from Philadelphia chromosome positive patients with chronic myeloid or acute lymphocytic leukemia (GROFFEN et al. 1984; HERMANS et al. 1987). These cells produce an *abl* protein with a constitutively active tyrosine kinase in which the N-terminus is replaced by sequences from the *bcr* gene (CLARK et al. 1987).

The *lck* gene is remarkable because of its high expression in cells of the T-lineage. Recent studies indicate that its expression and state of phosphorylation are modified during T-cell activation (MARTH et al. 1987; VEILETTE et al. 1988). However, there is no evidence that the $p56^{lcK}$ protein plays a role in growth control of cells from this lineage nor is it known whether it can be oncogenically activated.

III. The *ras* Family Oncogenes

The *ras* family consists of three closely related genes, H-*ras*, K-*ras*, and N-*ras*, each of which encodes a 21-kDa protein that can bind (FINKEL et al. 1984) and hydrolyze GTP (MCGRATH et al. 1984; SWEET et al. 1984). The v-H-*ras* and v-K-*ras* genes were originally discovered as the transforming genes of the Harvey and Kirsten strains of mouse sarcoma virus, respectively, while N-*ras* was detected as an NIH 3T3-cell-transforming oncogene from a human neuroblastoma (for review see BARBACID 1987). Oncogenic activation of *ras* results from a single-point mutation in amino acids 12, 13, or 61 and correlates with a decrease in the ability to hydrolyze GTP (for review see GIBBS et al. 1985; BARBACID 1987; BOS 1988). It has also been shown that overexpression of the normal H-*ras* gene can weakly activate its transforming capacity (CHANG et al. 1982).

K-*ras* and N-*ras* genes have been found to be frequently associated with human tumors of nonviral origin. The recent application of sensitive detection techniques has shown that up to 90% of human pancreas carcinomas contain K-*ras* with a mutation in position 12. There appears to be a certain cell type specificity since the N-*ras* gene is most frequently found in cells from human acute myeloid leukemia and in patients with myelodysplastic syndrome (For review see BOS 1988).

The most direct evidence which links the *ras* genes to growth control stems from experiments which demonstrated that microinjection of antibodies against $p21^{H\text{-}ras}$ into quiescent NIH 3T3 cells inhibit serum-induced stimulation of cell growth (MULCAHY et al. 1985). In addition, microinjection of $p21^{H\text{-}ras}$ or $p21^{H\text{-}ras}$ proteins into these cells induces a temporary morphological transformation and growth factor independence (FERAMISCO et al. 1984; STACEY and KUNG 1984). A possible role of the *ras* family proteins in signal transductions is suggested by their similarity to the family of G proteins. Like *ras* proteins, G proteins can bind and hydrolyze GTP and are located at the inner face of the plasma membrane. Indirect evidence indicates that G proteins are involved in coupling the activity of certain growth factor receptors with that of phospholipase C, a rate-limiting enzyme in the production of the second messengers diacylglycerol and inositol triphosphate (for review see BERRIDGE 1987; BERRIDGE and IRVINE 1984; HANLEY and JACKSON 1987). Thus, the mitogenic activities of bombesin and thrombin can be inhibited by pertussis toxin, a specific G protein inhibitor (LETTERIO et al. 1985; PARIS and POUYSSÉGUR 1986). Taken together, these observations raise the possibility that $p21^{ras}$ plays a role in the transduction of signals initiated by growth factor receptors and that mutations leading to its constitutive activation bypass the cell's requirement for growth factor stimulation. However, except for the detection of a gene *(gap)* whose product stimulates the GTP hydrolytic activity of $p21^{ras}$ and which also seems to be essential for the transforming activity of $p21^{ras}$ proteins (TRAHEY and MCCORMICK 1987; VOGEL et al. 1988; ADARI et al. 1988), little progress has been made in recent years which elucidates the precise role of *ras* in mammalian cells.

IV. Serine Threonine Kinase-Type Oncogenes

A protein with a central role in signal transduction during growth control is protein kinase C. This enzyme, which is activated after growth factor stimulation of quiescent fibroblasts, can also be activated by tumor promoters, such as phorbol esters (NISHIZUKA 1983; CASTAGNA et al. 1982). The immediate agonists of the enzyme are diacylglycerol, calcium ions, and phospholipids (for reviews see NISHIZUKA 1986; PARKER and ULLRICH 1987). The activated enzyme is capable of phosphorylating proteins on serine and threonine and its substrates include the EGF and insulin receptors (COCHET et al. 1984; BOLLAG et al. 1986). However, contrary to the view that protein kinase C has a positive role in growth control, phosphorylation of these receptors leads to decreased tyrosine kinase activities (DAVIS and CZECH 1985; HUNTER et al. 1984; BOLLAG et al. 1986). Circumstantial evidence suggests that protein kinase C, plays a role in the transformation of fibroblasts by the oncogenes *ras, sis, fms, fos,* and *src* (PREISS et al. 1986; JACKOWSKI et al. 1986; GOULD et al. 1985). In addition, overexpression of the protein kinase C-1 gene in NIH 3T3 cells leads to alterations that are typical of transformed cells (PERSONS et al. 1988; HOUSEY et al. 1988) although no example has been reported for the natural occurrence of an oncogenic version of the protein kinase C gene.

Two retroviral oncogenes have been found to encode proteins with a serine/threonine kinase activity, namely v-*mos* and v-*mil*/v-*raf*. v-*mos* was originally detected as the transforming gene of Moloney murine sarcoma virus (MAXWELL and ARLINGHAUS 1985a, b) and its product is related to the catalytic subunit of bovine cAMP-dependent protein kinase (BARKER and DAYHOFF 1982; VERMA et al. 1980). Besides its autophosphorylating capacity the v-*mos* protein exhibits a DNA-binding activity which is stimulated by the presence of nucleotide triphosphates (SETH et al. 1987). The v-*mil* gene was identified as one of two oncogenes contained in the avian retrovirus MH2 (COLL et al. 1983; JANSEN et al. 1983; KAN et al. 1983) while the mammalian homolog, v-*raf*, comes from the mouse sarcoma virus strain 3611 (RAPP et al. 1983). Based on its sequence similarities to v-*raf*, a gene termed A-*raf* has also been described (HULEIHEL et al. 1986). Whereas the murine c-*mos* gene can be oncogenically activated by simple overexpression (WOOD et al. 1984), c-*raf* appears to require a 5′ truncation of the gene (MÖLDERS et al. 1985; ISHIKAWA et al. 1987). The role in growth control of the *mos* and *mil*/*raf* proteins remains to be elucidated. It is possible that in normal cells they either act downstream of *ras* or in a pathway which is independent of *ras* gene function since microinjection of antibodies to $p21^{ras}$ appears to block DNA synthesis in v-*sis*, v-*src*, v-*fos*, and v-*fms*-transformed but not in v-*mos*- or v-*raf*-transformed fibroblasts grown in low serum (SMITH et al. 1986). An interesting property of the v-*mil*/v-*raf* gene is its ability to cooperate with v-*myc* in cell transformation (see below). In addition, it was recently discovered that v-*raf* facilitates a lineage switch of c-*myc*-overexpressing B cells to macrophages (KLINKEN et al. 1988) by a mechanism which is not yet understood.

A new potential member of the serine/threonine kinase oncogene family is the *pim*-1 gene, which was found to be activated by retroviral integration in a mouse T-cell lymphoma (CUYPERS et al. 1984). The protein encoded by this gene exhibits a significant similarity to the catalytic subunit of protein kinases as well as the products of the *mos* and *abl* genes (VAN BEVEREN and VERMA 1985).

In essentially all growth-factor-stimulated as well as in oncogene-transformed fibroblasts the ribosomal subunit S6 becomes phosphorylated by a specific kinase in certain specific serine and threonine residues (THOMAS et al. 1982). This raised the possibility that the gene encoding the kinase involved might itself be an oncogene. However, the recent characterization of the major mammalian S6 kinase indicates that it does not correspond to any of the known oncogene-encoded serine/threonine kinases (G. THOMAS, personal communication).

D. Genes Encoding Nuclear Proteins

To trigger cell proliferation, the mitogenic signal initiated by growth factors at the cell surface must eventually reach the nucleus and activate or suppress genes whose products are required for traversal of the cell cycle. A number of years ago it was discovered that *fos* and *myc*, two genes which were already known to act as oncogenes, when transduced by retroviruses, are activated following serum stimulation of quiescent fibroblasts (KELLY et al. 1983). These findings have been extended to a number of other cell types stimulated by a variety of different growth factors. In the past 5 years more than 80 genes have been identified whose expression is induced by serum stimulation of fibroblasts (COCHRAN et al. 1983; LINZER and NATHANS 1983; LAU and NATHANS 1985; LAU and NATHANS 1987; ALMENDRAL et al. 1988). These genes generally fall into two classes: "immediate early genes," like c-*fos* and v-*jun*, which reach a maximal expression about 15–30 min after stimulation; and "early genes," such as c-*myc*, which reach a peak about 2 h after stimulation. A common feature of most of the immediate early genes is that expression of their mRNAs can be further induced by cycloheximide treatment of the cells. Superinduction results from an increased mRNA stability (perhaps by the inhibition of the synthesis of short-live nucleases) and from a block in feedback inhibition of transcription (see below). These observations indicate that serum-inducible genes are tightly controlled (ALMENDRAL et al. 1988; GREENBERG et al. 1986; LAU and NATHANS 1987). Several serum-induced genes encode nuclear proteins which have a known role in transcriptional regulation which, due to their resemblance to genes encoding zinc-finger-type proteins (LEMAIRE et al. 1988; CHAVRIER et al. 1988; SUKHATME et al. 1988), are suspected of regulating transcription. Other serum-induced genes, however, encode proteins which are not nuclear and which have no known oncogenic counterparts. Of those with a recognized function, some play a role in wound healing while others seem to encode growth factors (PAULSON et al. 1987; ROLLINS et al. 1988; RYSECK et al. 1989). These genes will not be discussed further.

An apparently completely different class of nuclear proteins which possibly also play a role in growth control are the steroid and thyroid hormone receptors. These nuclear receptors act as transcriptional regulators following binding of their cognate hormone. Hormones, in contrast to peptide growth factors, can enter the cell without the need for cell surface receptors. The v-*erb*A oncogene, a mutated form of the thyroid hormone receptor, is so far the only example of a gene in this category. Finally, there are a several nuclear-

Table 5. Nuclear protein-type oncogenes

Oncogene	Original source	Properties
v-fos	Murine retroviruses	"Immediate early genes"
fra-1[a]	Serum-induced fibroblasts	"Immediate early genes"
fosB[a]	Serum-induced fibroblasts	"Immediate early genes"
jun/AP-1	Avian retrovirus	"Immediate early genes"
junB[a]	Serum-induced fibroblasts	"Immediate early genes"
junD[a]	Serum-induced fibroblasts	"Immediate early genes"
v-myc	Avian retroviruses	"Early gene"
N-myc	Human neuroblastoma	
L-myc[a]	Human lung carcinoma	
v-erbA	Avian retrovirus	c-erbA corresponds to thyroid hormone receptor gene; family of at least two genes
v-myb	Avian retroviruses	Maximally expressed in immature hematopoietic cells; family of three genes
v-ets	Avian retrovirus	Highly expressed in mitogen-stimulated T cells; family of three genes
p53	SV40-transformed cells	

[a] Not yet shown to be activatable as an oncogene.

type oncogenes (of which we will discuss only a few) whose connection to growth control is not clearly established. Some of the nuclear protein-type oncogenes are listed in Table 5.

I. Immediate Early Genes

1. The *fos* Gene Family

The v-*fos* oncogene was originally detected in the murine retroviruses FBJ-MSV and FBR-MSV (FINKEL et al. 1966; FINKEL et al. 1973; FINKEL et al. 1975). Both viruses induce osteosarcomas and fibrosarcomas when inoculated into newborn rodents (KELLOFF et al. 1969; CURRAN and VERMA 1984; CURRAN et al. 1984). A c-*fos* construct lacking 3′ untranslated sequences specifically induces bone hypertrophy when it is constitutively expressed in transgenic mice (RÜTHER et al. 1987). The FBR *fos* gene has a stronger in vitro fibroblast-transforming capacity than the FBJ *fos* gene and is also able to immortalize nonestablished rodent cells (LEE et al. 1979; LEVY et al. 1973, 1987; JENUWEIN et al. 1985). A comparative study of the two viral genes showed that two small in-frame deletions in the carboxy-terminal region of FBR-*fos* enhance its tumorigenicity and that a point mutation approximately in the middle of the protein activates its immortalizing function (JENUWEIN et al. 1985; JENUWEIN and MÜLLER 1987). However, c-*fos* can already be oncogenically activated without structural alterations of the coding region provided that its expression is driven by a strong promoter/enhancer and that negative regulatory sequences in the 3′ noncoding region of the gene are removed (MILLER et al. 1984).

The notion that *fos* is involved in growth control is supported by the observation that treating cultured cells either with serum or with various growth factors (GREENBERG and ZIFF 1984; MÜLLER et al. 1984; COCHRAN et al. 1984; KRUIJER et al. 1984; SKOUV et al. 1986) or microinjecting them with the transforming version of the *ras* protein (STACEY et al. 1987) leads to a transient induction of c-*fos* expression. In addition, activation of several pathways involved in growth factor signaling, such as protein kinase C activation or an elevation in the level of intracellular calcium, induce transient c-*fos* expression (BÜSCHER et al. 1988). The induction of c-*fos* is mediated by a serum-responsive element in the *fos* promoter region (TREISMAN 1985; TREISMAN 1986; PRYWES and ROEDER 1986). This element consists of 24 nucleotides exhibiting a dyad symmetry and binds a specific protein, the "serum response factor," whose gene has recently been cloned (NORMAN et al. 1988). The notion that *fos* expression is a key element in the control of proliferation is supported by the observation that inhibition of *fos* protein expression by antisense *fos* RNA blocks the G0–G1 transition of growth-factor-stimulated cells (HOLT et al. 1986; NISHIKURA and MURRAY 1987). On the other hand, transciption of the c-*fos* gene is not only induced during growth stimulation but also during growth inhibition (BRAVO et al. 1985) and during differentiation (MITCHELL et al. 1985; MÜLLER and WAGNER 1984; MÜLLER et al. 1986a). The transient induction of c-*fos* can be explained by both the instability of the RNA and by the fact that the gene negatively regulates itself (SASSONE CORSI et al. 1988c; SCHÖNTHAL et al. 1988). Parallel to the induction of c-*fos* by serum there are two *fos*-related genes, *fra*-1 (COHEN and CURRAN 1988) and *fos*-B (M. ZERIAL and R. BRAVO, submitted for publication), which are also transiently expressed.

The c-*fos* gene encodes a phosphoprotein of 55 kDa which is localized in the cell nucleus, is associated with chromatin, and binds to DNA in vitro (CURRAN et al. 1984; RENZ et al. 1987; SAMBUCETTI and CURRAN 1986). $p55^{c\text{-}fos}$ contains a zinc finger-like domain which is absent in the viral genes (VAN BEVEREN et al. 1983). The protein has a highly charged central region which might be involved in DNA-binding (VAN BEVEREN et al. 1981; VAN STRAATEN et al. 1983) as well as five regularly spaced leucines known as the "leucine zipper motif" (LANDSCHULZ et al. 1988). The leucine zipper is necessary for interaction with the transcription factor-*jun*/AP-1, which itself also contains a leucine zipper motif (see below) and is also required for cell transformation by v-*fos* (KOUZARIDES and ZIFF 1988). Participation of c-*fos* protein in transcriptional regulation has been directly demonstrated for the adipocyte-specific aP2 gene (DISTEL et al. 1987), the α_1 (III) collagen gene (SETOYAMA et al. 1986), and the collagenase gene (SCHÖNTHAL et al. 1988). In addition, *fos* protein acts as strong transcriptional activator when fused to the DNA-binding domain of the yeast Lex A protein (LECH et al. 1988).

2. The *jun* Gene Family

The v-*jun* oncogene was first detected as the transforming gene of avian sarcoma virus 17 (CAVALIERI et al. 1985; MAKI et al. 1987; BOS et al. 1988; VOGT and TJIAN 1988). Recently, two new members of the *jun* family have been isolated using v-*jun* as a probe. These genes are called *jun*-B (RYDER et al. 1988) and *jun*-D (M. YANIV, personal communication). In fibroblasts, c-*jun* and *jun*-B transcription are induced by serum, EGF (RYSECK et al. 1988; QUANTIN and BREATHNACH 1988; RYDER et al. 1988) and TPA (ANGEL et al. 1988a) with kinetics very similar to those of c-*fos*. A first clue to the function of *jun* came from the finding that the C-terminus of v-*jun* is related to the DNA-binding domain of the yeast transactivator protein GCN4 (VOGT et al. 1987). The idea that *jun* itself encodes a transactivator protein was supported by the finding that the DNA-binding domain of GCN4 can be functionally exchanged with the C-terminus of v-*jun*. Likewise, the N-terminus of GCN4, known to correspond to the transactivator domain, could be replaced by the equivalent domain of v-*jun* (STRUHL et al. 1987, 1988). In addition, the DNA sequence to which GCN4 specifically binds is identical to the binding site of AP-1, a family of transcription factors involved in mediating phorbol-ester-induced gene transcription (LEE et al. 1987; ANGEL et al. 1987). This suggested that the c-*jun* product is related to if not identical to an AP-1 transcription factor, a notion which was confirmed by sequence comparisons as well as by the finding that AP-1 and recombinant c-*jun* protein bind to the same DNA sequences (BOHMANN et al. 1987; ANGEL et al. 1988b).

v-*Jun* protein exhibits several structural differences in comparison to c-*jun*/AP-1. Besides a stretch of viral *gag* sequences which are fused to its N-terminus, it exhibits a 26-amino-acid N-terminal deletion as well as three nonconservative amino acid substitutions, two of them in the putative DNA-binding domain (VOGT and TIJAN 1988). These structural changes could alter the magnitude of the transcriptional activity, the range of genes v-*jun* interacts with, and the selection of protein/protein interactions during transcriptional regulation. The observation that c-*jun*/AP-1 is a mediator for phorbol-ester-induced gene activation (ANGEL et al. 1988b) provides a link between the activation of protein kinase C to a nuclear protooncogene/transcription factor. In contrast to the situation with c-*fos*, protein expression of c-*jun*/AP-1 is subject to positive regulation in both phorbol-ester-treated cells and in cells overexpressing c-*fos* (LAMPH et al. 1988; ANGEL et al. 1988a; WASYLYK et al. 1988). This autoregulation could mediate some of the long-term effects described for tumor promoters and growth factors but might also play a role in other processes such as in development and memory (CURRAN and MORGAN 1988) where transient signals are converted into long-term ones.

The DNA-binding and transactivating functions of c-*jun*/AP-1 are strongly enhanced after dimerization with c-*fos* protein (HALAZONETIS et al. 1988; NAKABEPPU et al. 1988; SASSONE-CORSI et al. 1988a, b; CHIU et al. 1988). This reaction requires intact leucine-zipper domains on both molecules (KOUZARIDES and ZIFF 1988).

In conclusion, the activation of the c-*fos* and c-*jun* genes might be a point of convergence where growth factors as well as other extracellular signals are translated into a transcriptional complex which regulates specific gene expression (for review see CURRAN and FRANZA 1988; CURRAN and MORGAN 1988).

Whether or not some of the genes regulated by this transcriptional complex are themselves potential oncogenes is an interesting question for future studies.

II. Early Genes

1. The *myc* Gene Family

The *myc* gene was originally found as the transforming gene of the MC29 virus as well as that of three other independently isolated avian retroviruses which induce myelocytomatosis, endotheliomas, and carcinomas of the liver and kidney (for review see GRAF and BEUG 1978; GRAF and STEHELIN 1982; BISTER et al. 1983). It was later also discovered in a feline retrovirus which causes lymphoid leukemia (NEIL et al. 1984). There are two other *myc*-related genes, N-*myc* and L-*myc*, of which N-*myc* has been shown to cooperate with activated *ras* in the transformation of rodent fibroblasts in vitro and to induce their tumorigenicity (SCHWAB et al. 1985; YANCOPOULOS et al. 1985; SMALL et al. 1987). N-*myc* has been found to be amplified in human neuroblastomas, retinoblastomas, and small lung cancers. In neuroblastomas the degree of amplification seems to correlate with an advanced stage of the tumor (SCHWAB et al. 1984). In contrast, L-*myc* is frequently amplified in small cell lung carcinomas but shows no correlation to the state of disease (for review see SCHWAB 1985, 1986). c-*myc* was the first oncogene shown to become oncogenically activated by retroviral enhancer/promoter insertion (HAYWARD et al. 1981; NEEL et al. 1981; PAYNE et al. 1981, 1982). Subsequently the c-*myc* locus was found to be rearranged in essentially all B-cell tumors (murine plasmacytomas and human Burkitt's lymphomas) tested (for review see KLEIN and KLEIN 1986). Likewise, expression of c-*myc* under the control of an immunoglobulin enhancer element induces B-cell neoplasms in transgenic mice (ADAMS et al. 1985). In this connection it is curious that infection of hematopoietic cells in vitro with v-*myc*-containing viruses leads to the selective proliferation of transformed macrophages rather than lymphoid cells (GRAF et al. 1981; GRAF 1973; GAZZOLO et al. 1979; BEUG et al. 1979).

The link between *myc* and growth control stems from the observation that treatment of quiescent fibroblast cultures with serum or growth factors leads to an increase of c-*myc* mRNA with a peak at about 2 h (KELLY et al. 1983; KRUIJER et al. 1984. While c-*myc* is regulated at the level of transcriptional initiation there is evidence that it can also be regulated posttranscriptionally at the level of RNA stability (GREENBERG and ZIFF 1984; MUELLER et al. 1984; BLANCHARD et al. 1985; DEAN et al. 1986) or mRNA elongation (BENTLEY and GROUDINE 1986, 1988; EICK et al. 1985, 1987). It has been reported that microinjection of recombinant *myc* protein into quescent NIH cells makes them more susceptible to the effects of growth factors such as PDGF and EGF (KACZMAREK et al. 1985). Also, overexpression of c-*myc* can reduce growth factor requirement of fibroblasts and hematopoietic cells (ARMELIN et al. 1984; SORRENTINO et al. 1986; CORY et al. 1987) and exposure of T cells to a *myc* antisense oligonucleotide inhibits their entry into S-phase (HEIKKILA et al.

1987). Circumstantial evidence links c-*myc* expression to suppression of differentiation. For example, the differentiation of murine erythroleukemia cells, which is accompanied by c-*myc* downregulation, can be blocked by overexpression of c-*myc* in these cells (LACHMAN and SKOULTCHI 1984; LACHMAN et al. 1985; COPPOLA and COLE 1986). Likewise, *myc* antisense oligonucleotides inhibit the proliferation of the human promyelocytic cell line HL60 (WICKSTROEM et al. 1988).

In spite of many years of intensive studies the function of *myc* is still not clear. Like *fos* and *jun*, *myc* encodes a nuclear phosphoprotein (EVAN and HANCOCK 1985) which exhibits a leucine zipper domain (LANDSCHULZ et al. 1988) and binds to DNA (BEIMLING et al. 1985; HANN and EISENMAN 1984; PERSSON and LEDER 1984). However, neither a DNA target sequence nor another protein interacting with *myc* protein has yet been identified.

III. Hormone Receptor Genes

1. The *erb*A Gene Family

The v-*erb*A oncogene was originally detected as one of the two oncogenes contained in the ES4 strain of AEV (VENNSTRÖM and BISHOP 1982). The first example of a link between oncogenes and transcriptional regulators was the demonstration that the v-*erb*A gene encodes a mutated version of the thyroid hormone receptor, a protein related to steroid hormone receptors (SAP et al. 1986; WEINBERGER et al. 1986). Interstingly, the $p75^{gag,\,v\text{-}erbA}$ protein encoded by AEV has lost its ability to bind the ligand but retained its ability to bind DNA (SAP et al. 1986). Although v-*erb*A has no or only a weakly transforming ability on its own (FRYKBERG et al. 1983; SEALY et al. 1983), it cooperates with tyrosine kinase-type oncogenes and the activated *ras* oncogene in the transformation of avian erythroid cells (see below). It can also enhance the transformed phenotype of primary chick embryo fibroblasts (GANDRILLON et al. 1987). The human cellular homologue of the v-*erb*A gene, c-*erb*Aα, is located on human chromsome 17 (SPURR et al. 1984). In addition, there is at least one other member of the family, c-*erb*Aβ, located on chromosome 3 (WEINBERGER et al. 1986). Only tenuous evidence exists which suggests an involvement of these genes in human neoplasia (for review see GREEN and CHAMBON 1986).

IV. Other Nuclear Oncogenes

1. The *myb* Gene Family

The v-*myb* oncogene is contained in the AMV and the E26 avian leukemia viruses, which cause an acute myeloblastosis and an erythroblastosis/myeloblastosis, respectively (for review see BEARD 1963; GRAF and BEUG 1978; NEDYALKOV et al. 1975). There are at least three members of the *myb* gene family, namely c-*myb*, from which v-*myb* was derived (ROUSSEL et al. 1979;

KLEMPNAUER et al. 1982), as well as *myb*-A and *myb*-B (NOMURA et al. 1988). c-*myb* is most highly expressed in immature lymphoid, myeloid, and erythroid cells (WESTIN et al. 1982; DUPREY and BOETTIGER 1985; KUEHL et al. 1988; REED et al. 1987), but has also been detected after serum stimulation of quiescent fibroblasts (THOMPSON et al. 1986). The notion that c-*myb* plays a role in hematopoietic growth control/differentiation is supported by the findings that incubation of bone marrow cells with an antisense oligonucleotide inhibits colony formation of myeloid, erythroid, and megakaryocytic cells (GEWIRTZ and CALABRETTA 1988) and that overexpression of the c-*myb* gene in mouse erythroleukemia cells inhibits their differentiation (CLARKE et al. 1988). In addition, expression of v-*myb* in mature macrophages induces their proliferation and dedifferentiation into myeloblast-like cells (BEUG et al. 1987; NESS et al. 1987) without abolishing their growth factor dependence (BEUG et al. 1982b; LEUTZ et al. 1984). It is intriguing that, while c-*myb* seems to act in all hematopoietic lineages and also blocks the differentiation of erythroid cells, v-*myb* specifically transforms myeloid cells.

The viral *myb* genes are altered at both their 5′ and 3′ ends relative to c-*myb* (KLEMPNAUER et al. 1982; GERONDAKIS and BISHOP 1986). In addition, several murine plasmacytomas exhibit c-*myb* genes with 5′ end truncations (SHENG-ONG et al. 1986; REED et al. 1987), suggesting that this is a prerequisite for oncogenic activation. A number of acute human myeloid leukemias have been found to contain rearranged and/or overexpressed c-*myb* genes (PELLICI et al. 1984; BARLETTA et al. 1987), possibly implying the *myb* oncogene in human neoplasia. The c-*myb* and v-*myb* proteins are nuclear phosphoproteins capable of binding DNA (KLEMPNAUER et al. 1984; BOYLE et al. 1984; MOELLING et al. 1985) and most likely act as transcriptional regulators (BIEDENKAPP et al. 1988).

2. The *ets* Gene Family

The v-*ets* oncogene was transduced as one of the two oncogenes of the avian acute leukemia virus E26 (LEPRINCE et al. 1983; NUNN et al. 1983a, b). It is derived from c-*ets*-1 (LEPRINCE et al. 1988), a gene that is highly expressed in avian lymphoid cells (GHYSDAEL et al. 1986; POGNONEC et al. 1988). Another related gene, c-*ets*-2 (WATSON et al. 1988; BOULUKOS et al. 1988), is predominantly expressed in macrophages and, like c-*ets*-1, encodes a nuclear phosphoprotein (BOULUKOS et al. 1988) which possibly acts in transcriptional control. Recently, a third member of the *ets* family was found, the *erg* gene (REDDY et al. 1987). The v-*ets* gene of E26 virus is fused to v-*myb* and is responsible for the erythroid and fibroblast-transforming capacities of the virus (BEUG et al. 1984; NUNN and HUNTER 1989; GOLAY et al. 1988; YUAN et al. 1988). It is, however, not clear whether it can act as an oncogene by itself and what link, if any, it has to growth control. The human *ets* genes have been shown to be occasionally associated with specific chromosomal breakpoints in human leukemias (for review see PAPAS et al. 1987).

3. The p53 Oncogene

p53 was originally discovered as a nuclear phosphoprotein complexed with the large T antigen of SV40 virus (LANE and CRAWFORD 1979; LINZER and LEVINE 1979). p53 cDNA was molecularly cloned from a number of different mouse and human cell lines (JENKINS et al. 1984b; for review see JENKINS and STURZBECHER 1987). The overexpressed gene cooperates with the *ras* oncogene in inducing anchorage-independent growth and reducing growth factor requirements (ELIYAHU et al. 1984; JENKINS et al. 1985; PARADA et al. 1984). In addition, expression of the gene under the control of both the SV40 and a retroviral promoter induces the immortalization of senescent fibroblasts (JENKINS et al. 1984a). p53 has been implicated in the control of the mammalian cell cycle based on the observation that microinjection of p53 antibodies into quiescent fibroblasts seems to inhibit the G0 to G1 transition after serum stimulation (MERCER et al. 1982). Until very recently the assumption was that the normal p53 gene encodes a regulatory protein which can act as an oncogene if constitutively expressed. It now appears that a p53 cDNA isolated from normal mouse tissues behaves like an anti-oncogene (see next section) in that it can inhibit transformation. The explanation for this apparent discrepancy is that most p53 cDNAs previously characterized exhibit point mutations which inactivated its function as an antioncogene. These mutated p53 forms can in turn neutralize the suppressing function of wild-type p53 and thus, operationally speaking, act as oncogenes. (A. LEVINE, personal communication).

E. Cooperation Between Oncogenes

Epidemological studies on human neoplasms suggest that tumors arise as a consequence of a stepwise accumulation of independent events in descendants of a single cell (for review see KNUDSON 1986; FARBER and CAMERON 1980; KLEIN and KLEIN 1985). What are these events? The key feature of malignant cell transformation is the escape from normal growth control mechanisms. However, cell transformation is a complex phenomenon which includes loss of growth factor dependence, changes in cell morphology, acquisition of anchorage-independent growth in vitro, loss of contact inhibition, and acquisition of the ability to self-renew. A different property which is often observed is the escape from cellular senescence, termed "immortalization" or "establishment." Finally, some cell types have to acquire the capacity of "invasiveness" to become fully malignant. These features do not necessarily apply to cells of all tissues in a given species or to the same tissue in different species. For example, normal fibroblastoid cells are already capable of self-renewal but not of anchorage-independent growth; in contrast, normal hematopoietic progenitors are capable of forming colonies in semisolid tissue culture medium but cannot self-renew. Accordingly, the acquisition of self-renewal (or proliferation) capacity is a criterion for the transformation of hematopoietic cells while the acquisition of anchorage-independent growth potential is a criterion for transformation of fibroblasts. Similarly, in contrast to sessile cell types, hematopoietic cells do not have to acquire the property of invasiveness to become malignant since they inherently possess the capacity to enter the bloodstream. Species differences also play a role: This is exemplified by the observation that in primary chick embryo fibroblast cultures the v-*myc* gene transforms but does not immortalize while in mammalian fibroblasts it immortalizes but does not transform (for review see COLE 1986). Also, for reasons which are unclear, avian hematopoietic cells are much more susceptible to the in vitro transforming effects of oncogenes than those of a mammalian origin (for review see GRAF and BEUG 1978).

Before discussing cooperativity one might first ask if oncogenes exist which induce neoplasms in a one-step process. At first sight, this is the case for a number of retrovirus-borne oncogenes. For example, the v-*sis*-containing simian sarcoma virus can induce sarcomas in marmosets (WOLFE et al. 1971) and the v-*src*-containing Rous sarcoma virus can do the same in chickens (ROUS 1911). However, the tumors induced in these animals are polyclonal and arise by reinfection of target cells. Therefore immortalization of the transformed cells, a potentially necessary event for tumor formation, is obviated under these circumstances. Another complication in interpreting the effects of retroviral oncogenes is the fact that they are not only highly and constitutively expressed but are also often mutated and truncated. A more direct approach to test for the ability of a single oncogene to induce tumors is its introduction into the germline of animals using an appropriate promoter. The clearest example so far, in what appears to be a one-step event, are transgenic mice harboring the activated *neu* oncogene which develop adenocarcinomas (MULLER et al. 1988). In these experiments the *neu* oncogene had been placed under the control of the mouse mammary tumor virus LTR, known to be an enhancer/promoter element which leads to a high expression of the transgene in epithelial tissues. Because of this experimental design it is still possible that under natural circumstances carcinogenesis mediated by the *neu* gene requires both an activating mutation in its coding region as well as another event, such as one that leads to its increased expression.

In contrast to the situation with the *neu* gene a large body of evidence indicates that most oncogenes are incapable of transforming or causing neoplasms if not coexpressed with another oncogene. Several approaches have been taken do demonstrate this point. First, some tumor cell lines were found to contain more than one oncogene. For example, the human promyelocytic leukemia cell line HL60 contains an amplified c-*myc* gene as well as an activated H-*ras* gene (MURRAY et al. 1983). It cannot be excluded, however, that these oncogenes became activated during the establishment or passaging of the cell line since the original tumor material is no longer available for comparative studies. Even in cases where multiple oncogenes are detectable in fresh human tumor material (SUÁREZ et al. 1987), it is difficult to assess which of them were causally involved in the development of the neoplasm.

Another approach to study oncogene cooperativity is the introduction of various pairs of oncogenes into the same cells. Such studies showed that while primary rat embryo fibroblasts transfected by the v-H-*ras* gene or the v-*myc* gene alone do not become transformed or tumorigenic, they do so if the two oncogenes are introduced together (LAND et al. 1983). Similarly, transgenic mice expressing both the v-H-*ras* and *myc* oncogenes develop neoplasms at a higher incidence and more rapidly (SINN et al. 1987; ANDRES et al. 1988) than transgenic animals expressing the *ras* or *myc* oncogene only (STEWART et al. 1984; LEDER et al. 1986; ADAMS et al. 1985; SINN et al. 1987; QUAIFE et al. 1987). However, even in mice expressing both oncogenes tumorigenesis is usually monoclonal or oligoclonal, suggesting that at least one additional event is required for malignant growth. These studies have helped to define two general categories of oncogenes capable of cooperating in fibroblast transformation: oncogenes whose products are located in the cytoplasm or at the cell membrane ("cytoplasmic oncogenes") such as *ras* and *src*; and oncogenes whose products are located in the nucleus ("nuclear oncogenes") such as *myc*, p53, and possibly *myb* (PARADA et al. 1984; WEINBERG 1985; LAND 1986). Even though the precise mechanism by which these two classes of oncogenes act and cooperate is still largely elusive it has been postulated that cytoplasmic oncogenes induce cell transformation in fibroblasts (morphological changes and serum independence) while nuclar oncogenes are responsible for immortalization. It should also be added here that several transforming genes from DNA tumor viruses fall into the same two categories of cytoplasmic and nuclear oncogenes. An example of a cytoplasmic oncogene of this type is the polyoma middle T antigen which is known to activate $pp60^{c\text{-}src}$ (see above). This protein resides in the cytoplasm and can functionally replace $p21^{ras}$ while the adenovirus E1a protein and the SV40 large T antigen (known to form a complex with p53 antigen) are nuclear and can functionally replace *myc* (LAND 1986).

A third approach to investigate how oncogenes cooperate is based on the observation that several of the acute avian leukemia viruses carry pairs of oncogenes. The fact that they have been originally selected for their exceedingly high oncogenic potential suggests that each of the transduced oncogene pairs represent distinctive steps in the development of leukemia which have become joined into a single virus. Three such strains exist, capable of causing erythroid or myeloid leukemias with latency periods as short as 2 weeks. These viruses carry six oncogenes in three different pairwise combinations (for review see GRAF and BEUG 1978; GRAF and STEHELIN 1982; BISTER 1984). They are the v-*erb*A, v-*erb*B-containing ES4 strain of avian erythroblastosis virus (AEV), the v-*mil*, v-*myc*-containing Mill Hill 2 (MH2) strain, and the v-*myb*, v-*ets*-containing E26 strain. The mechanisms by which these oncogene pairs cooperate have been elucidated in great detail for the AEV and MH2 viruses which will be used here as paradigms.

A scheme illustrating the mechanism of transformation by AEV in erythroid cells and by MH2 in myeloid cells is outlined in Fig. 2. AEV with its v-*erb*A and v-*erb*B oncogenes selectively transforms erythroid progenitor cells while MH2 with is v-*mil* and v-*myc* oncogenes selectively transforms myelomonocytic cells (GRAF et al. 1981; SAMARUT and GAZZOLO 1982; GAZ-

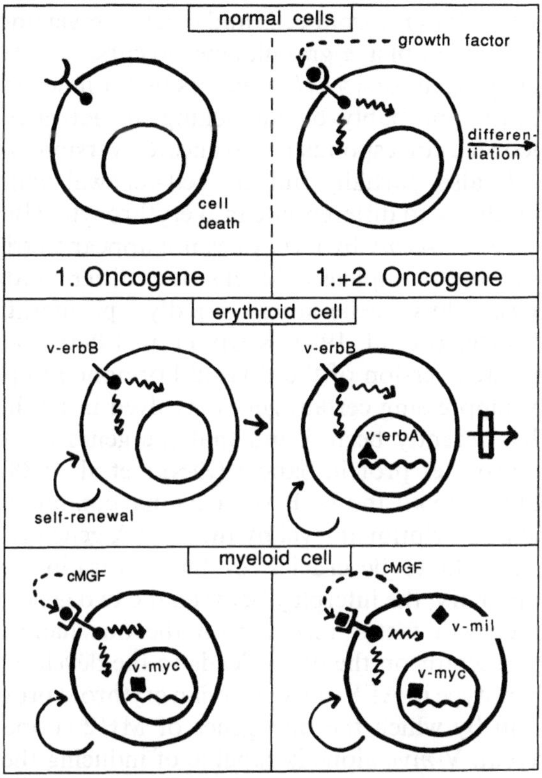

Fig. 2. Oncogene cooperation in avian hematopoietic cells. The *top panel* depicts a hematopoietic progenitor cell with and without stimulation by a growth factor. In the absence of an appropriate growth factor the cell dies. In its presence, it can divide and differentiate. The *middle panel* shows the effects of the two oncogenes of AEV (v-*erb*B and v-*erb*A) on an erythroid target cell. v-*erb*B acts as a primary oncogene in that it induces the cell's self-renewal, proliferation, and factor independence. v-*erb*A expressed in addition leads to a total block of differentiation. The *bottom panel* illustrates the effects of the two oncogenes of the MH2 virus (v-*myc* and v-*mil*) on a myelomonocytic target cell. The v-*myc* gene acts as a primary oncogene in that it induces the cell's proliferation without abolishing its growth factor requirement (in this case, cMGF). If v-*mil* is expressed in addition, the cells synthesize their own cMGF and grow in an autocrine fashion. *Symbols:* The *large circle* represents a cell, the *small circle* its nucleus. *Zigzag arrows,* signals emitted by an activated receptor; *straight arrows,* differentiation; *curved arrows,* self-renewal; *wavy line in the nucleus,* DNA

ZOLO et al. 1979; BEUG et al. 1979). The target cell specificity is not a result of the presence or absence of cell surface recepors for AEV and MH2 since the viruses can infect (but not transform) nontarget cells (see below). In both cases inactivation of one of the two oncogenes either abolishes or decreases the leukemogenic potential of the virus (FRYKBERG et al. 1983; GRAF et al. 1986).

How do these viral oncogenes transform their respective target cells? Normal hematopoietic progenitor cells require specific growth factors for their survival, proliferation, and differentiation. In chickens these factors are erythropoietin (EPO) for erythroid cells and chicken myelomonocytic growth factor (cMGF) for macrophage/granulocyte precursors (LEUTZ et al. 1984, 1989). In erythroid progenitors, v-*erb*B induces factor independence by a non-autocrine mechanism, probably by mimicking an activated growth factor receptor (this oncogene corresponds to a truncated version of the EGF receptor, see above). It also partially induces self-renewal without completely blocking the cell's ability to differentiate into erythrocytes (BEUG et al. 1982a; FRYKBERG et al. 1983). v-*erb*A by itself does not appear to trigger the growth of erythroid cells but completely blocks the differentiation of v-*erb*B-transformed cells, thus generating rapidly proliferating, immature erythroblasts (FRYKBERG et al. 1983; KAHN et al. 1986). v-*erb*A, which corresponds to a mutated version of the thyroid hormone receptor (see above), probably acts by suppressing certain genes involved in erythroid differentiation. Indeed it has recently been found that the genes for the erythrocyte-specific anion transporter protein band 3 (ZENKE et al. 1988) as well as carbonic anhydrase (M. ZENKE and H. BEUG, personal communication) are both suppressed at the transcriptional level by the v-*erb*A gene product. These cellular genes are thought to be crucial for the maturation of a functioning erythrocyte. In summary, the interplay between the two genes of AEV, encoding an aberrant growth factor receptor on the one hand and a aberrant nuclear hormone receptor on the other, leads to the development of a highly leukemogenic phenotype in AEV-infected avian erythroid progenitor cells.

The mechanism by which the oncogenes of MH2 cooperate in myeloid cells is very different. V-*myc* alone is capable of inducing the self-renewal of macrophage-like cells (BEUG et al. 1979; GAZZOLO et al. 1979; GRAF et al. 1986) which, however, remain dependent on cMGF. Introduction of v-*mil* into v-*myc*-transformed macrophages induces their factor independence via the induction of cMGF production and establishment of an autocrine system (GRAF et al. 1986; VON WEIZSÄCKER et al. 1986). This appears to be an important step in leukemogenesis since v-*myc* is weakly leukemogenic on its own but is highly oncogenic when expressed together with v-*mil* (GRAF et al. 1986). A hybrid v-*mil*/v *raf* gene has also been shown to cooperate with v-*myc* in the transformation of murine lymphoid cells (for review see RAPP et al. 1986).

These examples show that the mechanisms of action and cooperation of two different pairs of cytoplasmic and nuclear oncogenes varies with the lineage of the target cell. The two model viruses also provide a basis for the concept of "primary" and "auxiliary" oncogenes (GRAF et al. 1985). For example, v-*erb*B can be considered as the primary oncogene of AEV since it is capable of inducing self-renewal (cell transformation). In contrast, v-*erb*A can be considered as an auxiliary oncogene since it enhances the transformed phenotype of v-*erb*B-transformed erythroblasts. In support of this concept is the observation that several other virus isolates containing v-*erb*B exist although no others contain v-*erb*A, suggesting that v-*erb*B was acquired first during the evolution of AEV. Similarly, in this scheme, v-*myc* acts as the primary oncogene of MH2 virus since it is capable of transforming myelomonocytic cells on its own while v-*mil* has an auxiliary function since it cannot transform by itself but induces cMGF production and, thus, autocrine growth. Again, several independent virus isolates exist which contain v-*myc* but no others with v-*mil*, making it likely that v-*myc* is the oncogene which was acquired first from the host cell during the evolution of the MH2 virus. It is interesting to note that v-*erb*B (as well as other cytoplasmic oncogenes) can replace the function of v-*mil* in myeloid cells and therefore acts as an auxiliary oncogene in this particular cell type (Adkins et al. 1984; Graf et al. 1986). This suggests that the signals emitted from one and the same oncogene are interpreted in different ways by cells belonging to different lineages. The elucidation of the molecular basis of these cell-type-specific differences remains a challenging problem whose answer may provide a key to the understanding of the process of lineage determination during hematopoietic cell differentiation.

No models of oncogene function and cooperativity would be complete without considering the possible involvement of anti-oncogenes. A considerable body of evidence exists which implicates the retinoblastoma *(Rb)* gene as a negative regulator of epithelial cell growth whose inactivation contributes to oncogenesis (for recent reviews see Klein 1987; Ponder 1988). It is not clear, however, if the *Rb* gene is also active in other tissues and how many other suppressor-type genes exist. An unexpected new candidate for a suppressor gene is p53, a gene whose mutated from has long been known to act as an oncogene (see the previous section). Another interesting suppressor gene is K-*rev*-1. This gene suppresses cell transformation by the *ras* oncogene and exhibits a surprising similarity to the *ras* gene family (Kitayama et al. 1989). In conclusion, perhaps the most realistic scenario for the development of neoplasias is the stepwise accumulation in a single cell of mutations which on the one hand deregulate positively acting genes and on the other inactivate negatively acting genes. The resulting alterations in the balance between self-renewal and differentiation are a prerequisite for the onset of uncontrolled proliferation and may eventually result in oncogenesis.

Acknowledgements. We would like to thank S. Ness, C.H. Heldin, H. Land, and B. Vennström for discussions and H. Davies for typing the manuscript.

References

Adams JM, Harris AW, Pinkert CA, Corcoran LM, Alexander WS, Cory S, Palmiter RD, Brinster RL (1985) The c-*myc* oncogene driven by immunoglobulin enhancers induce lymphoid malignancy in transgenic mice. Nature 318:533–538

Adari H, Lowy DR, Willumsen BM, Der CJ, McCormick F (1988) Guanosine triphosphatase activating protein (GAP) interacts with the p21 *ras* effector binding domain. Science 240:518–521

Adkins B, Leutz A, Graf T (1984) Autocrine growth induced by *src*-related oncogenes in transformed chicken myeloid cells. Cell 39:439–445

Almendral JM, Sommer D, MacDonald-Bravo H, Burckhardt J, Perera J, Bravo R (1988) Complexity of the early genetic response to growth factors in mouse fibroblasts. Mol Cell Biol 8:2140–2148

Andres A-C, van der Valk MA, Schönenberger CA, Flückiger F, LeMeur M, Gerlinger P, Groner B (1988) Ha-*ras* and c-*myc* oncogene expression interferes with mor-

phological and functional differentiation of mammary epithelial cells in single and double transgenic mice. Genes Dev 2:1486–1495
Angel P, Imagawa M, Chiu R, Stein B, Imbra RJ, Rahmsdorf HJ, Jonat C, Herrlich P, Karin M (1987) Phorbol ester inducible genes contain a common *cis* element recognized by a TPA-modulated *trans*-acting factor. Cell 49:729–739
Angel P, Hattori K, Smeal T, Karin M (1988a) The *jun* proto-oncogene is positively autoregulated by its product, jun/AP1, Cell 55:875–885
Angel P, Allegretto EA, Okino ST, Hattori K, Boyle WJ, Hunter T, Karin M (1988b) Oncogene *jun* encodes a sequence-specific *trans*-activator similar to AP-1. Nature 332:166–171
Armelin HA, Armelin MCS, Kelly K, Stewart T, Leder P, Cochran BH, Stiles CD (1984) Functional role for c-*myc* in mitogenic response to platelet-derived growth factor. Nature 310:655–660
Assoian RK, Grotendorst GR, Miller DM, Sporn MB (1984) Cellular transformation by coordinated action of three peptide growth factors from human platelets. Nature 309:804
Balduzzi PC, Notter MDF, Morgan HR, Shibuya M (1981) Some biological properties of two new avian sarcoma viruses. J Virol 40:268–275
Barbacid M (1987) *ras* genes. Annu Rev Biochem 56:779–827
Bargmann CI, Hung M-C, Weinberg RA (1986a) Multiple independent activations of the *neu* oncogene by a point mutation altering the transmembrane domain of p 185. Cell 45:649–657
Bargmann CI, Hung M-C, Weinberg RA (1986b) The *neu* oncogene encodes an epidermal growth factor receptor-related protein. Nature 319:226–230
Barker WC, Dayhoff MO (1982) Viral *src* gene products are related to the catalytic chain of mammalian cAMP-dependent protein kinase. Proc Natl Acad Sci USA 68:1520–1524
Barletta C, Pellici P, Kenyon L, Smith SD, Dalla-Favera R (1987) Relationship between c-*myb* locus and the 6q-chromosomal aberration in leukemias and lymphomas. Science 287:1064–1067
Beard JW (1963) Avian virus growths and their etiological agents. Cancer Res 7:1–127
Begg AM (1927) A filterable endothelioma of the fowl. Lancet 1:912–915
Beimling P, Benter T, Sander T, Moelling K (1985) Isolation and characterization of he human cellular *myc* gene product. Biochemistry 24:6349–6355
Benedict SH, Maki Y, Vogt PK (1985) Avian retrovirus S13: properties of the genome and of the transformation-specific protein. Virology 145:154–164
Bentley DL, Groudine M (1986) A block to elongation is largely responsible for decreased transcription of c-*myc* in differentiated HL60 cells. Nature 321:702–706
Bentley DL, Groudine M (1988) Sequence requirements for premature termination of transcription in the human c-*myc* gene. Cell 53:245–256
Berridge MJ (1987) Inositol lipids and cell proliferation. Bioch Biophys Acta 907:33–45
Berridge MJ, Irvine RF (1984) Inositol trisphosphate, a novel second messenger in cellular signal transduction. Nature 312:315–321
Besmer P, Snyder HW Jr, Murphy JE, Hardy WD Jr, Parodi (1983) The Parodi-Irgens feline sarcoma virus and simian sarcoma virus have homologous oncogenes, but in different contexts of the viral genomes. J Virol 46:603–613
Besmer P, Murphy JE, George PC, Qiu F, Gergold PJ, Lederman L, Snyder HW Jr, Brodeur D, Zuckerman EE, Hardy WD (1986a) A new acute transforming feline retrovirus and relationship of its oncogene v-*kit* with the protein kinase gene family. Nature 320:415–421
Besmer P, Lader E, George PC, Bergold PJ, Qui F-H, Zuckerman EE, Hardy WD (1986b) A new acute transforming feline retrovirus with *fms* homology specifies a C-terminally truncated version of the c-*fms* protein that is different from SM-feline sarcoma virus v-*fms* protein. J Virol 60:194–203
Betsholtz C, Johnsson A, Heldin C-H, Westermark B (1985) Efficient reversion of simian sarcoma virus-transformation and inhibition of growth factor-induced mitogenesis by suramin. Proc Natl Acad Sci USA 83:6440–6444

Beug H, Hayman MJ (1984) Temperature-sensitive mutants of avian erythroblastosis virus: surface expression of the *erb*B product correlates with transformation. Cell 36:963–972

Beug H, Claviez M, Jockusch B, Graf T (1978) Differential expression of Rous sarcoma virus specific transformation parameters in enucleated cells. Cell 14:843–857

Beug H, v. Kirchbach A, Däderlein G, Conscience JF, Graf T (1979) Chicken hematopoietic cells transformed by seven strains of defective avian leukemia viruses display three distinct phenotypes of differentiation. Cell 18:375–390

Beug H, Palmieri S, Freudenstein C, Zentgraf H, Graf T (1982a) Hormone-dependent terminal differentiation in vitro of chicken erythroleukemia cells transformed by *ts* mutants of avian erythroblastosis virus. Cell 28:907–919

Beug H, Hayman MJ, Graf T (1982b) Myeloblasts transformed by the avian acute leukemia virus E26 are hormone-dependent for growth and for the expression of a putative *myb*-containing protein, p 135 E26. EMBO J 1:1069–1073

Beug H, Leutz A, Kahn P, Graf T (1984) *ts* mutants of E26 leukemia virus allow transformed myeloblasts, but not erythroblasts or fibroblasts, to differentiate at the nonpermissive temperature. Cell 39:579–588

Beug H, Blundell P, Graf T (1987) Reversibility of differentiation and proliferative capacity in avian myelomonocytic cells transformed by ts E26 leukemia virus. Genes Dev 1:277–286

Biedenkapp H, Borgmeyer U, Sippel AE, Klempnauer K-H (1988) Viral *myb* oncogene encodes a sequence-specific DNA-binding activity. Nature 335:835–837

Birchmeier C, Birnbaum D, Waitches G, Fasano O, Wigler M (1986) Characterization of an activated human c-*ros* gene. Mol Cell Biol 6:3109–3116

Bister K (1984) Molecular biology of avian acute leukaemia viruses. In Goldman JM, Jarrett O (eds) Mechanisms of viral leukaemogenesis. Churchill Livingstone, Edinburgh (Leukaemia and lymphoma research, vol 1)

Bister K, Enrietto P, Graf T, Hayman M (1983) The transforming gene of avian acute leukemia virus MC29. In: Neth R et al. (eds) Haematology and blood transfusion. Mechanism of malignant transformation, Springer, Berlin Heidelberg New York (Modern trends in human leukemia, vol 5)

Biswas R, Basu M, Sen-Majumdar A, Das M (1985) Intrapeptide autophosphorylation of the epidermal growth factor receptor: regulation of kinase catalytic function by receptor dimerization. Biochemistry 24:3795–3802

Blanchard J-M, Piechacyzk M, Dani C, Chambard J-C, Franchi A, Pouyssegur J, Jeanteur P (1985) C-*myc* gene is transcribed at high rate in Go-arrested fibroblasts and is post-transcriptionally regulated in response to growth factors. Nature 317:443–445

Bohmann D, Bos TJ, Nishimura T, Vogt PK, Tjian R (1987) Human protooncogene c-*jun* encodes a protein with antigenic and enhancer binding properties of transcription factor AP-1. Science 238:1386–1392

Bolen JB, Thiele CJ, Israel MA, Yonemoto W, Lipsich LA, Brugge JS (1984) Enhancement of cellular *src* gene product associated tyrosyl kinase activity following polyoma virus infection and transformation. Cell 38:767–777

Bollag GE, Roth RA, Beaudoin J, Mochly-Rosen D, Koshland DE Jr (1986) Protein kinase C directly phosphorylates the insulin receptor in vitro and reduces its protein-tyrosine kinase activity. Proc Natl Sci USA 83:5822–5824

Bos JL (1988) The *ras* gene family and human carcinogenesis. Mutat Res (Elsevier) 195:255–271

Bos T, Bohmann D, Tsuchie H, Tjian R, Vogt P (1988) v-*jun* encodes a nuclear protein with enhancer binding proteins of AP-1. Cell 52:705–712

Boulukos KE, Pognonec P, Begue A, Galibert F, Gesquiere D, Stehelin D, Ghysdael J (1988) Identification in chickens of an evolutionary conserved cellular *ets*-2 gene (c-*ets*-2) encoding nuclear proteins related to the products of the c-*ets* proto-oncogene. EMBO J 7:697–705

Boyle WJ, Lampert MA, Lipsick JS, Baluda MA (1984) Avian myeloblastosis virus and E26 virus oncogene products are nuclear proteins. Proc Natl Acad Sci USA 81:4265–4269
Bravo R, Burckhardt J, Curran T, Müller R (1985) Stimulation and inhibition of growth by EGF in different A431 cell clones is accompanied by the rapid induction of c-*fos* and c-*myc* proto-oncogenes. EMBO J 5:695–700
Brown AM, Papkoff J, Fung YK, Shackleford GM, Varmus HE (1987) Identification of protein products encoded by the proto-oncogene *int*-1. Mol Cell Biol 7:3871–3977
Brown AMC, Wildin RS, Prendergast TJ, Varmus HE (1986) A retrovirus vector expressing the putative mammary oncogene *int*-1 causes partial transformation of a mammary epithelial cell line. Cell 46:1001–1009
Browning PJ, Bunn HF, Cline A, Shuman M, Nienhuis AW (1986) Replacement of COOH-terminal truncation of v-*fms* with c-*fms* sequences markedly reduces transformation potential. Proc Natl Acad Sci USA 83:7800–7804
Büscher M, Rahmsdorf HJ, Liftin M, Karin M, Herrlich P (1988) Activation of the c-*fos* gene by UV and phorbol ester: different signal transduction pathways converge to the same enhancer element. Oncogene 3:301–311
Cabrera CV, Alonso MC, Johnston P, Phillips RG, Lawrence PA (1987) Phenocopies induced with anti-sense RNA identify the wingless gene. Cell 50:659–663
Cartwright CA, Kaplan PL, Cooper JA, Hunter T, Eckhart W (1986) Altered sites of tyrosine phosphorylation in pp60c-*src* associated with polyoma middle tumor antigen. Mol Cell Biol 6:1562–1570
Cartwright CA, Eckhart W, Simon S, Kaplan PL (1987) Cell transformation by pp60c-*src* mutated in the carboxy-terminal regulatory domain. Cell 49:83–91
Castagna M, Takai Y, Kaibuchi K, Sano K, Kikiwa V, Nishizuka Y (1982) Direct activation of calcium-activated, phospholipid-dependent protein kinase by tumor-promoting phorbol esters. J Biol Chem 257:7847–7851
Cavalieri P, Ruscio T, Tinoco R, Benedict S, Davis C, Vogt PK (1985) Isolation of three new avian sarcoma viruses: ASV 9, ASV 17, and ASV 25. Virology, 143:680–683
Chang EH, Gonda MA, Ellis RW, Scolnick EM, Lowy DR (1982) Human genome contains four genes homologous to transforming genes of Harvey and Kirsten murine sarcoma viruses. Proc Natl Acad Sci USA 79:4848–4852
Chavrier P, Zerial M, Lemaire P, Almendral J, Bravo R, Charnay P (1988) A gene encoding a protein with zinc fingers is activated during G0/G1 transition in cultured cells. EMBO J 7:29–35
Chiu R, Boyle W, Meek J, Smeal T, Hunter T, Karin M (1988) The c-*fos* protein interacts with c-*Jun*/AP-1 to stimulate transcription of AP-1 responsive genes. Cell 54:541–552
Clark SS, McLaughlin J, Crist WM, Champlin R, Witte O (1987) Unique forms of the *abl* tyrosine kinase distinguish Ph'-positive CML from Ph'-positive ALL. Science 235:85–88
Clarke ME, Kukowska-Latallo JF, Westin E, Smith M, Prochownik EV (1988) Constitutive expression of a c-*myb* cDNA blocks Friend murine erythroleukemia cell differentiation. Mol Cell Biol 8:884–892
Cochet C, Gill GN, Meisenhelder J, Cooper JA, Hunter T (1984) C-kinase phosphorylates the epidermal growth factor receptor and reduces its epidermal growth factor-stimulated tyrosine protein kinase activity. J Biol Chem 259:2553–2558
Cochran BH, Reffel AC, Stiles CD (1983) Molecular cloning of gene sequences regulatd by platelet-derived growth factor. Cell 33:939–947
Cochran BH, Zullo J, Verma IM, Stiles CD (1984) Expression of the c-*fos* oncogene and a newly discovered r-*fos* is stimulated by platelet-derived growth factor. Science 226:1080–1082
Cohen DR, Curran T (1988) *fra*-1: a serum-inducible, cellular immediate-early gene that encodes a *fos*-related antigen. Mol Cell Biol 8:2063–2069

Cole MD (1986) The *myc* oncogene: its role in transformation and differentiation. Annu Rev Genet 20:361–384

Coll J, Righi M, de Taisne C, Dissous C, Gegonne A, Stehelin D (1983) Molecular cloning of the avian acute transforming retrovirus MH2 reveals a novel cell-derived sequence (v-*mil*) in addition to the *myc* oncogene. EMBO J 2(12):2189–2194

Cooper CS, Park M, Blair DG, Tainsky MA, Huebener K, Croce CM, Vande Woude GF (1984) Molecular cloning of a new transforming gene from a chemically transformed human cell line. Nature 311:29–33

Cooper JA (in press) The *src*-family of protein-tyrosine kinases. In: Kemp B, Alewood PF (eds) Peptides and protein phosphorylation. CRC Pres, Boca Raton

Coppola JA, Cole MD (1986) Constituive c-*myc* oncogene expression blocks mouse erythroleukaemia cell differentiation but not commitment. Nature 320:760–763

Cory S (1986) Activation of cellular oncogenes in hemopoietic cells by chromosome translocations. Adv Cancer Res 47:189–234

Cory S, Harris AW, Langdon WY, Alexander WS, Corcoran LM, Palmiter RD, Pinkert CA, Brinster RL, Adams JM (1987) The *myc* oncogene and lymphoid neoplasia: from translocations to transgenic mice. Hamatol-Bluttransfus 31:248–251

Cotton PC, Brugge JS (1983) Neural tissues express high levels of the cellular *src* gene product pp60c-src. Mol Cell Biol 3:1157–1162

Courtneidge SA (1985) Activation of the pp60c-*src* kinase by middle T antigen binding or by dephosphorylation. EMBO J 4:1471–1477

Courtneidge SA, Smith AE (1983) Polyoma virus transforming protein associates with the product of the c-*src* cellular gene. Nature 303:435–439

Coussens L, Yang-Feng TL, Liao Y-C, Chen E, Gray A, McGrath J, Seeburg PH, Libermann TA, Schlessinger J, Francke U, Levinson A, Ullrich A (1985) Tyrosine kinase receptor with extensive homology to EGF receptors shares chromosomal location with *neu* oncogene. Science 230:1132–1139

Coussens L, Van Beveren C, Smith D, Chen E, Mitchell RL, Isacke CM, Verma IM, Ullrich A (1986) Structural alteration of viral homologue of receptor protooncogene *fms* at carboxyl terminus. Nature 320:277–280

Curran T, Franza BR Jr (1988) *Fos* and *jun*: the AP-1 connection. Cell 55:395–397

Curran T, Morgan JI (1988) Memories of *fos*. Bio Essays 7:255–258

Curran T, Verma IM (1984) The FBR murine osteosarcoma virus I. Molecular analysis and characterization of a 75,000 Da *gag-fos* fusion product. Virology 135:218–228

Curran T, Miller AD, Zokas L, Verma IM (1984) Viral and cellular *fos* proteins: a comparative analysis. Cell 36:257–268

Cuttitta F, Carney DN, Mulshine J, Moody TW, Fedorko J, Fischler A, Minna JD (1985) Bombesin-like peptides can function as autocrine growth factors in human small-cell lung cancer. Nature 316:823–826

Cuypers HT, Selten G, Quint W, Zijlstra M, Robanus-Maandag E, Boelens W, Van Wezenbeek P, Melief C, Berns A (1984) Murine leukemia virus-induced T-cell lymphomagenesis: integration of proviruses in a distinct chromosomal region. Cell 37:141–150

Davis RJ, Czech MP (1985) Platelet derived growth factor mimics phorbol diester action on epidermal growth factor receptor phosphorylation at threonine-654. Proc Natl Acad Sci USA 82:4080–4084

Dean M, Levine RA, Campisi J (1986) C-*myc* regulation during retinoic acid-induced differentiation of F9 cells is post-transcirptional and associated with growth arrest. Mol Cell Biol 6:518–524

Delli-Bovi P, Curatola AM, Kern FG, Greco A, Ittmann M (1987) An oncogene isolated by transfection of Kaposi's sarcoma DNA encodes a growth factor that is a member of the FGF family. Cell 50:729–737

Delli-Bovi P, Curatola AM, Newman KM, Sato Y, Moscatelli D, Hewick RM, Rifkin DB, Basilico C (1988) Processing, secretion, and biological properties of a novel growth factor of the fibroblast growth factor family with oncogenic potential. Mol Cell Biol 8:2933–2941

Dickson C, Peters G (1987) Potential oncogene product related to growth factors. Nature 326:833
Dickson C, Smith R, Brookes S, Peters G (1984) Tumorigenesis by mouse mammary tumour virus: proviral activation of a cellular gene in the common integration region *int*-2. Cell 37:529–536
DiFiore PP, Pierce JH, Fleming TP, Hazan R, Ullrich A, Richter-King C, Schlessinger J, Waterfield MD (1987) Overexpression of the human EGF receptor confers an EGF-dependent transformed phenotype to NIH 3T3 cells. Cell 51:1063–1070
Distel RJ, Ro HS, Rosen BS, Groves DL, Spiegelmann BM (1987) Nucleoprotein complexes that regulate gene expression in adipocyte differentiation: direct participation of c-*fos*. Cell 49:835–844
Donner L, Fedele LA, Garon CF, Anderson SJ, Sherr CJ (1982) McDonough feline sarcoma virus: characterization of the molecularly cloned provirus and its feline oncogene (v-*fms*). J Virol 41:489–500
Doolittle RF, Hunkapiller MW, Hood LE, Devare SG, Robbins KC, Aaronson SA, Antoniades HN (1983) Simian sacroma virus *onc* gene, v-*sis*, is derived from the gene (or genes) encoding a platelet-derived growth factor. Science 221:275–277
Downing JR, Rettenmier CW, Sherr CJ (1988) Ligand-induced tyrosine kinase activity of the colony-stimulating factor 1 receptor in a murine macrophage cell line. Mol Cell Biol 8:1795–1799
Downward J, Parker P, Waterfield MD (1984a) Autophosphorylation sites on the receptor for epidermal growth factor. Nature 311:483–485
Downward J, Yarden Y, Mayes E, Scrace G, Totty N, Stockwell P, Ullrich A, Schlessinger J, Waterfield MD (1984b) Close similarity of epidermal growth factor receptor and v-*erb*B oncogene protein sequences. Nature 307:521–527
Duprey SP, Boettiger D (1985) Developmental regulation of c-*myb* in normal myeloid progenitor cells. Proc Natl Acad Sci USA 82:6937–6941
Eick D, Piechazyk M, Henglein B, Blanchard J-M, Traub B, Kofler E, Wiest S, Lenoir GM, Kofler E, Wiest S, Lenoir GM, Bornkamm GW (1985) Aberrant c-*myc* of Burkitt's lymphoma cells have longer half-lives. EMBO J 4:3717–3725
Eick D, Berger R, Polack A, Bornkamm GW (1987) Transcription of c-*myc* in human mononuclear cells is regulated by an elongation block. Oncogene 2:61–65
Eliyahu D, Raz A, Gruss P, Givol D, Oren M (1984) Participation of p53 cellular tumour antigen in transformation of normal embryonic cells. Nature 312:646–649
Evan GI, Hancock DC (1985) Studies on the interaction of the human c-*myc* protein with cell nuclei: p62 c-*myc* as a member of a discrete subset of nuclear proteins. Cell 43:253–261
Farber E, Cameron R (1980) The sequential analysis of cancer development. Adv Cancer Res 31:125–226
Feramisco JR, Gross M, Kamota T, Rosenberg M, Sweet RW (1984) Microinjection of the oncogene form of the human H-*ras* (T-24) protein results in rapid proliferation of quiescent cells. Cell 38:109–117
Finkel MP, Biskis BO, Jinkins PB (1966) Virus induction of osteosarcomas in mice. Science 151:698–701
Finkel MP, Reilly CA Jr, Biskis BO, Grecco IL (1973) Bone tumor viruses. In: Saunders CL, Bush RH, Ballon JE, Mahlum CC (eds) Colston papers, Proceedings 24th Symposium Colston Research Society, vol 24. Butterworths, London, pp 353–366
Finkel MP, Reilly CA Jr, Biskis BO (1975) Viral etiology of bone cancer. Front Radiat Ther Oncol 10:28–39
Finkel T, Der JJ, Cooper GM (1984) Activation of *ras* genes in human tumors does not affect localization, modification or nucleotide binding properties of p21. Cell 37:151–158
Frankel AE, Neubauer RL, Fischinger PJ (1976) Fractionation of DNA nucelotide transcripts from Moloney sarcoma viruses and isolation of sarcoma virus-specific complementary DNA. J Virol 18:481–490

Frykberg L, Palmieri S, Beug H, Graf T, Hayman MJ, Vennström B (1983) Transforming capacities of avian erythroblastosis virus mutants deleted in the *erbA* or *erbB* oncogenes. Cell 32:227–238

Fung Y-KT, Shackleford GM, Brown AMC, Sanders G, Varmus HE (1985) Nucleotide sequence and expression in vitro of cDNA derived from mRNA of int-1, a provirally activated mouse mammary oncogene. Mol Cell Biol 5:3337–3344

Gamett D, Tracy S, Robinson H (1986) Differences in sequences encoding the carboxyl-terminal domain of the epidermal growth factor receptor correlate with differences in the disease potential of viral *erb*B genes. Proc Natl Acad Sci USA 83:6053–6057

Gandrillon O, Jurdic P, Benchaibi M, Xiao J-H, Ghysdael J, Samarut J (1987) Expression of the v-*erb*A oncogene in chicken embryo fibroblasts stimulates their proliferation in vitro and enhances tumor growth in vivo. Cell 49:687–697

Gazit A, Igarashi H, Chiu I-M, Srinivasan A, Yaniv A, Tronick SR, Robbins KC, Aaronson SA (1984) Expression of the normal human *sis*/PDGF-2 coding sequence induces cellular transformation. Cell 39:89–97

Gazzolo L, Moscovici C, Moscovici MG, Samarut J (1979) Response of hematopoietic cells to avian acute leukemia viruses: effects on the differentiation of the target cells. Cell 16:627–638

Geissler EN, Ryan MA, Housman DE (1988) The dominant-white spotting (W) locus of the mouse encodes the c-*kit* proto-oncogene. Cell 55:185–192

Gerondakis S, Bishop JM (1986) Structure of the protein encoded by the chicken protooncogene c-*myb*. Mol Cell Biol 6:3677–3684

Gewirtz AM, Calabretta B (1988) A c-*myb* antisense oligodeoxynucleotide inhibits normal human hematopoiesis in vitro. Science 242:1303–1306

Ghysdael J, Gegonne A, Pognonec P, Dermis D, LePrince D, Stehelin D (1986) Identification and preferential expression in thymic and bursal lymphocytes of a c-*ets* oncogene-encoded Mr 54,000 cytoplasmic protein. Proc Natl Acad Sci USA 83:1714–1718

Gibbs JB, Sigal IS, Scolnick EM (1985) Biochemical properties of normal and oncogenic *ras* p21. Trends Biochem Sci 9:350–353

Gill GN, Kawamoto T, Cochet C, Le A, Sato JD, Masui H, McLeod C, Mendelsohn J (1984) Monoclonal anti-epidermal growth factor receptor antibodies which are inhibitors of epidermal growth factor binding and antagonists of epidermal growth factor-stimulated tyrosine protein kinase activity. J Biol Chem 259:7755–7760

Golay J, Introna M, Graf T (1988) A single point mutation in the v-*ets* oncogene affects both erythroid and myelomonocytic cell differentiation. Cell 55:1147–1158

Golden A, Nemeth SP, Brugge JS (1986) Blood platelets express high levels of the pp60c-*src* specific tyrosine kinase activity. Proc Natl Acad Sci USA 83:852–856

Gould K, Hunter T (1988) Platelet-derived growth factor induced multisite phosphorylation of pp60c-*src* increases its tyrosine kinase activity. Mol Cell Biol 8:3345–3356

Gould KL, Woodgett JR, Cooper JA, Buss JE, Shalloway D, Hunter T (1985) Protein kinase C phosphorylates pp60-*src* at a novel site. Cell 42:849–857

Graf T (1973) Two types of target cells for transformation with avian myelocytomatosis virus. Virology 54:398–413

Graf T, Stehelin D (1982) Avian leukemia viruses. Oncogenes and genome structure. BBA Rev Cancer 651:245–271

Graf T, v. Weizsäcker F, Grieser S, Coll J, Stehelin D, Patschinsky T, Bister K, Bechade C, Calothy G, Leutz A (1986) V-*mil* induces autocrine growth and enhances tumorigenicity in v-*myc* transformed avian macrophages. Cell 100:357–364

Graf T, Beug H (1978) Avian leukemia viruses: interaction with their target cells in vivo and in vitro. BBA Revs Cancer, 516:269–299.

Graf T, Adkins B, Leutz A, Beug H, Kahn P (1985) Cooperativity between "primary" and "auxiliary" oncogenes of defective avian leukemia viruses. In: Marks PA (ed) Bristol-Myers Cancer Symposia, Genetics, Cell Differentiation, and Cancer, vol 7. Academic Press, New York, pp 171–182

Graf T, v. Kirchbach A, Beug H (1981) Characterization of the hematopoietic target cells of AEV, MC29 and AMV avian leukemia viruses. Exptl Cell Res 131:331–343
Green S, Chambon P (1986) A superfamily of potentially oncogenic hormone receptors. Nature 324:615–617
Greenberg ME, Ziff EB (1984) Stimulation of 3T3 cells induces transcription of the c-*fos* proto-oncogene. Nature 311:433–438
Greenberg ME, Hermanowski AL, Ziff EB (1986) Effect of protein synthesis inhibitors on growth factor activation of c-*fos*, and actin gene transcription. Mol Cell Biol 6:1050–1057
Groffen J, Stephenson JR, Heisterkamp N, de Klein A, Bartram CR, Grosveld G (1984) Philadelphia chromosomal breakpoints are clustered within a limited region, *bcr*, on chromosome 22. Cell 36:93–99
Halazonetis TD, Georgopoulos K, Greenberg ME, Leder P (1988) c-*jun* dimerizes withitself and with c-*fos*, forming complexes of different DNA binding affinities. Cell 55:917–924
Hampe A, Gobet H, Sherr CJ, Galibert F (1984) Nucleotide sequence of the feline retroviral oncogene v-*fms* shows unexpected homology with oncogenes encoding tyrosine-kinase specific protein kinases. Proc Natl Acad Sci USA 81:85–89
Hanley MR, Jackson T (1987) Transformer and transducer. Nature 328:668–669
Hann SR, Eisenman RN (1984) Proteins encoded by the human c-*myc* oncogene: differential expression in neoplastic cells. Mol Cell Biol 4:2486–2497
Hannink M, Donoghue DJ (1986) Cell surface expression of membrane-anchored v-*sis* gene products: glycosylation is not required for cell surface transport. J Cell Biol 103:2311–2322
Hayman MJ, Kitchener G, Vogt PK, Beug H (1985) The putative transforming protein of S13 avian erythroblastosis virus is a transmembrane glycoprotein with an associated protein kinase activity. Proc Natl Acad Sci USA 82:8237–8241
Hayman MJ, Kitchener G, Knight J, McMahon J, Watson R, Beug H (1986) Autophosphorylation of v-*erb*B does not correlate with cell transformation. Virology 150:270–275
Hayward WS, Neel BG, Astrin SM (1981) ALV-induced lymphoid leukosis: activation of a cellular *onc* gene by promoter insertion. Nature 290:475–480
Heikkila R, Schwab G, Wickstrom E, Loke SL, Pluznik DH, Watt R, Neckers LM (1987) A c-*myc* antisense oligodeoxynucleotide inhibits entry into S phase but not progress from G0 to G1. Nature 328:445–449
Heldin CH, Betsholtz C, Claesson-Welsh L, Westermark B (1987) Subversion of growth regulatory pathways in malignant transformation. Biochim Biophys Acta 907:219–244
Hermans A, Heisterkamp N, von Lindern M, van Baal S, Meijer D, van der Plas D, Wiedermann LM, Groffen J, Bootsma D, Grosveld G (1987) Unique fusion of *bcr* and c-*abl* genes in Philadelphia positive acute lymphoblastic leukemia. Cell 51:33–40
Holt JT, Gopal TV, Moulton AD, Nienhuis W (1986) Inducible production of c-*fos* antisense RNA inhibits 3T3 cell proliferation. Proc Natl Acad Sci USA 83:4794–4798
Housey GM, Johnson MD, Hsiao WLW, O'Brian CA, Murphy JP, Kirschmeier P, Weinstein IB (1988) Overproduction of protein kinase C causes disordered growth control in rat fibroblasts. Cell 52:343–354
Huang JS, Huang SS, Deuel TF (1984) Transforming protein of simian sarcoma virus stimulates autocrine growth of SSV-transformed cells through PDGF cell-surface receptors. Cell 39:79–87
Huleihel M, Goldsborough M, Cleveland JL, Gunnell M, Bonner T, Rapp UR (1986) Characterization of murine A-*raf*, a new oncogene related to the v-*raf* oncogene. Mol Cell Bil 6:3934–3938
Hunter T, Ling N, Cooper JA (1984) Protein kinase C phosphorylation of the EGF receptor at a threonine residue close to the cytoplasmic face of the plasma membrane. Nature 311:480–483

Iba H, Takeya T, Cross FR, Hanafusa T, Hanafusa H (1984) Rous sarcoma virus variants that carry the cellular *src* gene instead of the viral *src* gene cannot transform chicken embryo fibroblasts. Proc Natl Acad Sci USA 81:4424–4428

Ishikawa F, Takaku F, Nagao M, Sugimura T (1987) Rat c-*raf* oncogene activation by a rearrangement that produces a fused protein. Mol Cell Biol 7:1226–1232

Jackowski S, Rettenmier CW, Sherr CJ, Rock CO (1986) A guanine nucleotide-dependent phosphatidylinositol-4,5-diphosphate phospholipase C in cells transformed by the v-*fms* and v-*fes* oncogenes. J Biol Chem 261:4978–4985

Jackson TR, Blair LA, Marshall J, Goedert M, Hanley MR (1988) The *mas* oncogene encodes an angiotensin receptor. Nature 335:437–440

Jansen HW, Ruckert B, Lurz R, Bister K (1983) Two unrelated cell-derived sequences in the genome of avian leukemia and carcinoma inducing retrovirus. EMBO J 48:61–73

Jenkins JR, Sturzbecher HW (1988) The *p53* oncogene. In: Reddy EP, Skala AM, Curran T (eds) The oncogene handbook. Elsevier, pp 403–423

Jenkins JR, Rudge K, Curie GA (1984a) Cellular immortalization by a cDNA clone encoding the transformation-associated phosphoprotein p53. Nature 312:651–654

Jenkins JR, Rudge K, Redmond S, Wade-Evens A (1984b) Cloning and expression of full length mouse cDNA sequences encoding the transformation associated protein p53. Nucl Acids Res 12:5609–5626

Jenkins JR, Rudge K, Chumakov P, Curie GA (1985) The cellular oncogene p53 can be activated by mutagenesis. Nature 317:816–818

Jenuwein T, Muller R (1987) Structure-function analysis of *fos* protein: a single amino acid change activates the immortalizing potential of v-*fos*. Cell 48:647–657

Jenuwein T, Muller D, Curran T, Muller R (1985) Extended lifespan and tumorigenicity of nonestablished mouse connective tissue cells transformed by the *fos* oncogene of FBR-MSV. Cell 41:629–637

Johnsson A, Betsholtz C, Heldin C-H, Westermark B (1985b) Antibodies against platelet-derived growth factor inhibit acute transformation by simian sarcoma virus. Nature 317:438–440

Johnsson A, Betsholtz C, von der Helm K, Heldin C-H, Westermark B (1985) Platelet-derived growth factor agonist activity of a secreted form of the v-*sis* oncogene product. Proc Natl Acad Sci USa 82:1721–1725

Johnsson A, Heldin C-H, Wasteson Å, Westermark B, Deuel TF, Huang JS, Seeburg PH, Gray A, Ullrich A, Scrace G, Stroobant P, Waterfield MD (1984) The c-*sis* gene encodes a precursor of the B chain of platelet-derived growth factor. EMBO J 3:921–928

Josephs SF, Guo C, Ratner L, Wong-Staal F (1984) Human proto-oncogene nucleotide sequences corresponding to the transforming region of simian sarcoma virus. Science 223:487–490

Jurdic P, Benchaibi M, Gandrillon O, Samarut J (1987) Transforming and mitogenic effects of avian leukemia virus E26 on chicken hematopoietic cells and fibroblasts, respectively, correlate with level of expression of the provirus. J Virol 61:3058–3065

Kaczmareck L, Hyland JK, Watt R, Rosenberg M, Baserga R (1985) Microinjected c-*myc* as a competence factor. Science 228:1313–1315

Kahn P, Frykberg L, Graf T (1986) Cooperativity between v-*erb*A- and v-*src*-related oncogenes in erythroid cell transformation. In: Deinhardt F (ed) Proceedings. Hamburg XII Symp for Comp Res on Leukemia and Rel Diseases. Berlin Heidelberg New York, pp 41–50

Kan NC, Flordellis CS, Garon CF, Duesberg PH, Papas TS (1983) Avian carcinoma virus MH2 contains a transformation-specific sequence, MHT, and shares the *myc* sequence with MC29, CMII, and OK10 viruses. Proc Natl Acad Sci 80: 6566–6570

Katan M, Parker PJ (1988) Oncogenes and cell control. Nature 332:203–205

Kato J, Takeya T, Grandon C, Iba H, Levy JB, Hanafusa H (1986) Amino acid substitutions sufficient to convert the nontransforming pp60c-*src* protein to a transforming protein. Mol Cell Biol 6:4155–4160

Keating MT, Williams LT (1988) Autocrine stimulation of intracellular PDGF receptors in v-*sis*-transformed cells. Science 239:914–916

Kelloff GJ, Lane WT, Turner HC, Huebner RJ (1969) In vivo studies of the FBJ murine osteosarcoma virus. Nature 223:1379–1380

Kelly K, Cochran BH, Stiles CD, Leder P (1983) Cell-specific regulation of the c-*myc* gene by lymphocyte mitogens and platelet-derived growth factor. Cell 35:603–610

Khazaie K, Dull TJ, Graf T, Schlessinger J, Ullrich A, Beug H, Vennström B (198) Truncation of the human EGF receptor leads to differential transforming potentials in primary avian fibroblasts and erythroblasts. EMBO J 7:3061–3071

King CR, Kraus MH, Aaronson SA (1985a) Amplification of a novel v-*erb*B-related gene in a human mammary carcinoma. Science 229:974–976

King CR, Kraus MH, Aaronson SA (1985b) Human tumor cell lines with EGF receptor gene amplification in the absence of aberrant sized mRNAs. Nucl Acids Res 13:8447–8486

Kitayama H, Sugimoto Y, Matsuzaki T, Ikawa Y, Noda M (1989) A *ras*-related gene with transformation suppressor activity. Cell 56:77–84

Klein G (1987) The approaching era of the tumor suppressor genes. Science 238:1539–1545

Klein G, Klein E (1985) Evolution of tumors and the impact of molecular oncology. Nature 315:190–195

Klein G, Klein E (1986) Conditioned tumorigenicity of activated oncogenes. Cancer Res 46:3211:3224

Klempnauer K-H, Symonds G, Evans GI, Bishop JM (1984) Subcellular localization of proteins encoded by oncogenes of avian myeloblastosis virus and avian leukemia virus E26 and by the chicken c-*myb* gene. Cell 37:537–547

Klinken SP, Alexander WS, Adams JM (1988) Hematopoietic lineage switch: v-*raf* oncogene converts Eμ-*myc* transgenic B cells into macrophages. Cell 53:857–867

Kmiecik TE, Shalloway D (1987) Activation and suppression of pp60c-*src* transforming ability by mutation of its primary sites of tyrosine phosphorylation. Cell 49:65–73

Knight J, Zenke M, Disela C, Kowenz E, Vogt P, Engel JD, Hayman MJ, Beug H (1988) Temperature-sensitive v-*sea* transformed erythroblasts: a model system to study gene expression during erythroid differentiation. Genes Dev 2:247–258

Knudson AG Jr (1986) Genetics of human cancer. Annu Rev Genet 20:231–251

Koda T, Sasaki A, Matsushima S, Kakinuma M (1987) A transforming gene, *hst*, found in NIH 3T3 cells transformed with DNA from three stomach cancers and a colon cancer. J Cancer Res 78(4):325–328

Kouzarides T, Ziff E (1988) The role of the leucine zipper in the *fos-jun* interaction. Nature 336:646–651

Kozma SC, Redmond SM, Fu XC, Saurer SM, Groner B, Hynes NE (1988) Activation of the receptor kinase domain of the *trk* oncogene by recombination with two different cellular sequences. EMBO J 7:147–154

Kraus MH, Popescu NC, Amsbaugh SC, King CR (1987) Overexpression of the EGF receptor-related proto-oncogene *erb*B-2 in human mammary tumor cell lines by different molecular mechanisms. EMBO J 6:605–610

Kris RM, Lax I, Gullick W, Waterfield MD, Ullrich A, Fridkin M, Schlessinger J (1985) Antibodies against a synthetic peptide as a probe for the kinase activity of the avian EGF receptor and v-*erb*B protein. Cell 40:619–625

Kruijer W, Cooper JS, Hunter T, Verma IM (1984) Platelet-derived growth factor induces rapid but transient expression of the c-*fos* gene and protein. Nature 312:7330–7334

Kuehl WM, Bender TP, Stafford J, McClinton D, Segal S, Dmitrovsky E (1988) Expression and function of the c-*myb* oncogene during hematopoietic differention. Curr Top Microbiol Immunol 141:318–323

Lachman HM, Skoultchi AI (1984) Expression of c-*myc* changes during differentiation of mouse erythroleukemia cells. Nature 82:5323–5327

Lachman HM, Hatton KS, Skoultchi AI, Schildkraut CL (1985) C-*myc* mRNA levels in the cell cycle change in mouse erythroleukemia cells following inducer treatment. Proc Natl Acad Sci USA 82:5232–5327

Lamph WW, Wamsley P, Sassone-Corsi P, Verma IM (1988) Induction of protooncogene JUN/AP-1 by serum and TPA. Nature 334:629–631

Land H (1986) Oncogenes cooperate, but how? In: Kahn P, Graf T (eds) Oncogenes and growth control. Springer, Berlin Heidelberg New York, pp 304–311

Land H, Chen AC, Morganstern JP, Parada LF, Weinberg RA (1986) Behavior of *myc* and *ras* oncogenes in transformation of rat embryo fibroblasts. Mol Cell Biol 6:1917–1925

Land H, Parada LF, Weinberg RA (1983) Tumorigenic conversion of primary embryo fibroblasts requires at least two cooperating oncogenes. Nature 304:596–602

Landschulz WH, Johnson PF, McKnight SL (1988) The leucine zipper: a hypothetical structure common to a new class of DNA binding proteins. Science 240:1759–1764

Lane DP, Crawford LV (1979) T-antigen is bound to a host protein in SV40-transformed cells. Nature 278:261–263

Lang RA, Metcalf D, Gough NM, Dunn AR, Gonda TJ (1985) Expression of a hemopoietic growth factor cDNA in a factor-dependent cell line results in autonomous growth and tumorigenicity. Cell 43:531–542

Lau LF, Nathans D (1985) Identification of a set of genes expressed during the G0/G1 transition of cultured mouse cells. EMBO J 4:3145–3151

Lau LF, Nathans D (1987) Expression of a set of growth-related immediate early genes in BALB/c 3T3 cells: coordinate regulation with c-*fos* or c-*myc*. Proc Natl Acad Sci USA 84:1182–1186

Lax I, Kris R, Sasson I, Ullrich A, Hayman MJ, Beug H, Schlessinger J (1985) Activation of c-*erb*B in avian leukosis virus-induced erythroblastosis leads to the expresion of a truncated EGF receptor kinase. EMBO J 4:3179–3182

Leal F, William LT, Robbins KC, Aaronson SA (1985) Evidence that the v-*sis* gene product transforms by interaction with the receptor for platelet-derived growth factor. Science 230:327–330

Lech K, Anderson K, Brent R (1988) DNA-bound *fos* proteins activate transcription in yeast. Cell 52:179–184

Leder A, Pattengale PK, Kuo A, Stewart TA, Leder P (1986) Consequences of widespread deregulation of the c-*myc* gene in transgenic mice: multiple neoplasms and normal development. Cell 45:485–495

Lee CK, Chan EW, Reilly CA Jr, Pahnke VA, Rockus G, Finkel MP (1979) In vitro properties of FBR murine osteosarcoma virus. Proc Soc Exp Biol Med 162:214–220

Lee W, Mitchell P, Tjian R (1987) Purified transcription factor AP-1 interacts with TPA-inducible enhancer elements. Cell 49:741–752

Lemaire P, Révélant O, Bravo R, Charnay P (1988) Two mouse genes encoding potential transcription factors with identical DNA-binding domains are activated by growth factors in cultured cells. Proc Natl Acad Sci USA 85:4691–4695

LePrince D, Gegonne A, Coll J, de Taisne C, Schneeberger A, Lagrou C, Stehelin D (1983) A putative second cell-derived oncogene of the avian leukaemia retrovirus E26. Nature 306:395–397

LePrince D, Duterque-Coquillard M, Li P, Henry C, Flourens A, Debuire B, Stehelin D (1988) Alternative splicing within the chicken c-*ets*-1 locus: implications for transduction within the E26 retrovirus of the c-*ets* proto-oncogene. J Virol 62:3233–3241

Letterio JJ, Coughlin SR, Williams LT (1986) Pertussis toxin-sensitive pathway in the stimulation of c-*myc* expression and DNA synthesis by bombesin. Science 234:1117–1119

Leutz A, Beug H, Graf T (1984) Purification and characterization of cMGF, a novel chicken myelomonocytic growth factor. EMBO J 3:3191–3197

Leutz A, Damm K, Sterneck E, Kowenz E, Ness S, Frank R, Gausepohl H, Pan Y-CE, Smart J, Hayman MJ, Graf T (1989) Molecular cloning of the chicken

myelomonocytic growth factor (cMGF) reveals relationship to interleukin 6 and granulocyte colony stimulating factor. EMBO J 8:175–181
Levy JA, Hartley JW, Rowe WP, Huebner RJ (1973) Studies of FBJ osteosarcoma virus in tissue culture. I. Biological characteristics of the "C"-type viruses. J Natl Cancer Inst 51:525–539
Levy JA, Kazan PL, Reilly CA Jr, Finkel MP (1987) FBJ osteosarcoma virus in tissue culture III. Isolation and characterization of non-virus producing FBJ-transformed cells. J Virol 26:11–15
Libermann TA, Nusbaum HR, Razon N, Kris R, Lax I, Soreq H, Whittle N, Waterfield MD, Ullrich A, Schlessinger J (1985) Amplification enhanced expression and possible rearrangement of EGF receptor gene in primary human brain tumors of glial origin. Nature 313:144–147
Lidereau R, Callahan R, Dickson C, Peters G, Escot C, Ali IU (1988) Amplification of the *int*-2 gene in primary human breast tumours. Oncogene Res 2:285–291
Linzer DIH, Levine AJ (1979) Characterization of a 54K dalton cellular SV40 tumor antigen present in SV40 transformed cells and uninfected embryonal carcinoma cells. Cell 17:43–52
Linzer DIH, Nathans D (1983) Growth-related changes in specific mRNAs of cultivated mouse cells. Proc Natl Acad Sci USA 80:4271–4275
Lyman SD, Rohrschneider LR (1987) Analysis of functional domains of the v-*fms*-encoded protein of Susan McDonough strain sarcoma virus by linker insertion mutagenesis. Mol Cell Biol 7:3287–3296
Maki Y, Bos TJ, Davis C, Starbuck M, Vogt PK (1987) Avian sarcoma virus 17 carries the *jun* oncogene. Proc Natl Acad Sci USA 84:2848–2852
Marth JD, Lewis DB, Wilson CB, Gearn ME, Krebs EG, Perlmutter RM (1987) Regulation of *pp56lck* during T-cell activation: functional implications for the *src*-like protein tyrosine kinases. EMBO J 6:2727–2734
Martin GS (1986) The *erb*B gene and the EGF receptor. Cancer Surv 5:199–220
Martin-Zanca D, Hughes SH, Barbacid M (1986) A human oncogene formed by the fusion of truncated tropomyosin and protein tyrosine kinase sequences. Nature 319:743–748
Maxwell SA, Arlinghaus RB (1985) Use of site specific antipeptide antibodies to perturb the serine kinase catalytic activity of $p37^{mos}$. J Virology 55:874–876
Maxwell SA, Arlinghaus RB (1985) Serine kinase activity associated with Moloney murine sarcoma virus-124-encoded $p37^{mos}$. Virology 143:321–333
Mayer BJ, Hamaguchi M, Hanafusa H (1988) A novel viral oncogene with structural similarity to phospholipase C. Nature 332:272–275
McGrath JP, Capon DJ, Goeddel DV, Levinson AD (1984) Comparative biochemical properties of normal and activated human *ras* p21 protein. Nature 310:644–649
Mercer ME, Nelson D, DeLeo AB, Old LJ, Baserga R (1982) Microinjection of monoclonal antibody to protein p53 inhibits serum-induced DNA synthesis in 3T3 cells. Proc Natl Acad Sci USA 79:6309–6312
Merlino GT, Xu AH, Ishii S, Clark AJ, Semba K, Toyoshima K, Yamamoto T, Pastan I (1984) Amplification and enhanced expression of the epidermal growth factor receptor gene in A431 human carcinoma cells. Science 224:417–419
Miller AD, Curran T, Verma IM (1984) c-*fos* protein can induce cellular transformation: a novel mechanism of activation of a cellular oncogene. Cell 36:51–60
Mitchell RL, Henning-Chubb C, Huberman E, Verma IM (1986) c-*fos* protein expression is neither sufficient nor obligatory for differentiation of monomyelocytes to macrophages. Cell 45:497–504
Moelling K, Pfaff E, Beug H, Beimling P, Bunte T, Schaller HE, Graf T (1985) DNA-binding activity is associated with purified *myb* proteins from AMV and E26 viruses and is temperature-sensitive for E26 *ts* mutants. Cell 40:983–990
Molders H, Defesche J, Muller D, Bonner TI, Rapp UR, Muller R (1985) Integration of transfected LTR sequences into the c-*raf* proto-oncogene: activation by promoter insertion. EMBO J 4:693–698

Mulcahy LS, Smith MR, Stacey DW (1985) Requirement for *ras* proto-oncogene function during serum-stimulated growth of NIH-3T3-cells. Nature 313:241–243

Muller R, Wagner E (1984) Differentiation of F9 teratocarcinoma stem cells after transfer of c-*fos* proto-oncogenes. Nature 311:438–442

Muller R, Bravo R, Burckhardt J, Curran T (1984) Induction of c-*fos* gene and protein by growth factors precedes activation of c-*myc*. Nature 312:716–720

Muller R, Muller D, Guillbert L (1986a) Differential expression of c-*fos* in hematopoietic cells: correlation with differentiation of monomyelocytic cells in vitro. EMBO J 5:311:316

Muller R, Muller D, Verrier B, Bravo R, Herbst H (1986b) Evidence that expression of c-*fos* protein in amnion cells is regulated by external signals. EMBO 5:311–316

Muller WJ, Sinn E, Pattengale PK, Wallace R, Leder P (1988) Single-step induction of mammary adenocarcinoma in transgenic mice bearing the activated c-*neu* oncogene. Cell 54:105–115

Murray M, Cunningham J, Parada LF, Dautry F, Lebowitz P, Weinberg RA (1983) The HL-60 transforming sequence: a *ras* oncogene coexisting with altered *myc* genes in hemopoietic tumors. Cell 33:749–757

Mushinski JF, Potter M, Bauer SR, Reddy EP (1983) DNA rearrangement and altered RNA expression of the c-*myb* oncogene in mouse plasmacytoid lymphosarcomas. Science 220:795–798

Nakabeppu Y, Ryder K, Nathans D (1988) DNA binding activities of three murine *jun* proteins: stimulation by *fos*. Cell 55:907–915

Neckameyer WS, Shibuya M, Hsu MT, Wang LH (1986) Proto-oncogene c-*ros* codes for a molecule with structural features common to those of growth factor receptors and displays tissue specific and developmentally regulated expression. Mol Cell Biol 6:1478–1486

Nedyalkov ST, Bozhkov SP, Todorov G (1975) Experimental erythroblastosis in the Japanese quail (*Coturnix coturnix japonica*) induced by E26 leukosis strain. Acta Vet (Brno) 44:75–78

Neel BG, Hayward WS, Robinson HL, Fang J, Astrin S (1981) Avian leukosis virus induced tumors have common proviral integration sites and synthesize discrete new RNAs. Oncogenesis by promoter insertion. Cell 23:323–334

Neil JC, Hughes D, McFarlane R, Wilie NM, Onions DE, Lees G, Jarret WO (1984) Transduction and rearrangement of the *myc* gene by feline leukaemia virus in naturally occurring T-cell leukemias. Nature 308:814–820

Ness SA, Beug H, Graf T (1987) v-*myb* dominance over v-*myc* in doubly transformed chick myelomonocytic cells. Cell 51:41–50

Nilsen TW, Maroney PA, Goodwin RG, Rottman FM, Crittenden LB, Raines MA, Kung H (1985) c-*erb*B activation in ALV-induced erythroblastosis: novel RNA processing and promoter insertion result in expression of an amino-truncated EGF receptor. Cell 41:719–726

Nishizuka Y (1983) The role of protein kinase C in cell surface signal transduction and tumour promotion. Nature 308:693–698

Nishizuka Y (1986) Studies and perspectives of protein kinase C. Science 233:305–312

Nishikura K, Murray JM (1987) Antisense RNA of proto-oncogene c-*fos* blocks renewed growth of quiescent 3T3 cells. Mol Cell Biol 7:639–649

Nomura N, Takahashi M, Matsui M, Ishii S, Date T, Sasamoto S, Ishizaki R (1988) Isolation of human cDNA clones of *myb*-related genes, A-*myb* and B-*myb*. Nucl Acids Res 16:11 075–11 089

Norman C, Runswick M, Pollock R, Treisman R (1988) Isolation and properties of cDNA clones encoding SRF, a transcription factor that binds to the c-*fos* serum response element. Cell 55:989–1003

Nunn MF, Hunter H (1989) The *ets* sequence is required for induction of erythroblastosis in chickens by avian retrovirus E26. J Virol 63:398–402

Nunn MF, Seeburg PH, Moscovici C, Duesberg PH (1983a) Tripartite structure of the avian erythroblastosis virus E26 transforming gene. Nature 306:391–395

Nunn M, Weiher H, Bullock P, Duesberg P (1983b) Avian erythroblastosis virus E26: nucleotide sequence of the tripartite *onc* gene and of the LTR, and analysis of the cellular prototype of the viral *ets* sequence. Virology 133:330–339
Nusse R (1989) The *int* genes in mammary tumorigenesis and in normal development. Trends Genet 4:291–295
Nusse R, Varmus HE (1982) Many tumors induced by the mouse mammary tumour virus contain a provirus integrated in the same region of the host genome. Cell 31:99–109
Oker-Blom M, Hortling L, Kallio A, Nurmiaho EL, Westermarck H (1978) OK10 virus, an avian retrovirus resembling the acute leukaemia viruses. J Gen Virol 40:623–633
Padhy LC, Shih C, Cowing D, Finklestein R, Weinberg RA (1982) Identification of a phosphoprotein specifically induced by the transforming DNA of rat neuroblastomas. Cell 28:865–871
Papas TS, Bhat NK, Chen TT, Dubois G, Fisher RJ, Fujiwara S, Pribyl J, Saachi N, Seth A, Showalter SD, Watson DK, Zweig M, Ascione R (1987) The *ets* gene in cells and viruses: implications for leukemias and other human diseases. In: Peschle C (ed) Normal and neoplastic blood cells: from genes to therapy. New York Academy of Sciences, pp 171–191
Papkoff J, Brown AM, Varmus HE (1987) The *int*-1 proto-oncogene products are glycoproteins that appear to enter the secretory pathway. Mol Cell Biol 7:3978–3984
Parada LF, Land H, Weinberg RA, Wolf D, Rotter W (1984) Cooperation between gene encoding p53 tumour antigen and *ras* in cellular transformation. Nature 312:649–651
Paris S, Pouysségur J (1986) Pertussis toxin inhibits thrombin-induced activation of phosphoinositide hydrolysis and Na+/H+ exchange in hamster fibroblasts. EMBO J 5:55–60
Park M, Dean M, Cooper CS, Schmidt M, O'Brien SJ, Blair DG, Vande Woude GF (1986) Mechanism of *met* oncogene activation. Cell 45:895–904
Parker PJ, Ullrich A (1987) Protein kinase C. J Cell Physiol 5:53–56
Parsons SJ, Crentz CE (1986) p60src activity detected in the chromaffin granule membrane. Biochem Biophys Res Commun 134:736–742
Paulson Y, Hammacher A, Heldin CH, Westermark B (1987) Possible positive autocrine feedback in the prereplicative phase of human fibroblasts. Nature (London) 328:715–717
Payne GS, Courtneidge SA, Crittenden LB, Fadly AM, Bishop JM, Varmus HE (1981) Analysis of avian leukosis virus DNA and RNA in bursal tumors: viral gene expression is not required for maintenance of the tumor state. Cell 23:311–322
Payne GS, Bishop JM, Varmus HE (1982) Multiple arrangements of viral DNA and an activated host oncogene (c-*myc*) in bursal lymphomas. Nature 295:209–214
Pellici PG, Lanfrancone L, Brathwaite MD, Wolman SR, Dalla Favera R (1984) Amplification of the c-*myb* oncogene in a case of human acute myelogenous leukemia. Science 225:1117–1121
Persons DA, Wilkison WO, Bell RM, Finn OJ (1988) Altered growth regulation and enhanced tumorigenicity of NIH 3T3 fibroblasts transfected with protein kinase C-I DNA. Cell 52:447–458
Persson H, Leder P (1984) Nuclear localization and DNA binding properties of a protein expressed by human c-*myc* oncogene. Science 225:718–720
Peters G, Brookes S, Smith R, Dickson C (1983) Tumorigenesis by mouse mammary tumor virus: evidence for a common region for provirus integration in mammary tumours. Cell 38:369–377
Piwnica-Worms H, Saunders KB, Roberts TM, Smith AE, Cheng SH (1987) Tyrosine phosphorylation regulates the biochemical and biological properties of pp60c-*src*. Cell 49:75–82

Pognonec P, Boulukos KE, Gesquiere JG, Stehelin D, Ghysdael J (1988) Mitogenic stimulation of thymocytes results in the calcium-dependent phosphorylation of c-*ets*-1 proteins. EMBO J 7:977–983

Ponder B (1988) Gene losses in human tumours. Nature 335:400–402

Potter M, Reddy EP, Wivel NA (1978) Immunoglobulin production by lymphosarcomas induced by Abelson virus in mice. NCI Monogr 48:311–321

Preiss J, Loomis C, Bishop W, Stein R, Niedel J, Bell RM (1986) Quantitative measurement of sn-1,2-diacylglycerols present in platelets, hepatocytes and *ras*- and *sis*-transformed normal rat kidney cells. J Biol Chem 261:8597–8600

Prywes R, Roeder RG (1986) Inducible binding of a factor to the c-*fos* enhancer. Cell 47:777–784

Quantin B, Breathnach R (1988) Epidermal growth factor stimulates transcription of the c-*jun* proto-oncogene in rat fibroblasts. Nature 334:538–539

Quiafe CJ, Pinkert CA, Ornitz DM, Palmiter RD, Brinster R (1987) Pancreatic neoplasia induced by *ras* expression in acinar cells of transgenic mice. Cell 48:1023–1034

Rapp UR, Goldsborough MD, Mark GE, Bonner TI, Groffen J, Reynolds FH Jr, Stephenson J (1983) Structure and biological activity of v-*raf*, a unique oncogene in lung carcinogenesis. Proc Natl Acad Sci USA 80:4218–4222

Rapp UR, Cleveland JL, Storm SM, Beck TW, Huleihel M (1986) Transformation by *raf* and *myc* oncogenes. Int Symp Princess Takamatsu Cancer Res Fund 17:55–74

Rapp UR, Cleveland JL, Bonner TI, Storm SM (1988) The *raf* oncogene. In: Reddy EP, Skalka AM, Curran T (eds) The oncogene handbook, Elsevier, Amsterdam, pp 213–253

Reddy ES, Rao VN, Papas TS (1987) The *erg* gene: a human gene related to the *ets* oncogene. Proc Natl Acad Sci USA 84:6131–6135

Reddy EP, Skalka AM, Curran T (1988) The oncogene handbook. Elsevier Amsterdam

Reed JC, Alpers JD, Scherle PA, Hoover RG, Nowell PC, Prystowsky MB (1987) Proto-oncogene expression in cloned T lymphocytes: mitogens and growth factors induce different patterns of expression. Oncogene 1:223–228

Renz M, Verrier B, Kurz C, Muller R (1987) Chromatin association and DNA binding properties of the c-*fos* proto-oncogene product. Nucl Acids Res 15:277–292

Reynolds AB, Vila J, Lansing TJ, Potts WM, Weber MJ, Parsons JT (submitted) Activation of the oncogenic potential of the avian cellular *src* protein by specific structural alteration of the carboxy-terminus.

Rijsewijk F, Schuermann M, Wagenaar E, Parren C, Weidel D, Nusse R (1987a) The *Drosophila* homolog of the mouse mammary oncogene *int*-1 is identical to the segment polarity gene wingless. Cell 50:649–657

Rijsewijk F, van Deemter L, Wagenaar E, Sonnenberg A, Nusse R (1987b) Transfection of the *int*-1 mammary oncogene in cuboidal RAC mammary cell line results in morphological transformation and tumorgenicity. EMBO J 6:127–131

Robbins KC, Devare SG, Reddy EP, Aaronson SA (1982a) In vivo identification of the transforming gene product of simian sarcoma virus. Science 218:1131–1133

Robbins KC, Hill RL, Aaronson SA (1982b) Primate origin of the cell-derived sequences of simian sarcoma virus. J Virol 41:721–725

Robbins KC, Antoniades HN, Devare SG, Hunkapiller MW, Aaronson SA (1983) Structural and immunological similarities between simian sarcoma virus gene product(s) and human platelet-derived growth factor. Nature 305:605–608

Robbins KC, Leal F, Pierce JH, Aaronson SA (1985) The v-*sis*/*PDGF-2* transforming gene product localizes to cell membranes but is not a secretory protein, EMBO J 4:1783–1792

Rogelj S, Weinberg RA, Fanning P, Klagsbrun M (1988) Basic fibroblast growth factor fused to signal peptide transforms cells. Nature 331:173–175

Rollins BJ, Morrison ED, Stiles CD (1988) Cloning and expression of *JE*, a gene inducible by platelet-derived growth factor and whose product has cytokine-like properties. Proc Natl Acad Sci USA 85:3738–3742

Rosenthal A, Lindquist PB, Bringman TS, Goeddel DV, Derynck R (1986) Expression in rat fibroblasts of a human transforming growth factor-(alpha) cDNA results in transformation. Cell 46:301–309

Rous P (1911) A sarcoma of the fowl transmissible by an agent separable from the tumor cells. J Exp Med 13:397–411

Roussel MF, Saule S, Lagrou C, Rommens C, Beug H, Graf T, Stehelin D (1979) Three new types of viral oncogenes of cellular origin specific for haematopoietic cell transformation. Nature 281:425–455

Roussel MF, Rettenmier CW, Look AT, Sherr CJ (1984) Cell surface expression of v-*fms*-coded glycoproteins is required for transformation. Mol Cell Biol 4:1999–2009

Roussel MF, Downing JR, Rettenmier CW, Sherr CJ (1988) A point mutation in the extracellular domain of the human CSF-1 receptor (c-*fms* proto-oncogene product) activates its transforming potential. Cell 55:979–988

Royer-Pokora B, Beug H, Claviez M, Winkhardt H-J, Friis RR, Graf T (1978) Transformation parameters in chicken fibroblasts transformed by AEV and MC29 avian leukemia viruses. Cell 13:751–750

Rüther U, Garber C, Komitowski D, Muller R, Wagner EF (1987) Deregulated c-*fos* expression interferes with normal bone development in transgenic mice. Nature 325:412–416

Ryder K, Lau L, Nathans D (1988) A gene activated by growth factors is related to the oncogene v-*jun*. Proc Natl Acad Sci USA 85:1487–1491

Ryseck R-P, Hirai SI, Yaniv M, Bravo R (1988) Transcriptional activation of c-*jun* during the G0/G1 transition in mouse fibroblasts. Nature 334:535–537

Ryseck R-P, MacDonald-Bravo H, Mattei MG, Bravo R (1989) Cloning and sequence of a secretory protein induced by growth factors in mouse fibroblasts. Exp Cell Res 180:266–275

Sacca R, Stanley ER, Sherr CJ, Rettenmeier CW (1986) Specific binding of the mononuclear phagocyte colony-stimulating factor CSF-1 to the product of the v-*fms* oncogene. Proc Natl Acad Sci USA 83:3331–3335

Sakamoto H, Yoshida T, Nakakuki M, Odagiri H, Miyagawa K, Sugimura T, Terada M (1988) Cloned *hst* gene from normal human leukocyte DNA transforms NIH3T3 cells. Biochem Biophy Res Commun 151:965–972

Samarut J, Gazzolo L (1982) Target cells infected by avian erythroblastosis virus differentiate and become transformed. Cell 28:921–929

Sambucetti LC, Curran T (1986) The *fos* protein complex is associated with DNA in isolated nuclei and binds to DNA cellulose. Science 234:1417–1419

Sap J, Muñoz A, Damm K, Ghysdael J, Leutz A, Beug H, Vennström B (1986) The c-*erb*A protein is a high affinity receptor for thyroid hormone. Nature 324:635–640

Sariban E, Luebbers R, Kufe D (1988) Transcriptional and posttranscriptional control of c-*fos* gene expression in human monocytes. Mol Cell Biol 8:340–346

Sassone-Corsi P, Ransone LJ, Lamph WE, Verma IM (1988a) Direct interaction between *fos* and *jun* nuclear oncoproteins: role of the leucine zipper domain. Nature 336:692–695

Sassone-Corsi P, Lamph WW, Kamps M, Verma IM (1988b) *fos*-associated cellular p39 is related to nuclear transcription factor AP-1. Cell 54:553–560

Sassone-Corsi P, Sisson JC, Verma IM (1988c) Transcriptional autoregulation of the proto-oncogene *fos*. Nature 334:314–319

Schechter AL, Stern DF, Vaidyanathan L, Decker SJ, Drebin JA, Greene MI, Weinberg RA (1984) The *neu* oncogene: an *erb*B-related gene encoding a 185,000-Mr tumor antigen. Nature 312:513–516

Schönthal A, Herrlich P, Rahmsdorf HJ, Ponta H (1988) Requirement for *fos* gene expression in the transcriptional activation of collagenase by other oncogenes and phorbol esters. Cell 54:325–334

Schwab M (1985) Amplification of N-*myc* in human neuroblastomas. Trends Genet 1:271–275

Schwab M (1986) Amplification of proto-oncogenes and tumor progression. In: Kahn P, Graf T (eds) Oncogenes and growth control. Springer, Berlin Heidelberg New York, pp 332–339
Schwab M, Ellison J, Busch M, Rosenau W, Varmus HE, Bishop JM (1984) Enhanced expression of the human gene N-*myc* consequent to amplification of DNA may contribute to malignant progression of neuroblastoma. Proc Natl Acad Sci USA 81:4940–4944
Schwab M, Varmus HE, Bishop JM (1985) The human N-*myc* gene contributes to tumorigenic conversion of mammalian cells in culture. Nature 316:160–162
Sealy L, Privalsky ML, Moscovici G, Moscovici C, Bishop JM (1983) Site-specific mutagenesis of avian erythroblastosis virus: *erb*B is required for oncogenicity. Virology 130:155–178
Semba K, Kamata N, Toyoshima K, Yamamoto T (1985) A v-*erb*B related protooncogene, c-*erb*B-2 is distinct from the c-*erb*B-1/epidermal growth factor-receptor gene and is amplified in a human salivary gland adenocarcinoma. Proc Natl Acad Sci USA 82:6497–6501
Seth A, Priel E, Vande Woude GF (1987) Nucleoside triphosphate-dependent nucleic-acid-binding properties of *mos* protein. J Virol 56:144–152
Setoyama C, Frunzio R, Lian G, Mudryi M, De Combrugghe B (1986) Transcriptional activation encoded by the v-*fos* gene. Proc Natl Acad Sci USA 83:3213–3217
Shackleford GM, Varmus HE (1987) Expression of the proto-oncogene *int-1* is restricted to postmeiotic male germ cells and the neural tube of mid-gestational embryos. Cell 50:89–95
Shalloway D, Coussens PM, Yaciuk P (1984) Overexpression of the c-*src* protein does not induce transformation of NIH 3T3 cells. Proc Natl Acad Sci USA 81:7071–7075
Shen-Ong GLC, Morse HC, Potter M, Mushinski JF (1986) Two modes of c-*myb* activation in virus-induced mouse myeloid tumors. Mol Cell Biol 6:380–392
Sherr CJ (1987) Fibroblast and hematopoietic cell transformation by the *fms* oncogene (CSF-1 receptor). J Cell Physiol 5:83–87
Sherr CJ (in press) The *fms* oncogene. BBA Rev Cancer (in press)
Sherr CJ, Rettenmier CW, Sacca R, Roussel MF, Look AT, Stanley ER (1985) The c-*fms* proto-oncogene product is related to the receptor for the mononuclear phagocyte growth factor. Cell 41:665–676
Shibuya M, Hanafusa H, Balduzzi PC (1982) Cellular sequences related to three new *onc* genes of avian sarcoma virus (*fps, yes,* and *ros*) and their expression in normal and transformed cells. J Virol 42:143–152
Shih C, Padhy LC, Murray M, Weinberg RA (1981) Transforming genes of carcinomas and neuroblastomas introduced into mouse fibroblasts. Nature 290:261–264
Sinn E, Muller W, Pattengale P, Tepler I, Wallace R, Leder P (1987) Coexpression of *MMTV*/v-*Ha-ras* and *MMTV*/c-*myc* genes in transgenic mice: synergistic action of oncogenes in vivo. Cell 49:465–475
Skouv J, Christensen B, Skibski L, Autrup H (1986) The skin tumor-promoter 12-*O*-tetradecanoyolphorbol-13-acetate induces transcription of the c-*fos* proto-oncogene in human bladder epithelial cells. Carcinogenesis 7:331–333
Slamon DJ, Clark GM, Wong SG, Levin WJ, Ullrich A, McGuire WL (1987) Human breast cancer: correlation of relapse and survival with amplification of the HER-2/neu oncogene. Science 235:177–182
Small M, Hay N, Ramsay G, Schwab M, Bishop JM (1987) N-*myc* induces tumorigenic conversion of cultured rat cells. Mol Cell Biol 7:1638–1645
Smith MR, DeGudicibus SJ, Stacey DW (1986) Requirement for c-*ras* proteins during viral oncogene transformation. Nature 320:540–543
Sorrentino V, Drozdoff V, McKinney MD, Zeitz L, Fleissner E (1986) Potentiation of growth factor activity by exogenous c-*myc* expression. Proc Natl Acad Sci USA 83:8167–8171
Sporn MB, Todaro GJ (1980) Autocrine secretion and malignant transformation of cells. N Engl J Med 303:878–880

Spurr NK, Solomon E, Jansson M, Sheer D, Goodfellow PN, Bodmer WF, Vennström B (1984) Chromosomal localization of the human homologues to the oncogenes *erb*A and B. EMBO J 3:159–163

Stacey DW, Kung HF (1984) Transformation of NIH 3T3 cells by microinjection of Ha-*ras* p21 protein. Nature 310:508–511

Stacey DW, Watson T, Kung HF, Curran T (1987) Microinjection of transforming *ras* protein induces c-*fos* expression. Mol Cell Biol 7:523–527

Stern DF, Hare DL, Ceccini MA, Weinberg RA (1987) Construction of a novel oncogene based on synthetic sequences encoding epidermal growth factor. Science 235:321–324

Stewart TA, Pattengale PK, Leder P (1984) Spontaneous mammary adenocarcinomas in transgenic mice that carry and express MTV/*myc* fusion genes. Cell 38:627–637

Stocking C, Löliger C, Kawai M, Suciu S, Gough N, Ostertag W (1988) Identification of genes involved in growth autonomy of hematopoietic cells by analysis of factor-independent mutants. Cell 53:869–879

Struhl K (1987) The DNA-binding domains of the *jun* oncoprotein and the yeast GCN4 transcriptional activator proteins are functionally homologous. Cell 50:841–846

Struhl K (1988) The JUN oncoprotein, a vertebrate transcription factor, activates transcription in yeast. Nature 332:649–650

Suarez HG, Nardeux PC, Andeol Y, Sarasin A (1987) Multiple activated oncogenes in human tumors. Oncogene Res 1:201–207

Sukhatme VP, Cao X, Chang LC, Tsai-Morris C-H, Stamenkovich D, Ferreira PCP, Cohen DR, Edwards SA, Shows TB, Curran T, Le Beau MM, Adamson ED (1988) A zinc finger-encoding gene coregulated with c-*fos* during growth and differentiation, and after cellular depolarization. Cell 53:37–43

Sweet RW, Yokoyama S, Kamata T, Feramisco R, Rosenberg M, Gross M (1984) The product of *ras* is a GTPase and the T24 oncogenic mutant is deficient in this activity. Nature 311:273–275

Taira M, Yoshida T, Miyagawa K, Sakamoto H, Terada M, Sugimura T (1987) cDNA sequence of human transforming gene *hst* and identification of the coding sequence required for transforming activity. Proc Natl Acad Sci USA 84:2980–2984

Takahashi M, Cooper GM (1987) *ret* transforming gene encodes a fusion protein homologous to tyrosine kinases. Mol Cell Biol 7:1378–1385

Tam JP, Marquardt H, Rosberger DF, Wong TW, Todaro GJ (1984) Synthesis of biologically active rat transforming growth factor I. Nature 309:376–378

Tamura T, Simon E, Niemann H, Snoek GT, Bauer H (1986) gp140-v-fms molecules expressed at the surface of cells transformed by the McDonough strain of feline sarcoma virus are phosphorylated in tyrosine and serine. Mol Cell Biol 6:4745–4748

Tempest PR, Cooper CS, Major GN (1986) The activated human *met* gene encodes a protein tyrosine kinase. FEBS Lett 209:357–361

Thomas G, Martin-Perez J, Siegmann M, Otto AM (1982) The effect of serum, EGF, PGF2 alpha and insulin on S6 phosphorylation and the initiation of protein and DNA synthesis. Cell 30:235–242

Thompson CB, Challoner PB, Neiman PE, Groudine M (1986) Expression of the c-*myb* proto-oncogene during cellular proliferation. Nature 319:374–380

Trahey M, McCormick F (1987) A cytoplasmic protein stimulates normal N-*ras* p21 GTPase, but does not affect oncogenic mutants. Science 238:542–545

Treisman R (1985) Transient accumulation of c-*fos* RNA following serum stimulation requires a conserved 5′ element and c-*fos* 3′ sequences. Cell 42:889–902

Treisman R (1986) Identification of a protein-binding site that mediates transcriptional response of the c-*fos* gene to serum factors. Cell 46:567–574

Ullrich A, Coussens L, Hayflick JS, Dull TJ, Gray A, Tam AW, Lee J, Yarden Y, Libermann TA, Schlessinger J, Downward J, Mayes ELV, Whittle N, Waterfield MD, Seeburg PH (1984) Human epidermal growth factor receptor cDNA sequence and aberrant expression of the amplified gene in A431 epidermoid carcinoma cells. Nature 309:418–425

Ushiro H, Cohen S (1980) Identification of phosphotyrosine as a product of epidermal growth factor-activated protein kinase in A-431 cell membranes. J Biol Chem 255:8363–8365

Van Beveren C, Verma IM (1985) Homology among oncogenes. Curr Top Microbiol 123:73–98

Van Beveren C, Van Straaten F, Galleshaw JA, Verma IM (1981) Nucleotide sequence of the genome of a murine sarcoma virus. Cell 27:97–108

Van Beveren C, Van Straaten F, Curran T, Muller R, Verma IM (1983) Analysis of FBJ-MuSV provirus and c-*fos* (mouse) gene reveals that viral and cellular *fos* gene products have different carboxy-termini. Cell 32:1241–1255

Van Ooyen A, Nusse R (1984) Structure and nucleotide sequence of the putative mammary oncogene *int*-1; proviral insertions leave the protein-encoding domain intact. Cell 39:233–240

Van Ooyen A, Kwee V, Nusse R (1985) The nucleotide sequence of the human *int*-1 mammary oncogene; evolutionary conservation of coding and non-coding sequences. EMBO J 4:2905–2909

Van Straaten F, Muller R, Curran T, Van Beveren C, Verma IM (1983) Complete nucleotide sequence of human c-*onc* gene: deduced amino acid sequence of the human c-*fos* gene protein. Proc Natl Acad Sci USA 80:3183–3187

Veillette A, Horak I, Horak EM, Bookman MA, Bolen JB (1988) Alterations of the lymphocyte-specific protein tyrosine kinase (p56lck) during T-cell activation. Mol Cell Biol 8:4353–4361

Velu TJ, Beguinot L, Vass WS, Willingham MC, Merlino GT, Pastan I, Lowy DR (1987) Epidermal growth factor-dependent transformation by a human EGF receptor proto-oncogene. Science 238:1408–1410

Vennström B, Bishop JM (1982) Isolation and characterization of chicken DNA homologous to the two putative oncogenes of avian erythroblastosis virus. Cell 28:135–143

Verma IM, Lai MHT, Bosselman RA, McKennett MA, Fan H, Berns A (1980) Molecular cloning of unintegrated Moloney mouse sarcoma virus. Proc Natl Acad Sci USA 77:1773–1777

Vogel US, Dixon RAF, Schaber MD, Diehl RE, Marshall MS, Scolnick EM, Sigal IS, Gibbs JB (1988) Cloning of bovine GAP and its interaction with oncogenic *ras* p21. Nature 335:90–93

Vogt P, Tjian R (1988) *Jun*: a transcriptional regulator turned oncogenic. Oncogene 3:3–7

Vogt PK, Bos TJ, Doolittle RF (1987) Homology between the DNA-binding domain of the GCN4 regulatory protein of yeast and the carboxyl-terminal region of a protein coded for by the oncogene *jun*. Proc Natl Acad Sci USA 84:3316–3319

von Weizsäcker F, Beug H, Graf T (1986) Temperature-sensitive mutants of MH2 avian leukemia virus that map in the v-*mil* and the v-*myc* oncogene respectively. EMBO J 5:1521–1527

Wasylyk C, Imler JL, Wasylyk B (1988) Transforming but not immortalizing oncogenes activate the transcription factor PEA1. EMBO J 7:2475–2483

Watanabe S, Lazar E, Sporn MB (1987) Transformation of normal rat kidney (NRK) cells by an infectious retrovirus carrying a synthetic rat type (alpha) transforming growth factor gene. Proc Natl Acad Sci USA 84:1258–1262

Waterfield MD, Scrace GT, Whittle N, Stroobant P, Johnsson A, Wasteson A, Westermark B, Heldin C-H, Huang HS, Deuel TF (1983) Platelet-derived growth factor is structurally related to the putative transforming protein p28sis of simian sarcoma virus. Nature 304:35–39

Watson DK, McWilliams M, Papas TS (1988) Molecular organization of the chicken *ets* locus. Virology 164:99–105

Weinberg RA (1985) The action of oncogenes in the cytoplasm and nucleus. Science 230:770–776

Weinberger C, Thompson EC, Ong ES, Lebo R, Gruol DJ, Evans RM (1986) The c-*erb*A gene encodes a thyroid hormone receptor. Nature 324:641–646

Weinstein Y, Ihle JN, Lavu S, Reddy EP (1986) Truncation of the c-*myb* gene by a retroviral integration in an interleukin 3-dependent myeloid leukemia cell line. Proc Natl Sci USA 83:5010–5014

Westin EH, Gallo RC, Arya SK, Eva A, Souza LM, Baluda MA, Aaronson SA, Wong-Staal F (1982) Differential expression of the *amv* gene in human hematopoietic cells. Proc Natl Acad Sci USA 79:2194–2198

Wheeler EF, Rettenmier CW, Look AT, Sherr CJ (1986) The v-*fms* oncogene induces factor independence and tumorigenicity in a CSF-1 dependent macrophage cell line. Nature 234:377–380

Wickstrom EL, Bacon TA, Gonzales A, Freeman DL, Lyman GH, Wickstrom E (1988) Human promyelocytic leukemia HL-60 cell proliferation and c-*myc* protein expression are inhibited by a antisense pentadecadeoxynucleotide targeted against c-*myc* mRNA. Proc Natl Acad Sci USA 85:1028–1032

Wieschaus E, Riggelman R (1987) Autonomous requirements for the segment polarity gene armadillo during *Drosophila* embryogenesis. Cell 49:177–194

Wilkinson DG, Bailes JA, McMahon AP (1987) Expression of the proto-oncogene *int*-1 is restricted to specific neural cells in the developing mouse embryo. Cell 50:79–88

Wilkinson DG, Peters G, Dickson C, McMahon AP (1988) Expression of the FGF-related proto-oncogene *int*-2 during gastrulation and neurulation in the mouse. EMBO J 7:691–695

Witte ON (1986) Functions of the *abl* oncogene. Cancer Surv 5:183–197

Wolfe LG, Deinhardt F, Theilen GJ, Rabin H, Kawakami T, Bustad LK (1971) Induction of tumors in marmoset monkeys by simian sarcoma virus, type 1 (*Lagothrix*) a preliminary report. J Natl Cancer Inst 47:1115–1120

Wong PMC, Chung S-W, Nienhuis AW (1987) Retroviral transfer and expression of the interleukin-3 gene in hemopoietic cells. Genes Dev 1:358–365

Wood TG, McGeady ML, Baroudy BM, Blair DG, Vande Woude GF (1984) Mouse c-*mos* oncogene activation is prevented by upstream sequences. Proc Natl Acad Sci USA 81:7817–7821

Woolford J, McAuliffe A, Rohrschneider LR (1988) Activation of the feline c-*fms* proto-oncogene: multiple alterations are required to generate a fully transformed phenotype. Cell 55:965–977

Yamada G, Kitamura Y, Sonoda H, Harada H, Taki S, Mulligan RC, Osawa H, Diamantstein T, Yokoyama S, Taniguchi T (1987) Retroviral expression of the human IL-2 gene in a murine T cell line results in cell growth autonomy and tumorigenicity. EMBO J 6:2705–2709

Yamamoto T, Nishida T, Mitajimi N, Kawai S, Ooi T, Toyoshima K (1983) The *erb*B gene of avian erythroblastosis virus is a member of the *src* gene family. Cell 35:71–78

Yamamoto T, Hihara H, Nishida T, Kawai S, Toyoshima K (1983) A new avian erythroblastosis virus, AEV-H, carries *erb*B gene responsible for the induction of both erythroblastosis and sarcomas. Cell 34:225–232

Yancopoulos GD, Nisen PD, Tesfaye A, Kohl N, Goldfarb MP, Alt F (1985) N-*myc* can cooperate with *ras* to transform normal cells in culture. Proc Nat Acad Sci USA 82:5455–5459

Yarden J, Ullrich A (1988) Growth factor receptor tyrosine kinases. Ann Rev Biochem 57:443–478

Yarden Y, Schlessinger J (1987) Epidermal growth factor induces rapid, reversible aggregation of the purified epidermal growth factor receptor. Biochemistry 26:1443–1451

Yarden Y, Escobedo JA, Kuang W-J, Yang-Feng TL, Daniel TO, Tremble PM, Chen EY, Ando ME, Harkins RN, Francke U, Fried VA, Ullrich A, Williams LT (1986) Structure of the receptor for platelet-derived growth factor helps define a family of closely related growth factor receptors. Nature 323:226–232

Yoshida MC, Wada M, Satoh H, Yoshida T, Sakamoto H, Miyagawa K, Yokota J, Koda T, Kakinuma M, Sugimura T (1988) Human *HST1* (HSTF1) gene maps to chromosome band 11q13 and coamplifies with the *int*-2 gene in human cancer. Proc Natl Acad Sci USA 85:4861–4864

Yoshida T, Migyagawa K, Odagiri H, Sakamoto H, Little PF, Terada M, Sugimura T (1987) Genomic sequences of *hst*, a transforming gene encoding a protein homologous to fibroblast growth factors and the *int*-2 encoded protein. Proc Natl Acad Sci USA 84:305–309

Yu YH, Richert N, Ito S, Merlino GT, Pastan I (1984) Characterization of epidermal growth factor receptor gene expression in malignant and normal human cell lines. Proc Natl Acad Sci USA 81:7308–7312

Yuan CC, Kan N, Dunn KJ, Papas TS, Blair DG (1988) Properties of a murine retroviral recombinant of avian acute leukemia virus E26: a murine fibroblast assay for v-*ets* function. J Virol 63:205–215

Zhan X, Bates B, Hu X, Goldfarb M (1988) The human FGF-5 oncogene encodes a novel protein related to fibroblast growth factors. Mol Cell Biol 8:3487–3495

Zenke M, Kahn P, Disela C, Vennström B, Leutz A, Keegan K, Hayman MJ, Choi H-R, Yew N, Engel JD, Beug H (1988) v-*erb*A specifically suppresses transcription of the avian erythrocyte anion transporter (Band 3) gene. Cell 1988 52:107–119

Zerial M, Toschi L, Ryseck R-P, Schuermann M, Müller R, Bravo R (1989) The product of a novel growth factor-activated gene, *fos*B, interacts with JUN proteins enhancing their DNA binding activity (submitted for publication)

Appendix A. Alternate Names for Growth Factors

Activins (A, AB, B)
    FSH releasing protein (FRP)

Colony-stimulating factor 1 (CSF-1)
    Macrophage colony-stimulating factor (M-CSF)

Epidermal growth factor (EGF)
    Urogastrone

Erythropoietin (epo; EP)
    Hemopoietine
    Erythrocyte stimulating factor (ESF)

Fibroblast growth factors (FGF)
    Acidic fibroblast growth factor (aFGF)
    Basic fibroblast growth factor (bFGF)
    Brain-derived growth factor (BNDF; BDGF)
    Heparin binding growth factor (HBFG)
    Endothelial cell growth factor (ECGF)
    Retina-derived growth factor (RDGF)
    Eye-derived growth factor (EDGF)
    Kidney angiogenic factor (KAF)
    Adrenal growth factor (AGF)
    Corpus luteum angiogenic factor (CLAF)
    Ovarian growth factor (OGF)
    Placental angiogenic factor (PAG)
    Hepatocyte growth factor (HGF)
    Myogenic growth factor (MGF)
    Cartilage-derived growth factor (CDGF)
    Bone growth factor (BGF)
    Seminiferous growth factor (SGF)
    Prostatropin (PGF)
    Tumor-derived growth factor (TDGF)
    Hepatoma-derived growth factor (HDGF)
    Melanoma-derived growth factor (MDGF)
    Mammary tumor-derived growth factor (MTGF)

Gastrin-releasing peptide (GRP)
    Mammalian bombesin

Granulocyte colony-stimulating factor (G-CSF)
Macrophage/granulocyte inducer type 1, granulocyte (MG-1G)
Pluripotent colony-stimulating factor (pluripoetin)
Granulocyte-macrophage colony-stimulating factor-β (GM-CSFβ)

Granulocyte-macrophage colony-stimulating factor (GM-CSF)
Macrophage-granulocyte inducer
Colony-stimulating factor 2 (CSF-2)

Inhibin

Insulin-like growth factor I (IGF-I)
Somatomedin A
Somatomedin C
Basic somatomedin

Insulin-like growth factor II (IGF-II)
Multiplication stimulating activity (MSA)

Interferon-α (IFN-α)
Leukocyte (Le) interferon
Type I interferon

Interferon-β (IFN-β)
Fibroblast (F) interferon
Type I interferon

Interferon-γ (IFN-γ)
Immune interferon
T interferon
Type II interferon

Interleukin-1 (IL-1)
Interleukin-1 α
Interleukin-1 β
Lymphocyte activating factor (LAF)

Interleukin-2 (IL-2)
T-cell growth factor (TCGF)

Interleukin-3 (IL-3)
Mast cell growth factor
P-cell stimulating factor
Multi-colony-stimulating factor
Burst promoting activity
WEHI-3 hematopoietic growth factor
Thy-1 inducing factor
Histamine cell-producing stimulating factor
20α-dehydrogenase-inducing factor

Interleukin-4 (IL-4)
B-cell stimulatory factor 1 (BSF-1)
T-cell growth factor (TCGF)

B-cell growth factor I (BCGF-1)
Mast-cell growth factor (MCGF)

Interleukin-5 (IL-5)
T-cell replacing factor (TRF-1)
B-cell growth factor II (BCGF II)
B-cell differentiation factor μ (BCDFμ)
Eosinophil differentiation factor (EDF)
IgA enhancing factor (IgA-EF)
B-cell maturation factor (BMF)
B-cell growth and differentiation factor (BGDF)

Interleukin-6 (IL-6)
B-cell stimulatory factor 2 (BSF-2)
Interferon-β2
26-kDa protein
Hepatocyte stimulating factor
Hybridoma/plasmacytoma growth factor
Interleukin-HP1
Macrophage granulocyte inducer type 2

Lymphotoxin
Tumor necrosis factor-β (TNF-β)

Müllerian inhibiting substance (MIS)
Anti-Müllerian hormone (AMH)
Müllerian inhibiting factor (MIF)

Nerve growth factor (NGF)

Platelet-derived endothelial cell growth factor (PD-ECGF)

Platelet-derived growth factor (PDGF-AA, PDGF-BB, PDGF-AB)

Transforming growth factor-α (TGF-α)
Sarcoma growth factor (SGF)

Transforming growth factor-β (TGF-β)
Cartilage inducing factor-A (CIF-A)
Cartilage inducing factor-B (CIF-B)
BSC-1 growth inhibitor (BSC-1 GI)
Differentiation inhibitor (DI)
Polyergin

Tumor necrosis factor (TNF)
Cachectin
Tumor necrosis factor-α (TNF-α)

Appendix B. Chromosomal Locations of Growth Factors/Growth FactorReceptors

Human

chromosome 1	Nerve growth factor (p22.1) Transforming growth factor-β2 (q41)
chromosome 2	Inhibin-α (distal portion of long arm) Inhibin-β_B (near centromere on short arm) Interleukin-1β (2q13-2q21) Transforming growth factor-α (p11-p13)
chromosome 4	Basic fibroblast growth factor Epidermal growth factor (q25-q27)
chromosome 5	Acidic fibroblast growth factor Colony-stimulating factor 1 (q33.1) Colony-stimulating factor 1 receptor (q33.2-33.3) Granulocyte-macrophage colony-stimulating factor (q23-31) Interleukin-3 (q23-q31) Interleukin-4 (q23-31) Interleukin-5 (q23.3-32) Platelet-derived growth factor receptor (β-subunit)
chromosome 6	Inteferon-γ receptor
chromosome 7	Epidermal growth factor receptor (p14-p12) Erythropoietin (q11–q22) Inhibin-β_A Interleukin-6 (p21) Platelet-derived growth factor A chain
chromosome 9	Interferon-α Interferon-β
chromosome 11	Insulin-like growth factor II
chromosome 12	Insulin-like growth factor I Interferon-γ
chromosome 14	Transforming growth factor-β3 (q24)
chromosome 17	Granulocyte colony-stimulating factor (q21-q22) Nerve growth factor receptor (q12-q22)
chromosome 18	Gastrin-releasing peptide (q21)

chromosome 19	Müllerian inhibiting substance Transforming growth factor-β1 (q13)
chromosome 21	Interferon-α/β receptor
chromosome 22	Platelet-derived growth factor B chain

Mouse

chromosome 1	Transforming growth factor-β2
chromosome 2	Interleukin-1 α Interleukin-1β
chromosome 3	Epidermal growth factor Nerve growth factor
chromosome 4	Interferon-α Interferon-β
chromosome 5	Erythropoietin Interleukin-6
chromosome 7	Transforming growth factor-β1
chromosome 10	Interferon-γ Interferon-γ receptor
chromosome 11	Epidermal growth factor receptor Granulocyte colony-stimulating factor Granulocyte-macrophage colony-stimulating factor Interleukin-3 Interleukin-5 Nerve growth factor receptor
chromosome 12	Transforming growth factor-β3
chromosome 16	Interferon-α/β receptor
chromosome 18	Colony stimulating growth factor 1 receptor

Subject Index

MIX
Papier aus verantwortungsvollen Quellen
Paper from responsible sources
FSC® C105338
FSC
www.fsc.org

If you have any concerns about our products,
you can contact us on
ProductSafety@springernature.com

In case Publisher is established outside the EU,
the EU authorized representative is:
Springer Nature Customer Service Center GmbH
Europaplatz 3, 69115 Heidelberg, Germany

Printed by Libri Plureos GmbH
in Hamburg, Germany